no. of prot. = elec

at wt = prot + neutrons

elef. are min

The Essence of
Organic Chemistry

The Essence of
Organic

Jane M. Cram and Donald J. Cram

University of California, Los Angeles

Chemistry

ADDISON-WESLEY PUBLISHING COMPANY

Reading, Massachusetts • Menlo Park, California
London • Amsterdam • Don Mills, Ontario • Sydney

This book is in the

ADDISON-WESLEY SERIES IN CHEMISTRY

Second printing, June 1978

Copyright © 1978 by Addison-Wesley Publishing Company, Inc. Philippines
copyright 1978 by Addison-Wesley Publishing Company, Inc.

ISBN 0-201-01031-3
ABCDEFGHIJK-HA-798

Preface

No authors write a new textbook of elementary organic chemistry without the belief that they can improve on the many already available. We are no exception. Our major challenge in developing this book was to encompass the main themes of organic chemistry in the space of 500 pages. A second challenge was to develop a text which was as little dependent on the teacher and on ancillary material as possible. A third challenge has been to introduce enough biological chemistry to illustrate the analogies and dissimilarities between the compounds and reactions found in the laboratory and in the cell. Although the students to whom this book is addressed have different career objectives, enough of them are oriented toward the life sciences to justify our stress of those features of organic chemistry most relevant to life processes.

The enormous correlative power of the structural theory of organic chemistry finds its expression in the classes of organic compounds on the one hand, and the classes of organic reactions on the other. The organization of the text at the chapter level follows the traditional pattern in which compound classes are taken up in order of their increasing complexity. The first four chapters deal largely with the nature of the chemical bond as applied to the structures, reactions, and physical properties of hydrocarbons. In these chapters, the other classes of compounds are introduced as derivatives of the hydrocarbons and of simple inorganic compounds already familiar to the student. Chapter 5 treats alcohols, phenols, and ethers, which are the simple organic derivatives of water. Chapter 6 introduces the alkyl halides, sulfates, phosphates, thiols, and sulfides, which are organic derivatives of the simple inorganic acids. The compound classes treated through Chapter 6 can be interconverted largely through simple substitution, addition, and elimination reactions. Thus the reaction classes with their similar general mechanistic features are used to integrate otherwise individual reactions into families of reactions.

By the time Chapter 7 is reached, enough compounds and reactions have been discussed to allow a thorough treatment of chirality and of dynamic stereochemistry. Once developed, the stereochemical themes appear again and again in the subsequent discussions. Chapters 8 and 9 deal with carboxylic acids and their near relatives, the acid chlorides, esters, and fats. These compounds and their interconvertibility provide analogies for the amines, amides, and proteins of Chapter 10. In Chapter 11 on aldehydes and ketones, oxidation, reduction, and condensation reactions relate these compounds to those compound classes discussed previously. These middle chapters provide opportunities for comparing and contrasting the structures and reactions of organic and bio-chemistry. While the facts and theories of organic chemistry build on one another as the book progresses, the biochemical sections depend directly on the preceding organic sections rather than on preceding biochemical sections.

Chapters 12 through 15 in turn take up aromatic substitution, carbo-hydrates, carbon chain-building reactions, and organic spectroscopy. The unique character of each of these subjects and the absence of their interdependence allows them to be studied in any order.

In our attempt to make this book as self-contained as possible, subjects are not mentioned unless they are treated in depth. Problems are inter-spersed with the text so the students can test their knowledge section by section. The problems at the end of each chapter depend more on the chapter taken as a whole and on material found in previous chapters. The problems become progressively more difficult as the end of the problem section is approached. So that the students can learn whether they have been successful in answering our problems, *the answers to all problems are found at the end of the book*. The important topics and reactions summarized at the end of each chapter provide the students with an overview of the subject.

A separate Study Guide for Students has been written to help them learn the material in this text. The text and the Study Guide play reciprocal roles, not unlike those of participants in a dialogue. Thus the Study Guide poses a series of questions to which the continuity of the text provides the series of answers. Conversely, the Study Guide reveals the thought patterns required for solutions of the more difficult problems in the text. The questions of the Study Guide are Socratic in character and anticipate the development of the subject matter in the text. The questions of the text are designed to allow the students to measure their progress and to self-test their development.

We are indebted to Professor J. John Uebel of the University of New Hampshire and to Professor Kenneth G. Hancock of the University of California at Davis, who read the entire manuscript, for their many constructive comments and suggestions. The comments of a number of reviewers helped to shape the book in its early stages. It has been a pleasure to work with Miss Laura Rich, the editor, whose consistent interest helped to move the book through to completion. Lastly, we wish to acknowledge the impatience and lack of cooperation between the coauthors, without which this book would never have been completed.

Los Angeles, California
September 1977

J. M. C.
D. J. C.

Contents

CHAPTER 1

The Ingredients
of Organic Compounds:
Atoms, Orbitals,
and Bonds

Human beings are curious; otherwise the questions that organic chemistry answers would never have been asked. These questions originally centered around attempts to understand life processes and the operations of "organs," hence the term **organic chemistry**. Since compounds of carbon and hydrogen were always encountered in organs, organic chemistry became associated with the study of compounds of carbon and hydrogen. As the science developed, many compounds of carbon and hydrogen were synthesized in the laboratory which were identical to those isolated from "organic" sources. Many others were prepared that were unrelated to the *natural* compounds. Some of the syntheses involved inorganic starting materials, but most were transformations of one compound of carbon and hydrogen into a second, a third, etc. Since the properties of the compounds were found to be independent of their "natural" or "unnatural" origins, the term organic chemistry became more general, and now refers to the **chemistry of compounds that contain carbon and hydrogen**.

The intimate relationship between laboratory and cellular chemistry continues, and the term "organic" appropriately refers to that relationship. The evolutionary compounds of carbon produced in cells and those compounds prepared in the laboratory are governed by the same chemical rules. This book, then, describes the compounds of carbon and hydrogen and the rules that govern their transformations.

1.1 PERSPECTIVE   Matter exhibits different organizational characteristics at different temperatures. Above about 100,000°K, the electrons and nuclei that compose plasma are stable and associate to form atoms only below that temperature. Atoms, in turn, associate to form molecules only at temperatures below a few thousand degrees. The common chemical bonds of organic compounds are those between carbon and carbon, carbon and hydrogen, carbon and oxygen, carbon and nitrogen, carbon and sulfur, and carbon and the halogens. These bonds are stable only in the range between absolute zero and about 1000°K. Evolutionary organic compounds, those associated with living systems, are produced and used at temperatures that vary between about 260°K and 373°K, and are often unstable at higher temperatures. "Unnatural" organic compounds have been designed and synthesized which are stable to about 773°K, but these are noteworthy exceptions, since most organic compounds decompose at much lower temperatures.

What are the sizes of organic compounds? One of the smallest molecules is methane (the main component in natural gas), which is about 0.00000003 cm in diameter. One of the largest is polyethylene, which can be as long as 0.00009 cm. A small enzyme molecule—one of the machines of life—is about 0.0000003 cm in diameter. The size of objects in the world we perceive with the unaided eye ranges from 10^{-2} to 10^{10} cm (0.01 to 10 billion cm). Even with the help of the electron microscope, our sight can barely detect only the largest organic compounds, and then not with enough detail to provide much information about the molecular structure.

Where did organic compounds come from? Apparently hydrogen and helium were the "ultimate atomic starting materials" for the synthesis of the heavier elements. The atmospheres of most large planets are composed primarily of H_2, He, CH_4 (methane), CO, CO_2, N_2, NH_3, and H_2O. In laboratory experiments, when mixtures of these gases flow past a spark discharge (an artificial lightning), organic compounds are formed in profusion. The combination of lightning and ultraviolet light from the sun probably generated on this planet a "soup" of organic compounds, which accumulated during the first 2.5 billion years of its 4.5 billion-year existence. From this soup life evolved, which transformed the atmosphere from one which was reducing (without oxygen) to one that was oxidizing. The cells that developed utilized ultraviolet light to consume carbon dioxide, water, and organic compounds of the soup and to liberate oxygen. Oxygen in turn formed ozone O_3, which served as a filter for much of the ultraviolet radiation that had fueled the original formation of the organic soup.

This transformation from a reducing to an oxidizing atmosphere occurred gradually over a period extending from about 2.5 to 3 billion years in the age of the planet and was accompanied by the consumption of the organic soup by microorganisms. Both geological and fossil evidence points to the existence of life at the simple bacteria level as far back as 3.2 billion years ago. Since life and the orderly synthesis of organic compounds go hand in hand, evolutionary organic chemistry is about 3.2 billion years old. In sharp contrast, laboratory organic chemistry is only about 150 years old. The sources of organic compounds used in laboratory organic chemistry are petroleum, natural gas, and coal, which are the fossil fuels originally generated by life processes.

As we have just indicated, organic chemistry as a science is only about 150 years old. It is an empirical science whose development has depended on millions of separate laboratory observations. These observations have been rationalized in terms of the structural theory of organic chemistry, which makes the subject understandable and teachable. This theory accounts for more separate facts in a more consistent way than any other theory proposed up to now. Each structure of the millions of known organic compounds resembles a submicroscopic cathedral, each part of which contributes to a grand architectural masterpiece. The beauty of the theory is that a few simple ideas account for diverse complex phenomena and allow predictions to be made about millions of compounds that have never been, but might be, prepared. Many organic chemists, guided by the structural theory, have designed and synthesized hundreds of new organic compounds during their lifetimes.

The substances of greatest importance in our lives, with the exception of water and oxygen, are those containing carbon. They are the chief components of our bodies, our food and fuel, wood, natural and synthetic fibers for clothes, carpets, and furniture, natural and synthetic rubbers, plastics, paper, soap and detergents, dyes, and some pigments. These materials, which came into being either through the chemical reactions of evolution or by human design, all share the feature of possessing useful chemical and physical properties. People have changed the form of many naturally occurring substances to meet their needs. Examples are food, rubber, and the natural fibers, cotton, linen, wool, and silk, all of which are converted only in their physical forms for practical uses. By contrast, most dyes, drugs, plastics, synthetic fibers, and synthetic rubbers are constructed from smaller molecules by chemists.

No strict dividing line exists between organic chemistry and biochemistry. The changes which carbon compounds undergo are similar in the laboratory and in the cell. The biological compounds are frequently larger and more complex in structure, but the part of the large molecule which undergoes a change in any single reaction is no more complex than it is in small compounds. An object of this book is to present the structures of some organic compounds and to compare the reactions devised by chemists in the laboratory with those which have evolved in biological systems.

1.2 CARBON COMPOUNDS

Carbon atoms form strong carbon–carbon and carbon–hydrogen bonds. Carbon is unique among the elements in forming stable, continuous chains of identical atoms which are the backbones of durable compounds. These chains make possible the existence of tens of thousands of different compounds containing only carbon and hydrogen, which are called **hydrocarbons**. Hydrocarbons range in size from the gases methane and ethane to the plastic polyethylene.

$$CH_4 \qquad\qquad CH_3\!-\!CH_3 \qquad\qquad \text{etc.} \sim (CH_2\!-\!CH_2\!-\!CH_2\!-\!CH_2)_x \sim \text{etc.}$$

Methane
Molecular weight 16
Diameter 0.00000003 cm

Ethane
Molecular weight 28

Polyethylene
Molecular weight about 100,000
Maximum length about 0.00009 cm

Carbon also forms strong bonds with oxygen, nitrogen, sulfur, fluorine, chlorine, bromine, and iodine to give a wide variety of structures and properties.

The principles which govern the structures of carbon compounds are the same as those which apply to the structures of the inorganic compounds, such as water, ammonia, hydrogen sulfide, hydrogen chloride, sulfuric acid, and so on.

1. The electronic structure of each element determines the number and type of bonds it forms with other elements to make compounds.

2. A molecule has a definite arrangement in space for the atoms which compose it. This definite spatial arrangement gives the molecule a shape which can be predicted from the electronic structure of the elements.

We therefore start our study of organic chemistry with a review of the principles which apply to all compounds.

1.3 ELECTRON STRUCTURE AND COMBINING RULES FOR SMALL MOLECULES

The electron structure of a given element determines the number of bonds formed by each of its atoms in compounds with other elements. Each atom attempts to have enough electrons in its highest occupied energy level to match the stable electron structure of a rare gas.

Hydrogen fills the first energy level (which can hold a total of two electrons) by sharing a pair of electrons with another atom. It thereby achieves the stable electron configuration of helium. The chemical bond formed between two atoms which share one pair of electrons is called a **covalent bond**.

A dot represents one electron for the hydrogen atom.

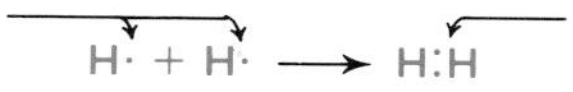

$$H\cdot + H\cdot \longrightarrow H{:}H$$

A pair of electrons (two dots) shared between two atoms constitutes a covalent bond.

The number of valence electrons (electrons in the highest occupied energy level) for the elements of the first three periods of the periodic table are shown in Fig. 1.1. From this figure, observe that oxygen needs two electrons to fill its highest occupied energy level with eight electrons. One way it can obtain these is by sharing a pair of electrons with two other atoms in two covalent bonds, as in water.

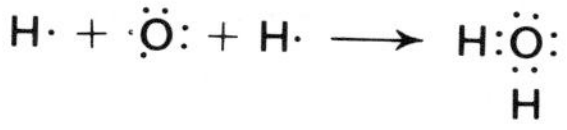

$$H\cdot + \cdot\ddot{O}{:} + H\cdot \longrightarrow H{:}\underset{H}{\ddot{O}}{:}$$

FIGURE 1.1 ELECTRON STRUCTURE OF ELEMENTS OF FIRST THREE PERIODS

Number of electrons in highest occupied level	1	2	3	4	5	6	7	8
First energy level	H	He						
Second energy level	Li	Be	B	C	N	O	F	Ne
Third energy level	Na	Mg	Al	Si	P	S	Cl	Ar
							Br	
							I	

An atom can form as many covalent bonds with other atoms as the maximum number of electrons needed to fill its highest occupied energy level. Nitrogen $:\dot{\ddot{N}}\cdot$ forms three bonds with three hydrogens, carbon $\cdot\dot{C}\cdot$ forms bonds with four hydrogens, and chlorine $:\dot{\ddot{C}l}:$ with only one. Formulas for the resulting compounds may be drawn to show both the number of atoms involved and the direct attachment of the hydrogens to the other kind of atom.

hydrogen chloride   water   ammonia   methane

In these formulas a line between symbols for the elements represents the pair of electrons which form the covalent bond between the two atoms. Pairs of electrons which are not shared—nonbonding pairs—are represented by two dots.

Throughout this book, there are problems inserted after many of the sections within the chapters. These problems test your understanding of the material in the preceding section by asking you to apply the principle to slightly different situations from the ones presented in the text. Be sure to work the problems as you finish studying each section.

Write your answers to all problems. Organic chemistry is a pencil-and-paper subject as well as a laboratory science. You must train your hand to write as you train your eye to read. Answers to all problems are at the end of this book. Check your answers with the printed ones in every detail.

Problem **1.1** Draw electron structures for the following compounds, using a dash for a pair of electrons in a bond and two dots for each pair of nonbonding electrons.

 a) hydrogen bromide   **b)** hydrogen fluoride
 c) hydrogen sulfide   **d)** phosphine PH_3
 e) carbon tetrachloride CCl_4 (similar to methane)

1.4 SHAPES OF SMALL MOLECULES IN THREE DIMENSIONS

The submicroscopic size of molecules does not relieve them of their necessity to possess shape. Like familiar objects on a macro scale, the molecules of most compounds occupy the three dimensions of space. The size and the shape of the molecule depend upon the distances between the atoms and the angles formed by two bonds to the same atom.

The distance between the nuclei of two atoms bonded to each other is the **bond length**. The stable bond length represents the lowest possible energy level in which the repulsion between positively charged nuclei is offset by the attraction of the positive nuclei for the negatively charged electrons. Fortunately, the bond length is remarkably constant between the same two atoms regardless of the compound in which the bond length is measured. The carbon–carbon bond length is 1.54 Å; carbon–hydrogen bond lengths are shorter, 1.09 Å, and carbon–oxygen bond lengths are approximately 1.43 Å.

The angle between two bonds to the same central atom is called the **bond angle**. Together, the sizes of the bond angles and the lengths of the bonds determine the spatial arrangement of the bonded atoms and the three-dimensional shape of the molecule. In methane all the H—C—H bond angles are 109.5°. In this arrangement the electron pairs of the bonds are as distant as possible from one another. The hydrogens are also as far apart as possible.

Bonds represented by solid lines are in the plane of the paper.
Heavy wedged bond comes above the paper.
Dashed bond extends below the paper.

A three-dimensional representation of methane

The 109.5° angle is the only value possible for four equal bond angles around a central atom. It is the characteristic angle between any two bonds originating at a carbon bonded to four other atoms. While bond angles deviate from the standard value much more than do bond lengths, values of about 109° are most often found for carbon.

A carbon bonded to four atoms is called a **tetrahedral carbon**, because the four bonds point toward the vertices of a regular tetrahedron with the carbon at the center of the figure. In Fig. 1.2, the three-dimensional formulas for the structure of the methane molecule are shown from three angles of view. Ball-and-stick models and space-filling models for the same angles are also shown.

two-dimensional formula   three-dimensional formulas

**FIGURE 1.2
MODELS OF METHANE
VIEWED FROM THREE
ANGLES**

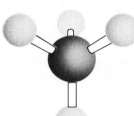

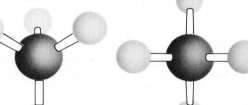

dark gray = carbon
light gray = hydrogen

Ball-and-stick models show clearly the angles formed by two C—H bonds (sticks), but the length of the bond is greatly exaggerated in relation to the size of the atoms. Dashed lines show the outlines of the regular tetrahedron with carbon at the center and a hydrogen at each vertex.

Space-filling models are made to give the relative sizes of atoms and the accurate distances between them, but the bond angles are less obvious.

The H—O—H bond angle in water (105°) and the H—N—H bond angle in ammonia (107°) are slightly smaller than the H—C—H bond angle in methane. The three atoms of the water molecule must lie in the same plane, but they are not in one line. The four atoms of ammonia form a three-dimensional figure like a three-legged stool. Note that methane, water, and ammonia, which have similar bond angles, all have four pairs of electrons around the central atom.

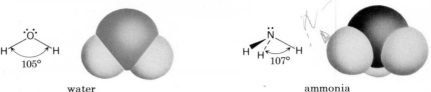

water   ammonia

It is important for your hand to learn to draw what your eye is learning to see. Practice in drawing simple formulas now will increase your ability to handle larger structures later.

Problem   **1.2** Draw the structure for each of these compounds in a two-dimensional form, showing all the bonds, and in a three-dimensional form in the manner shown above for methane.

  a) methane CH_4 **b)** ammonium ion NH_4^+ (nitrogen is tetrahedral)
  c) carbon tetrachloride CCl_4 **d)** chloroform $CHCl_3$

1.5 ATOMIC AND MOLECULAR ORBITAL VIEWS OF BONDING

The space around the nucleus of an atom that might be occupied by one electron or a pair of electrons is called an **atomic orbital**. The shape and size of an orbital depend upon the energy of the electrons occupying the orbital. The shapes, dimensions, and locations of the orbitals are derived from quantum mechanical calculations based on electron emission spectra.

The first electronic energy level has one orbital (occupied by one or two electrons) which is an s orbital, called $1s$. The second energy level has four orbitals, an s and three p's ($2s$, $2p_x$, $2p_y$, and $2p_z$). The shapes of s and p orbitals are depicted in Figs. 1.3 and 1.4.

FIGURE 1.3   SHAPES OF ATOMIC ORBITALS

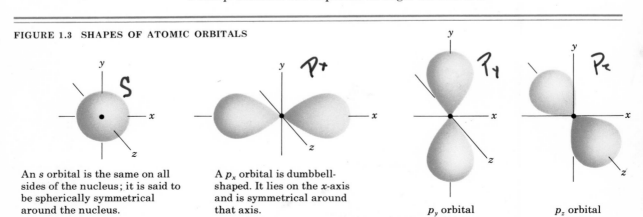

An s orbital is the same on all sides of the nucleus; it is said to be spherically symmetrical around the nucleus.

A p_x orbital is dumbbell-shaped. It lies on the x-axis and is symmetrical around that axis.

p_y orbital

p_z orbital

The three p orbitals (p_x, p_y, and p_z) have the same shape and size; they differ only in the orientation about the nucleus.

FIGURE 1.4
THREE *p* ORBITALS

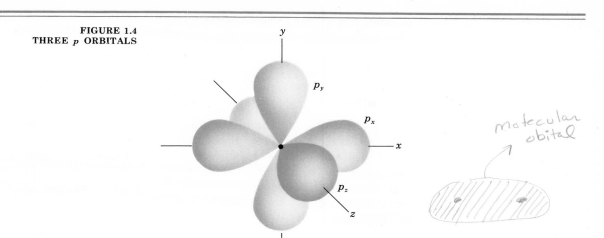

Two atoms are held close to each other by an attractive force called a bond. A **covalent bond** holds two atoms together in a molecule by the attraction of both nuclei for the same pair of electrons. The pair of electrons is said to be *shared* by the two nuclei.

The simplest model for a covalent bond is the overlap of an atomic orbital of one atom with an atomic orbital of the other atom. The greater the overlap of the two orbitals, the stronger is the binding force of the bond. Both of the electrons in the bond are attracted to both nuclei. The region around the two nuclei which is occupied by the pair of electrons is called a **molecular orbital** (see Fig. 1.5).

FIGURE 1.5
BONDING USING
ATOMIC ORBITALS

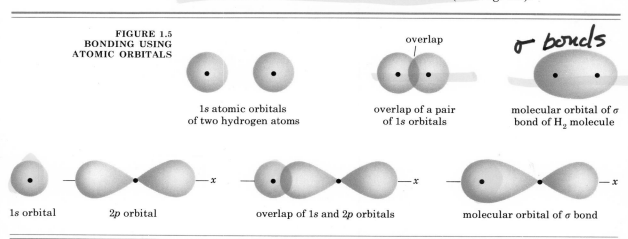

The preferred position of the hydrogen 1s orbital (Fig. 1.5) is at the end of the $2p_x$ orbital on the *x*-axis where overlap is greatest. Since both the 1s and $2p_x$ orbitals are symmetrical about the *x*-axis, the bonding orbital is also symmetrical about the *x*-axis. A bonding orbital which is symmetrical about the axis connecting the nuclei of the two atoms is called a sigma bonding orbital and the bond is a **sigma (σ) bond**.

Problem   **1.3** Draw the bond in terms of the overlap of atomic orbitals and the molecular orbital for each of the following.

    a) hydrogen fluoride HF   **b)** fluorine F_2

1.6 THE ATOMIC ORBITALS OF CARBON

The picture of covalent bonds formed by overlap of simple atomic orbitals does not account for the experimentally determined arrangement of atoms of most organic compounds. The three p orbitals are perpendicular to one another (90°). If these p orbitals were used in bonding carbon to hydrogen, the H—C—H bond angle should be 90°. But methane has *four* equivalent C—H bonds, not three, with bond angles of 109.5°.

The carbon atom has six electrons distributed in the first and second electron energy levels: $1s^2, 2s^2, 2p_x^1, 2p_y^1, 2p_z^0$. The two electrons of the first energy level do not participate in bonding. A carbon atom uses all four orbitals and all four valence electrons of the second energy level to form four equivalent bonds. The four orbitals used for these four bonds are equivalent and are a hybrid of the 2s and the three 2p orbitals. These hybridized orbitals are called **sp³ hybrid orbitals**.

Repulsion between electron pairs forces the pairs of electrons and the bonds they form to be distributed as far apart as possible around the central carbon atom. This distribution gives an angle between the sp^3 hybrid orbitals of 109.5°. The spatial representation of the four sp^3 hybrid orbitals for a carbon atom is shown in Fig. 1.6.

Hybridization of carbon's 2s and three 2p orbitals into four sp^3 orbitals, with one electron in each orbital, provides the means for forming four equivalent bonds at carbon. These bonds are much stronger and give a more stable compound than would be possible if only p orbitals could be used. The sp^3 hybrid orbitals afford the possibility of greater overlap with orbitals of other atoms, a factor which produces strong bonds. The overlap of another orbital with an sp^3 orbital is at the end of the large lobe of the sp^3 orbital. The atom bonded to the carbon is thus located beyond the end of the large lobe as shown in Fig. 1.7.

The bonding orbital formed by the overlap of s and sp^3 orbitals or by sp^3 and sp^3 orbitals is symmetrical around the line connecting the two nuclei, and is another example of a sigma bond orbital.

This concept of bonding with sp^3 hybridized orbitals is consistent with the experimentally determined spatial arrangement of the four hydrogen atoms around the carbon atom in methane. The calculated best angle of 109.5° for two sp^3 orbitals is the same value as that found for the H—C—H bond angle.

The electron pairs around the nitrogen and oxygen atoms in ammonia and in water are also distributed as far apart as possible in sp^3 orbitals. The nonbonding pairs of electrons also occupy sp^3 orbitals as shown in Fig. 1.8.

FIGURE 1.6 *sp³* HYBRID ORBITALS

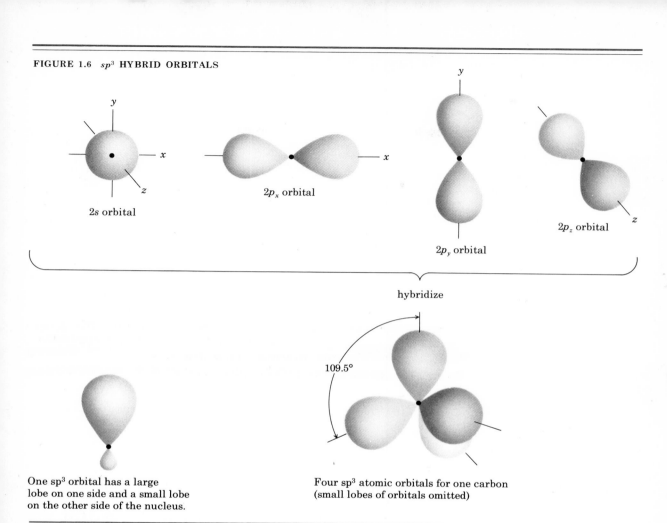

2s orbital

2p_x orbital

2p_y orbital

2p_z orbital

hybridize

109.5°

One sp³ orbital has a large
lobe on one side and a small lobe
on the other side of the nucleus.

Four sp³ atomic orbitals for one carbon
(small lobes of orbitals omitted)

FIGURE 1.7 BONDING WITH *sp³* ORBITALS

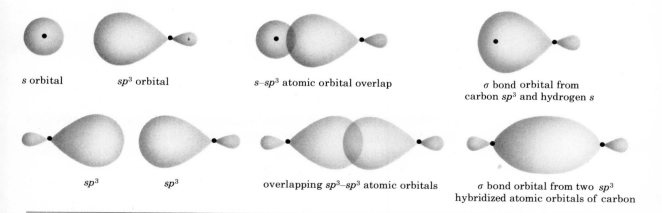

s orbital

sp³ orbital

s–sp³ atomic orbital overlap

σ bond orbital from
carbon sp³ and hydrogen s

sp³

sp³

overlapping sp³–sp³ atomic orbitals

σ bond orbital from two sp³
hybridized atomic orbitals of carbon

FIGURE 1.8 GEOMETRY OF MOLECULES USING sp^3 ORBITALS

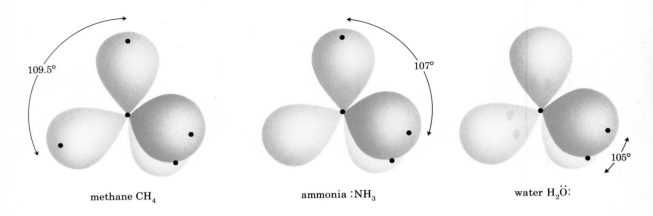

methane CH_4 ammonia $:NH_3$ water $H_2\ddot{O}:$

By comparison with bonding electron pairs, nonbonding pairs appear to exert greater repulsion, and angles involving their orbitals are larger than 109°. As a consequence the other angles in a molecule containing nonbonding electron pairs are slightly compressed and smaller (Fig. 1.8).

Problem **1.4** Draw the overlap of atomic orbitals and the molecular orbital for each of the following.

a) a carbon–nitrogen bond   **b)** a carbon–oxygen bond

1.7 FUNCTIONAL GROUPS

Most organic compounds contain atoms other than carbon and hydrogen, like oxygen or nitrogen. An example is methanol, in which one of the hydrogens of methane is replaced by a **hydroxyl group** —**OH**, with a carbon–oxygen bond.

methane   methyl group   hydroxyl group   methanol

The **CH_3** group of atoms, which is methane minus one of the hydrogens, is called the **methyl group**. It is an assembly of atoms which can form a bond with atoms like the hydroxyl group to make a compound. An atom or a group of atoms, such as —OH, which takes the place of a hydrogen in methane is called a **functional group**.

Think of methanol as being related to the hydrocarbon methane, from which it gets most of its name. Methanol is also related to water, with one of the hydrogens of water replaced by the methyl group. Water contains a hydroxyl group attached to a hydrogen. The presence of the hydroxyl group —OH in methanol is signified by the "ol" part of the methanol name.

Some other important functional groups which can be joined to the methyl group are **chloro** —Cl, **amino** —NH$_2$, and **sulfhydryl** —SH. The formulas for the compounds containing these functional groups are drawn here.

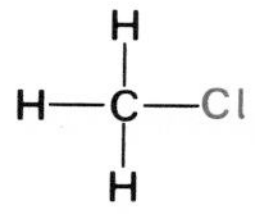

chloromethane	methylamine	methanethiol
(related to	(related to	(related to
hydrogen chloride)	ammonia)	hydrogen sulfide)

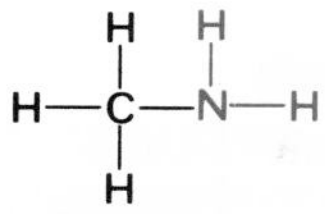

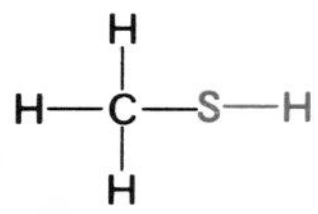

Each of the compounds shown above is derived from an inorganic compound in which one hydrogen has been replaced by the methyl group. Methylamine is derived from ammonia NH$_3$, methanethiol is derived from hydrogen sulfide H$_2$S, and chloromethane from hydrogen chloride HCl.

Problem **1.5** What would be the approximate values for the following bond angles in methylamine and in methanol, if you assume no variation in value from those formed by the same atoms in methane, water, and ammonia?

a) H—C—H   **b)** H—C—N   **c)** H—C—O   **d)** C—N—H

e) H—N—H   **f)** C—O—H

1.8 CLASSES OF ORGANIC COMPOUNDS

Compounds that contain the same functional group belong to the same **class** of organic compounds. Each of the carbon compounds that we have discussed so far belongs to a different class of compounds. Recall that methane is a hydrocarbon, since it has only carbon and hydrogen atoms. We can form another hydrocarbon similar to methane, but containing two carbons and six hydrogens, if we join together two methyl groups by a carbon–carbon bond. This hydrocarbon is called ethane. Propane and butane, the familiar heating fuels, also belong to this series of hydrocarbons. Propane has three carbons and butane has four.

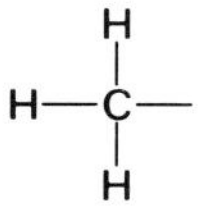

| methyl | methyl | ethane | propane | butane |

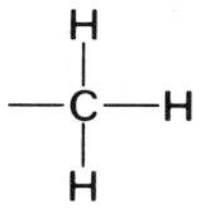

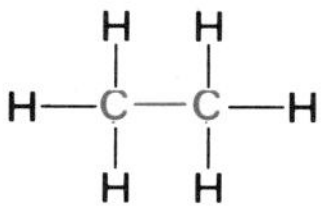

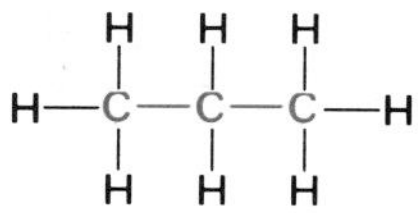

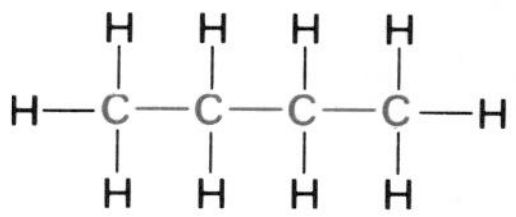

Methanol belongs to the class known as **alcohols**. All compounds which contain the hydroxyl functional group —OH are called alcohols. The structure of another alcohol can be formed by removing one hydrogen from ethane and attaching an —OH in its place. Because this alcohol is related to ethane and contains the —OH functional group, it is called **ethanol**, but is better known as the "alcohol" of whiskies, wines, and beer.

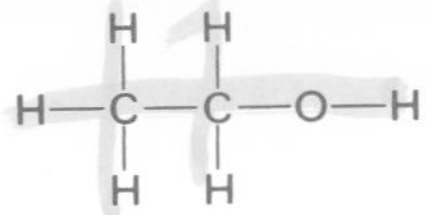

$$H-\underset{\underset{H}{|}}{\overset{\overset{H}{|}}{C}}-\underset{\underset{H}{|}}{\overset{\overset{H}{|}}{C}}-O-H$$

ethanol

Compounds in the same class have physical and chemical properties which are quite similar. This similarity is extremely fortunate, because it enables us to study a few compounds as representatives of a large number of others. In this manner we can learn the characteristics of many thousands of compounds by considering only a few.

Compounds like methylamine $CH_3—NH_2$, which contains the amino group —NH_2, are called **amines**. Next to oxygen, compounds containing nitrogen are the most important and numerous in organic chemistry.

Chloromethane and other compounds containing a halogen atom (F, Cl, Br, or I) are called **alkyl halides**. Methanethiol belongs to the class of compounds called **thiols**.

Problem **1.6** Draw the three-dimensional structures of chloromethane, methanol, and methylamine in the manner shown for methane in Section 1.4. (Make the C—Cl, C—O, or C—N bond in the plane of the paper first.)

1.9 THE POLAR COVALENT BOND In the preceding sections many examples were given of covalent bonding—the sharing of a pair of electrons—between unlike atoms: C—Cl, C—O, C—H, C—N, N—H. How is the sharing of the electrons affected by the different nuclei?

In the case of hydrogen H_2 or fluorine F_2, the bonding electrons are shared equally by the two nuclei.

H:H   :F:F:

At the other extreme is lithium fluoride. The electrons are so strongly attracted by the fluorine nucleus and so weakly held by the lithium nucleus that they become associated only with the fluorine. Fluorine, with an excess of one electron, now has a negative charge, and lithium, with a deficiency of one electron, has a positive charge. The electrons are *not shared*.

Li+   :F:−

In the crystal, oppositely charged independent ions are held together by an electrostatic attraction which is called an **ionic bond**.

What lies between the two extremes of electrostatic attraction of an ionic bond and equally shared electrons of a covalent bond? Some nuclei *attract electrons* more strongly than do others. These atoms are said to be more **electronegative** than the others. The electronegativity of

independent ions held together by electrostatic attraction what is called an ionic bond.

atoms increases as you pass from left to right in the periodic table, $C < N < O < F$, and from bottom to top, $P < N; S < O; Cl < F$. The values of the electronegativities of elements are shown in Fig. 1.9.

FIGURE 1.9
PAULING ELECTRO-
NEGATIVITY VALUES FOR
ELEMENTS

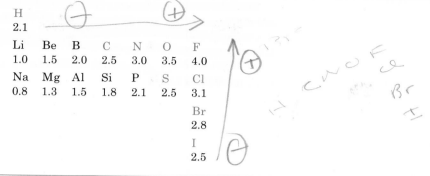

H 2.1						
Li	Be	B	C	N	O	F
1.0	1.5	2.0	2.5	3.0	3.5	4.0
Na	Mg	Al	Si	P	S	Cl
0.8	1.3	1.5	1.8	2.1	2.5	3.1
						Br 2.8
						I 2.5

The electrons in a covalent bond between two atoms of greatly different electronegativity are unequally shared by the two nuclei. In a covalent bond between carbon and oxygen, for example, the oxygen exerts a stronger pull on the electrons than does the carbon. The result of this unequal sharing is a **polarized covalent bond**. The carbon has become slightly positive, having *less* than its equal share of the pair of electrons, and the oxygen has become slightly negative, having *more* than its share of the electrons. This effect can be formulated in three ways.

C :O   C—O   C—O
          δ^+ δ^- $\longleftrightarrow$

The bond polarization is described by use of the symbols δ^+, δ^- (delta plus, delta minus), indicating partial plus and minus charges, or by $\longleftrightarrow$, describing the direction of electron attraction. In this arrow the cross identifies the partially positive end and the head the partially negative end of a bond.

The electrons in a covalent bond between carbon and hydrogen are very nearly equally shared by the two nuclei, and most C—H bonds are *nonpolar*. The direction of polarization of bonds in the classes of compounds previously introduced is shown in Fig. 1.10.

FIGURE 1.10   POLAR COVALENT BONDS

ethanol   ethylamine   fluoromethane

C—C	C—O	C—N	C—F
	$\longleftrightarrow$	$\longleftrightarrow$	$\longleftrightarrow$
equal	or	or	or
C—H	C—O	C—N	C—F
	δ^+ δ^-	δ^+ δ^-	δ^+ δ^-
nearly equal		polarized	

A polarized covalent bond really resembles a dipole, since one atom is a positive pole and the other is a negative pole. The polarity of unsymmetrical covalent bonds is associated with many of the reactions of carbon compounds. In general, most reactions involve changes in the polarized covalent bonds (C—O, C—Cl, etc.) and not in the nonpolar covalent bonds (C—C and C—H). For example, in ethanol it is the C—O or O—H bond which breaks during most reactions, and not the C—C or C—H bonds. The hydrocarbon part of the molecule is frequently undisturbed, and only the functional group and its attached carbon undergo changes.

Those atoms in the structures of Fig. 1.10 which are marked positive or negative are the sites of new bond making or bond breaking in the molecules. Thus, as we will see later, all alcohols react somewhat similarly and in a manner characteristic of an —OH group, even though they have different hydrocarbon parts. Such a generalization allows us to simplify the study of the physical and chemical properties of a large number of alcohols by the study of just a few alcohols.

Problem   **1.7** For the structure of the following compound, indicate the direction of the polarity of the bonds by the use of arrows and by the symbols δ^+ and δ^-.

$$\text{Cl}-\overset{\overset{\displaystyle H}{|}}{\underset{\underset{\displaystyle H}{|}}{C}}-\overset{\overset{\displaystyle H}{|}}{\underset{\underset{\displaystyle H}{|}}{C}}-\text{O}-\text{H}$$

1.10 POLAR COMPOUNDS   A compound containing polar covalent bonds has separated partial positive and negative charges which generate an electric dipole. The force of an electric field required to orient the molecules of the compound in the field depends on both the magnitude of the charges and the distance between them. The **dipole moment μ (mu)** is a function of both the amount of charge and the distance between charges.

The effect of individual bond dipoles on the dipole moment of the compound depends on the relative orientation of the polar bonds. The moments of bonds with dipoles oriented in opposite directions cancel each other. The dipole moment thus serves as a useful quantitative measure of the extent to which a compound is polar and electrically unsymmetrical.

The dipole moment of a compound also provides information concerning the shape of a molecule. The sizable dipole moment for water, $\mu = 1.84$ units, indicates the electrical dissymmetry of the molecule. If the two hydrogens lay in opposite directions equidistant from the oxygen, all three atoms coaxial, the moment for one O—H bond would cancel the moment for the other O—H bond. Such a compound would have a zero dipole moment even though it had two polar bonds.

$$\text{H}-\text{O}-\text{H}$$
$$\mu = 0$$

$$\mu = 1.84 \text{ units}$$

We conclude that the water molecule is bent, since its dipole moment is considerably greater than zero. On the other hand a dipole moment of zero for CCl_4 attests to the symmetry of the four C—Cl bonds around the carbon.

Dipole moments for some small compounds are given in Fig. 1.11.

FIGURE 1.11
DIPOLE MOMENTS

methylamine	methanol	chloromethane	carbon tetrachloride	water
$\mu = 1.33$	$\mu = 1.69$	$\mu = 1.86$	$\mu = 0$	$\mu = 1.84$ units

1.11 HYDROGEN BONDING

Hydrogen–oxygen, hydrogen–nitrogen, and hydrogen–fluorine bonds are the most highly polarized of ordinary covalent bonds. When a strongly electronegative atom such as oxygen attracts the bonding electrons away from an attached hydrogen, no other electrons shield the positive charge of the bare hydrogen nucleus, a proton. This hydrogen nucleus is very small and the positive charge is highly localized. The nonbonding electrons as well as the bonding electrons which are closely held give the oxygen a high electron density. In compounds containing hydroxyl groups a strong association occurs between the partial positive charge of the hydrogen nucleus of one molecule and the partial negative charge of an oxygen of another molecule. This association is called **hydrogen bonding**. In the structure of water the dotted lines represent hydrogen bonding.

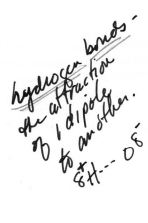

The hydrogen nucleus becomes associated with a particular pair of nonbonding electrons occupying an sp^3 orbital of oxygen. For this reason, hydrogen bonding has a specific directional character, which plays an important role in the structure of proteins (Chapter 10).

Strong hydrogen bonding is also found in ammonia and hydrogen fluoride, whose structures are shown here.

Hydrogen bonding between molecules is much stronger than most other *inter*molecular forces (about 5–8% as strong as an average covalent bond). It is responsible for many of the physical properties observed for compounds containing the —OH group (alcohols and acids) and the —NH group (amines and amides).

Melting and boiling points of compounds that form hydrogen bonds occur at much higher temperatures than would be expected in the absence of hydrogen bonding. Hydrogen bonds produce an association between molecules of water in the liquid or in the solid state, so that a greater amount of energy is required to separate molecules than the formula H_2O would suggest. Hydrogen bond associations can be overcome at moderate temperatures of 0–150°C, whereas direct cleavage of covalent bonds by heat requires 300–500°C. Some comparative melting and boiling points are given in Fig. 1.12.

FIGURE 1.12 TRANSITION TEMPERATURES FOR HYDROGEN-BONDED COMPOUNDS	Compound	MW	mp, °C	bp, °C	Compound	MW	mp, °C	bp, °C
	H_2O	18	0	100	CH_3OH	32	−98	65
	NH_3	17	−78	−33	CH_3NH_2	31	−94	−6
	CH_4*	16	−182	−161	CH_3CH_3*	30	−183	−89

* no hydrogen bond

Hydrogen bonding between solute and solvent, replacing solute–solute and solvent–solvent attractions, makes many compounds soluble and even miscible (soluble in all proportions) with water. As the number of carbons in the compound increases for the number of possible hydrogen bonds, solubility in water decreases. Low molecular weight carboxylic acids, alcohols, and amines (as well as liquid ammonia at −33°C) serve as moderately polar solvents with hydrogen bonding capacity.

1.12 EFFECT OF INTERMOLECULAR ATTRACTIONS ON PHYSICAL PROPERTIES

Intermolecular attractions exist between all molecules, polar and nonpolar, in the solid and liquid states. Energy must be supplied to overcome these associations to allow molecules in a crystal to move past each other (melting), and to escape from contact altogether (boiling).

The weakest of the intermolecular interactions is known as a **van der Waals force**. This force is most simply described as the attraction between two temporarily polarized bonds. The more intermolecular contacts there are between two molecules, the more bond polarizations there are, and the stronger are the van der Waals forces. The difference in the contacts between methane molecules and octane molecules is depicted in Fig. 1.13.

In a series of hydrocarbons, the melting and boiling points become higher with increasing molecular weight, due to an accumulation of van der Waals forces, as shown in Fig. 1.14.

increased MW means mp + bp become higher to overcome Van's

FIGURE 1.13
INTERACTIONS BETWEEN
HYDROCARBON MOLECULES

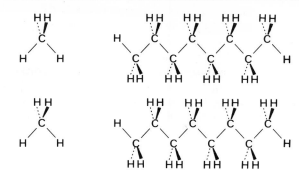

methane
(similar alignment)

octane
(two molecules aligned for maximum
contact in the crystal structure)

FIGURE 1.14
TRANSITION TEMPERATURES
FOR SATURATED
HYDROCARBONS

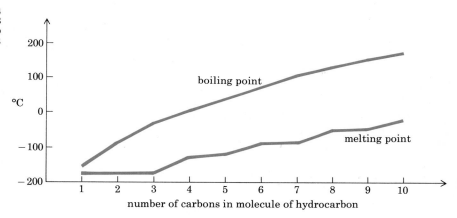

The property of solubility of one substance in another depends on the interactions of molecules of the two compounds involved. The attraction between two different nonpolar compounds closely resembles the interaction of the molecules of each with its kind. A solution of one solid nonpolar compound (the solute) in a liquid nonpolar compound (the solvent) is easily produced. The molecules of the solid slip past each other as in melting, and their neighbors in the crystal are replaced by molecules of the solvent. The solvent molecules have an intermolecular attraction for the solute molecules similar to the solute-for-solute attraction. Because of the different shapes of the solvent and solute molecules, the two seldom pack into the same crystal, and the crystal structure of the solid cannot be maintained. Saturated hydrocarbons are found to be miscible in all proportions with other saturated hydrocarbons.

Polar molecules cling together more strongly than do nonpolar substances. Polar substances are, however, readily soluble in polar solvents. The strong intermolecular interactions, particularly those due to hydrogen bonding, that hold the molecules of a polar solute together are replaced by similar polar interactions between solvent and solute. *Like dissolves like* is an adage of considerable utility. For example, ethanol and methanol are miscible with water, an alcohol with six carbons has a limited solubility in water, and the hydrocarbon with six carbons has a very low solubility in water.

New Terms and Topics

Temperature range for organic compounds to be stable and for biological compounds to be made and used (Section 1.1)

Size of organic compounds (Section 1.1)

Sources of organic compounds (Section 1.1)

Common elements which bond with carbon (Section 1.2)

Hydrocarbons (Section 1.2)

Electron structure of molecules (Section 1.3)

Three-dimensional shape of molecules (Section 1.4)

Tetrahedral carbon (Section 1.4)

Bond angles and bond lengths (Section 1.4)

Molecular models (Section 1.4)

Covalent bond (Sections 1.3 and 1.5)

Sigma orbital and sigma bond (Section 1.5)

Atomic orbital overlap and molecular orbital (Section 1.5)

Hybridized atomic orbital: sp^3 (Section 1.6)

Functional groups of carbon compounds: —OH, —NH_2, —SH, Cl (Section 1.7)

Classes of organic compounds: hydrocarbons, alcohols, amines, thiols, and alkyl halides (Section 1.8)

Electronegativity (Section 1.9)

Polarity of unsymmetrical covalent bonds (Section 1.9)

Dipole moment of molecules (Section 1.10)

Hydrogen bonding (Section 1.11)

Intermolecular attractions; van der Waals force (Section 1.12)

Problems

Brief answers to all problems are given at the end of this book. Further discussions of solutions to problems are given in the Study Guide.

1.8 Define by words (and drawings if appropriate) the following terms.

a) *s* atomic orbital

b) covalent bond

c) sp^3 hybridized orbital

d) molecular orbital

e) sigma bond

f) bond angle

g) bond length

h) tetrahedral carbon

i) hydrocarbon

j) methyl group

k) functional group

l) polar covalent bond

1.9 a) What is the size of the H—C—H bond angle in methane?

   b) How do the H—O—H and H—N—H angles of water and ammonia compare with the H—C—H angle of methane?

   c) What would you expect the size of the Cl—C—Cl angle in carbon tetrachloride CCl_4 to be?

1.10 Write formulas for the following groups of atoms.

a) methyl CH_3- **b)** hydroxyl $R-OH$ **c)** amino $R-NH_2$ **d)** sulfhydryl $R-SH$

1.11 Give the formula and name of an organic compound related to methane and each of the following inorganic compounds.

a) water   **b)** hydrogen bromide   **c)** ammonia   **d)** hydrogen sulfide

1.12 Name the class to which each compound in Problem 1.11 belongs.

The following questions require a logical extension of the information contained in this chapter.

1.13 a) Write the formula of a compound in which methyl groups have replaced both hydrogens of water. CH_3-O-CH_3 dimethyl ether

b) Write the formula of a compound in which methyl groups have replaced two hydrogens of ammonia. $R-N\langle {CH_3 \atop CH_3}$

1.14 a) Indicate the polarity of bonds in each of the following structures by the use of arrows.

i)

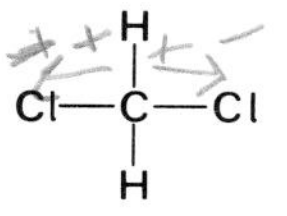

Cl—C—Cl with H above and H below

dichloromethane

ii)

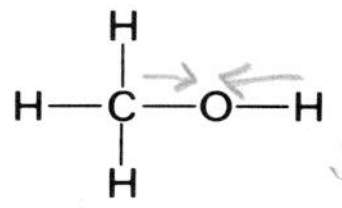

H—C—O—H with H above and H below

methanol

iii)

H—C—O—C—H with H atoms

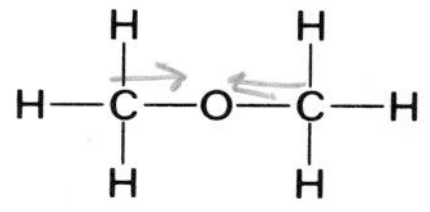

dimethyl ether

b) Would you expect (i) and (iii) above to have a dipole moment other than zero? Justify your answer.

1.15 Depict the hydrogen bonding between two methanol molecules and between one methanol molecule and one water molecule.

CH_2-O-H

$O-H-H_2C$

$N'-CH_3 \atop CH_3$

CH_2OH

CH_2OH

CH_2OH ... $O-H$ ... $H-C-O-H$ ... $O-H-H$ ...

CHAPTER 2

Structural Theory Applied to Saturated Hydrocarbons and Their Derivatives

The structural theory of organic chemistry in its original form was based on a concept little more complicated than that of balls and sticks. Balls, which symbolized atoms, were held together in molecules by sticks, which symbolized bonds. Balls that stood for carbon carried four sticks, those for hydrogen or halogen, one stick, those for oxygen, two sticks, and those for nitrogen, three sticks.

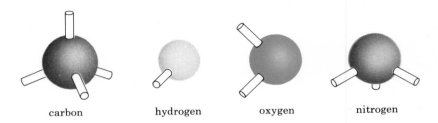

carbon   hydrogen   oxygen   nitrogen

In what were termed **saturated** molecules, only one stick joined two balls. In **unsaturated** molecules, two balls of the appropriate type could be bound together by two and occasionally even three sticks. In the ball-and-stick models of saturated compounds, the sticks were attached to carbon in the unique manner that placed them as far from one another as possible (tetrahedral arrangement).

This remarkably simple, homespun model for organic compounds correlated structural information gathered for thousands of organic compounds. Imaginative scientists, using only this model, predicted the structures of tens of thousands of compounds that might be prepared, many of which were later either synthesized or found in nature.

The structural theory in this rudimentary form served the science well for years before the electron was discovered, and for many more years before it was realized that the bonds were actually electron pairs. Never in the history of science have so few correlated and predicted so much with so little. Although the two-electron bond gave great depth and breadth to the structural theory, the primitive ball-and-stick model still stares up from any page on which a structural formula is written.

As newcomers to the world of organic molecules, you are somewhat in the same position as the early chemists in discovering the structures of carbon compounds. For many working chemists and for you, a ball-and-stick model of an organic molecule is a reasonable representation of the molecule, one which can be held in your hand and manipulated. Above all the model provides a concrete image that can be brought to mind when the compound is discussed. **Organic chemistry is a visual art**. The ball-and-stick model still depicts the three-dimensional quality of the molecule, the order in which the atoms are bonded together, and the bond angles. It does, however, *misrepresent* the distance between nuclei and the sizes of the atoms in relation to their bonds—a failure which the more recent and refined space-filling models are designed to overcome.

2.1 HYDROCARBONS SIMILAR TO METHANE

Previously, in Section 1.7, we saw that the methyl group, methane minus one hydrogen, could join other groups of atoms, such as the hydroxyl group —OH, to form many different classes of compounds. In a formal way, two methyl groups can be joined by a carbon–carbon bond to make a new hydrocarbon, called **ethane** (Section 1.8).

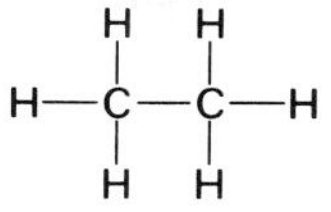

$$\begin{array}{c} \quad\ \ \, H\quad\ H \\ \ \ \ \ | \quad\ \ | \\ H-C-C-H \\ \ \ \ \ | \quad\ \ | \\ \quad\ \ H\quad\ H \end{array}$$

ethane

The structure of ethane, as represented above, shows clearly the order of linkage of the two carbons and the six hydrogens in the molecule as well as the four bonds for each carbon. The two- and three-dimensional structural formulas and both models of ethane are shown below.

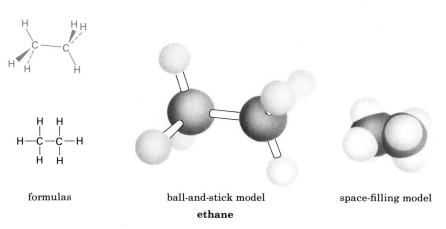

formulas   ball-and-stick model   space-filling model
ethane

Methane and ethane are the two smallest representatives of a series of hydrocarbons called **alkanes**. The molecular formulas of the first six members of the alkane series are listed below.

CH_4 (methane) C_3H_8 (propane) C_5H_{12} (pentane)
C_2H_6 (ethane) C_4H_{10} (butane) C_6H_{14} (hexane)

In each of these hydrocarbons the carbon atoms are bonded to other carbon atoms to form a continuous chain of carbons. The remaining bonding sites for each carbon are occupied by hydrogens. In our study of larger compounds, we will observe the great importance and necessity for knowing the *sequence of atoms* in the structure—exactly what atoms are bonded to what other atoms.

The three-dimensional structures and models for propane and butane show clearly the zigzag nature of the carbon chain, with the C—C—C bond angles maintained at 109.5° (Fig. 2.1).

(handwritten margin notes)
meth–1
eth–2
pro–3
but–4
pent–5
hex–6
hept–7
oct–8
non–9
dec–10

C—C – 109.5°

FIGURE 2.1
STRUCTURE OF SOME
SMALL ALKANES

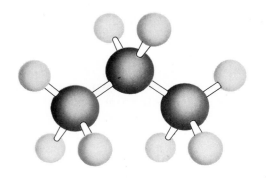

methane ethane

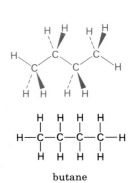

propane

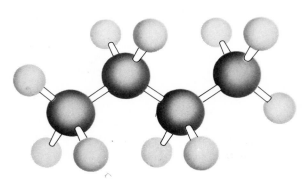

models of propane

butane

models of butane

In the next section, when we introduce isomers, it will become obvious why structural formulas of compounds are always given, and why most molecular formulas, such as C_5H_{12}, are *never* used alone. In order to save space and time, the structure of a compound is usually written linearly as a detailed or a condensed formula, as illustrated below for pentane.

condensed formulas

$CH_3-CH_2-CH_2-CH_2-CH_3$ $CH_3CH_2CH_2CH_2CH_3$

structural formula
showing all bonds

structural formula
showing C—C bonds only

structural formula
showing no bonds

two types of condensed formulas

In the **condensed formulas** some or all of the covalent bonds are understood to be present without being indicated. The same information, however, is conveyed in these shorthand notations as in the detailed structural formula.

Like ordinary chains, chains of carbon atoms are not always stretched out to the fullest extent. They coil in a variety of directions, but each arrangement keeps the bond angles and bond lengths constant.

$$
\begin{array}{ccc}
& \text{CH}_2 \quad \text{CH}_2 & \\
\text{H}_3\text{C} \quad \text{CH}_2 \quad \text{CH}_3 & \quad \text{H}_3\text{C} \quad \text{CH}_2 \ \text{CH}_2 \ \text{CH}_2 \ \text{CH}_3 & \quad \text{H}_3\text{C} \quad \text{CH}_2 \ \text{CH}_2 \ \text{H}_3\text{C—CH}_2
\end{array}
$$

pentane

These representations of pentane all show the same sequence of atomic linkages. All three formulas contain two terminal carbons, each bonded to one carbon and three hydrogens. They also contain three internal carbons, each bonded to two carbons and two hydrogens. When you write the condensed formulas for pentane, you should keep in mind the compound's three-dimensional nature.

The names and structures of the alkane hydrocarbons containing up to ten carbons are given in Fig. 2.2. These names and structures should be learned immediately, because they provide the basis for the names and structures of many other compounds. Note that after the first four alkanes, the names of the compounds use a counting system to tell you the number of carbons—**pent** = five, **hex** = six, etc.—**pentane** (5), **hexane** (6), **heptane** (7), **octane** (8), **nonane** (9), and **decane** (10).

The general formula for the alkane series is C_nH_{2n+2}. Alkanes are called *saturated* hydrocarbons, because they have the maximum number of hydrogens attached to the carbons in the chain.

By dropping one hydrogen from the end of an alkane chain, we obtained what is called an **alkyl group** (Section 1.7). The name of the alkyl group is derived from the name of the alkane by substituting the ending "yl" in place of "ane."

methyl group   ethyl group   propyl group

These groups appear over and over again as the hydrocarbon parts of the compounds discussed in future chapters. The names and structures of these alkyl groups, given in Fig. 2.2, are as familiar to organic chemists as the words "house," "father," or "book" are to people learning the English language.

FIGURE 2.2 THE ALKANE HYDROCARBONS C_1 THROUGH C_{10}

Molecular formula	Name	Structural formula	Condensed formula	Alkyl group
CH_4	Methane	H—C—H (with H above and below C)	CH_4	CH_3— methyl
C_2H_6	Ethane	structural formula	CH_3CH_3	CH_3CH_2— ethyl
C_3H_8	Propane	structural formula	$CH_3CH_2CH_3$	$CH_3CH_2CH_2$— propyl
C_4H_{10}	Butane	structural formula	$CH_3CH_2CH_2CH_3$	$CH_3CH_2CH_2CH_2$— butyl
C_5H_{12}	Pentane	structural formula	$CH_3CH_2CH_2CH_2CH_3$	$CH_3CH_2CH_2CH_2CH_2$— pentyl
C_6H_{14}	Hexane	structural formula	$CH_3CH_2CH_2CH_2CH_2CH_3$	$CH_3(CH_2)_4CH_2$— hexyl
C_7H_{16}	Heptane	structural formula	$CH_3CH_2CH_2CH_2CH_2CH_2CH_3$ or $CH_3(CH_2)_5CH_3$	$CH_3(CH_2)_5CH_2$— heptyl
C_8H_{18}	Octane	structural formula	$CH_3(CH_2)_6CH_3$	$CH_3(CH_2)_6CH_2$— octyl
C_9H_{20}	Nonane	structural formula	$CH_3(CH_2)_7CH_3$	$CH_3(CH_2)_7CH_2$— nonyl
$C_{10}H_{22}$	Decane	structural formula	$CH_3(CH_2)_8CH_3$	$CH_3(CH_2)_8CH_2$— decyl

Problem **2.1** Give the molecular formulas for ethane, propane, butane, and pentane. By what number of carbons and hydrogens does each of these formulas differ from the one smaller and the one larger? Write the structural formula for each of these molecules showing all C—C and C—H bonds and then write the condensed formula for each.

2.2 STRUCTURAL ISOMERS

Two compounds with the molecular formula C_4H_{10} are known. One, **butane**, boils at 0°C. The second has a boiling point of −12° and is known as **isobutane** (or **methylpropane**). Thus the formula C_4H_{10} actually stands for two separate compounds. The difference in boiling point and other physical properties displayed by isobutane must be due to a different order of linkage of the atoms from that which we have already seen for butane.

A little reflection reveals that it is possible to have two arrangements of four carbons—a continuous chain or a branched chain of carbons.

$$CH_3-CH_2-CH_2-CH_3 \qquad CH_3-\underset{\underset{CH_3}{|}}{CH}-CH_3$$

butane
(continuous chain)   isobutane
(branched chain)

Upon examining these structures we see that the sequence of atoms in the two is different. Butane has two methyl groups (carbon bonded to three hydrogens and one carbon) and two internal carbons with bonds to two carbons and two hydrogens. Isobutane has three methyl groups and one internal carbon with bonds to three carbons and one hydrogen. A CH_2 group, carbon bonded to only two hydrogens, is known as a **methylene group**.

Two compounds which have the same molecular formula but have different structural formulas are called **isomers**. The isomers we have just considered are known as **structural** or **constitutional isomers**, because their atoms are arranged in different sequences. Structural isomers have physical properties that differ from one another. You will encounter many examples of isomers in your study of organic compounds.

Since isomers are distinct compounds, they must be assigned separate names. Notice that both the structures and the names of compounds are important for you to learn. Your ability to grasp the material of Chapter 3 will depend on your command of material in Chapters 1 and 2, and so on throughout this book.

Three compounds with the molecular formula C_5H_{12} are shown in Fig. 2.3.

There are two ways to distinguish structural isomers from each other: (1) by drawing the structures so that the order of the linkages is obviously different, as in Fig. 2.3, and (2) by giving them separate names. Pentane has its carbons in a continuous chain (the "straight-chain" isomer), while the other two isomers, methylbutane and

FIGURE 2.3
THREE ISOMERS WITH
THE FORMULA C_5H_{12}

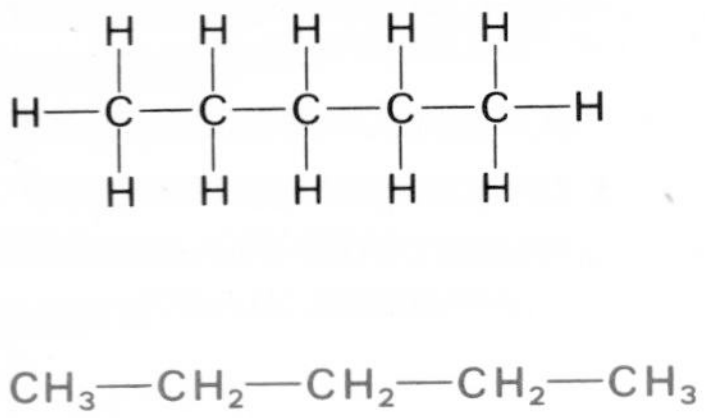

$$CH_3-CH_2-CH_2-CH_2-CH_3$$

pentane
bp 36°
2 methyl groups

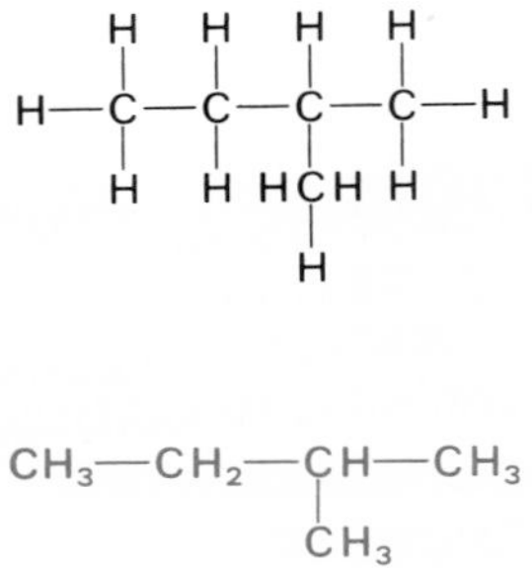

$$CH_3-CH_2-\underset{\underset{CH_3}{|}}{CH}-CH_3$$

methylbutane (isopentane)
bp 28°
3 methyl groups

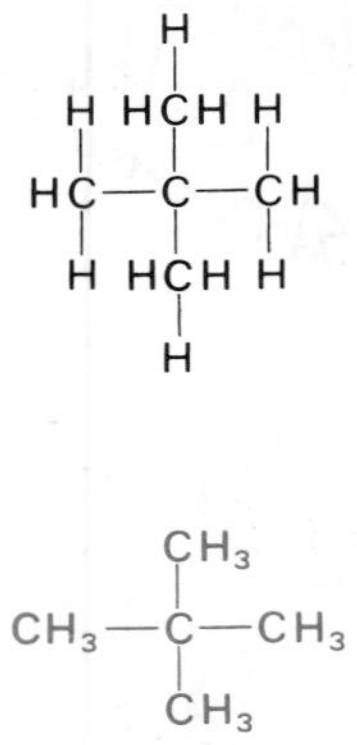

$$CH_3-\underset{\underset{CH_3}{\overset{\overset{CH_3}{|}}{|}}}{C}-CH_3$$

dimethyl propane (neopentane)
bp 9.5°
4 methyl groups and
1 carbon bonded to 4 carbons

dimethylpropane, have carbon side-chains attached to a main chain with fewer than five carbons (the "branched-chain" isomers).

Problems

2.2 Give the molecular formulas for hexane, octane, and decane. Write the condensed structures of these compounds. Check to make sure your formulas contain the appropriate number of bonds for each carbon and the appropriate number of hydrogens.

2.3 Write structures for the five isomers of C_6H_{14}. How many methyl groups does each isomer have?

2.3 SYSTEMATIC NOMENCLATURE FOR ALKANES—IUPAC

Although there are only three isomers of C_5H_{12}, there are five isomers of C_6H_{14}. As we continue to add CH_2 units to the molecule, we find the number of isomers increasing rapidly. The names can be quite awkward if we have to invent more and more prefixes to distinguish them. Obviously a different system of nomenclature is needed to simplify the task.

The most complete system of nomenclature for organic compounds is that devised by the International Union of Pure and Applied Chemistry, **IUPAC**. The aim is to have an international system in which each name applies clearly and unambiguously to one, and only one, compound. Knowledge of the rules allows a chemist from any country to translate formula into name or name into formula.

The IUPAC rules for naming alkanes are as follows.

1. Find the longest continuous carbon chain in the molecule. The name of the alkane with this number of carbons is the name of the parent compound.

2. Number the carbons of this longest chain starting at the end closest to a carbon having an attached group.

$$CH_3 \overset{③}{-} \overset{④}{CH} \overset{⑤}{-} \overset{⑥}{CH_2} - CH_2 - CH_3$$
$$\underset{①}{CH_3} - \underset{②}{CH_2}$$

3-methylhexane

3. Groups attached to the chain of the parent compound are called **substituents**. List the carbon substituents by their alkyl names (Figs. 2.2 and 2.7) in alphabetical order and add them as prefixes to the name of the parent alkane.

4. Assign each alkyl prefix the number of the carbon of the parent compound to which it is attached. If two alkyl groups have the same number, use the number twice. The number precedes the name of the alkyl group, and is separated by a hyphen from any letters preceding or following it.

$$CH_3 - CH_2$$
$$\underset{⑥}{CH_3} - \underset{⑤}{CH_2} - \underset{④}{CH} - \underset{③}{CH} - \underset{②}{CH_2} - \underset{①}{CH_3}$$
$$CH_3$$

3-ethyl-4-methylhexane

5. If two alkyl groups are identical, use the prefix di- before the alkyl name (tri = three, tetra = four), as in dimethyl. Give a number for each substituent. If two have the same number, repeat the number, using a comma between digits.

$$CH_3$$
$$\underset{①}{CH_3} - \underset{②}{C} - \underset{③}{CH_2} - \underset{④}{CH} - \underset{⑤}{CH_2} - \underset{⑥}{CH_2} - \underset{⑦}{CH_2} - \underset{⑧}{CH_2} - \underset{⑨}{CH_2} - \underset{⑩}{CH_3}$$
$$CH_3 \qquad CH_2 - CH_3$$

2,2-dimethyl-4-ethyldecane

In this example the three alkyl side chains are each given a number. Two methyl groups are indicated by the prefix di-. Substituents are placed in alphabetical order with a hyphen between letters and number and a comma between numbers.

Problems **2.4** Write structures for these compounds.

    a) 2,2-dimethylpentane   **b)** 4-ethyl-2-methylhexane
    c) 3-ethyl-4-propyloctane

2.5 Name the five C_6H_{14} isomers written in Problem 2.3.

2.4 CONFORMATIONS OF LARGE MOLECULES

Ethane and larger hydrocarbons, such as butane, contain two or more tetrahedral carbons. What is the spatial relationship of the atoms when there are so many in sequence?

$H-C=109.5°$

Imagine that we assemble the ethane molecule by attaching three hydrogens to each carbon and making one carbon–carbon bond. The bond angles are all approximately 109.5°. How are the hydrogens of carbon #1 arranged in relation to the hydrogens of carbon #2? A family of geometric arrangements can be imagined. Such arrangements of atoms are called the **conformations** of the molecules. Stable conformations are called **conformers**. The conformations of ethane are shown in Fig. 2.4.

FIGURE 2.4
CONFORMATIONS OF ETHANE

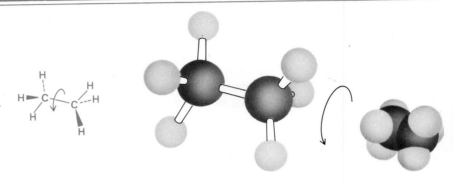

staggered conformation —stable, a conformer

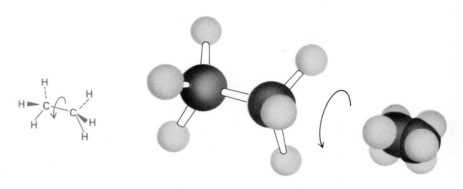

eclipsed conformation —unstable, not a conformer

Two arrangements which show hydrogens in extreme positions.

Another method of depicting molecular arrangements to emphasize the relationship of atoms attached to adjacent carbons is called a Newman projection. The molecule is viewed with one carbon directly behind the other carbon. The eye looks down the central C—C bond. Groups *a* attached to the front carbon are shown as radiating from a point of intersection. Groups *b* attached to the back carbon are shown as radiating from the circle.

A staggered conformation of ethane can be converted to an eclipsed conformation by turning one methyl group through 60° with respect to the other methyl, as indicated by the arrows in Fig. 2.4. A further turn through 60° brings it back to a staggered conformation again. Such a motion is referred to as **rotation** of atoms around the carbon–carbon bond. At room temperature the ethane molecule rotates easily through the staggered and eclipsed conformations.

Rotation of groups around a C—C bond is possible because the bond is composed of two coaxial overlapping sp^3 orbitals and is symmetrical around the line connecting the two nuclei. Remember that a bond which has this kind of symmetry is called a sigma bond (Section 1.5). Rotation around a sigma bond carries the molecule through a series of high-energy and low-energy conformations.

 [*The staggered conformation of ethane is the more stable, lower-energy conformation, because it places the electron pairs as far from one another as possible.*

The relationship of Newman projections to models is illustrated in Fig. 2.5 by views of ethane looking down the C—C bond.

In butane, the greater number of carbon–carbon bonds makes more conformations possible. Consider the C_2—C_3 bond and the possible

**FIGURE 2.5
ARRANGEMENT OF ATOMS
IN ETHANE (END-ON VIEW)**

Newman
projection

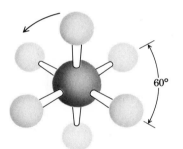

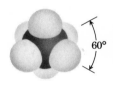

staggered conformation—lower energy

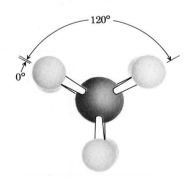

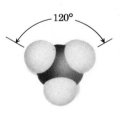

eclipsed conformation—higher energy

FIGURE 2.6 CONFORMATIONS OF BUTANE ABOUT THE C_2–C_3 BOND

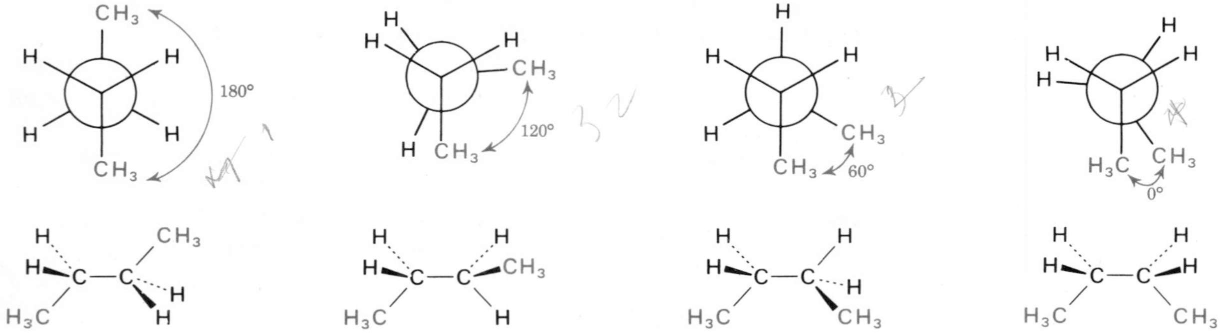

(a)
bonds staggered,
methyl groups as far
apart as possible

(b)
bonds eclipsed,
H and methyl
eclipsed

(c)
bonds staggered,
methyl groups close
together (gauche)

(d)
bonds eclipsed,
methyls eclipsed

arrangements of the attached methyl groups and hydrogens, which are pictured in Fig. 2.6. Since the methyl group occupies more space and has more repelling electrons than does hydrogen, we can understand why the conformation having the methyls staggered (180°) is the most stable one.

Problem **2.6 a)** On the basis of the repulsion by electrons between groups or atoms, how would you rank the conformations of butane shown in Fig. 2.6 for stability?

b) Draw the entire butane molecule in the most favored conformation for all bonds. The sketch should use dashed bonds for those behind the paper and wedged bonds for those above.

2.5 SOURCES AND USES OF ALKANES

The smaller alkanes, up to C_4, are gases at room temperature, those from C_5 to about C_{20} are liquids, and those above C_{20} are waxy solids. Paraffin wax is a mixture of solid alkanes.

All of the alkanes have a very low solubility in water. They are all lighter than water, floating on top in a mixture with water.

Natural gas and petroleum are the chief sources of alkanes at the present time. The principal component of natural gas is methane, but the other gaseous alkanes—ethane, propane, and butane—are also present.

Liquid petroleum is a complex mixture in which alkanes predominate. Distillation separates petroleum into its principal fractions: gasoline (C_6 through C_9, bp 40–200°); kerosene (C_8 through C_{14}, bp 175–325°); gas oil (C_{12} through C_{18}, bp above 275°); and lubricating oils and greases (above C_{18}).

The yield and quality of natural gasoline are improved by processes known as isomerization, cracking, and alkylation. Isomerization changes straight-chain alkanes into branched-chain hydrocarbons which make a better fuel. The "octane" rating for gasoline involves a scale on which 2,2,4-trimethylpentane ("isooctane") is the standard for good antiknock behavior (100) and heptane is the standard for poor fuel (0). Cracking breaks large alkanes into smaller ones, and alkylation builds larger and more branched alkanes from smaller ones.

Most of the alkanes produced commercially are used as fuels. They also serve as solvents for certain compounds because of their low cost.

The reaction of most organic compounds with molecular oxygen is an oxidation reaction which produces carbon dioxide and water with the evolution of large amounts of heat. The oxidation or burning of methane, propane, or butane furnishes heat for homes and appliances. Gasoline and kerosene are hydrocarbon mixtures whose combustion in engines is turned into work by expansion of the carbon dioxide and water vapor produced.

Combustion of hydrocarbons with molecular oxygen occurs only at high temperatures and is initiated by heat from a match or an ignition spark. Once ignited, the combustion of hydrocarbons is spontaneous, since the heat evolved during the reaction sustains the reactants at the high temperature necessary for the reaction to proceed. Excess heat evolved by the reaction is lost by the compounds to the surroundings. The quantity of heat lost by the compounds is called the **heat of reaction**, ΔH (the change in the heat, read "delta H"). The equation for the reaction of propane with oxygen is given as an example.

$$CH_3CH_2CH_3 + 5\,O_2 \longrightarrow 3\,CO_2 + 4\,H_2O \qquad \Delta H = -489\ \text{kcal}$$

The source of the heat change in the reaction is the difference in the stability of the bonds in the reactants and the products. In order to break a covalent bond and separate the two nuclei, energy must be supplied by heat or light. When two nuclei come together and a bond is formed between them, energy is released, usually in the form of heat. The amount of heat required or released depends upon the strengths or bond energies of the bonds involved. A few values of bond energies are given here. They represent the energy needed to break the bond, dividing the bonding electrons equally between the two atoms.

C—H	99 kcal/mole	C=O (in CO_2)	192 kcal/mole
C—C	83 kcal/mole	H—O	111 kcal/mole
O—O (O_2)	119 kcal/mole		

The difference in the sum of the bond energies of the reactants and of the products of a reaction defines the change in the heat content of the system (the heat of the reaction ΔH). This change in heat content, the **heat of combustion**, can be measured by the use of a calorimeter or calculated from bond energies.

The calculated heat of combustion of one mole of propane with 5 moles of oxygen to give 4 moles of water and 3 moles of carbon dioxide is shown below.

8 C—H bonds × 99 = 792	6 C=O bonds × 192 = 1152
2 C—C bonds × 83 = 166	8 H—O bonds × 111 = $\underline{888}$
5 O—O bonds × 119 = $\underline{595}$	2040 kcal
1553 kcal	

$$1553 - 2040 = -487 \text{ kcal/mole of propane}$$

Hydrocarbons are highly flammable in the presence of oxygen. Otherwise alkanes are inert to most chemical reagents. The saturated hydrocarbon portion of other classes of compounds is usually inert except for C—H and C—C bonds adjacent to functional groups.

Some small compounds containing halogens are nonflammable. Compounds of virtually all other classes burn readily, some explosively because of low ignition temperatures. Besides CO_2 and H_2O, the fully oxidized products of organic compounds containing sulfur or nitrogen are SO_3 or NO_2.

Biological oxidation is a controlled reaction which allows the energy given up to be used to run other biological reactions. The stepwise oxidations lead ultimately to CO_2 and H_2O, though many products are formed in between. The body actions such as brain functions and muscle contractions (which are produced by reactions of organic compounds) obtain a supply of energy from these highly controlled oxidations. Some of these reactions will be discussed in later sections.

A few alkanes are found in small quantities in living organisms. Long-chain alkanes occur in the waxy leaves of evergreens, probably serving to reduce the loss of water. A rather exotic use of alkanes has been discovered as the chemical sex attractant of a species of mushroom fly. The fluid secreted by the female fly was found to be a mixture of unbranched alkanes, from C_{15} to C_{30}, with heptadecane $C_{17}H_{36}$ as the most active compound. Such chemicals which serve as communication between individual insects are called **pheromones**. Other examples are given in later chapters.

Problem **2.7 a)** Write the equation for the combustion of butane with oxygen.
b) Calculate the heat of combustion for one mole of butane using bond energies.

2.6 SIMPLE NAMES FOR SMALL ALKYL GROUPS

Alkyl groups containing three and four carbons are frequently encountered both in the structures and names of many organic compounds. The names of the simple alkyl groups given in Fig. 2.7 should be learned at this time. These groups are all derived by removing a hydrogen from a parent hydrocarbon. The general symbol for any alkyl substituent is R—.

Note that the two propyl groups shown in Fig. 2.7 are isomeric, as are the four butyl groups. By removal of a hydrogen from a methyl group of

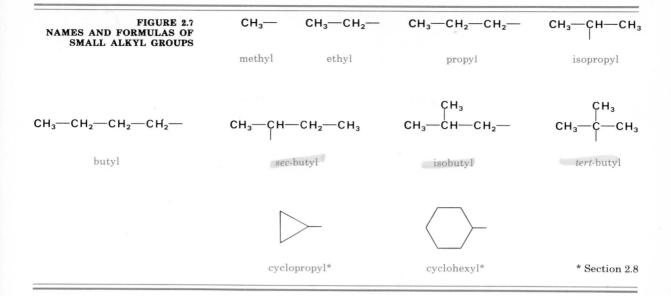

FIGURE 2.7
NAMES AND FORMULAS OF
SMALL ALKYL GROUPS

CH₃—   CH₃—CH₂—   CH₃—CH₂—CH₂—   CH₃—CH—CH₃

methyl   ethyl   propyl   isopropyl

CH₃—CH₂—CH₂—CH₂—   CH₃—CH—CH₂—CH₃   CH₃—CH—CH₂—   CH₃—C—CH₃

butyl   sec-butyl   isobutyl   tert-butyl

cyclopropyl*   cyclohexyl*   * Section 2.8

propane, the propyl group is derived. The isopropyl group involves the removal of a hydrogen from the central carbon of propane.

The tertiary (*tert*-) prefix refers to an alkyl group whose position of attachment is at a carbon that is already bonded to three other carbons. A secondary (*sec*-) prefix is applied to a carbon already bonded to two other carbons, as in *sec*-butyl. A primary carbon has the attachment to a carbon that is bonded to not more than one other carbon. No prefix is needed for unbranched primary groups, but sometimes *n*, for "normal," is used. The prefix *iso* is ambiguous, for isobutyl is a primary group while isopropyl is a secondary group.

Problems

2.8 Specify how each of the butyl groups is obtained by the removal of a hydrogen from a parent alkane.

2.9 Draw the structures of these compounds.

a) 4-*tert*-butyloctane   **b)** 3-isopropylheptane   **c)** 2,2-dimethylpentane

2.10 Give the names of these compounds.

a) CH₃—CH₂—CH₂—CH—CH₂—CH—CH₃
            | |
            CH₃—C—CH₃   CH₂—CH₃
            |
            CH₃

b) CH₃—CH—CH—CH₂—CH₂—CH₃
            | |
            H₃C CH₂—CH₂—CH₃

c) CH₃—CH—CH₂—CH₂—CH—CH₃
            | |
            CH₃   CH₃

Next to the C—C and C—H bonds, the C—O bond occurs most frequently in organic compounds. Methanol and ethanol, the simplest examples of compounds containing the C—O bond, were introduced in the first chapter. These alcohols are called derivatives of the alkanes methane and ethane, because the hydrocarbon portions of the two molecules are the methyl and ethyl groups respectively. They also contain the hydroxyl group —OH, which is the characteristic functional group for alcohols.

CH₃—OH   CH₃—CH₂—OH

methanol   ethanol

In a two-carbon compound the —OH group may be attached to either carbon to give the same substance, ethanol. Compounds containing the hydroxyl group may also be derived from larger alkanes, and in these cases attachment of the —OH to different carbons may produce isomeric structures. For example, let us look at propanol, which has three carbons.

Where is the —OH group attached in propanol? How many isomeric alcohols are there with the formula C_3H_8O? You can decide by numbering the carbons.

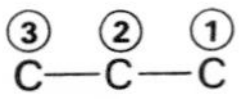

③ ② ①
C—C—C

If the —OH is attached to carbon #2, is that compound different from the one with the —OH attached to carbon #1? Test the atomic sequence after you have filled in all the other bonds with hydrogens. The structures you get depict different compounds. The names 1-propanol and 2-propanol designate the carbon to which the oxygen is bonded.

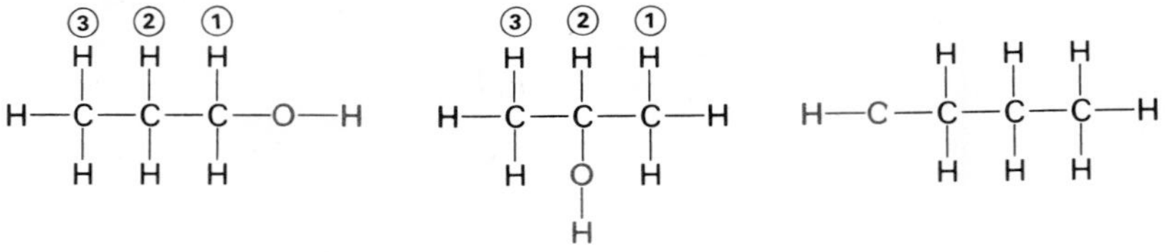

1-propanol
mp −127°, bp 97°

2-propanol
mp −89°, bp 84°

3-propanol (?)

Examine the structure of 3-propanol. Is this compound distinct from the others? To decide whether 3-propanol is possible, examine the sequences of the atoms of all three structures. The third structure is identical to 1-propanol. The carbons should be numbered to give the lowest number in the name.

Problem

2.11 Draw the structures for the four butanols. This can be done best by using the structures of both butane and methylpropane (isobutane) as parent compounds. Replace one H by an —OH to form an alcohol, but do not duplicate structures. There are *only* four butanols.

Other derivatives of alkanes—alkyl halides, amines, and thiols—were shown in Section 1.7. Like alcohols, the alkyl halides, amines, and thiols having three or more carbons have constitutional isomers. Just as there are two propanols, there are two chloropropanes, two amino ($-NH_2$) and two sulfhydryl ($-SH$) derivatives of propane. There are four butane derivatives for each class of compound. (See Problem 2.11.)

The common names for the three- and four-carbon derivatives are most often used. These names are formed from the names of the alkyl groups to which the functional group is attached. These alkyl groups are named in the preceding section. The common names and formulas of some of these alkane derivatives are shown in Fig. 2.8.

FIGURE 2.8
FORMULAS AND NAMES OF DERIVATIVES OF C₁ THROUGH C₄ ALKANES

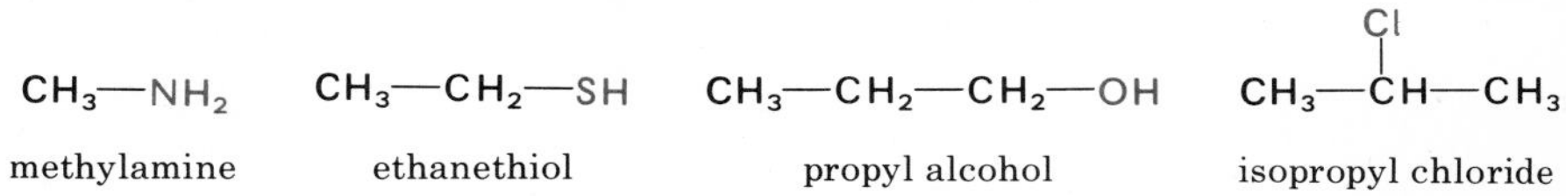

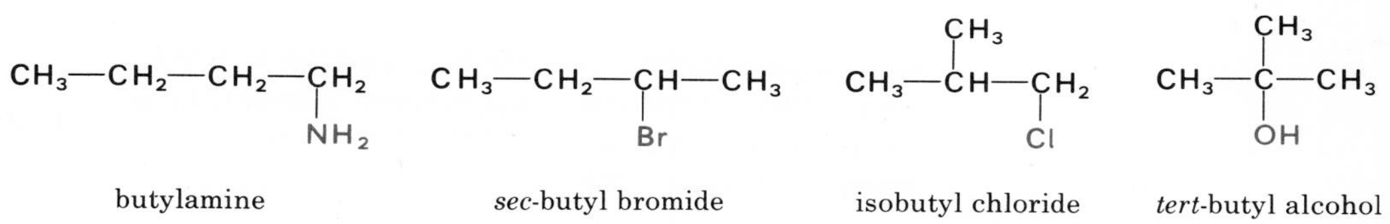

CH_3-NH_2
methylamine

CH_3-CH_2-SH
ethanethiol

$CH_3-CH_2-CH_2-OH$
propyl alcohol

$CH_3-\overset{\overset{\displaystyle Cl}{|}}{CH}-CH_3$
isopropyl chloride

$CH_3-CH_2-CH_2-\underset{\underset{\displaystyle NH_2}{|}}{CH_2}$
butylamine

$CH_3-CH_2-\underset{\underset{\displaystyle Br}{|}}{CH}-CH_3$
sec-butyl bromide

$CH_3-\overset{\overset{\displaystyle CH_3}{|}}{CH}-\underset{\underset{\displaystyle Cl}{|}}{CH_2}$
isobutyl chloride

$CH_3-\overset{\overset{\displaystyle CH_3}{|}}{\underset{\underset{\displaystyle OH}{|}}{C}}-CH_3$
tert-butyl alcohol

Alcohols are designated as primary, secondary, or tertiary alcohols depending on the state of substitution of the carbon to which the $-OH$ is attached. If the $-OH$ is attached to a tertiary carbon, a carbon bonded to three other carbons (Section 2.6), then the alcohol is called a tertiary alcohol.

Problem **2.12** For all the alcohols containing three or four carbons, specify which are primary, secondary, or tertiary.

While only one alcohol has the formula C_2H_6O, another sequence of atoms is possible. If the two carbons are both bonded to the oxygen, as in C—O—C, and the hydrogens are divided between the two carbons, a new class of compound called an **ether** is formed. In an ether both of the hydrogens of water have been replaced by hydrocarbon groups. Dimethyl ether (shown below) and ethanol are constitutional or structural isomers.

CH_3-O-CH_3

dimethyl ether

Problem **2.13 a)** Draw the structure of the ether which is isomeric with the two propanols.
b) Draw the structures of three ether isomers of $C_4H_{10}O$.

The order of linkages of atoms is clearly important and fundamental to the structural theory of organic chemistry. The organic chemist *always* writes the structures of the compounds between which he wishes to differentiate. The task of distinguishing between isomers is sometimes too great to be left to names alone.

2.8 CYCLIC HYDRO-CARBONS

If carbon can form a long chain of carbons, can the chain also be closed into a ring? **Cyclohexane** C_6H_{12} has such a ring structure and meets all the requirements of four bonds for each carbon.

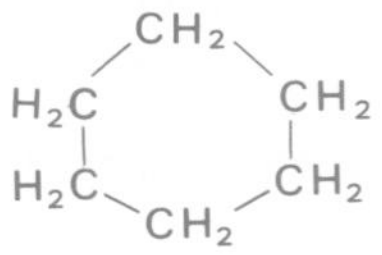

cyclohexane
bp 81°

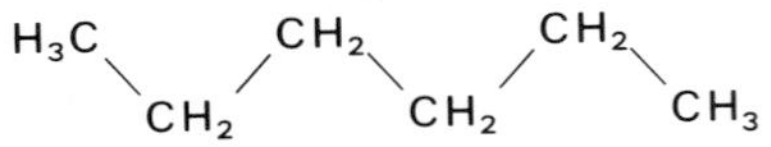

hexane
bp 68°

Cyclohexane, however, is not a simple alkane, since its formula C_6H_{12} does not fit the general formula for alkanes (C_nH_{2n+2}). Cyclohexane belongs to another series of hydrocarbons called **cycloalkanes**, whose general formula is C_nH_{2n}. The properties of the cycloalkanes both resemble and in some important respects differ from those of the alkanes, which are *open-chain* compounds. The bonds for the two missing hydrogens are accounted for by the additional C—C bond, since cyclohexane has six C—C and hexane only five C—C bonds.

All simple ring compounds containing three to thirty ring carbons are known. Structures of the four smallest rings are shown in Fig. 2.9.

FIGURE 2.9
CYCLOALKANES

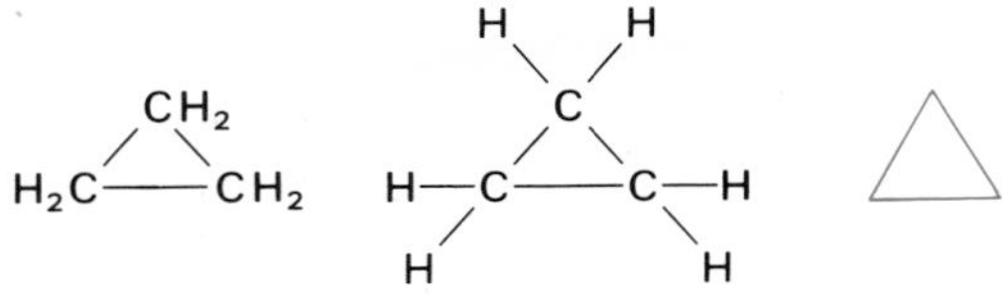

cyclopropane

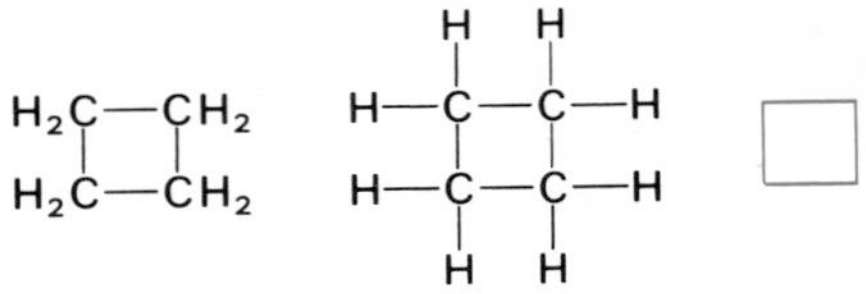

In these line representations each vertex of the figure stands for a CH$_2$ group.

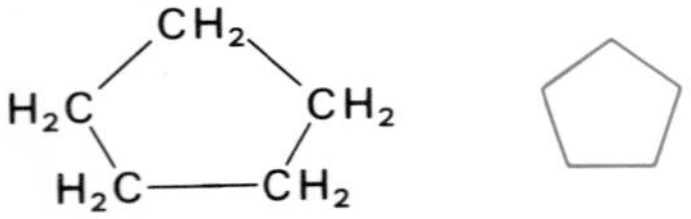

cyclobutane

cyclopentane

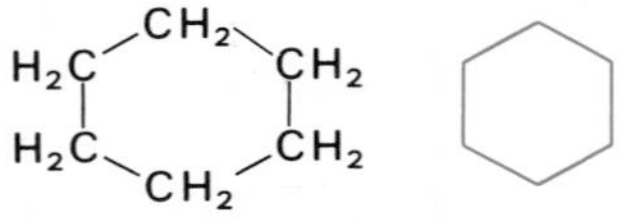

cyclohexane

The nomenclature for cycloalkanes follows the general rules for alkanes. The ring is the parent structure. The carbons of the ring are numbered starting with a carbon having a substituent and proceeding in the direction of the nearest second substituent. The following compounds illustrate the system.

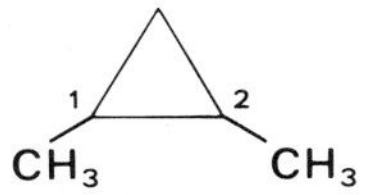

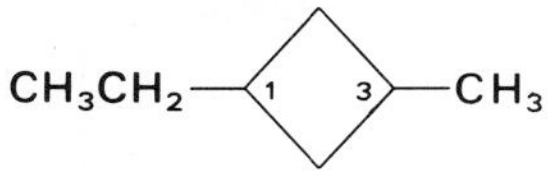

1,2-dimethylcyclopropane   1-ethyl-3-methylcyclobutane

Problem   **2.14** Name these compounds.

a)

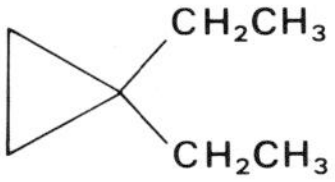

11 di ethyl cyclopropane

b)

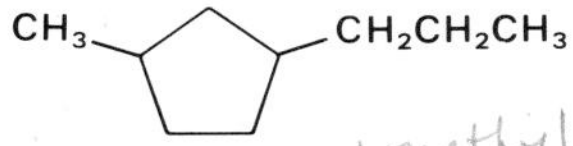

1 methyl 3 propyl cyclopentane

c)

1 butyl, 1,4 dimethyl cyclohexane

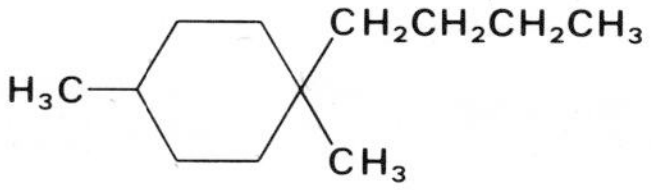

2.9 ISOMERISM IN CYCLIC COMPOUNDS

How many structural isomers of dimethylcyclopentane can you draw?

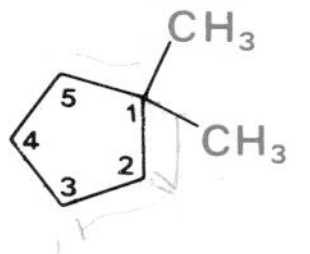

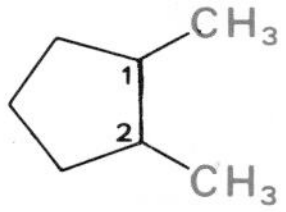

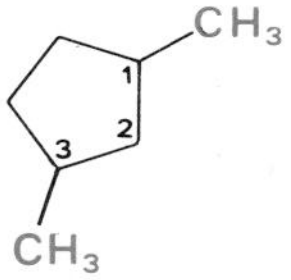

1,1-dimethylcyclopentane   1,2-dimethylcyclopentane   1,3-dimethylcyclopentane

There are three carbons on which the methyl groups can be distributed without duplicating structures.

The three-dimensional character of cycloalkanes provides the basis for a new question of possible isomerism. When two alkyl groups are attached to the ring on separate carbons, they can be on the same side of the ring or on opposite sides. In Fig. 2.10, why do the two right-hand

FIGURE 2.10
***CIS-TRANS* ISOMERS OF 1,2-DIMETHYLCYCLOPENTANE**

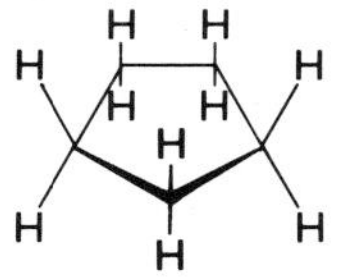

cyclopentane

five hydrogens above the ring, five below

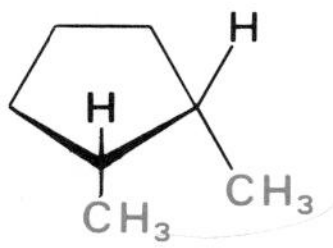

cis-1,2-dimethylcyclopentane
mp −50°, bp 130°
two methyl groups on same side of ring

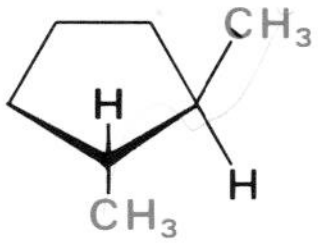

trans-1,2-dimethylcyclopentane
mp −89°, bp 124°
two methyl groups on opposite sides of ring

structures represent two isomers of 1,2-dimethylcyclopentane, and not merely two different conformations of the same compound?

To answer the question of isomers versus conformers one must know whether an atom (H) attached to an sp^3 carbon can pass through the plane of the carbon and two attached groups (A and D). In other words, can carbon turn inside out?

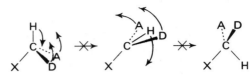

The answer is that bond angles cannot be distorted more than a few degrees without incurring a large amount of strain. The motion required to have one group pass through the plane defined by carbon and the other two groups is not possible without breaking bonds. Consequently the two dimethylcyclopentane structures are not interconvertible as conformations, but represent separate compounds.

Isomers which differ from each other only in the geometry of the molecule and not in the order of linkage of the atoms are called **geometric isomers**. They also are called *cis-trans* isomers. To distinguish between the two compounds, the isomer with substituents on the same side of the ring is designated by the prefix *cis* (meaning "on the same side"). The prefix *trans* (meaning "across from") indicates the two methyls are on opposite sides of the ring. A more specific method of naming such isomers with no ambiguity is given in Section 4.2.

Thus there are three structural isomers of dimethylcyclopentanes. In addition, 1,2-dimethylcyclopentane and 1,3-dimethylcyclopentane also exist as geometric isomers (*cis* and *trans* forms) for a total of five isomers.

Problem **2.15 a)** Draw the structures of *cis*- and *trans*-1-ethyl-2-propylcyclopentane.
b) How many dimethylcyclobutanes are there? How many dimethylcyclopropanes?

Cyclopropane and cyclobutane cannot have the ordinary bond angles found in open-chain alkanes. With three carbons in the ring, cyclopropane must have a C—C—C angle of 60°. It is surprising that such a ring can be made and is stable enough to be easily isolated and manipulated. The molecule is under much strain due to the distortion in the bond angle. Cyclobutane, whose C—C—C angles are 90°, is also strained. Besides angle strain, small rings have considerable strain introduced by the forced eclipsing of C—C and C—H bonds.

Cycloalkanes with five or more carbons have C—C—C, C—C—H, and H—C—H bond angles of approximately 109°. Cyclopentane and cyclohexane are the most common ring structures. The carbons of cyclopentane lie close to one plane. Planarity does not produce angle strain, because a regular planar pentagon has angles of 108°.

2.10 CONFORMATIONS OF CYCLOHEXANE

Six-membered ring systems are very common in both natural products and important synthetic materials. Six-membered rings are easily constructed with ball-and-stick models which have normal carbon bond angles. In such models the six carbons do not lie in a single plane, but follow the usual zigzag pattern of a carbon chain around the ring. Should all carbon atoms of cyclohexane be forced into the same plane, the C—C—C angle would have to be 120°, an unnecessary spreading of the bond angles which would introduce structural strain. In the nonplanar cyclohexane ring (a puckered ring) 109° bond angles are maintained.

cyclohexane

Cyclohexane can pucker to give two types of conformations, as shown in Fig. 2.11. The chair form of cyclohexane has *no* eclipsed bonds,

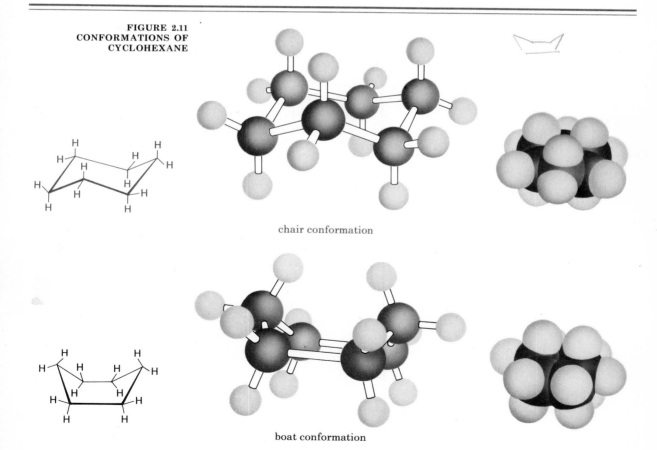

**FIGURE 2.11
CONFORMATIONS OF
CYCLOHEXANE**

chair conformation

boat conformation

Two conformations which show hydrogens at extreme positions.

and has the smallest atom and bond repulsions possible for the structure. All C—C bonds are gauche to each other. The boat form has both C—C and C—H bonds in eclipsed conformations and is less stable. The chair form is predominant in most substituted cyclohexanes.

In the chair conformation of cyclohexane the twelve hydrogens are distributed in two sets of six, which are geometrically different.

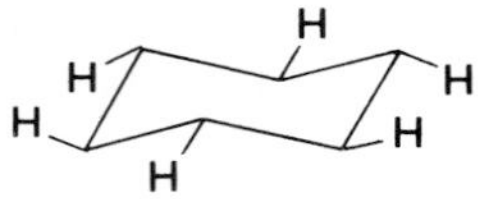

equatorial hydrogens
(bonds radiating outward from ring)

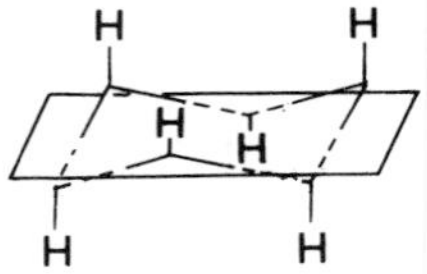

axial hydrogens
(bonds perpendicular to ring)

Cyclohexane interconverts between the two chair conformers. A simple flip of the model serves to interconvert one chair form of cyclohexane into the other by small rotations about all C—C bonds. Such a movement changes all the axial H's into equatorial H's, and all the equatorial H's into axial H's.

The two chair conformers are more easily seen with methylcyclohexane, which also flips rapidly. Although there are two conformers in the chair form of methylcyclohexane, one chair form is predominant over the other (Fig. 2.12). The three axial hydrogens on one side of the chair are quite close together. When a larger group like the methyl replaces one of these axial hydrogens, the atomic interactions are increased considerably. Consequently the conformation with methyl in the equatorial position is preferred. Interactions between nonbonded groups which are great enough to affect the conformation are called **steric hindrance**.

**FIGURE 2.12
TWO CHAIR CONFORMATIONS
OF METHYLCYCLOHEXANE**

minor repulsions
methyl in equatorial position
(more stable)

repulsions stronger
methyl in axial position
(less stable)

Problem **2.16** Draw structures of these compounds in a chair conformation. In drawing a chair conformation, make the opposite sides of the ring parallel, and the bonds to the axial positions parallel to each other.

    a) cyclohexane
    b) cis-1,3-dimethylcyclohexane (both methyls equatorial)
    c) cis-1,2-dimethylcyclohexane

New Terms and Topics

Saturated hydrocarbons—alkanes (Section 2.1)
Systematic nomenclature—IUPAC for alkanes (Section 2.3)
Structural isomers (Section 2.2)
Conformations of ethane, butane: staggered and eclipsed groups (Section 2.4)
Rotation about bonds (Section 2.4)
Primary, secondary, and tertiary alcohols (Section 2.7)
Bond energy (Section 2.5)

Heat of reaction; heat of combustion (Section 2.5)
Combustion of hydrocarbons (Section 2.5)
Pheromones (Section 2.5)
Cycloalkanes (Section 2.8)
Cis-trans isomerism of substituted cycloalkanes (Section 2.9)
Chair conformation of cyclohexane; equatorial and axial positions (Section 2.10)
Steric hindrance (Section 2.10)

Summary of Reactions

Combustion of Alkanes (Section 2.5)

alkane + O_2 $\longrightarrow$ CO_2 + H_2O

CH_3—CH_2—CH_3 + 5 O_2 $\longrightarrow$ 3 CO_2 + 4 H_2O
   propane

Problems

2.17 Name the following compounds by IUPAC nomenclature.

a) 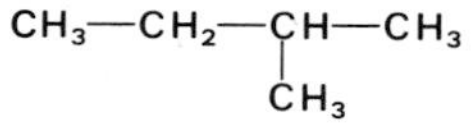 CH_3—CH_2—$\overset{\underset{\displaystyle CH_3}{|}}{CH}$—$CH_3$

b) CH_3—CH_2—OH

c) 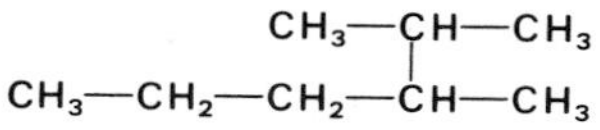 CH_3—CH_2—CH_2—$\overset{\underset{\displaystyle CH_3}{|}}{CH}$—$CH_3$ CH_3—CH—CH_3

d) 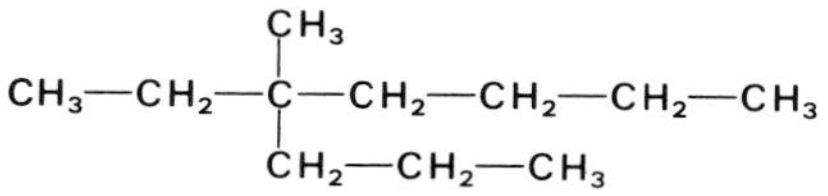 CH_3—CH_2—$\overset{\underset{\displaystyle CH_2—CH_2—CH_3}{|}}{\overset{\displaystyle |}{\underset{}{C}}}$—$CH_2$—$CH_2$—$CH_2$—$CH_3$ with CH_3 top

e) 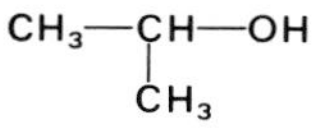 CH_3—$\overset{\underset{\displaystyle CH_3}{|}}{CH}$—OH

f) CH_3—CH_2—CH_2—CH_2—OH

g) 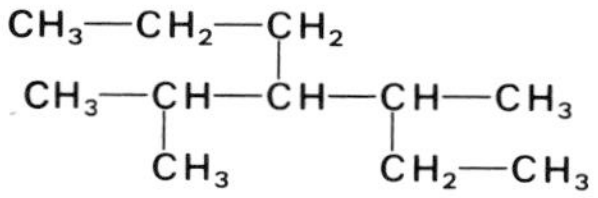 CH_3—CH_2—CH_2
   CH_3—CH—CH—CH—CH_3
        $|$ $|$
        CH_3 CH_2—CH_3

h)

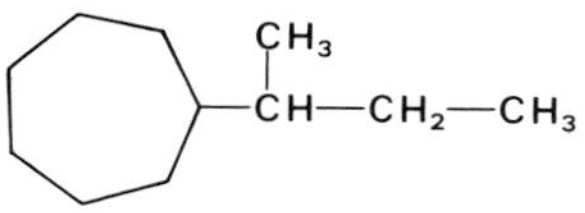

i)

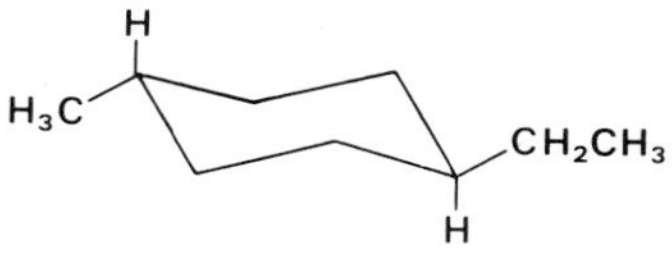

2.18 Draw structures for the following compounds.

   a) cyclopentane
   d) *sec*-butyl alcohol
   g) isobutyl chloride
   j) cyclobutane
   m) ethylamine

   b) 4-ethyl-2-methylhexane
   e) octane
   h) *tert*-butyl alcohol
   k) 4-propylheptane
   n) *cis*-1,3-dibromocyclohexane

   c) 2-pentanol
   f) 3,3-diethylpentane
   i) *cis*-1-ethyl-2-isopropylcyclopropane
   l) dimethyl ether

2.19 Using five carbons and the necessary number of hydrogens and other atoms, write a structure for each of the following.

a) an open-chain primary alcohol

b) an open-chain secondary alcohol

c) an open-chain tertiary alcohol

d) a cyclic secondary alcohol

e) a cyclic tertiary alcohol

f) a cyclic compound that is also a primary alcohol

2.20 For the formula $C_5H_{12}O$, draw the structures of eight alcohols and six ethers.

2.21 Specify which of the following alcohols are primary, secondary, and tertiary.

a) CH_3—CH_2—OH   ethanol

b) CH_3—$\underset{\underset{CH_3}{|}}{CH}$—OH   isopropyl alcohol

c) ▷—OH   cyclopropanol

d) CH_3—$\underset{\underset{CH_2-OH}{|}}{CH}$—$CH_3$ isobutyl alcohol

e) CH_3—$\underset{\underset{CH_3}{|}}{\overset{\overset{CH_3}{|}}{C}}$—$CH_2$—OH   2,2-dimethyl-1-propanol (neopentyl alcohol)

f) 1-methylcyclohexanol

2.22 There are seven isomeric dichlorocyclohexanes, including *cis-trans* isomers, having a molecular formula $C_6H_{10}Cl_2$. Draw formulas for these compounds using geometrical figures for the rings.

2.23 For the structures in each set, identify the structural isomers, the geometric isomers, and those structures that represent the same compound. (Some structures may not fit into any of these categories.)

Set 1

a) $CH_3CH_2\underset{\underset{CH_3}{|}}{CH}CH_3$

b) $CH_3CH_2CH_2CH_2CH_3$

c) $\underset{\underset{CH_2-CH_2}{|}}{CH_2}$—$\overset{}{CH}$—$CH_3$

Set 2

d) (cyclopropane with CH₃ and Cl)

e) (cyclopropane with Cl and CH₃)

f) (cyclopropane with Cl and CH₃)

g) (cyclopropane with CH₃ and Cl)

Set 3

h) $CH_3\underset{\underset{OH}{|}}{CH}CH_2\underset{\underset{CH_3}{|}}{CH}CH_3$

i) $CH_3\underset{\underset{CH_3}{|}}{CH}CH_2\underset{\underset{CH_3}{|}}{CH}$—OH

j) $CH_3\underset{\underset{OH}{|}}{\overset{\overset{CH_3}{|}}{C}}CH_2CH_2CH_3$

k) $CH_3\underset{\underset{OCH_3}{|}}{CH}CH_2CH_2CH_3$

Set 4

l) (cyclohexane with Cl and OH)

m) (cyclohexane with OH and Cl)

n) (cyclohexane with Cl and OH)

o) (cyclohexane with Cl and OH)

p) (oxane ring with Cl and CH₃)

2.24 a) Draw two chair conformations for *cis*-1,2-dichlorocyclohexane and two for *trans*-1,2-dichlorocyclohexane.

b) Which conformation in each pair would you expect to be the more stable?

2.25 These two compounds are isolated from plant materials and belong to a numerous and important group of natural products called terpenes. Other members of this group are natural rubber, camphor, and Vitamin A.

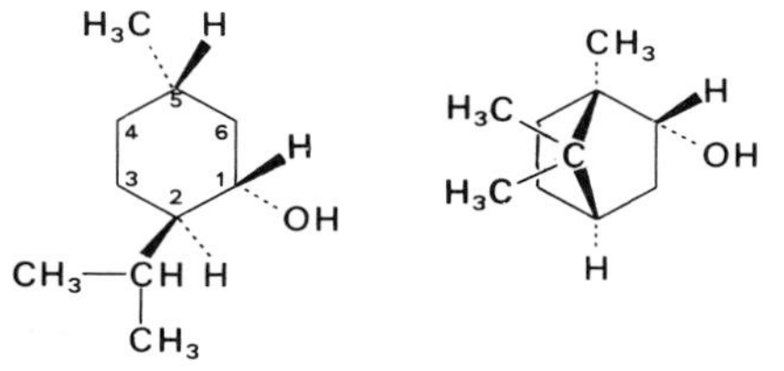

menthol   borneol

a) Give the molecular formula for each compound.
b) Are these molecules isomers of each other?
c) For menthol, describe the structure in terms of ring size, hydrocarbon substituents, functional group, and relationship of the three groups to each other on the ring.
d) What similar features do these molecules have? What differences?
e) Draw a *cis-trans* isomer of menthol.
f) Draw menthol using a chair conformation of cyclohexane.
g) Draw borneol using the boat conformation of cyclohexane.

2.26 Draw the atomic orbital overlap picture for each of the following.

a) methanol   **b)** fluoromethane

2.27 How many six-membered carbon ring isomers of the following structure are there? Draw them in the manner shown.

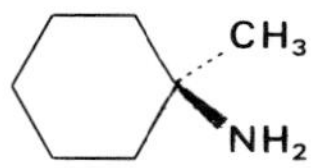

2.28 Inositol, a substance that is widespread in living material, is an important component of some phospholipids (Section 9.15). How many *cis-trans* isomers are there? Sketch them in the manner shown.

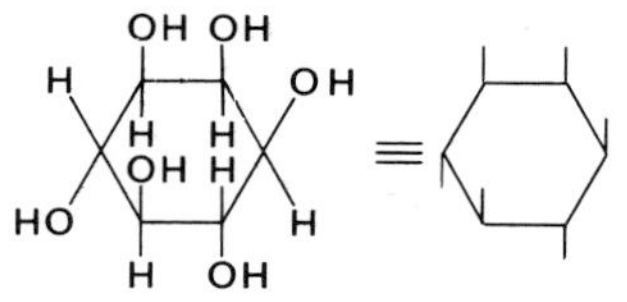

inositol

CHAPTER 3

Structural Theory Applied to Unsaturated Compounds

The primitive ball-and-stick model of the early structural theory also allowed the structures of "unsaturated" compounds to be visualized. For example, the structure of the hydrocarbon ethylene was deduced as follows. A weighed amount of this gas was burned to give carbon dioxide and water, which were also weighed. From these observed weights and the atomic weights of carbon, hydrogen, and oxygen was deduced the empirical formula CH_2. From the experimentally determined density of the gas, the molecular weight of ethylene was found to be about 28. Thus the substance possessed the molecular formula C_2H_4, with two fewer hydrogens than ethane C_2H_6.

It was also known that ethylene could be converted to ethane by adding one mole of hydrogen for each mole of ethylene in the presence of a catalyst, but that ethane resisted further addition of hydrogen, and was therefore "saturated." The theory required each carbon (ball) to carry four bonds (sticks), and each hydrogen one. A unique ball-and-stick structure for ethylene was provided by the idea that the two carbons of ethylene were held together not by one but by two bonds. Thus was born the idea of the carbon–carbon double bond (two balls held together by two sticks). Since a hydrogen molecule (two balls and one stick) could be added to ethylene, ethylene was said to be "unsaturated."

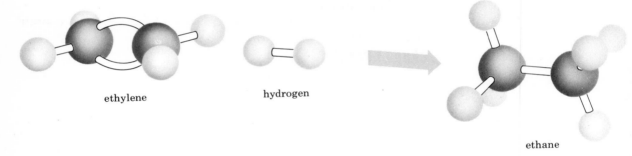

ethylene   hydrogen   ethane

The results of many similar experiments were interpreted in terms of multiple bonds between atoms. For example, compounds were found with carbon–carbon triple bonds C≡C, with carbon–oxygen double bonds C=O, and carbon–nitrogen double and triple bonds C=N and C≡N. When one mole of the compound reacted with just one mole of hydrogen H_2, researchers deduced the presence of a double bond. If two moles of hydrogen reacted, they deduced the presence of a triple bond (or perhaps two double bonds). A few examples of unsaturated compounds are shown which by addition of hydrogen formed compounds you have already encountered.

$$\text{H}-\text{C}\equiv\text{C}-\text{H} + 2\ H_2 \xrightarrow{\text{catalyst}} \text{H}-\underset{\underset{\text{H}}{|}}{\overset{\overset{\text{H}}{|}}{\text{C}}}-\underset{\underset{\text{H}}{|}}{\overset{\overset{\text{H}}{|}}{\text{C}}}-\text{H}$$

acetylene   ethane

$$\underset{\text{formaldehyde}}{\overset{\displaystyle H}{\underset{\displaystyle H}{>}}C{=}O + H_2} \xrightarrow{\text{catalyst}} \underset{\text{methanol}}{H{-}\overset{\displaystyle H}{\underset{\displaystyle H}{C}}{-}O{-}H}$$

$$\underset{\text{acetonitrile}}{H{-}\overset{\displaystyle H}{\underset{\displaystyle H}{C}}{-}C{\equiv}N + 2\,H_2} \xrightarrow{\text{catalyst}} \underset{\text{ethylamine}}{H{-}\overset{\displaystyle H}{\underset{\displaystyle H}{C}}{-}\overset{\displaystyle H}{\underset{\displaystyle H}{C}}{-}\overset{\displaystyle H}{N}{-}H}$$

In this chapter the structures of unsaturated compounds are presented. Besides the two series of hydrocarbons exemplified by ethylene ($H_2C{=}CH_2$) and acetylene ($H{-}C{\equiv}C{-}H$), the structures of aromatic hydrocarbons, such as benzene (C_6H_6) will be considered along with the unusual bonding which characterizes aromatic compounds. Finally, structures of the many classes of compounds containing $C{=}O$ double bonds will be introduced here.

3.1 UNSATURATED HYDROCARBONS— ALKENES AND ALKYNES

Alkenes and **alkynes** are two series of unsaturated hydrocarbons which, in contrast to the saturated alkanes and cycloalkanes, contain fewer than the maximum number of hydrogens per carbon.

The simplest alkene is **ethylene**, whose molecular formula is C_2H_4. Ethylene has two fewer hydrogens than its alkane counterpart ethane C_2H_6.

$$\underset{\displaystyle H}{\overset{\displaystyle H}{>}}C{=}C\underset{\displaystyle H}{\overset{\displaystyle H}{<}}$$

ethylene (common name)
ethene (IUPAC)

Each carbon in ethylene is bonded to one carbon and two hydrogens, and each carbon has one electron which in a saturated compound would be used in a bond to a hydrogen. However, in ethylene, the two carbons share the extra two electrons as well as those of the C—C bond. Thus in all, the two carbons share *four* electrons to form a carbon–carbon double bond.

The monoalkenes are defined as compounds that contain one double bond; their general formula is C_nH_{2n}.

In the systematic nomenclature (IUPAC) for alkenes, the "ane" ending of alkanes is replaced by "ene." This ending is attached to the stem name of the alkane having the same number of carbons as the *longest chain containing the double bond* in the alkene.

The carbon chain is numbered starting from the end closest to the double bond. Although the double bond involves two carbons, only the lower number is used to designate the position of the double bond.

These rules are illustrated by three alkenes—**1-hexene**, **2-hexene**, and **3-hexene**.

1-hexene

2-hexene

3-hexene

The stem **hex-** indicates a chain of six carbons, and the ending **ene** indicates one carbon–carbon double bond. The position of the double bond is given by the number of the carbon where it begins. These three hexenes provide another example of structural or constitutional isomerism, which was introduced in the previous chapter. Note that the hexenes are also isomeric with cyclohexane, but the double bond gives the alkenes very different chemical properties from those of cyclohexane. The nomenclature and chemical properties will be treated in detail in the chapter on alkenes (Chapter 4).

A hydrocarbon containing a carbon–carbon triple bond —C≡C— is a member of the **alkyne series**. The general formula for the alkynes is C_nH_{2n-2}. **Acetylene** C_2H_2 is the smallest member of this series.

$$H-C\equiv C-H \qquad\qquad CH_3-CH_2-C\equiv C-CH_3$$

acetylene (common name)   2-pentyne
ethyne (IUPAC)

The IUPAC ending for compounds containing a C≡C bond is "yne." The longest chain containing the triple bond becomes the stem name with "yne" added. The position of the triple bond is designated, as in alkenes, by the lower-numbered carbon of the C≡C bond.

For acetylene, ethylene, and the smallest representative of many other classes, common names are deeply entrenched in everyday usage and should be learned. For the larger compounds, IUPAC nomenclature is used, for the obvious reason that the systematic names and structures can be intertranslated with a knowledge of a few simple rules.

Other unsaturated hydrocarbons contain two or more multiple bonds (double or triple), and cyclic as well as open-chain compounds can contain double bonds. The terms *di-*, *tri-*, *tetra-*, etc. preceding *ene* refer to the number of double bonds. Of the many possibilities, these compounds are illustrative.

$$CH_2{=}CH-CH{=}CH_2 \qquad CH_2{=}CH-CH_2-CH{=}CH_2$$

1,3-butadiene   1,4-pentadiene

cyclohexene

1,3-cyclohexadiene

Problems   **3.1** Write structures for the following compounds.

    a) propene   **b)** 1-butene   **c)** 2-pentene

    d) cyclopentene   **e)** 1,3-pentadiene   **f)** 1,4-cyclohexadiene

3.2 Name the following compounds.

    a) $CH_3-CH=CH-CH_2-CH=CH-CH_2-CH=CH-CH_3$

    b) $CH_3-CH=CH-CH_3$ **c)** $CH_2=C=CH_2$ **d)**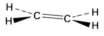

3.2 GEOMETRY OF UNSATURATED HYDROCARBONS

Carbon bonded to only three other atoms is no longer a tetrahedral carbon. Instead, the carbon and its three bonded atoms lie in the same plane, and the bond angles have been found to be approximately 120°.

Ethylene is a planar molecule; that is, all the nuclei lie in the same plane.

ethylene

The carbon and the two hydrogens attached to carbon #1 define one plane, which is also occupied by carbon #2 and its two attached hydrogens. The models of ethylene are shown in Fig. 3.1.

FIGURE 3.1 ETHYLENE

ethylene—side view

How do we describe ethylene in terms of the carbon atomic orbitals used for bonding? Carbon bonded to three other atoms is a trigonal carbon which uses three hybridized sp^2 orbitals and one p orbital to form the four carbon bonds. These three sp^2 orbitals are a hybridization of the one 2s orbital and the $2p_x$ and $2p_y$ orbitals, leaving the $2p_z$ unhybridized. The four carbon electrons occupy the three sp^2 orbitals and the one pure p orbital.

The axes of the three sp^2 orbitals lie in one plane and radiate from the central carbon atom, forming angles of 120°. The axis of the p orbital is perpendicular to the plane containing the axes of the hybrid orbitals as shown in Fig. 3.2.

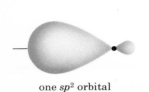

one *sp²* orbital

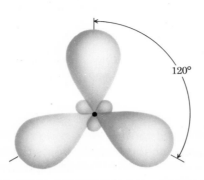

120°

three *sp²* orbitals lying in one plane
top view

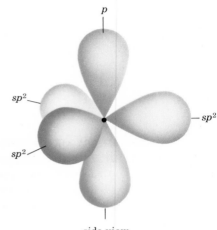

p

sp²

sp²

sp²

sp²

side view

Bonds between carbon and its three attached atoms are formed by overlap of the three *sp²* orbitals with orbitals of the attached atoms. The overlapped atomic orbitals thus form molecular orbitals which constitute sigma bonds.

overlapping atomic orbitals for σ bonds

molecular orbitals for σ bonds

ethylene

The orbital picture developed thus far does not provide for the planarity of ethylene, nor for the two $2p_z$ carbon orbitals, each containing one electron. The second bond of the carbon–carbon double bond is provided by overlap of these two p_z orbitals. The axis of each *p* orbital is perpendicular to the plane of the *sp²* orbitals. The *p* orbitals can overlap effectively *only when their axes are parallel to each other*. Overlap of the two *p* orbitals must occur side-to-side, since the axes are not directed toward each other for end-to-end overlap. No side-to-side overlap is possible when the two *p* orbital axes are perpendicular to each other. The atomic orbital overlap and the molecular orbital are depicted here.

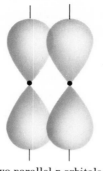 

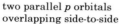

two parallel *p* orbitals
overlapping side-to-side

a π bonding
molecular orbital

The side-to-side overlap of *p* orbitals forms a **pi bond**. A pi bond is obviously not symmetrical about the line connecting the two nuclei and does not lie in the plane of the sigma bonds. In fact, the molecular orbital of the pi bond extends above and below the plane which the atoms and the sigma bonds occupy (Fig. 3.3).

**FIGURE 3.3
PI BONDING ORBITAL OF
ETHYLENE**

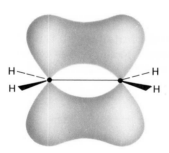

A result of the side-to-side overlap of *p*-orbitals in the pi bond is responsible for an important fact about ethylene, or any other compound containing a carbon–carbon double bond: at ordinary temperatures there is no rotation about this bond. A double bond is said to have *restricted rotation*.

The geometry of acetylene is as simple as the formula suggests. The four atoms lie in a straight line, $H-C\equiv C-H$.

What orbitals are needed for the carbons of acetylene? Once again hybridization is necessary for each carbon to form four bonds to two other atoms. There must be one sigma bond connecting carbon to carbon and a sigma bond connecting carbon to hydrogen. These sigma bonds use two hybrid *sp* orbitals of each carbon which are formed by mixing the 2s and $2p_x$ orbitals. The two *sp* orbitals extend in opposite directions along the *x*-axis with an angle of 180° between them.

One *sp* orbital extends
primarily in one direction.

Two *sp* orbitals extend in
opposite directions along
the same axis.

The carbon–carbon sigma bond is formed by end-to-end overlap of an *sp* orbital from each carbon, as shown below. The carbon–hydrogen bond involves overlap of the opposite *sp* orbital of carbon with the 1*s* orbital of hydrogen.

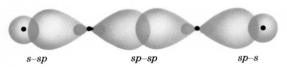

s–sp   *sp–sp*   *sp–s*

overlap of atomic orbitals

acetylene

molecular orbitals of σ bonds

The $2p_y$ and $2p_z$ orbitals are still lying on the *y* and *z* axes perpendicular to each other and to the line of the two *sp* orbitals. The second and third bonds of the triple bond are pi bonds formed by side-to-side overlap of parallel *p* orbitals of each carbon, as shown in Fig. 3.4.

FIGURE 3.4
PI BONDING ORBITALS OF
ACETYLENE

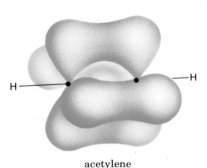

acetylene

The carbon–carbon distance is shorter for multiple bonds than for single bonds. In double bonds C=C it is 1.34 Å, and in triple bonds C≡C it is 1.20 Å, as compared with the single bond C—C of 1.54 Å.

3.3 STEREOISOMERISM OF ALKENES—*CIS-TRANS* ISOMERISM

For many alkenes, such as 1,2-dichloroethene, two structures may be drawn, one with the two chlorines on the same side of the C═C bond and one with the chlorines on opposite sides of the double bond.

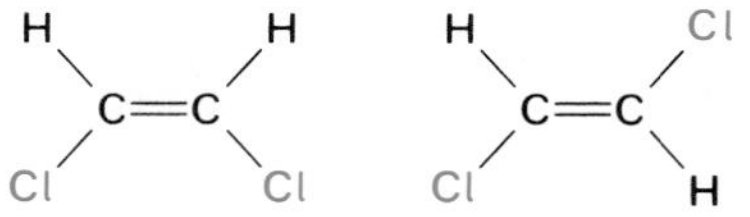

same side　　　　opposite sides

1,2-dichloroethene

The question arises whether these structures represent separate compounds or different conformations of the same compound. A significant feature of the bonding and geometry of alkene linkages is the restriction of rotation of attached groups around the C═C double bond (preceding section). Restricted rotation *prevents* the interconversion of the two 1,2-dichloroethenes into each other, without the breaking of the pi bond of the double bond. The two structures represent two isolable, stable compounds with different physical properties. The isomers are designated as *cis* (on the same side) and *trans* (on the other side).

cis-1,2-dichloroethene　　　　*trans*-1,2-dichloroethene

mp −80°, bp 60°　　　　mp −50°, bp 47°

These compounds provide another example of *cis-trans* isomerism, observed with substituted cycloalkanes in Section 2.9. *The simple requirement for cis-trans isomerism in alkenes is that each unsaturated carbon must have two different groups attached to it.* The groups on carbon #1 may be the same as or different from the groups attached to carbon #2, as in these pentenes.

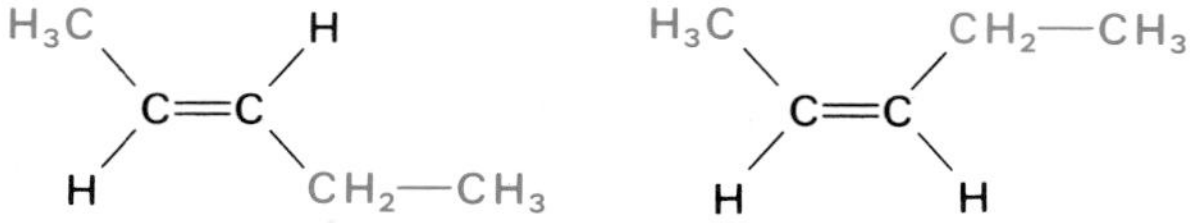

trans-2-pentene　　　　*cis*-2-pentene

If all four groups attached to the doubly bonded carbons are different, there are still two isomers, but for nomenclature purposes some further designation of the relationship of the groups is required. A method for such designation is given in Section 4.2.

Problem **3.3** Which of the following compounds have *cis-trans* isomers? Draw the structures of the isomeric compounds.

a) $CH_2{=}CH{-}CH_2{-}CH_3$ **b)** $CH_3{-}CH{=}CH{-}CH_3$

c) $CH_3{-}CH_2{-}CH{=}\underset{\underset{CH_3}{|}}{C}{-}CH_2{-}CH_3$ **d)** $CH_3{-}CH{=}CCl_2$

e) $CH_3{-}CH_2{-}CH{=}CH{-}CH{=}CH_2$

f) $CH_3{-}CH{=}CH{-}CH{=}CH{-}CH_2{-}CH_3$

Some of the spatial relationships of *cis-trans* isomers are demonstrated by an interesting and important pair of isomers, maleic and fumaric acids, shown in Fig. 3.5.

FIGURE 3.5
CIS-TRANS ISOMERS

maleic acid
(*cis*-butenedioic acid)
mp 130°

fumaric acid
(*trans*-butenedioic acid)
mp 286°

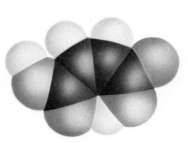

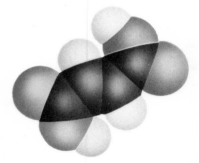

Space is limited for groups on the same side of the double bond. Some steric interference of atoms is evident.

Groups on opposite sides of the double bond are far apart.

In a long carbon chain containing a $C{=}C$ bond, the shape of the molecule as a whole depends on whether the chains attached to the double bond are *cis* or *trans* to one another. The isomeric 5-dodecenes of Fig. 3.6 illustrate this dependence.

FIGURE 3.6
GEOMETRY OF LARGE
MOLECULES CONTAINING
C=C IN CARBON CHAINS

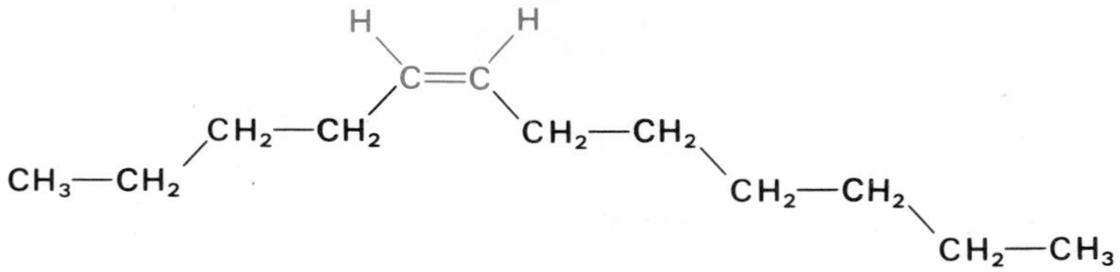

Cis C=C prevents the carbon chain
from becoming roughly linear.

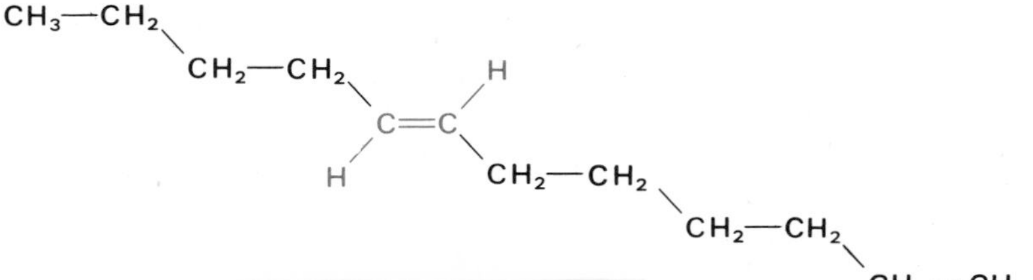

Trans C=C enables the molecule
to stretch out.

Two *cis* C=C

Smallest ring to accommodate a
trans double bond is eight.

Problem **3.4** Rings smaller than C_8 containing a *trans* C—CH=CH—C structural unit
have not been isolated. Give an explanation.

The lack of regularity in the shapes of alkenes and dienes causes them
to have slightly lower melting points than the corresponding saturated
compounds (Fig. 3.7).

FIGURE 3.7
COMPARATIVE MELTING
POINTS OF HYDROCARBONS

Compound	mp, °C	Compound	mp, °C
hexane	−95	*cis*-3-hexene	−135
		trans-3-hexene	−113

If the substituents on the carbon–carbon double bond are strongly polar, *cis* and *trans* isomers will have widely different dipole moments (Section 1.10). See Fig. 3.8.

FIGURE 3.8
DIPOLE MOMENTS DEPEND ON GEOMETRY OF MOLECULES

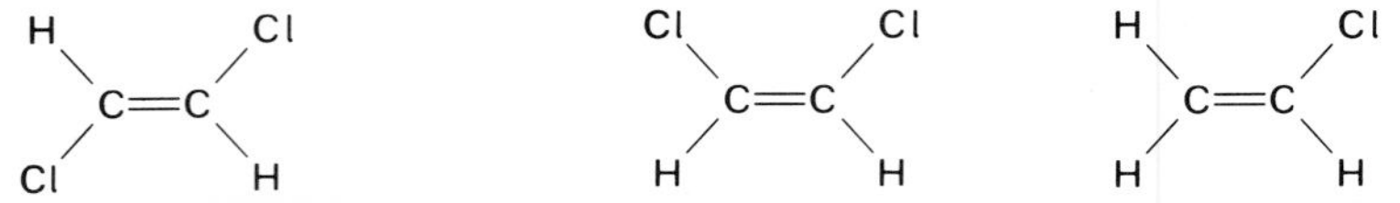

trans-1,2-dichloroethene
$\mu = 0$ units

cis-1,2-dichloroethene
$\mu = 1.9$ units

vinyl chloride
(chloroethene)
$\mu = 1.5$ units

Moments of C—Cl bonds
cancel each other.

Moments of C—Cl bonds
enhance each other.

Problem   **3.5** How do you expect the dipole moment of 1,1,2-trichloroethene to compare with the values for compounds in Fig. 3.8?

3.4 AROMATIC COMPOUNDS   A large number of organic compounds have a ring of six carbons containing three double bonds. This kind of structure is important enough to be distinguished from all the structures we have so far examined. It is called an **aromatic ring**. What is so important and special about this arrangement of six carbons and three double bonds? The structure of **benzene C_6H_6**, the simplest compound having this ring, is shown here.

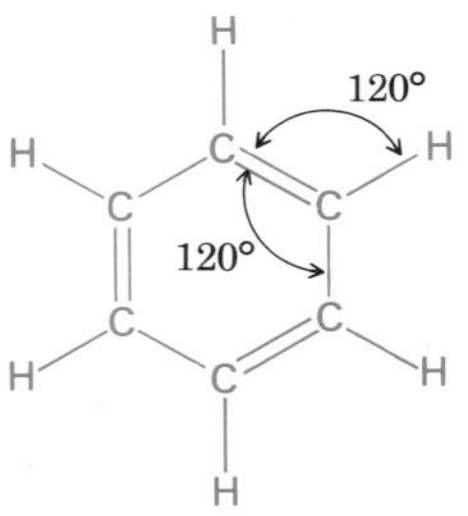

In this representation each vertex stands for a C—H.

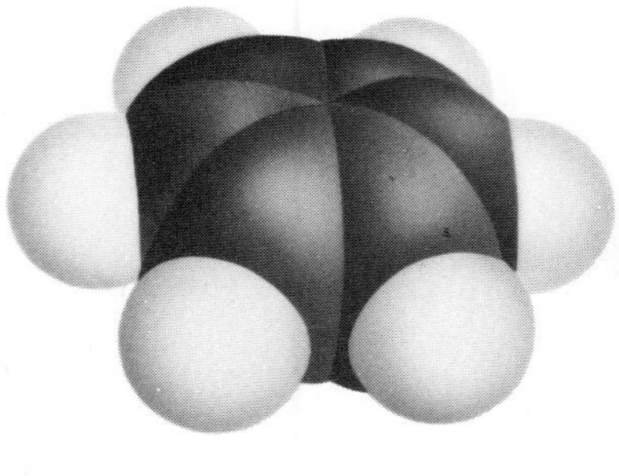

benzene

Each carbon in benzene is bonded to three atoms. The bond angles are the expected 120° for a trigonal carbon, and all six carbons and six hydrogens lie in one plane. The unusual fact about the aromatic ring in benzene is that it has the shape of a regular hexagon, with all bond lengths equal (1.39 Å). The structure we have shown should have three C—C bonds of 1.54 Å and three C=C bonds of 1.34 Å, the C=C bond length in the usual alkene.

How do we account for the unusual bond lengths in benzene? The answer lies in the disposition of the fourth orbital for each carbon, the *p* orbital containing one electron. Three sp^2 orbitals of each carbon form sigma bonds, all lying in the plane formed by the six carbon nuclei. One *p* orbital at each carbon extends above and below the carbon ring with its axis perpendicular to the ring. Side-to-side overlap of *p* orbitals on adjacent carbons forms pi bonds. Such an arrangement provides three double bonds. If you were constructing this molecule, would you form a double bond between carbons 1 and 2, 3 and 4, 5 and 6? Or would you use 2 and 3, 4 and 5, 6 and 1?

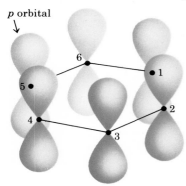
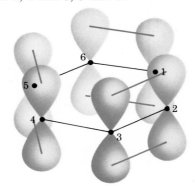
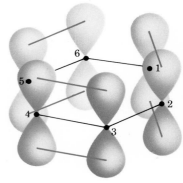

This ambiguity is associated with the unusual properties of benzene and other compounds containing the aromatic ring. Both ways of forming the double bonds are equally proper, and there is no means to choose between them. As a result, *each p orbital overlaps both of the adjacent p orbitals at the same time*. In this way no choice can or need be made. A pi bond molecular orbital for benzene is depicted in Fig. 3.9.

FIGURE 3.9
ONE PI BOND MOLECULAR
ORBITAL FOR BENZENE

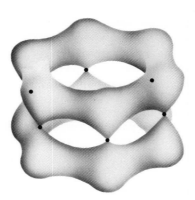

In the molecular pi orbitals for benzene, the electrons of the pi bonds are associated with more than two carbon nuclei—in fact with all six carbon nuclei at once. They are *delocalized electrons*, since they cannot be assigned to one specific bond between two specific atoms. The pi bonds also are *delocalized bonds*. The phenomenon of delocalized electrons is called **resonance**.

To depict completely this delocalization of bonds in benzene, two formulas are drawn. The actual structure of benzene is a **resonance hybrid** of these two structures, which are called **resonance structures**. The resonance structures are connected by a double-headed arrow to indicate that *together* they represent the compound benzene. The double-headed arrow ↔ must be carefully distinguished from two oppositely headed arrows ⇌ which indicate a reaction that goes in both directions—an equilibrium.

benzene   benzene

(one compound—two resonance structures)

When only one structure is written for benzene, the intention is that it stands for both. Sometimes a circle inside the hexagon is used to show the delocalization of bonds.

The equality of the carbon–carbon bond lengths of benzene provides the most compelling evidence supporting the delocalization of the pi electrons in benzene. The chemical properties of benzene differ from those of alkenes, and the differences are associated with this electron delocalization. These unique chemical properties are described in Chapter 12.

Delocalization of electrons or of charge stabilizes molecules and ions. A resonance hybrid has a lower energy than any of the individual resonance structures taken separately. *When delocalization is possible, it occurs.* In the next section we will see that a resonance hybrid imposes certain restrictions on the spatial arrangements of atoms and the freedom of rotation about bonds in molecules and ions.

Problem **3.6** Only one compound exists (no isomers) of 1,2-disubstituted benzenes. With 1,2-dibromobenzene as your example, write formulas for isomers which might be possible if the ring contained ordinary localized double and single bonds. Explain why these isomers don't exist.

As a substituent, the benzene ring minus one hydrogen C_6H_5— is called the **phenyl group**, as in **2-phenylbutane** (also called *sec*-butylbenzene).

2-phenylbutane

Examples of derivatives of benzene are **chlorobenzene, phenol** (hydroxybenzene), and **aniline** (aminobenzene). Benzene and all of these derivatives are called aromatic compounds. As a family of substituents they are known as aryl groups (as distinguished from alkyl groups). The general symbol for an aryl substituent is Ar—.

Problems   **3.7** Write two resonance structures for each of these compounds.

a) methylbenzene   b) 2-phenylbutane   c) chlorobenzene
d) phenol   e) aniline

3.8 Which of these molecular formulas might be associated with an aromatic compound?

a) C_8H_{12} b) C_8H_{10} c) $C_{10}H_{10}$

Draw a reasonable structure for those that are aromatic.

A compound whose structure closely resembles that of benzene is **pyridine**. Pyridine C_5H_5N has a six-membered ring, with three double bonds, in which one carbon–hydrogen unit of benzene has been replaced by a nitrogen. Like benzene, pyridine is a resonance hybrid.

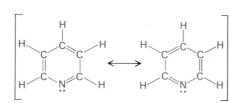

pyridine

The three double bonds of pyridine afford the same delocalization of electrons in this cyclic amine as was observed in benzene. The unshared electron pair on the nitrogen is *not part of the pi-system*, but is *localized on nitrogen* and occupies one of the three sp^2 orbitals and the same region of space that a C—H bond would. Like ammonia and saturated amines, pyridine is a base. Many derivatives of pyridine are found in compounds produced in nature.

3.5 CARBONATE AND GUANIDINIUM IONS

Many familiar ions and molecules have bonding arrangements that involve electron delocalization. The structure of the carbonate ion is a classic example. One carbon is attached to three oxygens by four pairs of bonding electrons. All atoms occupy the same plane, all three carbon-oxygen distances are equal, and all O—C—O bond angles are 120°. These facts indicate that the fourth bond to carbon is delocalized equally among the three oxygens, so that each carbon–oxygen bond has some pi bond character.

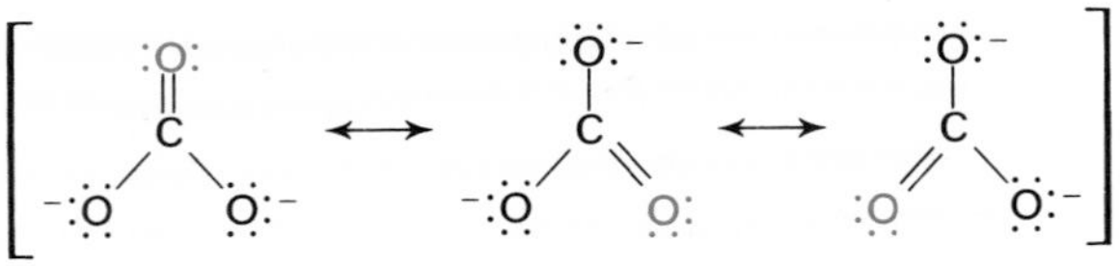

$$\left[\;\; \overset{\ddots\,\ddot{O}\,\ddots}{\underset{\,^{-}\!:\ddot{O}\ddot{O}:^{-}}{C}} \;\;\longleftrightarrow\;\; \overset{\ddots\,\ddot{O}:^{-}}{\underset{\,^{-}\!:\ddot{O}\ddot{O}:}{C}} \;\;\longleftrightarrow\;\; \overset{\ddots\,\ddot{O}:^{-}}{\underset{\,:\ddot{O}\ddot{O}:^{-}}{C}} \;\;\right]$$

carbonate ion

The two negative charges are also *delocalized* in such a way as to place an equal amount of negative charge on each of the three oxygen atoms. Thus each oxygen carries approximately two-thirds of a negative charge. These facts are symbolized by the three equivalent structures and the two double-headed arrows separating them, as shown above.

The delocalized electron pair of the carbonate ion occupies a molecular orbital which includes all four nuclei, as in this diagram.

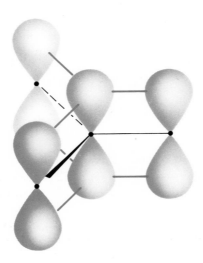

The guanidium ion, a positively charged resonance hybrid of importance in biological systems, illustrates the restrictions imposed by electron and charge delocalization.

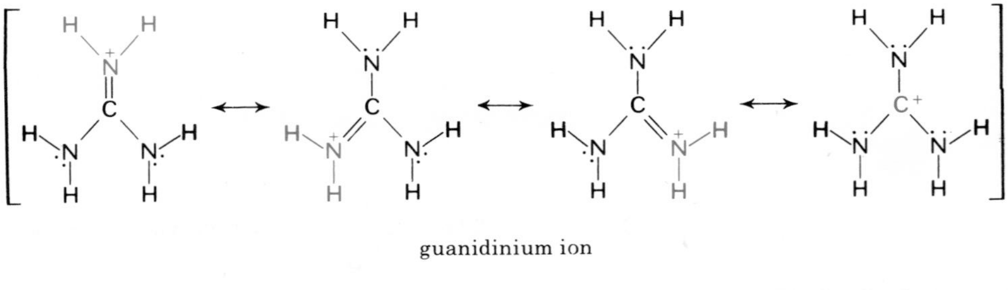

guanidinium ion

In the guanidinium ion the spatial relationships of the hydrogens, nitrogens, and carbon are very nearly what would be expected if all the C—N bonds were double (C=N). All of the carbon and nitrogen bond angles are approximately 120°, and the ion is planar. All of the C—N bonds are the same length—between that of a single and a double bond. All three of the nitrogen atoms carry the same amount of positive charge, which is also shared with the carbon.

One of the requirements and characteristics imposed on molecules or ions by resonance is that the bonded atoms and their attached atoms occupy fixed positions determined by the partial double-bond nature of the bonds. As in structures containing full double bonds, rotation about the bonds involved in delocalization is restricted and the bond angles approach the 120° value.

Problem **3.9** The following ions are resonance hybrids. Draw the resonance structures for each.

a) CH_3-C with $\overset{..}{O}:$ (double bond) and $\overset{..}{O}:^-$

b) pyrrole-type ring with $\overset{..}{N}^-$

c) $CH_3-CH=CH-\overset{+}{C}H_2$

d) $H_2N-\underset{\overset{||}{+NH_2}}{C}-NH-CH_2-CH_2-OH$

e) cyclopentadienyl ring with $\overset{..}{C}^-$ and H

3.6 CARBON–OXYGEN DOUBLE BONDS— ALDEHYDES AND KETONES

Carbon forms a double bond with oxygen and nitrogen and a triple bond with nitrogen. Compounds containing the C=O group, the **carbonyl group**, are among the most important organic compounds, both biologically and in the laboratory. They are divided into two large classes of compounds according to the type of atom Z which is attached to the carbon O=C—Z. The remaining bond to carbon is either to a hydrogen or to a carbon of a hydrocarbon group.

The two classes containing C=O are as follows.

1. Aldehydes, where Z = H, and **ketones**, where Z = C.
2. Carboxylic acids and acid derivatives, where Z = OH, OR, NH_2, Cl, or SR (see Section 3.7).

Aldehydes *always* have Z = H and the aldehyde functional group is —CH=O. Since three of the carbon bonds are accounted for in the aldehyde functional group, it must always be the beginning of the carbon chain, the #1 carbon, as shown in butanal.

$$CH_3-CH_2-CH_2-C\overset{O}{\underset{H}{\diagup}}$$

Aldehyde name = alkane name + *al* ending.

butanal

The two smallest aldehydes are known by their common names, **formaldehyde** $CH_2{=}O$ and **acetaldehyde** $CH_3{-}CH{=}O$.

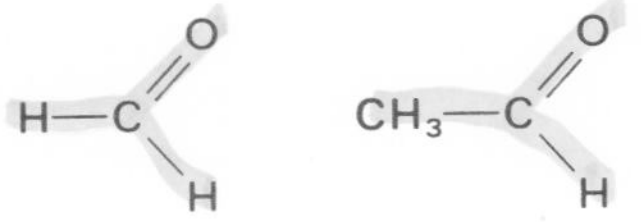

formaldehyde   acetaldehyde

Ketones have the ${=}O$ on an internal carbon as in **butanone**. In this case it is not necessary to write "2-butanone," since there is no other possible four-carbon ketone structure.

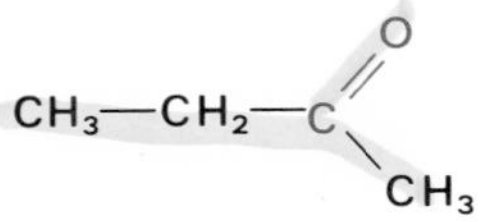

Ketone name = alkane name + *one* ending.

butanone

The simplest ketone, propanone, is better known by its common name **acetone**.

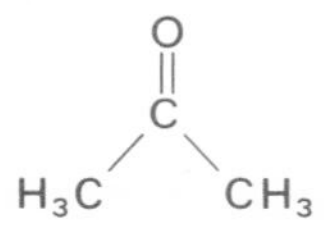

acetone

Problems   **3.10** What is the minimum number of carbons needed to produce open-chain isomeric ketones? Draw the structures of the three isomers.

3.11 Draw the structures for the four aldehydes which are isomers of the ketones in Problem 3.10.

3.7 THE CARBOXYLIC ACID FAMILY OF COMPOUNDS

Carboxylic acids contain the **carboxyl functional group** $-CO_2H$, as in the familiar compound **acetic acid**.

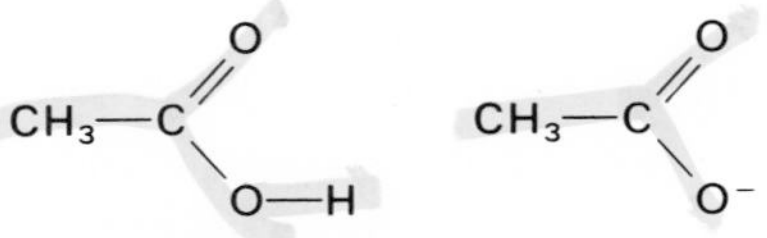

acetic acid   acetate ion

Acetic acid gives up a hydrogen ion, H^+, the one bonded to the oxygen, to a hydroxide ion. The **acetate ion** is left. Ions formed from carboxylic acids are called **carboxylate ions**.

Once again the two smallest carboxylic acids are known by their historical, common names, which reflect the origin of the acid. **Formic acid** was obtained from the distillate of red ants (L., *formica* = ant), and **acetic acid** is the essential constituent of vinegar (L., *acetum* = vinegar). Note that the common name of the aldehyde (previous section) is related to the common name of the acid.

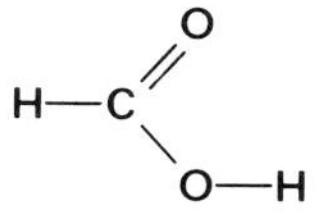

formic acid 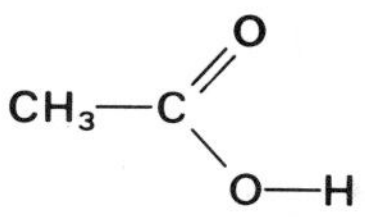   acetic acid

Butanoic acid is another example of carboxylic acid.

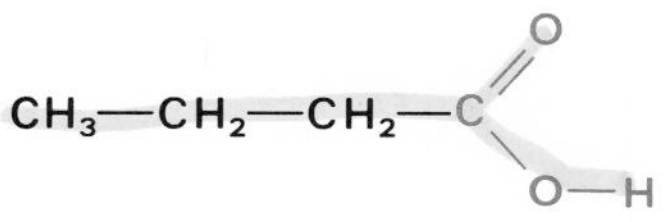

butanoic acid

Acid name = alkane name + *oic acid* ending.

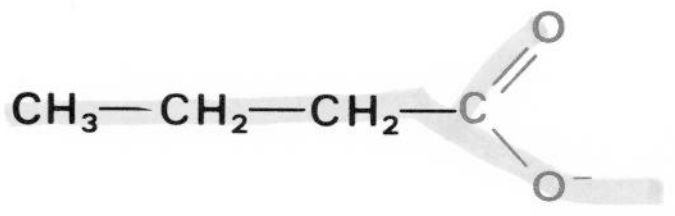

butanoate ion

Ion name = alkane name + *oate ion* ending.

The simple carboxyl functional group is converted into a **carboxylic ester functional group** by changing the —OH to —O—CH$_3$, or to any other —O—alkyl or —O—aryl group, as in **methyl butanoate**. An ester is a derivative of both the carboxylic acid and of the alcohol.

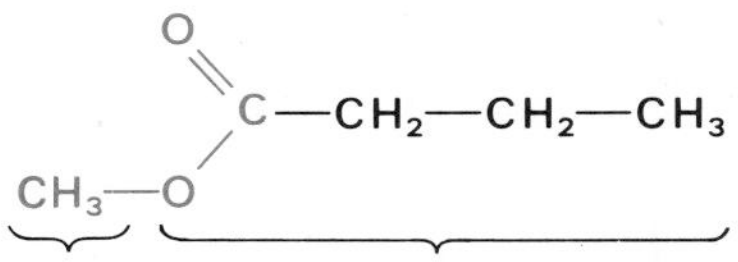

from alcohol   from acid

methyl butanoate

When the —OH of the acid is replaced by —NH$_2$, an **amide** is formed, as in **butanamide**.

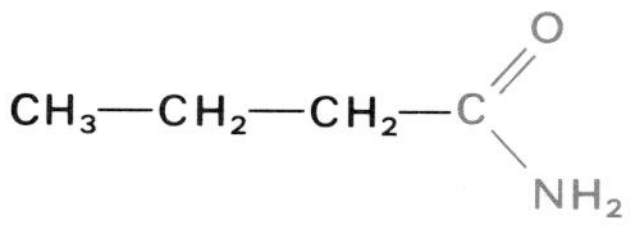

butanamide

Esters and amides are two examples of carboxylic acid derivatives which will be considered in detail in Chapters 8–10.

The classes of compounds containing the carbonyl group, C=O, are summarized in Fig. 3.10.

FIGURE 3.10 CLASSES OF COMPOUNDS CONTAINING THE CARBONYL GROUP, C=O

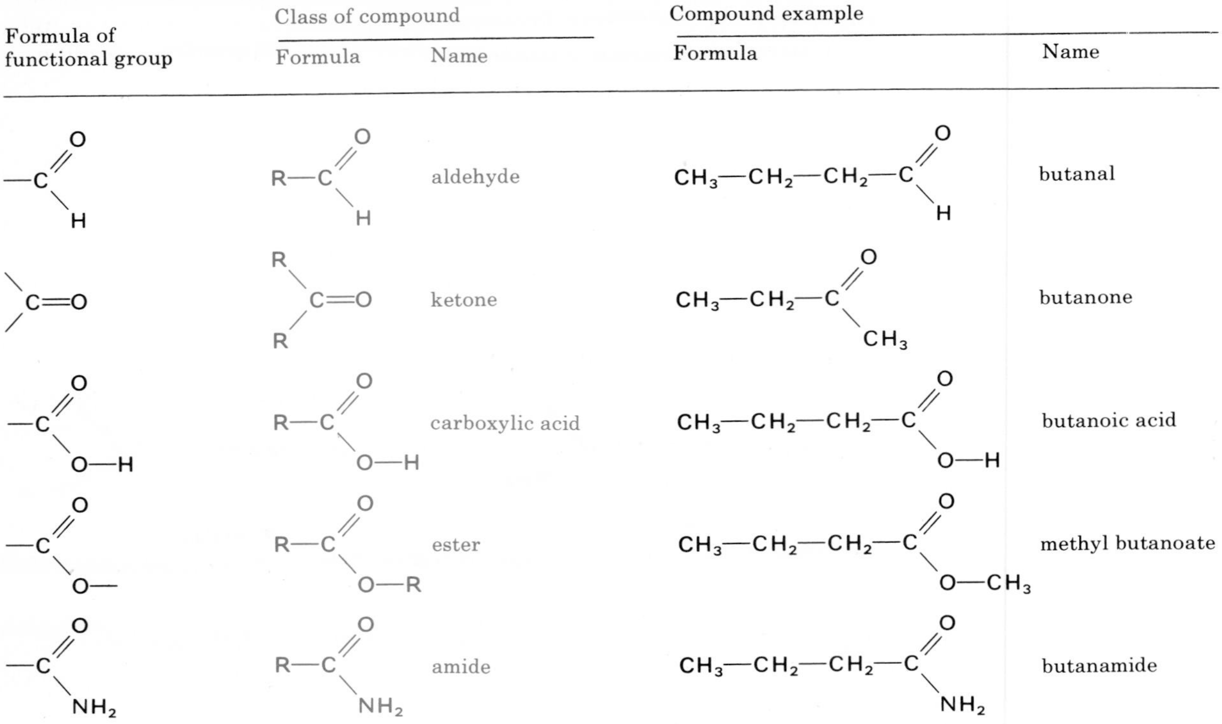

Formula of functional group	Class of compound		Compound example	
	Formula	Name	Formula	Name
$-C\overset{O}{\underset{H}{\big\Vert}}$	$R-C\overset{O}{\underset{H}{\big\Vert}}$	aldehyde	$CH_3-CH_2-CH_2-C\overset{O}{\underset{H}{\big\Vert}}$	butanal
$\overset{\diagdown}{\underset{\diagup}{C}}=O$	$\underset{R}{\overset{R}{\diagdown}}C=O$	ketone	$CH_3-CH_2-C\overset{O}{\underset{CH_3}{\big\Vert}}$	butanone
$-C\overset{O}{\underset{O-H}{\big\Vert}}$	$R-C\overset{O}{\underset{O-H}{\big\Vert}}$	carboxylic acid	$CH_3-CH_2-CH_2-C\overset{O}{\underset{O-H}{\big\Vert}}$	butanoic acid
$-C\overset{O}{\underset{O-}{\big\Vert}}$	$R-C\overset{O}{\underset{O-R}{\big\Vert}}$	ester	$CH_3-CH_2-CH_2-C\overset{O}{\underset{O-CH_3}{\big\Vert}}$	methyl butanoate
$-C\overset{O}{\underset{NH_2}{\big\Vert}}$	$R-C\overset{O}{\underset{NH_2}{\big\Vert}}$	amide	$CH_3-CH_2-CH_2-C\overset{O}{\underset{NH_2}{\big\Vert}}$	butanamide

Problems **3.12** Identify the class of compound for each of the following.

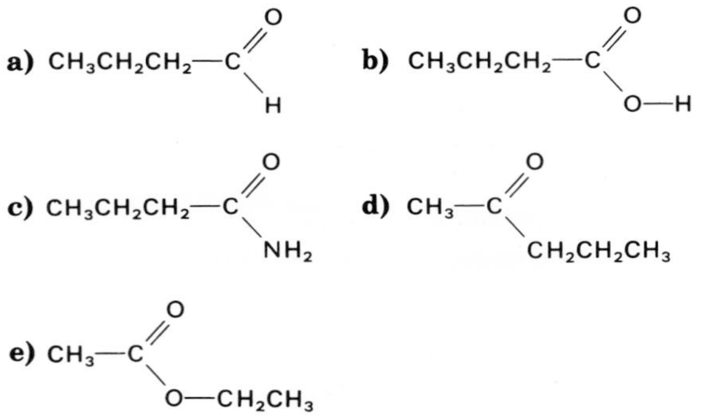

a) $CH_3CH_2CH_2-C\overset{O}{\underset{H}{\big\Vert}}$ b) $CH_3CH_2CH_2-C\overset{O}{\underset{O-H}{\big\Vert}}$

c) $CH_3CH_2CH_2-C\overset{O}{\underset{NH_2}{\big\Vert}}$ d) $CH_3-C\overset{O}{\underset{CH_2CH_2CH_3}{\big\Vert}}$

e) $CH_3-C\overset{O}{\underset{O-CH_2CH_3}{\big\Vert}}$

3.13 Write the structures for hexane, hexanal, 2-hexanone, hexanoic acid, sodium hexanoate, methyl hexanoate, and hexanamide.

3.8 ORGANIC ACIDS AND BASES The chemical reaction that everybody knows and uses is that of an acidic substance with a basic or alkaline substance. It is impossible to watch television without learning that acid in the stomach (actually

HCl) can be neutralized by sodium bicarbonate (in baking soda and "Alkaseltzer"), calcium carbonate ("Tums"), and dihydroxyaluminum sodium carbonate ("Rolaids"). The reactions most thoroughly covered in the first chemistry course are those involving the transfer of a proton from an acid to a base.

$$HCl + H_2O \longrightarrow H_3O^+ + Cl^-$$

acid   base

$$HCl + NH_3 \longrightarrow NH_4^+ + Cl^-$$

acid   base

The emphasis on acid-base reactions is well founded, for the transfer of a proton from one substance to another is without doubt the most important type of reaction. It occurs essentially instantaneously and is an integral part of nearly all biological reactions and of a large proportion of laboratory reactions.

It is important, then, that we be able to recognize acids and bases among the classes of organic compounds. Since the transfer of a proton is the theme of our story, *we define acids as those classes of compounds which donate a proton. Those classes which accept a proton are bases.*

Familiar examples of acidic substances are sulfuric acid, hydrogen chloride, hydronium ion, acetic acid, ammonium ion, and hydrogen sulfide.

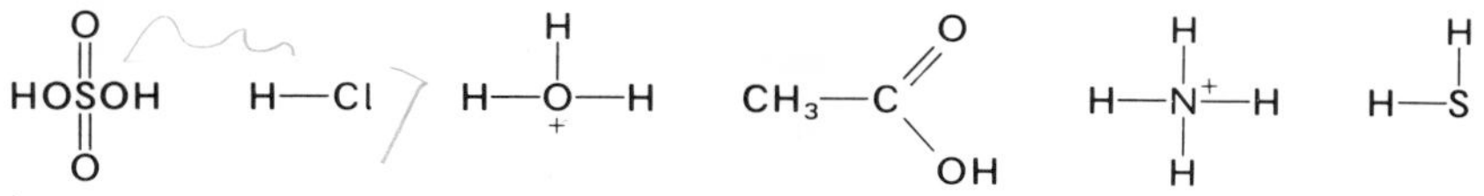

Acetic acid reacts with an aqueous solution of sodium hydroxide to form water and an aqueous solution of sodium acetate, a salt. Hydroxide ion is the base, the proton acceptor. Only the hydrogen attached to oxygen in the acetic acid is transferred, while those hydrogens attached to carbon in the methyl group remain unaffected.

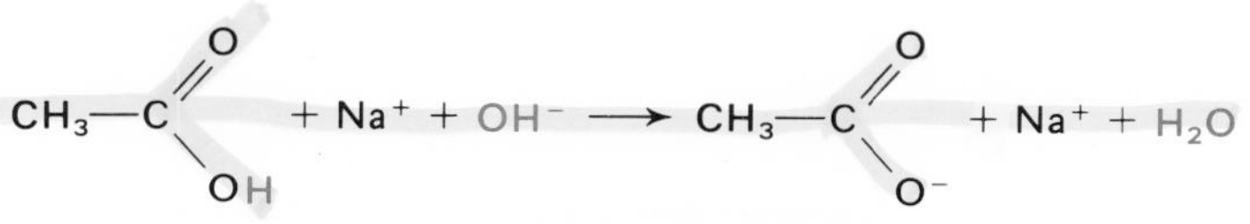

acetic acid   hydroxide ion   acetate ion   water
(acid)   (base)

In another reaction acetic acid donates a proton to water, which acts as a base, to form acetate and hydronium ions. Here acetate ion must act as a base and hydronium ion as an acid.

$$\text{CH}_3-\overset{\displaystyle \text{O}}{\underset{\displaystyle \text{OH}}{\text{C}}} + \text{H}-\overset{\displaystyle \text{H}}{\text{O}} \longrightarrow \text{CH}_3-\overset{\displaystyle \text{O}}{\underset{\displaystyle \text{O}^-}{\text{C}}} + \overset{\displaystyle \text{H}}{\underset{+}{\text{H}-\text{O}-\text{H}}}$$

acid　　　　　　base

$$\text{CH}_3-\overset{\displaystyle \text{O}}{\underset{\displaystyle \text{O}^-}{\text{C}}} + \overset{\displaystyle \text{H}}{\underset{+}{\text{H}-\text{O}-\text{H}}} \longrightarrow \text{CH}_3-\overset{\displaystyle \text{O}}{\underset{\displaystyle \text{OH}}{\text{C}}} + \text{H}-\overset{\displaystyle \text{H}}{\text{O}}$$

base　　　　　　acid

Acetate ion is called the **conjugate base** of acetic acid, and hydroxide ion is called the **conjugate base** of water. Hydronium ion is the **conjugate acid** of water.

Carboxylic acids are the most obvious organic acids. Any compound, however, which has a hydrogen bonded to an oxygen, sulfur, or nitrogen in some cases is capable of reacting as an acid to donate the proton to a base. Other organic compounds which can serve as acids include alcohols, phenols, thiols, and alkylammonium ions, which will be discussed shortly.

As you know from your previous study of acids, all acids are not of equal strength in donating protons. Sulfuric acid and hydrogen chloride are much stronger acids than acetic acid. The hydronium ion in aqueous solution is a much stronger acid than acetic acid. For our purposes here, we need an idea of the order of the strength of the classes of organic compounds as acids. Many reactions of alcohols are initiated by proton transfer, and a knowledge of the relative strengths of acids is important so that you will know how strong an acid or a base is needed for a particular reaction.

Some simple generalizations can be made. The acidity of a hydrogen increases in the order C—H < N—H < O—H. Note that this order correlates with the position of the atoms attached to hydrogen across the period in the periodic table. The acidity of protons attached to atoms increases as those atoms proceed down a group in the periodic table, e.g., O—H < S—H. Hydrogen chloride is a strong acid. Water is more acidic than ammonia, and hydronium ion is a stronger acid than ammonium ion. Hydrogen sulfide is more acidic than water.

As acids:　　$\text{H}-\text{Cl} > \text{H}_2\text{O} > \text{NH}_3$;　　$\text{H}_3\text{O}^+ > \text{NH}_4^+$;　　$\text{H}_2\text{S} > \text{H}_2\text{O}$

The strength of a base is determined by its relative tendency to accept a proton. The strongest base that is easily available is amide ion, which is stronger than hydroxide ion; hydrogen sulfide ion is a much weaker base. Ammonia is more basic than water. Chloride ion is a very weak base.

As bases:　　$\text{NH}_2^- > \text{OH}^- > \text{SH}^- > \text{Cl}^-$;　　$\text{NH}_3 > \text{H}_2\text{O}$

Note that the order of base strength is the reverse of the order of acid strength of the conjugate acids. This is a natural outcome of the method of determining relative strengths. For example, hydrogen chloride is said to be a stronger acid than water, because it gives a proton to hydroxide ion to form water and chloride ion.

$$HCl + OH^- \longrightarrow H_2O + Cl^-$$

acid 1   base 2   acid 2   base 1

The chloride ion is recognized as a weaker base than the hydroxide ion because the chloride ion does not remove a proton from the water to form the hydroxide ion.

This observation leads us to the generalization that an acid will donate its proton to the conjugate base of any acid weaker than itself. It will not donate a proton to the conjugate base of an acid much stronger than itself.

The proton transfer involving two acids and two bases reaches equilibrium. For two acids of widely different strengths, the equilibrium lies extremely far on the side toward the formation of the weaker acid, as in this example.

$$HCl \quad + \quad CH_3CO_2^- \longrightarrow CH_3CO_2H \quad + \quad Cl^-$$

hydrogen chloride   acetate ion   acetic acid   chloride ion

For two acids of somewhat the same strength, the equilibrium lies near the middle.

$$CH_3CO_2H + CH_3CH_2CO_2^- \rightleftharpoons CH_3CH_2CO_2H + CH_3CO_2^-$$

acetic acid   propanoate ion   propanoic acid   acetate ion

Problem   **3.14** Which is the stronger acid?

a) H_3O^+ or H_2O b) NH_3 or NH_4^+ c) OH^- or H_2O
d) NH_4^+ or H_3O^+ e) NH_3 or H_2S

Besides carboxylic acids, other organic compounds with a hydrogen attached to an oxygen or nitrogen show a tendency to donate a proton. The same order of acid strength and base strength we know for inorganic compounds applies to compounds in which a hydrogen has been replaced by a carbon group, like an alkyl or an aryl. Thus alcohols are more acidic than amines, protonated alcohols are stronger acids than alkylammonium ions, and all are much stronger than most hydrocarbons.

As acids: $R—OH > R—NH_2 \gg R—CH_3$; $R—OH_2^+ > R—NH_3^+$

These comparisons of acid strength are summarized in Fig. 3.11. The order of strength of the conjugate bases is listed in Fig. 3.12.

FIGURE 3.11   RELATIVE STRENGTH OF ACIDS

Increasing acid strength	Mineral acids H_2SO_4 and HCl	(very strong)
	Hydronium ion and protonated alcohols H_3O^+ and $R-OH_2^+$	(strong)
	Carboxylic acids $R-CO_2H$	(moderately strong)
	Phenols $Ar-OH$, ammonium and alkylammonium ions NH_4^+ and $R-NH_3^+$	(weak)
	Water and alcohols H_2O and $R-OH$	(very weak)
	Ammonia and amines NH_3 and $R-NH_2$	(very, very weak)
	Hydrocarbons	(extremely weak)

FIGURE 3.12   ORDER OF BASE STRENGTH

Increasing base strength	Amide and alkylamide ions NH_2^- and $R-NH^-$	(very strong)
	Alkoxide and hydroxide ions $R-O^-$ and OH^-	(strong)
	Ammonia, amines and phenoxide ion NH_3, $R-NH_2$ and $Ar-O^-$	(moderately strong)
	Carboxylate ions $R-CO_2^-$	(weak)
	Water and alcohols H_2O and $R-OH$	(weak)
	Chloride and hydrogen sulfate ions Cl^- and HSO_4^-	(very weak)

Problem   3.15 Draw the structures for the following compounds which may act as proton donors.

a) formic acid   b) 2-propanol   c) water
d) dimethylamine   e) methylammonium ion   f) phenol
g) 2-methyl-2-pentanol   h) sulfuric acid   i) propanoic acid
j) propylamine

3.9 A LOOK AT A LARGE MOLECULE

Biologically important compounds frequently are large molecules and contain many functional groups. Antibiotics often are such large molecules which substitute in a biochemical process for a similar molecule necessary to bacterial reproduction, and in that way they inhibit bacterial growth. Aureomycin (also called chlortetracycline) is an antibiotic which is active against many bacteria. This is its structure.

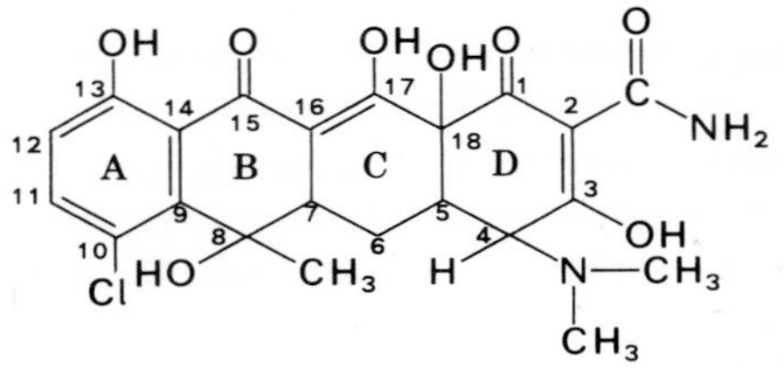

aureomycin
(chlortetracycline)

Problem   3.16 Examine the structure of aureomycin carefully, atom by atom.

a) What is the molecular formula? What is the molecular weight?

b) Why is it called a tetracycline? How many carbons are in each ring? What is the degree of unsaturation of each ring?

c) How many functional groups (not counting carbon–carbon double bonds) are there? How many different kinds of functional groups are represented?

d) How many molecules of hydrogen (H_2) would be needed to saturate all the double bonds in one molecule of aureomycin?

e) In the structure above, the carbons are given the numbers shown. List the functional groups by name and by formula according to the carbon to which they are attached.

f) Are all the hydroxyl functions attached to similar carbons? If not, describe the difference, identifying each function by carbon number.

New Terms and Topics

Multiple bonds—double and triple bonds (Section 3.1)

Alkenes and alkynes (Section 3.1)

Geometry of double and triple bonds (Section 3.2)

Stereoisomerism of alkenes—*cis-trans* isomers (Section 3.3)

Aromatic compounds—benzene and pyridine (Section 3.4)

Delocalization of electron pairs, bonds, and charges —resonance structures (Section 3.4)

Geometry of resonance structures (Section 3.5)

Carbonyl compounds—aldehydes and ketones (Section 3.6)

Carboxylic acids and derivatives—esters and amides (Section 3.7)

Organic acids and bases; relative strength (Section 3.8)

Problems

3.17 There are nine isomeric open-chain compounds exclusive of stereo-isomers with the molecular formula C_5H_8. Draw the formulas for the following.

    a) three alkynes   **b)** six dienes

3.18 Indicate the kind of orbitals carbon uses for each bond in the following compounds.

  a) $CH_2=CH-CH_2-OH$ **b)** $CH_3-CH=CH-C\equiv C-H$ **c)**

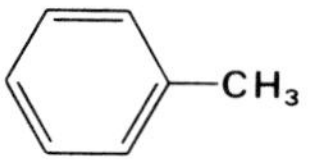

3.19 Name the class for each compound.

  a) $CH_3-CH_2-\underset{\underset{\textstyle CH_3}{|}}{CH}-CH=O$ **b)** $CH_3-CH_2-\overset{\overset{\textstyle O}{\|}}{C}-O-CH_2-CH_2-CH_3$ **c)** $CH_3-CH_2-O-CH_2-CH_3$

  d) $CH_3-\underset{\underset{\textstyle CH_3}{|}}{CH}-CH_2-\overset{\overset{\textstyle O}{\|}}{C}-CH_3$ **e)** 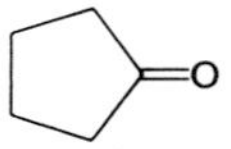   **f)** 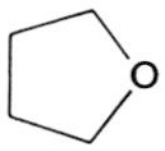   **g)**

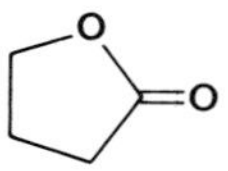

  h) $CH_3-CH_2-\overset{\overset{\textstyle O}{\|}}{C}-NH_2$ **i)** $CH_3-CH_2-CH_2-NH_2$

3.20 In the following groups which structure is not an isomer of the other two?

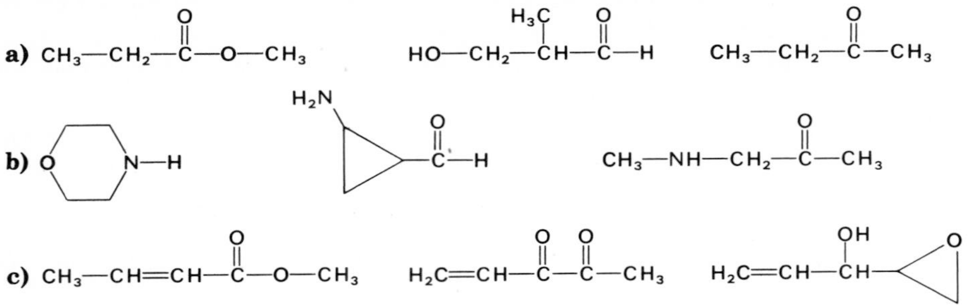

a) $CH_3—CH_2—\overset{\overset{O}{\|}}{C}—O—CH_3$ $HO—CH_2—\overset{\overset{H_3C}{|}}{CH}—\overset{\overset{O}{\|}}{C}—H$ $CH_3—CH_2—\overset{\overset{O}{\|}}{C}—CH_3$

b) [morpholine structure: O···N—H ring]   [cyclopropane structure with H_2N and $\overset{\overset{O}{\|}}{C}—H$] $CH_3—NH—CH_2—\overset{\overset{O}{\|}}{C}—CH_3$

c) $CH_3—CH=CH—\overset{\overset{O}{\|}}{C}—O—CH_3$ $H_2C=CH—\overset{\overset{O}{\|}}{C}—\overset{\overset{O}{\|}}{C}—CH_3$ $H_2C=CH—\overset{\overset{OH}{|}}{CH}—$[cyclopropane with O]

3.21 Which of the following compounds exist as *cis-trans* isomers? Draw the structures of those that do.

a) $CH_2=CCl_2$ **b)** $C_6H_5CH=CHC_6H_5$ **c)** $CH_3CH_2CH=CHCH=CHCH_3$ **d)** $CH_3C≡CCH_2CH_3$

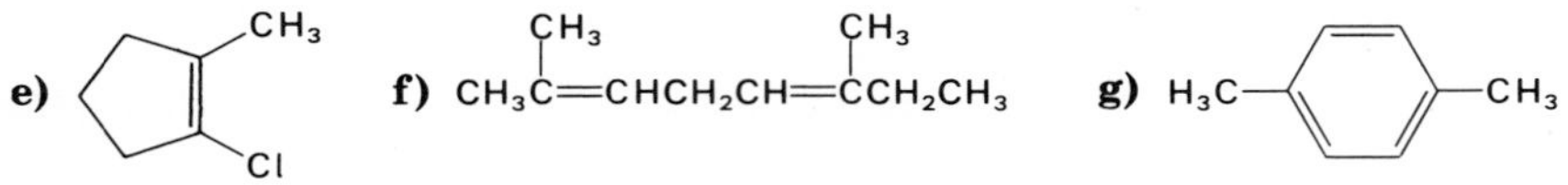

e) [cyclopentene with CH_3 and Cl]   **f)** $CH_3\overset{\overset{CH_3}{|}}{C}=CHCH_2CH=\overset{\overset{CH_3}{|}}{C}CH_2CH_3$ **g)** H_3C—[benzene ring]—CH_3

3.22 Draw the resonance structures for the following compounds or ions.

a) 1,3-dimethylbenzene   **b)** pyridine   **c)** acetate ion   **d)** guanidinium ion

3.23 Rank the compounds in each group in order of decreasing acid strength.

a) $CH_3CH_2CH_2—NH_2$ $CH_3CH_2CH_2—SH$ $CH_3CH_2CH_2—OH$

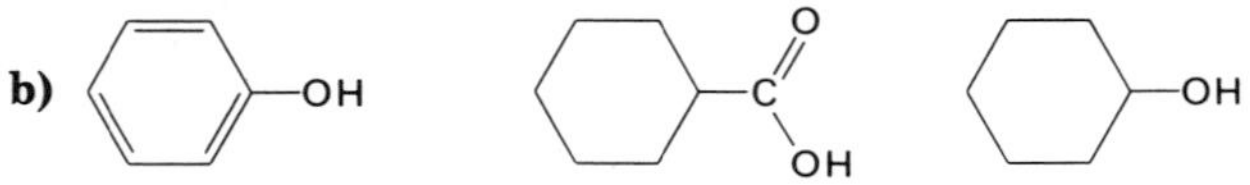

b) [benzene ring]—OH [cyclohexane ring]—$\overset{\overset{O}{\|}}{C}\overset{OH}{}$ [cyclohexane ring]—OH

c) $CH_3CH_2CH_2—\overset{+}{O}H_2$ $CH_3CH_2CH_2—OH$ $CH_3CH_2CH_2—\overset{+}{N}H_3$

3.24 Draw the structure of the conjugate acid of each of these compounds.

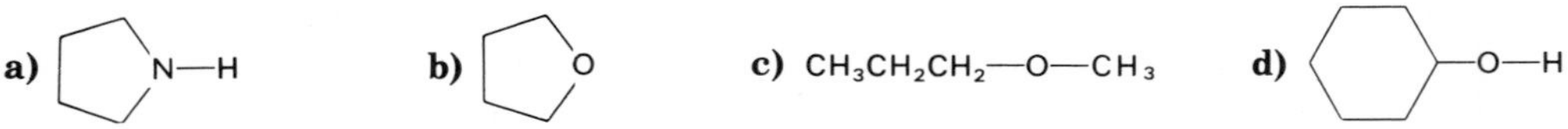

a) [pyrrolidine ring]$N—H$ **b)** [tetrahydrofuran ring]O **c)** $CH_3CH_2CH_2—O—CH_3$ **d)** [cyclohexane ring]—$O—H$

e) $CH_3CH_2—S—CH_3$ **f)** $CH_3C\overset{\overset{O}{\diagup}}{\diagdown}_{O—H}$

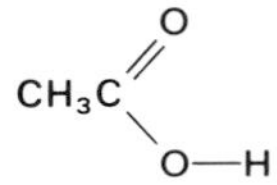

3.25 The following compounds are representatives of carbohydrates, fats, and proteins. What are the functional groups in each compound?

a) 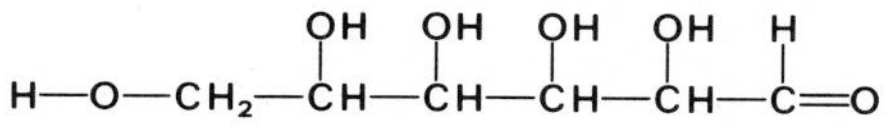 $H{-}O{-}CH_2{-}\overset{\displaystyle OH}{\underset{\displaystyle |}{CH}}{-}\overset{\displaystyle OH}{\underset{\displaystyle |}{CH}}{-}\overset{\displaystyle OH}{\underset{\displaystyle |}{CH}}{-}\overset{\displaystyle OH}{\underset{\displaystyle |}{CH}}{-}\overset{\displaystyle H}{\underset{\displaystyle |}{C}}{=}O$ glucose—a carbohydrate

b) 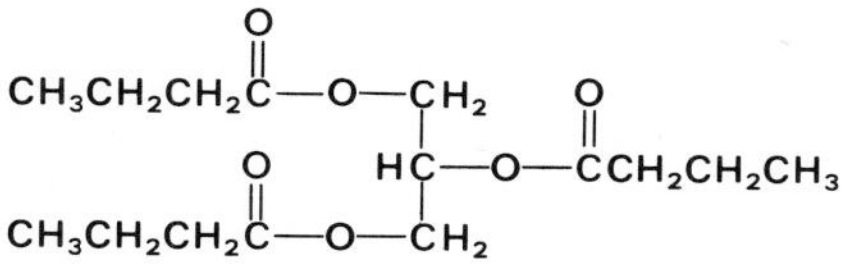 $CH_3CH_2CH_2\overset{\displaystyle O}{\overset{\displaystyle \|}{C}}{-}O{-}CH_2$ O

$\quad H\overset{}{C}{-}O{-}\overset{\displaystyle \|}{C}CH_2CH_2CH_3$ glyceryl tributanoate—a fat from butter

$\overset{\displaystyle O}{\overset{\displaystyle \|}{}}$

$CH_3CH_2CH_2\overset{\displaystyle O}{\overset{\displaystyle \|}{C}}{-}O{-}CH_2$

c) $NH_2{-}\underset{\displaystyle \underset{\displaystyle CH_3}{|}}{CH}{-}\overset{\displaystyle O}{\overset{\displaystyle \|}{C}}{-}NH{-}\underset{\displaystyle \underset{\displaystyle CH_2{-}OH}{|}}{CH}{-}\overset{\displaystyle O}{\overset{\displaystyle \|}{C}}{-}NH{-}\underset{\displaystyle \underset{\displaystyle CH_2{-}SH}{|}}{CH}{-}\overset{\displaystyle O}{\overset{\displaystyle \|}{C}}{-}OH$ a tripeptide—a short version of a protein

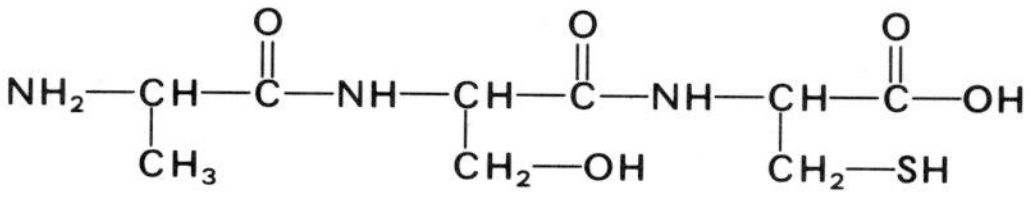

3.26 Write four structures for $C_5H_{10}O$ all of which have a different functional group. Name the functional groups you have used in each.

CHAPTER 4

Alkenes and
Alkynes

Some 280 to 350 million years ago, in shallow seas, countless microorganisms used solar energy to convert carbon dioxide and water into organic compounds. When these organisms died, some of the organic compounds they had synthesized survived and accumulated. As a result of geological events, these residues were trapped in shales and in limestone and sandstone deposits. The effects of time, pressure, and heat converted the compounds largely into hydrocarbons. These liquids and dissolved gases, when compacted in porous materials, poured into subterranean pools which formed the great petroleum reserves of our planet.

By similar processes fueled by sunlight, plants in profusion lived and died, and some of their organic remains accumulated in the form of peat. Again the effects of time, heat, and pressure tended to drive the organic material toward compounds poorer in oxygen and richer in carbon, and coal was produced.

Coal and petroleum are the fossil fuels, which are part of our evolutionary inheritance from the distant past. Just as the industrial revolution was fueled by coal, the "chemical revolution" of the 20th century has been fueled by petroleum. Petroleum contains several hundred organic compounds that have been isolated and characterized. Even so, it is simpler than coal, which is a chemical mess! Although both serve as fuels, petroleum is the main source of the organic compounds that have changed the character of our environment so drastically in the past 50 years.

Petroleum consists largely of alkanes, cycloalkanes, alkenes, and aromatic compounds. The first two classes might be called the sigma-hydrocarbons, because their atoms are held together solely by sigma bonds. Ordinarily they are referred to as the saturated hydrocarbons. The alkenes, alkynes, and aromatic hydrocarbons might be called the pi-hydrocarbons, because their distinguishing feature is the presence of carbon–carbon double or triple bonds. They have been called the unsaturated hydrocarbons. Petroleum is much richer in saturated than in unsaturated hydrocarbons. The saturated compounds serve as fuels, whereas the unsaturated compounds are the primary starting materials for the plastics, rubbers, fibers, and drugs on which we have grown so dependent. Fortunately, the saturated compounds are also readily converted by various processes to the alkenes, alkynes, and aromatic hydrocarbons.

In this chapter we describe the chemistry of alkenes and alkynes, particularly their most important reactions in the laboratory and their related transformations in biological processes. These unsaturated hydrocarbons are the first classes of compounds treated because their structures are simple, and their reactions link petroleum and coal to the classes of compounds that have functional groups containing oxygen, sulfur, and nitrogen.

4.1 SOURCES AND USES OF ALKENES, DIENES, AND POLYENES Alkenes, dienes, and polyenes are unsaturated hydrocarbons which contain, respectively, one, two, or many carbon–carbon double bonds $C=C$. The simple alkenes and polyenes are mainly used as starting materials for the preparation of other compound classes.

Simple alkenes are made from alkanes industrially by processes which remove hydrogens from adjacent carbons. Carbon–hydrogen bonds are broken when ethane is passed quickly through a tube at 800°. Ethylene and hydrogen are the products. Millions of tons of ethylene are produced in the United States every year by this process.

$$CH_3CH_3 \xrightarrow{\ 800°\ } CH_2{=}CH_2 + H_2$$

Ethylene (ethene) $CH_2{=}CH_2$, the simplest alkene, is consumed in large quantities as the basic starting material in the production of ethanol, diethyl ether, ethylene glycol (the main ingredient in antifreeze), acetaldehyde, acetic acid, and the plastic polyethylene.

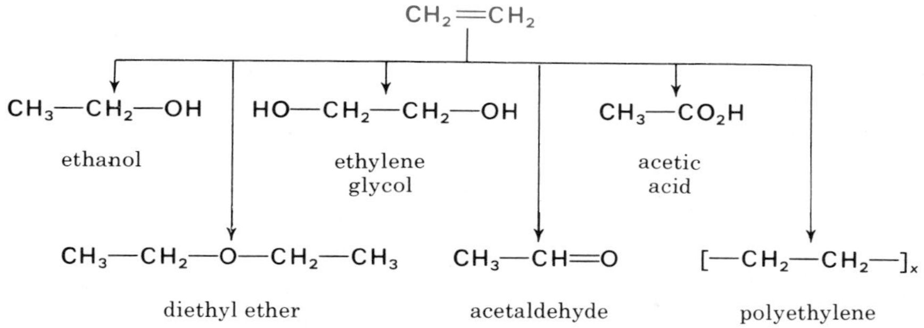

Ethylene, vinyl acetate, and vinyl chloride are the starting materials for the plastics polyethylene, polyvinyl acetate, and polyvinyl chloride. Vinyl acetate and vinyl chloride are made from acetylene (Section 4.9).

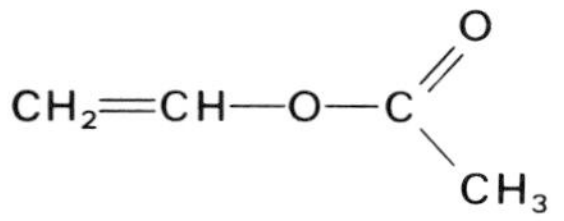

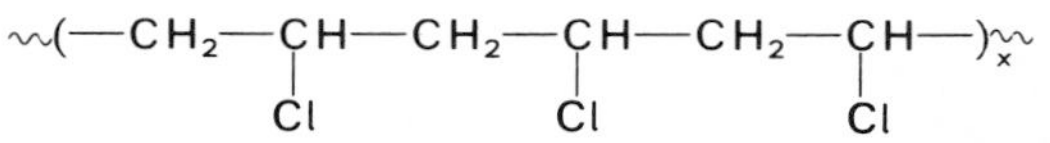

Other alkenes and dienes also find use in the production of polymers. A **polymer** is a very large molecule made up of repeating units of small molecules designated as monomers. Besides the monomers mentioned above, others that give important polymers are propylene, isobutylene, butadiene, isoprene, chloroprene, and styrene.

$$CH_2=CH-CH_3 \qquad CH_2=\overset{\overset{\displaystyle CH_3}{|}}{C}-CH_3 \qquad CH_2=CH-CH=CH_2$$

propylene   isobutylene   butadiene

$$CH_2=\overset{\overset{\displaystyle CH_3}{|}}{C}-CH=CH_2 \qquad CH_2=\overset{\overset{\displaystyle Cl}{|}}{C}-CH=CH_2 \qquad CH_2=CH-$$

isoprene   chloroprene   styrene

Natural rubber, obtained from the latex of the rubber tree, is a polymer of isoprene.

$$\left[-CH_2-CH=\overset{\overset{\displaystyle}{|}}{\underset{\underset{\displaystyle CH_3}{|}}{C}}-CH_2\!\mid\!CH_2-CH=\overset{\overset{\displaystyle}{|}}{\underset{\underset{\displaystyle CH_3}{|}}{C}}-CH_2- \right]_x$$

natural rubber
(showing two isoprene units joined together)

Some alkenes, dienes, and polyenes are found in plants, though seldom in animals in any quantity. Many belong to the family of natural products called **isoprenoids** (as does natural rubber), which are composed of isoprene units. The volatile isoprenoids have interesting and often pleasant odors, like those from the rose and from lemon peel. (See Section 4.10 for more of these compounds.)

Many compounds containing C=C bonds also have other groups such as the hydroxyl, carbonyl, and carboxyl functional groups. The chemistry of the C=C bond in such compounds is sometimes modified by the presence of the additional group and is described in appropriate later chapters of the book.

In the laboratory alkenes are prepared from saturated compounds having the same carbon skeleton and a functional group which may be eliminated by an appropriate reaction. The removal of water from alcohols (Section 5.6) and the removal of hydrogen halide from alkyl halides (Section 6.5) are two examples of reactions which give alkenes. More complex alkenes are built up from aldehydes, ketones, and esters by the Grignard reaction (Section 14.3).

4.2 NOMENCLATURE OF ALKENES

In our introduction to alkenes in Section 3.1, some of the rules of IUPAC nomenclature were described. A more complete set of rules is given here.

1. The longest carbon chain containing the double bond is selected for the base name. The stem of the alkane having this number of carbons is used with the ending "ene" to show one double bond. For two double bonds the ending is "diene," etc.

2. The carbon chain is numbered starting from the end closest to the double bond. The location of the double bond is given by the lower number of the two doubly bonded carbons.

3. Hydrocarbon groups, such as alkyl or aryl, halogens, and amino groups are named as substituents. They are assigned the number of the carbon to which they are attached. Substituents are listed alphabetically as prefixes to the base name. Numbers are separated from names by hyphens.

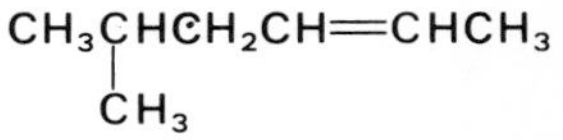

[handwritten: 5 methyl 2 hexene] *[handwritten: 1 butene]* *[handwritten: 3 bromo cyclo hexene]* *[handwritten: 4 phenyl 1,3 pentadiene]*

$CH_3CHCH_2CH=CHCH_3$ $CH_2=CHCH_2CH_3$
$\quad\;\; |$
$\quad\;\; CH_3$

[Br, 3, 2, 1 labeling on cyclohexene ring] $CH_3C=CHCH=CH_2$

5-methyl-2-hexene   1-butene   3-bromocyclohexene 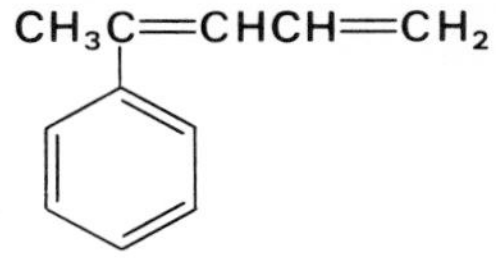   4-phenyl-1,3-pentadiene

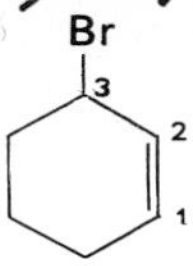

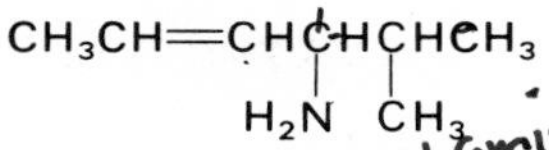

$CH_3CH=CHCHCHCH_3$
$\qquad\qquad |\quad |$
$\qquad\; H_2N\;\; CH_3$

[handwritten: 4 amino] *[handwritten: 5 methyl]* *[handwritten: 2 hexene]*

4-amino-5-methyl-2-hexene

4. Geometric isomers or stereoisomers of alkenes are identified by a notation which is based on a system of priority assigned to groups attached to the C=C. The isomer with the higher priority groups on the same side of the double bond is given the prefix *Z* (*zusammen* = together). The isomer with the higher priority groups on opposite sides of the double bond has the prefix *E* (*entgegen* = opposite).

Priority of groups is based on the **atomic number** of atoms attached to the C=C and is determined according to these rules.

a) The higher the *atomic number* of the atom attached to the unsaturated carbon, the higher the priority of that group.

Br > Cl > S > F > O > N > C > H

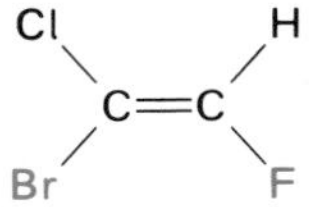

[structure of]
$$\text{Cl} \quad\quad\quad\quad \text{H}$$
$$\diagdown\qquad\qquad\diagup$$
$$\text{C}=\text{C}$$
$$\diagup\qquad\qquad\diagdown$$
$$\text{Br}\quad\quad\quad\quad \text{F}$$

Z-1-bromo-1-chloro-2-fluoroethene

Bromine is higher in priority than chlorine. Fluorine is higher in priority than hydrogen. The isomer with Br and F on same side is *Z*.

b) For two atoms of the same atomic number, the selection devolves upon the next attached atoms, and so on until a difference is found.

$—CH_2—Cl > —CH_2—OH > —CH_2—CH_3 > —CH_3$

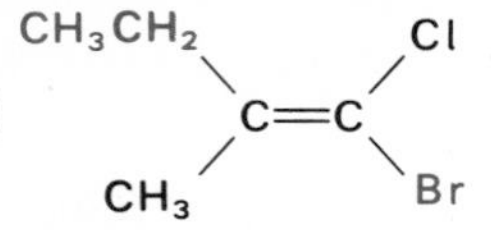

E-1-bromo-1-chloro-2-methyl-1-butene

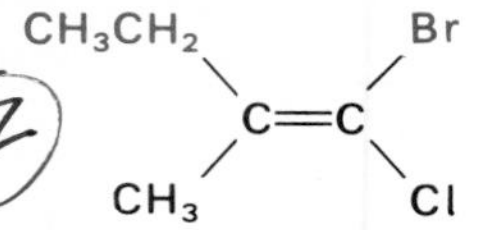

Z-1-bromo-1-chloro-2-methyl-1-butene

The bromo and the ethyl groups have higher priority than do the chloro and the methyl groups, respectively. Therefore the name is based on the relative positions of the bromo and the ethyl groups.

c) Double bonds are counted as two single bonds, i.e., C=O as O—C—O.

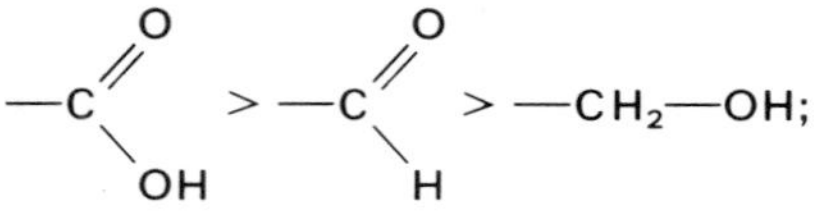

$—CH=CH_2 > —CH_2—CH_3$

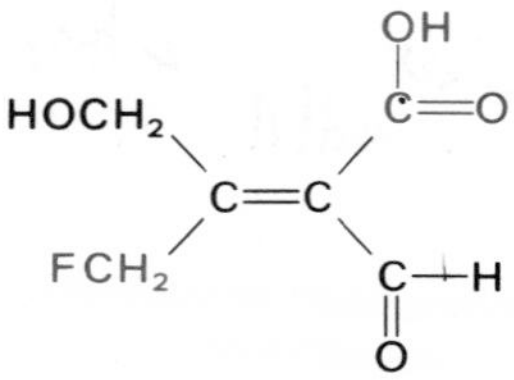

the *E*-isomer

Fluorine has higher priority than OH. Carboxylic acid group has higher priority than the aldehyde group. Fluorine and carboxylic acid are on opposite sides. Thus the structure shown is the *E*-isomer.

This is an *E*-*Z* way to distinguish geometric isomers of alkenes and of cyclic compounds.

Problems **4.1** Name the following compounds. For the structures where geometric isomers exist, identify the isomer shown as *E* or *Z*.

a) 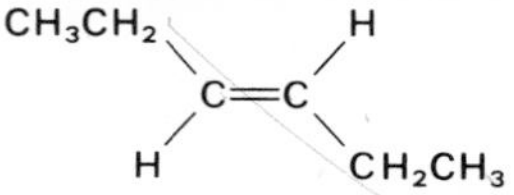

b) 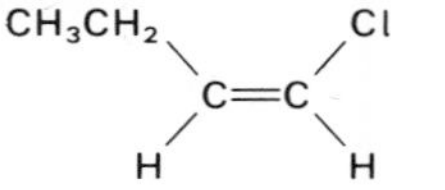

c) 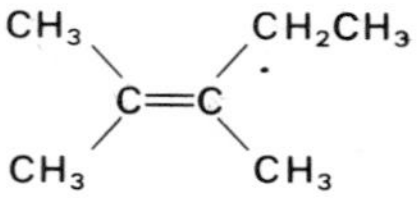

d) 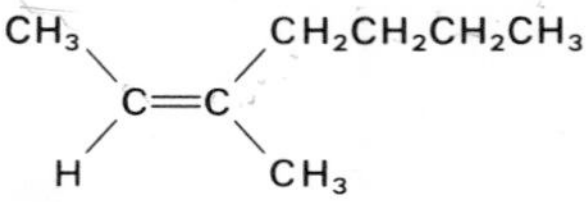

e) 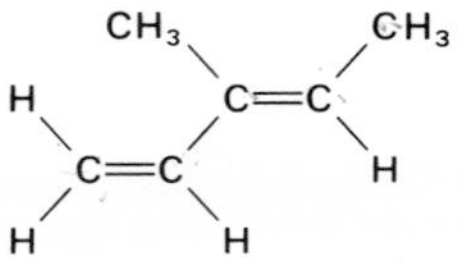

f) 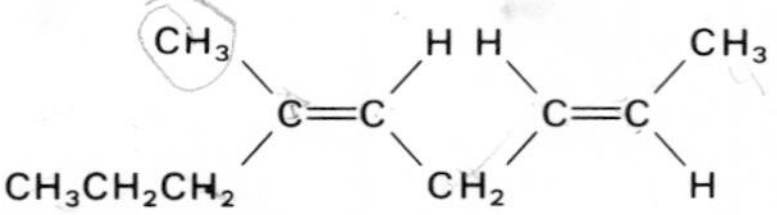

4.2 Name the following compounds and identify the isomer shown as *E* or *Z* when appropriate.

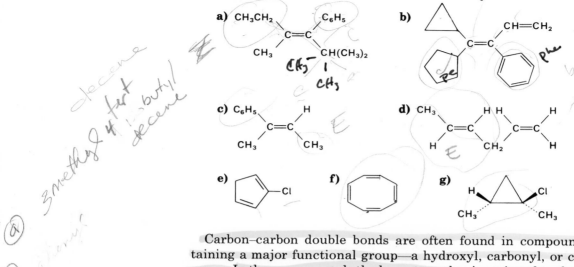

a) CH₃CH₂, C₆H₅, C=C, CH₃, CH(CH₃)₂

b)

c) C₆H₅, H, C=C, CH₃, CH₃

d) CH₃, H H, H, C=C, H, CH₂, C=C, H

e) Cl

f)

g) H, Cl, CH₃, CH₃

Carbon–carbon double bonds are often found in compounds containing a major functional group—a hydroxyl, carbonyl, or carboxyl group. In these compounds the lowest number is assigned to the carbon of the functional group, and the number for the double bond follows from that. (Nomenclature of these classes of compounds is given in Chapters 5, 8, and 11.)

⑤ ④ ③ ② ①
CH₂=CH—CH₂—CH₂—CO₂H

4-pentenoic acid

For the smallest alkenes, common names are still used (IUPAC names in parentheses).

$$CH_2=CH_2 \qquad CH_3—CH=CH_2 \qquad CH_3—\overset{\overset{\displaystyle CH_3}{|}}{C}=CH_2$$

ethylene   propylene   isobutylene
(ethene)   (propene)   (methylpropene)

Alkenes were originally known as "olefins." The name survives in the term **olefinic linkage** for a C=C bond, **olefinic carbons** for the unsaturated carbons, and **olefinic hydrogens**, those attached to the unsaturated carbons.

Two unsaturated groups have common names which designate special attachments. The **vinyl** group is attached at the olefinic carbon, and the **allyl** group is attached at the carbon adjacent to the olefinic carbon.

CH₂=CH—   CH₂=CH—Cl   CH₂=CH—CH₂—   CH₂=CH—CH₂—Cl

vinyl group   vinyl chloride   allyl group   allyl chloride
(chloroethene)   (3-chloro-1-propene)

The most general type of reaction which alkenes undergo is the addition of two atoms or groups of atoms to the two unsaturated carbons. The double bond is converted to a single bond as one new bond is made between each carbon and a new atom.

$$\begin{array}{c}\diagup\\\diagup\end{array}C{=}C\begin{array}{c}\diagdown\\\diagdown\end{array} + A{-}B \longrightarrow \begin{array}{cc}|&|\\-C-C-\\|&|\\A&B\end{array}$$

A reaction in which two compounds join to produce one compound is called an addition reaction. In an addition reaction an unsaturated compound is converted into a saturated compound. The pi electrons of the C=C bond and electrons from the adding reagent are used to make the new bonds. As a result of this addition, the geometry of groups around the two originally unsaturated carbons changes from a planar to a tetrahedral arrangement.

Unsaturated hydrocarbon groups are easily converted to saturated hydrocarbons when shaken under a slight pressure of hydrogen gas in the presence of a metal or metal oxide acting as a catalyst. A **catalyst** is a substance which, when added to the reaction, increases the rate of the reaction without itself being consumed. Without the catalyst the addition of hydrogen to an alkene is too slow to observe.

The addition of molecular hydrogen in the presence of a metal catalyst is called **catalytic hydrogenation**. The usual hydrogenation catalysts are finely divided platinum or palladium, or a finely divided form of nickel known as "Raney nickel."

Propene reacts with hydrogen gas with a platinum catalyst to give propane.

$$CH_3{-}CH{=}CH_2 + H_2 \xrightarrow[\text{1 atm}]{\text{Pt } 25°} CH_3{-}CH_2{-}CH_3$$

propene   propane

The particular conditions for the reaction are placed above and below the arrow in the equation. In this instance, the hydrogen gas was kept at a pressure of one atmosphere (1 atm, 760 torr) at 25°C over a platinum catalyst.

Two very important facts are expressed by the equation just given. One is that propene and hydrogen give propane. The other is that *by learning that propene and hydrogen form propane, we also learn that all alkenes react with hydrogen gas in the presence of catalysts to give alkanes.*

Remember that in studying the reactions of a particular functional group represented by one compound, we can expect that other compounds having the same functional group will react in the same way. By such generalizations, the reactions of many organic compounds become reduced to the reactions of a few compounds. We could always select one or two compounds for examples, but we use a wider selection so that you will become familiar with more compounds. Here we use 1,2-dimethylcyclohexene.

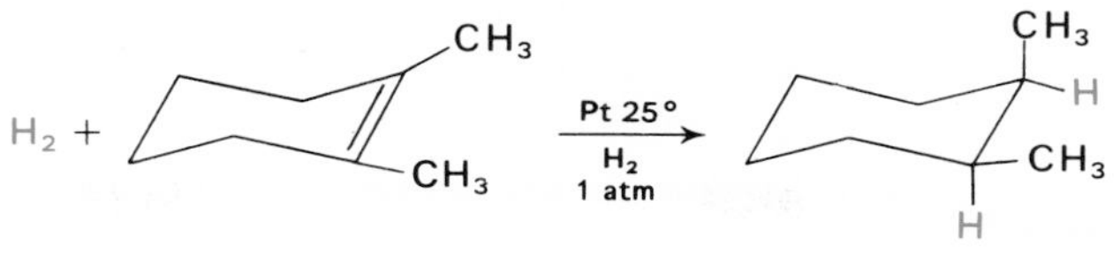

1,2-dimethylcyclohexene   *cis*-1,2-dimethylcyclohexane

The particular isomer of dimethylcyclohexane which is produced in this reaction indicates that the two hydrogen atoms were added on the same side of the double bond. Such reactions are called *cis*-additions. A reaction that gives only one of a number of possible isomers is said to be **stereospecific** (stereo = three-dimensional). This reaction followed a single stereochemical pathway which did not allow the hydrogens to add on opposite sides of the double bond with the formation of the other geometric isomer—*trans*-1,2-dimethylcyclohexane.

The pathway of catalytic hydrogenation is determined by the action of the catalyst. The metal adsorbs hydrogen molecules and alkene molecules on its surface. In this manner the H—H bond is weakened and bond formation occurs readily between each of the hydrogens and the two unsaturated carbons of a nearby alkene molecule.

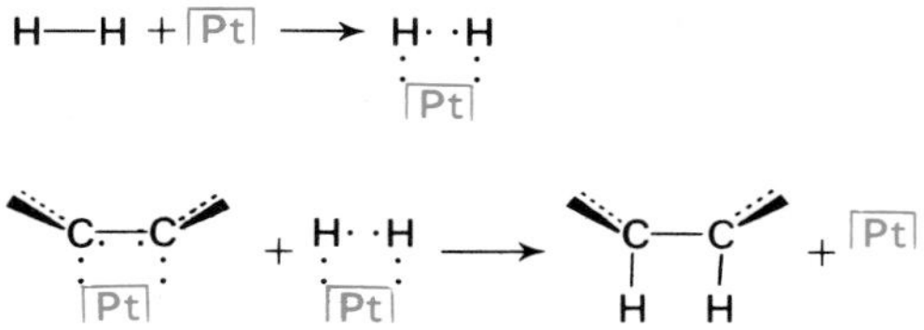

Another example of hydrogenation is that of 3-phenyl-1-propene to give 1-phenylpropane.

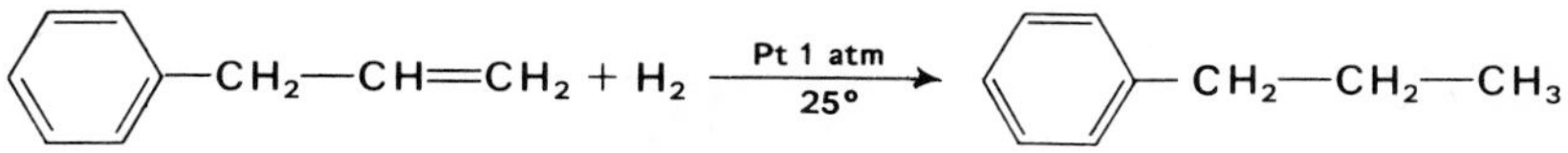

3-phenyl-1-propene   1-phenylpropane (propylbenzene)

In this reaction, note that the benzene ring is not hydrogenated under the conditions for which propene adds hydrogen readily. The benzene ring does not act as an alkene. This difference is typical of the difference in the reactions of alkenes and aromatic compounds.

4.4 ADDITION OF POLAR REAGENTS TO ALKENES

Alkenes react with many reagents in an addition reaction. The alkene and the reagent, usually in solution, need only to be mixed to show evidence of combining. Evolution of heat, a change in color, or a separation into two liquid layers sometimes provides perceptible evidence of the transformation. These addition reactions of alkenes occur very rapidly at room temperature.

The additions of some strongly polar reagents are treated here.

a) Acidic reagents Strong halogen acids H—X add readily to alkenes, the hydrogen going to one carbon, the halogen (X) to the other. The product is an **alkyl halide**.

$$\text{C}=\text{C} + \text{H--X} \longrightarrow \text{H--C--C--X}$$

With a concentrated solution of hydrogen chloride or hydrogen bromide in acetic acid (as solvent), 2-butene gives *sec*-butyl halide.

$$\text{H--Br} + \text{CH}_3\text{--CH}=\text{CH--CH}_3 \xrightarrow{\text{acetic acid}} \text{CH}_3\text{--CH--CH--CH}_3$$
$$\overset{|}{\text{H}} \quad \overset{|}{\text{Br}}$$

2-butene   2-bromobutane
                                      (*sec*-butyl bromide)

Hydrogen chloride, in concentrated aqueous solution, adds to isobutylene (methylpropene) to produce *tert*-butyl chloride.

$$\overset{\text{CH}_3}{\underset{|}{\text{CH}_3\text{--C}}}=\text{CH}_2 + \text{H--Cl} \longrightarrow \text{CH}_3\text{--}\overset{\text{CH}_3}{\underset{\text{Cl}}{\overset{|}{\underset{|}{\text{C}}}}}\text{--CH}_3$$

isobutylene   *tert*-butyl chloride

The detailed steps of the process by which reagents combine to give products is called the **reaction mechanism**. A polar reagent such as hydrogen bromide HBr adds to an alkene by a two-step mechanism. The HBr is strongly polar, with the hydrogen positive and the bromine negative. The hydrogen as a hydrogen ion, a proton H$^+$, is transferred to the alkene, in this example, 2-butene.

$$\text{CH}_3\text{--CH}=\text{CH--CH}_3 + \overset{\delta^+}{\text{H}}\text{--}\overset{\delta^-}{\text{Br}} \Longrightarrow \text{CH}_3\text{--CH--}\overset{+}{\text{CH}}\text{--CH}_3 + \text{Br}^-$$

2-butene   *sec*-butyl cation
                                      (a carbonium ion)

In the addition of the hydrogen ion, the C—H bond is formed using the pi electrons of the double bond. A reagent like the hydrogen ion which is deficient in electrons and is seeking electrons to use in bonding is called an **electrophilic** ("electron-loving") reagent. The bromide ion is also formed.

The second carbon of the original double bond is now deficient in electrons and possesses a full positive charge. A positive ion in which carbon bears the positive charge is an alkyl cation or a **carbonium ion**. A carbonium ion is an extremely reactive, high-energy electrophilic species. It reacts immediately after its formation with the negatively charged bromide ion to give the final product, the alkyl halide.

$$CH_3-\underset{\underset{H}{|}}{CH}-\underset{+}{CH}-CH_3 + Br^- \longrightarrow CH_3-\underset{\underset{H}{|}}{CH}-\underset{\underset{Br}{|}}{CH}-CH_3$$

the reaction
intermediate

2-bromobutane
(*sec*-butyl bromide)

The carbonium ion is an example of a **reaction intermediate**, an unstable high-energy species which reacts immediately with a reagent until a stable product is formed. Molecules or ions that react with positive carbon are called **nucleophilic** ("nucleus-loving") reagents. In the carbonium ion reaction with bromide ion, the Br^- acts as a nucleophile. In the overall process, HBr furnished an electrophile (H^+) and a nucleophile (Br^-) to the double bond. The evidence for this mechanism is discussed in the next section (Section 4.5).

b) Hydroxylated reagents —water and alcohols

Water is a very polar compound, but it is not a strong enough acid to transfer a proton to an alkene. The addition of water, called **hydration**, is therefore catalyzed by the presence of a strong acid, usually dilute sulfuric acid. Hydration is a very important commercial process for the preparation of alcohols from alkene petroleum products, such as ethylene and propylene.

$$CH_2{=}CH_2 + H_2O \xrightarrow[140°]{H_2SO_4} CH_3-CH_2-OH$$

ethylene   ethanol

$$CH_3-CH{=}CH_2 + H_2O \xrightarrow{H_2SO_4} CH_3-\underset{\overset{|}{OH}}{CH}-CH_2-H$$

propene   isopropyl alcohol

This reaction is very similar to the addition of hydrogen halides. The first step of the two-step mechanism is the same, the transfer of a hydrogen ion from the strong acid to the alkene. In this case the strong acid is the hydronium ion, H_3O^+, which serves as the electrophilic reagent. Once again a carbonium ion is formed as a reaction intermediate.

$$H_2O + H_2SO_4 \longrightarrow H_3O^+ + HSO_4^-$$

$$CH_2{=}CH-CH_3 + H_3O^+ \longrightarrow \underset{\underset{H}{|}}{CH_2}-\underset{+}{CH}-CH_3 + H_2O$$

propene   isopropyl cation

In the second step the carbonium ion forms a C—O bond with water to give a protonated alcohol. This combination is very much like the action of H_2O in taking on an electron-deficient H^+ to give H_3O^+.

$$CH_3—\underset{+}{CH}—CH_3 + H_2O \longrightarrow CH_3—\underset{\overset{|}{{}^+OH_2}}{CH}—CH_3 \xrightarrow{H_2O} CH_3—\underset{\overset{|}{OH}}{CH}—CH_3 + H_3O^+$$

isopropyl cation   protonated alcohol   2-propanol (isopropyl alcohol)

The protonated alcohol transfers a hydrogen ion to another molecule of water, which is present as solvent. This H_2O is shown in the equation over the second arrow. In this manner the catalyst, H_3O^+, is *regenerated* to be used again.

Carbonium ions are captured competitively by water and any negative ions present in the solution. Hydrogen halides are not used as catalysts, since we saw that halide ions are quite successful nucleophiles in combining with the carbonium ions (previous section). On the other hand, the hydrogen sulfate ion, HSO_4^-, is not a good competitor against other species like water, and sulfuric acid serves well as a catalyst.

In a similar reaction, ethanol is added to ethylene to produce diethyl ether (ordinary "ether"). This reaction is used in the commercial preparation for diethyl ether, but is seldom used for other alkenes and alcohols.

$$CH_2{=}CH_2 + CH_3—CH_2—OH \xrightarrow[180°]{H_2SO_4} CH_3—CH_2—O—CH_2—CH_3$$

ethylene   ethanol   diethyl ether

Problems   **4.3** Write the equation for each of these reactions.

     a) 3-hexene + hydrogen chloride

     b) 1-chloro-2-methylcyclopentene + hydrogen with platinum

     c) cyclohexene + water with sulfuric acid

4.4 Devise a mechanism based on mechanisms already discussed in this section for the acid-catalyzed addition of ethanol to ethylene.

4.5 ORIENTATION IN ADDITION OF UNSYMMETRICAL REAGENTS

Unsymmetrical alkenes can react with water (or hydrogen halide) to give two isomers, whose structures depend on which of the unsaturated carbons becomes attached to the hydroxyl group. For example, water and 1-pentene could give 1-pentanol or 2-pentanol.

$$CH_2{=}CH—CH_2CH_2CH_3 + H_2O \xrightarrow{H_2SO_4} CH_3\underset{\overset{|}{OH}}{C}HCH_2CH_2CH_3 \quad or \quad \underset{\overset{|}{OH}}{C}H_2CH_2CH_2CH_2CH_3$$

OH on which carbon?   2-pentanol   1-pentanol

Is the product a mixture of the two alcohols, or is there a structural guide which steers the reaction toward one or the other? In most cases both alcohols are produced, but one is predominant. The next question is, can you predict in advance of experiment the structure of the dominant product?

A rule based on the observed products from such reactions, first stated by Markownikoff (a Russian chemist) in 1871, predicts the direction of addition of acidic reagents (such as HX) to alkenes and vinyl halides (haloalkenes). **Markownikoff's rule** states that the *hydrogen* of the reagent will go to that carbon of the double bond which *already has the greater number of hydrogens.* These equations indicate the predominant products.

$$CH_2\!\!=\!\!CHCH_2CH_2CH_3 + H_2O \xrightarrow{\ H_2SO_4\ } \underset{\underset{\displaystyle H \quad OH}{|\quad\ |}}{CH_2CHCH_2CH_2CH_3} + \underset{\underset{\displaystyle OH\ H}{|\ \ \ |}}{CH_2CHCH_2CH_2CH_3}$$

1-pentene	2-pentanol (major)	1-pentanol (minor)

$$\underset{\text{isobutylene}}{\overset{\overset{\displaystyle CH_3}{|}}{CH_3C}\!\!=\!\!CH_2 + HCl} \longrightarrow \underset{\text{\textit{tert}-butyl chloride (100\%)}}{\overset{\overset{\displaystyle CH_3}{|}}{\underset{\underset{\displaystyle Cl}{|}}{CH_3CCH_2}}\!\!-\!\!H} \quad \left(\underset{\text{no trace}}{\overset{\overset{\displaystyle CH_3}{|}}{\underset{\underset{\displaystyle H}{|}}{CH_3CCH_2}}\!\!-\!\!Cl}\right)$$

$$CH_2\!\!=\!\!CH\!\!-\!\!Cl + HCl \longrightarrow \underset{\underset{\displaystyle H \quad Cl}{|\quad\ |}}{CH_2\!\!-\!\!CH\!\!-\!\!Cl}$$

vinyl chloride   1,1-dichloroethane (nearly 100%)

The direction of addition which the empirical rule predicts can be accounted for by the formation of a carbonium ion in the two-step mechanism. The positive hydrogen ion goes to the carbon which will give the more stable of the two possible carbonium ion intermediates.

Carbonium ion stability has been found to be in the order of tertiary > secondary > primary. Recall that alkyl groups were labeled tertiary, secondary, or primary according to the number of carbons attached to that carbon which is bonded to a functional group or to a carbon chain (Section 2.6). The same nomenclature system applies to carbonium ions. The positive charge on carbonium ions polarizes all of the bonds to the positive carbon. Tertiary carbonium ions are more stable than others, because the alkyl groups attached to the electron-deficient carbon are more polarizable and electron-releasing than are hydrogens. Movement of electrons toward a carbon bearing a positive charge helps to disperse the positive charge over more atoms. Reducing the concentration of charge at one point by spreading it over many carbons and hydrogens gives greater stability to the ion.

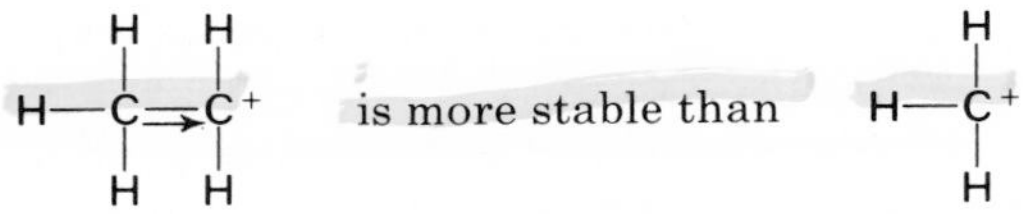

$$H-\underset{\underset{H}{|}}{\overset{\overset{H}{|}}{C}}=\underset{\underset{H}{|}}{\overset{\overset{H}{|}}{C}}{}^{+} \quad \text{is more stable than} \quad H-\underset{\underset{H}{|}}{\overset{\overset{H}{|}}{C}}{}^{+}$$

The addition of a hydrogen ion to isobutylene, $(CH_3)_2C{=}CH_2$, could give a tertiary carbonium ion or a primary carbonium ion. The tertiary cation is the more stable reaction intermediate. In the addition of hydrogen chloride to isobutylene, the tertiary cation is formed and reacts with the chloride ion to produce the tertiary halide, *tert*-butyl chloride.

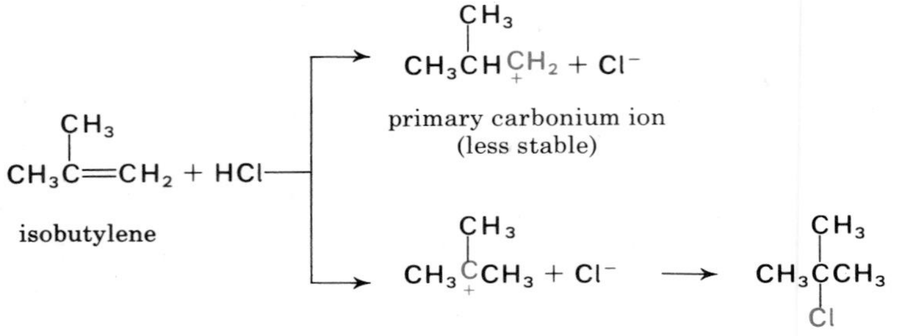

$$\underset{\text{isobutylene}}{CH_3\overset{\overset{CH_3}{|}}{C}{=}CH_2 + HCl}$$

$$CH_3CH\underset{+}{CH_2} + Cl^-$$
primary carbonium ion
(less stable)

$$CH_3\underset{+}{\overset{\overset{CH_3}{|}}{C}}CH_3 + Cl^- \longrightarrow CH_3\underset{\underset{Cl}{|}}{\overset{\overset{CH_3}{|}}{C}}CH_3$$

tertiary carbonium ion   *tert*-butyl chloride
(more stable)   (100% of product)

Problems **4.5** Write the structures of the two carbonium ions which could be formed by the addition of a proton to the following alkenes. Designate which cation you expect to be the more stable intermediate.

a) $CH_2{=}CHCH_2\underset{\underset{CH_3}{|}}{C}HCH_3$ b) 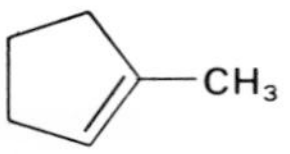 ⬠—CH_3 c) $CH_3CH{=}CHCH_2CH_2CH_3$

4.6 Write equations for the following reactions, labeling both major and minor products.

a) 1-hexene + dilute aqueous sulfuric acid

b) 2-phenyl-2-butene + hydrogen chloride

4.6 ADDITION OF HALOGENS TO ALKENES   Molecular halogens add to alkenes to produce dihaloalkanes. Bromine and 2-butene give 2,3-dibromobutane.

$$CH_3CH{=}CHCH_3 + Br_2 \longrightarrow CH_3\underset{\underset{Br}{|}}{C}H{-}\underset{\underset{Br}{|}}{C}HCH_3$$

2-butene   2,3-dibromobutane

The reaction occurs very rapidly and is used as a quick diagnostic test for an alkene or an alkyne. The alkene and the alkyl dihalide are colorless compounds. The solution of bromine in carbon tetrachloride is added one drop at a time to a solution containing the unknown compound. The disappearance of the red-brown color of bromine indicates the probable presence of carbon–carbon unsaturation. (Aromatic hydrocarbons do not react under these conditions.)

Molecular bromine and chlorine are nonpolar but highly polarizable compounds. They become polarized by the electron-rich double bond. The bromine–bromine bond breaks, both electrons going to one bromine, as new bonds form between the electron-deficient bromine and the unsaturated carbons. A stable bromide ion is left and a positive ion is formed. These two ions then react in a second step to form the dihalide.

$$CH_3CH\!=\!CHCH_3 + :\!\overset{..}{Br}\!\!-\!\!\overset{..}{Br}: \longrightarrow Br^- + CH_3\overset{+}{C}H\!-\!CHCH_3 \longrightarrow CH_3CH\!-\!CHCH_3$$

A *curved arrow* indicates the fate of an electron pair. Here the pair of pi electrons of the double bond forms the first bond between carbon and bromine. The bridged cation with bromine bonded to two carbons, called a bromonium ion, is more stable than the ordinary carbonium ion would be.

4.7 OTHER ADDITION REACTIONS OF ALKENES

In the reaction of an alkene with molecular bromine, the addition product possesses two new functional groups, two bromines attached to adjacent carbons (Section 4.6). This product differs from that obtained by the addition of HX or H_2O, in which only one of the added groups produces a functional group, while the second is a hydrogen. Other reactions of alkenes give two types of products having an oxygen functional group attached to each carbon.

a diol or glycol	an oxirane (a cyclic ether)	two carbonyl products from one alkene

a) Diols from action of neutral potassium permanganate on alkenes

A neutral aqueous solution of potassium permanganate is a mild reagent. Its reaction with an alkene is rapid, nevertheless, and produces a diol having the two hydroxyl groups on adjacent carbons.

$$3\,CH_3\!-\!CH\!=\!CH\!-\!CH_3 + 2\,K^+MnO_4^- + 4\,H_2O \longrightarrow 3\,CH_3\!-\!\underset{HO}{\overset{H}{C}}\!-\!\underset{OH}{\overset{H}{C}}\!-\!CH_3 + 2\,MnO_2 + 2\,K^+OH^-$$

2-butene

2,3-butanediol

Neutral potassium permanganate is dark purple in solution. The marked change in color upon reaction makes it a good diagnostic test for alkenes and alkynes. If the permanganate solution is added slowly to a solution of the alkene, the color disappears and a brown suspension of solid manganese dioxide is formed.

b) Oxiranes from alkenes with peracids Peracids possess a functional group in which a hydroperoxide group, —O—O—H, is attached to the carbonyl group, C=O. These acids bear the same relationship to carboxylic acids that hydrogen peroxide H—O—O—H does to water. This class of compound is important mainly because members act as oxygen atom donors to alkenes. **Peracetic acid** is an example.

$$CH_3-C \overset{O}{\underset{O-O-H}{\big<}}$$

peracetic acid

Peracetic acid reacts with an alkene to form an oxirane, a three-membered-ring ether. The peracetic acid is converted into acetic acid. The reaction of 2-pentene illustrates the reaction.

$$CH_3-C\overset{O}{\underset{O-O-H}{\big<}} \;+\; CH_3CH{=}CHCH_2CH_3 \;\longrightarrow\; CH_3\overset{O}{\overset{\triangle}{CH{-}CH}}CH_2CH_3 \;+\; CH_3-C\overset{O}{\underset{OH}{\big<}}$$

peracetic acid   2-pentene   2-ethyl-3-methyloxirane   acetic acid

The oxygen atom bonds with both of the unsaturated carbons from the same side of the double bond to give the three-membered ring.

Oxiranes are useful as intermediate compounds for synthetic sequences, because they react in predictable ways to give products which are otherwise difficult to prepare. (See Section 5.11.)

Problems **4.7** Draw the structure of the product of each of these reactions.

    a) 1,5-hexadiene + neutral potassium permanganate

    b) cyclopentene + peracetic acid

4.8 Draw the structure of the geometric isomer of the oxirane formed by the reaction of peracetic acid with each of the following.

    a) *cis*-2-pentene   **b)** *trans*-2-pentene

c) Reaction of alkenes with ozone—ozonolysis Some strong oxidizing agents are capable of breaking both carbon–carbon bonds of a double bond. Before X-ray crystal structure and spectroscopic techniques (Chapter 15) were available, the structures of new compounds isolated from natural sources had to be determined by their degradation to smaller known compounds. One process used extensively for degradation is **ozonolysis**, which is a cleavage of the C=C by reaction with ozone. This reaction indicates the exact position of each C=C in the original alkene.

A cyclic peroxide called an ozonide is formed initially in the ozonolysis reaction.

$$O_3 + \quad \overset{\diagdown}{\underset{\diagup}{C}}{=}\overset{\diagup}{\underset{\diagdown}{C}} \quad \longrightarrow \quad \overset{\diagup}{\underset{\diagdown}{C}}\overset{O-O}{\underset{O}{\diagdown \diagup}}\overset{\diagup}{\underset{\diagdown}{C}}$$

an ozonide

Because of its tendency to explode, the ozonide is not isolated. The ozonide is cleaved by treatment with zinc and water to give two compounds, each with C=O O=C where the original C=C was. On ozonolysis and cleavage an olefinic carbon with two carbon substituents gives rise to a ketone. If the olefinic carbon has one hydrogen and one carbon substituent, the product for that carbon is an aldehyde.

$$\underset{H}{\overset{R}{\diagdown}}C{=}C\underset{R}{\overset{R}{\diagup}} \quad \xrightarrow{O_3} \quad \xrightarrow{H_2O,\ Zn} \quad \underset{H}{\overset{R}{\diagdown}}C{=}O + O{=}C\underset{R}{\overset{R}{\diagup}}$$

The products from ozonolysis of 2-methyl-2-pentene are propanal and acetone.

$$\underset{\text{2-methyl-2-pentene}}{CH_3CH_2CH{=}\overset{\overset{\displaystyle CH_3}{|}}{C}CH_3} + O_3 \longrightarrow CH_3CH_2\overset{O-O}{\underset{O}{CH\diagdown\diagup C(CH_3)_2}} \xrightarrow{Zn,\ H_2O} \underset{\text{propanal}}{CH_3CH_2\overset{H}{\overset{|}{C}}{=}O} + \underset{\text{acetone}}{O{=}\overset{\overset{\displaystyle CH_3}{|}}{C}CH_3}$$

Problems   **4.9** Draw the structures of the products formed in these reactions.

a) $CH_3CH{=}CHCH_2CH_2CH{=}\overset{\overset{\displaystyle}{\underset{\displaystyle CH_3}{|}}}{C}CH_2CH_3 + O_3$ followed by Zn + H₂O

b) CH₃— ⬠ $\xrightarrow{O_3}$ $\xrightarrow{Zn,\ H_2O}$

4.10 Write a possible structure for a compound which on ozonolysis and cleavage by zinc and water gives the products listed.

a) two equivalents of $CH_3\overset{\overset{\displaystyle}{\underset{\displaystyle O}{||}}}{C}CH_2CH_3$ + one equivalent of $H\overset{\overset{\displaystyle}{\underset{\displaystyle O}{||}}}{C}CH_2CH_2\overset{\overset{\displaystyle}{\underset{\displaystyle O}{||}}}{C}H$

b) $CH_3\overset{\overset{\displaystyle}{\underset{\displaystyle O}{||}}}{C}CH_3 + CH_3\overset{\overset{\displaystyle}{\underset{\displaystyle O}{||}}}{C}CH_2CH_2\overset{\overset{\displaystyle}{\underset{\displaystyle O}{||}}}{C}CH_2\overset{\overset{\displaystyle}{\underset{\displaystyle O}{||}}}{C}H$

4.8 ALKYNES Hydrocarbons containing a carbon–carbon triple bond are called **alkynes**. The simplest alkyne is **acetylene**, $H-C\equiv C-H$, known by its common name. The names of other alkynes follow the IUPAC system.

1. The longest carbon chain containing the $C\equiv C$ bond is selected as the base structure. Its name has the stem of the name of the alkane having the same number of carbons. The ending "yne" is added to the stem.

2. The carbons are numbered from the end nearest the $C\equiv C$ bond.

3. Hydrocarbon, halogen, and amino groups are named as substituents with the number of the carbon to which the substituent is attached.

$$CH_3-C\equiv C-CH_3 \qquad\qquad H-C\equiv C-H$$

2-butyne ethyne (IUPAC)
acetylene (common)

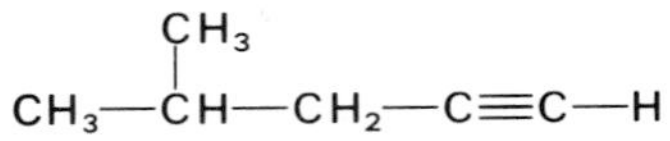

$$\underset{\text{4-methyl-1-pentyne}}{CH_3-\overset{\overset{\displaystyle CH_3}{|}}{C}H-CH_2-C\equiv C-H}$$

Recall that the geometry of the alkyne group is linear, the four atoms $C-C\equiv C-C$ lying in one line (Section 3.2). Since only this linear arrangement of atoms around the triple bond is possible, no stereoisomers exist.

Acetylene is obtained in large quantities from coke, the solid residue from the distillation of volatile compounds of coal. Coke is heated with calcium oxide CaO at 2000° to give calcium carbide and carbon monoxide. Calcium carbide and water react to give acetylene and CaO.

$$3\ C + CaO \xrightarrow{2000°} CaC_2 + CO; \qquad CaC_2 + H_2O \longrightarrow HC\equiv CH + CaO$$

coke calcium
carbide
acetylene

Most other alkynes must be prepared indirectly, usually from acetylene. The acetylenic hydrogen on a terminal alkyne group is weakly acidic. It can be removed by an exceedingly strong base such as the amide ion NH_2^-.

$$H-C\equiv C-H + Na^+NH_2^- \longrightarrow H-C\equiv C^-Na^+ + NH_3$$

sodium sodium ammonia
amide acetylide

The acetylide ion is used in the preparation of longer chain alkynes by reaction with alkyl halides. See Section 6.3(c) for a similar reaction.

4.9 REACTIONS OF ALKYNES

The principal type of reaction of alkynes, like that of alkenes, is addition. The same reagents add to the triple bond as to the double bond with some variations in conditions. The reactions of alkynes can usually be stopped after the first stage of addition to give alkene products, or they can be carried on to a second addition to give saturated products.

a) Catalytic hydrogenation

The addition of hydrogen gas to an alkyne by catalytic hydrogenation leads to alkenes. A continuation of hydrogenation leads to the alkane.

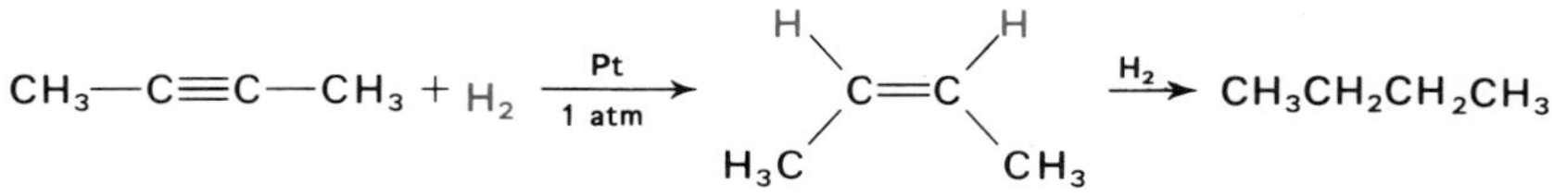

$$CH_3-C\equiv C-CH_3 + H_2 \xrightarrow[\text{1 atm}]{\text{Pt}} \underset{H_3C}{\overset{H}{\diagdown}}C=C\underset{CH_3}{\overset{H}{\diagup}} \xrightarrow{H_2} CH_3CH_2CH_2CH_3$$

2-butyne   cis-2-butene   butane

Catalysts for hydrogenation of alkynes are the same metals as for alkenes, platinum and palladium (Section 4.3). The two hydrogens add from the same side of the triple bond and the alkene has the two hydrogens in the cis position.

b) Addition of hydrogen halide

Hydrogen chloride adds once to an alkyne to form a haloalkene. Further addition gives a dihaloalkane.

$$H-C\equiv C-H + HCl \longrightarrow CH_2=CH-Cl \xrightarrow{HCl} CH_3-CHCl_2$$

acetylene   vinyl chloride   1,1-dichloroethane

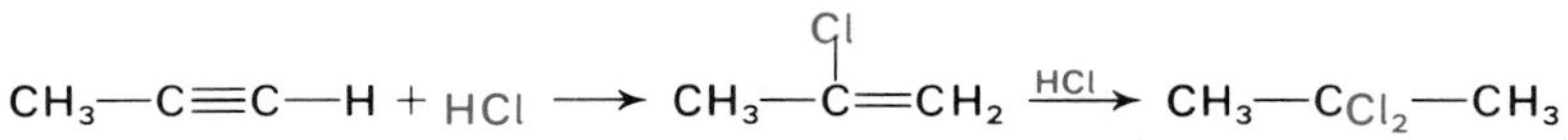

$$CH_3-C\equiv C-H + HCl \longrightarrow CH_3-\overset{Cl}{\underset{|}{C}}=CH_2 \xrightarrow{HCl} CH_3-CCl_2-CH_3$$

propyne   2-chloropropene   2,2-dichloropropane

The direction of the addition of HCl to an alkyne follows Markownikoff's rule (Section 4.5). The carbonium ion intermediate formed is the more stable one. This cation reacts quickly with the negative chloride ion left from the HCl to give a chloroalkene in which the halogen is bonded to an unsaturated carbon.

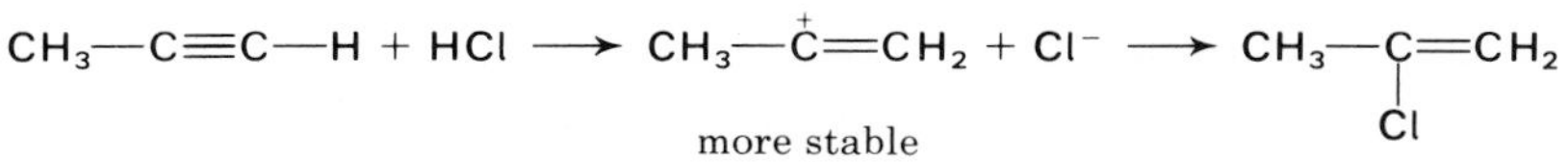

$$CH_3-C\equiv C-H + HCl \longrightarrow CH_3-\overset{+}{C}=CH_2 + Cl^- \longrightarrow CH_3-\underset{Cl}{\overset{|}{C}}=CH_2$$

more stable
cation

c) Addition of water   Hydration of alkynes requires catalysis by both sulfuric acid and mercuric sulfate. Mercuric ion (Hg^{2+}) is known to form an association with an alkyne whereby the triple bond is weakened. The addition product first formed with water is an unsaturated alcohol, an **enol** ("ene + ol"), which is an unstable structure. The enol quickly isomerizes by movement of a proton from oxygen to carbon to form a stable compound, the ketone (or acetaldehyde from acetylene).

$$HC\equiv CH + H_2O \xrightarrow[HgSO_4]{H_2SO_4} CH_2\!=\!CH\!-\!OH \longrightarrow CH_3\!-\!CH\!=\!O$$

vinyl alcohol   acetaldehyde
(an unstable enol)

$$CH_3C\equiv CH + H_2O \xrightarrow[HgSO_4]{H_2SO_4} CH_3\!-\!\underset{\underset{OH}{|}}{C}\!=\!CH_2 \longrightarrow CH_3\!-\!\underset{\underset{O}{\|}}{C}\!-\!CH_3$$

propyne   enol of acetone   acetone

The addition of water to an alkyne follows Markownikoff's rule. Thus for all alkynes (except acetylene) the hydration product is a ketone and not an aldehyde, e.g., acetone and not propanal, $CH_3CH_2CH\!=\!O$. The addition of water does not go beyond the ketone stage.

Acetic acid adds to acetylene under conditions similar to hydration to give vinyl acetate, a monomer useful for making polymers.

$$HC\equiv CH + CH_3CO_2H \longrightarrow CH_3C\!\!\overset{\displaystyle O}{\underset{\displaystyle O-CH=CH_2}{<}}$$

vinyl acetate

Note that in this reaction the first addition product is stable. No isomerization of the vinyl acetate occurs, since there is no H on the oxygen to move.

d) Halogens   Bromine and chlorine add to alkynes to form dihaloalkenes and tetrahaloalkanes.

$$H\!-\!C\equiv C\!-\!H + Br_2 \longrightarrow CHBr\!=\!CHBr \xrightarrow{Br_2} CHBr_2\!-\!CHBr_2$$

1,2-dibromoethene   1,1,2,2-tetrabromoethane

Problem   **4.11** Write the equations for the following reactions.

a) 3-hexyne + water + sulfuric acid + mercuric sulfate catalysts

b) 1,5-hexadiyne + hydrogen chloride (excess)

4.10 TERPENES   Many plants produce hydrocarbons as metabolic products. Alkenes and cycloalkenes are found in great abundance, including such examples as natural rubber, turpentine, some colored plant pigments, and some components of the essential oils of fragrant plants. The essences of the volatile oils of many plants have been in great demand since antiquity for perfumes—rose, lavender—and flavorings—lemon and peppermint. Volatile oils of other plants have been used for a variety of medicinal purposes, such as antiseptics and emollients—oil of eucalyptus, turpentine, camphor.

The great interest in essential oils very early challenged organic chemists to establish the structures of the components and to synthesize the compounds. They found that many of these compounds have formulas containing a multiple of five carbons. Their carbon skeletons are composed of isoprene units.

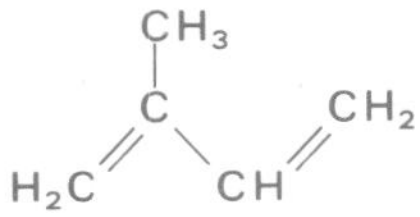

isoprene 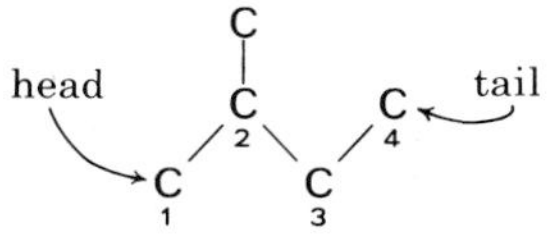   isoprenoid unit

The isoprene units are usually found attached to each other in a *head-to-tail* fashion, although some exceptions are known.

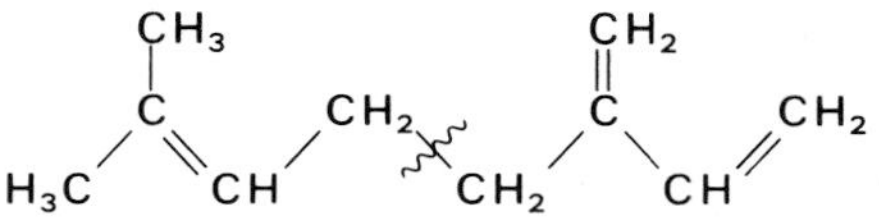

myrcene   head-to-tail
(oil of bayberry)   isoprene units

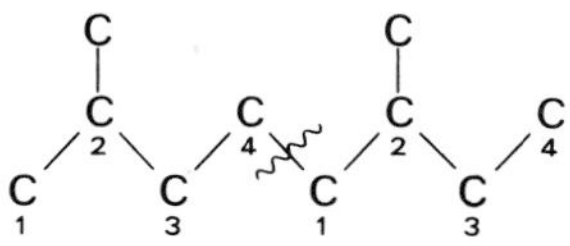

These isoprenoid compounds are called **terpenes**, and are grouped according to the number of isoprene units in the molecule.

monoterpenes	C_{10}	two isoprene units
sesquiterpenes	C_{15}	three isoprene units
diterpenes	C_{20}	four isoprene units
triterpenes	C_{30}	six isoprene units
tetraterpenes	C_{40}	eight isoprene units

The monoterpenes (C_{10}) are found as open-chain and cyclic compounds. Examples contain double bonds and hydroxyl or carbonyl functional groups. The three types of carbon structures are here represented by geraniol, limonene, and pinene, compounds which are found in the volatile oils of many plants.

Geraniol has a sweet rose odor. Besides being the chief constituent of attar of roses, the oil of East Indian geranium, and citronella, geraniol is found in oils of lemon grass, orange blossoms, nutmeg, and lavender. It is used in perfumery.

Geraniol has a typical monoterpene carbon skeleton, two isoprene units joined head-to-tail. It retains some of the isoprene unsaturation (two double bonds) and also contains one hydroxyl group.

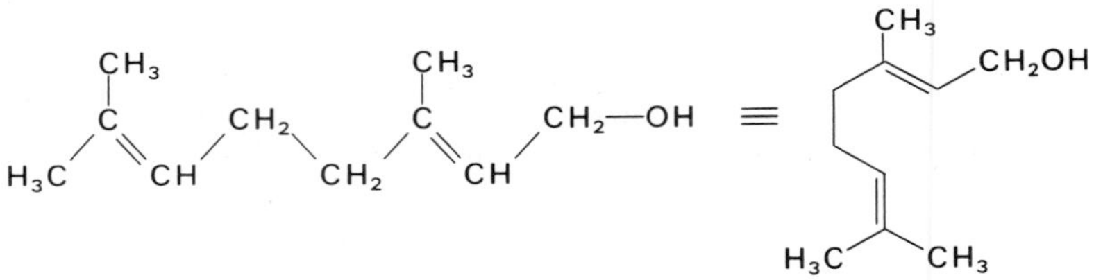

$$CH_3 \quad\quad CH_2 \quad\quad CH_3 \quad\quad CH_2{-}OH \;\equiv\; CH_3 \;\; CH_2OH$$

geraniol

Other examples of open-chain monoterpenes are the hydrocarbon myrcene (oil of bayberry), the alcohol citronellol (rose and citronella), and the aldehyde geranial (lemon grass, lemon and orange oils). Structures of these compounds are shown in Fig. 4.1.

The characteristic odor of the oils from lemon and orange peels is due to an oily liquid called **limonene**, which makes up 80–90% of these volatile oils. Limonene is a lesser constituent of the essential oils of many other plants, notably the herbs spearmint, peppermint, and lavender. It is a hydrocarbon, a monocyclic diene of ten carbons.

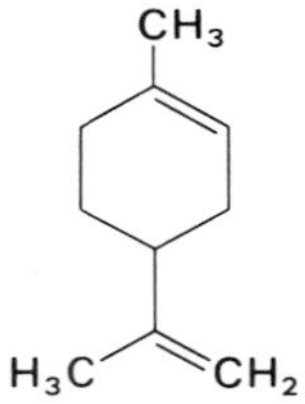

$$CH_3$$
$$H_3C \quad CH_2$$

limonene

Limonene is representative of a large number of terpenes with the same carbon skeleton, a six-membered ring with methyl and isopropyl substituents in the 1,4-positions of the ring. Besides double bonds in and out of the ring, terpenes with the same carbon skeleton have hydroxyl, ether, and carbonyl functional groups. The structures of menthol, an alcohol (from peppermint), of eucalyptol, an ether (from eucalyptus), of thymol, a phenol (from thyme), and of carvone, a ketone (from caraway and dill) are shown in Fig. 4.1.

Among the many familiar compounds, **pinene** is the most important hydrocarbon. It is the major constituent of turpentine, the volatile

FIGURE 4.1   REPRESENTATIVE MONOTERPENES

Open-chain monoterpenes

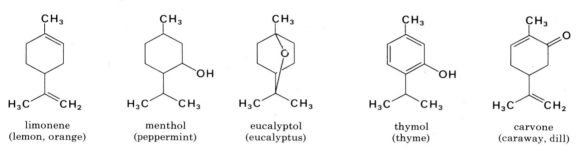

| myrcene
(bayberry) | geraniol
(rose) | citronellol
(rose) | geranial
(lemon grass) |

Monocyclic monoterpenes

limonene
(lemon, orange) menthol
(peppermint) eucalyptol
(eucalyptus) thymol
(thyme) carvone
(caraway, dill)

Bicyclic monoterpenes

α-pinene
(pine, cypress) borneol
(many plants) camphor
(camphor)

component of pine resins. Pinene is also widely found as a lesser constituent of many oils.

The most unusual feature of pinene is the four-membered ring in a bicyclic structure. (The double bond may be in other positions.)

α-pinene
(other isomers have
C=C in other
positions)

A bicyclic carbon skeleton is a structure which has two or more carbon atoms common to two rings. *All* of the ring atoms of the bicyclic structure are counted in tracing only *two* rings. A third ring is obvious, but the carbons in it are *all duplicated* in one of the first two rings.

Another well-known bicyclic skeleton appears in camphor, a ketone (from the camphor tree), and in borneol, an alcohol constituent of many oils. These compounds are shown in Fig. 4.1.

Most essential oils are a mixture of many terpenes, which either are quite volatile (have low boiling points) or sublime easily, as does camphor. For example, the oil of rosemary, an extremely aromatic herb, contains borneol, camphor, eucalyptol, pinene, and camphene, a bicyclic monoterpene hydrocarbon (formulas in Fig. 4.1).

It is believed that in plants the role of many terpenes, particularly turpentine, is to provide a defense against pests. Certainly very few insects can penetrate pine bark or needle, a fact probably due to the resin present.

Larger terpenes are important in the chemistry of vertebrates. **Vitamin A**, or **retinol**, is a diterpene alcohol, C_{20}, which is essential in very small quantity for proper cell reactions. Its best-known role is in the cycle of reactions of vision.

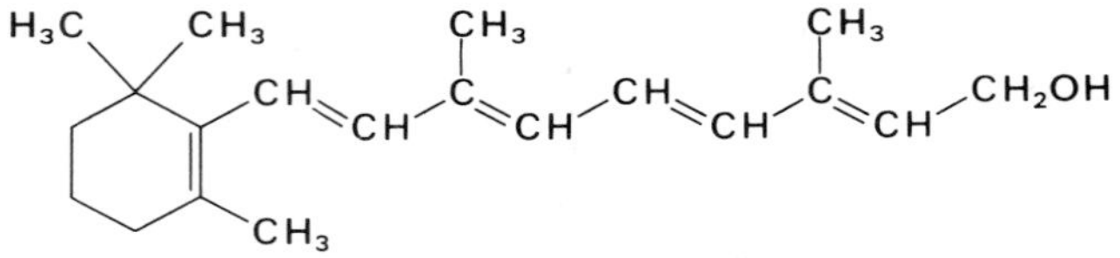

vitamin A
(retinol)

Some plant terpenes are an important part of the human diet because they are converted in the liver to vitamin A. Such is the case with a group of carotenoids. Those carrots that we are encouraged to eat contain the yellow pigment **β-carotene**, which is a tetraterpene hydrocarbon (C_{40}). The colors of carrots, tomatoes, and many other red, orange, and yellow plant products are due to the presence of carotenoids similar to carotene. Color of organic compounds is associated with long chains of alternating double and single bonds. (See Section 15.2.)

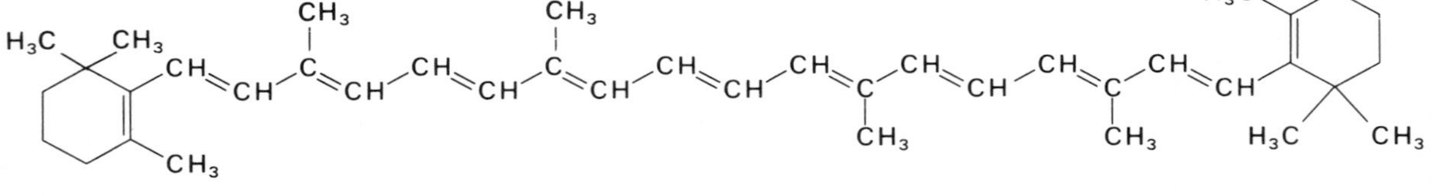

β-carotene
(yellow, carrot)

Problem **4.12** Examine carefully the structures of β-carotene and of vitamin A shown above and answer these questions.

a) Note that vitamin A has a carbon structure that is identical to one-half of the carbon structure of β-carotene. What is the structure of the cyclic portion of each molecule—ring size, unsaturation, and substitution on the ring?

b) What is the number of C=C bonds in each compound? How are they spaced in the molecule?

c) Indicate on a drawing of each structure the isoprene units. Note any isoprene units that are *not* joined head-to-tail.

New Terms and Topics

Alkene, diene, polyene—olefinic linkage (Section 4.1)

Polymer (Section 4.1)

Nomenclature of alkenes—*E–Z* isomers (Section 4.2)

Addition reaction (Section 4.3)

Catalytic hydrogenation; catalyst; *cis*-addition (Section 4.3)

Electrophilic reagent, nucleophilic reagent (Section 4.4)

Reaction mechanism; reaction intermediate (Section 4.4)

Two-step mechanism of addition of polar reagents (Section 4.4)

Carbonium ions—primary, secondary, tertiary (Section 4.4)

Orientation of addition of unsymmetrical reagents—Markownikoff's rule (Section 4.5)

Relative stability of carbonium ion reaction intermediates (Section 4.5)

Mechanism of halogen addition (Section 4.6)

Hydroxylation; 1,2-glycols (Section 4.7)

Peracetic acid; oxirane (Section 4.7)

Ozonolysis (Section 4.7)

Alkynes—structure and nomenclature (Section 4.8)

Terpenes—essential oils (Section 4.10)

Vitamin A, carotene, geraniol, limonene (Section 4.10)

Summary of Reactions

PREPARATION OF SMALL ALKENES FROM ALKANES (Section 4.1)

$$CH_3-CH_3 \xrightarrow{800°} CH_2=CH_2 + H_2$$

ethane · · · · · · · · · ethylene

ADDITION REACTIONS OF ALKENES

1. Hydrogen—catalytic hydrogenation (Section 4.3)

$$H_2 + R_2C=CR_2 \xrightarrow{Pt} R_2CH-CHR_2$$

$$H_2 + CH_3-CH=CH-CH_2CH_3 \xrightarrow[25°\ 1\ atm]{Pt} CH_3-CH_2-CH_2-CH_2-CH_3$$

2-pentene · · · · · · · · · · · · pentane

2. Bromine (Section 4.6)

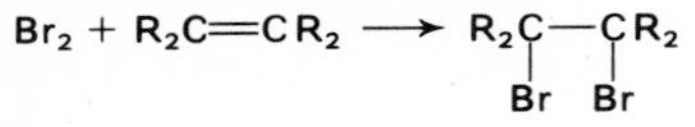

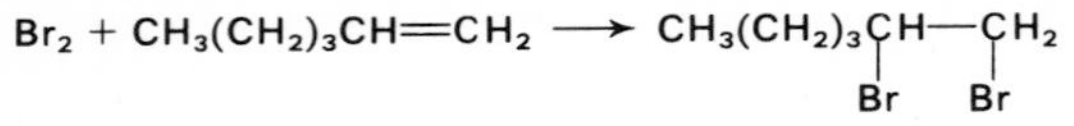

$$Br_2 + R_2C{=}CR_2 \longrightarrow R_2\underset{\displaystyle Br}{C}-\underset{\displaystyle Br}{C}R_2$$

$$Br_2 + CH_3(CH_2)_3CH{=}CH_2 \longrightarrow CH_3(CH_2)_3\underset{\displaystyle Br}{C}H-\underset{\displaystyle Br}{C}H_2$$

1-hexene   1,2-dibromohexane

3. Hydrogen bromide (Sections 4.4 and 4.5)

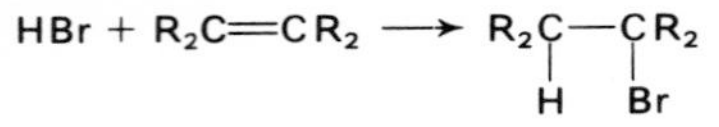

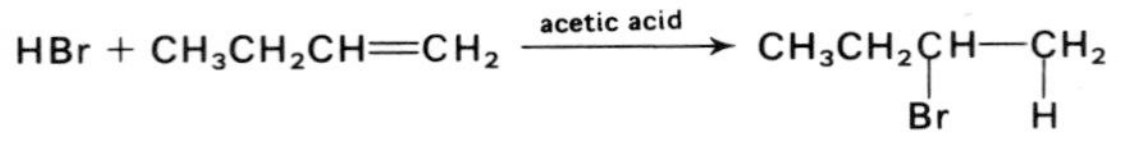

$$HBr + R_2C{=}CR_2 \longrightarrow R_2\underset{\displaystyle H}{C}-\underset{\displaystyle Br}{C}R_2$$

$$HBr + CH_3CH_2CH{=}CH_2 \xrightarrow{\text{acetic acid}} CH_3CH_2\underset{\displaystyle Br}{C}H-\underset{\displaystyle H}{C}H_2$$

1-butene   2-bromobutane
(*sec*-butyl bromide)

4. Water—hydration (Sections 4.4 and 4.5)

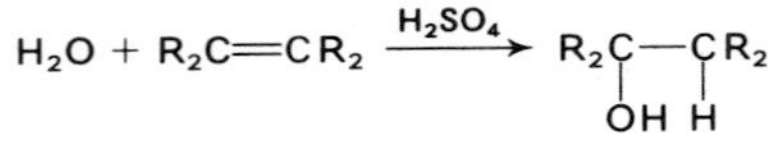

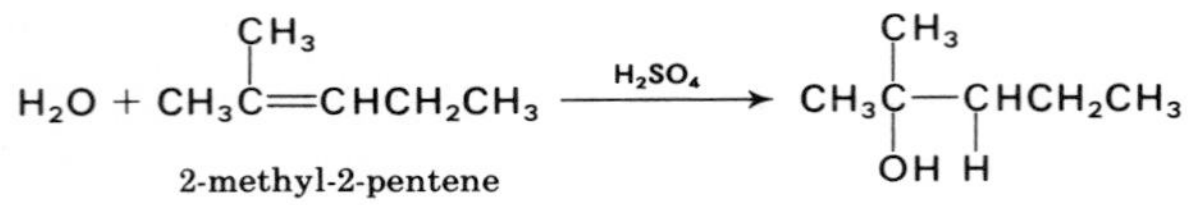

$$H_2O + R_2C{=}CR_2 \xrightarrow{H_2SO_4} R_2\underset{\displaystyle OH}{C}-\underset{\displaystyle H}{C}R_2$$

$$H_2O + CH_3\underset{}{\overset{\displaystyle CH_3}{C}}{=}CHCH_2CH_3 \xrightarrow{H_2SO_4} CH_3\underset{\displaystyle OH}{\overset{\displaystyle CH_3}{C}}-\underset{\displaystyle H}{C}HCH_2CH_3$$

2-methyl-2-pentene   2-methyl-2-pentanol

5. Reaction with potassium permanganate (neutral solution) (Section 4.7)

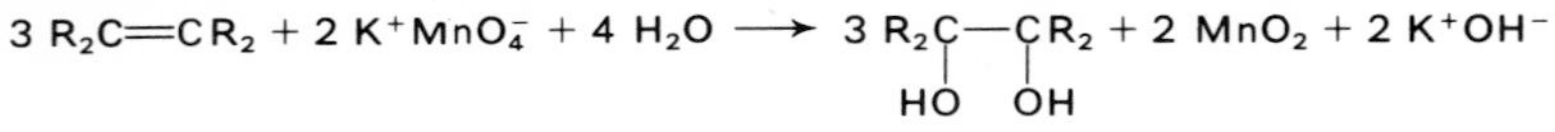

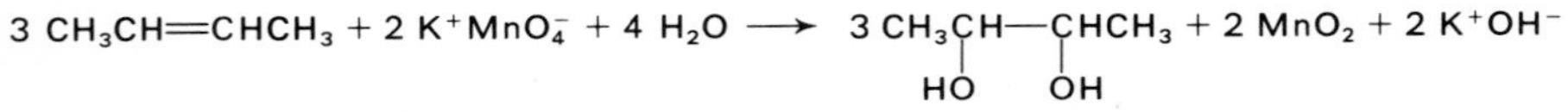

$$3\ R_2C{=}CR_2 + 2\ K^+MnO_4^- + 4\ H_2O \longrightarrow 3\ R_2\underset{\displaystyle HO}{C}-\underset{\displaystyle OH}{C}R_2 + 2\ MnO_2 + 2\ K^+OH^-$$

$$3\ CH_3CH{=}CHCH_3 + 2\ K^+MnO_4^- + 4\ H_2O \longrightarrow 3\ CH_3\underset{\displaystyle HO}{C}H-\underset{\displaystyle OH}{C}HCH_3 + 2\ MnO_2 + 2\ K^+OH^-$$

2-butene   2,3-butanediol

6. Reaction with peracetic acid (Section 4.7)

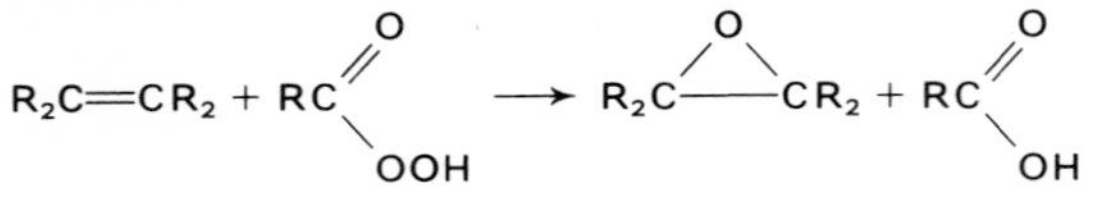

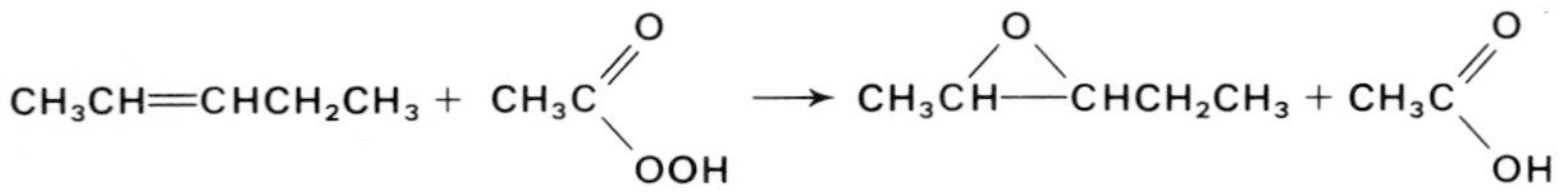

$$R_2C{=}CR_2 + RC\underset{\displaystyle OOH}{\overset{\displaystyle O}{\diagdown}} \longrightarrow R_2C\overset{\displaystyle O}{\diagup\diagdown}CR_2 + RC\underset{\displaystyle OH}{\overset{\displaystyle O}{\diagdown}}$$

$$CH_3CH{=}CHCH_2CH_3 + CH_3C\underset{\displaystyle OOH}{\overset{\displaystyle O}{\diagdown}} \longrightarrow CH_3CH\overset{\displaystyle O}{\diagup\diagdown}CHCH_2CH_3 + CH_3C\underset{\displaystyle OH}{\overset{\displaystyle O}{\diagdown}}$$

2-pentene   peracetic acid   2-ethyl-3-methyl-oxirane   acetic acid

7. Ozonolysis (Section 4.7)

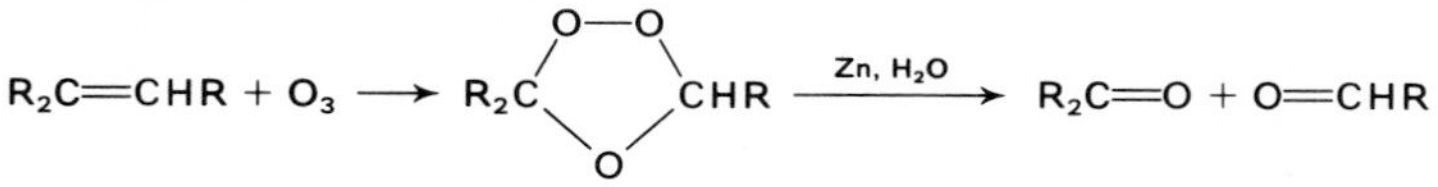

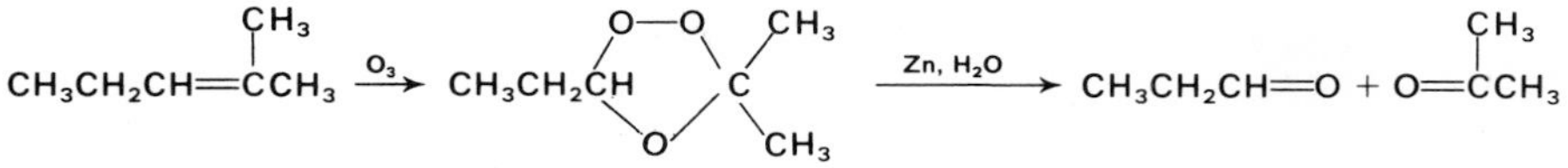

$$R_2C{=}CHR + O_3 \longrightarrow R_2C\underset{O}{\overset{O-O}{\diagdown \diagup}}CHR \xrightarrow{\text{Zn, H}_2O} R_2C{=}O + O{=}CHR$$

$$CH_3CH_2CH{=}\underset{\underset{CH_3}{|}}{C}CH_3 \xrightarrow{O_3} CH_3CH_2\underset{O}{\overset{O-O}{CH \diagdown \diagup}}\underset{\underset{CH_3}{|}}{C}{-}CH_3 \xrightarrow{\text{Zn, H}_2O} CH_3CH_2CH{=}O + O{=}\underset{\underset{CH_3}{|}}{C}CH_3$$

2-methyl-2-pentene   an ozonide   propanal   acetone

PREPARATION OF ACETYLENE FROM COKE (Section 4.8)

$$3\ C + CaO \longrightarrow CaC_2\ +\ CO; \qquad CaC_2 + H_2O \longrightarrow HC{\equiv}CH + CaO$$

coke calcium   calcium   carbon   acetylene
    oxide   carbide   monoxide

ADDITION REACTIONS OF ALKYNES (Section 4.9)

1. Water—hydration

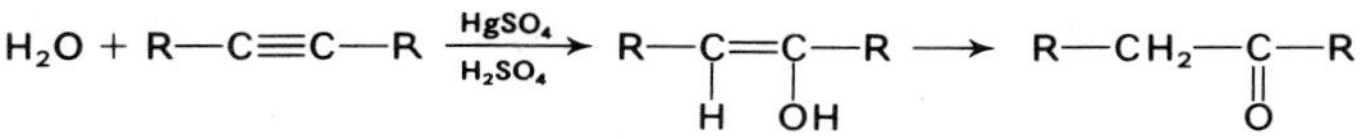

$$H_2O + R{-}C{\equiv}C{-}R \xrightarrow[\text{H}_2\text{SO}_4]{\text{HgSO}_4} R{-}\underset{\underset{H}{|}}{C}{=}\underset{\underset{OH}{|}}{C}{-}R \longrightarrow R{-}CH_2{-}\underset{\underset{O}{\|}}{C}{-}R$$

             enol   ketone

$$H_2O + H{-}C{\equiv}C{-}CH_3 \xrightarrow[\text{H}_2\text{SO}_4]{\text{HgSO}_4} H{-}\underset{\underset{H}{|}}{C}{=}\underset{\underset{OH}{|}}{C}{-}CH_3 \longrightarrow CH_3{-}\underset{\underset{O}{\|}}{C}{-}CH_3$$

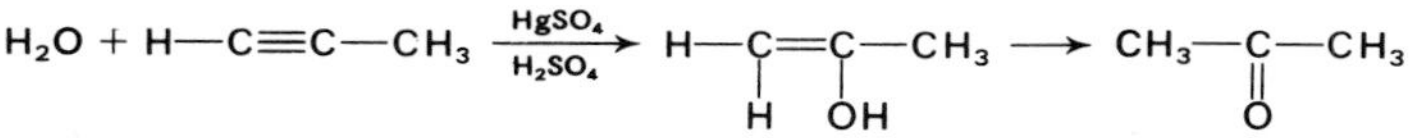

   propyne   enol of acetone   acetone

$$HC{\equiv}CH + H_2O \xrightarrow[\text{HgSO}_4]{\text{H}_2\text{SO}_4} CH_2{=}CH{-}OH \longrightarrow CH_3{-}CH{=}O$$

acetylene   vinyl alcohol   acetaldehyde
          unstable enol

2. Hydrogen halide

$$H{-}X + R{-}C{\equiv}C{-}R \longrightarrow RCH{=}CXR$$

$$H{-}Cl + H{-}C{\equiv}C{-}H \longrightarrow H_2C{=}CHCl \xrightarrow{\text{HCl}} CH_3{-}CHCl_2$$

           vinyl chloride   1,1-dichloroethane

3. Halogen

$$R{-}C{\equiv}C{-}R + Cl_2 \longrightarrow RCCl{=}CRCl \xrightarrow{\text{Cl}_2} RCCl_2{-}CRCl_2$$

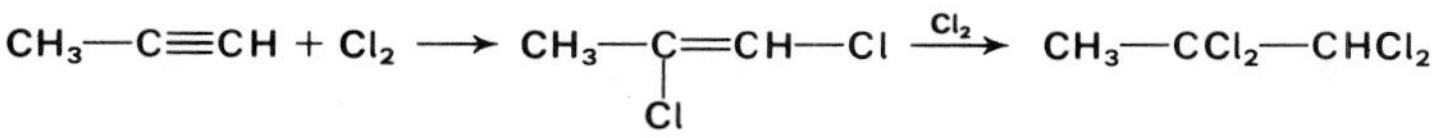

$$CH_3{-}C{\equiv}CH + Cl_2 \longrightarrow CH_3{-}\underset{\underset{Cl}{|}}{C}{=}CH{-}Cl \xrightarrow{\text{Cl}_2} CH_3{-}CCl_2{-}CHCl_2$$

   propyne   1,2-dichloropropene   1,1,2,2-tetrachloropropane

4. Hydrogen

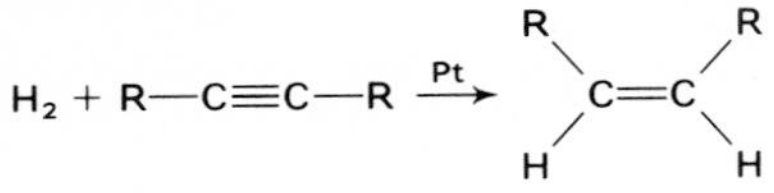

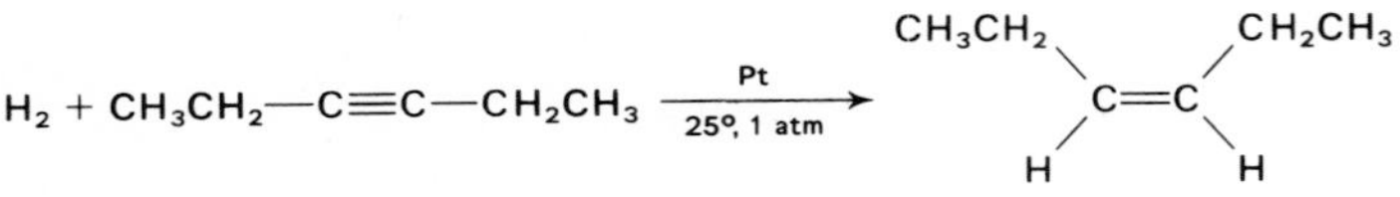

$$H_2 + R-C\equiv C-R \xrightarrow{Pt} \underset{H}{\overset{R}{C}}=\underset{H}{\overset{R}{C}}$$

$$H_2 + CH_3CH_2-C\equiv C-CH_2CH_3 \xrightarrow[25°,\ 1\ atm]{Pt} \underset{H}{\overset{CH_3CH_2}{C}}=\underset{H}{\overset{CH_2CH_3}{C}}$$

3-hexyne   *cis*-3-hexene

Problems

4.13 Draw structures for the following compounds.

a) 2-pentene

b) 2-ethyl-1,3-butadiene

c) 3-penten-1-yne

d) 1,4-hexadiyne

e) *cis*-1,2-diphenylethene

f) 3-phenyl-1-butene

g) *E*-1-phenyl-1-propene

h) isoprene

i) vinyl chloride

j) polyvinyl chloride

k) allyl chloride

l) fumaric acid

m) *trans*-cyclooctene

n) *Z*-3-*tert*-butyl-2-pentene

o) *E*-1-methyl-2-bromocyclopentanol

p) *E,E*-2-chloro-2,4-hexadiene

4.14 Give IUPAC names for the following.

a) 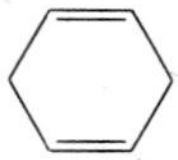 (cyclohexadiene ring structure)

b) 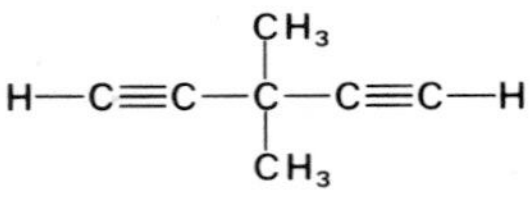 $H-C\equiv C-\underset{CH_3}{\overset{CH_3}{C}}-C\equiv C-H$

c) $CH_3-CH_2-\underset{}{\overset{CH_3}{CH}}-\underset{}{\overset{Br}{CH}}-CH=CH_2$

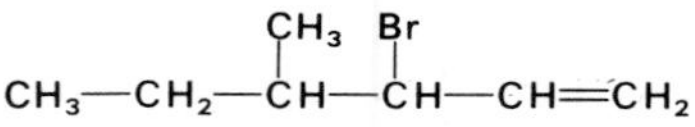

d) 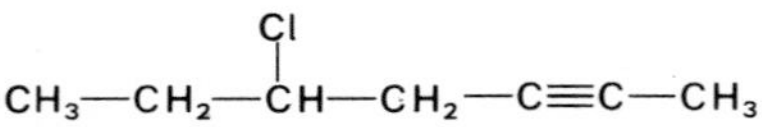 $CH_3-CH_2-\underset{}{\overset{Cl}{CH}}-CH_2-C\equiv C-CH_3$

e) $CH_3-CH_2-CH=C=CH-CH_2-CH_3$

4.15 Draw structures for the products of the following reactions. (Give the major product when two are possible.)

a) $CH_3CH_2CH=C(CH_3)_2 + H_2O + H_2SO_4$

b) 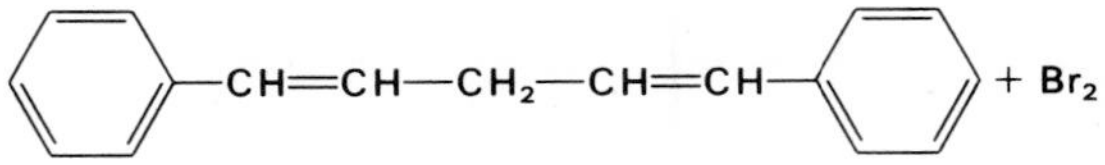 (phenyl)$-CH=CH-CH_2-CH=CH-$(phenyl) $+ Br_2$

c) 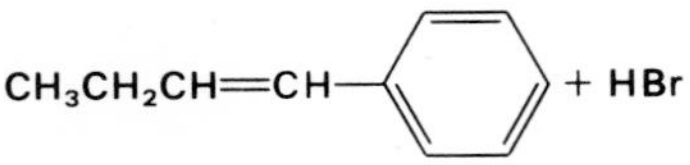 $CH_3CH_2CH=CH-$(phenyl) $+ HBr$

d) $CH_3CH_2CH_2C\equiv CH + Na^+NH_2^-$

e) $CH_3C\equiv CCH_3 + H_2 + Pt$

f) 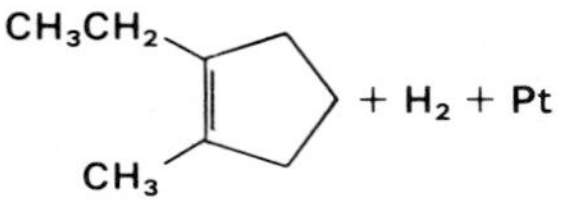 (cyclopentene with CH_3CH_2 and CH_3 substituents) $+ H_2 + Pt$

g) $HC\equiv CCH_2CH_2CH_3 + HCl$ (unlimited supply)

h) $CH_3CH_2CH=CHCH_3 +$ neutral $KMnO_4$

i) $CH_3CH_2CH=CHCH_2CH_3 + CH_3CO_3H$

j) $CH_3CH_2C\equiv CH + H_2O + H_2SO_4 + HgSO_4$

4.16 a) The addition of hypochlorous acid HOCl to an alkene produces a chloroalcohol.

$$CH_2{=}CH_2 + H{-}O{-}Cl \longrightarrow Cl{-}CH_2{-}CH_2{-}OH$$

On the basis of relative electronegativity of oxygen and chlorine, predict and draw the structure for the major product of addition of HOCl to 1-butene.

b) If the addition of HOCl to 1-butene is a two-step process like that of other electrophilic reagents, what is the most likely structure for the intermediate cation?

4.17 Give the structures of the products formed from each of the following compounds with ozone followed by zinc and water cleavage.

a) limonene   **b)** geraniol (formulas in Fig. 4.1)

4.18 Explain why *cis*-cyclohexene is a known, stable compound, but *trans*-cyclohexene and cyclohexyne are not known.

cis-cyclohexene   *trans*-cyclohexene   cyclohexyne

4.19 Allene (propadiene $CH_2{=}C{=}CH_2$) has the molecular formula C_3H_4. All three carbons lie on one axis. The central carbon atom is *sp* hybridized with two unhybridized *p* orbitals. The terminal carbons are sp^2 hybridized with one *p* orbital unhybridized. The three atoms of each CH_2 group form a plane. Are the two CH_2 planes perpendicular to each other, do they coincide (coplanar), or do they rotate with respect to each other? Justify your answer in terms of a molecular orbital picture. Think of the orbital drawing of ethylene and acetylene in developing your answer. Draw your best structure for allene showing the relative positions of the four hydrogens.

4.20 Bromine in methanol (CH_3OH) solution adds to ethylene to give 1,2-dibromoethane and $Br{-}CH_2CH_2{-}OCH_3$. How do you account for the formation of the bromoether? Write equations for the reaction steps. (*Hint:* Remember the formation of diethyl ether from the acid-catalyzed reaction of ethanol with ethylene.)

4.21 a) What is the major product of this reaction? What is a possible minor product?

$$\text{(cyclohexane ring)}{=}CH_2 + H_2O + H_2SO_4$$

Account for the preferential formation of the major product.

b) Which compound of the pair given would you predict to be the more reactive with aqueous sulfuric acid? Justify your choice.

(i) 1-pentene and 2-methyl-1-butene

(ii) cyclopentene and 1-methylcyclopentene

4.22 Offer a simple explanation of the following facts.

a) Polyisoprene (rubber) that possesses the all *cis* structure stretches, but the material that possesses the all *trans* structure stretches very little.

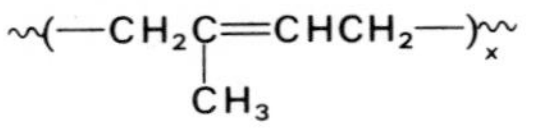

$$\sim\!\!(\!-CH_2\underset{\underset{\displaystyle CH_3}{|}}{C}\!=\!CHCH_2\!-\!)_x\!\!\sim$$

b) The melting point of 2,4-hexadiyne is much higher (64°) than that of 1,5-hexadiyne (−6°).

4.23 Identify the isoprene units in the following naturally occurring compounds by drawing dashed lines as is done for limonene, etc., in Section 4.10.

a)

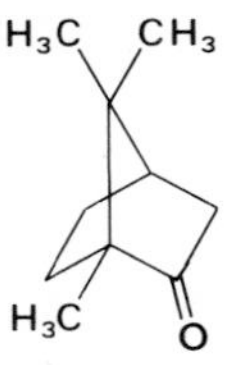

camphor

b)

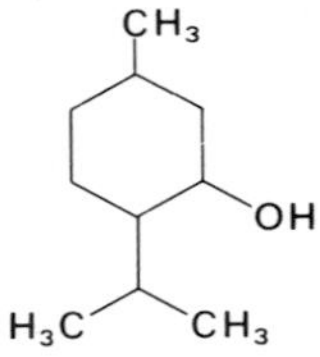

menthol

c)

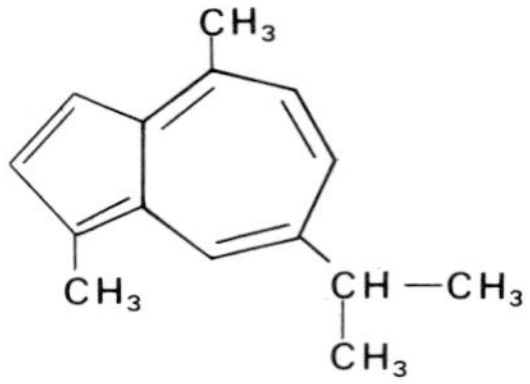

guaiazulene

d)

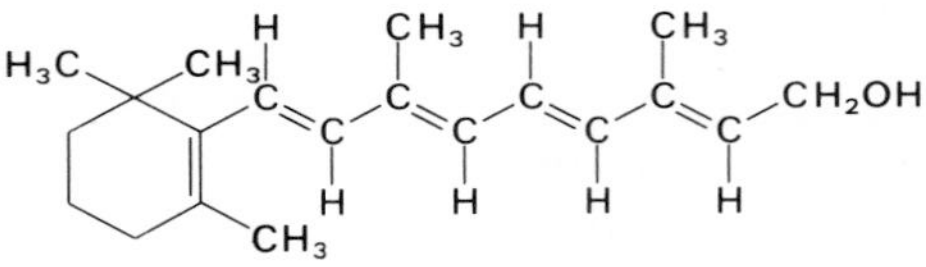

vitamin A

e)

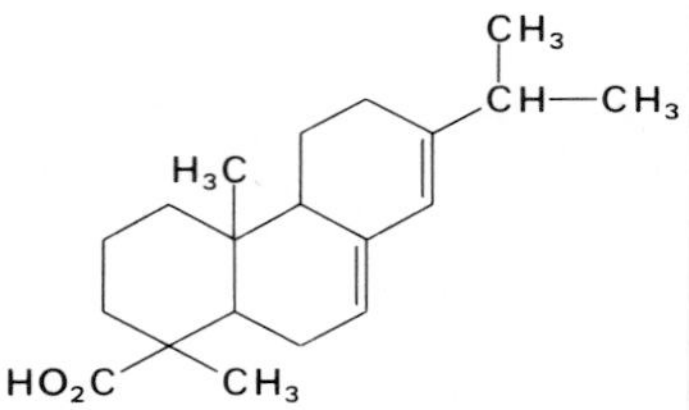

abietic acid
(chief constituent of
pine rosin)

4.24 Draw eucalyptol in a boat conformation (Fig. 4.1).

4.25 Draw the structure of the product of geraniol with each of these reagents.

a) dry hydrogen bromide   **b)** peracetic acid

CHAPTER 5

Alcohols, Phenols, and Ethers

Every fact and concept in this book was an exciting discovery at some time. Ancient peoples undoubtedly discovered, forgot, and rediscovered certain facts many times before records were kept. Now novelty in science can be identified and is reported in a cumulative scientific literature.

The discovery of ethanol as the component common to a variety of attractive beverages took place hundreds of years after the discovery of the beverages themselves. The compound is a volatile liquid of simple molecular formula, C_2H_6O, that historically was difficult to separate completely from water, a fact that probably delayed its characterization as a pure compound.

Unlike ethanol, which has been associated with sin or pleasure, cholesterol has usually been related to pain. Gallstones are almost pure crystalline cholesterol, and this fact led to its early isolation and recognition as a large compound of molecular formula $C_{27}H_{46}O$. The structure of ethanol, CH_3CH_2OH, was elucidated shortly after its purification because of its simplicity. The relatively complicated structure of cholesterol (Section 5.13) took investigators almost half a century to determine fully.

The features the two compounds share are that both were known in antiquity, both frequently are implicated in the disease of hypertension, and both contain a hydrocarbon joined to hydroxyl groups. The last property defines each compound as an alcohol, and alcohols are the subject of this chapter.

5.1 SOURCES AND USES OF ALCOHOLS

Ethanol is one of the few alcohols available in any quantity as a natural product. The properties of "alcohol" or ethyl alcohol have been known since people first collected fruits and grapes and allowed them to ferment. Fermentation is the breakdown of carbohydrates in the presence of an enzyme catalyst into smaller compounds. Ethanol is the product of the fermentation of plant materials which contain carbohydrates—fruits and grains. The natural fermentation solution from grapes achieves a maximum alcohol content of about 12% in wine. On exposure to oxygen, the ethanol is oxidized to acetic acid and the solution becomes vinegar.

$$HOCH_2(CHOH)_4CH{=}O \xrightarrow{\text{enzyme}} 2\ CH_3CH_2OH + 2\ CO_2 \qquad CH_3CH_2OH + O_2 \xrightarrow{\text{enzyme}} CH_3C\overset{\displaystyle O}{\underset{\displaystyle OH}{\big\backslash\!\!\big/}} + H_2O$$

glucose   ethanol   acetic acid (vinegar)

Whiskeys and brandies are distilled from fermentation solutions and have a much higher alcohol content—40–45%, or "80–90 proof." Enzyme catalysts are not active in solutions of such high alcohol concentration, and the alcohol is safe from reaction with oxygen.

For commercial chemical use, ethanol and other small alcohols are obtained from petroleum alkenes. Acid-catalyzed hydration of the alkenes ethylene, propene, 1- and 2-butene, and isobutylene (Section 4.4) gives ethyl, isopropyl, *sec*-butyl, and *tert*-butyl alcohols, respectively.

$$CH_2{=}CH_2 + H_2O \xrightarrow[140°]{H_2SO_4} CH_3CH_2OH$$

Protonated

$$CH_3\overset{CH_3}{\underset{}{C}}{=}CH_2 + H_2O \xrightarrow{H_2SO_4} CH_3\overset{CH_3}{\underset{OH}{C}}CH_3$$

| ethylene | ethanol | isobutylene | *tert*-butyl alcohol |

Due to the direction of addition to alkenes (Markownikoff's rule, Section 4.5), except for ethanol only secondary and tertiary alcohols are obtained by hydration of alkenes. Primary alcohols other than ethanol require special syntheses (see Section 5.7).

Methanol was originally known as "wood alcohol," because it was a product of the destructive distillation of wood. Methanol is extremely poisonous, its first action being destruction of the optic nerves, leading to blindness. Further poisoning ultimately causes death. Methanol is often added to ethanol to make commercial ethanol unsafe for human consumption and therefore free of a beverage tax. (The product is known as "denatured alcohol.")

Commercial methanol, one of the cheapest chemicals, is prepared by hydrogenation of carbon monoxide at a high temperature and pressure with copper chromite catalyst. The carbon monoxide is made from coke and steam.

$$C + H_2O \longrightarrow CO + H_2; \quad CO + 2\,H_2 \xrightarrow[2\ atm]{CuCrO_2,\ 400°} CH_3OH$$

The lower molecular weight alcohols are used extensively as low-priced, moderately polar solvents for organic compounds.

Small diols, such as ethylene glycol and propylene glycol, find commercial use in the manufacture of polymers, the polyesters (Section 9.1). Ethylene glycol is also the main ingredient in antifreeze.

$$\underset{\underset{OH}{|}}{CH_2}{-}\underset{\underset{OH}{|}}{CH_2} \qquad \underset{\underset{OH}{|}}{CH_2}{-}\underset{\underset{OH}{|}}{CH}{-}CH_3 \qquad \underset{\underset{OH}{|}}{CH_2}{-}\underset{\underset{OH}{|}}{CH}{-}\underset{\underset{OH}{|}}{CH_2}$$

| ethylene glycol (1.2-ethanediol) | propylene glycol (1,2-propanediol) | glycerol (1,2,3-propanetriol) |

Glycerol occurs as the alcohol component of the esters, fats, and vegetable oils (Section 9.13). Glycerol finds extensive use as a "wetting agent," a substance to absorb moisture in lotions. This property of absorbing water is due to the miscibility of glycerol with water and to the strong hydrogen bonds between water and glycerol.

Another important use of glycerol is in the manufacture of glyceryl trinitrate, the triester of nitric acid better known as nitroglycerin. The ester is used as an active ingredient in dynamite and as a vasodilator to prevent coronary spasm.

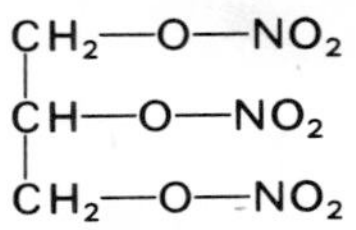

$$CH_2—O—NO_2$$
$$CH—O—NO_2$$
$$CH_2—O—NO_2$$

glyceryl trinitrate

Other important biological alcohols which are found free or as parts of esters include ethanolamine, choline cation, and glucose.

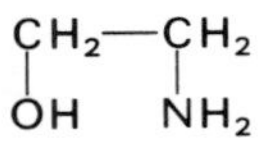

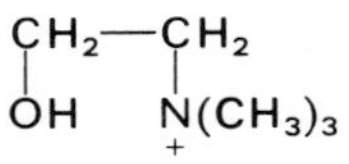

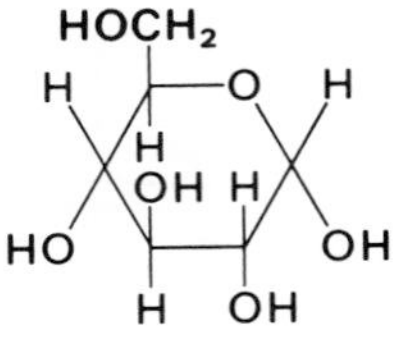

$$CH_2—CH_2$$
$$OH \quad NH_2$$

ethanolamine

$$CH_2—CH_2$$
$$OH \quad N(CH_3)_3$$
$$+$$

choline cation

glucose

As you already know, the functional group for an alcohol is the hydroxyl group —OH. The oxygen–hydrogen bond and the bond attaching the group to the carbon chain are both strongly polar, with the carbon to which the hydroxyl is attached having a partial positive charge. Reactions of these compounds involve changes in the polar bonds.

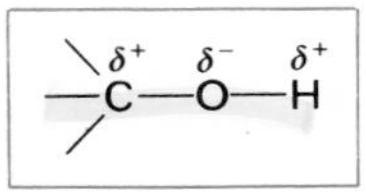

$$\overset{\delta^+}{\underset{}{}}\ \overset{\delta^-}{\underset{}{}}\ \overset{\delta^+}{\underset{}{}}$$
$$—C—O—H$$

Both the chemical and physical properties of alcohols are strongly determined by the polarity of the functional group and of the compounds themselves. One of the important factors contributing to the polarity of an alcohol is the bond angle at the oxygen atom. In general this bond angle C—O—H is close to 105°, as in H—O—H of water, and the C—O and H—O bond moments add to give alcohols a dipole moment of about 1.70 units.

The shape of a molecule of ethanol is shown in the models depicted here.

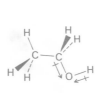

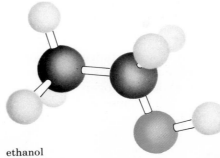

ethanol

5.2 NOMENCLATURE OF ALCOHOLS

In IUPAC nomenclature, major functional groups, such as hydroxyl groups, are included in the name of the parent compound. The parent name for a compound having two or more major functional groups is determined by the following order of precedence: carboxylic acid > carboxylic acid derivatives > aldehyde > ketone > alcohol. Thus a compound which has no functional group of greater precedence than the hydroxyl group has the name of the alcohol as the parent compound. The IUPAC rules for alcohols follow.

1. The longest carbon chain containing the hydroxyl group is used as the parent compound. The name of the alcohol substitutes "ol" as an ending for the "e" of the alkane corresponding to this longest chain.

2. When C=C and C≡C are present in the compound, the parent compound is chosen to include the unsaturated groups if possible, even if this does not make the longest chain the parent compound.

3. The carbons are numbered beginning at the end closest to the hydroxyl group. For alcohols with the —OH attached to a ring, numbering starts with the carbon bearing the —OH.

$$CH_3-\underset{\underset{OH}{|}}{CH}-CH_2-CH_2-CH_3$$

2-pentanol

cyclopentanol

$$CH_3-CH{=}CH-CH_2-OH$$

2-buten-1-ol

4. Hydrocarbon, halogen, amino, and alkoxyl groups are named as substituents of the parent alcohol. They are given the number of the carbon to which they are attached.

$$HO-CH_2-\underset{\underset{Cl}{|}}{CH}-\underset{\underset{Cl}{|}}{CH_2}$$

2,3-dichloro-1-propanol

$$NH_2-CH_2-CH_2-OH$$

2-amino-1-ethanol
(ethanolamine)

$$CH_3-\underset{\underset{OH}{|}}{CH}-CH_2-\underset{\underset{CH_2CH_3}{|}}{C}{=}CH_2$$

4-ethyl-4-penten-2-ol
(longest chain including the C=C)

$$CH_3-CH_2-\underset{\underset{CH_3}{|}}{\overset{\overset{CH_3}{|}}{C}}-CH_2-\underset{\underset{CH_2-OH}{|}}{CH}-CH_2-CH_3$$

2-ethyl-4,4-dimethyl-1-hexanol
(longest chain containing the hydroxyl group)

5. For compounds which contain more than one hydroxyl group, the number of —OH groups is denoted by the insertion before "ol" of the appropriate syllable: *di, tri, tetra*.

$$CH_3-\underset{\underset{OH}{|}}{CH}-\underset{\underset{OH}{|}}{CH}-\underset{\underset{OH}{|}}{CH_2}$$

1,2,3-butanetriol

1,2-cyclohexanediol

Compounds having hydroxyl groups on adjacent carbons have special properties in addition to those of the usual alcohols. They are called **1,2-glycols**.

Compounds that have a hydroxyl group attached to a C=C or C≡C carbon, or to a carbon bearing an amino, halogen, alkoxyl, or another hydroxyl group are generally unstable and cannot be isolated except in special circumstances.

$$CH_3—CH(OH)_2 \qquad CH_2=CH—OH$$

1,1-ethanediol   vinyl alcohol (an enol)
(unstable)   (unstable)

Some other alcohols that are known by their common names are important, such as **benzyl alcohol** and **allyl alcohol**. Note that both of these alcohols have the —OH on a carbon adjacent to an unsaturated carbon.

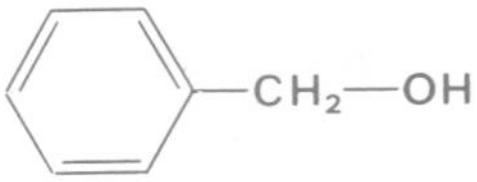

$—CH_2—OH$ $CH_2=CH—CH_2—OH$

benzyl alcohol   allyl alcohol
(phenylmethanol)   (2-propen-1-ol)

Recall that alcohols are designated as primary, secondary, or tertiary, according to the number of carbon groups attached to the carbon bearing the hydroxyl group (Section 2.7). Benzyl and allyl alcohols are both primary.

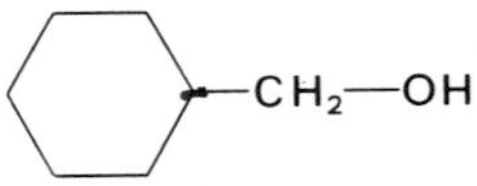

cyclohexylmethanol   cyclopropanol   1-methylcyclohexanol
(primary)   (secondary)   (tertiary)

Problems   **5.1** Name the following compounds by the IUPAC system.

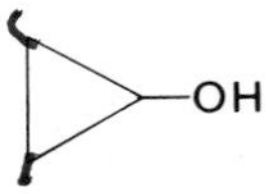

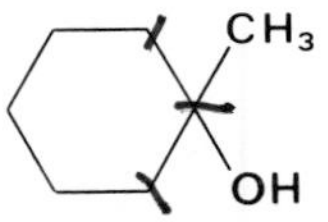

a) 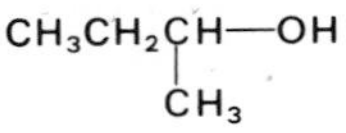 $CH_3CH_2\underset{\underset{CH_3}{|}}{CH}—OH$
     b) 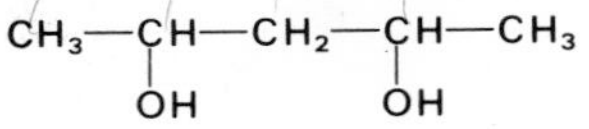 $CH_3—\underset{\underset{OH}{|}}{CH}—CH_2—\underset{\underset{OH}{|}}{CH}—CH_3$
     c) 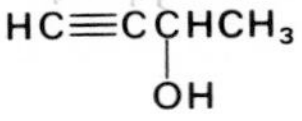 $HC\equiv C\underset{\underset{OH}{|}}{C}HCH_3$

d) 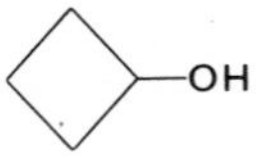   —OH
     e) 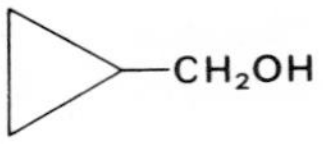 $—CH_2OH$
     f)

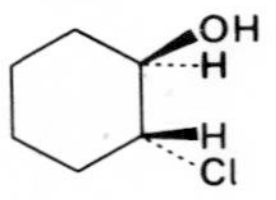

g)

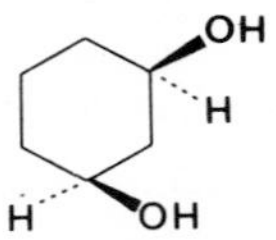

h) 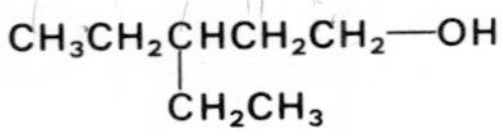 $CH_3CH_2\underset{\underset{CH_2CH_3}{|}}{C}HCH_2CH_2—OH$

i) 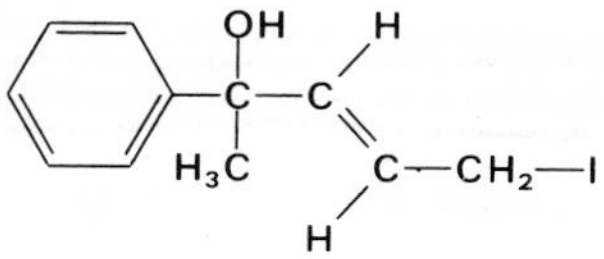 (structure)

OH H
C—C
H₃C   C—CH₂—I
H

j) 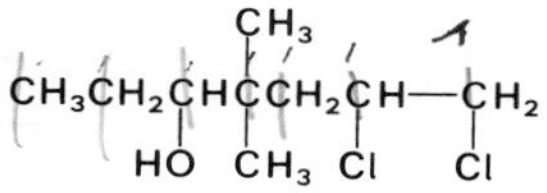 $CH_3CH_2CHCCH_2CH-CH_2$
HO CH₃ Cl   Cl

(handwritten: 3 heptanol / 4,4 dimethyl / 6,7 dichloro)

k) 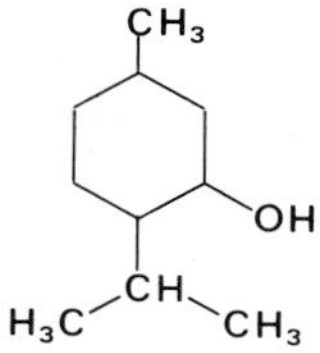 (structure)

CH₃ (cyclohexane ring) OH
H₃C—CH—CH₃

(handwritten: 1 cyclohexanol / 2 isopropyl / 5 methyl)

5.2 Write structures for the following compounds.

    a) isopropyl alcohol   **b)** glycerol   **c)** allyl alcohol

    d) benzyl alcohol   **e)** phenol   **f)** isobutyl alcohol

    g) *E*-4-aminocyclohexanol   **h)** *cis*-3-hexen-2-ol

    i) 1,2,3,4,5,6-hexanehexol

    j) five saturated isomeric alcohols of formula C_4H_8O

    k) 1-phenyl-1-propanol

5.3 PHYSICAL PROPERTIES OF ALCOHOLS

The simple alcohols, compounds that contain the hydroxyl as the main functional group, have higher boiling points and greater solubility in water than hydrocarbons or alkyl halides of comparable molecular weight. Both of these properties are due to the hydrogen bonding between —OH groups of the alcohol and of the alcohol with the water. The values are given in Fig. 5.1.

FIGURE 5.1 COMPARISON OF PHYSICAL PROPERTIES OF ALCOHOLS AND OTHER COMPOUNDS

Compound	MW	mp, °C	bp, °C
CH_3CH_3	30	−172	−89
CH_3—OH	32	−98	+65
CH_3—Cl	50	−97	−24
$CH_3CH_2CH_3$	44	−189	−42
CH_3CH_2—OH	46	−117	+78
CH_3CH_2—Cl	64	−139	+13
$CH_3CH_2CH_2CH_3$	58	−138	−0.5
$CH_3CH_2CH_2$—OH	60	−127	+97

Problem   **5.3** Sketch a diagram showing the hydrogen bonding between an alcohol ROH with itself and with water.

5.4 ALCOHOLS AS ACIDS AND BASES

The hydroxyl group gives alcohols many properties that are similar to those of water. Strong acids donate a proton to water.

$$H_2SO_4 + H_2O \longrightarrow H_3O^+ + HSO_4^-$$

With strong mineral acids, an alcohol also acts as a base accepting the proton. The protonated alcohol cation is called an **oxonium ion**.

$$H_2SO_4 + CH_3\!-\!OH \longrightarrow CH_3\!-\!\overset{\overset{+}{|}}{\underset{|}{O}}\!-\!H \;+\; HSO_4^-$$
$$H$$

| methanol (as a base) | methyloxonium ion | hydrogen sulfate ion |

The reaction of sulfuric acid with alcohols, as with water, goes essentially to completion.

An alcohol can act as an acid and donate a proton to a base which is stronger than the conjugate base of the alcohol, its anion. In the IUPAC nomenclature, the anion of an alcohol is called an alkoxide ion. An example is the reaction of 1-propanol with sodium amide, in which the equilibrium lies far to the side of forming the propoxide ion and ammonia.

$$CH_3CH_2CH_2\!-\!OH + Na^+NH_2^- \longrightarrow CH_3CH_2CH_2\!-\!O^-Na^+ + NH_3$$

| 1-propanol | sodium amide | sodium propoxide | ammonia |

Alcohols are slightly weaker acids than water, and alkoxide ions are slightly stronger bases than hydroxide ion. The equilibrium of methoxide ion–water and methanol–hydroxide ion in a methanol–water solvent is very nearly 1:1.

$$CH_3\!-\!O^- + H_2O \; \rightleftarrows \; CH_3\!-\!OH + OH^-$$

methoxide ion   methanol

Other alkoxide ions are slightly stronger bases than methoxide ion. Sodium hydroxide is thus not a strong enough base to generate a large concentration of alkoxide ion from alcohol. At the same time any water present in a solution of alkoxide salts, such as sodium ethoxide in ethanol, will reduce the concentration of the ethoxide ion by forming hydroxide ion and ethanol.

$$CH_3CH_2\!-\!O^- + H_2O \; \rightleftarrows \; CH_3CH_2\!-\!OH + OH^-$$

Two practical methods of preparing solutions of metal alkoxides involve reactions which produce hydrogen gas as a product and go to completion. One is the use of an ionic compound containing hydride ions, $H\!:^-$, as the base, such as sodium hydride. The products of the reaction of sodium hydride with *tert*-butyl alcohol are hydrogen gas and sodium *tert*-butoxide.

$$CH_3-\underset{\underset{CH_3}{|}}{\overset{\overset{CH_3}{|}}{C}}-OH + Na^+H^- \longrightarrow CH_3-\underset{\underset{CH_3}{|}}{\overset{\overset{CH_3}{|}}{C}}-O^-Na^+ + H_{2(g)}$$

tert-butyl sodium   sodium *tert*-butoxide
alcohol   hydride

Like water, alcohols react with very active metals such as metallic sodium or potassium to form hydrogen gas and the metal alkoxide.

$$2\ CH_3CH_2-OH + 2\ Na° \longrightarrow 2\ CH_3CH_2-O^-Na^+ + H_{2(g)}$$

sodium ethoxide

The reaction of ethanol with sodium is quite vigorous and gives off much heat. The reaction must be controlled by slow addition of pieces of metal, and by cooling the mixture. Otherwise the hydrogen gas–air mixture will ignite and then the alcohol will ignite.

Reactions which form either the conjugate acid or the conjugate base of an alcohol, the alkyloxonium or alkoxide ions, are important, since one of them is usually the first step of a reaction involving alcohols. The reactions of Section 5.5(a) are ones in which oxonium ions are first formed and then proceed to react further. Reactions of alkoxide ions with alkyl halides are considered in Section 6.3.

Problem   **5.4** Write equations for these acid–base reactions.

  a) sulfuric acid + isobutyl alcohol   **b)** *tert*-butyl alcohol + potassium
  c) cyclohexanol + sodium amide

5.5 REACTIONS OF ALCOHOLS THAT CLEAVE THE C—O BOND

An alcohol may be readily converted to an alkyl halide if the alcohol is converted first to its oxonium ion. The proton transfer is then followed by a reaction in which the halogen takes the place of the oxygen functional group in the oxonium ion. The process involves the breaking of the C—O bond and the forming of a new C—Cl bond.

$$HCl + R-OH \longrightarrow R-\overset{+}{O}H_2 + Cl^- \longrightarrow R-Cl + H_2O$$

A reaction in which one functional group replaces another functional group in a compound by a process of breaking and making bonds between carbon and the functional groups is called a **substitution reaction**. The conversion of an alcohol to an alkyl halide is a type of substitution reaction in which one electronegative group replaces another. The substitution reaction is shown in this general manner, where Nu: is the incoming electronegative group and L: is the leaving group.

$$Nu: + \quad -\overset{|}{\underset{|}{C}}-L \longrightarrow Nu-\overset{|}{\underset{|}{C}}- \quad + :L$$

entering
group

leaving
group

In the neutral alcohol, the C—O bond does not break readily to give an ordinary substitution reaction. Several methods are available to modify the alcohol so that the C—O bond is more easily broken. Two methods of making the alcohol more reactive are used in the following conversions of alcohols to alkyl halides. In Section 5.5(a) a strong acid forms the alkyloxonium ion, and in Section 5.5(b), the hydroxyl group is changed first to an inorganic ester group.

a) Reactions of alcohols with hydrogen halides The overall reaction of an alcohol with a hydrogen halide is a sequence of two reactions which gives an alkyl halide as the final product. In the reaction below of 1-propanol and hydrogen bromide, the first reaction is the formation of the propyloxonium ion and bromide ion. The bromide ion then displaces water to give 1-bromopropane.

$$CH_3CH_2CH_2-OH + HBr \longrightarrow CH_3CH_2CH_2-\overset{+}{O}H_2 + Br^-$$

1-propanol   propyloxonium ion

$$CH_3CH_2CH_2-\overset{+}{O}H_2 + Br^- \longrightarrow CH_3CH_2CH_2-Br + H_2O$$

1-bromopropane

The second equation illustrates a substitution in which Nu is Br^- and L is H_2O.

$$Nu: + -\overset{|}{\underset{|}{C}}-L \longrightarrow Nu-\overset{|}{\underset{|}{C}}- + :L \qquad :\overset{..}{Br}:^- + -\overset{|}{\underset{|}{C}}-\overset{+}{\underset{\underset{H}{|}}{\overset{..}{O}}}-H \longrightarrow :\overset{..}{Br}-\overset{|}{\underset{|}{C}}- + :\overset{..}{\underset{\underset{H}{|}}{O}}-H$$

The old bond C—L is strongly polarized, with L having the greater share of the electrons. The bond breaks to give both of the electrons to the group that is leaving (L:). The incoming group (Nu:) brings an unshared pair of electrons to form the new C—Nu bond. Remember that a reagent which furnishes a pair of electrons and is seeking a nucleus with which to share the electrons is called a **nucleophilic reagent** or **nucleophile** (Section 4.4). The substance that the nucleophile acts on is the **substrate.** A substitution reaction that is initiated by a nucleophilic reagent acting on the substrate compound is called a **nucleophilic substitution.**

With primary oxonium ions, the bromide ion pushes the water molecule out by simultaneously forming the C—Br bond as the C—O bond breaks. The bromide ion is an electron-rich reagent which uses a pair of its electrons to form a bond with an electron-deficient nucleus. The carbon atom bearing the protonated hydroxyl group is partially positive and is becoming electron-deficient as the departing water takes the C—O bonding electrons. The reaction occurs in one step.

$$Br^- + CH_3CH_2CH_2—\overset{+}{O}H_2 \longrightarrow CH_3CH_2CH_2—Br + H_2O \quad \text{(one step)}$$

Although benzyl and allyl alcohols are primary alcohols, they react very rapidly with hydrogen halides to give benzyl and allyl halides, respectively.

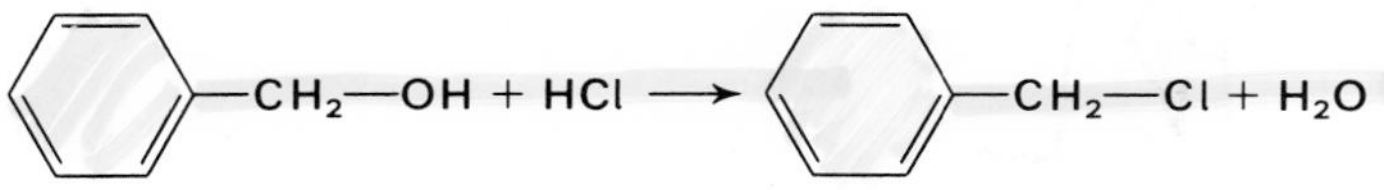

| benzyl alcohol | benzyl chloride |

Tertiary alcohols react rapidly with concentrated hydrochloric, hydrobromic, and hydriodic acids to give the tertiary halides and some alkene.

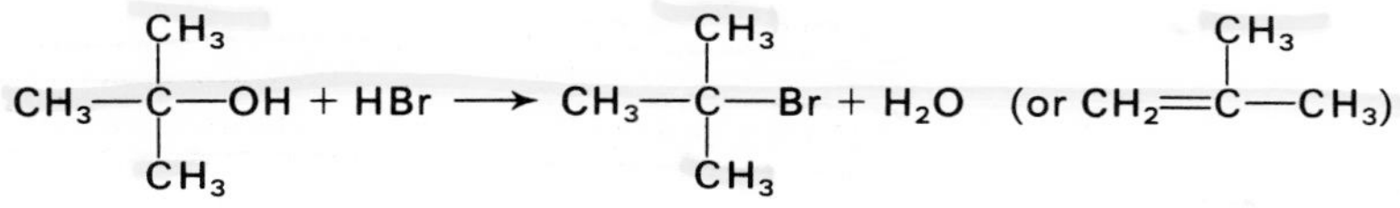

tert-butyl alcohol   *tert*-butyl bromide   isobutylene

The *tert*-butyloxonium ion first formed by proton transfer cleaves at the C—O bond, giving water and *tert*-butyl cation, a carbonium ion. Water takes the C—O bonding electrons, leaving a carbon with only six electrons and a positive charge. The carbonium ion then reacts rapidly with bromide ion to give the *tert*-butyl bromide.

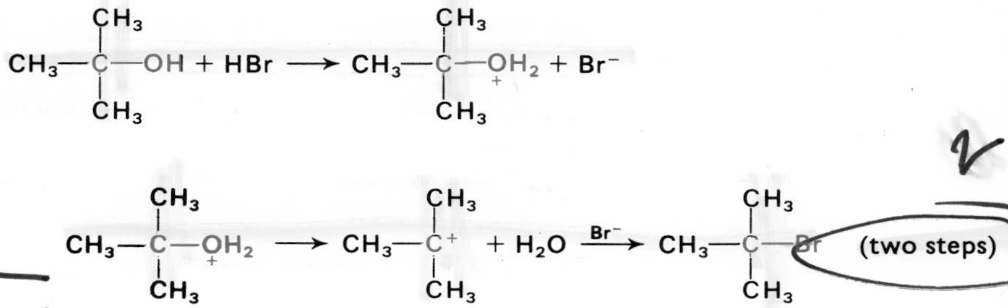

tert-butyloxonium ion   *tert*-butyl cation   *tert*-butyl bromide

Tertiary alcohols react rapidly because they form the tertiary carbonium ion. Formation of the carbonium ion is slower from secondary alcohols. Primary alcohols probably do not form carbonium ions at all, but react directly as noted above. The ease of reaction through carbonium ion formation corresponds to the order of carbonium ion stability—tertiary > secondary > primary—which was observed in the reactions of alkenes (Section 4.5).

b) Reactions of alcohols with halides of sulfur and phosphorus

A more effective method of converting primary and secondary alcohols to alkyl halides involves the use of the sulfur or phosphorus halides, thionyl chloride $SOCl_2$, phosphorus trichloride PCl_3, and phosphorus tribromide PBr_3.

$$CH_3CH_2-OH + Cl-\overset{Cl}{\underset{}{S}}=O \longrightarrow CH_3CH_2-Cl + SO_2 + HCl$$

ethanol   thionyl chloride   chloroethane

$$3\ CH_3\overset{CH_3}{\underset{}{CH}}CH_2-OH + PBr_3 \longrightarrow 3\ CH_3\overset{CH_3}{\underset{}{CH}}CH_2-Br + P(OH)_3$$

isobutyl alcohol   phosphorus tribromide   isobutyl bromide   phosphorous acid

All of these inorganic nonmetal halides convert the alcohol to an ester of an inorganic acid and a halide ion in the first of the two reactions. In the second reaction the C—O bond of the ester breaks as the inorganic acid group is easily displaced by the halide.

$$CH_3CH_2-OH + Cl-\overset{Cl}{\underset{}{S}}=O \longrightarrow CH_3CH_2-O-\overset{Cl}{\underset{}{S}}=O + HCl \longrightarrow CH_3CH_2-Cl + SO_2 + HCl$$

The ester can be isolated after the first reaction, but usually the second reaction is brought about by heating immediately. In these conversions the poor leaving group, the —OH, has been changed into a good leaving group, the inorganic sulfur dioxide or phosphorous acid.

5.6 DEHYDRATION OF ALCOHOLS TO ALKENES

Alcohols undergo another type of reaction called an elimination. An **elimination reaction** is one in which a compound loses two groups on adjacent atoms to form an alkene. It is the reverse of an addition reaction.

An alcohol loses water (dehydrates) to form an alkene in the presence of a strong acid and at temperatures higher than those normally used for substitution reactions. Reaction conditions vary widely with the structure of the alcohol. Elimination reactions always compete with substitution reactions, but some control is possible in many examples.

Tertiary alcohols are dehydrated readily even at room temperature with concentrated sulfuric acid.

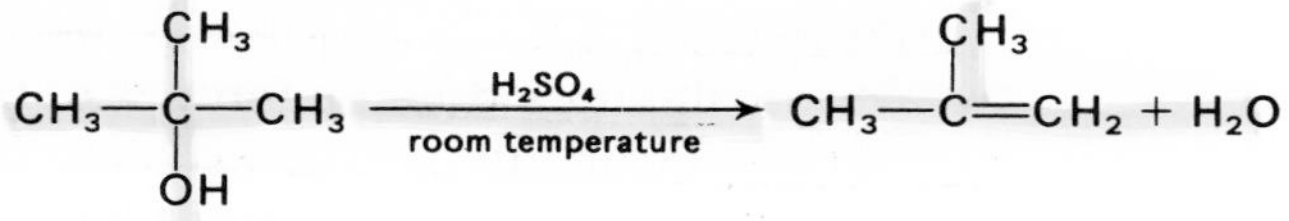

$$CH_3-\underset{\underset{OH}{|}}{\overset{\overset{CH_3}{|}}{C}}-CH_3 \xrightarrow[\text{room temperature}]{H_2SO_4} CH_3-\overset{\overset{CH_3}{|}}{C}=CH_2 + H_2O$$

tert-butyl alcohol   isobutylene

The reaction proceeds through the carbonium ion, which loses a proton to form a double bond.

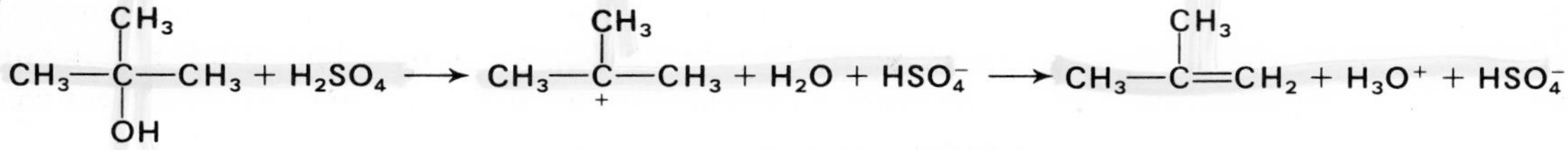

$$CH_3-\underset{\underset{OH}{|}}{\overset{\overset{CH_3}{|}}{C}}-CH_3 + H_2SO_4 \longrightarrow CH_3-\overset{\overset{CH_3}{|}}{\underset{+}{C}}-CH_3 + H_2O + HSO_4^- \longrightarrow CH_3-\overset{\overset{CH_3}{|}}{C}=CH_2 + H_3O^+ + HSO_4^-$$

Secondary alcohols require somewhat higher temperatures for dehydration than do tertiary alcohols. Primary alcohols require much more drastic conditions, as in this example of ethanol.

$$CH_3CH_2-OH \xrightarrow[180°]{H_2SO_4} CH_2=CH_2 + H_2O$$

Problem **5.5** Write equations for the following reactions.

  a) 3-methyl-3-hexanol + concentrated HCl

  b) 2-chlorocyclopentanol + $SOCl_2$ heated

  c) isopropyl alcohol + concentrated HBr

  d) 1-hexanol + PBr_3

  e) cyclohexanol + H_2SO_4 (100°)

5.7 METHODS OF PREPARATION OF ALCOHOLS

Alcohols are the products of many reactions of other classes of compounds. Some of these reactions are discussed here. Other examples are only cited here and will be treated in detail in later chapters.

Reactions which give alcohols can be divided into two categories:

1. Those in which only the functional group is changed, with the number of carbons remaining the same.

2. Those in which two, or occasionally three, organic compounds are combined to make an alcohol larger than the starting compounds.

Some examples of the first category are given here. Reactions belonging to the second category are treated in Chapter 14.

a) Hydration of alkenes

The acid-catalyzed addition of water to alkenes has been considered in detail in the previous chapter (Section 4.4). The hydration of 2-methyl-1-pentene serves as an example.

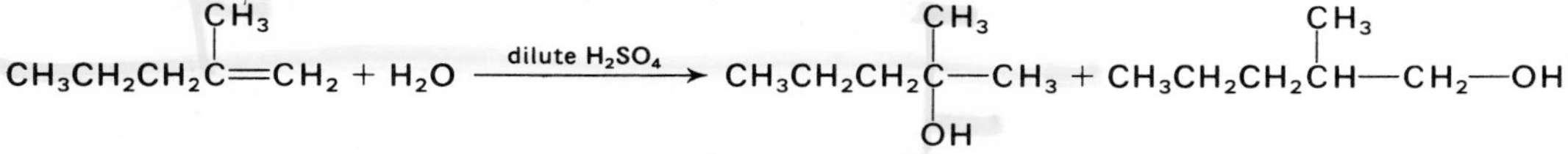

$$CH_3CH_2CH_2\overset{\overset{CH_3}{|}}{C}=CH_2 + H_2O \xrightarrow{\text{dilute } H_2SO_4} CH_3CH_2CH_2\underset{\underset{OH}{|}}{\overset{\overset{CH_3}{|}}{C}}-CH_3 + CH_3CH_2CH_2\overset{\overset{CH_3}{|}}{C}H-CH_2-OH$$

2-methyl-1-pentene   2-methyl-2-pentanol   2-methyl-1-pentanol
                                       (major product, 98%)   (minor product, 2%)

Addition of the water according to Markownikoff's rule gives the tertiary alcohol in strong preference to the primary alcohol. Predominant products are in the order of tertiary > secondary > primary alcohols, corresponding to the order of stability of carbonium ion intermediates (Section 4.5).

b) Hydroboration of alkenes

Frequently there are two reaction sequences which can be used to convert one class of compound into the same new class. In some examples the isomer produced in the first sequence is different from the isomer produced in the second sequence. The difference is very important, because it provides some choice in the kind of product the chemist can make from the same starting material.

An outstanding example of a reaction which can produce a different set of isomers is the conversion of alkenes to alcohols via a reaction called **hydroboration**. The overall result of this sequence of two reactions is that the alcohol produced is the least substituted alcohol ("anti-Markownikoff" product) in contrast to the acid-catalyzed hydration. The product which is highly predominant occurs in the order primary > secondary > tertiary alcohol.

Hydroboration is the addition of the elements of borane BH_3 to an alkene to form a trialkylborane. The hydrogen goes to one olefinic carbon, the boron to the other.

$$\diagdown{B}{-}H + \diagup{C}{=}C\diagdown \longrightarrow \diagup{B}{-}\overset{|}{C}{-}\overset{|}{C}{-}H$$

Borane BH_3 does not exist under ordinary conditions. The actual reagent used is diborane B_2H_6, which in the reaction acts as BH_3. Each of the H—B units adds to an alkene, in this example propene, giving tripropylborane as the product.

$$B_2H_6 + 6\ CH_3{-}CH{=}CH_2 \longrightarrow 2\ (CH_3{-}\overset{\overset{\text{H}}{|}}{C}H{-}CH_2{-})_3B$$

propene   tripropylborane

Trialkylboranes are versatile compounds which can be converted to many other classes of compounds. The conversion to alcohols is their best known reaction. Tripropylborane is treated with hydrogen peroxide and NaOH. The final product is an alcohol and sodium borate.

$$(CH_3CH_2CH_2{-})_3B + 3\ H_2O_2 + 3\ OH^- \longrightarrow 3\ CH_3CH_2CH_2{-}OH + BO_3^{3-} + 3\ H_2O$$

tripropylborane   1-propanol

The predominant alcohol product (greater than 95%) is the primary alcohol, 1-propanol. The direction of addition of diborane to an alkene is opposite to Markownikoff's rule, because the addition of H and B is simultaneous and does not proceed through a carbonium ion. Primary alcohols are formed in preference to secondary or tertiary alcohols, as in this example.

$$\underset{\text{2-methyl-1-pentene}}{CH_3CH_2CH_2\overset{\overset{\displaystyle CH_3}{|}}{C}=CH_2} \xrightarrow{B_2H_6} \xrightarrow{H_2O_2,\ OH^-} \underset{\text{2-methyl-1-pentanol (99\%)}}{CH_3CH_2CH_2\overset{\overset{\displaystyle CH_3}{|}}{C}H-CH_2-OH}$$

c) Catalytic hydrogenation of aldehydes and ketones

In a reaction similar to the addition of hydrogen to the C=C bond of alkenes, hydrogen also adds to the C=O bond of aldehydes and ketones. Hydrogenation of aldehydes produces primary alcohols, while ketones give secondary alcohols. Obviously, tertiary alcohols cannot be obtained this way.

$$\underset{\text{2-methylbutanal}}{CH_3CH_2\overset{\overset{\displaystyle CH_3}{|}}{C}HCH=O} + H_2 \xrightarrow{Ni} \underset{\text{2-methyl-1-butanol}}{CH_3CH_2\overset{\overset{\displaystyle CH_3}{|}}{C}HCH_2-OH}$$

cyclohexanone   cyclohexanol

The same hydrogenation catalysts—platinum, palladium, and nickel—are used for the addition to carbonyl compounds and to alkenes.

Certain metal hydrides can also convert carbonyl compounds as well as esters to alcohols. These reactions are considered in Sections 9.7 and 11.11.

More complex alcohols are built from smaller compounds by the action of organomagnesium halides (Grignard reagents) with aldehydes, ketones, or esters. These reactions are treated in Chapter 14, on carbon chain-building reactions (Section 14.3 and 14.4).

Problem

5.6 Write the structures for the alcohols produced by the sequence of reactions of diborane, hydrogen peroxide, and sodium hydroxide with the following alkenes.

  a) 3-methyl-3-hexene   **b)** 3-phenyl-1-butene

  c) 1-methylcyclopentene

5.7 Write the structures for the products A to F of the reactions indicated.

  a) $CH_3CH_2CH_2CH_2CH=O \xrightarrow{H_2,\ Ni} A \xrightarrow{PBr_3} B$

  b) $CH_3CH_2CH_2CH=CH_2 \xrightarrow{H_2O,\ H_2SO_4} C \xrightarrow{HBr} D$

  c) $CH_3CH_2CH_2CH=CH_2 \xrightarrow[\text{2) } H_2O_2,\ OH^-]{\text{1) } B_2H_6} E \xrightarrow{SOCl_2} F$

5.8 PHENOLS Compounds having a hydroxyl group attached to an aromatic ring belong to the class of compounds called **phenols**. The chemical properties of phenols and alcohols are sufficiently different to warrant their having separate names. The simplest member of the phenols is also named **phenol** or hydroxybenzene.

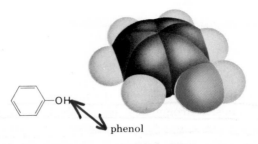

phenol

Phenol is used in large quantities commercially in the preparation of dyes, antioxidants, phenolic resins, and many pharmaceutical compounds.

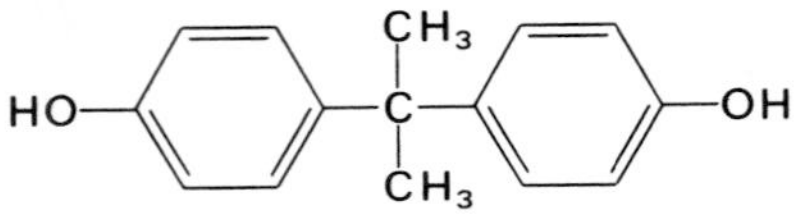

bis-phenol A
(used to form epoxy resins)

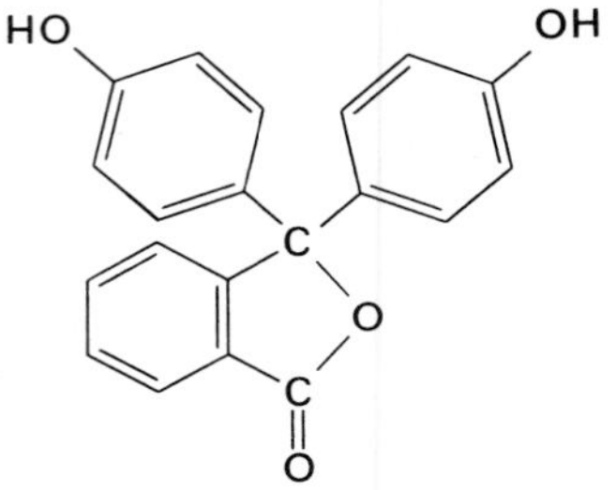

phenolphthalein
(an indicator dye)

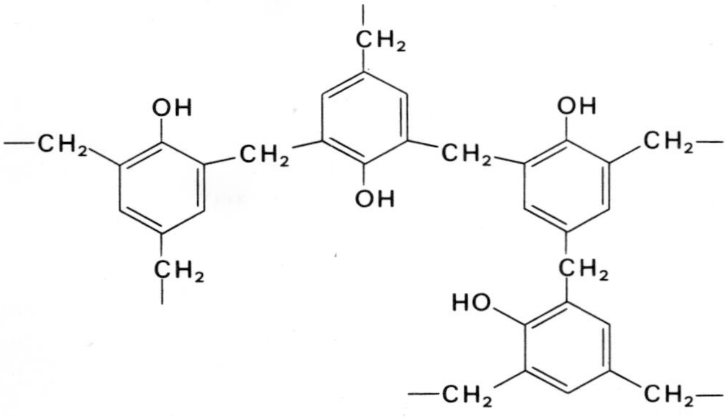

phenol-formaldehyde resin
("Bakelite")

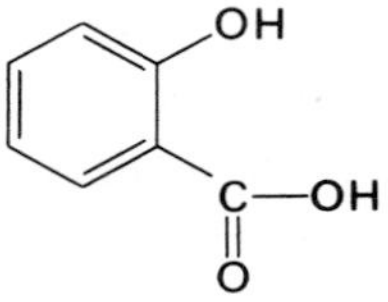

salicylic acid
(used to make
aspirin)

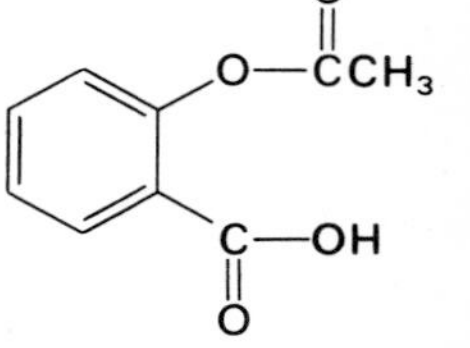

acetylsalicylic acid
(aspirin)

Under the name carbolic acid, phenol was used as an antiseptic. Because of its toxicity to tissues, phenol is seldom used as an antiseptic today. Many substituted phenols are effective antiseptics, such as hexachlorophene and two active ingredients in "Lysol," a commercial disinfectant.

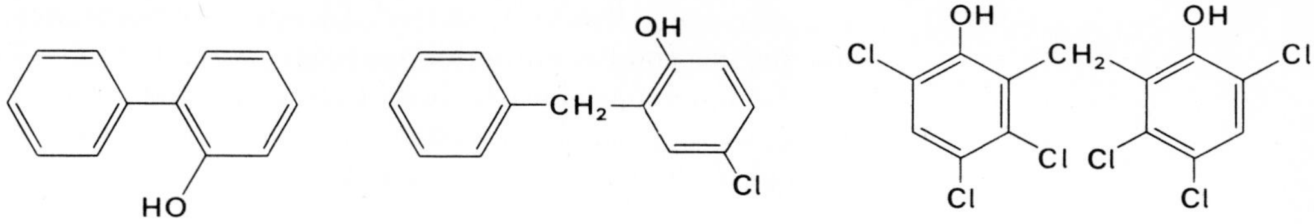

2-phenylphenol
(in "Lysol")

2-benzyl-4-chlorophenol

hexachlorophene

The indiscriminate use of hexachlorophene with possible toxicity effects has caused it to be placed under severe restrictions in spite of its very effective control against some diseases such as infant diarrhea.

Among the three dihydroxybenzenes and their substituted compounds there are many familiar products. **Hydroquinone (1,4-dihydroxybenzene)** finds extensive use as a photographic developer and an antioxidant. An antiseptic ingredient of some mouthwashes is 4-hexylresorcinol (4-hexyl-1,3-dihydroxybenzene).

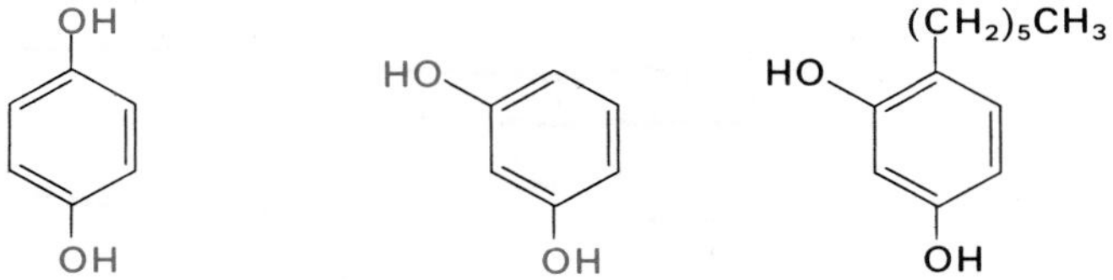

1,4-dihydroxybenzene
(hydroquinone)
(developer)

1,3-dihydroxybenzene
(resorcinol)

4-hexylresorcinol
(antiseptic)

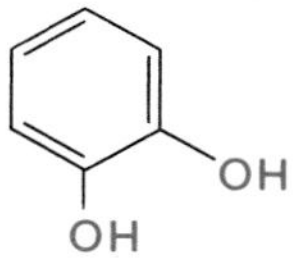

1,2-dihydroxybenzene
(catechol)

Many mono-methyl ethers of substituted **catechols** are found in essential oils, for example vanillin and eugenol.

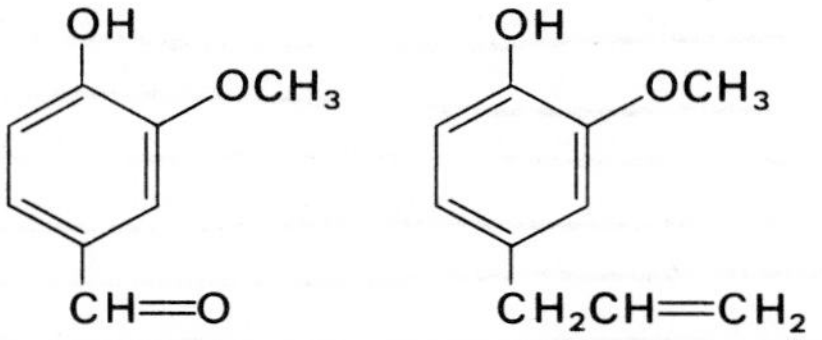

vanillin
(vanilla bean)

eugenol
(oil of cloves)

In the body the very important adrenal hormones, epinephrine or adrenaline and norepinephrine, are substituted catechols. In the event of a sudden demand by the body for energy and strength, the emission of epinephrine from the adrenal gland stimulates the breakdown of liver glycogen, the energy reservoir, to blood glucose.

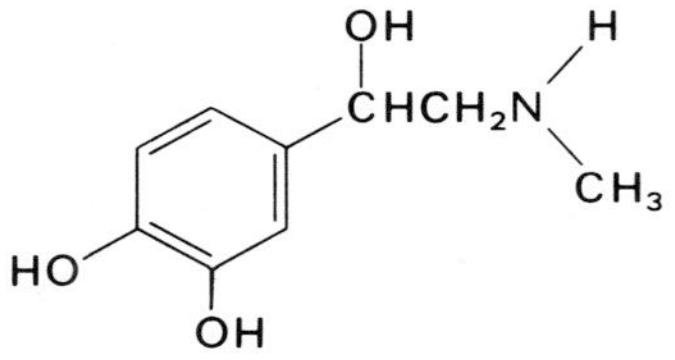

epinephrine
(adrenaline)

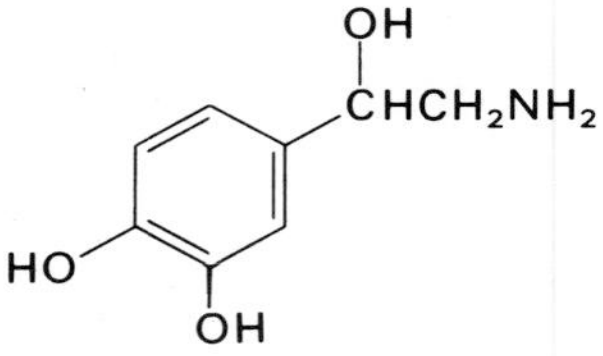

norepinephrine
(missing a CH_3 on N of epinephrine)

Many phenolic compounds are known by their common names as shown above. In the IUPAC nomenclature:

1. The parent compound is phenol, the benzene ring with the attached hydroxyl group.
2. The numbering of carbons in the ring starts with the carbon bearing the hydroxyl group.
3. Hydrocarbon, halogen, and amino substituents are indicated as prefixes.

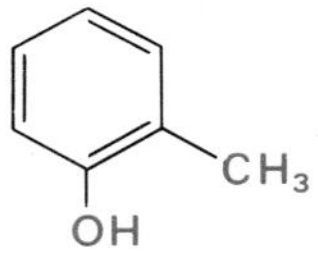

2-methylphenol
o-methylphenol
(o-cresol)

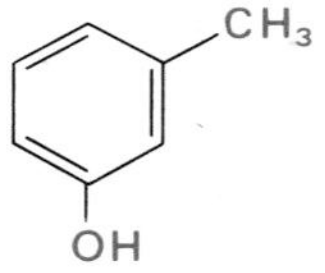

3-methylphenol
m-methylphenol
(m-cresol)

4-methylphenol
p-methylphenol
(p-cresol)

The naming of the three methylphenols above illustrates an important method of designating the relative positions of two groups attached to a benzene ring. The prefix *ortho-* (*o-*) indicates groups on adjacent (1,2) carbons. Groups 1,3 to each other have the prefix *meta-* (*m-*), while groups directly opposite on the ring (1,4) are said to be *para-* (*p-*) to each other. This system of designation is convenient because it can be used without reference to any official order of numbering of substituents. For example, in epinephrine, shown above, the hydroxyl groups are *ortho* to each other and the carbon side-chain is *para* to one hydroxyl and *meta* to the other.

Problem   **5.8** Name the following compounds by any convenient name.

a) 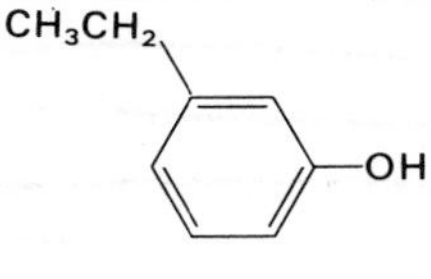 CH₃CH₂—⟨benzene ring⟩—OH

b) 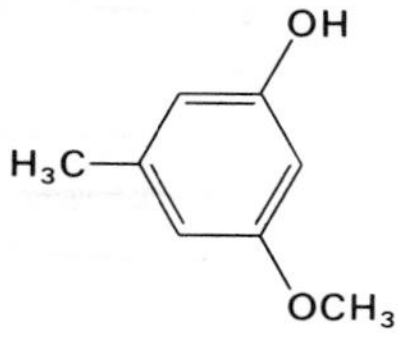 H₃C—⟨benzene ring with OH and OCH₃⟩

c)  H₂N—⟨benzene ring⟩—OH

d) 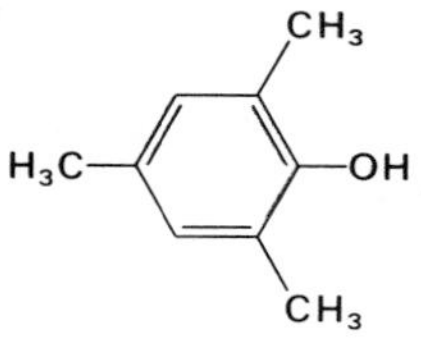 H₃C—⟨benzene ring with CH₃, OH, CH₃⟩

e) 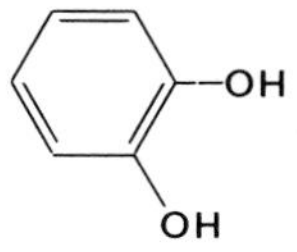 ⟨benzene ring⟩—OH with OH

f) 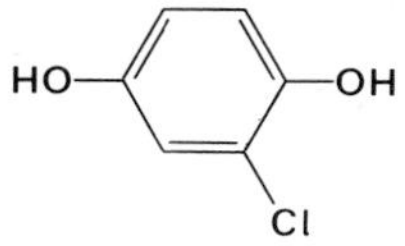 HO—⟨benzene ring with OH and Cl⟩

5.9 ACIDITY OF PHENOLS

One of the most striking properties of phenols is the degree of acidity. The acidity of a hydroxyl group is much greater on an aromatic ring than in saturated alcohols. The strength of phenols as acids falls between that of carboxylic acids and that of alcohols, as shown in Fig. 3.11.

The difference in strength of acids is shown by the action of sodium hydroxide, which converts phenol to sodium phenoxide and water. Recall that sodium hydroxide does not remove the proton from alcohols (Section 5.4).

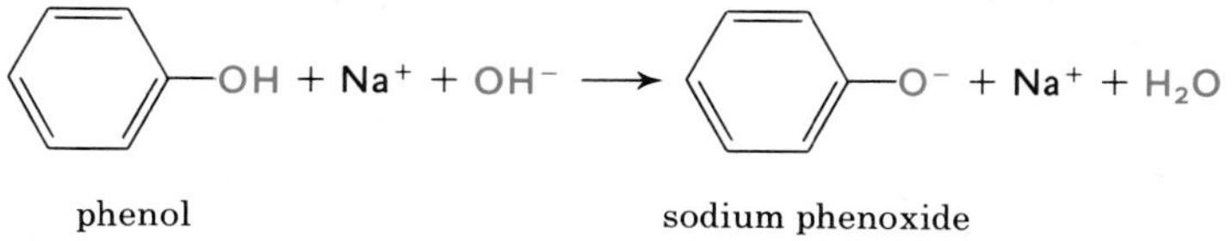

$$\langle C_6H_5\rangle\text{—OH} + Na^+ + OH^- \longrightarrow \langle C_6H_5\rangle\text{—O}^- + Na^+ + H_2O$$

phenol   sodium phenoxide

The greater acidity of phenols as compared with alcohols is due to the increased delocalization of electrons and charge in phenoxide ion over phenol, which is absent in the alkoxide ion. The resonance structures of both phenol and phenoxide ion are shown here.

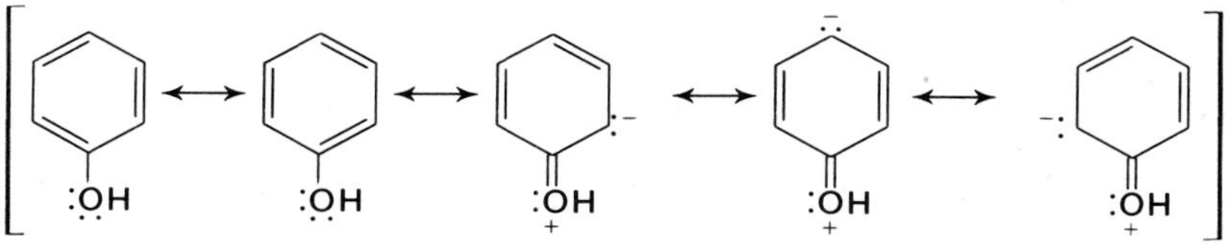

resonance structures of phenol

In three of the resonance structures of phenol, a pair of the oxygen's nonbonding electrons are delocalized into the ring, resulting in a positive charge on the oxygen atom—as in the hydronium ion—and a negative charge distributed on either of the two *ortho* or one *para* carbons of the ring. These structures with a separation of charge provide a small amount of added stability to phenol.

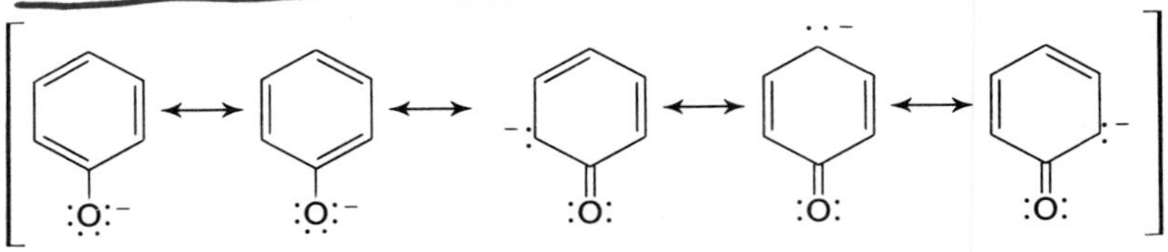

resonance structures of phenoxide ion

Three of the structures in the phenoxide ion also have a pair of nonbonding electrons from oxygen delocalized into the ring. However, in the ion the delocalization of negative charge into the ring does not involve separation of positive and negative charges. In the phenoxide ion the three structures having charge delocalized into the ring provide much greater stabilization to the resonance hybrid than do the corresponding structures in phenol.

The position of equilibrium between the neutral hydroxyl compound and the negative ion is the measure of acidity of the hydroxyl group. Since the phenoxide ion is more stable relative to phenol than is an alkoxide ion (which has no delocalization stabilization) relative to the alcohol, phenols are stronger acids than alcohols.

Phenol does not undergo the usual nucleophilic substitution and elimination reactions which are characteristic of alcohols. As we can see from the resonance structures for phenol shown above, the aromatic C—O bond is not an ordinary polarized C—O single bond. The three resonance structures which have separated positive and negative charges put a positive charge on the oxygen (making it electron-deficient instead of partially negative as in an alcohol), and the bond to carbon has some double-bond character C=O, which makes the carbon–oxygen bond more difficult to cleave.

Problems   **5.9** Carbonic acid H_2CO_3 (CO_2 + H_2O) is a stronger acid than phenols and a weaker acid than carboxylic acids. Write an acid–base reaction which you would expect to occur for the following compounds.

   a) phenol, carbonic acid, and their conjugate bases

   b) acetic acid, carbonic acid, and their conjugate bases

5.10 Draw the resonance structures for *p*-cresol (4-methylphenol) and for the 4-methylphenoxide ion.

5.10 ETHERS   **Ethers** are a class of compounds in which both hydrogens of water have been replaced by an alkyl or aryl group. They are relatively inert liquids of low polarity and high volatility. They serve primarily as low-boiling solvents, particularly **diethyl ether, tetrahydrofuran, and dioxane**.

$$CH_3CH_2—O—CH_2CH_3$$

diethyl ether
("ether")
bp 35°

tetrahydrofuran
("THF")
bp 65°

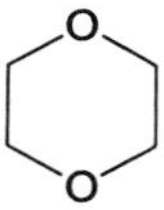

dioxane

bp 101°

Diethyl ether has a low solubility in water and can be dried easily. It provides a slightly polar solvent, which is nonproton-donating and reasonably unreactive. Tetrahydrofuran and dioxane are miscible with water, and each mixed with water becomes a solvent which dissolves ionic and nonionic compounds in the same medium.

Ether and THF pose a hazard when used as solvents because of their great volatility and low ignition temperatures. They should never be used near flames or exposed heating elements. During storage in the absence of antioxidants, they react with oxygen to form unstable and explosive peroxides.

Simple ethers are commonly named by their carbon groups plus the word "ether."

$$CH_3—O—CH_2CH_2CH_3$$

methyl propyl ether
1-methoxypropane (IUPAC)

More complicated structures are named according to the IUPAC rules, which use the alkoxy group as a prefix to the name of the parent hydrocarbon, as in **1-methoxypropane**.

Ethers are generally prepared by the reaction of sodium alkoxide or sodium phenoxide with alkyl halides (Section 6.3). Hydroxyl groups are converted to ether groups primarily for the purpose of protecting the hydroxyl group from reaction with other reagents, since ether groups are ordinarily unreactive.

5.11 OXIRANES (EPOXIDES)   The **oxiranes** are three-membered ring cyclic ethers. They are an extremely important type of ether because they are highly reactive compounds, in contrast to the unreactive open-chain ethers.

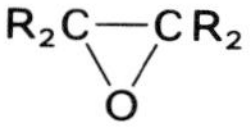

$$R_2C—CR_2$$
$$\diagdown O \diagup$$

an oxirane

$$H_2C—CH_2$$
$$\diagdown O \diagup$$

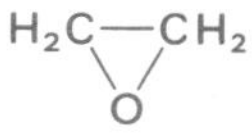

oxirane (ethylene oxide)

The three-membered ring in cyclopropanes and oxiranes (epoxides) is under a large amount of strain because of the large distortion of the

C—C—C and C—O—C bond angles from the normal 109° to about 60°. This makes the C—O bond in oxiranes considerably weaker than the usual one in alcohols or regular ethers. The relief of strain obtainable by breaking the C—O bond and opening the oxirane ring provides the driving force for reactions of oxiranes.

The simplest member of the class is **oxirane** or **ethylene oxide**. It is made commercially in large quantities from ethylene by various industrial processes. One method is the reaction of ethylene with *tert*-butyl hydroperoxide. (The peroxide is prepared by the addition of hydrogen peroxide to isobutylene.) Both the ethylene oxide and *tert*-butyl alcohol products are useful.

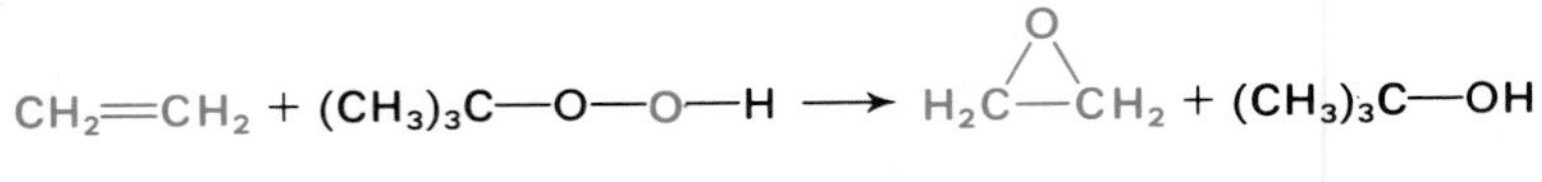

$$CH_2{=}CH_2 + (CH_3)_3C{-}O{-}O{-}H \longrightarrow H_2C{-}CH_2 + (CH_3)_3C{-}OH$$

| ethylene | *tert*-butyl hydroperoxide | ethylene oxide | *tert*-butyl alcohol |

Ethylene oxide is used to make ethylene glycol for antifreeze and other compounds with a functional group on each carbon.

Substituted oxiranes are prepared from alkenes directly by oxidation using percarboxylic acids, such as peracetic acid (Section 4.7).

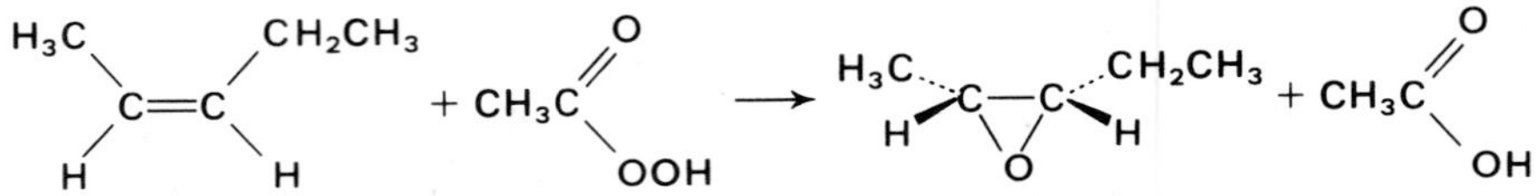

| *cis*-2-pentene | peracetic acid | *cis*-2-ethyl-3-methyloxirane (*cis*-2,3-epoxypentane) | acetic acid |

In the IUPAC name, the oxygen in the ring is the #1 atom. Common names for oxiranes are very often used. They are derived from the alkene name by adding the word oxide, as in ethylene oxide. In others the oxygen is named as a prefix, "epoxy," as shown below.

Two oxiranes which have been found in nature have quite different roles. Squalene epoxide, a triterpene, is the noncyclic compound which forms the rings in the biosynthesis of cholesterol and other steroids.

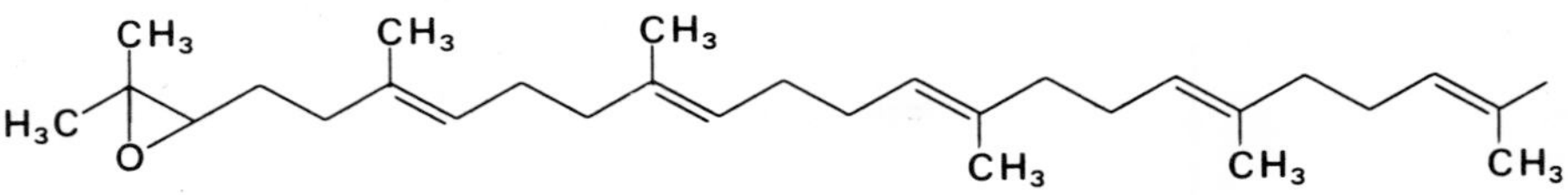

2,3-epoxysqualene

A nineteen-carbon oxirane has been isolated as the active component in the sex-attractant of the female gypsy moth, *cis*-7,8-epoxy-2-methyl-octadecane.

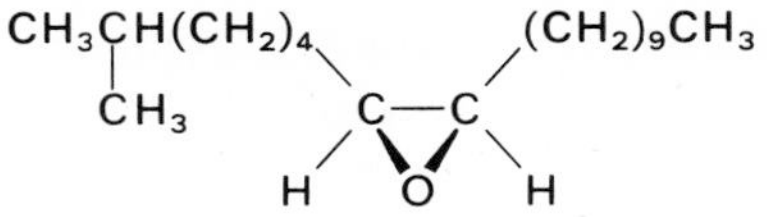

$$CH_3CH(CH_2)_4 \diagdown \qquad \diagup (CH_2)_9CH_3$$

$$CH_3 \qquad C-C$$

$$\diagup \diagdown \diagdown$$

$$H \quad O \quad H$$

sex-attractant of female gypsy moth

Although the molecule has no other functional groups, the male gypsy moth responds to a compound having the methyl group and cyclic ether only in the positions shown. In pest control experiments that are now underway, the compound is used to attract male gypsy moths to traps containing the sex-attractant compound and poison. A phenomenally small quantity of material, a matter of molecules per acre, evaporating into the air from a trap is effective in attracting moths over an area of many acres. Thus the gypsy moth's own pheromone can be used to control its population.

5.12 REACTIONS OF OXIRANES (EPOXIDES)

Oxiranes react under acidic conditions with a nucleophile in a typical substitution reaction. The unusual feature of oxirane reactions is the ring opening by strong nucleophiles in the absence of acid.

a) Acidic reagents

Oxirane rings are opened easily under mild acidic conditions. The reaction of ethylene oxide with aqueous sulfuric acid yields ethylene glycol.

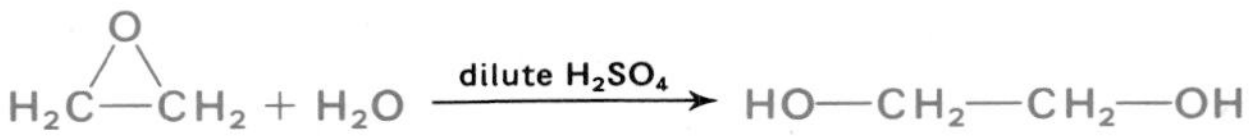

$$H_2C \overset{O}{\overset{\diagup \diagdown}{-}} CH_2 + H_2O \xrightarrow{\text{dilute } H_2SO_4} HO-CH_2-CH_2-OH$$

ethylene glycol

The reaction proceeds through the protonated oxirane, which is attacked by water. In the nucleophilic substitution one C—O bond breaks and a new one forms.

Ethylene oxide also reacts with hydrogen halides and with hydrogen sulfide under acidic conditions.

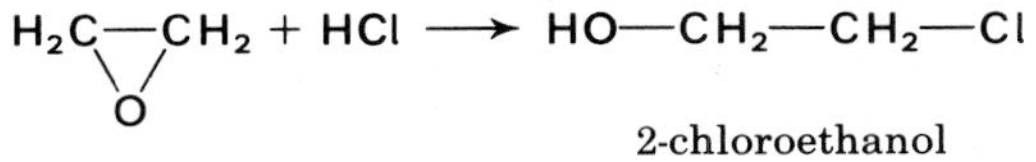

$$H_2C \overset{\diagdown \diagup}{\underset{O}{-}} CH_2 + HCl \longrightarrow HO-CH_2-CH_2-Cl$$

2-chloroethanol

$$H_2C \overset{\diagdown \diagup}{\underset{O}{-}} CH_2 + H_2S \xrightarrow{H_2SO_4} HO-CH_2-CH_2-SH$$

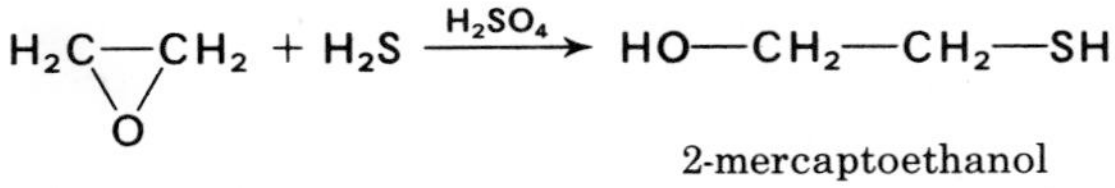

2-mercaptoethanol

b) Substitution without acid

The driving force in the relief of ring strain as the oxirane ring opens is shown by the reaction of oxiranes in the absence of acids. Ammonia and amines are stronger bases than is ethylene oxide and are therefore protonated in acidic solution. The ammonium and alkylammonium ions can no longer act as nucleophiles in acidic solution. The reaction of ethylene oxide with ammonia is a simple nucleophilic substitution without benefit of acid catalyst.

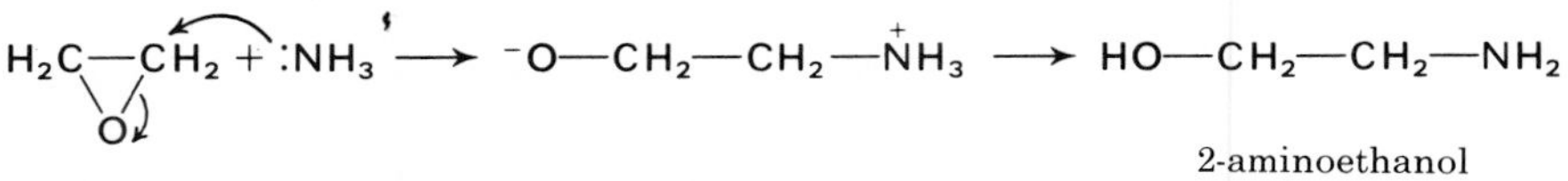

$$H_2C—CH_2 + :NH_3 \longrightarrow {}^-O—CH_2—CH_2—\overset{+}{N}H_3 \longrightarrow HO—CH_2—CH_2—NH_2$$

2-aminoethanol

Problem **5.11** Write formulas for the starting materials and for compounds A through F.

a) 1-hexene $\xrightarrow{\text{CH}_3\text{CO}_3\text{H}}$ A $\xrightarrow{\text{dil. H}_2\text{SO}_4}$ B

b) ethylene oxide $\xrightarrow{\text{H}_2\text{SO}_4 \text{ in CH}_3\text{OH}}$ C

c) 2-butene $\xrightarrow{\text{CH}_3\text{CO}_3\text{H}}$ D $\xrightarrow{\text{H}_2\text{S, H}_2\text{SO}_4}$ E

d) cyclopentene oxide $\xrightarrow{\text{CH}_3\text{NH}_2}$ F

5.13 STEROIDS   A group of familiar and important hormones and the well-known cholesterol belong to a family of compounds called **steroids**. Steroids can have alcoholic or phenolic hydroxyl groups, carbonyl groups, or carboxylic acid groups. They all have in common a basic tetracyclic carbon structure, with three six-membered rings and one five-membered ring fused into what is called the **steroid nucleus**.

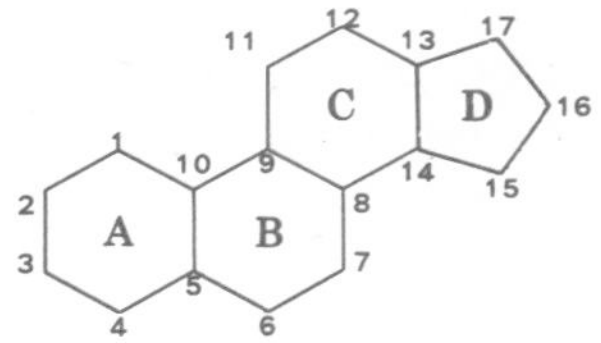

steroid nucleus showing numbering system

All of the steroids found in vertebrates are derived from the straight chain triterpene, 2,3-epoxysqualene (for the formula of this compound see Section 5.11).

   Cholesterol is the most abundant steroid of mammals. It is found in almost all tissues, particularly in the brain and spinal column. It is both a component in the diet and a product synthesized in the liver (via squalene). It is sometimes deposited on the inner walls of blood vessels of human beings, causing high blood pressure and hardening of the arteries. For this reason the biosynthesis and regulation of the transport and deposition of cholesterol are the subjects of extensive study. Most of the cholesterol in tissues appears as an ester.

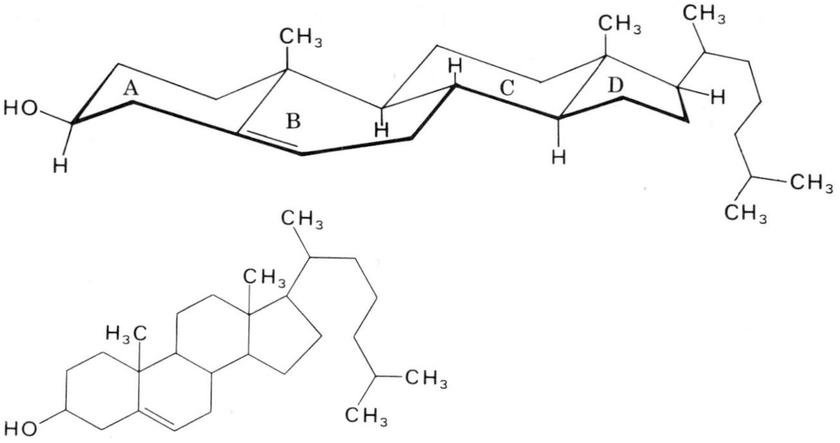

cholesterol

Cholesterol is the biological precursor of the steroid hormones and bile acids. Cholic acid, a bile acid, aids in the digestion of fats.

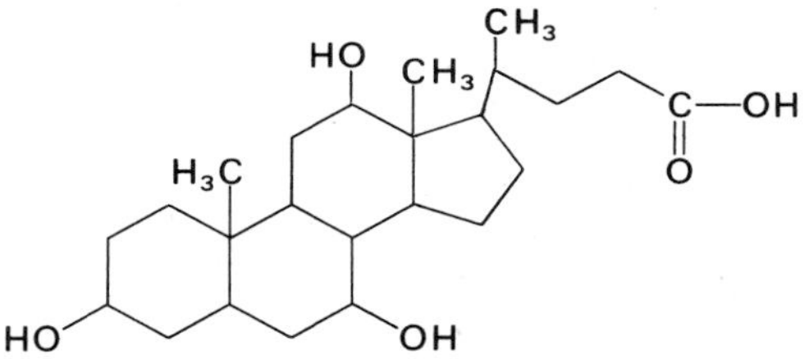

cholic acid

Hormones are compounds secreted by various endocrine glands in trace amounts. They function as chemical messengers carried by the blood to various target organs, where they regulate a variety of physiological and metabolic activities in vertebrates. Many hormones have a steroidal structure, including female and male sex hormones and adrenal corticosteroids. Their structures are shown below.

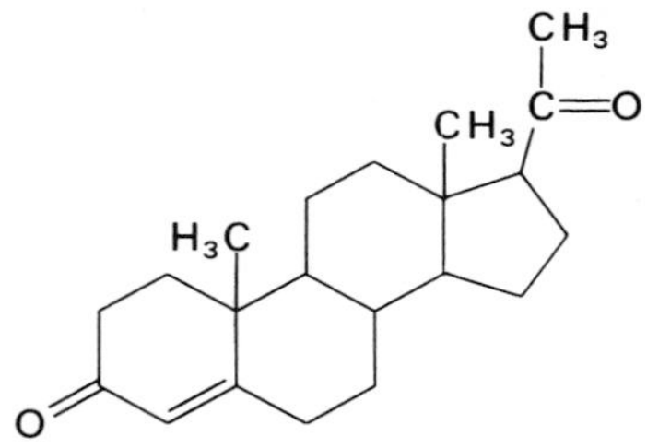

progesterone
(pregnancy hormone)

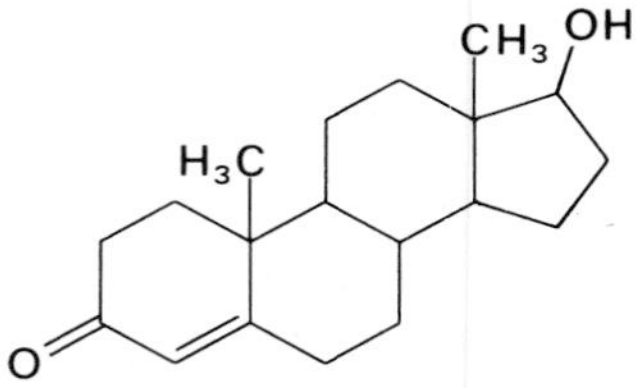

testosterone
an androgen
(male hormone)

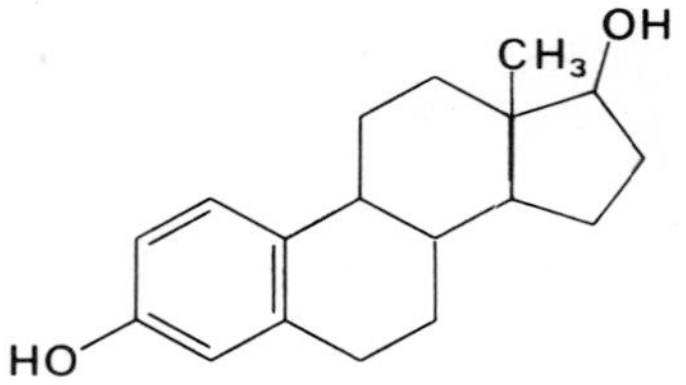

estradiol

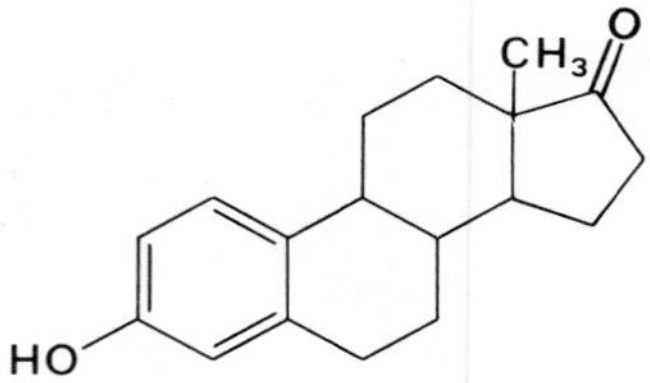

estrone

the major estrogens (female hormones)

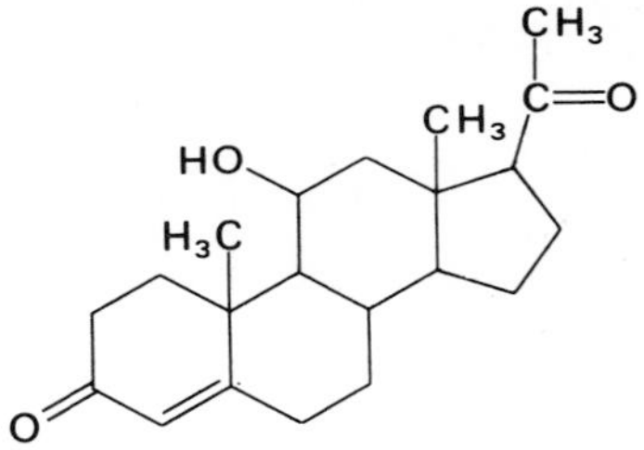

corticosterone

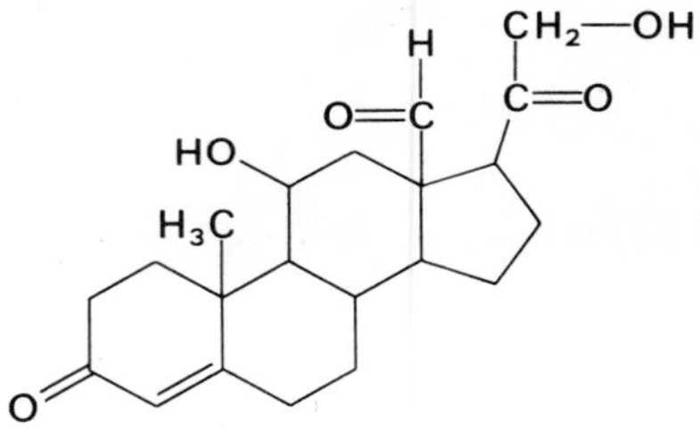

aldosterone

the major adrenal corticosteroids (adrenal hormones)

Problem **5.12** The similarity of the structures of compounds with such different biological roles is astonishing. Examine the formulas for cholesterol and the six steroid hormones carefully and answer these questions about them.

a) How does the unsaturation of the rings vary in the different structures? See steroid nucleus above for designation of rings.

b) How many substituents attached to the steroid nucleus does each compound have?

c) Each compound has an oxygen group in the same position on ring A (see designation of rings in the steroid nucleus above), yet they are not the same group. Identify the functional group on ring A in each compound and the number of the carbon carrying the group.

d) Other similarities of substituent positions exist. Note that a methyl group appears on carbon #10 at the ring junction between rings A and B in all compounds except the ones with ring A aromatic (where no substituent is possible). In what other two positions is there some substituent in every structure?

e) Cholesterol is the precursor of progesterone. What structure changes are needed for this conversion?

f) Progesterone is the precursor of testosterone and corticosterone. In each case what changes occur?

g) Testosterone is the precursor of estradiol. What change in structure is needed?

New Terms and Topics

Nomenclature of alcohols (Section 5.2)
Substitution reaction at a saturated carbon (Section 5.5)
Nucleophilic reagent and leaving group (Section 5.5)
Carbonium ion intermediates from alcohols (Section 5.5)
Elimination reaction (Section 5.6)
Hydroboration (Section 5.7)

Structure of phenols (Section 5.8)
Acidity of phenols (Section 5.9)
Ethers—inert solvents (Section 5.10)
Oxiranes—ring strain and reactivity (Section 5.11)
Ethylene oxide (Section 5.11)
Steroids—cholesterol and steroid hormones (Section 5.13)

Summary of Reactions

REACTIONS OF ALCOHOLS

1. **Reactions of alcohols as acids and bases** (Section 5.4)

   a) **With active metals**

   $$2 \text{ RO—H} + 2 \text{ Na}° \longrightarrow 2 \text{ RO}^-\text{Na}^+ + \text{H}_2$$

   $$2 \text{ CH}_3\text{CH}_2\text{—O—H} + 2 \text{ Na} \longrightarrow 2 \text{ CH}_3\text{CH}_2\text{—O}^-\text{Na}^+ + \text{H}_2$$

   ethanol · · · · · · · sodium ethoxide

   b) **As an acid**

   $$\text{RO—H} + \text{B}^- \longrightarrow \text{RO}^- + \text{HB}$$

   $$(\text{CH}_3)_3\text{C—O—H} + \text{Na}^+\text{H}^- \longrightarrow (\text{CH}_3)_3\text{C—O}^-\text{Na}^+ + \text{H}_2$$

   tert-butyl · · sodium · · · · · sodium *tert*-butoxide
   alcohol · · · · hydride

   c) **As a base** (Section 5.4)

   $$\text{R—OH} + \text{HA} \longrightarrow \text{R—}\overset{+}{\text{O}}\text{H}_2 + \text{A}^-$$

   $$\text{CH}_3\text{—OH} + \text{H}_2\text{SO}_4 \longrightarrow \text{CH}_3\text{—}\overset{+}{\text{O}}\text{H}_2 + \text{HSO}_4^-$$

   methanol · · · · · · · methyloxonium ion

2. Substitution reactions of alcohols

a) With hydrogen halides (Section 5.5)

$$R{-}OH + HBr \longrightarrow R{-}Br + H_2O$$

$$CH_3CH_2CH_2{-}OH + HBr \longrightarrow CH_3CH_2CH_2{-}Br + H_2O$$

    1-propanol   1-bromopropane

b) With nonmetal halides (Section 5.5)

$$R{-}OH + SOCl_2 \longrightarrow R{-}Cl + SO_2 + HCl$$

$$CH_3CH_2{-}OH + SOCl_2 \longrightarrow CH_3CH_2{-}Cl + SO_2 + HCl$$

ethanol   thionyl chloride   chloroethane

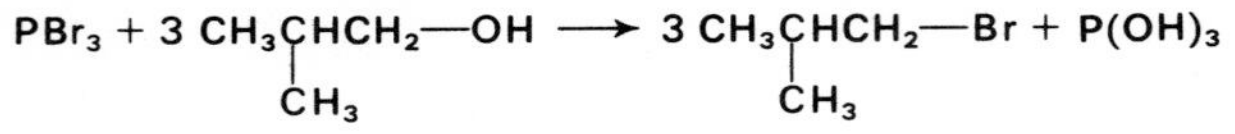

$$PBr_3 + 3\ CH_3\underset{\underset{CH_3}{|}}{C}HCH_2{-}OH \longrightarrow 3\ CH_3\underset{\underset{CH_3}{|}}{C}HCH_2{-}Br + P(OH)_3$$

    phosphorus   isobutyl   isobutyl   phosphorous
    tribromide   alcohol   bromide   acid

3. Dehydration of alcohols to alkenes (Section 5.6)

$$RCH_2{-}CHR{-}OH \xrightarrow{H_2SO_4} RCH{=}CHR + H_2O$$

$$(CH_3)_3C{-}OH \xrightarrow[\text{room temperature}]{H_2SO_4} (CH_3)_2C{=}CH_2 + H_2O$$

tert-butyl alcohol   isobutylene

PREPARATION OF ALCOHOLS

1. By hydroboration of an alkene (Section 5.7)

$$RCH{=}CH_2 \xrightarrow{B_2H_6} (RCH_2{-}CH_2{-})_3B \xrightarrow{H_2O_2,\ OH^-} RCH_2CH_2{-}OH + BO_3^{-3}$$

$$CH_3CH_2CH{=}CH_2 \xrightarrow{B_2H_6} (CH_3CH_2CH_2CH_2{-})_3B \xrightarrow{H_2O_2,\ OH^-} CH_3CH_2CH_2CH_2OH + BO_3^{-3}$$

    1-butene   tributylborane   1-butanol

2. By hydrogenation of carbonyl compounds (Section 5.7)

$$R_2C{=}O + H_2 \xrightarrow{Ni} R_2CH{-}OH$$

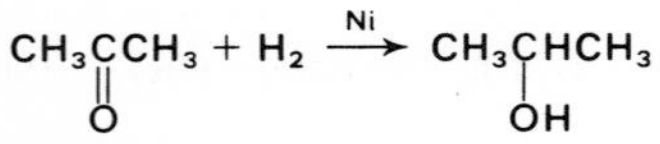

$$CH_3\underset{\underset{O}{\|}}{C}CH_3 + H_2 \xrightarrow{Ni} CH_3\underset{\underset{OH}{|}}{C}HCH_3$$

    acetone   isopropyl alcohol

PHENOL AS AN ACID (Section 5.9)

$$C_6H_5-OH + Na^+OH^- \longrightarrow C_6H_5-O^-Na^+ + H_2O$$

phenol   sodium phenoxide

RING OPENING REACTIONS OF OXIRANES (Section 5.12)

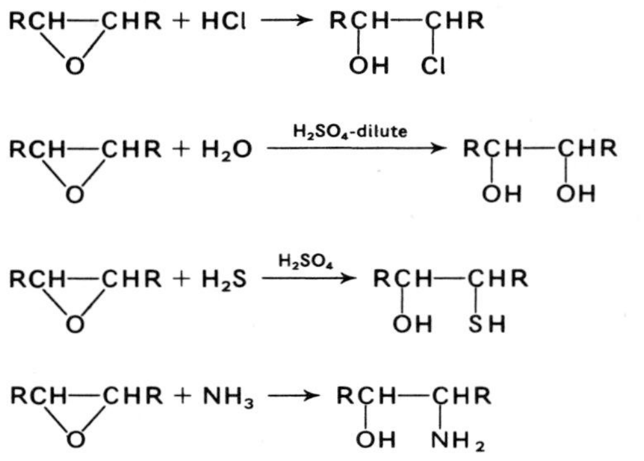

$$\underset{\underset{O}{\diagup\diagdown}}{RCH-CHR} + HCl \longrightarrow \underset{\underset{OH \quad Cl}{|\quad|}}{RCH-CHR}$$

$$\underset{\underset{O}{\diagup\diagdown}}{RCH-CHR} + H_2O \xrightarrow{H_2SO_4\text{-dilute}} \underset{\underset{OH \quad OH}{|\quad|}}{RCH-CHR}$$

$$\underset{\underset{O}{\diagup\diagdown}}{RCH-CHR} + H_2S \xrightarrow{H_2SO_4} \underset{\underset{OH \quad SH}{|\quad|}}{RCH-CHR}$$

$$\underset{\underset{O}{\diagup\diagdown}}{RCH-CHR} + NH_3 \longrightarrow \underset{\underset{OH \quad NH_2}{|\quad|}}{RCH-CHR}$$

Problems

5.13 Write structures for the following compounds.

a) 1,2,3-cyclohexanetriol   **b)** 2-penten-1-ol   **c)** potassium *tert*-butoxide

d) 3-chloro-4-methyl-1,2-hexanediol   **e)** ethylene oxide   **f)** methyl 3-phenyl-2-butyl ether

g) 2,4,6-tribromophenol   **h)** tetrahydrofuran   **i)** *m*-methoxyphenol

j) hydroquinone   **k)** catechol   **l)** *cis*-2,3-dimethyloxirane

5.14 Write the structures for the products of the following reactions.

a) 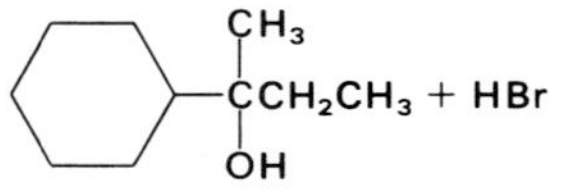 cyclohexyl$-\underset{\underset{OH}{|}}{\overset{\overset{CH_3}{|}}{C}}CH_2CH_3$ + HBr

b) 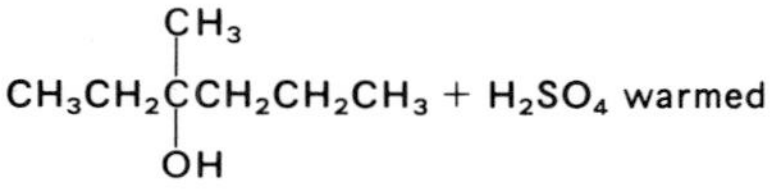 $CH_3CH_2\underset{\underset{OH}{|}}{\overset{\overset{CH_3}{|}}{C}}CH_2CH_2CH_3$ + H_2SO_4 warmed

c) $CH_3CH_2CH_2CH_2CH{=}CH_2$ + B_2H_6

d) 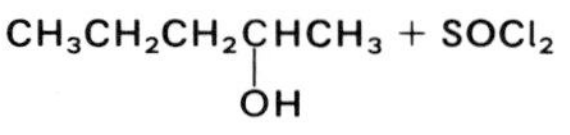 $CH_3CH_2CH_2\underset{\underset{OH}{|}}{C}HCH_3$ + $SOCl_2$

e) 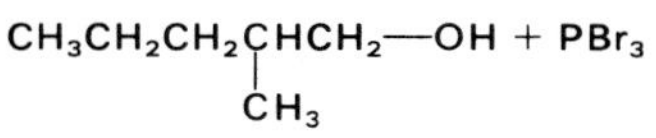 $CH_3CH_2CH_2\underset{\underset{CH_3}{|}}{C}HCH_2{-}OH$ + PBr_3

f) 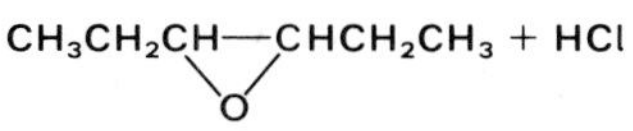 $CH_3CH_2CH\underset{O}{\overset{\diagup\diagdown}{-}}CHCH_2CH_3$ + HCl

g) 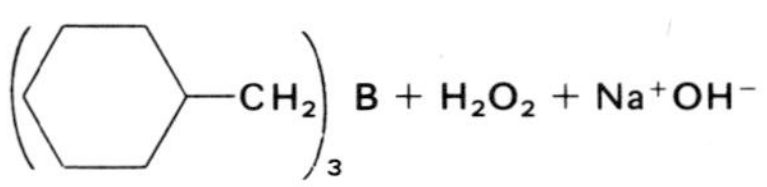 $\left(\text{cyclohexyl}{-}CH_2 \right)_3 B$ + H_2O_2 + Na^+OH^-

h) $CH_3CH_2CH_2\overset{\overset{H}{|}}{C}{=}O$ + H_2 + Pt

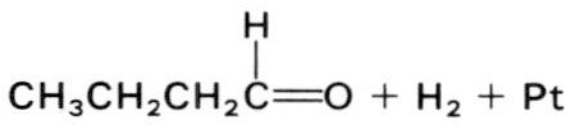

5.15 Write structures for starting materials and products A through L.

a) cyclohexanol $\xrightarrow[100°]{H_2SO_4}$ A $\xrightarrow[Pt]{H_2}$ B

b) 2-heptene $\xrightarrow{CH_3CO_3H}$ C $\xrightarrow{H_2S, H_2SO_4}$ D

c) 2-butanol $\xrightarrow[100°]{H_2SO_4}$ E $\xrightarrow{neut.\ KMnO_4}$ F

d) decanal $\xrightarrow{H_2,\ Pt}$ G $\xrightarrow{PCl_3}$ H

e) 2-methyl-1-pentene $\xrightarrow{B_2H_6}$ I $\xrightarrow{H_2O_2,\ OH^-}$ J

f) 2-phenyl-2-propanol $\xrightarrow{H_2SO_4}$ K $\xrightarrow[2)\ H_2O_2,\ OH^-]{1)\ B_2H_6}$ L

5.16 Draw the resonance structures for phenoxide ion.

5.17 In which of the following compounds would you expect *intramolecular* hydrogen bonding? Draw an appropriate structure for each that would.

a) 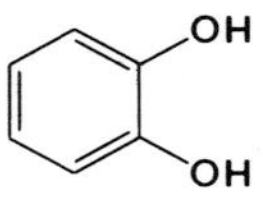 **b)** 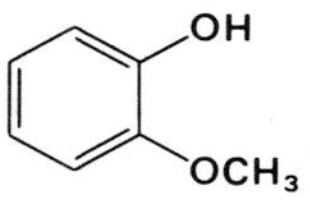 **c)** 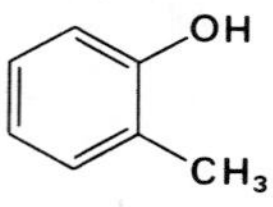 **d)**

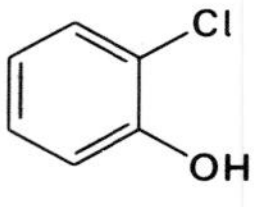

e) 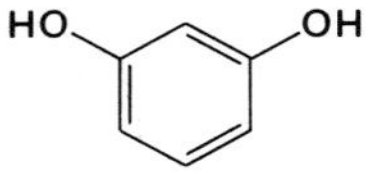 **f)**

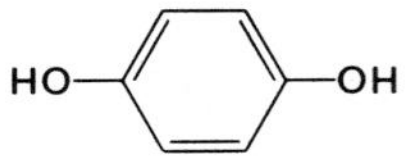

5.18 Draw the structures of the (major) products of these reactions.

a) 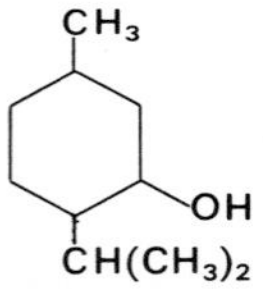 + H₂SO₄

menthol

b) 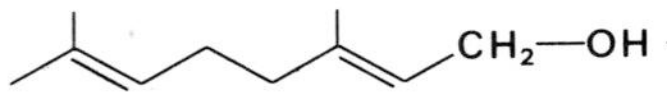 CH₂—OH + PCl₃

geraniol

c) 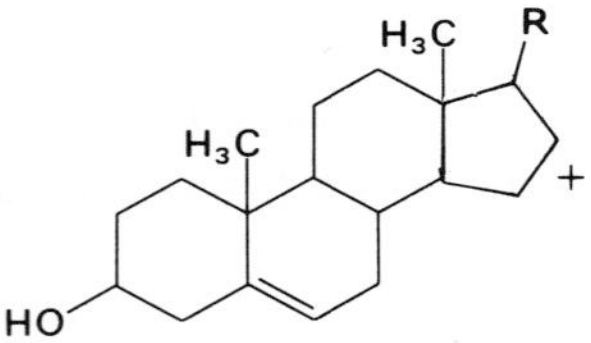 + SOCl₂

cholesterol

5.19 Choose an alkene and the appropriate reagent to produce each of the following alcohols.

a) $CH_3CH_2\overset{\overset{\displaystyle CH_3}{|}}{C}HCH_2{-}OH$

b) $CH_3CH_2\overset{\overset{\displaystyle CH_3}{|}}{\underset{\underset{\displaystyle OH}{|}}{C}}CH_3$

c)

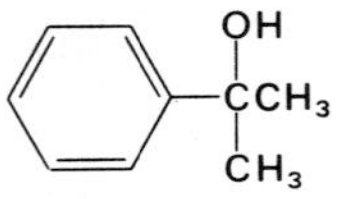

5.20 Choose an alcohol and the appropriate reagent to produce each of the following compounds.

a) $CH_3CH{=}\overset{\overset{\displaystyle CH_3}{|}}{C}CH_2CH_2CH_3$

b) $CH_2{=}CHCH_2CH_3$

CHAPTER 6

Alkyl Halides and
Their Relatives

In the molecular evolution that produced life, organohalides must have been rejected repeatedly as a class of nature's intermediates. Most of the other classes of organic compounds are well represented in nature's storeroom. Only a few compounds produced by lower organisms, such as molds and actinomyces, contain halogen, an example being griseofulvin, a useful fungicide. Another example is the antibiotic aureomycin (Section 3.9). Both compounds contain a halogen attached to the aromatic ring.

griseofulvin

The virtual absence of organohalides in nature is probably related to their natural use. When found, organohalides often appear to have evolved as a biological weapon that one microorganism uses to guard itself from another microorganism.

Compounds that perturb the normal chemistry of microorganisms are of enormous interest and importance to us. In the battle for food and good health, scientists have discovered and doctors and farmers have used a variety of organohalides. Vast numbers of people are alive today and have food to eat because organohalides have destroyed the bacteria and nematodes, the spirochetes, beetles, mosquitoes, and weeds that are the competitors of the human species. In this chapter we discuss the chemistry of the alkyl halides and their natural counterparts, the alkyl phosphates.

6.1 SOURCES AND USES OF ALKYL HALIDES

Most of the common alkyl halides are liquids that are heavier than water or other organic compounds. Methyl fluoride, chloride, and bromide, however, are gases. Most alkyl halides are insoluble in water, but are miscible with liquid hydrocarbons. Dichloromethane CH_2Cl_2, chloroform $CHCl_3$, and carbon tetrachloride CCl_4 are important laboratory and commercial solvents for nonpolar substances and their reactions. Many of the stable and cheap chlorides are used as solvents for dry cleaning, for extraction of fatty materials from animal tissues, and for separation of substances by liquid–liquid extraction (e.g., a water–carbon tetrachloride combination). An increase in the halogen content of organic compounds decreases their flammability and therefore increases their usefulness for many purposes.

Some small alkyl chlorides are made by reaction of the alkane directly with chlorine gas, using ultraviolet light or a high temperature.

$$CH_4 \xrightarrow[-HCl]{Cl_2} CH_3Cl \xrightarrow[-HCl]{Cl_2} CH_2Cl_2 \xrightarrow[-HCl]{Cl_2} CHCl_3 \xrightarrow[-HCl]{Cl_2} CCl_4$$

A mixture of all possible chlorides, mono-, di-, tri-, etc., is obtained which must be separated by careful distillation. The reaction is feasible only for hydrocarbons which have only one or two carbons with hydrogens for substitution, e.g., methane, ethane.

Polyhalogenated compounds have many specific uses. While carbon tetrachloride and chloroform are toxic when inhaled over a period of time, other similar compounds are nontoxic. "Freon," dichlorodifluoromethane CCl_2F_2, is nontoxic and nonflammable. It is used as a condensable gas in cooling systems, such as refrigerators and automobile air conditioning. A compound known as halothane, $F_3CCHBrCl$, is widely used as an anesthetic.

The carbon–fluorine bond is very strong. Its formation is accompanied by the evolution of much heat, and once formed it is not easily broken. The C–F bond energy is about 116 kcal/mole. The great strength of the C–F bond gives perfluorocarbons (compounds in which all the bonds of carbon are attached to fluorines, no hydrogens) unique properties of stability against heat and light deterioration. The polymer "Teflon" is a compound containing a chain of hundreds of carbons, all of which have only C—C and C—F bonds. Teflon is extraordinarily slippery and finds uses as a self-lubricating plug in valves and stopcocks and as a coating for cooking utensils.

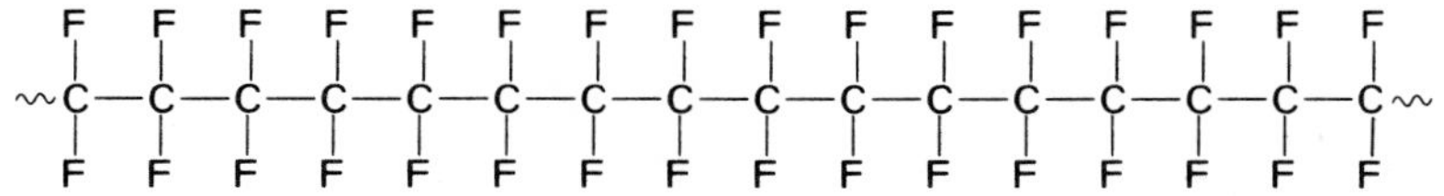

Teflon

Some multichloro compounds are potent insecticides, of which DDT, dieldrin, and lindane are examples.

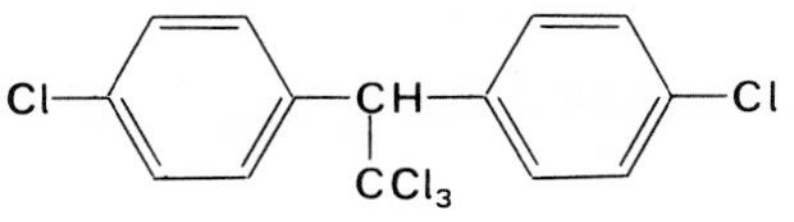

DDT 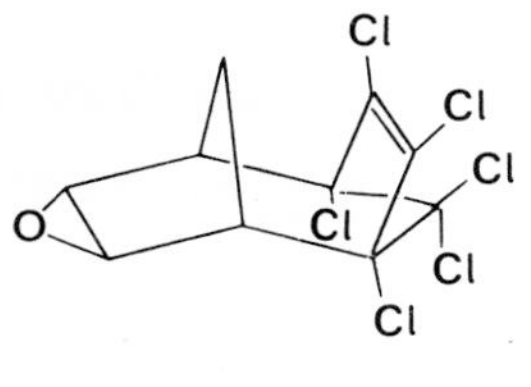   dieldrin 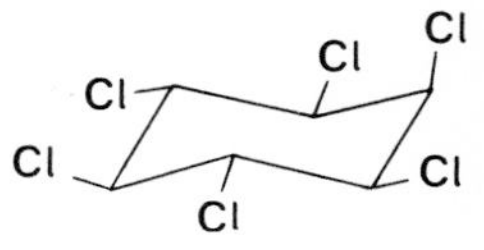   lindane

The persistence of these compounds, without further reaction, through the life cycles of plants, small animals, fish, and larger mammals has led to partial or total bans on their use. However, DDT, as used in the control of the mosquito as a carrier of malaria and encephalitis, has probably saved more lives than has any other synthetic compound.

Alkyl halides are rare in nature. A few naturally occurring halides are found as products of molds or bacteria. In the laboratory alkyl halides as a class of compounds serve chiefly as valuable synthetic intermediates through which alcohols are converted to ethers and nitriles.

6.2 NOMENCLATURE OF ALKYL HALIDES

Systematic IUPAC nomenclature for organic halides uses the halogen as a substituent whose name appears as a prefix to that of the parent hydrocarbon. Carbons of the hydrocarbon are numbered starting from the end closest to the first substituent.

$CH_3CH_2CH_2$—Br   ⬡—Br $(C_6H_5)_3C$—Cl

1-bromopropane   bromocyclohexane   triphenylchloromethane

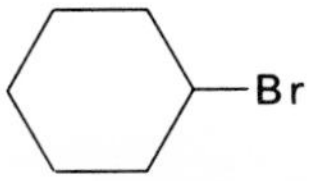

Compounds having two, three, or four halogens bonded to the same carbon are stable. This stability contrasts with that of compounds having a combination of halogen, hydroxyl, or amino groups attached to the same carbon, which are not stable (Section 5.2). The best known of the polyhalogenated compounds are called by their common names, carbon tetrachloride and chloroform.

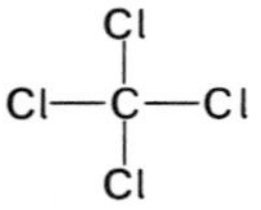

$$Cl-\underset{\underset{Cl}{|}}{\overset{\overset{Cl}{|}}{C}}-Cl \qquad Cl-\underset{\underset{Cl}{|}}{\overset{\overset{H}{|}}{C}}-Cl \qquad Cl-\underset{\underset{H}{|}}{\overset{\overset{H}{|}}{C}}-Cl$$

IUPAC = tetrachloromethane   trichloromethane   dichloromethane
common = carbon tetrachloride   chloroform

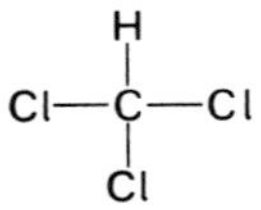

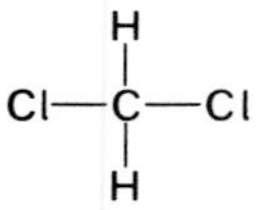

$CH_3CH_2CH_2CH_2CCl_2CH_3$ ⬡—CF_3

2,2-dichlorohexane   trifluoromethylbenzene

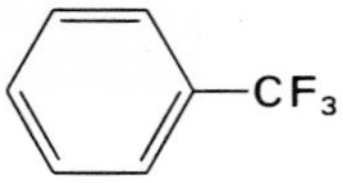

Other halides are also best known by their common names, such as allyl, vinyl, and benzyl halides.

CH_2=CH—CH_2—Cl CH_2=CH—Cl ⬡—CH_2—Br

allyl chloride   vinyl chloride   benzyl bromide

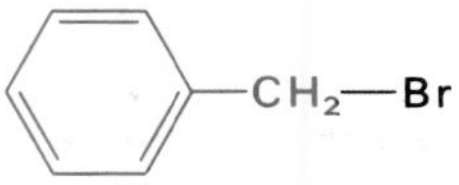

Alkyl halides are classified as primary, secondary, or tertiary, according to the number of carbon groups attached to the carbon bearing the halogen, in the manner used for alcohols.

Problems

6.1 Draw structures for the following compounds.

   a) 2-bromo-2-chloro-1,1,1-trifluoroethane   **b)** 3-bromocyclohexene

   c) perchlorocyclopentadiene (1,2,3,4,5,5-hexachlorocyclopentadiene)

   d) 2,2,4-tribromopentane   **e)** iodoform   **f)** allyl bromide

6.2 Identify each of these halides as primary, secondary, or tertiary.

a) ⬡—Cl   **b)** CH₃CHCH₂CH₃ with Cl   **c)** ⬡—CH₂—Cl

d) CH₃C with CH₃ and Cl attached to benzene ring

6.3 REACTIONS OF ALKYL HALIDES—SUBSTITUTION

The principal reaction of alkyl halides is the substitution of other electronegative atoms or groups of atoms for the halogen.

$$Nu^- + R—Cl \longrightarrow Nu—R + Cl^-$$

Alkyl halides are known as alkylating agents, since in the substitution reaction an alkyl group is transferred from one atom to another.

The carbon–halogen bond is both polar and polarizable, with the carbon having a partial positive charge.

$$\overset{\delta^+}{\underset{|}{-C-}}\overset{\delta^-}{Cl}$$

The carbon–fluorine bond has a large bond energy (116 kcal/mole) and is not easily broken. Consequently, alkyl fluorides do not react in ordinary substitution reactions. The other halogens with weaker C—X bonds are readily displaced in nucleophilic substitutions. In order of reactivity, the iodide ion is more easily displaced than the bromide ion, which in turn is more reactive than the chloride.

Alkyl halides react with a variety of reagents in the substitution reaction to produce many classes of compounds. Among the most useful are those reactions with alkoxide and phenoxide ions to give ethers and with cyanide ion to form nitriles, R—C≡N.

a) Alkyl halides with alkoxide and phenoxide ions

The reaction of alkyl halides with alkoxide or phenoxide ions, RO⁻ or ArO⁻, provides a simple procedure for making ethers. The two carbon fragments may be the same or they may be quite different, as in this reaction of sodium propoxide with ethyl chloride. The reaction is called the **Williamson synthesis** of ethers.

$$CH_3CH_2CH_2—O^-Na^+ + CH_3CH_2—Cl \longrightarrow CH_3CH_2CH_2—O—CH_2CH_3 + Na^+Cl^-$$

sodium propoxide   ethyl chloride   ethyl propyl ether

The same ethyl propyl ether may be made by reversing the roles played by the propyl and ethyl parts, using 1-chloropropane and sodium ethoxide.

$$CH_3CH_2CH_2\!-\!Cl + CH_3CH_2\!-\!O^-Na^+ \longrightarrow CH_3CH_2CH_2\!-\!O\!-\!CH_2CH_3 + Na^+Cl^-$$

1-chloropropane   sodium ethoxide   ethyl propyl ether

As a rule, either carbon portion can be halide or alkoxide, but there are two general exceptions. Ethers with a tertiary carbon group must be made from the tertiary alkoxide and another halide. When treated with an alkoxide ion, tertiary halides give only alkenes as products (see Section 6.5).

$$(CH_3)_3CO^-K^+ + CH_3CH_2\!-\!Br \longrightarrow (CH_3)_3C\!-\!O\!-\!CH_2CH_3 + K^+Br^-$$

potassium   ethyl   <i>tert</i>-butyl
tert-butoxide   bromide   ethyl ether

Bromobenzene does not undergo a nucleophilic substitution under these conditions; therefore phenyl ethers may be made only from a phenoxide ion and an alkyl halide.

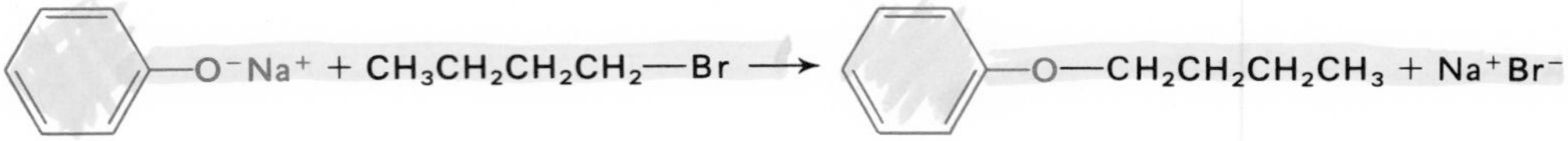

sodium   1-bromobutane   butyl phenyl ether
phenoxide

Recall that phenol is a strong enough acid to react with sodium hydroxide to give sodium phenoxide and water (Section 5.9).

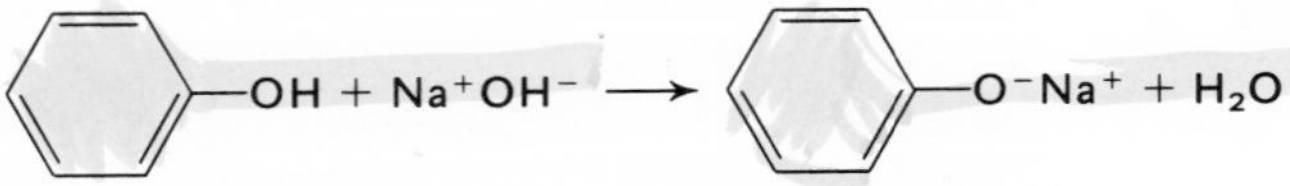

phenol   sodium
phenoxide

In contrast to phenol, the alcohols are not acidic enough to form their salts when treated with sodium hydroxide, and the sodium alkoxides must be prepared by the reaction of sodium metal or sodium hydride with the alcohol (Section 5.4).

$$2\,CH_3\!-\!OH + 2\,Na \longrightarrow 2\,CH_3\!-\!O^-Na^+ + H_2$$

b) Alkyl halides with sodium hydroxide   Primary halides are reactive toward hydroxide ion to form alcohols by substitution of a hydroxide for a halide ion.

$$CH_3CH_2CH_2-Br + OH^- \longrightarrow CH_3CH_2CH_2-OH + Br^-$$

1-bromopropane   1-propanol

For most saturated alkyl halides the conversion to alcohols is not important, since the source of the halide itself is usually the alcohol or the alkene. Conversion of the alkene directly to the alcohol by the addition of water is more usual than the sequence using the alkyl halide.

Benzyl and allyl halides are especially reactive with hydroxide ion.

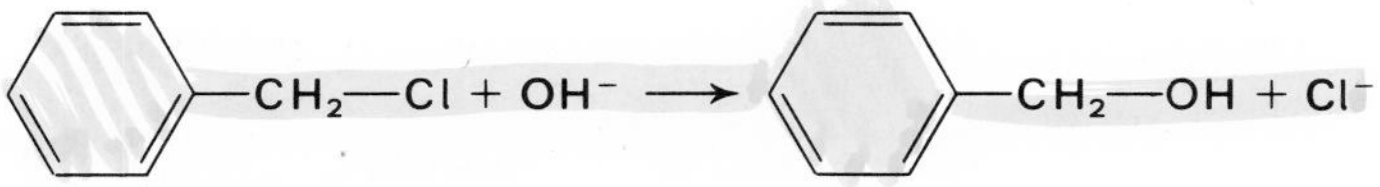

benzyl chloride   benzyl alcohol

$$CH_2{=}CHCH_2-Cl + OH^- \longrightarrow CH_2{=}CHCH_2-OH + Cl^-$$

allyl chloride   allyl alcohol

c) Primary halides with sodium cyanide

Substitution of primary alkyl halides with carbon reagents which can form new carbon–carbon bonds provides a method of increasing the carbon chain length. In Chapter 14 there are many examples of such reactions of carbanions, reagents with a negative charge on carbon. The simplest example of such a reagent is the cyanide ion, $C{\equiv}N^-$.

The cyanide ion, $:C{\equiv}N:^-$, is a simple carbon nucleophile that is readily available as sodium cyanide, Na^+CN^-. The displacement of the halide in a primary alkyl halide by the cyanide ion produces a **nitrile**.

$$R-C{\equiv}N \qquad CH_3-C{\equiv}N$$

a nitrile   acetonitrile

Nitriles are synthetically useful intermediates that can be either hydrolyzed to acids (Section 8.10) or hydrogenated to amines (Section 10.6). The reaction of benzyl bromide with cyanide ion gives phenylacetonitrile. The nitrile contains one more carbon than the original halide.

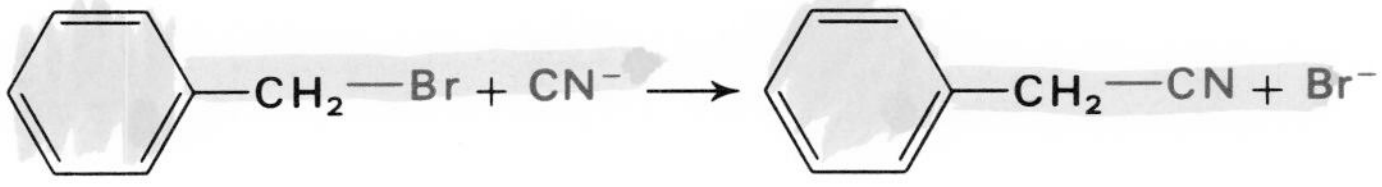

benzyl bromide   phenylacetonitrile

Secondary halides give poor yields of nitriles; tertiary halides give only alkenes by elimination.

6.3 Write equations for the following reactions.

    a) methyl iodide + sodium isopropoxide

    b) sodium phenoxide + 2-chloropentane

    c) 1-bromopropane + sodium cyanide

6.4 MECHANISMS OF NUCLEOPHILIC SUBSTITUTION AND OF ELIMINATION

How do reactions occur? In what order are bonds made and broken during reactions? Do short-lived reaction intermediates that can't be isolated intervene between starting material and product in a reaction? These questions define the subject of reaction mechanisms.

In the substitution reactions of alcohols with hydrogen halides, two mechanisms were depicted for the conversion of alcohols to alkyl halides (Section 5.5). As with alcohols, alkyl halides and alkyl sulfonates may undergo nucleophilic substitution by two different mechanisms—one-step and two-step processes.

A one-step mechanism is depicted as one in which the nucleophile attacks the substrate and the leaving group departs in one concerted operation. A reaction which occurs by this mechanism is called a **bimolecular nucleophilic substitution** (S_N2) because the *two species*, nucleophile and substrate, are involved at the time the leaving group departs.

nucleophile   primary or secondary carbon attached to leaving group   detached leaving group

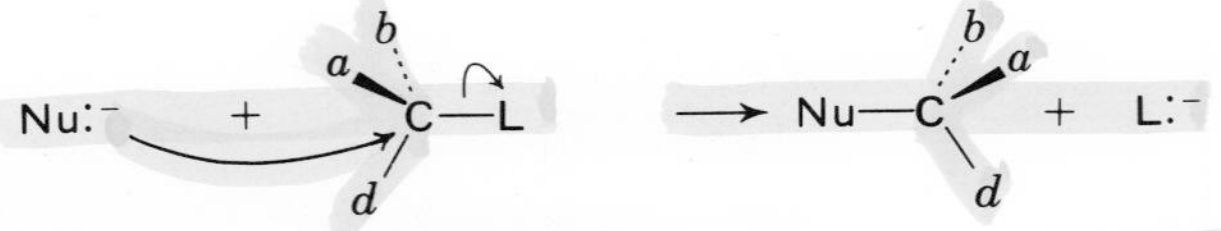

C—Nu bond formation occurs simultaneously with C—L bond cleavage.

In the one-step mechanism, the nucleophile with its high electron density, and often a negative charge, is attracted to the carbon with a partial positive charge. During the substitution, *five* groups are gathered around the central carbon atom. The aggregate is either neutral or negatively charged, usually the latter.

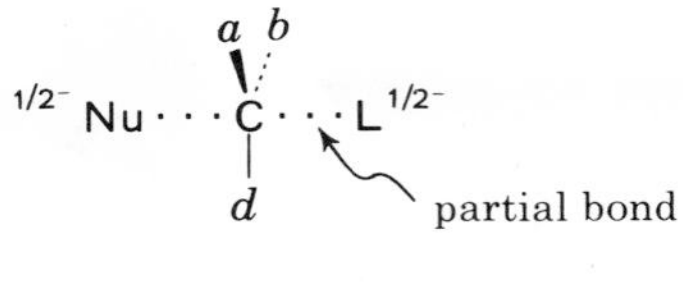

The one-step mechanism is generally characteristic of primary alcohols and primary alkyl halides. Substitutions which proceed by this mechanism are useful reactions which give a good yield of the expected product.

Access to the carbon for the reagent Nu is easiest on the side opposite the leaving group. The leaving group departs with its pair of bonding electrons.

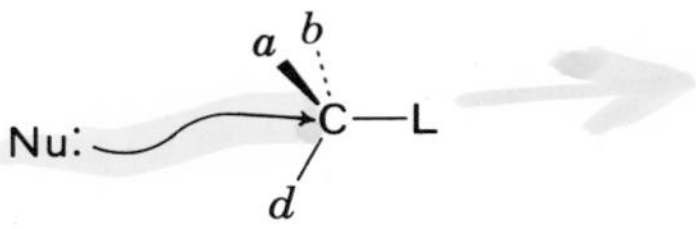

During the bimolecular nucleophilic substitution a carbon which starts with four substituents temporarily passes through a state in which it has five substituents. What is the effect on this reaction if the carbon groups attached to the carbon being attacked are large groups? Large alkyl groups occupy much space, and they hinder the approach of the attacking group (they create *steric hindrance*), making the reaction proceed more slowly. The reactivity order of alkyl halides in reaction with the alkoxide ion is consistent with the postulated one-step bimolecular mechanism (S_N2). Methyl halides, the least hindered, react the most rapidly. Primary halides react more rapidly than secondary. Tertiary halides do not give a substitution reaction with alkoxide ion; instead only alkenes are produced.

By contrast with the substitution of alkyl halides with alkoxide ion described above, tertiary halides react very rapidly with nucleophilic solvents to give substituted products. The mechanism is depicted as a two-step process with the solvent becoming a reactant only in the second step, after the leaving group has departed.

The first of the two steps of this mechanism is the *ionization* of the substrate to form a carbonium ion, which is captured almost immediately by the solvent in the second step. This mechanism is called a **unimolecular nucleophilic substitution** (S_N1).

Step 1 Bond cleavage—ionization of the substrate

carbonium ion of   leaving
reasonable stability   group

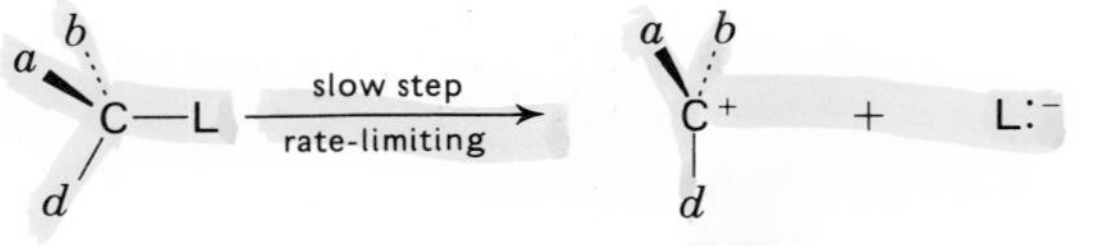

Step 2 Bond formation

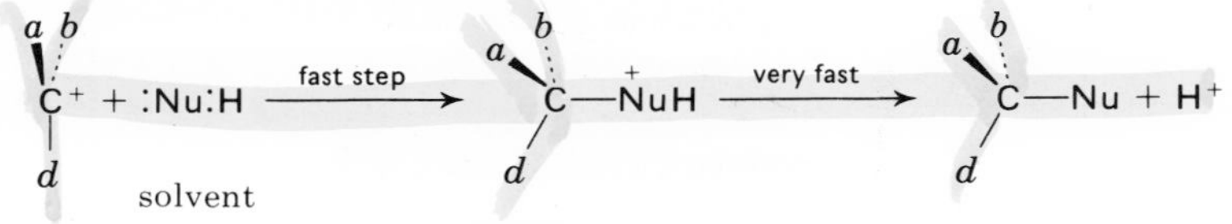

solvent

The first step, ionization, is the slower, rate-limiting step. The second step, the reaction of the carbonium ion with the solvent, occurs very much faster than the first step. The final step, deprotonation of the product, is extremely rapid.

The characteristics of the solvolysis reactions are consistent with the order of stability of carbonium ions, that is, tertiary faster than secondary and very much faster than primary halides.

$$\begin{matrix} & \overset{\displaystyle C}{|} & & \overset{\displaystyle C}{|} & & & \\ C\!-\!\underset{\displaystyle \underset{|}{C}}{\overset{+}{C}} & > & C\!-\!\underset{+}{C}H & \gg & C\!-\!\underset{+}{C}H_2 & > & \underset{+}{C}H_3 \end{matrix}$$

The two-step mechanism of ionization of the substrate to the carbonium ion and the leaving group is characteristic of tertiary halides, which form the most stable carbonium ions and are least subject to concerted displacement of the halide. Solvolysis reactions of tertiary halides give moderate yields of substitution products.

In discussing the one-step reaction mechanism characteristic of primary halides and the ionization mechanism of tertiary halides, we have said very little about secondary halides. The choice of mechanism for secondary halides becomes a matter of the constitution of the halide itself, the stability of the carbonium ion which could be formed, the ionizing power of the solvent, and almost every other possible factor. In general, negatively charged nucleophiles favor S_N2, and neutral nucleophiles and ionizing solvents favor S_N1.

Nucleophilic reagents also have properties of bases, and some, like hydroxide, alkoxide, phenoxide, and cyanide ions, are quite good bases. If the reagent forms a new bond with a hydrogen of the substrate— primary alkyl halide or tertiary carbonium ion—the product formed is an alkene instead of the substitution product.

$$B:^- + H\!-\!\overset{|}{\underset{|}{C}}\!-\!\overset{|}{\underset{|}{C}}\!-\!X \longrightarrow ^{\backslash}\!\!\!\!\underset{/}{C}\!\!=\!\!\overset{\backslash}{\underset{/}{C}} + HB + X^-$$

The alkene is produced in an elimination of the elements of H—X from the alkyl halide. In nearly every nucleophilic substitution, elimination is a competing process between the same two reactants. The production of alkene is favored over the substitution product by an increase in basicity of the reagent, an increase in temperature of the reaction, and by conditions of the reaction which encourage ionization to carbonium ions.

Carbonium ions are excellent proton donors, if there is a hydrogen on a carbon adjacent to the electron-deficient carbon. The reaction of strong bases with a carbonium ion is proton transfer to give an alkene, in almost total preference to the bond formation with carbon to give the substitution product.

$$CH_3-\underset{+}{\overset{\overset{\displaystyle CH_3}{|}}{C}}-CH_2-H + OH^- \longrightarrow CH_3-\overset{\overset{\displaystyle CH_3}{|}}{C}=CH_2 + H_2O$$

With primary halides, the base must remove the proton from the neutral alkyl halide. The proton removed is one attached to a carbon adjacent to the carbon bearing the halogen, so that a new carbon–carbon bond can be formed as the halogen leaves, as shown in the next section.

Problems

6.4 Draw *cis*-4-bromo-1-methylcyclohexane in two chair conformations. Which of these conformations should be best suited for a rear-side attack by hydroxide ion? Draw the structure of the alcohol produced by the rear-side attack. Is the product *cis*- or *trans*-4-methylcyclohexanol?

6.5 Arrange the following compounds in decreasing order of reactivity toward CH_3O^- in a bimolecular nucleophilic substitution.

$$CH_3-\underset{\overset{\displaystyle |}{CH_3}}{\overset{\overset{\displaystyle CH_3}{|}}{C}}-Br \qquad CH_3CH_2CH_2CH_2-Br \qquad CH_3CH_2\overset{\overset{\displaystyle Br}{|}}{C}HCH_3 \qquad CH_3-I \qquad CH_3\overset{\overset{\displaystyle CH_3}{|}}{C}HCH_2-Br$$

6.6 Draw the structures of the products of the unimolecular nucleophilic substitution of *cis*-4-bromo-1-methylcyclohexane with water and with ethanol in an ethanol–water solvent.

6.7 Given the relative stabilities of the carbonium ions formed as reaction intermediates, arrange the following compounds in decreasing order of reactivity with water by a unimolecular substitution mechanism.

$$CH_2=CHCH_2-Br \qquad CH_3CH_2CH_2-Br \qquad CH_3\overset{\overset{\displaystyle |}{Br}}{C}HCH_3 \qquad CH_3\overset{\overset{\displaystyle CH_3}{|}}{\underset{\underset{\displaystyle CH_3}{|}}{C}}Br \qquad CH_3-Br$$

6.5 ALKENES FROM ALKYL HALIDES

Nucleophiles which are strong bases may attack a hydrogen on a carbon adjacent to the carbon bearing the halogen (X). The elements of H—X are removed from adjacent carbons to produce an alkene. A reaction which produces alkenes from saturated compounds is called an *elimination*.

$$B^- + H-\overset{|}{\underset{|}{C}}-\overset{|}{\underset{|}{C}}-X \longrightarrow \overset{}{\underset{}{C}}=\overset{}{\underset{}{C} } + X^- + H-B$$

This elimination reaction of alkyl halides with strong bases proceeds in good yields and is useful for converting alcohols to alkenes via the synthetic sequence alcohol → alkyl halide → alkene.

For tertiary halides this elimination is the *only* reaction with strong bases; no substitution occurs.

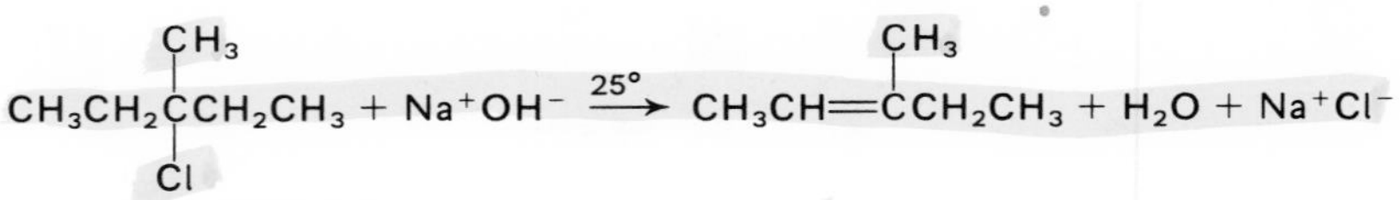

$$CH_3CH_2\overset{\overset{\displaystyle CH_3}{|}}{\underset{\underset{\displaystyle Cl}{|}}{C}}CH_2CH_3 + Na^+OH^- \xrightarrow{25°} CH_3CH{=}\overset{\overset{\displaystyle CH_3}{|}}{C}CH_2CH_3 + H_2O + Na^+Cl^-$$

3-chloro-   3-methyl-2-pentene
3-methylpentane   (predominant alkene)

With primary and secondary halides, elimination competes with substitution. Higher temperatures and bulkier bases, such as *tert*-butoxide ion, encourage elimination at the expense of substitution.

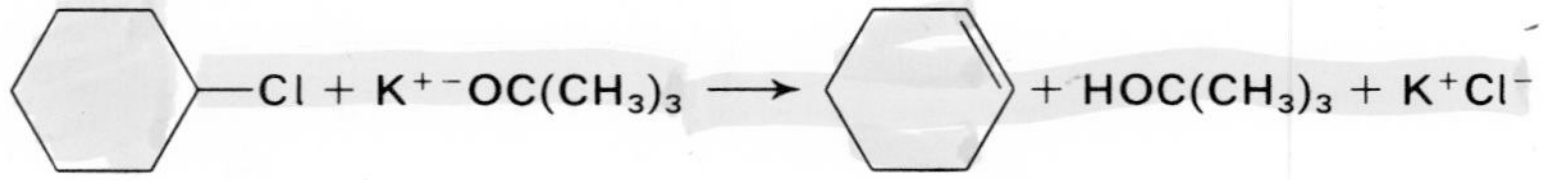

$$\text{chlorocyclohexane} \qquad\qquad\qquad\qquad \text{cyclohexene}$$

The alkene product which predominates when there is a choice is the *more highly substituted ethylene*.

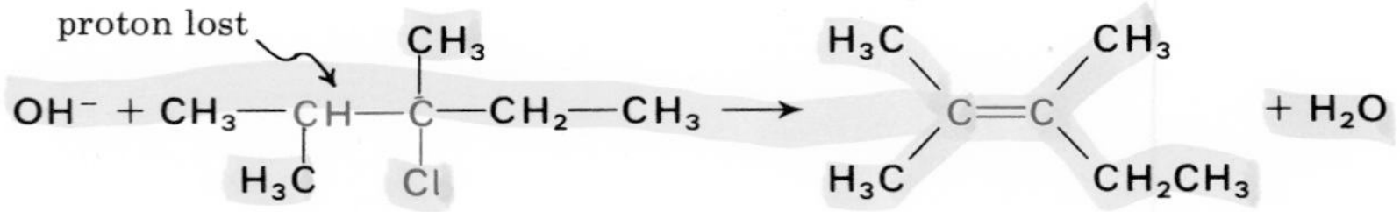

major product

Loss of the hydrogen from the methyl group CH_3 would give $R_2C{=}CH_2$, while loss of a hydrogen from the ethyl group $-CH_2CH_3$ would give $R_2C{=}CHCH_3$. Each of these alkenes is less stable and thus less favored in the reaction than the major alkene product formed by loss of hydrogen from the isopropyl group.

Problem   **6.8** Write the structure for the major product of each of these reactions.

a) 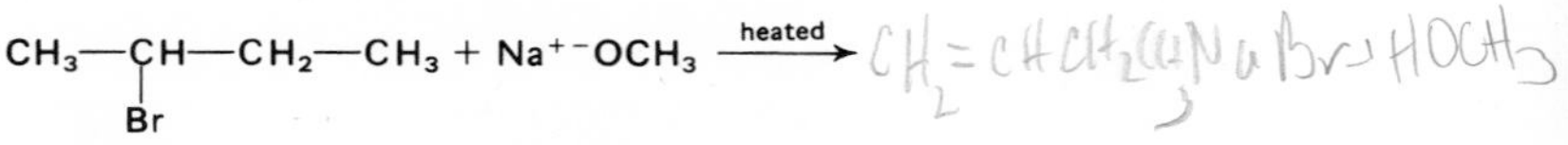 $CH_3-\underset{\underset{\displaystyle Br}{|}}{CH}-CH_2-CH_3 + Na^+{}^-OCH_3 \xrightarrow{\text{heated}}$ *[handwritten:] $CH_2{=}CHCH_2CH_3$ Na Br $\rightsquigarrow$ HOCH$_3$*

b) 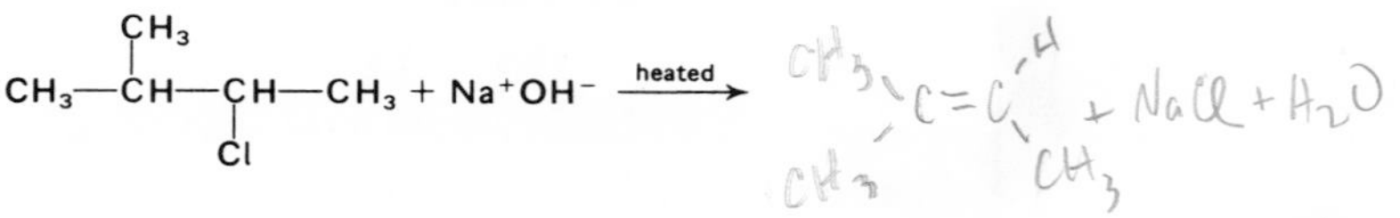 $CH_3-\overset{\overset{\displaystyle CH_3}{|}}{CH}-\underset{\underset{\displaystyle Cl}{|}}{CH}-CH_3 + Na^+OH^- \xrightarrow{\text{heated}}$ *[handwritten:] $\overset{CH_3}{\underset{CH_3}{{>}}}C{=}C\overset{H}{\underset{CH_3}{{<}}} + NaCl + H_2O$*

c) $CH_3-\overset{\overset{\displaystyle CH_3}{|}}{CH}-\underset{\underset{\displaystyle C_6H_5}{|}}{\overset{\overset{\displaystyle Cl}{|}}{C}}-CH_2-CH_3 + Na^+OH^- \xrightarrow{\text{heated}}$

[handwritten:] $\overset{CH_3}{\underset{CH_3}{{>}}}C{=}C\overset{CH_2CH_3}{\underset{C_6H_5}{{<}}} + NaCl + H_2O$

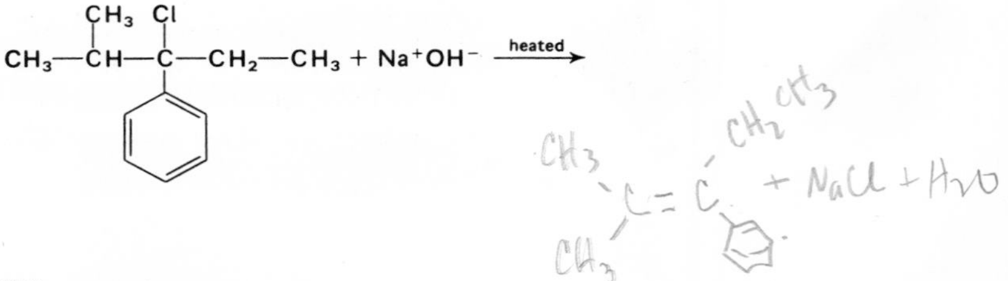

6.6 PREPARATION OF ALKYL HALIDES

The chief reactions by which alkyl halides are obtained have been discussed in the previous chapters on alkenes and on alcohols. From alkenes, alkyl halides are made by the addition of hydrogen halide HX or halogen X_2 (Section 4.4 and 4.6).

$$CH_2{=}CH_2 + Cl_2 \longrightarrow Cl{-}CH_2{-}CH_2{-}Cl$$
1,2-dichloroethane

$$CH_3CH{=}CH_2 + HCl \longrightarrow CH_3\underset{\underset{Cl}{|}}{C}HCH_3$$
propene   2-chloropropane

The addition of HCl according to Markownikoff's rule does not give primary alkyl halides, only secondary and tertiary halides.

Two reagents can be used to convert alcohols to alkyl halides. From primary and secondary alcohols, the best method to prepare a halide is by the use of thionyl chloride or a phosphorus trihalide (Section 5.5).

$$3\ CH_3CH_2\underset{\underset{CH_3}{|}}{C}HCH_2{-}OH + PBr_3 \longrightarrow 3\ CH_3CH_2\underset{\underset{CH_3}{|}}{C}HCH_2{-}Br + P(OH)_3$$
2-methyl-1-butanol   1-bromo-2-methylbutane

$$CH_3(CH_2)_4CH_2{-}OH + SOCl_2 \longrightarrow CH_3(CH_2)_4CH_2{-}Cl + SO_2 + HCl$$
1-hexanol   1-chlorohexane

The substitution of a halogen for the hydroxyl group by hydrogen halide works best with tertiary alcohols and some very reactive primary alcohols, such as benzyl alcohol (Section 5.5).

$$CH_3CH_2\underset{\underset{CH_3}{|}}{\overset{\overset{CH_3}{|}}{C}}{-}OH + HCl \longrightarrow CH_3CH_2\underset{\underset{CH_3}{|}}{\overset{\overset{CH_3}{|}}{C}}{-}Cl + H_2O\ (+\ CH_3CH{=}\overset{\overset{CH_3}{|}}{C}CH_3)$$

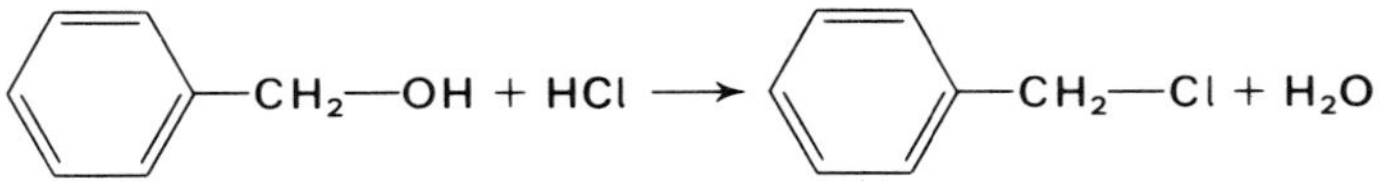

6.7 ALCOHOL DERIVATIVES—ESTERS OF SULFONIC AND PHOSPHORIC ACIDS

A common means of turning the hydroxyl into a good leaving group is to convert it into an ester of sulfonic or phosphoric acid. In general, sulfonate esters are used in the laboratory and diphosphate or triphosphate esters occur in biological sequences. Substitution and elimination reactions of these two classes give products under less acidic conditions than are needed for alcohols to undergo the same changes.

The ester of a *sulfonic acid*, R—SO$_3$H or Ar—SO$_3$H, such as **benzenesulfonic acid** or *p*-**toluenesulfonic acid** is easy to make and to handle.

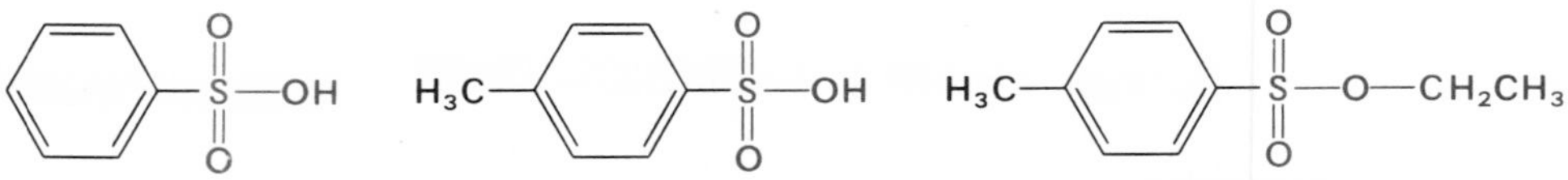

benzenesulfonic
acid

p-toluenesulfonic acid
(an arenesulfonic acid)

ethyl *p*-toluenesulfonate
(ethyl tosylate)

Remember from the nomenclature of phenols (Section 5.8) that *p*- (*para*-) as a prefix specifies the 1,4 relationship between groups attached to the benzene ring. Thus *p*-toluenesulfonic acid has the methyl in the #4 position and the sulfonic acid group in the #1 position.

Arenesulfonic acids are very strong acids, rather comparable to sulfuric acid, whose structure they resemble. The anion, arenesulfonate ion, being the conjugate base of a strong acid like halide ions, is an excellent leaving group in substitution reactions.

The *p*-toluenesulfonates are generally used because they are more often solids and thus are more easily handled than the benzene-sulfonates. The name *p*-toluenesulfonyl group (Ar—SO$_2$—) is shortened to the "tosyl" group, and the ester is ethyl tosylate. The ester is prepared by the reaction of the alcohol with the chloride of the sulfonic acid, *p*-toluenesulfonyl chloride (tosyl chloride). The reaction is rapid and evolves both heat and hydrogen chloride gas. The acidity of the reaction mixture is controlled by using pyridine as a base to react with the HCl as it forms. Pyridine used in large quantity serves as the solvent for the reaction as well.

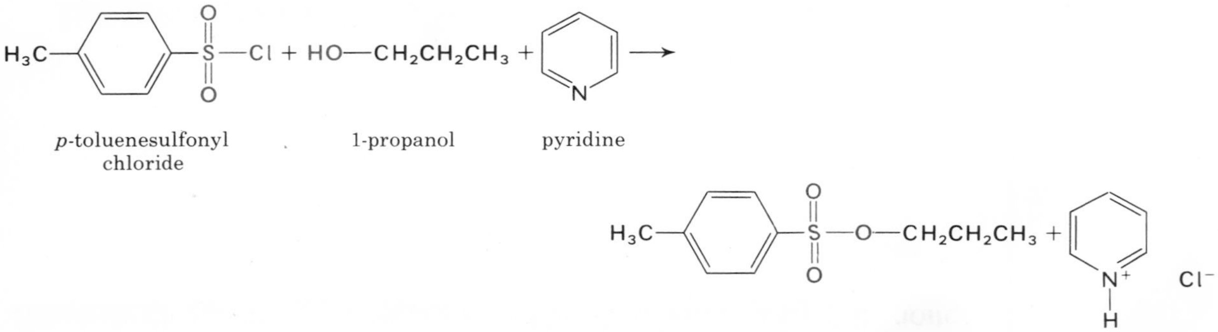

p-toluenesulfonyl
chloride

1-propanol

pyridine

propyl *p*-toluenesulfonate
(propyl tosylate)

pyridinium
chloride

The sulfonate group can be displaced by methanol to give an ether. Bond cleavage occurs at the C—O bond of the ester.

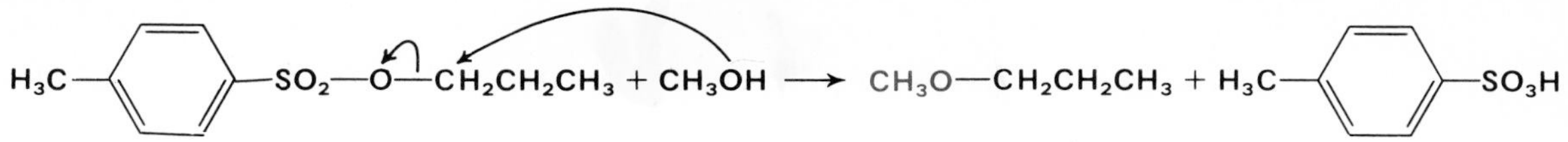

propyl tosylate   1-methoxypropane   toluenesulfonic acid

The combined two steps are a means of converting the alcohol to the ether without the use of an alkyl halide. The ester is made in a weakly basic medium, pyridine, and the substitution is carried out in a neutral, polar solvent.

The reaction of a sulfonate ester with a strong or moderate base does not give a substitution product. Instead an alkene is formed by elimination of the elements of the sulfonic acid.

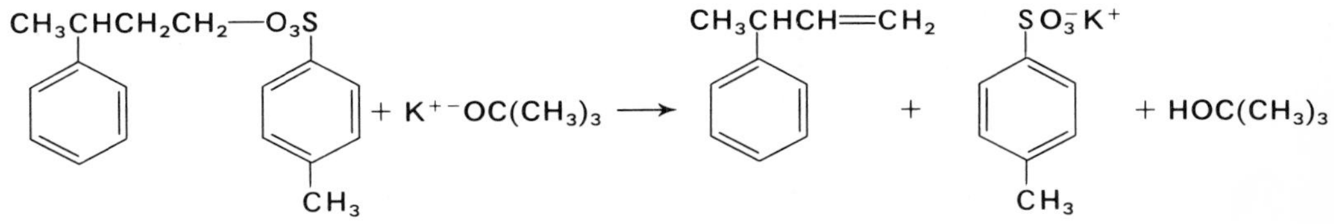

| 3-phenyl-1-butyl tosylate | potassium tert-butoxide | 3-phenyl-1-butene | potassium tosylate | tert-butyl alcohol |

The series of alcohol → sulfonate → alkene is particularly useful with primary alcohols, which are somewhat difficult to dehydrate directly by the use of strong acid. The sequence also eliminates the use of sulfuric acid for compounds that may be sensitive to strong acid at other functional groups.

Problem   6.9 Draw the structures of compounds A through E.

a) p-toluenesulfonyl chloride (A) + cyclohexylmethanol (B) $\xrightarrow{\text{pyridine}}$ C

b) C + K^+ $^-OC(CH_3)_3$ $\longrightarrow$ D   c) C + CH_3OH $\longrightarrow$ E

Esters of phosphoric acid have an easily displaced group in biological reactions, just like the sulfonate esters in laboratory reactions. We will consider the biological reactions of these phosphates in the next section.

There are several forms of phosphoric acid and the corresponding anions. Since the esters of three of the phosphoric acids are important biological compounds, we shall examine carefully the structures of the acids and anions. Orthophosphoric acid H_3PO_4 is usually called simply phosphoric acid, and its anion PO_4^{-3} is the phosphate ion. The structures of these species and of diphosphate ion $P_2O_7^{-4}$ and triphosphate ion $P_3O_{10}^{-5}$ are shown below.

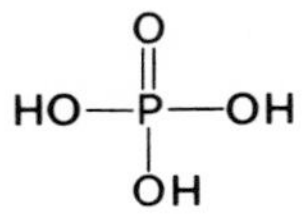

orthophosphoric acid 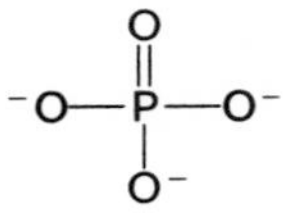   orthophosphate ion 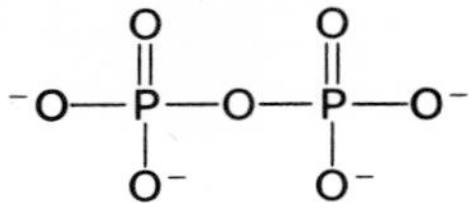   pyrophosphate ion (diphosphate ion) 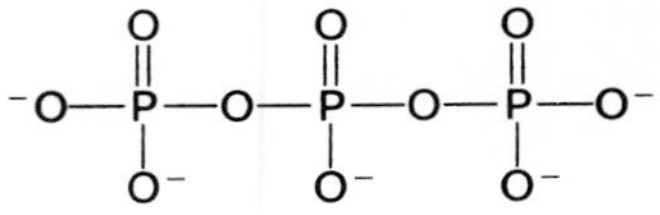   metaphosphate ion (triphosphate ion)

Note in the structure of the pyrophosphate ion that the two phosphorus atoms are bonded to the same oxygen. In the triphosphate ion another unit of PO_3^- is added by making another P—O bond.

Alkyl esters of the phosphoric acids have these structures.

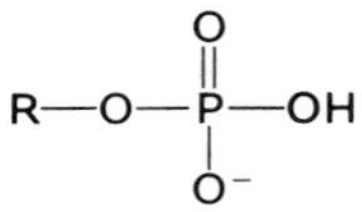

alkyl phosphate ion 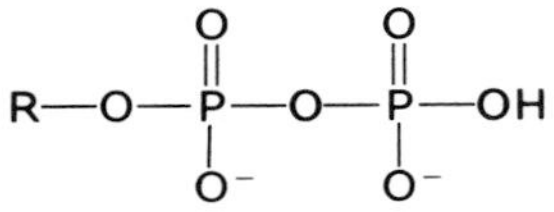   alkyl pyrophosphate ion (alkyl diphosphate ion) 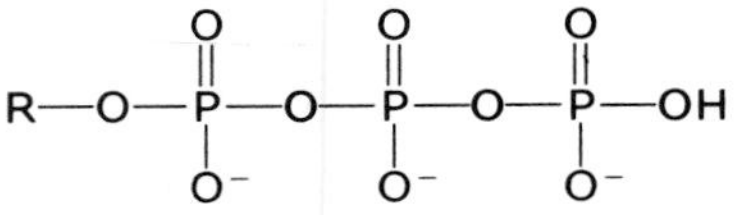   alkyl triphosphate ion

The first hydrogen of phosphoric acid H_3PO_4 is moderately acidic, stronger than acetic acid. The last hydrogens of HPO_4^{-2} and the multiple ions are very weakly acidic, less than phenol or ammonium ion. In the water solution of a cell this last hydrogen remains attached. Here and in further treatment of these structures, we will have to be somewhat vague as to the number of protons actually attached to the structures we use and the number of negative charges. The presence and absence of the intermediate protons depends upon the acidity of the medium, which varies over a wide range even in biological systems and is sometimes not known with certainty in the cell.

Problem   **6.10** Draw the structures of these phosphate esters. For simplicity use one negative charge per phosphate unit.

     a) isobutyl phosphate anion

     b) 3-methyl-2-butyl pyrophosphate anion

     c) cyclohexylmethyl triphosphate anion

     d) diethyl phosphate anion

Besides the ease of displacement of the various phosphate groups, these phosphate esters have two other important advantages. First, the large number of oxygens carrying negative charges makes these esters compatible with water, the cell medium. Carboxylate and sulfonate esters are insoluble in water. Second, the P—O bond in the P—O—P sequence has a large bond energy. When the P—O—P linkage is broken this large amount of energy is liberated for use in making other bonds. These esters thus serve as an energy reservoir for metabolic reactions.

Some examples of biological substitution of phosphate esters are given in Sections 6.8 and 6.10.

6.8 SUBSTITUTION IN THE BIOSYNTHESIS OF TERPENES

A major contribution to our knowledge of biological pathways has been the elucidation of the conversion of C_5 isoprenoid units to C_{10} and C_{15} units in the biosynthesis of steroids and terpenes. Lengthening of the carbon chain C_5 to C_{10} to C_{15} involves the substitution of a pyrophosphate group (diphosphate group) by an alkene acting as the nucleophilic reagent.

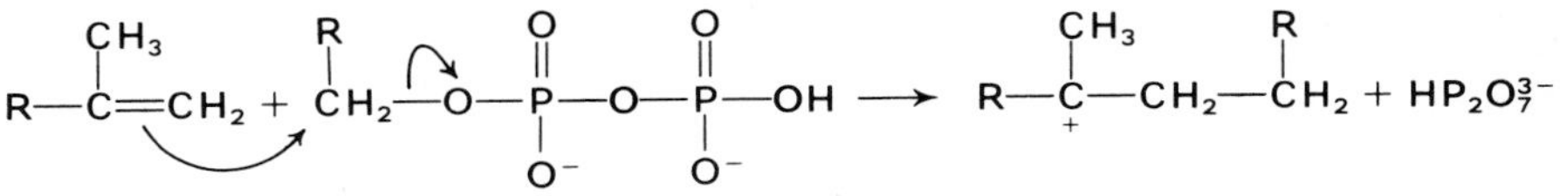

In the course of the biosynthesis of terpenes and steroids a C_5 isoprenoid compound, 3-methyl-3-butenyl pyrophosphate, is produced. Some of this compound is isomerized to 3-methyl-2-butenyl pyrophosphate by the removal of a hydrogen and its return to a new position which moves the position of the double bond.

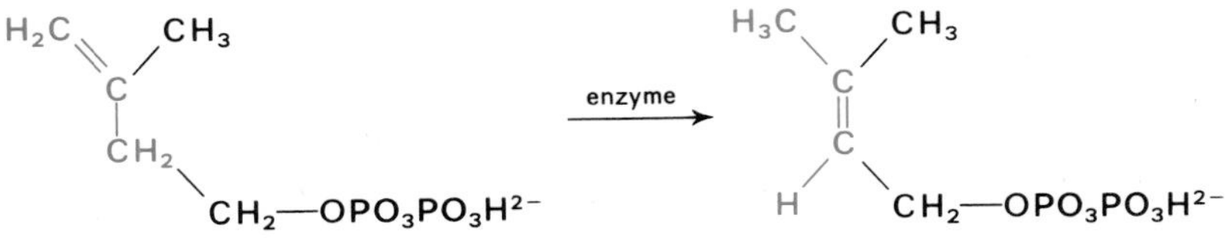

3-methyl-3-butenyl pyrophosphate
(isopentenyl pyrophosphate)

3-methyl-2-butenyl pyrophosphate
(dimethylallyl pyrophosphate)

The next step in the synthesis is the reaction of these two isomers with each other. The old and the new isomers have structural features to encourage such a combination. The terminal double bond of the isomer on the left (isopentenyl ester) acts as the nucleophilic reagent furnishing the pair of electrons for the new C—C bond. The allylic pyrophosphate group of the second isomer (dimethylallyl ester) is an exceptionally reactive group for substitution. Recall that allyl chloride reacts rapidly in a substitution reaction (Section 6.3).

All of the biogenetic reactions are enzyme catalyzed and controlled. The reaction between the two isomeric pyrophosphates proceeds as shown in Fig. 6.1 to give the pyrophosphate of the monoterpene geraniol (Section 4.10). The two units are joined head-to-tail.

dimethylallyl
pyrophosphate (C_5)

isopentenyl
pyrophosphate (C_5)

geranyl
pyrophosphate (C_{10})
(for geraniol, see Section 4.10)

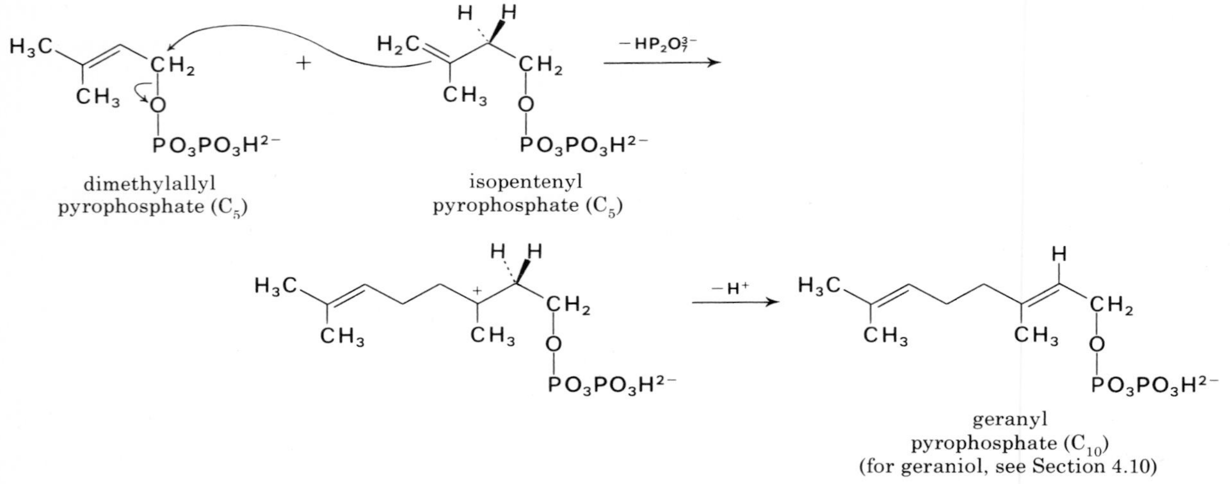

Note that the C_{10} product in Fig. 6.1 is again an allylic pyrophosphate. This compound is also capable of reacting with the terminal double bond of another isopentenyl pyrophosphate. In the second substitution reaction a sesquiterpene (C_{15}) compound is formed, farnesyl pyrophosphate.

Problem **6.11** Write reaction steps similar to those shown in Fig. 6.1 for the reaction of geranyl pyrophosphate (the C_{10} compound) with isopentenyl pyrophosphate (the original C_5 compound) to give a C_{15} pyrophosphate called farnesyl pyrophosphate.

By somewhat similar reactions of isomerization and substitution, two farnesyl pyrophosphates (C_{15}) join tail-to-tail. The C_{30} product of this substitution ultimately generated is squalene, the triterpene precursor of the steroid. (See Section 5.13 for introduction to the steroids.)

6.9 THIOLS, SULFIDES, AND DISULFIDES

Sulfur is in the same group in the periodic table as oxygen, and there are sulfur-containing analogs of most of the classes of compounds with an oxygen-containing functional group. Thiols, sulfides, and disulfides are the sulfur compounds corresponding to alcohols, ethers, and peroxides (R—O—O—R). The sulfonium halide is similar to alkyl-ammonium halides.

| thiol | sulfide | disulfide | sulfonium halide |

Thiols and sulfides are low boiling liquids. The compounds of low molecular weight have quite offensive odors reminiscent of the skunk, which owes its protective odor to a mixture of sulfur compounds including these structures.

$$\underset{\text{\textit{trans}-2-butene-1-thiol}}{\overset{H_3C}{\underset{H}{>}}C=C\overset{H}{\underset{CH_2-SH}{<}}}$$

$$\underset{\text{3-methyl-1-butanethiol}}{CH_3\overset{CH_3}{\underset{|}{C}}HCH_2CH_2-SH}$$

$$\underset{\substack{\text{\textit{trans}-2-butenyl-1} \\ \text{methyl disulfide}}}{\overset{H_3C}{\underset{H}{>}}C=C\overset{H}{\underset{CH_2-S-S-CH_3}{<}}}$$

Sulfur functional groups are found in many biologically important compounds. For example, three of the amino acids obtained from proteins have sulfur groups. **Methionine** has a methyl sulfide group which is involved in substitution reactions of methyl transfer to other compounds (see Section 6.10). **Cysteine** has a sulfhydryl group (SH) and **cystine** has a disulfide group (RSSR). Both of these latter amino acids are important in the protein of hair and wool (hence the distinctive odor of burning hair) and play a role in determining the shape of many proteins. We will learn more about all the amino acids in Chapter 10.

$$\underset{\text{cysteine}}{H-S-CH_2\overset{+NH_3}{\underset{|}{C}}HCO_2^-}$$

$$\underset{\text{cystine}}{\begin{array}{c} +NH_3 \\ | \\ S-CH_2CHCO_2^- \\ S-CH_2CHCO_2^- \\ | \\ +NH_3 \end{array}}$$

$$\underset{\text{methionine}}{CH_3-S-CH_2CH_2\overset{+NH_3}{\underset{|}{C}}HCO_2^-}$$

Two coenzymes have sulfur functional groups as the working portion of the large molecules: coenzyme A is a thiol, and lipoic acid has a disulfide group.

$$\underset{\text{part structure of coenzyme A}}{R-\overset{O}{\overset{||}{C}}-NH-CH_2CH_2-SH}$$

lipoic acid

A coenzyme is a compound whose presence is essential for some enzyme-catalyzed reactions to occur. The coenzyme functional group may be changed in the process, but some subsequent reaction regenerates the original coenzyme structure for use again.

In IUPAC nomenclature the name of the thiol is formed by addition of the suffix "thiol" to the alkane name. The position of the SH is given by number.

$$\underset{\text{2-propanethiol}}{\overset{\displaystyle \overset{\text{SH}}{|}}{CH_3\text{—}CH\text{—}CH_3}} \qquad \underset{\text{1,3-propanedithiol}}{HS\text{—}CH_2\text{—}CH_2\text{—}CH_2\text{—}SH}$$

The anion RS^- is the conjugate base of the thiol RSH. It carries the name of the thiol with "ate" ending added.

$$CH_3\text{—}S^-\,Na^+$$

sodium methanethiolate

Sulfides and disulfides are named according to the rules for ethers (Section 5.10), as diethyl sulfide. For a substituent $CH_3\text{—}S\text{—}$, "thio" is used in place of "oxy" to denote the sulfur atom, as "methylthio" instead of "methyloxy" or "methoxy."

$$CH_3CH_2\text{—}S\text{—}CH_2CH_3 \qquad CH_3CH_2\text{—}S\text{—}S\text{—}CH_2CH_3 \qquad CH_3\text{—}S\text{—}CH_2CH_3$$

$$\begin{array}{ccc} \text{diethyl sulfide} & \text{diethyl disulfide} & \text{methylthioethane} \\ \text{(common name)} & \text{(common name)} & \text{(IUPAC name)} \end{array}$$

Thiols are easily prepared from alkyl halides and sodium hydrogen sulfide Na^+HS^- in a substitution reaction.

$$\underset{\text{bromoethane}}{CH_3CH_2\text{—}Br} + HS^- \longrightarrow \underset{\text{ethanethiol}}{CH_3CH_2\text{—}SH} + Br^-$$

Thiols are more reactive than alcohols, their oxygen counterparts, but find little general use in the laboratory. The biological occurrence of the sulfur functional groups, however, is of great importance in metabolism.

In acid–base reactions, thiols are stronger acids than are alcohols, just as H_2S is a stronger acid than H_2O (Section 3.8). The thiolate ions RS^- and sulfides $R\text{—}S\text{—}R$ are weaker bases than their oxygen analogs RO^- and $R\text{—}O\text{—}R$. As a consequence of the acidity of thiols, their salts may be prepared by reaction of thiols with sodium hydroxide.

$$\underset{\text{ethanethiol}}{CH_3CH_2\text{—}S\,H} + OH^- \longrightarrow \underset{\substack{\text{ethanethiolate} \\ \text{ion}}}{CH_3CH_2\text{—}S^-} + H_2O$$

As nucleophiles in substitution reactions, thiols, thiolate ions, and sulfides are more reactive than the corresponding oxygen compounds. These nucleophilic substitutions occur easily under mild conditions with the sulfur group displacing a halide or p-toluenesulfonate ion.

$$CH_3CH_2{-}Br + HS{-}CH_2CH_3 \longrightarrow CH_3CH_2{-}S{-}CH_2CH_3 + HBr$$

bromoethane   ethanethiol   diethyl sulfide

$$CH_3CH_2{-}Br + (CH_3CH_2)_2S \longrightarrow (CH_3CH_2)_3S^+Br^-$$

bromoethane   diethyl sulfide   triethylsulfonium bromide

On the other hand, the alkyl groups of positive sulfonium ions undergo attack by a nucleophile. A dialkyl sulfide is the leaving group. The reaction is so indiscriminate as to which alkyl group undergoes substitution that it is seldom used in organic synthesis. Nucleophilic substitution of sulfonium ions is, however, the method used in biological processes for transferring a methyl group from one compound to another in reactions closely controlled by enzymes. Transmethylations of the kind exemplified by the reaction of propylamine and a methyl-sulfonium ion play an important role in the biosynthesis of many compounds (see next section).

$$CH_3CH_2CH_2{-}NH_2 + CH_3{-}\overset{+}{S}R_2 \longrightarrow CH_3CH_2CH_2{-}\overset{+}{N}H_2{-}CH_3 + R_2S$$

propylamine   methylsulfonium   methylpropylammonium   dialkyl
                              ion   ion   sulfide

Thiols are rapidly dehydrogenated to disulfides by a reaction with oxygen, which does not have a counterpart in reactions of alcohols. Hydrogenation can cleave the S—S bond of disulfides to form two thiols again. In this example cysteine is converted to cystine.

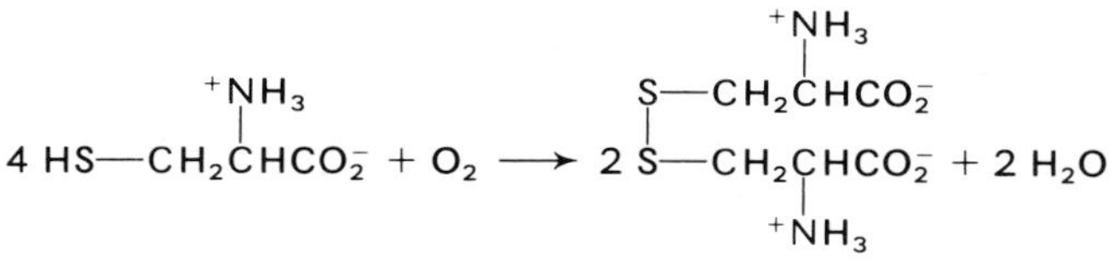

$$4\ HS{-}CH_2\overset{\overset{\displaystyle +NH_3}{|}}{C}HCO_2^- + O_2 \longrightarrow 2\ \underset{\underset{\displaystyle +NH_3}{|}}{\overset{\overset{\displaystyle +NH_3}{|}}{S{-}CH_2CHCO_2^-}}{S{-}CH_2CHCO_2^-} + 2\ H_2O$$

cysteine   cystine

Problem   **6.12** Write an equation for each of the following reactions.

a) 2-butanethiol + methyl tosylate

b) ethylthiocyclopentane + ethyl iodide

6.10 BIOLOGICAL TRANS-METHYLATION REACTIONS

Few biological examples of nucleophilic substitution reactions at a saturated carbon are known. Transmethylations all appear to occur by this process, however. As in so many cases, biological reactions accomplish a special conversion which is not practical in the laboratory.

In the synthesis of many biological compounds, a methyl group is transferred from one electronegative atom to another—sulfur to oxygen or nitrogen. In such transmethylations, a sulfonium ion of the amino acid methionine is the methyl donor and functional groups of nitrogen or oxygen are the acceptors. As we have noted in the previous section, the reactions are enzyme-catalyzed nucleophilic substitutions on a methylsulfonium ion with a sulfide as leaving group.

$$R'NH_2 + CH_3-\overset{+}{S}R_2 \longrightarrow R'\overset{+}{N}H_2-CH_3 + R_2S$$

One of the most commonly encountered coenzymes, adenosine triphosphate (ATP), takes part in the transmethylation reaction. Adenosine triphosphate is a large molecule which undergoes many different types of reaction in the biosyntheses of many kinds of compounds, so much so that its structure will soon become familiar to you. The particular functional group which interests us now is the alkyl triphosphate group.

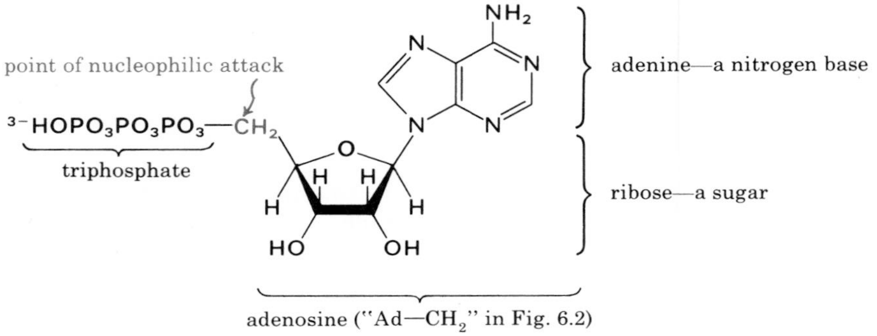

adenosine triphosphate (ATP)

The transmethylation reaction actually involves two nucleophilic substitutions.

1. The first reaction is the displacement of the triphosphate group of ATP, a good leaving group like the sulfonate group, by the sulfide of methionine. A methylsulfonium ion is formed.
2. In the second reaction an amino or hydroxyl group as nucleophile displaces a dialkyl sulfide from the methyl group of the methyl-sulfonium ion. The transmethylation reaction is illustrated in Fig. 6.2 by the conversion of the adrenal hormone norepinephrine to the active adrenal hormone epinephrine (adrenaline), whose formulas we first saw as phenolic compounds in Section 5.8.

Problem **6.13** Ethanolamine $(HO-CH_2CH_2-NH_2)$ is converted to choline ion $[HO-CH_2CH_2-N^+(CH_3)_3]$ by the transmethylation process shown in Fig. 6.2. Write the equations for the sequence of reactions needed to transfer one methyl group to ethanolamine.

FIGURE 6.2
METHYLATION OF
NOREPINEPHRINE TO
EPINEPHRINE—A
BIOLOGICAL
TRANSMETHYLATION

Reaction 1: Formation of the sulfonium ion Methionine displaces the triphosphate group of ATP (adenosine triphosphate).

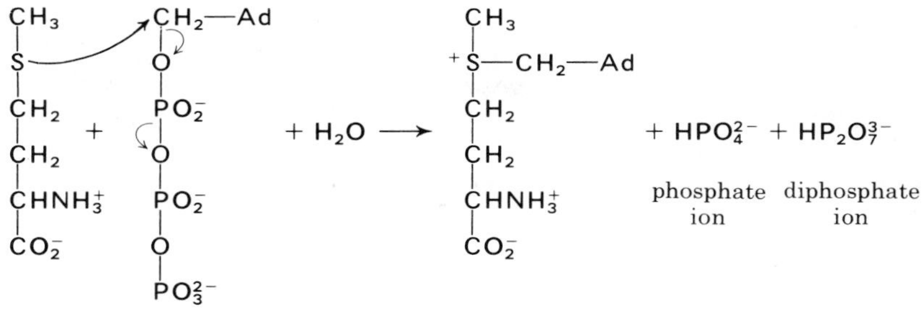

methionine   ATP   S-adenosyl-
                                      methionine

Reaction 2: Transmethylation Norepinephrine displaces the sulfide from methyl sulfonium ion.

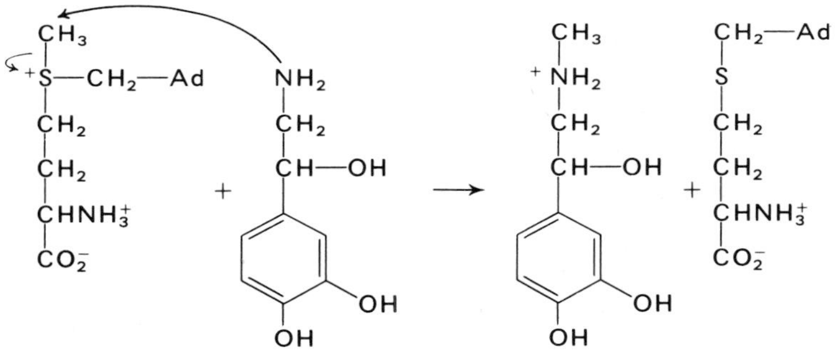

norepinephrine   epinephrine
(an adrenal hormone)   (adrenaline)

$Ad—CH_2—$ = adenosine = sugar–nitrogen base

New Terms and Topics

Alkyl halides (Section 6.1)

Nucleophilic substitution of alkyl halides (Section 6.3)

Bimolecular and unimolecular mechanisms of substitution (Section 6.4)

Elimination reaction of halides (Section 6.5)

Nitriles (Section 6.4)

Thiols, sulfides, disulfides, sulfonium ions (Section 6.9)

Arenesulfonic acid and esters (Section 6.7)

Substitution and elimination reactions of arenesulfonates (Section 6.7)

Esters of phosphoric acid—phosphate, pyrophosphate, and triphosphate esters (Section 6.7)

Adenosine triphosphate, ATP (Section 6.10)

Coenzymes—coenzyme A, lipoic acid, and ATP (Section 6.9)

Biological substitutions of pyrophosphate and triphosphate esters (Sections 6.8 and 6.10)

Summary of Reactions

REACTIONS OF PRIMARY AND SECONDARY ALKYL HALIDES

1. Substitution reactions

a) With alkoxide and phenoxide ions to form ethers (Section 6.3)

$$R—Cl + R'—O^-Na^+ \longrightarrow R—O—R' + Na^+Cl^-$$

$$CH_3CH_2CH_2—Cl + CH_3CH_2—O^-Na^+ \longrightarrow CH_3CH_2CH_2—O—CH_2CH_3 + Na^+Cl^-$$

1-chloropropane   sodium ethoxide   ethyl propyl ether

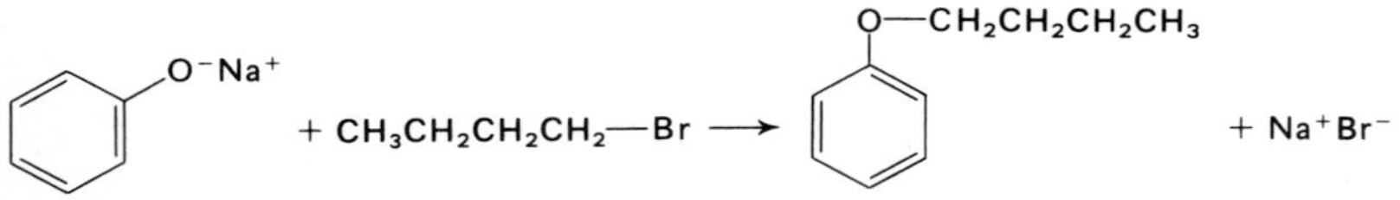

sodium phenoxide   1-bromobutane   butyl phenyl ether

$$(CH_3)_3C—O^-K^+ + CH_3I \longrightarrow (CH_3)_3C—O—CH_3 + K^+I^-$$

potassium   methyl   *tert*-butyl
tert-butoxide   iodide   methyl ether

b) With hydroxide ion to form alcohol (Section 6.3)

$$R—Cl + OH^- \longrightarrow R—OH + Cl^-$$

$$CH_3CH_2CH_2—Br + OH^- \longrightarrow CH_3CH_2CH_2—OH + Br^-$$

1-bromopropane   1-propanol

c) With cyanide ion (Section 6.6)

$$R—Cl + CN^- \longrightarrow R—CN + Cl^-$$

$$CH_3CH_2CH_2—Br + CN^- \longrightarrow CH_3CH_2CH_2—CN + Br^-$$

propyl bromide   butanonitrile

d) With HS⁻, thiols, and sulfides (Section 6.9)

$$R—Cl + HS^- \longrightarrow R—SH + Cl^-$$

$$Br—CH_2CH_2CH_2—Br + 2 HS^- \longrightarrow HS—CH_2CH_2CH_2—SH + 2 Br^-$$

1,3-dibromopropane   1,3-propanedithiol

$$R—S—H + R—Cl \longrightarrow R_2S + HCl$$

$$CH_3I + CH_3CH_2—SH \longrightarrow CH_3CH_2—S—CH_3 + HI$$

ethanethiol   ethyl methyl sulfide

2. Elimination reactions (Section 6.5)
With strong base

$$\underset{\underset{\displaystyle Cl}{|}}{RCH_2CHR} + B^- \longrightarrow RCH{=}CHR + HB + Cl^-$$

$$CH_3CH_2CH_2CH_2{-}Br + Na^+NH_2^- \longrightarrow CH_3CH_2CH{=}CH_2 + NH_3 + Na^+Br^-$$

    1-bromobutane   1-butene

PREPARATION OF ARENESULFONATE ESTERS (Section 6.8)
Alcohol with arenesulfonyl chloride

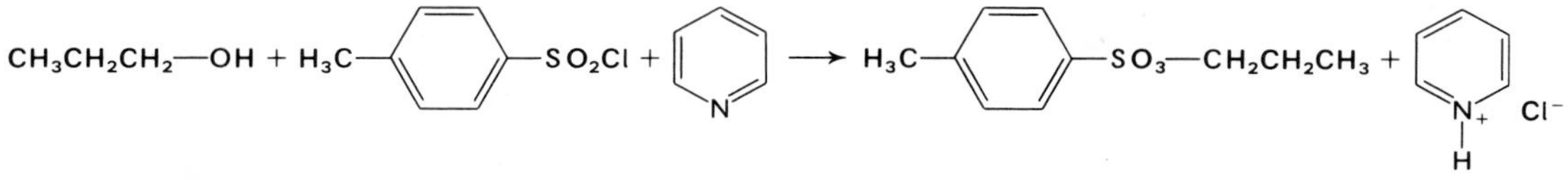

   1-propanol   p-toluenesulfonyl   pyridine   propyl p-toluenesulfonate   pyridinium
                         chloride   chloride

REACTIONS OF ARENESULFONATE ESTERS

1. Substitution with weak bases (Section 6.7)

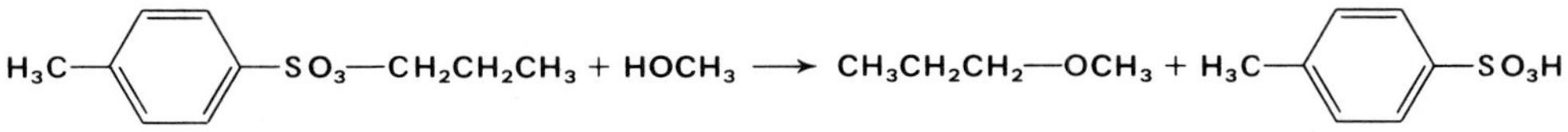

      propyl tosylate   1-methoxypropane   toluenesulfonic acid

2. Elimination with strong bases (Section 6.7)

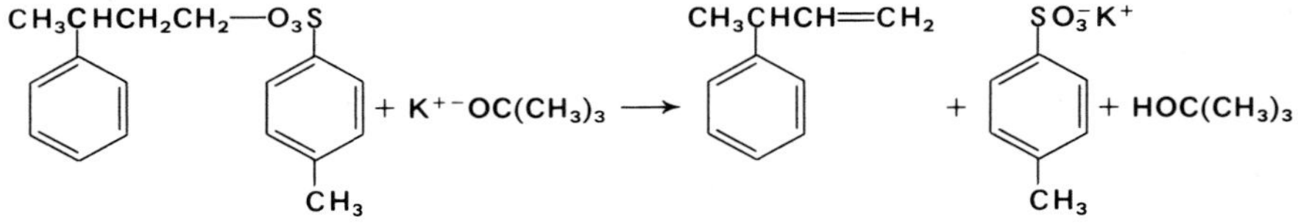

   3-phenyl-1-butyl   potassium   3-phenyl-1-butene   potassium   tert-butyl
       tosylate   tert-butoxide   tosylate   alcohol

REACTIONS OF THIOLS AND SULFIDES

1. Thiols with bases (Section 6.9)

$$R{-}SH + B^- \longrightarrow RS^- + BH$$

$$CH_3CH_2{-}SH + OH^- \longrightarrow CH_3CH_2{-}S^- + H_2O$$

   ethanethiol   ethanethiolate ion

2. Thiols and sulfides with alkyl halides (Section 6.9)

$$R-Br + R-SH \longrightarrow HBr + R-S-R \xrightarrow{R-Br} R_3S^+Br^-$$

$$CH_3CH_2-SH + CH_3CH_2-Br \xrightarrow{-HBr} (CH_3CH_2)_2S \xrightarrow{CH_3CH_2-Br} (CH_3CH_2)_3S^+Br^-$$

diethyl sulfide   triethylsulfonium bromide

3. Oxidation of thiols (Section 6.9)

$$2 R-SH + O_2 \longrightarrow R-S-S-R + H_2O$$

$$2 HS-CH_2\underset{\overset{|}{{}^+NH_3}}{CH}CO_2^- + O_2 \longrightarrow {}^-O_2C\underset{\overset{|}{{}^+NH_3}}{CH}CH_2-S-S-CH_2\underset{\overset{|}{{}^+NH_3}}{CH}CO_2^- + H_2O$$

     cysteine   cystine

Problems

6.14 Write equations for the following reactions using complete structures for all reactants and products.

a) 1-bromo-3-phenylpentane + sodium cyanide

b) allyl bromide + sodium ethoxide

c) 2-bromobutane + sodium hydroxide at 100°

d) sodium phenoxide + isobutyl chloride

e) 1-phenylethanol + p-toluenesulfonyl chloride in pyridine

f) cyclopentyl p-toluenesulfonate + methanol

g) sec-butyl bromide + sodium hydrogen sulfide

6.15 Draw the structures for the starting materials and compounds A through P.

a) 1-propanol $\xrightarrow{SOCl_2}$ A $\xrightarrow{Na^+ \, {}^-OCH(CH_3)_2}$ B

b) 3-methyl-1-pentanol $\xrightarrow{PCl_3}$ C $\xrightarrow{Na^+CN^-}$ D

c) 1-pentene $\xrightarrow{1) \, B_2H_6,}$ $\xrightarrow{2) \, H_2O_2, \, OH^-}$ E $\xrightarrow{PBr_3}$ F

d) tert-butyl alcohol $\xrightarrow{K°}$ G $\xrightarrow{benzyl \, chloride}$ H

e) isopropyl chloride $\xrightarrow{Na^+HS^-}$ I $\xrightarrow{methyl \, iodide}$ J

f) 3-methyl-1-butanol $\xrightarrow{p\text{-toluenesulfonyl chloride}}$ K $\xrightarrow{K^+ \, {}^-OC(CH_3)_3}$ L

g) cyclopentanol $\xrightarrow{Na°}$ M $\xrightarrow{tert\text{-butyl bromide}}$ N

6.16 Give three series of reactions to convert 2-phenylethanol into styrene (phenylethylene).

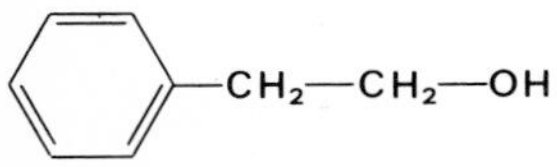

2-phenylethanol 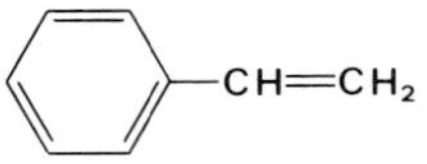   styrene

6.17 The benzyl cation is stabilized by resonance structures which involve the participation of the electron pairs of the aromatic ring. (The aromatic ring electronic structure is disrupted in these structures.) Draw the three contributing structures which have the positive charge on different ring carbons and ring electrons forming a double bond with the methylene group.

$$\left[\bigcirc\!\!-\overset{+}{C}H_2 \longleftrightarrow \bigcirc\!\!-\overset{+}{C}H_2\right]$$

6.18 Nicotinamide is methylated to methylnicotinamide cation by the biological transmethylation process. Sketch the process using the partial structure for ATP, $Ad-CH_2-OPO_3PO_3PO_3^{4-}$.

nicotinamide
(niacin)

methylnicotinamide
cation

6.19 You have cyclohexanol and methanol. Describe the reaction scheme which would give you the best yield of methyl cyclohexyl ether. You may use any other reagents you need.

6.20 By analogy with bimolecular (S_N2) and unimolecular (S_N1) mechanisms of nucleophilic substitution, describe two mechanisms of elimination which could be labeled bimolecular (E_2) and unimolecular (E_1). Give examples of compounds which should proceed by each mechanism.

6.21 Draw the resonance structures for the allyl cation.

Answers to the remaining problems are not directly stated in the chapter, but should be based on information given there.

6.22 Arrange each series of carbonium ions in order of decreasing stability.

a) $\overset{+}{C}H_3$,　$CH_3\overset{+}{C}HCH_3$,　$CH_3\overset{+}{\underset{\underset{\textstyle CH_3}{|}}{C}}CH_3$

b) $CH_2{=}CH\overset{+}{C}H_2$,　$CH_3\overset{\underset{\textstyle CH_3}{|}}{C}{=}CH\overset{+}{C}H_2$,　$CH_3CH{=}CH\overset{+}{C}H_2$

c) $CH_3CH{=}CH\overset{+}{C}H_2$,　$CH_3\overset{+}{C}HCH_2CH_3$,　$CH_2{=}CHCH_2\overset{+}{C}H_2$

6.23 Give an explanation for the fact that these neopentyl compounds are extremely unreactive in either a bimolecular or a unimolecular substitution reaction.

$CH_3\overset{\overset{\textstyle CH_3}{|}}{\underset{\underset{\textstyle CH_3}{|}}{C}}CH_2{-}Cl$　　　$CH_3\overset{\overset{\textstyle CH_3}{|}}{\underset{\underset{\textstyle CH_3}{|}}{C}}CH_2{-}OH$

2,2-dimethyl-1-chloropropane
(neopentyl chloride)

2,2-dimethyl-1-propanol
(neopentyl alcohol)

6.24 Give an explanation for the fact that allyl compounds react rapidly under conditions for unimolecular or bimolecular substitution—e.g., allyl chloride reacts as rapidly as 1-chloropropane with methoxide ion by a bimolecular (one-step) mechanism, while allyl alcohol reacts as rapidly as *tert*-butyl alcohol with conc. hydrobromic acid by a unimolecular (two-step) mechanism.

CHAPTER 7

Shells, Shoes, Screws, and Chiral Molecules

You can travel by either boat or airplane, but these vehicles are constitutionally very different! You can burn the organic compounds butane or methanol, but they are constitutionally dissimilar. Automobiles and bicycles provide alternative vehicles of transportation on land, but they are constitutionally different, perhaps to the same degree as an alcohol and an ether which have the same molecular formula. A Cadillac and a Volkswagen are both cars, but the capabilities are different—like a primary and a tertiary alcohol. A sports car and a station wagon with the same engine have the same working parts but different exteriors—like a *cis*-alkene and a *trans*-alkene. A Rolls Royce made for American consumption (steering wheel on the left) and the same model made for English use (steering wheel on the right) have identical parts, but a few of the parts are differently arranged. One car is a reflection, the image seen in a mirror, of the other.

Each Rolls Royce works equally well in its own environment (on the designated side of the road), but each in the opposite environment becomes a problem. A car is constructed to have the driver near the middle of the road for ease in seeing ahead. If you try to maintain this objective and drive an American car in the right-hand lane of an English road, you are in real trouble. The road signs are turned with their backs to you, and an approaching car wants to use the same lane as you. There is no means of accommodating all of the conventions with an American car on English roads.

This chapter is concerned with organic compounds which come in right-hand and left-hand models. Enzymes in living systems resemble cars and highways, in that only one of the two possible molecular structures of a compound (the car) fits the enzyme catalyst (the highway) and can be utilized in metabolism.

In this chapter we first explore a large number of mirror-image relationships in objects with which you are familiar. Second, we describe the occurrence and the structural requirements for the existence of molecules that have nonidentical mirror images. Last, we illustrate the importance of mirror-image relationships in compounds undergoing reactions, both in the laboratory and in the cell. We will see how these relationships contributed to our knowledge of reaction mechanisms and to the biological selection of materials for metabolism.

7.1 SHELLS, SHOES, AND SCREWS

A small child knows a shoe goes on a foot, but at first a right shoe and a left shoe are indistinguishable. With experience and training the child learns that for a "best fit" the left shoe goes on the left foot. With socks, however, either one of a pair will fit on the left foot.

The basic difference in the success of the fit of shoes and socks lies in the relationship of the two shoes to each other and the two socks to each other. The two socks are identical in every respect, that is, they are *superimposable* on each other. One shoe is *not superimposable* on the other. With both shoes sole down pointing left, the buckle of one shoe is on the side nearer the viewer and the buckle of the other is on the far side. No manner of turning will make all the parts of both shoes coincide—buckle to buckle, heel to heel, both soles down, etc.

What is the relationship between a left and a right shoe? The reflection of the left shoe, the image seen in a mirror, is not superimposable on the real left shoe. The mirror image of the left shoe is, however, identical and superimposable on the right shoe. The shoes are mirror images of each other and neither is superimposable on the other shoe or on its own mirror image, as shown in Fig. 7.1.

**FIGURE 7.1
NONSUPERIMPOSABLE
MIRROR IMAGE**

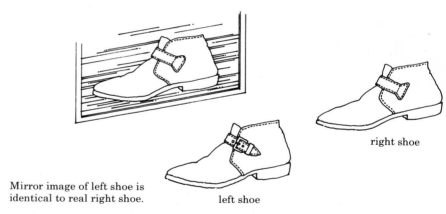

Mirror image of left shoe is identical to real right shoe.

left shoe

right shoe

An object which is not superimposable on its mirror image is said to have *handedness* or to be **chiral** (from the Greek word **cheir** for hand). An **achiral** object is one that *is* superimposable on its own mirror image. While chiral objects obviously can exist as nonidentical pairs, achiral objects can't and are singles.

Though chiral objects may exist in pairs, very often only one of the pair is desirable. An example is a screw or a threaded bolt. To insert the usual bolt into a nut one turns it to the right, clockwise. A turn to the left reverses the procedure. There is nothing wrong with a reverse thread and a counterclockwise turning, but by convention most screws and bolts are threaded in this one direction for convenience. For safety, valves on a hydrogen gas cylinder are threaded in the counterclockwise direction to prevent the careless use of a valve previously used for oxygen, thus avoiding a possible explosion.

The list of chiral objects could go on, but examples of two kinds of seashells tell the story. For no obvious reason, a single shell of many kinds of sea animals is chiral. Amazingly, in a collection of the same kind of shells, one structure is strongly predominant over that of its mirror image. Why? The inherited characteristic is not absolute but is dominant. Figure 7.2 contains pictures of several shells.

FIGURE 7.2 SEASHELLS—CHIRAL AND ACHIRAL

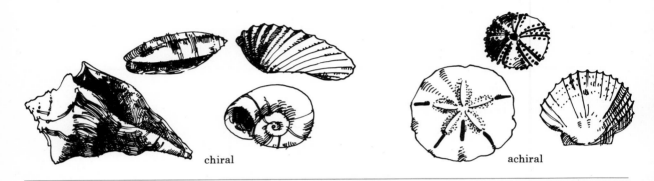

chiral   achiral

7.2 THE PLANE OF SYMMETRY

A shell which is composed of two parts, a bivalve, has another interesting characteristic. The two halves of a clam shell are mirror images of each other. Each half-shell is chiral and the left halves are easily distinguished from the right halves. However, the complete shell is composed of the two halves attached along one side in a hinge. While each half is chiral and is the mirror image of the other, the whole shell is superimposable on its own mirror image and is thus achiral, as shown in Fig. 7.3.

**FIGURE 7.3
PLANE OF SYMMETRY**

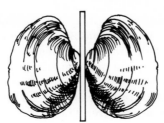

Whole shell is bisected by mirror plane.
Two half shells are mirror images.

How is it that an object with two chiral parts can itself be achiral? The answer lies in the symmetry of the object and the relationship of symmetry to chirality.

Symmetry can be described in terms of a number of characteristics, only one of which, **the plane of symmetry**, will be considered here. The plane of symmetry, the reflection plane, and the mirror plane are interchangeable terms and are illustrated by the whole clam shell in Fig. 7.3. An object has a plane of symmetry if it can be divided by a plane, acting as a mirror, that gives a reflection of one half of the object which is superimposable on the other half of the object. Chairs, purses, and rings are usually symmetrical objects with one or more mirror planes. *An object containing a mirror plane, or plane of symmetry, is identical with its mirror image, and therefore is achiral.*

Problems   **7.1** Which of the following objects are chiral (exclusive of markings)?

a) a man pitching a ball	**b)** a man up to bat	**c)** a baseball bat
d) a tennis racket	**e)** a golf club	**f)** a pair of pliers
g) a pair of scissors	**h)** a cup	**i)** a 3-legged stool
j) a square	**k)** a strung guitar	**l)** a symbol of a heart
m) one of a pair of dice	**n)** a paper clip	**o)** a coiled door spring
p) two links of a chain linked		

7.2 For all two-dimensional objects, the plane in which they lie is a plane of symmetry. Among the letters of the alphabet, identify those which contain another plane of symmetry.

A B C D E F G H I J K L M N O P Q R S T U V W X Y Z

7.3 A REVIEW OF THE TYPES OF ISOMERISM

In our study of organic compounds many aspects of structure have emerged which divide into two large categories. First, compounds have a definite atomic sequence of bonded atoms. Differences in atomic sequence of the same number and kind of atoms lead to **constitutional** or **structural isomers**, e.g., ethanol and dimethyl ether (Section 2.7) or 1-propanol and 2-propanol (Section 2.7). Second, given a defined sequence of bonded atoms, isomers arise because of different possible relationships between atoms which are not bonded to each other. The relationships of nonbonded atoms to each other are described in terms of their relative locations in space, and are called **stereochemical relationships**. Compounds that contain the same atomic sequence and differ only in the spatial relationship of their parts are called **stereoisomers**. Stereoisomers are nonsuperimposable on one another in spite of their containing the same atomic sequences.

Conformational isomers (or conformers) are one form of stereoisomers. They are distinguished by different arrangements of atoms that at ordinary temperature can interconvert by rotations about single bonds. The anti and gauche forms of butane and the axial and equatorial forms of methylcyclohexane (Sections 2.4 and 2.10) exemplify conformational isomers that equilibrate rapidly. A second example of stereoisomers is found in the structures of alkenes that are differentiated by *cis* and *trans* arrangements of groups about carbon–carbon double bonds, e.g., *cis*- and *trans*-2-butenes (Section 3.3). The *cis-trans* isomers of alkenes interconvert only at temperatures of several hundred degrees, since pi bonds must be broken. The *cis-trans* isomers of substituted cycloalkanes differentiate between *cis* and *trans* arrangements of groups attached to the ring, as in the 1,4-dimethylcyclohexanes (Section 2.10). These isomers can be interconverted only by chemical reactions that make and break sigma bonds.

A third example of stereoisomers, called **enantiomers**, is found in *chiral compounds*, such as (*R*)-2-butanol and (*S*)-2-butanol, described in the next section. Enantiomers can interconvert only as a result of sigma bond making and breaking processes. Other isomers which are also involved with chiral structures are diastereoisomers, which are described in Section 7.8.

Examples of all of these types of isomers (except diastereoisomers) are shown in Fig. 7.4.

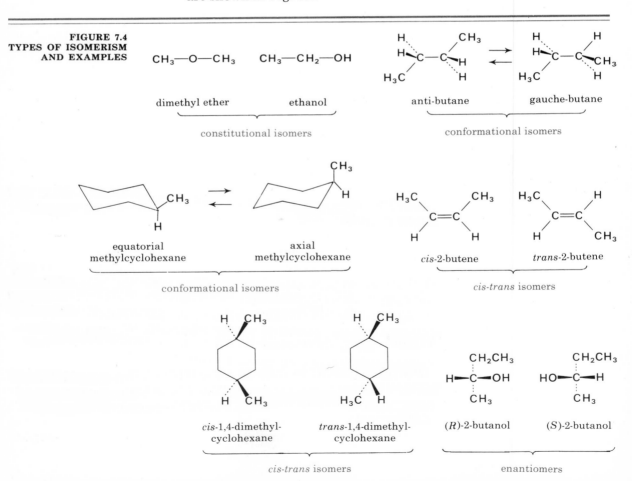

FIGURE 7.4
TYPES OF ISOMERISM AND EXAMPLES

CH_3—O—CH_3 CH_3—CH_2—OH

dimethyl ether   ethanol

constitutional isomers

anti-butane   gauche-butane

conformational isomers

equatorial methylcyclohexane   axial methylcyclohexane

conformational isomers

cis-2-butene   *trans*-2-butene

cis-trans isomers

cis-1,4-dimethyl-cyclohexane   *trans*-1,4-dimethyl-cyclohexane

cis-trans isomers

(*R*)-2-butanol   (*S*)-2-butanol

enantiomers

Some compounds, like some of the familiar objects mentioned before, are chiral. An example of such a compound is 2-butanol. The two structures that can be written for 2-butanol are shown here.

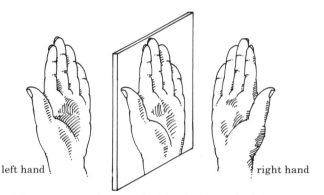

The molecules of 2-butanol having the first structure are *not identical* to the molecules having the second structure. The structures are not interconvertible and represent separate isomers. The two isomers of 2-butanol differ from each other in the same way that a right hand differs from a left hand. The two hands are not superimposable on each other, but are mirror images of each other, as shown in Fig. 7.5.

FIGURE 7.5
LEFT HAND, RIGHT HAND—
NONSUPERIMPOSABLE AND
MIRROR IMAGES

left hand   right hand

Mirror image of right hand is identical to real left hand, whose thumb is on left side and small finger on right.

Hands are chiral objects and the 2-butanols are chiral molecules. The chirality of compounds is of particular interest to us because a large number of natural compounds are chiral and often exist in only one of the two structures. Enzymes are chiral compounds. Nature makes use of chirality to control reactions by requiring that molecules fit into enzymes which catalyze their reactions, in the same way that a right hand fits into a right glove.

Let us examine chirality more closely. Two molecules which are mirror images of each other and are nonsuperimposable are called **enantiomers**. The characteristics of enantiomers and of *cis-trans* isomers, two kinds of stable stereoisomers, are different. Stereoisomers have the same atomic sequences. In enantiomers the distances between the same two groups are equal, whereas in *cis-trans* isomers distances between the same two groups are different.

Several types of molecular structures produce mirror images which are nonsuperimposable. The most common type involves the tetrahedral carbon atom with four unlike groups attached. Such a carbon atom is asymmetrically substituted and is called an **asymmetric carbon**, illustrated in Fig. 7.6.

**FIGURE 7.6
ENANTIOMERS WITH AN
ASYMMETRIC CARBON**

Four different groups attached to a carbon produce two chiral structures—enantiomers.

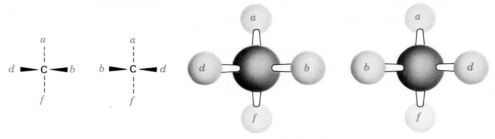

ball-and-stick models

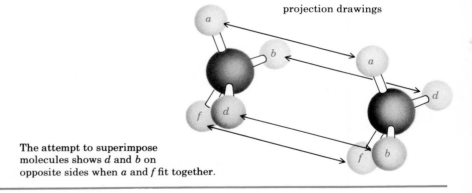

projection drawings

The attempt to superimpose molecules shows *d* and *b* on opposite sides when *a* and *f* fit together.

**FIGURE 7.7
REPRESENTATIONS
OF ENANTIOMERS**

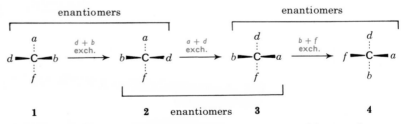

Exchange of any two groups produces the enantiomer of the starting compound. Structures 1 and 2 are enantiomeric compounds. Structures 1 and 3 represent the same compound, as do structures 2 and 4.

Only the breaking of bonds at an asymmetric carbon can convert one enantiomer into the other. There are only two possible enantiomers for any given sequence of atoms. *The exchange of bonding positions of any two attached groups in one enantiomer produces the other enantiomer*, as shown in Fig. 7.7.

Two compounds which are particularly important in metabolism are glycerol and **glyceraldehyde** (2,3-dihydroxypropanal). Glyceraldehyde exists as enantiomers, but glycerol does not, as shown in Fig. 7.8.

FIGURE 7.8
GLYCEROL AND THE
ENANTIOMERS OF
GLYCERALDEHYDE

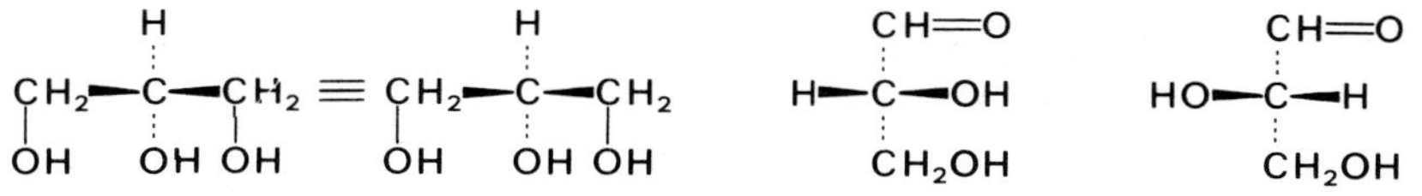

glycerol   mirror image R-glyceraldehyde S-glyceraldehyde

superimposable mirror images   nonsuperimposable mirror images

Problems

7.3 Draw the two enantiomers of the following compounds, each of which has one asymmetric carbon.

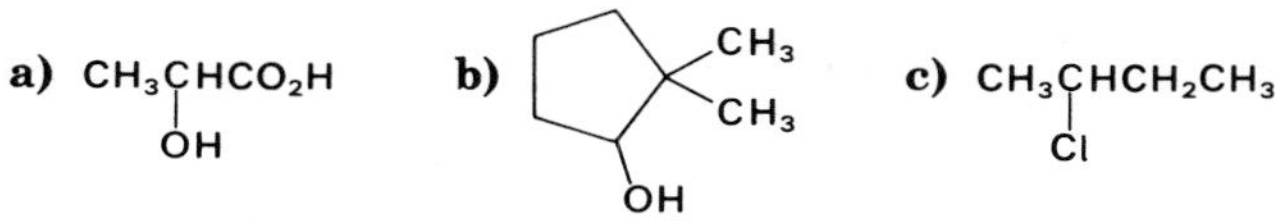

a) CH_3CHCO_2H
     OH

b)

c) $CH_3CHCH_2CH_3$
     Cl

7.4 Which of the following compounds exist as enantiomers? Draw both enantiomers of those that have them.

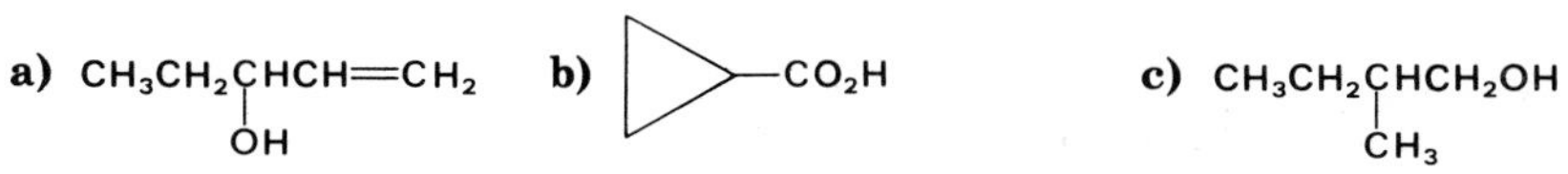

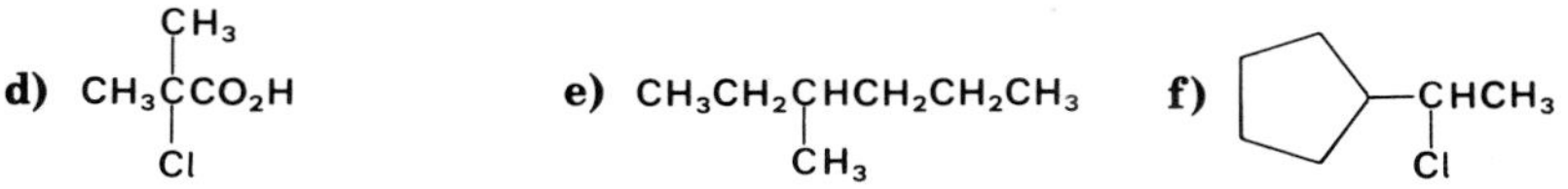

a) $CH_3CH_2CHCH=CH_2$
     OH

b) $-CO_2H$

c) $CH_3CH_2CHCH_2OH$
     CH_3

d) CH_3CCO_2H
     CH_3 (above), Cl (below)

e) $CH_3CH_2CHCH_2CH_2CH_3$
     CH_3

f) $-CHCH_3$
     Cl

What effect does the chirality have on the properties of these compounds? Enantiomers have identical physical properties with one exception which is directly related to the chirality. They interact with plane-polarized light in opposite ways. This property of optical activity is described in the next section.

7.5 OPTICAL ACTIVITY

The specific distinguishing physical property of enantiomers is the rotation of the plane of plane-polarized light as the light passes through a liquid or a solution. For this reason enantiomers were historically called **optical isomers**. This rotation by optically active compounds is easily measured by a polarimeter, and the degree of the rotation has a specific value for each compound.

Ordinary light has components that oscillate in all directions perpendicular to the direction in which the beam is propagated. Plane-polarized light oscillates also in a direction that is perpendicular to the direction of propagation, but the oscillation is confined to one plane, as depicted in Fig. 7.9.

FIGURE 7.9
PLANE-POLARIZED LIGHT WAVE

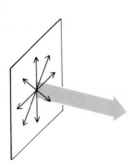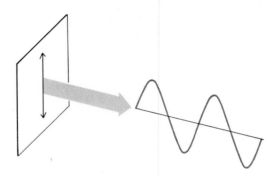

Components of ordinary light oscillate in all planes perpendicular to the direction of propagation of the beam.

Plane-polarized light wave oscillates in only one plane perpendicular to the direction of propagation of the beam.

FIGURE 7.10
SKETCH OF THE ESSENTIAL PARTS OF A POLARIMETER

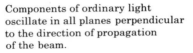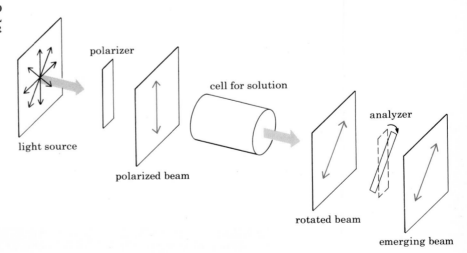

If the plane-polarized beam passes through a transparent substance, the plane of the oscillation may be changed. It may be rotated through an angle α which is characteristic of the substance causing the rotation.

Isomers which are nonsuperimposable, but are mirror images, when interposed in a beam of plane-polarized light cause a rotation of the plane through *equal angles in opposite directions.* A solution of a single enantiomer which transmits the light with the plane rotated clockwise is called *dextrorotatory* and the angle of rotation is given a positive ($+$) sign. The other enantiomer in a solution of the same concentration transmits the light with the plane rotated an equal number of degrees in a counterclockwise direction. It is the *levorotatory* isomer and the angle of rotation is given a negative ($-$) sign.

The angle of rotation of a solution can be measured quite precisely with a polarimeter. The important parts of a polarimeter are shown in Fig. 7.10.

The magnitude of the rotation of an enantiomer is reported as the specific rotation $[\alpha]$. This rotation is dependent on the wavelength (used as a subscript), the temperature (a superscript), often the concentration, and the solvent. All of these variables should be given when a rotation is reported. The monochromatic light most often used is the light emitted by the sodium lamp at 589 nm (called the sodium D line).

The specific rotation is calculated according to this equation.

$$[\alpha]_{\text{wavelength}}^{\text{temp}} = \frac{\text{observed rotation,}^\circ}{\text{length of sample tube (dm)} \times \text{conc. (g/ml)}}$$

7.6 ABSOLUTE CONFIGURATION AND NOTATION

The arrangement of substituents around a carbon atom is called the **configuration**. Enantiomers have opposite configurations. An enantiomer may easily assume various conformations by rotation of groups around the single bonds, but a change in the configuration requires the breaking of bonds at the asymmetric carbon.

Experimentally the separate enantiomers are distinguished by the opposite optical rotations they give. The enantiomer in bottle #1 is labeled ($+$)-2-butanol, and the enantiomer in bottle #2 is labeled ($-$)-2-butanol. An older method of designating enantiomers uses the small letter *d* for the *dextro*rotatory isomer ($+$), as in *d*-2-butanol, and the small letter *l* for the *levo*rotatory or ($-$) isomer, as in *l*-2-butanol.

The $+$ sign of rotation of the enantiomer in bottle #1 does not tell us which of the two configurations the sample actually possesses. For nearly a century the actual or **absolute configuration** of any enantiomer remained unknown. The first experimental determination of the absolute configuration of an enantiomer, sodium rubidium ($+$)-tartrate, was made in 1951 by X-ray crystallography. Finally the actual configuration of enantiomer #1 was known with certainty. Since that time it has become a regular procedure, though in many cases a lengthy one, to determine the absolute configuration of enantiomers of interest by X-ray crystallography.

The ability to determine the absolute configuration of compounds brought the need for a simple nomenclature system. The configuration of an enantiomer can be specified by placing a single letter R or S as a prefix to the name of the compound. The notation is based on the same system of priority of substituents* that is used in the E–Z designation of *cis-trans* isomers (Section 4.2).

For assignment of a configurational name to a compound, its structure is viewed with the group of lowest priority away from the eye. The other three substituents on the carbon project toward the viewer. If the priority order of these three substituents decreases in a clockwise direction, the configuration is designated as R, and if in a counterclockwise direction, the configuration is designated as S (Fig. 7.11).

FIGURE 7.11   CONFIGURATION DESIGNATION BY PRIORITY RULES

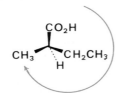

(*R*)-2-methylbutanoic acid

The isomer is R (*rectus* = right) if groups decrease in priority in a clockwise order.

(*S*)-2-methylbutanoic acid

The isomer is S (*sinister* = left) if groups decrease in priority in a counterclockwise order.

Another way of visualizing the configuration is to imagine the four bonds of the molecule to be arranged like the column and three spokes for the steering wheel of a car. The bond to the lowest priority is the column and the other three bonds are the spokes, with their three attached substituents lying on the wheel. The wheel is always turned in the direction of passing from the substituent of highest toward that

* Priority is established by the difference in *atomic number* of the atoms nearest to the asymmetric carbon.

1. The higher the **atomic number** of the atom bonded directly to the asymmetric carbon, the higher the priority of the substituent.

$Cl > S > F > O > N > C > H$

2. For two atoms of the same atomic number, the priority depends on the next attached atoms until a difference between groups is found.

$—CH_2Cl > —CH_2OH > —CH_2CH_3 > CH_3$

3. Double bonds are counted as two single bonds, i.e., C=O as O—C—O.

of next priority. If the resulting turn is to the right, the configuration is R (for *rectus*, meaning right). If the turn is to the left, the configuration is S (for *sinister*, meaning left).

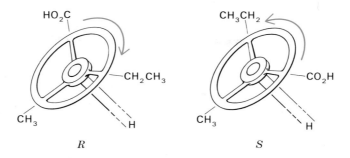

R   S

Problem   **7.5** Draw the configuration specified for each compound.

a) (R)-CH₃CHCO₂H
            |
            OH

b) (S)-CH₃CH₂CCH₂OH   with CH₃ up and NH₂ down

c) (S)-ClCH₂CHCO₂H
            |
            OH

d) (R)-HC≡CCHCH₂CH₃   with CH₃

7.7 RACEMIC FORMS   An equimolar mixture of two enantiomers in any phase is a racemic form, also called a *d,l* or a (±) mixture. The racemic form differs from each enantiomer in the value of all physical properties that depend on molecular associations, such as boiling point, melting point, and solubility. In a crystal the racemic form packs differently from either enantiomer alone. The (±) mixture and pure enantiomers, when crystalline, possess different crystal structures, lattice energies, and solubilities. Enantiomers have identical values for all physical constants except their optical rotations, which are of equal value and of opposite sign. The (±) mixture has an optical rotation of zero. Figure 7.12 lists some physical constants for one pair of enantiomers.

Chemical reactions of enantiomers and racemic forms *are identical*, except in environments that are themselves asymmetric. Asymmetric environments might be created by carrying out a reaction with an

FIGURE 7.12
PHYSICAL CONSTANTS
OF RACEMIC FORM
AND ENANTIOMERS

equal $(R),(S)$-isomer mixture
mp 120.5°
$[\alpha]_D^{20} = 0.0$

(S)-isomer
mp 133.8°
$[\alpha]_D^{20} = +156.6°$

(R)-isomer
mp 133.8°
$[\alpha]_D^{20} = -156.9°$ (H_2O)

hydroxyphenylacetic acid
(mandelic acid)

asymmetric catalyst, with an asymmetric reagent, or in an asymmetric solvent. In general, laboratory syntheses occur in *symmetric environments*, and enzyme-catalyzed reactions involve *asymmetric environments*.

Louis Pasteur discovered isomers that rotate light oppositely by finding crystals of sodium ammonium tartrate that were themselves mirror images of each other. So distinct were the crystal shapes that Pasteur was able to separate the crystals into two piles using a microscope and tweezers. Most ($\pm$) mixtures form crystals containing both enantiomers. Less laborious and more universal methods of resolving racemic forms into the enantiomers are discussed in Section 7.9.

Problem   **7.6** A sample of hydroxyphenylacetic acid (Fig. 7.12) gave a rotation $[\alpha]$ of $-120°$. What is the percent of R,S racemic form in the sample? What is the total percent of (R)-isomer in the sample?

7.8 DIASTEREOISOMERS

For compounds that contain more than one asymmetric carbon, there are more than two stereoisomers. If there are n asymmetric carbons, the *maximum number* of stereoisomers is 2^n. Compounds in which no two asymmetric carbons possess the same set of four attached groups reach this maximum number of stereoisomers. Some other compounds which have fewer than the maximum number will be considered in Section 7.10.

When $n = 1$, as in 2-butanol, the number of stereoisomers is two and the stereoisomers are enantiomerically related. When $n = 2$ the maximum number of stereoisomers is four. Since enantiomers or mirror-image relationships come only in pairs, some of the relationships between two of the stereoisomers must not be enantiomeric. Stereoisomers which exist because of asymmetric carbons and which are not enantiomers are **diastereoisomers** of one another. Figure 7.13 contains

FIGURE 7.13 DIASTEREOISOMERS— STEREOISOMERS WITH TWO OR MORE ASYMMETRIC CARBONS

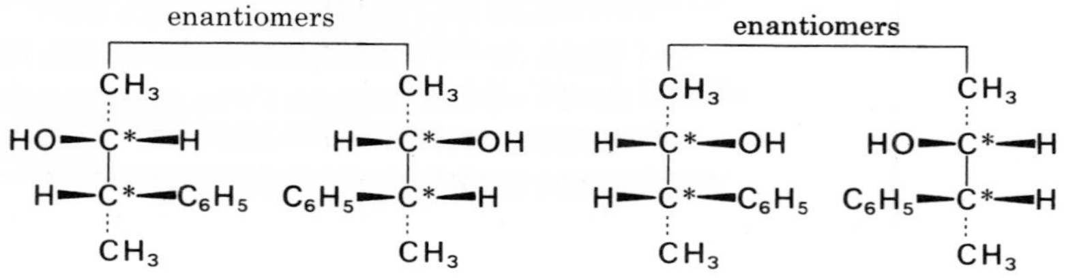

Asymmetric carbons are marked by an asterisk. A is diastereoisomerically related to C and D, and B is diastereoisomerically related to C and D. The racemic form composed of $A \cdot B$ is diastereoisomerically related to the racemic form composed of $C \cdot D$.

three-dimensional formulas for the four stereoisomers of 3-phenyl-2-butanol, whose two asymmetric carbons each have a different set of four attached groups. As the figure shows, among the four 3-phenyl-2-butanols, there are two sets of enantiomeric relationships and four sets of diastereoisomeric relationships.

Problems **7.7** The R or S designation of configuration of each asymmetric carbon is incorporated directly into the systematic name of each isomer. Thus isomer A of 3-phenyl-2-butanol (Fig. 7.13) is named 3-(S)-phenyl-2-(R)-butanol.

a) Provide names for isomers B, C, and D in Fig. 7.13.

b) With a general statement, indicate how enantiomeric relationships can be identified from the configurational names of the four isomers.

c) With a general statement, indicate how diastereoisomeric relationships can be identified from the configurational names of the four isomers.

7.8 How many stereoisomers and how many racemic forms can be composed from a compound which has four different asymmetric centers?

7.9 How many asymmetric centers are found in the following compounds, and how many stereoisomers will each have?

a) 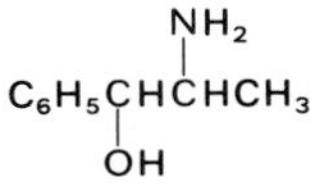
$$\underset{\underset{OH}{|}}{\overset{\overset{NH_2}{|}}{C_6H_5CHCHCH_3}}$$

b)
$$H_2N-\underset{}{\overset{\overset{CH_3}{|}}{CH}}-\underset{\underset{O}{\|}}{C}-NH-\underset{}{\overset{\overset{CH_3}{|}}{CH}}-\underset{\underset{O}{\|}}{C}-NH-\underset{}{\overset{\overset{CH_3}{|}}{CH}}-CO_2H$$

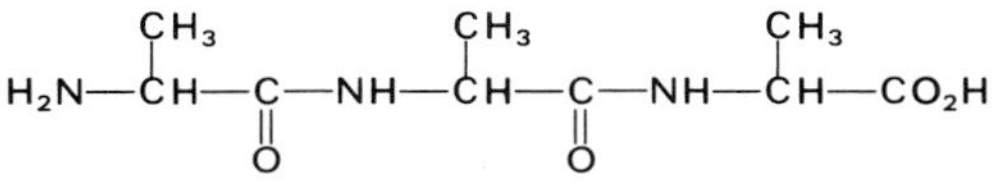

c)
$$H_2N-\underset{}{\overset{\overset{CH_2C_6H_5}{|}}{CH}}-\underset{\underset{O}{\|}}{C}-NH-\underset{}{\overset{\overset{CH_3}{|}}{CH}}-\underset{\underset{O}{\|}}{C}-NH-CH_2-CO_2H$$

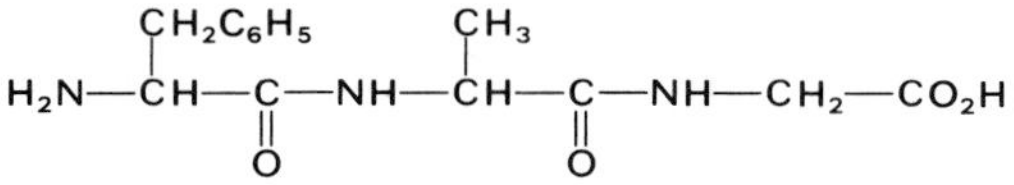

7.9 RESOLUTION OF RACEMIC FORMS

Diastereoisomers, unlike enantiomers, have different physical properties. For example, the magnitudes of the optical rotations of the diastereoisomerically related stereoisomers of 3-phenyl-2-butanols are much different (see Fig. 7.13), and the boiling points are very slightly different. The p-toluenesulfonate ester (tosylate ester) derivatives of the four isomers are all solids, and have the following melting points: A, 62–63°; B, 62–63°; C, 46–47°; D, 46–47°. Differences in crystal character and solubility serve as the most useful basis for separating diastereoisomers.

The two enantiomers that compose a racemic form of a compound have the same physical and chemical properties except in their interaction with plane-polarized light and their interaction with chiral reagents or catalysts. By a suitable reaction with an optically active reagent, the enantiomers of a racemic form can be converted to a mixture of crystalline diastereoisomers incorporating the chiral centers of both reactants. The diastereoisomers thus formed may be separated through the differences in their solubility. The less soluble diastereoisomer crystallizes from solution at a suitable concentration and is removed by filtration. The solution is then concentrated by evaporation of solvent and the more soluble diastereoisomer crystallizes and is obtained by filtration. Finally, the separated diastereoisomers are subjected to reactions that reconvert them to the *separated enantiomers* of the original racemic form.

The reaction of an organic acid and an organic base to form an organic salt is both rapid and easily reversed. Resolutions (the separation of a racemic form into its two enantiomers) usually involve salt formation between a racemic amine and an optically pure acid, or between an optically pure amine and a racemic acid, as illustrated in Fig. 7.14.

FIGURE 7.14 RESOLUTION OF A RACEMIC CARBOXYLIC ACID BY SALT FORMATION IN AN ACID–BASE REACTION

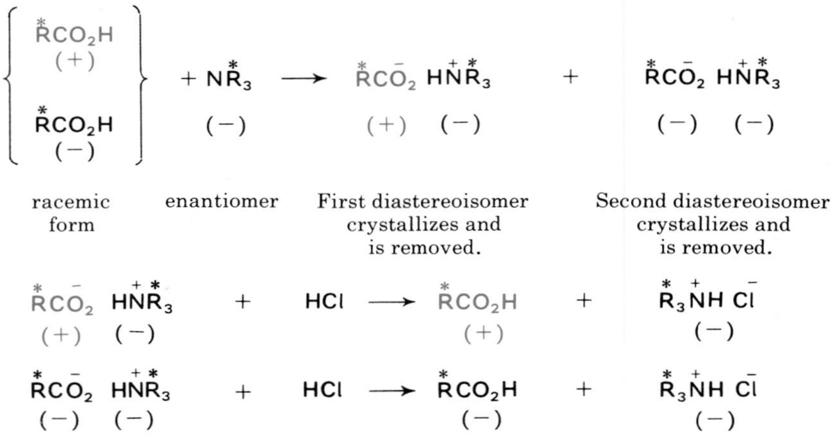

$$\left\{ \begin{array}{c} \overset{*}{R}CO_2H \\ (+) \\ \\ \overset{*}{R}CO_2H \\ (-) \end{array} \right\} \ + \ N\overset{*}{R}_3 \ \longrightarrow \ \overset{*}{R}C\bar{O}_2 \ HN^{+}\overset{*}{R}_3 \ + \ \overset{*}{R}C\bar{O}_2 \ HN^{+}\overset{*}{R}_3$$

racemic form	enantiomer	First diastereoisomer crystallizes and is removed.	Second diastereoisomer crystallizes and is removed.

(with labels: $(-)$ under $N\overset{*}{R}_3$; $(+)$ $(-)$ under first diastereoisomer; $(-)$ $(-)$ under second diastereoisomer)

$$\overset{*}{R}C\bar{O}_2 \ HN^{+}\overset{*}{R}_3 \ + \ HCl \ \longrightarrow \ \overset{*}{R}CO_2H \ + \ \overset{*}{R}_3\overset{+}{N}H \ \bar{Cl}$$
$$(+) \quad (-) \qquad\qquad\qquad\qquad (+) \qquad\qquad (-)$$

$$\overset{*}{R}C\bar{O}_2 \ HN^{+}\overset{*}{R}_3 \ + \ HCl \ \longrightarrow \ \overset{*}{R}CO_2H \ + \ \overset{*}{R}_3\overset{+}{N}H \ \bar{Cl}$$
$$(-) \quad (-) \qquad\qquad\qquad\qquad (-) \qquad\qquad (-)$$

Optically pure acids and amines are abundantly available in nature. Naturally occurring, optically active amines (the alkaloids) such as brucine, strychnine, quinine, and cinchonine (Section 10.1) and the two optically active tartaric acids are most used for resolution. Salts prepared from these compounds are easily decomposed by treatment with a strong acid or base. If the desired enantiomer is a carboxylic acid, treatment of its salt with a strong acid such as hydrochloric acid will form the carboxylic acid. If the desired enantiomer is an amine, treatment of its ammonium salt with a strong base such as sodium hydroxide will form the neutral amine.

Problem **7.10** Diagram the resolution steps for a racemic amine, such as 1-phenylethylamine, by the use of (−)-tartaric acid.

7.10 MESO FORMS Molecules exhibit symmetry properties, just as do familiar objects. Molecules that contain one asymmetric carbon contain no element of symmetry and are nonsuperimposable on their mirror images. The chirality of these asymmetric compounds has been demonstrated by many examples.

What of the compounds that have a plane of symmetry? Just as the objects in Section 7.2 that have a plane of symmetry are achiral, so we should expect such molecules to be superimposable on their mirror images, to be achiral, and to have an optical rotation of zero. With some exceptions due to restricted rotation, molecules without an asymmetric carbon possess a mirror plane and are achiral and optically inactive. There are structures, however, which contain *more than one* asymmetric carbon and yet possess a plane of symmetry. Like the clam shell in Fig. 7.3, these compounds are superimposable on their mirror images and are optically inactive.

When at least one isomer of a set of stereoisomers possesses a plane of symmetry, the total number of stereoisomers is less than the maximum number of 2^n. An example is provided by the stereoisomers of tartaric acid, 2,3-dihydroxybutanedioic acid, which have two asymmetric carbons with the same set of four attached groups. Only three stereoisomers of tartaric acid are known. A striking feature of the acids is that one of the three isomers has a plane of symmetry and the other two acids do not. The two acids which do not have a mirror plane are enantiomers. The relative configurations of the two asymmetric carbons are the determining factor, as is demonstrated by the structures of the three tartaric acids shown in Fig. 7.15.

	eclipsed conformations		superimposable mirror images	
FIGURE 7.15 **THREE STEREOISOMERS OF TARTARIC ACID**				

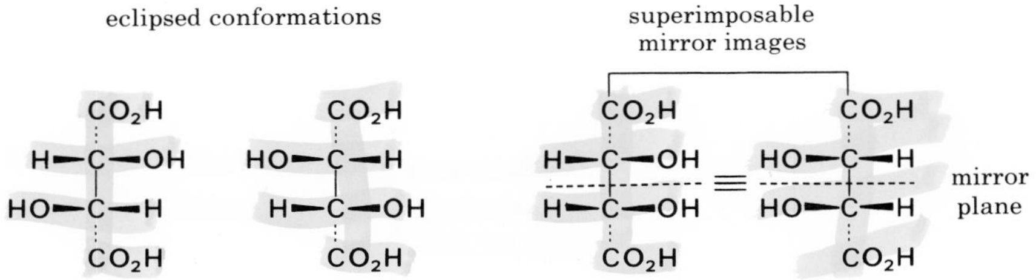

Configuration	RR	SS	RS	$\equiv$	SR
Melting point, °C	170	170	140		
$[\alpha]_D^{20}$ in water	$+12°$	$-12°$	$0°$		
Symmetry planes	none	none	one		one

A molecule may rotate to assume many different conformations. To be optically inactive, an isomer must show an obvious plane of symmetry in only one conformation, if that conformation can be achieved by the allowable rotations. Thus the RS isomer of tartaric acid is optically inactive because, in the eclipsed conformation, it has a plane of symmetry.

The isomers of tartaric acid that possess the RR and SS configurations are optically active and are enantiomers. The RS isomer and others like it that contain asymmetric carbons but are optically inactive because they contain a plane of symmetry are said to possess **meso configurations**.

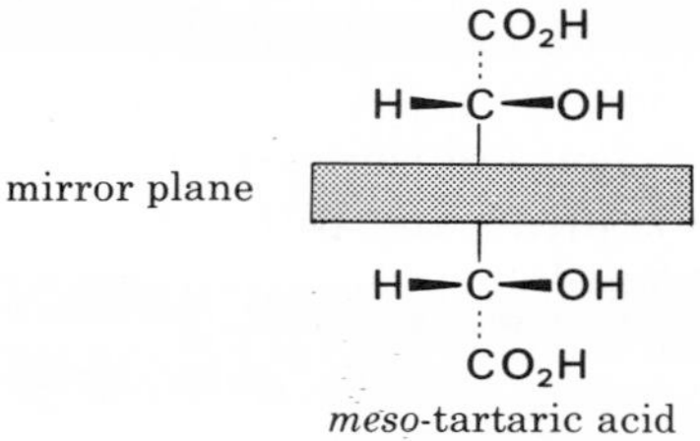

$$CO_2H$$
$$H \blacktriangleright C \blacktriangleright OH$$

mirror plane

$$H \blacktriangleright C \blacktriangleright OH$$
$$CO_2H$$

meso-tartaric acid

Many cyclic compounds have stereoisomers that contain mirror planes, several of which are exemplified below. Both of the *cis-trans* isomers of 1,4-dimethylcyclohexane contain mirror planes (plane of the page), do not contain asymmetric carbon atoms, and are optically inactive.

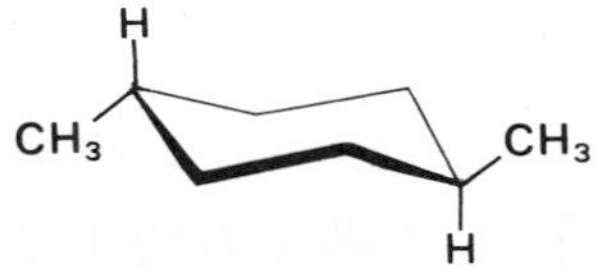

trans-1,4-dimethylcyclohexane    *cis*-1,4-dimethylcyclohexane

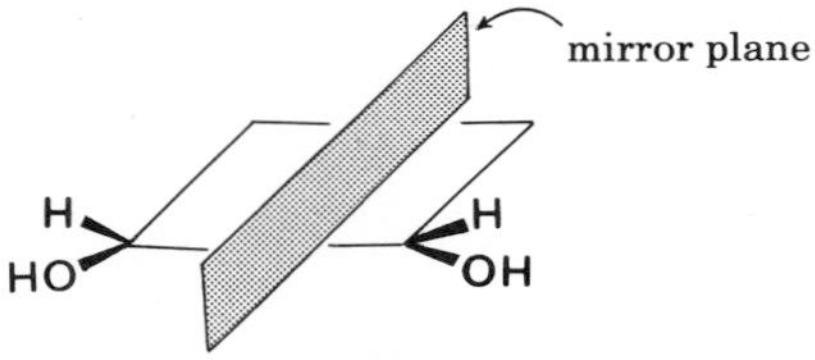

cis-1,2-cyclobutanediol (*meso*) 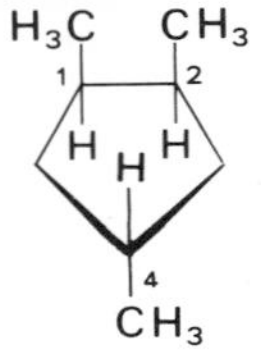   a *meso*-1,2,4-trimethylcyclopentane

The other examples of stereoisomers depicted are optically inactive because they contain mirror planes and asymmetric centers in *meso* configurations. The mirror plane of *cis*-1,2-cyclobutanediol bisects the ring between the C-1 and C-2 carbons (see diagram above). The mirror plane of 1,2-Z-4-E-trimethylcyclopentane is perpendicular to the plane of the page, bisects the bond between carbons #1 and #2 (C-1 and C-2) and passes through the C-4 carbon and its bonded groups.

Problems **7.11** Write three-dimensional structures for the two optically active and the two *meso* stereoisomers of 1,2,4-trimethylcyclopentane.

7.12 Write three-dimensional structures for all stereoisomers of the following compounds. Identify which are optically active and which are *meso*.

a) $CH_3 \!-\! \overset{\displaystyle H}{\underset{\displaystyle C_6H_5}{C}} \!-\! CH_2 \!-\! CH_3$ **b)** $C_6H_5 \!-\! \overset{\displaystyle HO}{\underset{\displaystyle H}{C}} \!-\! \overset{\displaystyle OH}{\underset{\displaystyle H}{C}} \!-\! C_6H_5$

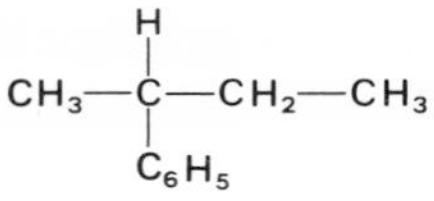

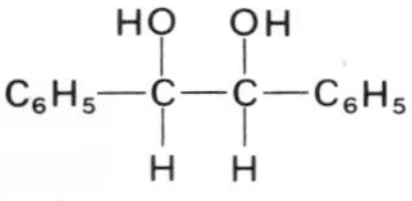

c)

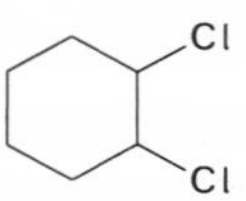

d)

7.13 Indicate whether the following compounds possess an R, an S, an RS, an SS, or an RR configuration.

a) 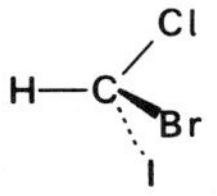
$$H-C\overset{Cl}{\underset{I}{\overset{|}{\diagdown}}}Br$$

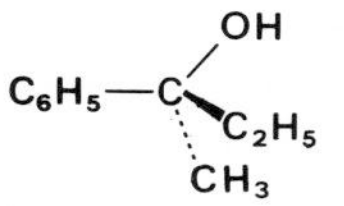

b)
$$C_6H_5-C\overset{OH}{\underset{CH_3}{\overset{|}{\diagdown}}}C_2H_5$$

c)
$$\overset{H_3C}{\underset{C_6H_5}{\diagup}}H-C-C-H\overset{CH_3}{\underset{C_6H_5}{\diagdown}}$$

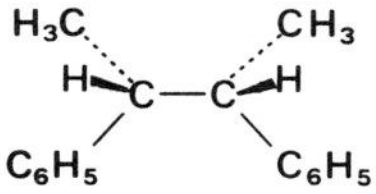

d) 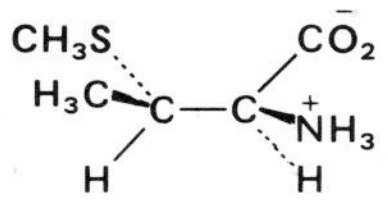
$$\overset{CH_3S}{\underset{H}{\diagup}}H_3C-C-C\overset{\overset{-}{CO_2}}{\underset{H}{\overset{\diagup}{\underset{NH_3}{+}}}}$$

e) 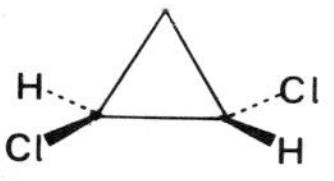
$$H\cdots\overset{\triangle}{}\cdots Cl \quad Cl\cdots\cdots H$$

7.14 Assume you prepared an ether from racemic 2-chloropentane and racemic sodium 2-pentoxide. Write three-dimensional structures for the three ethers you would get, and indicate which would be optically active. Indicate their configurations with the R, S notations.

7.11 DYNAMIC STEREOCHEMISTRY

The above sections have dealt with stereochemistry as applied to structures. Equally interesting are the many questions about what happens to configurations at a given carbon when reactions are carried out that involve breaking bonds to that carbon. These questions are important to the chemistry of biological systems, since most of the natural compounds contain asymmetric centers, and are constantly being synthesized and destroyed in the cell.

In the laboratory, hydration of an olefinic linkage in the absence of chiral centers provides a racemic secondary alcohol. For example, the hydration of the unsaturated acid fumaric acid produces racemic malic acid, as shown in Fig. 7.16. A carbonium ion intermediate is formed by the acid catalyst. The water adds to the two faces of the planar carbonium ion with equal probability (Fig. 7.16).

The mechanism of catalysis of hydration of the alkene by the enzyme is similar to that of the laboratory reaction. A proton donor, H_3O^+ in the regular reaction and HB^+ on the enzyme, reacts with the double bond of the fumaric acid to form a carbonium ion intermediate. The proton to be donated by the enzyme is positioned correctly to form a bond with only one of the two olefinic carbons, and only from the top face of the fumarate ion. The water molecule is in the niche between the positively charged carbon and a basic center on the enzyme A^- to which it is hydrogen bonded. Water adds to the positive carbon from the lower side only, while the A^- removes the proton from the water. In the laboratory reaction a water molecule can add to the positive carbon from either side and another water molecule removes the proton.

In Fig. 7.16 we show the enzyme in a schematic diagram with only the working parts specified. The site of the enzyme molecule which participates as a catalyst in the hydration contains four specific units. Two positive charges on the enzyme are countercharges for the two carboxylate ion ends of the fumarate ion. Two steric barriers hold the

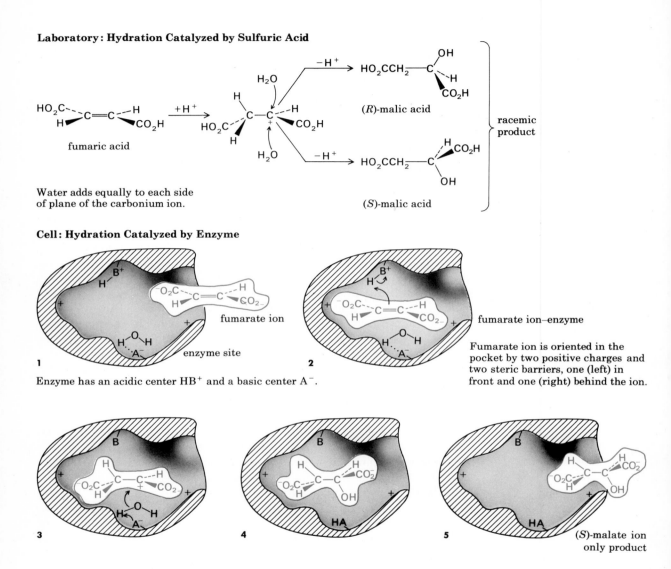

Laboratory: Hydration Catalyzed by Sulfuric Acid

fumaric acid

(R)-malic acid

(S)-malic acid

racemic product

Water adds equally to each side of plane of the carbonium ion.

Cell: Hydration Catalyzed by Enzyme

fumarate ion

enzyme site

1

Enzyme has an acidic center HB$^+$ and a basic center A$^-$.

fumarate ion–enzyme

2

Fumarate ion is oriented in the pocket by two positive charges and two steric barriers, one (left) in front and one (right) behind the ion.

3

4

5

(S)-malate ion only product

fumarate ion in the correct geometry for the acidic and basic centers to operate. In this way only the desired enantiomer of the product can form. In the enzyme the acidic center BH$^+$ is a protonated amine, R—NH$_3^+$, and the base, A$^-$, is a carboxylate ion, R—CO$_2^-$. The structure of a different enzyme is discussed in more detail in Section 10.18.

Problem

7.15 One of the experiments which contributed to the elucidation of the mechanism of the enzyme-catalyzed hydration of fumaric acid was done using deuterium oxide, D_2O, instead of water, H_2O. The malic acid produced contained deuterium on the carbon where D^+ added and OD where D_2O added. A second asymmetric carbon was created by the use of D_2O. Using the diagram in Fig. 7.16 as the model, draw the diastereoisomer produced in this reaction with D^+ and D_2O. Name the configuration at each asymmetric carbon by R,S nomenclature. Draw the enantiomer of this product and name its configurations.

7.12 STEREOCHEMISTRY OF NUCLEOPHILIC SUBSTITUTION

Interesting questions arise about the stereochemical course of reactions when the starting material and/or the product has an asymmetric center at the site of reaction. The most obvious example that comes to mind is the nucleophilic substitution at a saturated carbon—as in an alkyl halide in which the halogen is attached to an asymmetric carbon.

The reaction of optically active 1-phenyl-1-chloroethane with water gives 1-phenylethanol which is nearly completely racemic. This transformation is an example of the unimolecular nucleophilic substitution at a saturated carbon (S_N1 reaction, Section 6.4). The carbonium ion intermediate which is formed in the first step of the reaction becomes planar (with 120° angles) and therefore loses the original configuration of the starting alkyl halide. Water may then attack the planar carbonium ion on either face with equal probability and the two enantiomeric products are formed in nearly equal amounts. The stereochemical course of the reaction is formulated in Fig. 7.17.

FIGURE 7.17 LOSS OF CONFIGURATION IN THE S_N1 REACTION

(R)-1-phenyl-1-chloroethane   planar carbonium ion intermediate   (R)-1-phenyl-ethanol 49%   (S)-1-phenyl-ethanol 51%

almost racemic

Dynamic stereochemistry has proved very useful in elucidating reaction mechanisms. The observed racemic product in Fig. 7.17 suggests the reaction went through an intermediate that possessed a plane of symmetry. The sp^2-p hybridized carbonium ion as that intermediate provides an explanation for the experimental results.

One of the most beautiful examples of dynamic stereochemistry involves the proof that the bimolecular nucleophilic substitution reaction (S_N2 reaction, Section 6.4) occurs with a backside attack and produces an *inversion of configuration*. The reaction sequence contains three reactions.

Reaction 1

R—OH + Ts—Cl ⟶ HCl + R—OTs   (Ts = H_3C—⟨benzene ring⟩—SO_2)

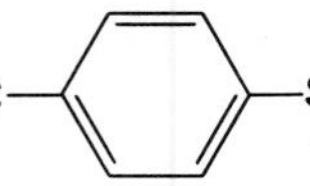

alcohol tosyl chloride   tosylate ester

Reaction 2   (an S_N2 reaction)

R—OTs + ⁻OAc ⟶ R—OAc + ⁻OTs

tosylate ester acetate ion   acetate ester   tosylate ion

Reaction 3

R—OAc + ⁻OH ⟶ R—OH + ⁻OAc

acetate ester   alcohol   acetate ion

The middle reaction is a bimolecular nucleophilic substitution reaction. The actual sequence is shown in Fig. 7.18 with the optically active alcohol (*S*)-1-phenyl-2-propanol.

FIGURE 7.18
INVERSION OF
CONFIGURATION IN THE
S_N2 REACTION

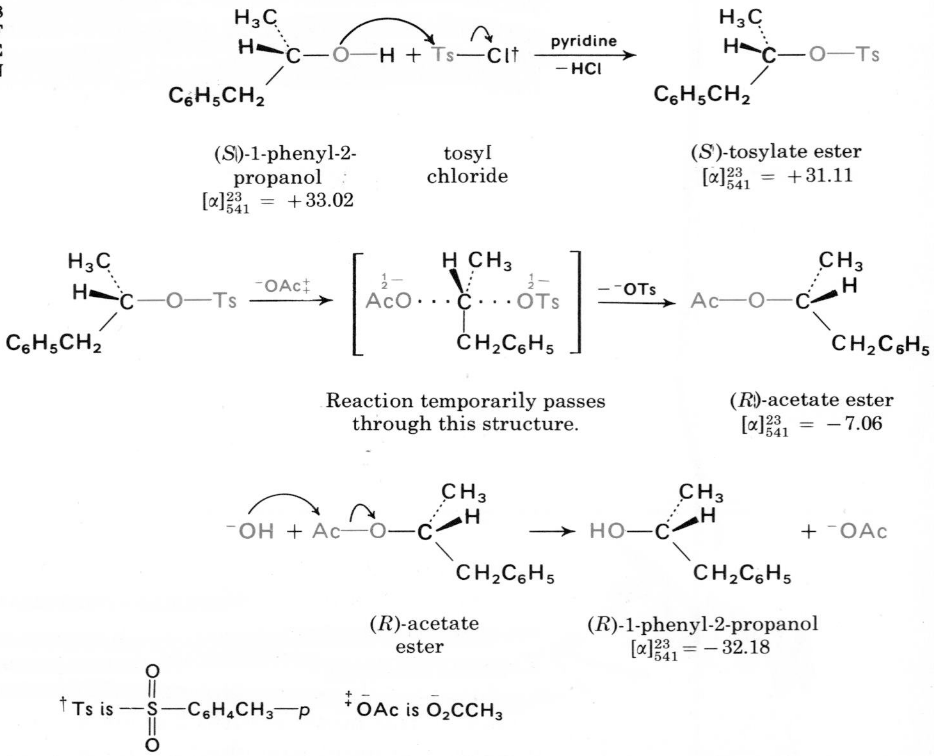

(*S*)-1-phenyl-2-
propanol
$[\alpha]_{541}^{23} = +33.02$

tosyl
chloride

(*S*)-tosylate ester
$[\alpha]_{541}^{23} = +31.11$

Reaction temporarily passes
through this structure.

(*R*)-acetate ester
$[\alpha]_{541}^{23} = -7.06$

(*R*)-acetate
ester

(*R*)-1-phenyl-2-propanol
$[\alpha]_{541}^{23} = -32.18$

† Ts is —S(=O)(=O)—$C_6H_4CH_3$—*p*   ‡ ⁻OAc is $\bar{O}_2CCH_3$

The alcohol (S)-1-phenyl-2-propanol is converted to its tosylate ester, a reaction that does not involve breaking the alcohol C—O bond, and therefore preserves the configuration at the asymmetric carbon. The (S)-tosylate ester is then treated with potassium acetate in acetone. The acetate ion acts as a nucleophile and displaces the tosylate group (TsO⁻) from the asymmetric carbon in a bimolecular nucleophilic substitution reaction. The acetate ester produced is hydrolyzed with a base in a reaction known to break only the C—O bond on the acid side without breaking the C—O bond on the alcohol side. Therefore the configuration at the asymmetric center is unaltered in this step. The optically pure 1-phenyl-2-propanol thus produced possesses the (R) configuration, enantiomeric to the starting alcohol. An *inversion of configuration* must have occurred during the bimolecular nucleophilic substitution reaction.

During the S_N2 reaction the H, CH_3, and $C_6H_5CH_2$ groups start on one side of the molecule, but end up on the other. Their relocation resembles that of the ribs of an umbrella that turns inside out in a wind.

Problems   **7.16** Draw the structures of (R)-2-bromopentane and of the ether formed when it reacts with sodium ethoxide. Which ether enantiomer is formed, (R) or (S)?

7.17 The conversion of 2-octanol into 2-chlorooctane is accomplished by several methods. Explain why with (R)-2-octanol the three methods produce the enantiomer of the halide shown.

(R)-2-octanol

+ HCl $\xrightarrow{\text{H}_2\text{O, heat}}$ 2-chlorooctane nearly racemic

+ H_3C—⟨benzene ring⟩—SO_2Cl ⟶ $\xrightarrow{\text{KCl}}$ (S)-2-chlorooctane

+ $SOCl_2$ with pyridine ⟶ (S)-2-chlorooctane

7.13 STEREOCHEMISTRY OF ELIMINATION INITIATED BY BASE

An elimination reaction which is initiated by a base to produce an alkene is a bimolecular process. Both the base and the substrate are involved in the rate-limiting step, which passes through a structure of high energy.

$$\overset{}{\underset{^{1/2^-}\text{B}\cdots\overset{|}{\text{H}}}{-\overset{|}{\text{C}}\cdots\overset{\overset{\displaystyle X^{1/2^-}}{\vdots}}{\underset{|}{\text{C}}}-}}$$

The elimination reaction occurs faster when the departing atoms or groups are in an anti conformation. In the drawing above, the H and X groups are shown anti to each other. All bonds are made and broken simultaneously.

The base furnishes electrons to form a bond with the proton being removed from the substrate. The pair of electrons of the C—H bond is used to push the halide X out and to generate a C=C double bond.

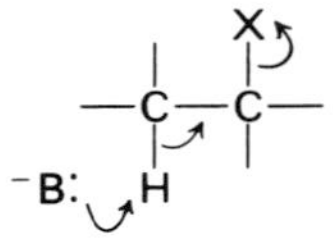

The electrons moving into the C=C bond are acting as a nucleophilic reagent to push the halide ion out. Just as in an ordinary nucleophilic substitution, the incoming electron pair approaches from the side of the carbon opposite to the halogen.

In the reaction of 1-bromo-1,2-diphenylpropane with hydroxide ion, the anti conformation required for reaction determines the particular *cis-trans* isomer of the alkene which is formed. The stereochemical course of the reaction of one of the stereoisomers is shown in Fig. 7.19.

FIGURE 7.19 STEREOCHEMISTRY OF BASE-INITIATED ELIMINATION

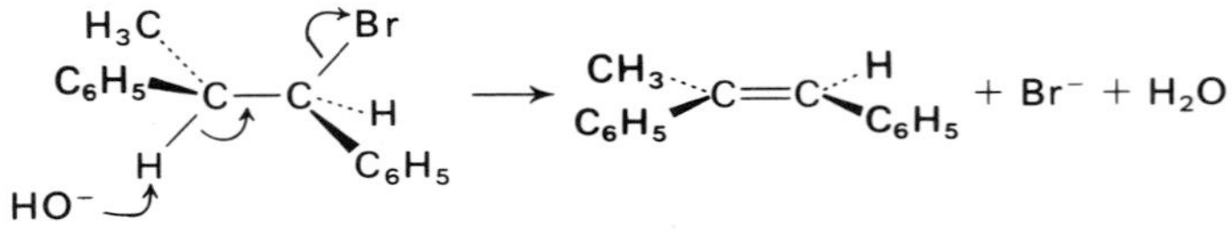

1-bromo-1,2-diphenylpropane

Z-1,2-diphenylpropene

The H and Br are in an anti relationship and thus depart from opposite sides of the molecule.

Both the stereoisomer of 1-bromo-1,2-diphenylpropane depicted in Fig. 7.19 and its enantiomer give the same alkene with the two phenyl groups *cis* to each other.

Bimolecular eliminations from compounds which cannot assume a conformation in which the departing groups are anti to each other require much higher temperatures and much stronger bases to be effected.

Problem 7.18 a) Draw the structure of one of the diastereoisomers of the 1-bromo-1,2-diphenylpropane isomer shown in Fig. 7.19 with the departing H and Br in an anti relationship.

b) Draw the structure of the isomer of 1,2-diphenylpropene which would be formed from the halide drawn in (a).

c) Name the configuration of the stereoisomer of the halide used in (a) and of the alkene formed in (b).

New Terms and Topics

Mirror image, reflection (Section 7.1)
Nonsuperimposable mirror image (Section 7.1)
Chiral object; achiral object (Section 7.1)
Plane of symmetry (Section 7.2)
Stereoisomers (Section 7.3)
Enantiomers; asymmetric carbon (Section 7.4)
Optical rotation; optically active compound (Section 7.5)
Dextrorotatory, levorotatory isomers (Section 7.5)
Configuration of a compound (Section 7.5)
R,S notation of configuration; priority rules (Section 7.6)

Racemic form (Section 7.7)
Diastereoisomers; number of stereoisomers (Section 7.8)
Meso forms (Section 7.10)
Resolution of racemic mixture (Section 7.9)
Stereochemical course of a reaction determined by enzyme catalysis (Section 7.11)
Stereochemical course of reactions involving making and breaking bonds at asymmetric carbons—substitution and elimination (Sections 7.12, 7.13)

Problems

7.19 How many stereoisomers does each of the following structures have? Draw the structure of any *meso* isomer.

a) 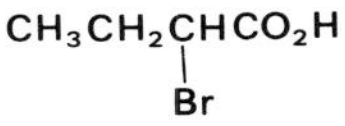 $CH_3CH_2CHCO_2H$
$\quad\quad\quad\quad\quad |$
$\quad\quad\quad\quad\quad Br$

b) 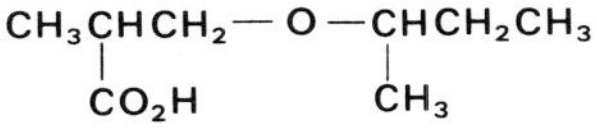 $CH_3CHCH_2-O-CHCH_2CH_3$
$\quad\quad\quad | \quad\quad\quad\quad\quad\quad |$
$\quad\quad\quad CO_2H \quad\quad\quad\quad CH_3$

c) 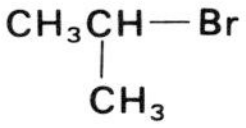 CH_3CH-Br
$\quad\quad\quad |$
$\quad\quad\quad CH_3$

d)

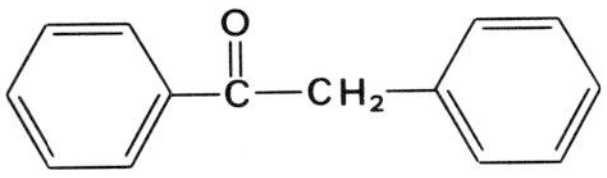

e)

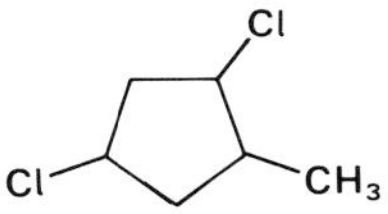

f)

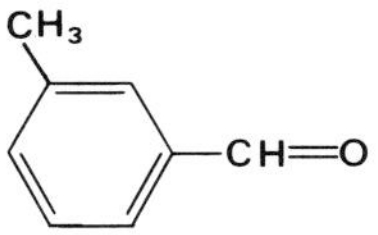

g) 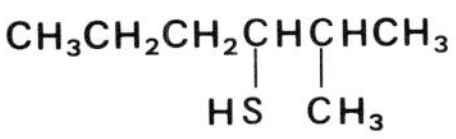 $CH_3CH_2CH_2CHCHCH_3$
$\quad\quad\quad\quad\quad\quad | \quad |$
$\quad\quad\quad\quad\quad\quad HS \quad CH_3$

h)

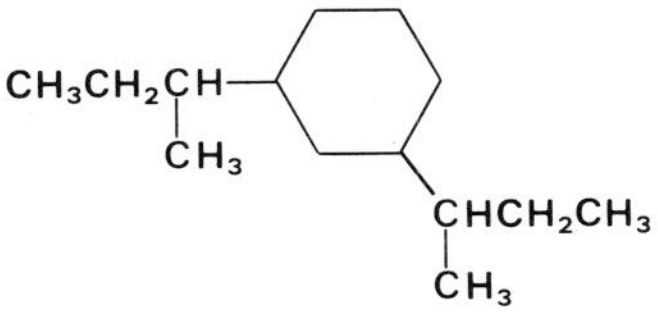

i) 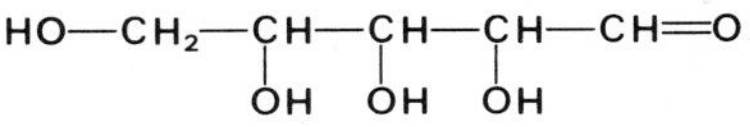 $HO-CH_2-CH-CH-CH-CH=O$
$\quad\quad\quad\quad\quad\quad | \quad\ | \quad\ |$
$\quad\quad\quad\quad\quad\quad OH \ OH \ OH$

j) 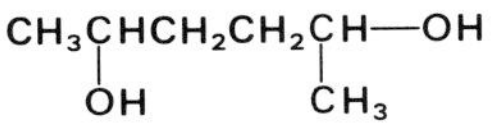 $CH_3CHCH_2CH_2CH-OH$
$\quad\quad\ \ | \quad\quad\quad\quad |$
$\quad\quad\ \ OH \quad\quad\quad CH_3$

7.20 Draw the indicated enantiomer for each compound.

a) (S)-cysteine $\quad (HS-CH_2CH-CO_2^-)$
$\quad\quad\quad\quad\quad\quad\quad\quad\quad\quad\quad |$
$\quad\quad\quad\quad\quad\quad\quad\quad\quad\quad\quad {}^+NH_3$

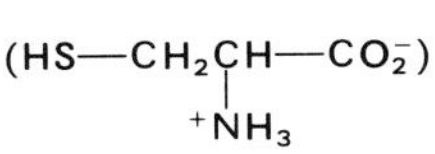

b) (R)- $CH_3CH_2-\overset{\overset{\textstyle CH_2OH}{|}}{CH}-CO_2H$

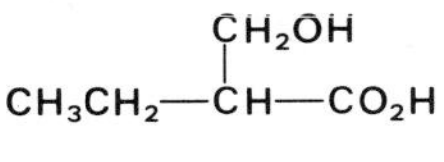

c) (R,R)- $CH_3\overset{\overset{\textstyle}{|}}{\underset{\underset{\textstyle Br}{|}}{CH}}-\overset{\overset{\textstyle CH_2CH_3}{|}}{\underset{\underset{\textstyle OH}{|}}{C}}CH(CH_3)_2$

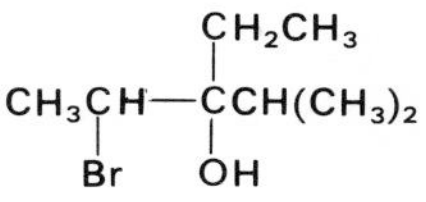

7.21 Write three-dimensional structures for all four stereoisomers of

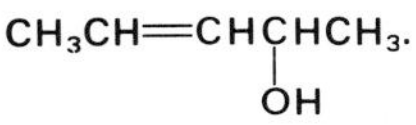

$CH_3CH=CHCHCH_3.$
$\quad\quad\quad\quad\quad\ |$
$\quad\quad\quad\quad\quad\ OH$

Among these stereoisomers, identify two sets of enantiomers, two sets of diastereoisomers, and two *cis-trans* isomers.

7.22 With three-dimensional formulas, trace the following conversions.

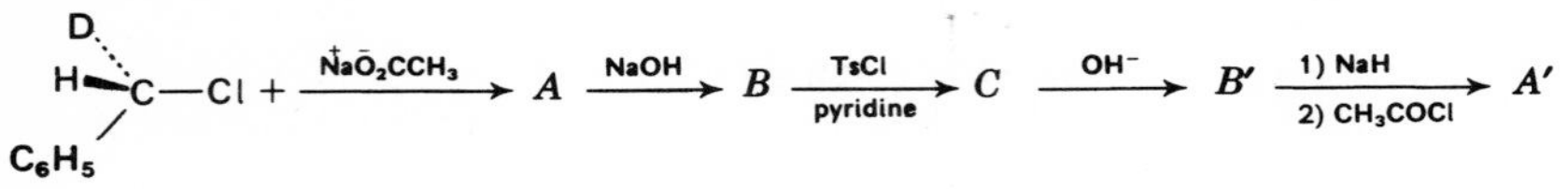

$$\underset{C_6H_5}{\overset{D}{\underset{|}{H}}}\!\!\!\!C\!-\!Cl\ +\ \xrightarrow{\text{Na}\bar{\text{O}}_2\text{CCH}_3}\ A\ \xrightarrow{\text{NaOH}}\ B\ \xrightarrow[\text{pyridine}]{\text{TsCl}}\ C\ \xrightarrow{\text{OH}^-}\ B'\ \xrightarrow[\text{2) CH}_3\text{COCl}]{\text{1) NaH}}\ A'$$

7.23 Indicate whether the following structures are enantiomeric, diastereomeric, or identical.

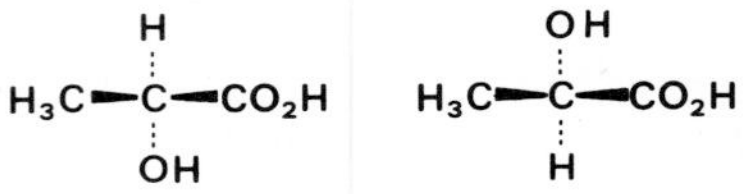

a) 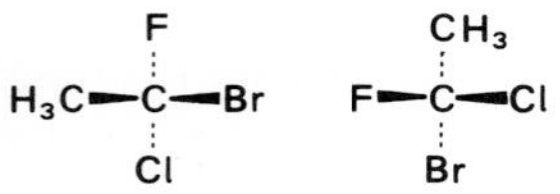 $H_3C\!-\!\underset{Cl}{\overset{F}{\underset{|}{\overset{|}{C}}}}\!-\!Br \qquad F\!-\!\underset{Br}{\overset{CH_3}{\underset{|}{\overset{|}{C}}}}\!-\!Cl$

b) $H_3C\!-\!\underset{OH}{\overset{H}{\underset{|}{\overset{|}{C}}}}\!-\!CO_2H \qquad H_3C\!-\!\underset{H}{\overset{OH}{\underset{|}{\overset{|}{C}}}}\!-\!CO_2H$

c) $CH_3CH_2\!-\!\underset{OH}{\overset{H}{\underset{|}{\overset{|}{C}}}}\!-\!CH_2\!-\!\underset{OH}{\overset{H}{\underset{|}{\overset{|}{C}}}}\!-\!CH_2CH_3 \qquad CH_3CH_2\!-\!\underset{H}{\overset{OH}{\underset{|}{\overset{|}{C}}}}\!-\!CH_2\!-\!\underset{H}{\overset{OH}{\underset{|}{\overset{|}{C}}}}\!-\!CH_2CH_3$

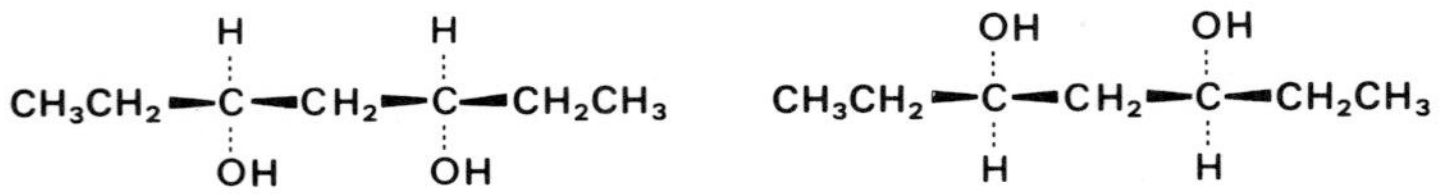

d)

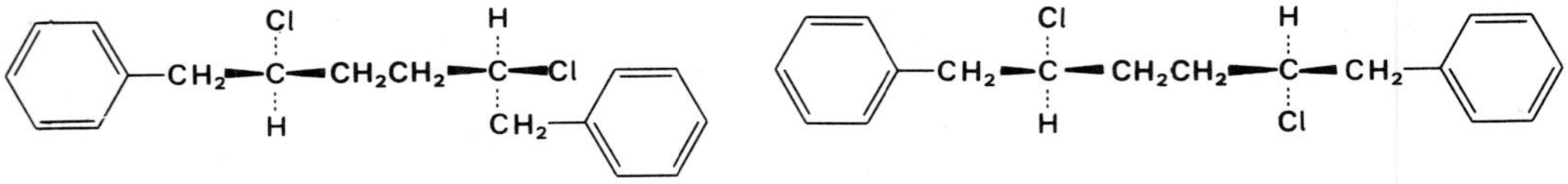

e)

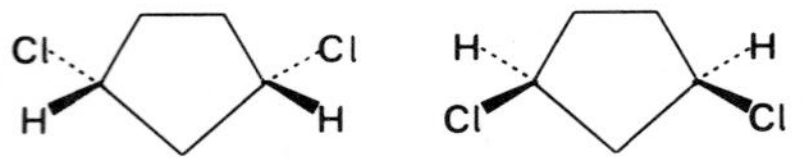

f)

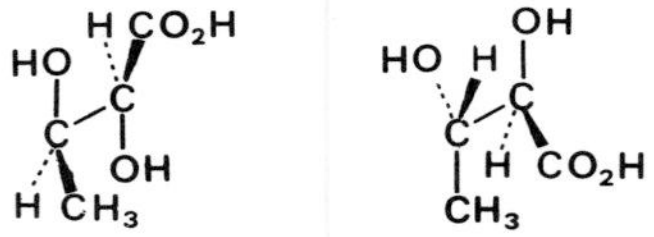

7.24 Draw two conformations of 2-chlorobutane in which a hydrogen on carbon #3 is in an anti position to the Cl. Account for the fact that 2-chlorobutane gives predominantly *trans*-2-butene over *cis*-2-butene (6:1) when treated with strong base.

Answers to the remaining problems require a combination of information from this chapter with information from previous chapters.

7.25 The reactions of *cis*- and *trans*-2,3-dimethyloxirane (2,3-epoxybutane) with aqueous sulfuric acid give 2,3-butanediol products which have zero rotation. The reaction is stereospecific. The product from the *cis*-oxirane can be resolved to (+) and (−)-enantiomers, while the product from the *trans*-oxirane cannot be resolved. Account for these findings by devising a mechanism which gives these stereochemical results.

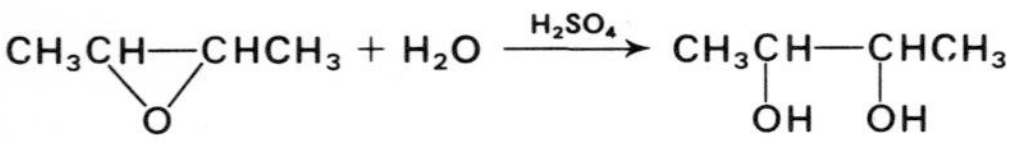

$$CH_3CH\!-\!CHCH_3\ +\ H_2O\ \xrightarrow{H_2SO_4}\ CH_3\underset{OH}{\overset{}{\underset{|}{C}}}H\!-\!\underset{OH}{\overset{}{\underset{|}{C}}}HCH_3$$

7.26 Review the series of biological reactions in Problem 6.18, in which nicotinamide was converted to methylnicotinamide cation using methionine and adenosine triphosphate (Section 6.11). Making use of the isotopes of hydrogen (hydrogen H, deuterium D, and tritium T) where needed, show what you expect the stereochemical course of each nucleophilic substitution reaction to be in this sequence.

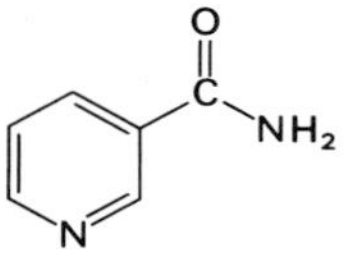

nicotinamide 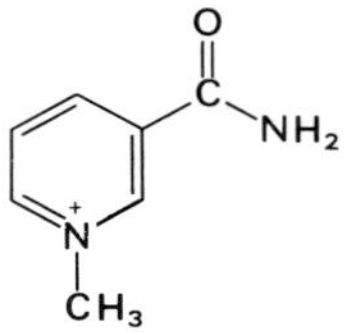   methylnicotinamide cation

7.27 In the presence of bromide ion, $(-)$-sec-butylbromide racemizes. (The rotation slowly diminishes to zero.) Describe the reaction which occurs and its stereochemical course which produces a racemized product.

7.28 Radioactive isotopes of halogens, carbon (^{14}C) and hydrogen (tritium ^{3}H), as well as the nonradioactive deuterium are often used to elucidate mechanisms of reactions. Account for the following observations.

$$CH_3CH_2CH_2CHDBr + *Br^- \longrightarrow CH_3CH_2CH_2CHD*Br + Br^-$$

When an optically active deuterated bromobutane (shown in the equation) is heated with radioactive bromide ion ($*Br^-$), the radioactive bromine is slowly incorporated into the sample. At the same time the rotation of the bromo-butane sample diminishes exactly twice as fast as the radioactive bromine is incorporated.

Carboxylic Acids and Acyl Chlorides

Of all the types of compounds, those containing a carbon–oxygen double bond enter into the widest variety of reactions of both the laboratory and biological type. The carboxylic acids and acid derivatives—esters, amides, acid halides, and anhydrides—plus the aldehydes and ketones all possess the carbon–oxygen double bond. The natural products whose principal functional groups belong to these classes include all the proteins, fats, and carbohydrates. The majority of metabolic reactions occur at these functional groups or at adjacent sites strongly influenced by them. In the laboratory, functional groups composed of or containing the carbon–oxygen double bond provide most of the chief methods of building large compounds from smaller ones.

We begin our study of compounds containing the C=O group with carboxylic acids. The carboxylic acids contain the carboxyl group in which the hydroxyl (OH) and carbonyl (C=O) groups are merged. Through the acid–base reactions, carboxylic acids control the reactions of many other functional groups. Through the reactions of acid derivatives, carboxylic acids participate in a large selection of important reactions. The patterns of physical and chemical properties produced by this combination of hydroxyl and carbonyl functional groups illustrate beautifully how simple structural theory embraces and integrates a diversity of chemical phenomena.

8.1 SOURCES AND USES OF CARBOXYLIC ACIDS

Carboxylic acids are stable compounds found in all living organisms. Acetic acid is the simplest compound that is common to the degradative and synthetic processes catalyzed by enzyme systems. It is the starting material for biosyntheses of terpenes, steroids, and long-chain fatty acids (Section 14.8); it is formed in the enzyme-catalyzed oxidation of ethanol, the product of carbohydrate fermentation (Section 5.1); and its ester is an intermediate product in the metabolism of fatty acids and of carbohydrates to carbon dioxide and water (Sections 14.9 and 14.12). The formula and models of acetic acid are shown here.

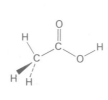
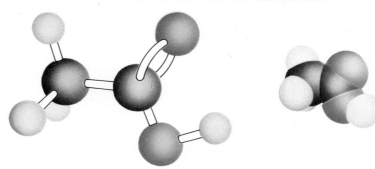

acetic acid

Large straight-chain carboxylic acids, saturated and unsaturated, e.g., octadecanoic and *cis*-9-octadecenoic acids, are obtained from

Carboxylic Acids
R—C(=O)—OH

animal and vegetable fats by hydrolysis of the ester groups (Section 9.4).

$$CH_3(CH_2)_{16}C\overset{O}{\underset{OH}{\diagup}}$$

$$CH_3(CH_2)_7{-}\overset{H}{\underset{}{C}}{=}\overset{H}{\underset{}{C}}{-}(CH_2)_7C\overset{O}{\underset{OH}{\diagup}}$$

octadecanoic acid
(stearic acid, from animal fat)

cis-9-octadecenoic acid
(oleic acid, from olive oil)

Some amino acids with the amino group in the 2-position are obtained from the hydrolysis of the amide groups of proteins (Section 10.14).

$$\underset{{}^+NH_3}{CH_2{-}CO_2^-}$$

$$\underset{{}^+NH_3}{CH_3{-}CH{-}CO_2^-}$$

glycine
(aminoacetic acid)

alanine
(2-aminopropanoic acid)

Other carboxylic acids which play an important role in metabolism are hydroxyacids, ketoacids, and dicarboxylic acids, such as lactic, pyruvic, and fumaric acids.

$$CH_3{-}\underset{OH}{CH}{-}C\overset{O}{\underset{OH}{\diagup}}$$

$$\underset{O}{\overset{H_3C}{\diagdown}}C{-}C\overset{O}{\underset{OH}{\diagup}}$$

fumaric acid structure

lactic acid
(2-hydroxypropanoic acid)

pyruvic acid
(2-oxopropanoic acid)

fumaric acid
(*trans*-butenedioic acid)

In the laboratory carboxylic acids are readily converted to all the other oxygen- and nitrogen-containing classes of compounds.

8.2 THE CARBOXYL FUNCTIONAL GROUP

The carboxyl functional group $-CO_2H$ is a combination of the carbonyl C=O and the hydroxyl OH groups. The two parts combine to produce a functional group with characteristics different from those of the carbonyl or hydroxyl groups taken alone.

The carbon–oxygen double bond is a strongly polar bond in which the carbon is electron-deficient and the oxygen is a center of high electron density and negative charge. The bond polarity can be depicted in several ways.

$$\underset{\delta^+ \quad \delta^-}{\diagup}C{=}O \qquad \underset{\longmapsto}{\diagup}C{=}O \qquad \left[\diagup C{=}O \longleftrightarrow \overset{+}{\diagup}C{-}\overset{-}{O}\right] \qquad \underset{\delta^+ \quad \delta^-}{H{-}O}\diagup\overset{C}{\underset{\delta^+}{}} \qquad \underset{\longmapsto}{H{-}O}\diagup\overset{C}{}$$

The O—H bond is strongly polar, with the oxygen partially negative and the hydrogen partially positive. The carboxyl group is an extremely polar group with four atoms having a partial charge.

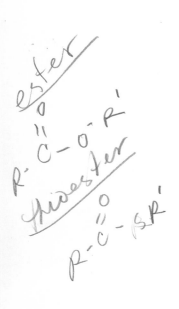

Reactions of carboxylic acids occur at these centers of charge.

Besides the carboxylic acid, other classes of compounds have an electronegative group attached to the carbonyl group. If the —OH of an acid is replaced by a —Cl, the compound is an **acid chloride**. An **ester** has an alkoxy, —OR, or aryloxy, —OAr, group in place of the —OH, and a **thioester** has an —SR. A **carboxylic anhydride** has a carboxylate group, —O$_2$CR, replacing the —OH. If an amino group, —NH$_2$ or —NHR, or —NR$_2$, replaces the —OH, the compound produced is an **amide**. These classes of compounds are known collectively as carboxylic acid derivatives, because of the similar features of their structures. They are all obtained from acids and are easily converted to acids. They all contain the group R—C=O, called an **acyl group**. General formulas for these classes are given in Fig. 8.1.

FIGURE 8.1 FUNCTIONAL GROUPS OF CARBOXYLIC ACID AND CARBOXYLIC ACID DERIVATIVES

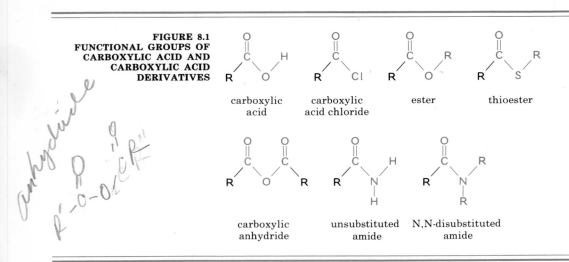

carboxylic acid

carboxylic acid chloride

ester

thioester

carboxylic anhydride

unsubstituted amide

N,N-disubstituted amide

Problem 8.1 Name and draw the structure of each functional group contained in the following compounds, whose uses are familiar.

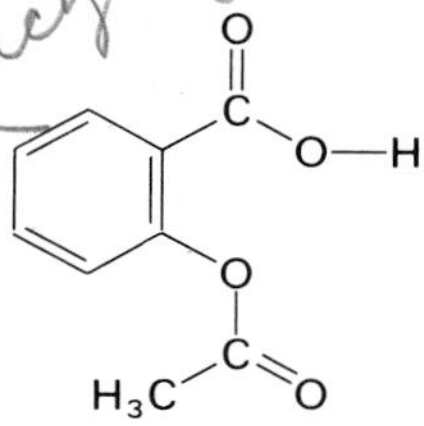

(marginal handwritten notes:) R—C (acyl group) ; alkane + oic ; -o-ic ; (2) ; unsaturated ; alkene + oic ; (3)

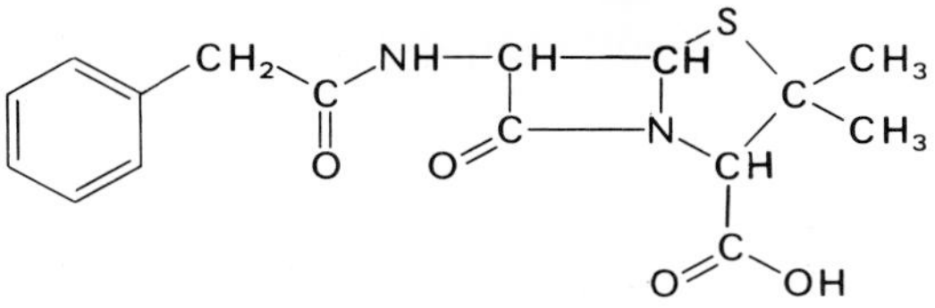

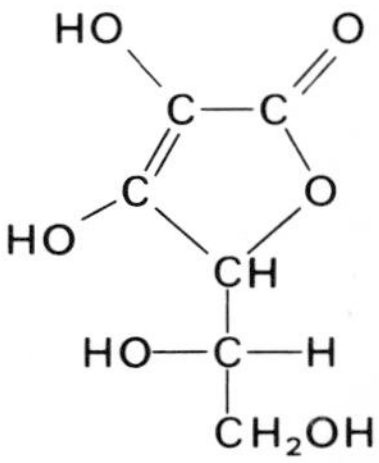

8.3 NOMENCLATURE OF CARBOXYLIC ACIDS

By the IUPAC system, the carboxyl group is the principal group in the parent compound, taking precedence over most other substituents in the name. The following rules are used in developing names for acids.

1. The names of straight-chain carboxylic acids are derived from the names of the corresponding alkanes by dropping the "e" ending and adding "oic acid" (two words). The carboxyl group must be at the beginning of the chain and no number is needed for the group.

$CH_3CH_2CH_3$ $CH_3CH_2C{\small\overset{O}{\underset{OH}{}}}$ $CH_3(CH_2)_{16}CH_3$ $CH_3(CH_2)_{16}C{\small\overset{O}{\underset{OH}{}}}$

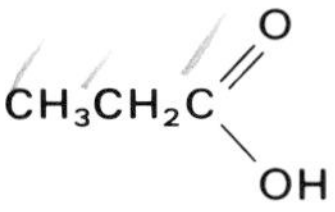

propane   propanoic acid   octadecane   octadecanoic acid

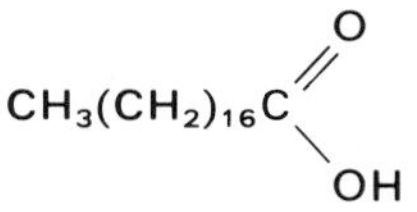

2. Other functional groups are named as substituents. The longest carbon chain containing the carboxyl group is the parent compound. The carboxyl carbon is #1. Each substituent on the chain is identified by its simplest name and a number is used to indicate its position on the chain.

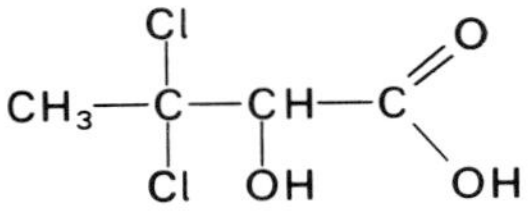

3,3-dichloro-2-hydroxybutanoic acid 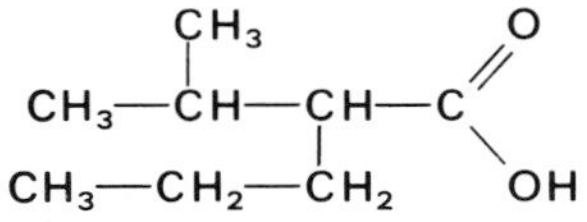   2-isopropylpentanoic acid

3. Names of unsaturated carboxylic acids are derived from the names of the corresponding alkenes and alkynes. The longest carbon chain containing both the carboxyl and the multiple bond is chosen as the parent compound. A number indicates the lower-numbered carbon of the multiple bond, with carbon #1 being the carboxyl group.

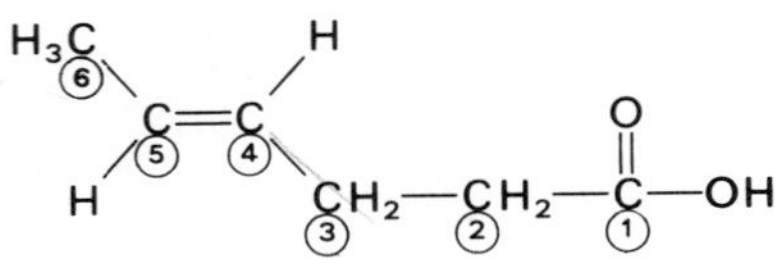

trans-4-hexenoic acid 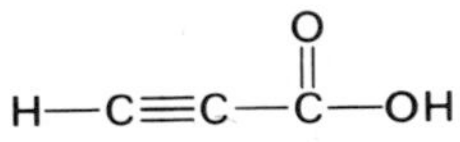   propynoic acid 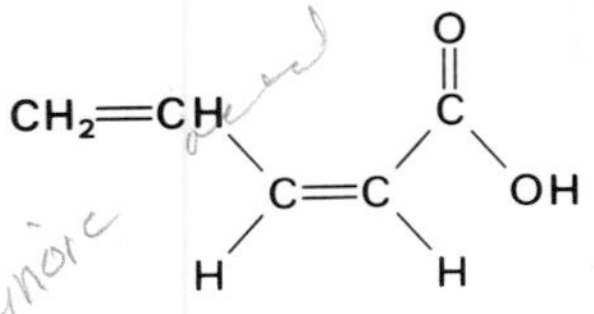   cis-2,4-pentadienoic acid

4. Cyclic acids add "carboxylic acid" to the name of the cyclic hydrocarbon. Numbering of the ring carbons starts with the carbon bearing the carboxyl group.

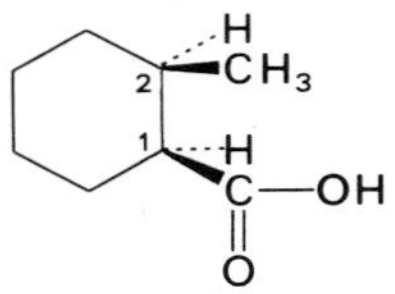

cis-2-methylcyclohexanecarboxylic acid

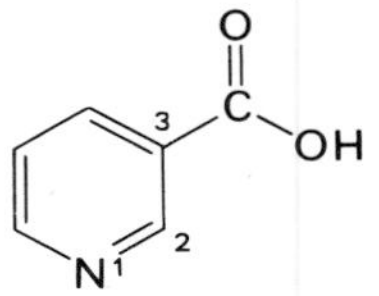

3-pyridinecarboxylic acid
(nicotinic acid)
(N is the #1 atom in the ring.)

5. Dicarboxylic acids have the ending "dioic acid." The longest chain is selected which includes both of the carboxyl carbons. Numbering starts with one of the carboxyl carbons and ends with the other. No numbers for these groups are needed since carboxyl groups must terminate chains. (Why?)

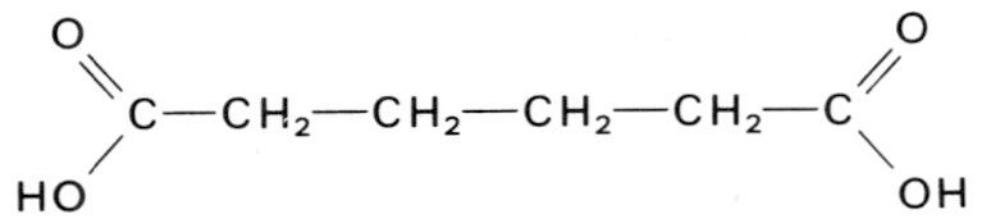

hexanedioic acid

Many small carboxylic acids are known by their common names, notably benzoic, formic, and acetic acids, and those which have biological origins such as lactic and pyruvic acids and the amino acids.

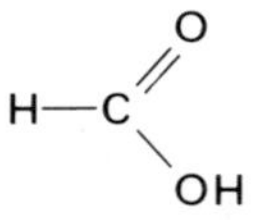

formic acid 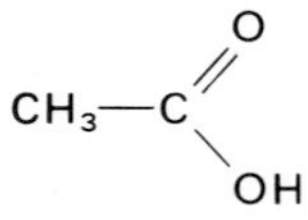   acetic acid 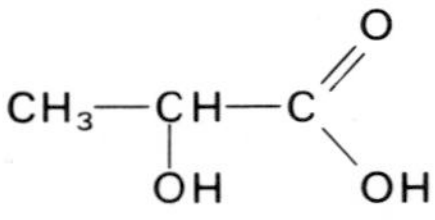   lactic acid 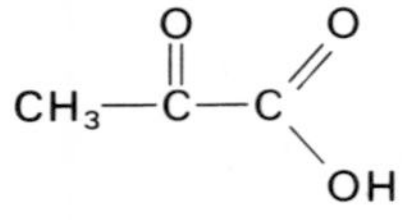   pyruvic acid

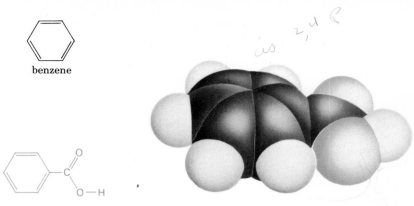

benzene

benzoic acid

While propanoic and butanoic acids are the preferred names, the names propionic and butyric are still used as well. The common and IUPAC names for some acids up to twenty carbons are given in Fig. 8.2.

	Alkane	Carboxylic acid		
FIGURE 8.2 **NAMES FOR CARBOXYLIC** **ACIDS UP TO C$_{20}$**	IUPAC name	Formula	IUPAC name	Common name
	Methane	H—CO_2H	Methanoic	Formic
	Ethane	CH_3—CO_2H	Ethanoic	Acetic
	Propane	CH_3—CH_2—CO_2H	Propanoic	Propionic
	Butane	$CH_3(CH_2)_2CO_2H$	Butanoic	Butyric
	Pentane	$CH_3(CH_2)_3CO_2H$	Pentanoic	Valeric
	Hexane	$CH_3(CH_2)_4CO_2H$	Hexanoic	Caproic
	Heptane	$CH_3(CH_2)_5CO_2H$	Heptanoic	
	Octane	$CH_3(CH_2)_6CO_2H$	Octanoic	
	Decane	$CH_3(CH_2)_8CO_2H$	Decanoic	
	Dodecane	$CH_3(CH_2)_{10}CO_2H$	Dodecanoic	
	Tetradecane	$CH_3(CH_2)_{12}CO_2H$	Tetradecanoic	
	Hexadecane	$CH_3(CH_2)_{14}CO_2H$	Hexadecanoic	Palmitic
	Octadecane	$CH_3(CH_2)_{16}CO_2H$	Octadecanoic	Stearic
	Eicosane	$CH_3(CH_2)_{18}CO_2H$	Eicosanoic	Arachidic

Problems   **8.2** Draw structures for the following compounds.

a) formic acid   b) dichloroacetic acid

c) 2-ethylbutanoic acid   d) 2,3-dihydroxypropanoic acid

e) benzoic acid

8.3 The longest carbon chain in the molecule may not contain the carboxyl group. Draw the structure for such an acid and name it.

8.4 Give systematic names for the following compounds.

a) $CH_3-\overset{\underset{|}{CH_3}}{CH}-\overset{\overset{O}{\|}}{C}-OH$

b) $CH_3-\overset{\underset{|}{OH}}{CH}-\overset{\underset{|}{CH_2CH_3}}{CH}-\overset{\overset{O}{\|}}{C}-OH$

c) $CH_3-(CH_2)_5-\overset{\underset{|}{OH}}{CH}-CH_2-\overset{\underset{|}{OH}}{CH}-(CH_2)_8-\overset{\overset{O}{\|}}{C}-OH$

d) (phenyl)$\overset{H}{\underset{}{C}}=\overset{H}{\underset{\overset{|}{\underset{\|}{C}-OH}\,O}{C}}$

e) $\overset{H}{\underset{CH_3-CH_2}{C}}=\overset{\overset{O}{\underset{\|}{C}-OH}}{\underset{H}{C}}$

8.5 Write structures for the following.

a) *cis*-2-butenoic acid

b) 2-isopropylhexanoic acid

c) 3-phenyl-2-bromopropanoic acid

d) *trans*-4-*tert*-butylcyclohexanecarboxylic acid

e) Because of its size the *tert*-butyl group is nearly always in the equatorial position of cyclohexane. Draw the chair conformation of (d).

<table>
<tr><td>**8.4 PHYSICAL
PROPERTIES OF
CARBOXYLIC ACIDS**</td><td>Aliphatic carboxylic acids are colorless liquids or low-melting white, waxy solids. Simple aromatic acids are white crystalline solids which sublime at moderate temperatures.</td></tr>
</table>

Aliphatic carboxylic acids are colorless liquids or low-melting white, waxy solids. Simple aromatic acids are white crystalline solids which sublime at moderate temperatures.

The boiling points of acids are higher than those of most other compounds of comparable molecular weight, because of carboxyl-to-carboxyl group hydrogen bonding. (See next section.) The high polarity of the functional group and its potentialities for hydrogen bonding make the small carboxylic acids soluble in water. Pure acetic acid (mp 17°, bp 118°) is a good solvent of medium polarity. It is also mildly corrosive to the skin.

The more volatile acids, acetic and propanoic, have a pungent, biting odor. The C_4 to C_6 acids have a rancid butter odor, and the phenyl-acetic acid odor is characteristic of horses. Carboxylic acids provide the sour taste associated with certain foods. Acetic acid is the chief component of vinegar, citric acid is found in citrus fruits, and tartaric acid in grapes.

(handwritten margin note: b.p. higher than same m.w. that do not hydrogen bond)

$HO-\overset{\underset{|}{CH_2-CO_2H}}{\underset{\overset{|}{CH_2-CO_2H}}{C}}-CO_2H$

citric acid

$\overset{HO-CH-CO_2H}{\underset{HO-CH-CO_2H}{}}$

tartaric acid

$C_6H_5-CH_2-CO_2H$

phenylacetic acid

long chain insoluble because of long C chain

Acids having more than four carbons for each carboxyl group are only slightly soluble in water. The nonpolar hydrocarbon portion of the molecule is too incompatible with water for easy dissolution.

8.5 HYDROGEN BONDING For carboxylic acids and amides, hydrogen bonding is a dominant feature in their physical properties. The data for boiling points given in Fig. 8.3 show the high values for acids and amides as compared with compounds of similar molecular weight in other classes.

FIGURE 8.3 COMPARISON OF BOILING POINTS FOR SELECTED COMPOUNDS, SHOWING DEPENDENCE ON DIPOLAR INTERACTIONS INCLUDING HYDROGEN BONDING

Class	Compound	Formula	MW	bp, °C
Amide	Acetamide	$CH_3-\overset{\overset{O}{\|\|}}{C}-NH_2$	59	221
Acid	Acetic acid	$CH_3-\overset{\overset{O}{\|\|}}{C}-OH$	60	118
Alcohol	1-Propanol	$CH_3-CH_2-CH_2-OH$	60	97
Ketone	Acetone	$CH_3-\overset{\overset{O}{\|\|}}{C}-CH_3$	58	56
Aldehyde	Propanal	$CH_3-CH_2-\overset{\overset{O}{\|\|}}{C}-H$	58	49
Ester	Methyl formate	$CH_3-O-\overset{\overset{O}{\|\|}}{C}-H$	60	31
Hydrocarbon	Butane	$CH_3-CH_2-CH_2-CH_3$	58	−0.5

Two hydrogen bonds often hold two molecules of acid together. In benzene, a solvent of low polarity, acetic acid exists as a dimer. It also is dimeric in the absence of a solvent. Pure acetic acid is called glacial acetic acid, because it crystallizes just below room temperature (17°).

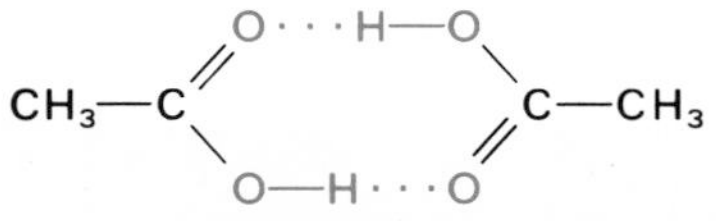

In dilute solution in water or ethanol, hydrogen bonding of acetic acid with the solvent replaces the carboxyl–carboxyl interaction.

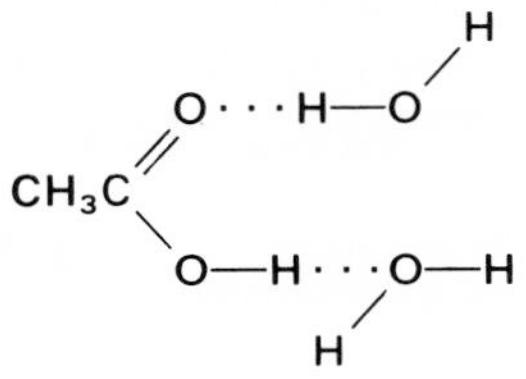

The amide–amide hydrogen bonding involves more than two molecules which accounts for the much higher melting and boiling points of acetamide (mp 82° and bp 221°) as compared with those of acetic acid (17° and 118°).

Problem 8.6 Draw a series of amide groups H—N—C=O, with each hydrogen bonded to the nitrogen of one molecule and to the oxygen of a second molecule.

8.6 SALT FORMATION OF CARBOXYLIC ACIDS

The carboxyl group is a proton donor to ammonia, amines, and all bases stronger than these, e.g., hydroxide and alkoxide ions.

$$CH_3-C\underset{OH}{\overset{O}{<}} + NH_3 \rightleftharpoons CH_3-C\underset{O^-}{\overset{O}{<}} + NH_4^+$$

Recall that the acid–base proton exchange is a rapidly reversible reaction, with the position of equilibrium between reactants and products being dependent on the relative strengths of the acids and bases involved (Section 3.8).

Proton transfers between oxygen and nitrogen atoms of functional groups are exceedingly rapid. The carboxylic acids and their conjugate bases, the carboxylate ions, play an important role in determining and maintaining the proper acidity in certain reaction media, particularly those of a biological nature. This role would be impossible unless proton transfers were extremely fast.

In the laboratory, strong acids like sulfuric acid and hydronium ion and moderately strong acids like acetic acid catalyze reactions. Nature's catalysts, enzymes, have side-chain ammonium ion groups which serve as proton donors and carboxylate ion groups which act as proton acceptors. The involvement of carboxyl and carboxylate groups in enzyme activity was illustrated in the enzyme-catalyzed hydration of fumaric acid (Section 7.11), where the $-CO_2^-$ ion accepted a proton from the protonated alcohol formed in the reaction.

8.7 THE ACIDITY OF CARBOXYLIC ACIDS

Since proton transfers from oxygen and nitrogen to oxygen and nitrogen are extremely fast and reversible, the tendency to donate and accept protons is expressed in terms of the position of equilibrium established and not in terms of the rate of proton transfer. The quantitative measure of the acidity of carboxylic acids is stated in terms of the equilibrium constant (K_a) for the proton transfer from the acid to the solvent. The standard solvent and proton acceptor is water.

$$RCO_2H + H_2O \rightleftharpoons RCO_2^- + H_3O^+ \qquad K_a = \frac{[RCO_2^-][H_3O^+]}{[RCO_2H](1)}$$

The equilibrium constant K_a is expressed in terms of the concentrations of products over concentrations of reactants. The concentration of the solvent (in this case water) is essentially constant and is set equal to unity. With K_a values we are able to make comparisons of the acidity of various carboxylic acids and determine the effect of changes in molecular structure on this property.

The acidity equilibrium constant for acetic acid in water at 25° is $K_a = 1.8 \times 10^{-5}$. In any other solvent the equilibrium constant changes depending upon the base strength of the solvent. Since H^+ itself is of very high energy, it must be covalently bound to the solvent, hence the basicity of the solvent greatly affects the K_a. Because both the carboxylic acid and its anion are solvated, the polarity of the solvent and its solvating properties also seriously affect K_a.

A more convenient means for expressing the strength of acids is in terms of pK_a units, defined as the negative of the logarithm of the equilibrium constant.

$$pK_a = -\log_{10} K_a$$

This expression gives a pK_a value for acetic acid of 4.8. On this scale a lower pK_a value represents a stronger acid (a greater tendency to donate a proton), and a higher value indicates a weaker acid.

Carboxylic acids are stronger proton donors than are alcohols. The great effect of the $=O$ on the acidity of the OH group attached to the same carbon is due to the delocalization of electrons and charge of the carboxylate ion, the product of the proton transfer.

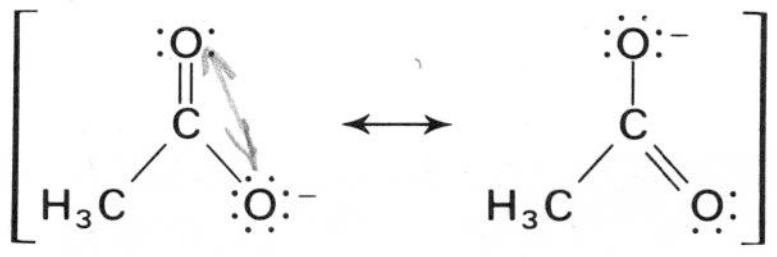

acetate ion

An ion or molecule whose electrons or charge can delocalize has a lower energy than would be expected from the calculated energy of one of its contributing resonance structures alone. In an equilibrium reaction, the lowering of the energy of one of the components due to delocalization of electrons and charge results in a larger equilibrium concentration of that component than would otherwise exist. For this reason the stability afforded by the electron delocalization of the acetate ion is the dominant factor that makes acetic acid a stronger acid than ethanol. The pK_a of ethanol against water as base is 16, as compared with 4.8 for acetic acid.

Problem **8.7** Write the equations for the following reactions.

    a) aqueous hydrochloric acid and sodium acetate

    b) formic acid and sodium methoxide

    c) butanoic acid and ethylamine

Acid equilibrium constants are easily determined by instrumental methods, and small differences due to structural changes can be measured.

The effect of substituting an electronegative group, such as chlorine, for hydrogen on the carbon chain in acetic acid is shown in the pK_a values given in Fig. 8.4.

FIGURE 8.4
VALUES OF pK_a FOR
CARBOXYLIC ACIDS

Acid	pK_a	Acid	pK_a
CH_3CO_2H	4.75	HCO_2H	3.75
$ClCH_2CO_2H$	2.85	$CH_3CH_2CO_2H$	4.87
Cl_2CHCO_2H	1.48	$CH_3CHClCO_2H$	2.83
Cl_3CCO_2H	0.70	$ClCH_2CH_2CO_2H$	3.98

The effect of chlorine substitution in increasing the ability of the acid to donate a proton is accounted for by the polarity of the Cl—C bond, which is in opposition to the direction of polarity of the CO_2H group in chloroacetic acid.

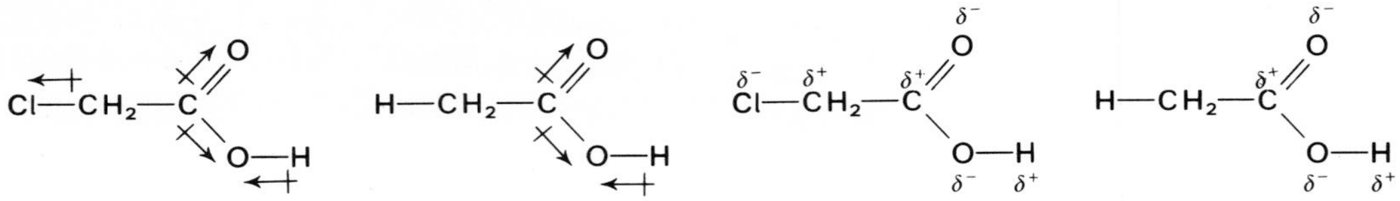

[whole molecule decreased stability] (handwritten margin note)

[ions increased stability] (handwritten margin note)

Two partially positive centers on adjacent carbon atoms destabilize a molecule and make it more reactive. This effect partially explains why chloroacetic acid is a slightly stronger acid than acetic acid.

The complete picture requires consideration of the effect of Cl on the chloroacetate ion as well, since the acid and its ion are in equilibrium with one another. The partially positive carbon attached to chlorine attracts electrons and thus compensates for some of the negative charge of the carboxylate group. As a result, the stability of the anion as a whole increases.

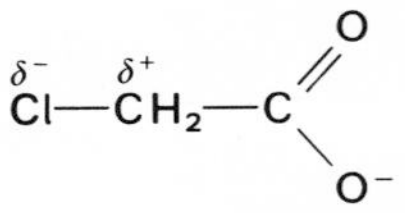

The increased stability of the anion coupled with the decreased stability of the acid results in an overall shift in equilibrium for chloroacetic acid that favors the anion. In other words, the chloro acid is stronger and has a lower pK_a (2.85) than the unsubstituted acid (4.75).

A strongly polarized bond to carbon affects the polarity of other bonds to the same atom, and even of bonds to carbon further along the chain. This phenomenon is called the **inductive effect**. Bond polarity in one place of a molecule "induces" bond polarity elsewhere in the molecule. The effect drops off rapidly with an increase in the number of bonds it must pass through.

Electronegative groups such as Cl, OH, NH$_2$, have an electron-withdrawing inductive effect. This effect weakens in the carboxylic acid as a Cl is moved away from the functional group. Additional chlorines increase the effect. Carboxylic acids carrying electron-withdrawing substituents are stronger than their parent unsubstituted acids. For example, trifluoroacetic acid is a very strong acid whose pK_a is less than zero (K_a is greater than 1).

An inductive effect that operates in the opposite direction is visible in the pK_a values for formic acid (3.75) and acetic acid (4.75). The methyl group of acetic acid is more electron-releasing than the hydrogen of formic acid. Thus the formate ion is more stable relative to formic acid than is acetate ion relative to acetic acid. Although acetic acid is slightly stronger than propanoic acid (pK_a = 4.87), the pK_a difference, 0.19, is much less than the difference between formic and acetic acids. This fact illustrates the decrease in inductive effect with increased distance. The inductive effect of the CH$_3$ group is opposite to that of chlorine. Alkyl groups show an electron-releasing inductive effect relative to hydrogen, resulting in a slight destabilization of the anions. The inductive effects of various alkyl groups are nearly equal to one another and are very small (Fig. 8.4).

Problem **8.8** Arrange the following bases in order of decreasing base strength.

Br$^-$ OH$^-$ CH$_3$CH$_2$CO$_2^-$ Cl$_3$CCO$_2^-$ NH$_3$ F$_3$CCO$_2^-$ CH$_3$CO$_2^-$ HCO$_2^-$

8.9 Arrange the following acids in order of decreasing acid strength in each group.

a) CH$_3$—CH—CO$_2$H   CH$_3$—CH—CO$_2$H   CH$_3$—CH—CO$_2$H
          | | |
          OH   F   Br

CH$_2$—CH$_2$—CO$_2$H
|
Br

b) H$_3$N$^+$—CH$_2$—CO$_2$H   HO—CH$_2$—CO$_2$H   HS—CH$_2$—CO$_2$H

8.9 CONVERSION OF CARBOXYLIC ACIDS TO ACID CHLORIDES

The conversion of carboxylic acids to their acid chlorides is accomplished through the use of inorganic acid chlorides. The two most often used are thionyl chloride SOCl$_2$ and phosphorus trichloride PCl$_3$, the same reagents that convert alcohols to alkyl chlorides (Section 5.5). The reaction of propanoic acid is given with each reagent.

[handwritten margin notes: RXNs w/ an O=C—OH needs SOCl$_2$ PCl$_3$]

$$CH_3CH_2\overset{\displaystyle O}{\overset{\|}{C}}\!-\!OH + SOCl_2 \longrightarrow CH_3CH_2\overset{\displaystyle O}{\overset{\|}{C}}\!-\!Cl + SO_2 + HCl$$

$$3\,CH_3CH_2\overset{\displaystyle O}{\overset{\|}{C}}\!-\!OH + PCl_3 \longrightarrow 3\,CH_3CH_2\overset{\displaystyle O}{\overset{\|}{C}}\!-\!Cl + P(OH)_3$$

propanoic acid   propanoyl chloride   phosphorous acid

In these reactions the bond attaching the hydroxyl group to the carbonyl carbon is broken and is replaced by a new carbon–chlorine bond.

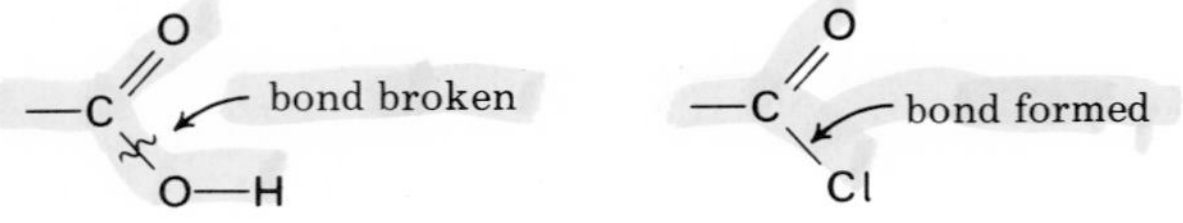

Another reaction of carboxylic acids, the acid-catalyzed reaction of carboxylic acids with alcohols to give esters, is treated in Section 9.11.

Problem   **8.10** Write the equations for these reactions.

     a) benzoic acid and phosphorus trichloride

     b) 2-hydroxypentanoic acid and thionyl chloride

8.10 METHODS OF SYNTHESIS OF CARBOXYLIC ACIDS

Carboxylic acids may be obtained from many other classes of compounds using a variety of types of reagents. The most obvious are the hydrolyses of naturally occurring acid derivatives, esters and amides, to give the corresponding acids.

Straight-chain fatty acids of even-numbered carbons, between 14 and 18 particularly, are obtained by treatment of the animal and vegetable fats with sodium hydroxide (Section 9.14). Saturated and certain unsaturated acids having the double bonds in the center of the chains are available from this source.

Nitriles react with water in an acid-catalyzed reaction to yield the corresponding acid.

$$CH_3CH_2C\!\equiv\!N + 2\,H_2O \xrightarrow{H_3O^+} CH_3CH_2CO_2H + NH_4^+$$

propanonitrile   propanoic acid

The hydrolysis of nitriles actually consists of two reactions, the first an acid-catalyzed addition of water to the nitrile to give the amide, followed in a second reaction by hydrolysis of the amide to the acid and ammonium ion (Section 10.11).

$$CH_3CH_2C{\equiv}N + H_2O \xrightarrow{H_3O^+} CH_3CH_2\overset{\overset{\displaystyle O}{\|}}{C}-NH_2 \xrightarrow{H_3O^+} CH_3CH_2CO_2H + NH_4^+$$

propanonitrile   propanamide   propanoic acid

In the reactions of alkyl halides you learned that nitriles are obtained from the substitution of a primary alkyl halide by sodium cyanide (Section 6.3). By this route it is possible to change the alkyl halide to an acid having one more carbon in the chain. Since the alkyl halide is made from the alcohol, this becomes a route by which an alcohol is converted to a carboxylic acid of one more carbon.

$$R-OH \xrightarrow{SOCl_2} R-Cl \xrightarrow{NaCN} R-CN \xrightarrow{H_3O^+} R-CO_2H$$

Another means of extending the carbon chain is discussed in detail in Chapter 14, on carbon chain-building reactions (Section 14.3). It involves the conversion of an alkyl halide (primary, secondary, or tertiary) to an **alkylmagnesium halide**, a reactive intermediate compound.

$$\overset{\overset{\displaystyle CH_3}{|}}{CH_3CH}-Br + Mg \xrightarrow{dry\ ether} \overset{\overset{\displaystyle CH_3}{|}}{CH_3CH}-Mg-Br$$

isopropyl   metallic   isopropylmagnesium
bromide   magnesium   bromide

The alkylmagnesium halide (called a **Grignard reagent**) reacts with carbon dioxide to form the magnesium salt of the carboxylic acid. The new carbon–carbon bond is formed where the carbon–bromine bond was in the original alkyl bromide. The carboxylate salt is acidified to give the acid.

$$\overset{\overset{\displaystyle CH_3}{|}}{CH_3CH}-Mg-Br + CO_2 \longrightarrow \overset{\overset{\displaystyle H_3C}{|}}{CH_3CH}-\overset{\overset{\displaystyle O}{\|}}{C}-O-Mg-Br \xrightarrow{H_3O^+} \overset{\overset{\displaystyle H_3C}{|}}{CH_3CH}-\overset{\overset{\displaystyle O}{\|}}{C}-OH + Mg^{2+} + Br^-$$

isopropylmagnesium   (dry ice)   bromomagnesium   2-methylpropanoic
bromide   2-methylpropanoate   acid

Common strong inorganic oxidizing agents, such as acidified sodium dichromate and potassium permanganate, oxidize primary alcohols and aldehydes to carboxylic acids having the same carbon skeleton.

$$3\ CH_3CH_2CH_2-OH + 2\ Cr_2O_7^{2-} + 16\ H^+ \longrightarrow 3\ CH_3CH_2CO_2H + 4\ Cr^{3+} + 11\ H_2O$$

1-propanol   propanoic acid

Problem 8.11 Draw the structures of each starting compound and product and give the reagents needed for each step of the following syntheses of a carboxylic acid.

a) benzyl alcohol → benzyl chloride → phenylacetonitrile → phenylacetic acid

b) *tert*-butyl chloride → *tert*-butylmagnesium chloride → 2,2-dimethyl-propanoic acid

c) 1-pentanol → pentanoic acid

8.11 GREEK-LETTER NOTATION AND NOMENCLATURE

For convenience, a Greek-letter notation frequently is used to designate the relative positions of two functional groups in a molecule. The letters α alpha, β beta, γ gamma, δ delta, ε epsilon, and ω omega (last) indicate that one functional group is located on the first, second, third, fourth, or last carbon from the other functional group. Some examples are given below.

$$\overset{\beta}{CH_3}-\overset{\alpha}{CH}-CO_2H$$
$$\underset{NH_2}{|}$$

α-amino acid

$$\overset{\gamma}{CH_3}-\overset{\beta}{C}-\overset{\alpha}{CH_2}-CO_2H$$
$$\underset{O}{\parallel}$$

β-ketoacid

$$\overset{\beta}{CH_3}-\overset{\alpha}{CH}-\overset{}{C}-\overset{\alpha}{CH_3}$$
$$\underset{OH}{|}\ \underset{O}{\parallel}$$

α-hydroxyketone

$$\overset{\gamma}{CH_3}-\overset{\beta}{CH}=\overset{\alpha}{CH}-\overset{}{C}-CH_3$$
$$\underset{O}{\parallel}$$

α,β-unsaturated ketone

Note that the Greek letters *do not correspond* to the numbering system for systematic nomenclature. For a carboxylic acid where the carboxyl carbon is #1, the α-carbon is #2.

An older system of nomenclature for acids and aldehydes uses Greek letters instead of numbers for substituents.

$$CH_3-\overset{OH}{\underset{|}{CH}}-CH_2-CO_2H$$

3-hydroxybutanoic acid
(β-hydroxybutyric acid)

The Greek letter notation is the most useful in conversational nomenclature and in making generalizations about the reactivity of certain carbons and substituents located in positions relative to the functional groups. For example, β-ketoacids readily lose CO_2 (Section 8.20).

8.12 HYDROXYACIDS

Carboxylic acids with hydroxyl substituents possess the properties of both alcohols and acids. Where the hydroxyl group is in a position close to the carboxyl, the hydroxyacids undergo unique reactions.

In the presence of a strong acid, 4-hydroxy- and 5-hydroxyacids form internal esters, called **lactones**. These stable five- and six-membered rings form readily whenever a catalytic amount of acid is present.

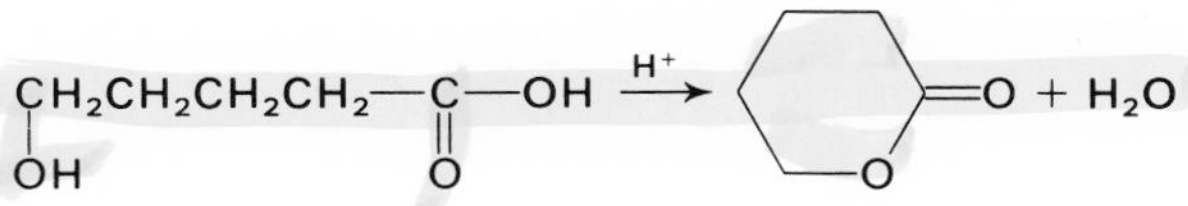

$$CH_2CH_2CH_2CH_2-C-OH \xrightarrow{H^+} \text{(ring)} =O + H_2O$$

with OH below and =O below the C.

5-hydroxypentanoic acid   δ-valerolactone
(tetrahydro-2-pyrone)

Under acidic conditions a 3-hydroxyacid (β-hydroxyacid) loses a molecule of water to form an unsaturated acid with a double bond primarily in the 2,3-position (α,β-unsaturated acid). This type of reaction is encountered frequently and is called a **β-elimination**.

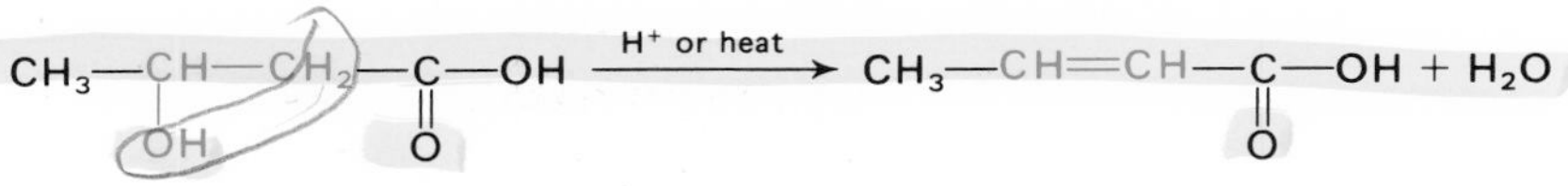

$$CH_3-CH-CH_2-C-OH \xrightarrow{H^+ \text{ or heat}} CH_3-CH=CH-C-OH + H_2O$$

with OH below the CH and O below the C.

3-hydroxybutanoic acid   2-butenoic acid
(β-hydroxybutyric acid)

The reverse of β-elimination is also a reaction which occurs readily. The α,β-unsaturated ester or acid is converted to a β-hydroxyester or acid by an acid-catalyzed addition of water.

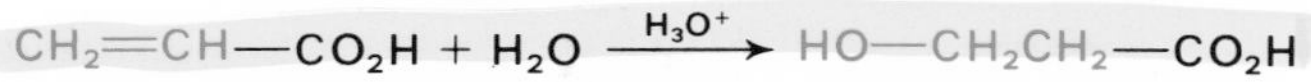

$$CH_2=CH-CO_2H + H_2O \xrightarrow{H_3O^+} HO-CH_2CH_2-CO_2H$$

propenoic acid   3-hydroxypropanoic acid
(acrylic acid)   (β-hydroxypropionic acid)

A high temperature encourages the elimination of water to give the unsaturated acid. A large quantity of water encourages the addition.

Some 2- and 3-hydroxyacids that are important natural products are shown here.

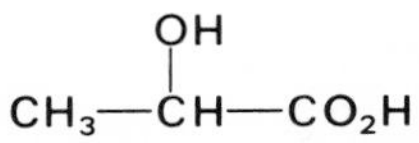

$$CH_3-CH-CO_2H \qquad HO-CH-CO_2H \qquad \begin{matrix} CH_2-CO_2H \\ HO-C-CO_2H \\ CH_2-CO_2H \end{matrix}$$

with OH above the CH in lactic acid; HO—CH—CO₂H above HO—CH—CO₂H in tartaric acid.

lactic acid 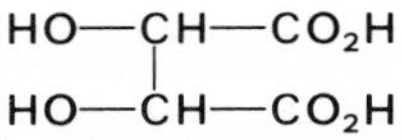   tartaric acid 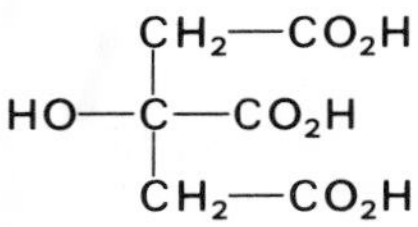   citric acid

Citric acid is the starting compound for a series of reactions by which small metabolic compounds are converted to CO_2 and H_2O. The series is known as the **citric acid cycle** (see Section 14.12).

8.13 PROSTAGLANDINS   In recent years a group of compounds called **prostaglandins** have been isolated and synthesized, and their functions have been investigated. The prostaglandins are a family of 20-carbon carboxylic acids with hydroxyl and sometimes carbonyl groups and a five-membered ring in the middle of the chain.

prostaglandin E$_2$ prostaglandin F$_{2\alpha}$

Prostaglandins are found in minute quantities in nearly all mammalian tissues and fluids. Although the mechanism of their action is not well understood, it is clear that they are involved at the cellular level in regulating many functions, including gastric acid secretion, contraction and relaxation of smooth muscles, inflammation and vascular permeability, body temperature, food intake, and blood platelet aggregation. For example, the antiinflammatory action of aspirin has been related to the reduction of prostaglandin, which causes the inflammation. One prostaglandin induces labor, one relieves breathing difficulties caused by asthma, another lowers blood pressure and reduces gastric acid formation.

The development of modified prostaglandins shows promise for drug use. The synthetic compounds are not inactivated as quickly and show greater tissue specificity, thus producing fewer undesirable side effects.

The biosynthesis of prostaglandins is from the C$_{20}$ unsaturated fatty acid, 8,11,14-eicosatrienoic acid (arachidonic acid).

arachidonic acid

8.14 ACID CHLORIDES   Acid chlorides, RCOCl, can be converted easily into esters and amides. They are valuable only because they show great reactivity in these transformations, and can serve as intermediates in the conversion of the less reactive carboxylic acids to more reactive classes of compounds. The driving force in making the highly reactive acid chlorides from carboxylic acids is the very high reactivity of thionyl chloride relative to its reaction products, SO$_2$ and HCl (shown in Section 8.9).

Acid chlorides cannot be readily prepared from any compounds other than their parent carboxylic acids. Although the simple acid chlorides can be purchased, the more complex compounds are prepared and used directly. Formyl chloride is inherently unstable, and has not been prepared. Acid halides are also known as **acyl halides**, and are referred to as **acylating agents**, since they transfer acyl groups R—C=O, from one electronegative group to another.

Because of their ready reaction with water, acid chlorides must be protected from moisture in the air during preparation and manipulation. Due to the reaction with water to produce carboxylic acids and hydrogen chloride, acid chlorides have a sharp, unpleasant odor, and their fumes irritate the skin and eyes.

Since there is usually no advantage in the use of acid fluorides or bromides, the cheaper chlorides are preferred.

8.15 NOMENCLATURE OF ACID CHLORIDES

The name of an acid chloride is based directly on the name of the corresponding acid. In the IUPAC system, the "oic acid" is dropped from the name of the acid and "oyl chloride" is added. The carboxyl contains the #1 carbon and the substituents are numbered in the same manner as for acids.

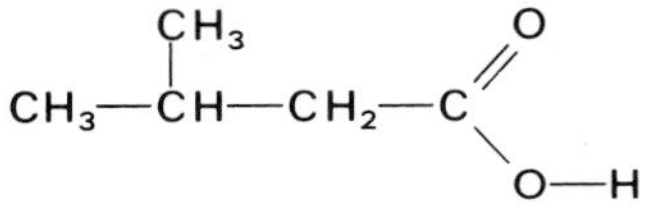

oic dropped
oyl chloride added

CH₃—CH(CH₃)—CH₂—C(=O)—O—H

3-methylbutanoic acid

CH₃—CH(CH₃)—CH₂—C(=O)—Cl

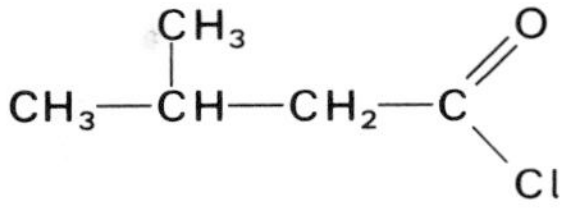

3-methylbutanoyl chloride

For acids which are usually called by their common names, the acid chloride name is derived from the common name by dropping the "ic acid" and adding "yl chloride."

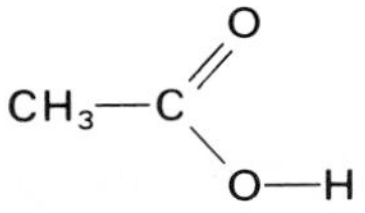

CH₃—C(=O)—O—H

acetic acid

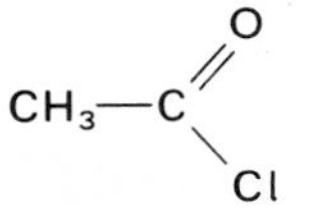

CH₃—C(=O)—Cl

acetyl chloride

CH₃—C(=O)—

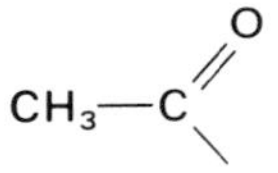

acetyl group
(an acyl group)

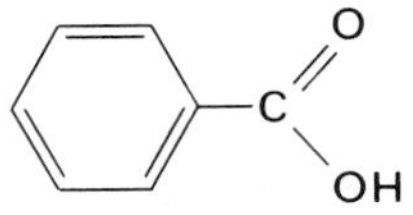

C₆H₅—C(=O)—OH

benzoic acid

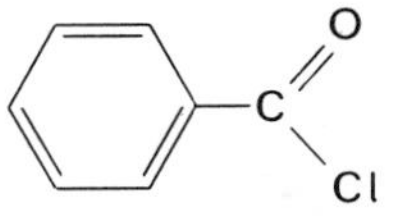

C₆H₅—C(=O)—Cl

benzoyl chloride

C₆H₅—C(=O)—

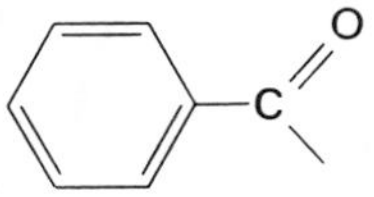

benzoyl group

Note the distinction between an acid chloride and an alkyl chloride. In an acid or *acyl chloride*, the chlorine atom is bonded to a carbonyl group, while in an *alkyl chloride*, the bond is to a saturated hydrocarbon group. It is this difference in structure that accounts for their difference in reactivities.

8.16 REACTIONS OF ACID CHLORIDES TO FORM ESTERS

Acid chlorides are very reactive toward nucleophilic reagents. The reaction of an acid chloride with an alcohol gives an ester and hydrogen chloride. The carbon–chlorine bond of the acid chloride and the oxygen–hydrogen bond of the alcohol are broken and a new carbon–

O‖ȱ-Cl + COH → ester + HCl

oxygen bond is made joining the two reactants into one compound, the ester.

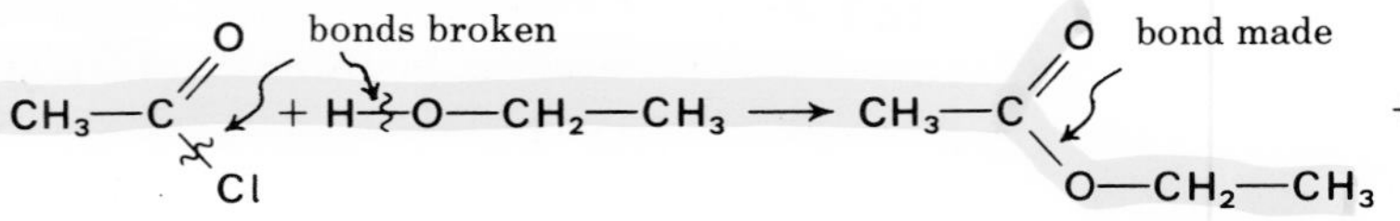

$$CH_3-C\overset{O}{\underset{Cl}{\big\backslash}} + H\overset{\textstyle |}{-}O-CH_2-CH_3 \longrightarrow CH_3-C\overset{O}{\underset{O-CH_2-CH_3}{\big\backslash}} + HCl$$

bonds broken   bond made

acetyl chloride   ethanol   ethyl acetate   hydrogen chloride

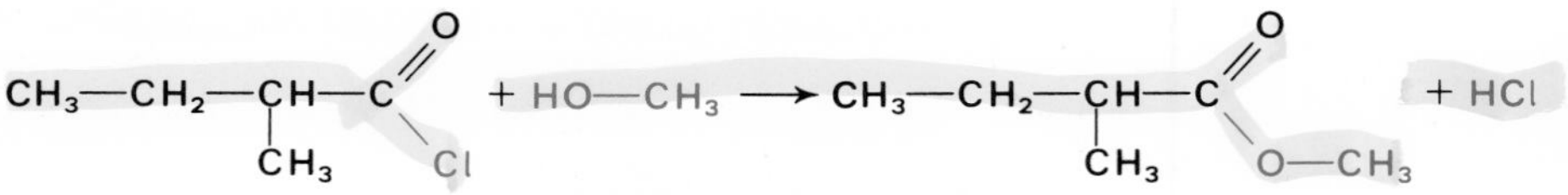

$$CH_3-CH_2-\underset{\underset{CH_3}{|}}{CH}-C\overset{O}{\big\backslash}_{Cl} + HO-CH_3 \longrightarrow CH_3-CH_2-\underset{\underset{CH_3}{|}}{CH}-C\overset{O}{\big\backslash}_{O-CH_3} + HCl$$

2-methylbutanoyl chloride   methanol   methyl 2-methylbutanoate

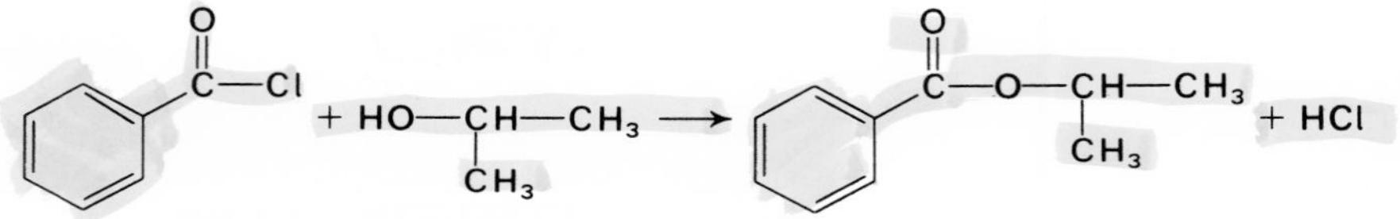

$$\underset{}{\bigcirc}\overset{O}{\underset{}{C}}-Cl + HO-\underset{\underset{CH_3}{|}}{CH}-CH_3 \longrightarrow \underset{}{\bigcirc}\overset{O}{\underset{}{C}}-O-\underset{\underset{CH_3}{|}}{CH}-CH_3 + HCl$$

benzoyl chloride   2-propanol   isopropyl benzoate

The yield of ester from acid chloride in these reactions is almost 100%, if water is carefully excluded. The hydrogen chloride escapes as a gas after the solution is quickly saturated. If triethylamine is also added to the reaction mixture, it neutralizes the hydrogen chloride produced.

$$(CH_3CH_2)_3N + HCl \longrightarrow (CH_3CH_2)_3\overset{+}{N}H \quad Cl^-$$

triethylamine   triethylammonium chloride

Phenol, behaving like an aromatic alcohol, reacts easily with 2-methylpropanoyl chloride to produce phenyl 2-methylpropanoate and hydrogen chloride.

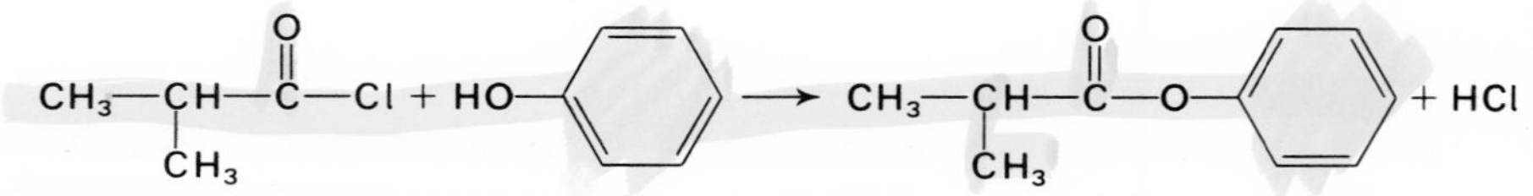

$$CH_3-\underset{\underset{CH_3}{|}}{CH}-\overset{O}{\underset{}{C}}-Cl + HO-\bigcirc \longrightarrow CH_3-\underset{\underset{CH_3}{|}}{CH}-\overset{O}{\underset{}{C}}-O-\bigcirc + HCl$$

2-methylpropanoyl chloride   phenol   phenyl 2-methylpropanoate

8.12 Draw the structures and name the alcohols and acid chlorides needed to make these esters.

a) 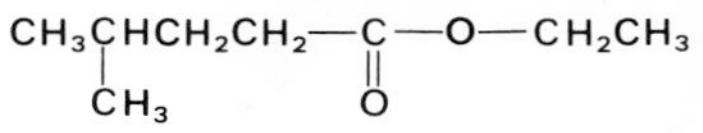 CH₃CHCH₂CH₂—C—O—CH₂CH₃   **b)**
       | ‖
       CH₃   O

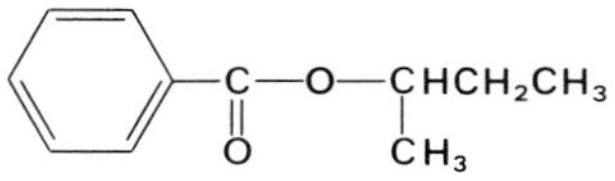

c) CH₃C—O—⬡
      ‖
      O

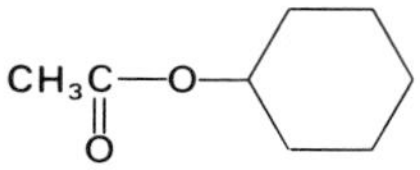

8.17 MECHANISM OF THE SUBSTITUTION REACTIONS OF ACID CHLORIDES AND OTHER ACID DERIVATIVES

The conversion of one acid derivative into another is a nucleophilic substitution reaction. The overall reaction involves an entering group Nu: which substitutes for a leaving group L:, both of which have electronegative atoms. Although both the entering and leaving groups possess unshared electron pairs and are therefore nucleophiles, the entering group is usually the stronger nucleophile.

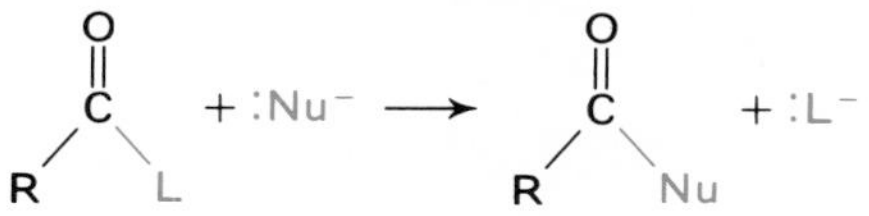

$$
\underset{R}{\overset{O}{\underset{\|}{C}}}\diagdown L + :Nu^- \longrightarrow \underset{R}{\overset{O}{\underset{\|}{C}}}\diagdown Nu + :L^-
$$

The polar bond of the carbonyl group C=O gives a carbon with partial positive charge.

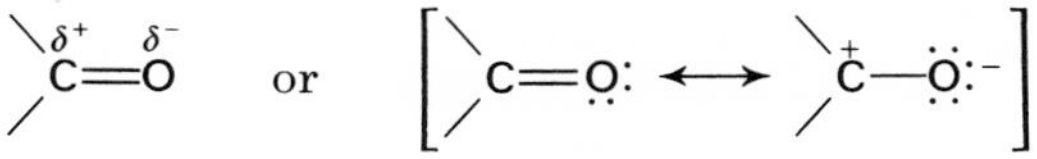

$$
\overset{\delta^+}{\underset{}{C}}=\overset{\delta^-}{O} \quad \text{or} \quad \left[\diagup\!\!\!\!C=\ddot{\underset{..}{O}}: \longleftrightarrow \overset{+}{\underset{}{C}}-\ddot{\underset{..}{O}}:^- \right]
$$

The electron-deficient carbon attracts a nucleophile with a pair of electrons available to form a new bond. The actual substitution reaction proceeds in two steps, an addition to the unsaturated carbon and an elimination to reform the carbon–oxygen double bond.

Step 1   Addition of the nucleophile to C=O

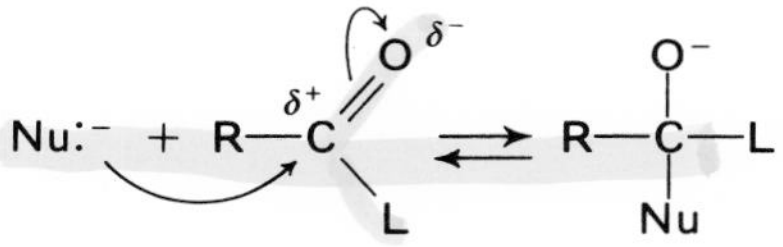

$$
\text{Nu:}^- + R\overset{\delta^+}{-}C\underset{L}{\overset{\overset{\displaystyle O}{\|}^{\delta^-}}{}} \rightleftharpoons R-\underset{\underset{Nu}{|}}{\overset{\overset{O^-}{|}}{C}}-L
$$

The tetrahedral adduct contains both the nucleophile and the leaving group.

nucleophile   acid derivative   the adduct (a short-lived reaction intermediate)

The first step is the formation of a new bond between the nucleophilic reagent and the electron-deficient carbon of the carbonyl group. One pair of electrons of the C=O bond is shifted to the oxygen, which then has a negative charge.

Step 2 Elimination of the leaving group

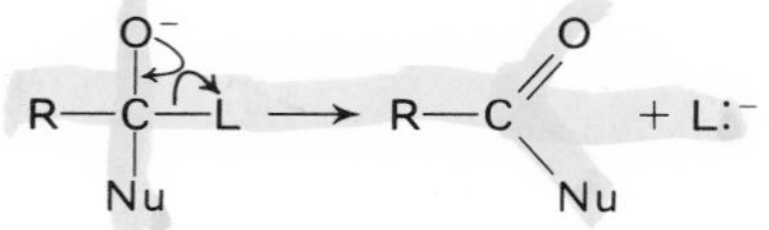

intermediate	new acid
adduct	derivative

The tetrahedral carbon is converted back to a trigonal carbon (carbon with bonds to three atoms only).

The bond between carbon and the leaving group is broken, with the bonding electrons going with the leaving group. The carbon–oxygen double bond is reformed. The negative charge on the oxygen in the adduct is carried away by L^-.

The first step of the reaction, in many cases, has been demonstrated to be reversible. The reaction intermediate has only a short lifetime and usually cannot be isolated or detected by direct means. Note that the adduct has the structure of a compound or ion which you learned earlier is unstable, that is, one having a combination of halogen, hydroxyl, alkoxyl, or amino groups bonded to the same carbon (Section 5.2). The unstable intermediate adduct forms a more stable compound by losing either the attacking nucleophile (reverse of step 1) or the group originally present (step 2).

The mechanism above is formulated using a negatively charged nucleophile which displaces a negatively charged leaving group. Most reactions of acyl chlorides, however, proceed with the less reactive neutral nucleophilic reagents displacing the negative chloride ion. The adduct intermediate formed is electrically neutral, but with a separation of plus and minus charges. Loss of the negative leaving group in step 2 leaves a protonated product which must lose a proton.

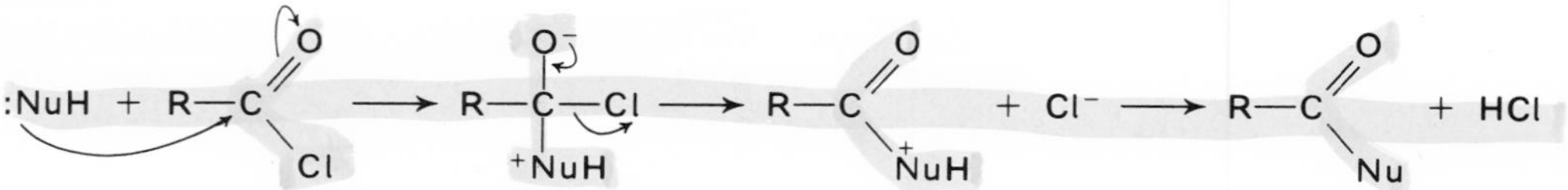

	neutral adduct	protonated product	neutral product

The steps of the mechanism of the reaction of acetyl chloride with ethanol are described below.

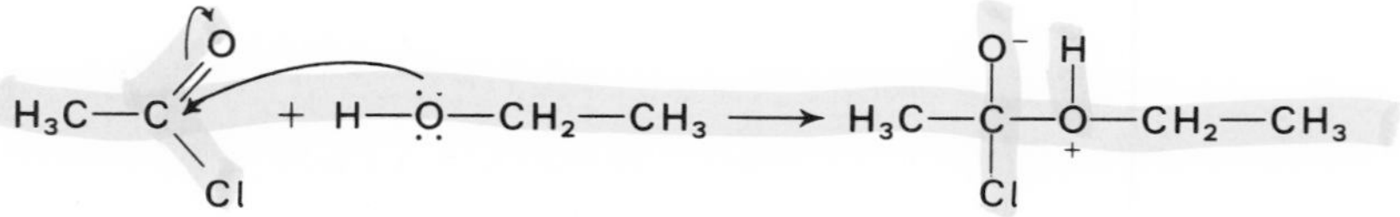

acetyl chloride	ethanol (neutral nucleophile)	reaction intermediate (electrically neutral)

$CH_3-\overset{\overset{\displaystyle O^-}{|}}{\underset{\underset{\displaystyle Cl}{}}{C}}-\overset{+}{O}-CH_2-CH_3 \longrightarrow Cl^- + H_3C-\overset{\overset{\displaystyle O}{\parallel}}{C}-\overset{+}{\underset{\underset{\displaystyle H}{}}{O}}-CH_2-CH_3 \longrightarrow H_3C-\overset{\overset{\displaystyle O}{\parallel}}{C}-O-CH_2-CH_3 + HCl$

leaving group   protonated ester   ethyl acetate
departs   (neutral ester)

Problem 8.13 a) Write the overall equation for the reaction of phenol with hexanoyl chloride.

b) Write the steps for the mechanism of the reaction.

8.18 REACTIONS OF ACID CHLORIDES TO FORM AMIDES

Ammonia and amines are excellent nucleophiles. When ammonia or an amine that contains at least one hydrogen on nitrogen reacts with an acid chloride, an amide or a substituted amide is formed. Ammonia and 2-methylbutanoyl chloride yield 2-methylbutanamide and hydrogen chloride.

$CH_3-CH_2-\overset{\overset{\displaystyle}{}}{\underset{\underset{\displaystyle CH_3}{|}}{CH}}-\overset{\overset{\displaystyle O}{\parallel}}{C}-Cl + 2\ NH_3 \longrightarrow CH_3-CH_2-\underset{\underset{\displaystyle CH_3}{|}}{CH}-\overset{\overset{\displaystyle O}{\parallel}}{C}-NH_2 + NH_4^+Cl^-$

2-methylbutanoyl chloride   2-methylbutanamide

The hydrogen chloride reacts with ammonia also, forming ammonium chloride. In order to have enough ammonia for amide formation, two moles of ammonia must be used for each mole of acid chloride.

The reaction of the acid chloride with ammonia is frequently carried out by careful addition of the acid chloride to a concentrated aqueous solution of ammonia. Although the acid chloride can also react with water, its reaction with the ammonia is so much faster that the yield of amide is high.

The reaction of an amine with an acid chloride gives an amide whose nitrogen has an attached carbon group. The amine also reacts with the hydrogen chloride to form a salt. The reaction of 2-butenoyl chloride with *tert*-butylamine gives N-*tert*-butyl-2-butenamide and *tert*-butyl-ammonium chloride.

$CH_3-CH=CH-\overset{\overset{\displaystyle O}{\parallel}}{C}-Cl + 2\ NH_2-C(CH_3)_3 \longrightarrow$

2-butenoyl   *tert*-butyl-
chloride   amine

$CH_3-CH=CH-\overset{\overset{\displaystyle O}{\parallel}}{C}-NH-C(CH_3)_3 + (CH_3)_3C-NH_3^+ \quad Cl^-$

N-*tert*-butyl-   *tert*-butyl-
2-butenamide   ammonium
                                  chloride

The amide can have one or two alkyl or aryl substituents on the nitrogen. This means that amines having one or two alkyl or aryl groups can react to produce stable amides—a process which requires the loss of one hydrogen from nitrogen. An amine with three alkyl groups and no hydrogen cannot form a stable amide, only an amide salt which cannot be isolated.

$$CH_3-\overset{\overset{\textstyle O}{\|}}{C}-Cl + 2\underset{\substack{\text{N-methyl-}\\ \text{aniline}}}{\langle\text{phenyl}\rangle-NH-CH_3} \longrightarrow CH_3-\overset{\overset{\textstyle O}{\|}}{C}-\underset{\underset{\textstyle CH_3}{|}}{N}-\langle\text{phenyl}\rangle + \langle\text{phenyl}\rangle-\overset{\overset{\textstyle H}{|}}{\underset{\underset{\textstyle CH_3}{|}}{\overset{+}{N}}}-H\ Cl^-$$

| acetyl chloride | N-methyl-aniline | N-methyl-N-phenylacetamide | N-methylanilinium chloride |

$$CH_3-\overset{\overset{\textstyle O}{\|}}{C}-Cl + (CH_3CH_2CH_2CH_2)_3N \rightleftharpoons CH_3-\overset{\overset{\textstyle O}{\|}}{C}-\overset{+}{N}(CH_2CH_2CH_2CH_3)_3\ Cl^-$$

tributylamine   (unstable species)

This difference in behavior (and others to be considered later) requires that amines be distinguished according to the number of carbon groups attached to nitrogen. An amine with one alkyl or aryl group and two hydrogens on the nitrogen is a **primary amine**, e.g., ethylamine. A **secondary amine** has two carbon groups on nitrogen and only one hydrogen, such as N-methylaniline. A compound with three alkyl groups and no hydrogen on the nitrogen is a **tertiary amine**, tributylamine.

$$CH_3CH_2-NH_2 \qquad \langle\text{phenyl}\rangle-NH-CH_3 \qquad (CH_3CH_2CH_2CH_2)_3N$$

| ethylamine | N-methylaniline | tributylamine |
| (a primary amine) | (a secondary amine) | (a tertiary amine) |

Note carefully that this use of *primary*, *secondary*, and *tertiary* for amines is quite different from the designation of primary, secondary, and tertiary alcohols and alkyl halides. For alcohols and alkyl halides, the distinction rests on the structure of *the one alkyl* group in the compound, because the reactivity of the alcohol or halide depends upon the number of carbon groups bonded to the carbon bearing the OH or Cl. In the case of amines, the reaction depends upon the number of groups attached to the nitrogen and very little on the structure of these groups.

Problem **8.14** Write the structures for the products of these reactions.

a) 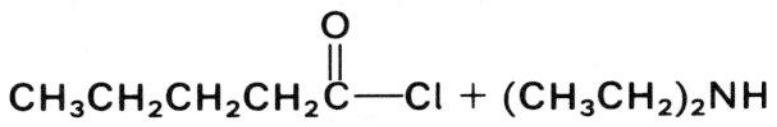 $CH_3CH_2CH_2CH_2\overset{\overset{\displaystyle O}{\|}}{C}-Cl + (CH_3CH_2)_2NH$

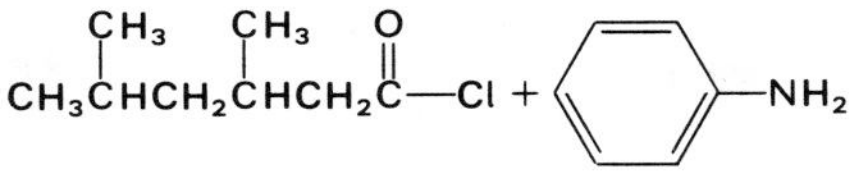

b) $CH_3\overset{\overset{\displaystyle CH_3}{|}}{C}HCH_2\overset{\overset{\displaystyle CH_3}{|}}{C}HCH_2\overset{\overset{\displaystyle O}{\|}}{C}-Cl +$ $-NH_2$

c) $\overset{\overset{\displaystyle O}{\|}}{C}-Cl + CH_3\overset{\overset{\displaystyle CH_3}{|}}{C}H-NH_2$

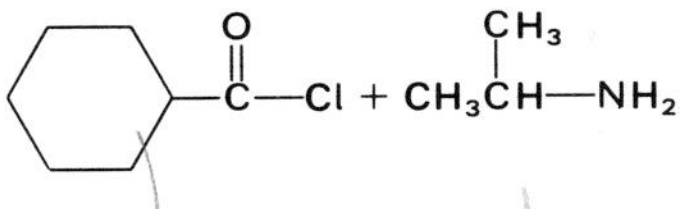

8.19 OTHER SUBSTITUTION REACTIONS OF ACYL CHLORIDES

The reaction of the acid chloride with a thiol to give a thioester is quite similar to the ester formation discussed above. Butanoyl chloride and 2-propanethiol produce isopropyl butanethioate and hydrogen chloride.

$$CH_3CH_2CH_2\overset{\overset{\displaystyle O}{\|}}{C}-Cl + HS-\overset{\overset{\displaystyle CH_3}{|}}{C}HCH_3 \longrightarrow CH_3CH_2CH_2\overset{\overset{\displaystyle O}{\|}}{C}-S-\overset{\overset{\displaystyle CH_3}{|}}{C}HCH_3 + HCl$$

butanoyl chloride   2-propanethiol   isopropyl butanethioate

Thioesters are not ordinarily made and used in the laboratory since their properties offer little advantage over those of oxygen esters. In biological systems, however, the formation of oxygen esters nearly always involves the use of thioesters of the thiol called coenzyme A. These reactions are described in Section 9.16. Other biological reactions in which the carboxyl functional group of a compound is present as a thioester of coenzyme A are illustrated in Chapter 14.

Acid anhydrides are formed when acid chlorides react with carboxylate ions. For example, propanoyl chloride and sodium propanoate give propanoic anhydride and sodium chloride.

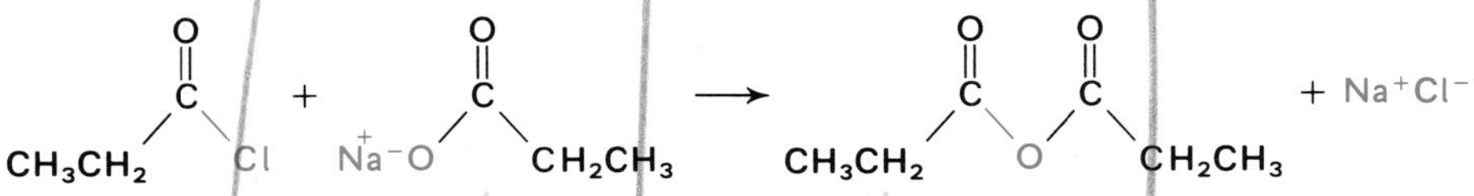

propanoyl chloride   sodium propanoate   propanoic anhydride

Carboxylic acid anhydrides are similar in their reactions to acid chlorides. Some special examples of acid anhydrides are described in Section 8.21.

The reactions of acyl chlorides are summarized in Fig. 8.5.

FIGURE 8.5
SCHEME OF REACTIONS
OF ACYL CHLORIDES

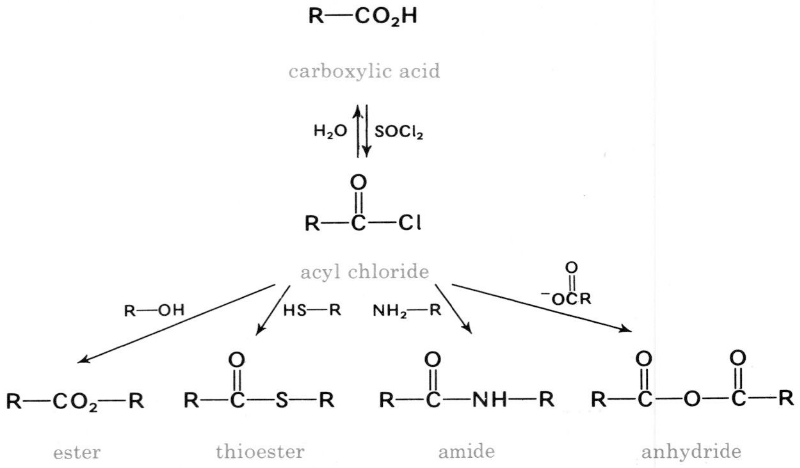

$$R—CO_2H$$

carboxylic acid

$$H_2O \updownarrow SOCl_2$$

$$R—\overset{\displaystyle O}{\overset{\|}{C}}—Cl$$

acyl chloride

$$R—OH \qquad HS—R \quad NH_2—R \qquad {}^-O\overset{\displaystyle O}{\overset{\|}{C}}R$$

$$R—CO_2—R \quad R—\overset{\displaystyle O}{\overset{\|}{C}}—S—R \quad R—\overset{\displaystyle O}{\overset{\|}{C}}—NH—R \quad R—\overset{\displaystyle O}{\overset{\|}{C}}—O—\overset{\displaystyle O}{\overset{\|}{C}}—R$$

ester   thioester   amide   anhydride

Problems

8.15 Write equations for reactions of the following compounds.

   a) 4-chlorobutanoyl chloride + 1-butanethiol

   b) benzoyl chloride + *tert*-butyl alcohol + triethylamine

   c) acetyl chloride + sodium acetate

8.16 Write the stepwise mechanism for the reaction of acetyl chloride with sodium acetate.

8.17 Write the equation for each of the following reactions.

   a) 2,3-diethylpentanoyl chloride + aniline, $C_6H_5—NH_2$

   b) 3-methylbutanoyl chloride + ammonia

   c) propanoyl chloride + N-methylpentylamine

8.20 DICARBOXYLIC ACIDS

Compounds containing two carboxyl groups are known as dicarboxylic acids. The IUPAC names of dicarboxylic acids take the name of the hydrocarbon chain containing both carboxyl groups and substitute "dioic acid" for the "e" ending of the hydrocarbon. The common names of some of these acids are more generally used than the systematic ones. The common names should be learned, since the acids occur frequently in the laboratory syntheses and biological reactions. The ancient mnemonic, "Oh, my, such good apple pie," is derived from the first letters of the common names of the saturated dicarboxylic acids—oxalic, malonic, succinic, glutaric, adipic, and pimelic.

$$HO—\overset{\displaystyle O}{\overset{\|}{C}}—OH \qquad\qquad HO—\overset{\displaystyle O}{\overset{\|}{C}}—\overset{\displaystyle }{\underset{\underset{\displaystyle O}{\|}}{C}}—OH$$

carbonic acid   oxalic acid
(unstable)   (ethanedioic)

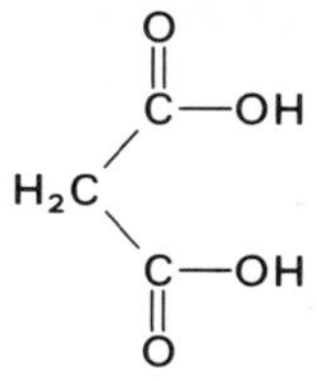

malonic
acid
(propanedioic)

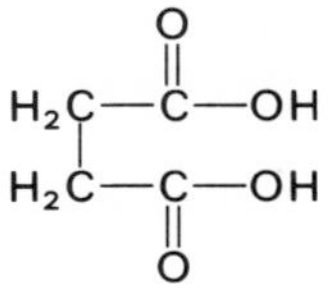

succinic acid
(butanedioic)

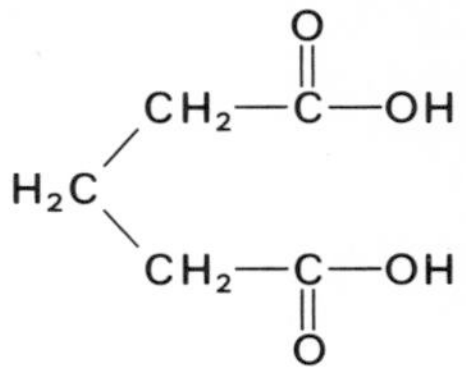

glutaric acid
(pentanedioic)

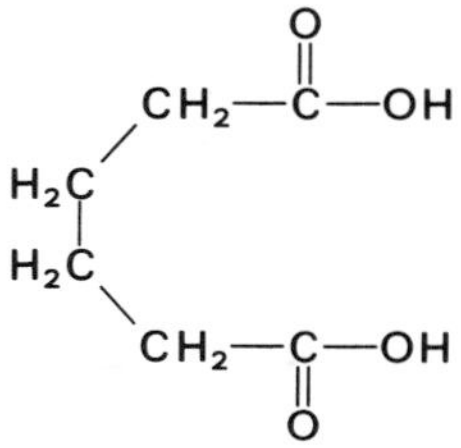

adipic acid
(hexanedioic)

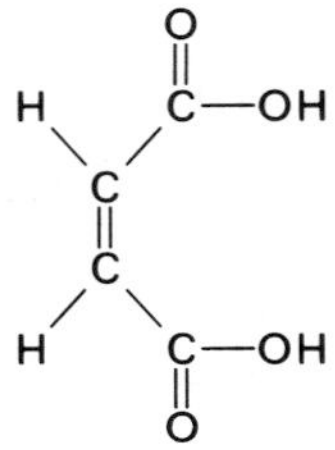

maleic acid
(*cis*-butenedioic)

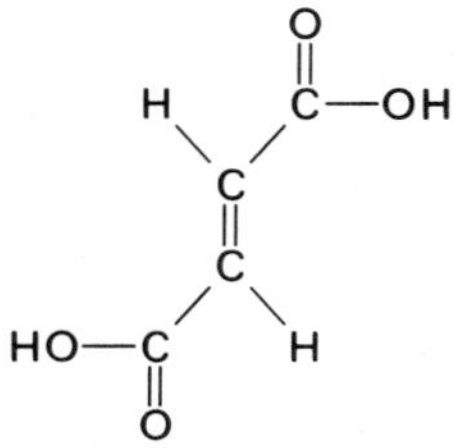

fumaric acid
(*trans*-
butenedioic)

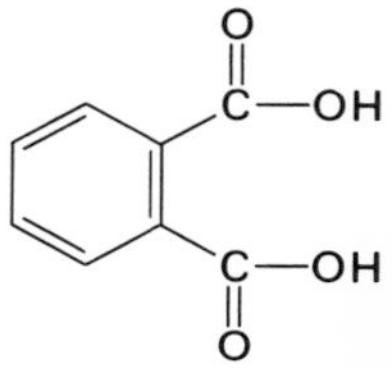

phthalic acid
(1,2-benzene-
dicarboxylic)

In these small molecules, the two carboxyl groups are not sufficiently isolated to act independently. Compounds with two functional groups on adjacent or nearby carbons frequently have properties that result from the interaction of the two groups. A striking example is the behavior of two of the dicarboxylic acids when heated. Succinic and glutaric acids lose a molecule of water when heated to form cyclic anhydrides containing five- and six-membered rings. The temperatures are high enough to drive the reactions to completion by loss of the water as a gas.

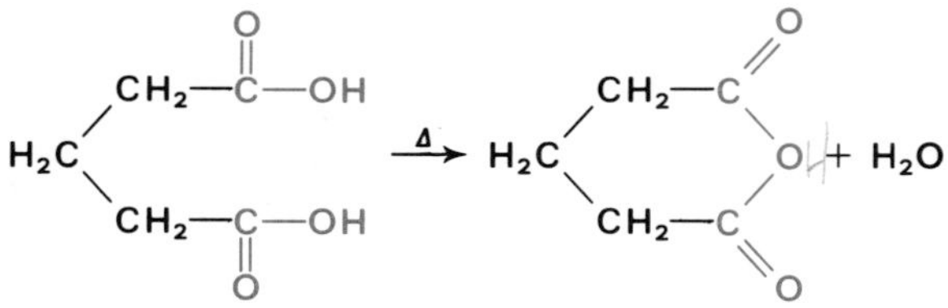

glutaric acid

glutaric anhydride

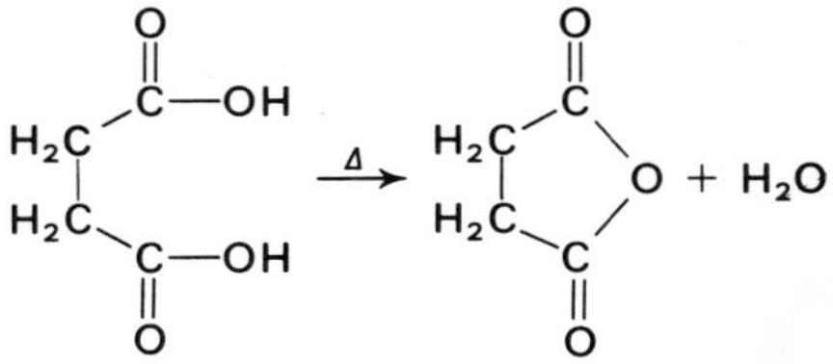

succinic acid

succinic anhydride

Five- and six-membered rings are unstrained (have bond angles similar to open-chain compounds) and form readily. Such rings close when conditions favor the interaction of two groups connected by a chain of proper length.

Problem   **8.18** Explain why maleic acid (*cis*-butenedioic acid) forms a cyclic anhydride when heated to 140°, whereas fumaric acid (*trans*-butenedioic acid) is stable at that temperature and does not lose water.

Carboxylic acids which have another carbonyl group as CO_2H or $C=O$ on the β-carbon are unstable and lose carbon dioxide when heated. Malonic acid and substituted malonic acids are easily decarboxylated to monocarboxylic acids.

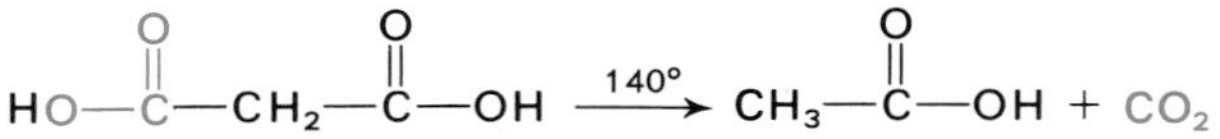

$$HO-\overset{O}{\underset{||}{C}}-CH_2-\overset{O}{\underset{||}{C}}-OH \xrightarrow{140°} CH_3-\overset{O}{\underset{||}{C}}-OH + CO_2$$

     malonic acid   acetic acid

Ketone groups in the β-position to a carboxyl group also destabilize the acid, and β-ketoacids are easily decarboxylated to methyl ketones.

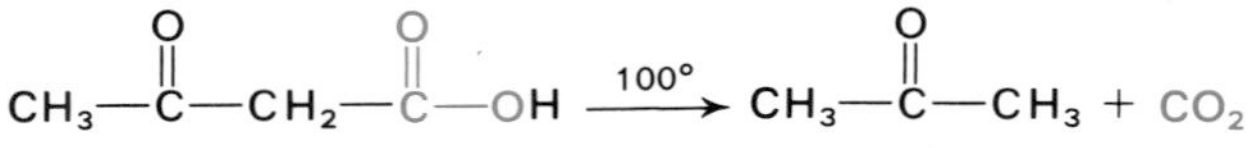

$$CH_3-\overset{O}{\underset{||}{C}}-CH_2-\overset{O}{\underset{||}{C}}-OH \xrightarrow{100°} CH_3-\overset{O}{\underset{||}{C}}-CH_3 + CO_2$$

    acetoacetic acid   acetone
    (a β-keto acid)

In the oxidation of biological compounds in metabolism to carbon dioxide and water, the decarboxylation of substituted malonic acids and of β-ketoacids is the principal method of producing carbon dioxide from a larger compound. Many of the metabolic reactions have as their final stage the production of β-carbonyl acids which lose CO_2 readily by enzyme catalysis, as illustrated in Section 14.10.

8.21 CARBOXYLIC ACID ANHYDRIDES

The name anhydride means "without water." When carboxylic acid anhydrides react with water, they give only carboxylic acids as products.

Carboxylic acid anhydrides of dicarboxylic acids and aromatic acids are solids. Those of other monocarboxylic acids are liquids. Acetic anhydride is a corrosive liquid.

Only acetic anhydride and certain cyclic anhydrides find general use. Cyclic anhydrides with five- or six-membered rings form readily when the diacids are heated and give off water (Section 8.20). The important cyclic anhydrides are succinic, glutaric, phthalic, and maleic anhydrides.

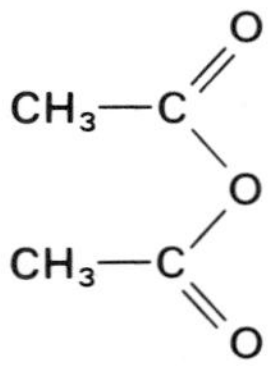

    acetic anhydride 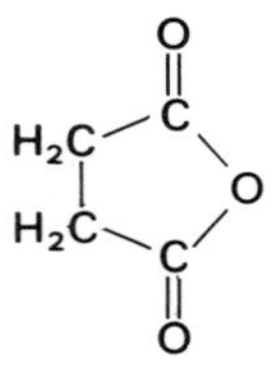   succinic anhydride 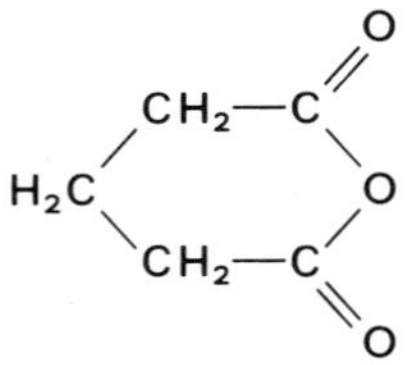   glutaric anhydride phthalic anhydride   maleic anhydride

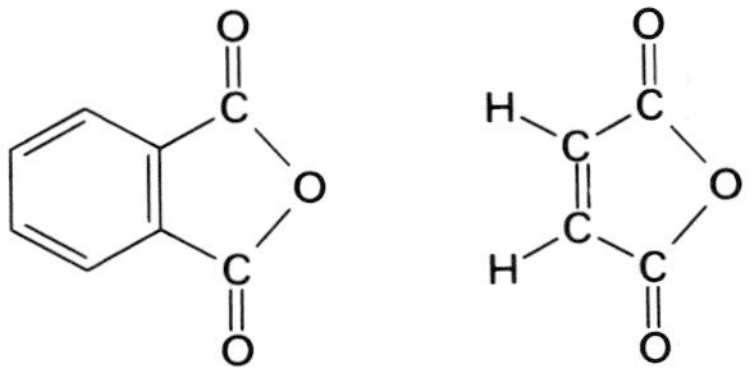

The reactions of nucleophiles with anhydrides produce the same compounds they give with acid chlorides. The carboxylate ion is the leaving group and the carboxylic acid is the other product. The reactivity of anhydrides is somewhat less than that of acid chlorides, though they are still the second most reactive acid derivative.

$$CH_3-\overset{\overset{O}{\|}}{C}-O-\overset{\overset{O}{\|}}{C}-CH_3 + CH_3-OH \longrightarrow CH_3-\overset{\overset{O}{\|}}{C}-O-CH_3 + HO-\overset{\overset{O}{\|}}{C}-CH_3$$

acetic anhydride   methanol   methyl acetate   acetic acid

Problem   8.19 Write the equation for each of the following reactions.

a) maleic anhydride + 1-propanol   **b)** succinic anhydride + ammonia

New Terms and Topics

Nomenclature of substituted carboxylic acids, unsaturated acids, and acid chlorides (Sections 8.3, 8.15)

Hydrogen bonding of acids (Section 8.5)

Scale of acidity of acids; pK_a (Section 8.7)

Effect of structure on acidity; electron-withdrawing inductive effect (Section 8.8)

General mechanism of nucleophilic substitution reactions of acid derivatives (Section 8.17)

Alkylmagnesium halides (Section 8.10)

Reactions of acid chlorides; transacylation (Section 8.16–8.19)

Cyclic anhydrides (Section 8.21)

Dicarboxylic acids (Section 8.20)

Prostaglandins (Section 8.13)

Summary of Reactions

REACTIONS OF CARBOXYLIC ACIDS

1. Salt formation (Section 8.6)

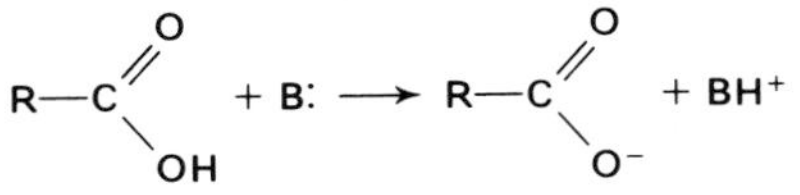

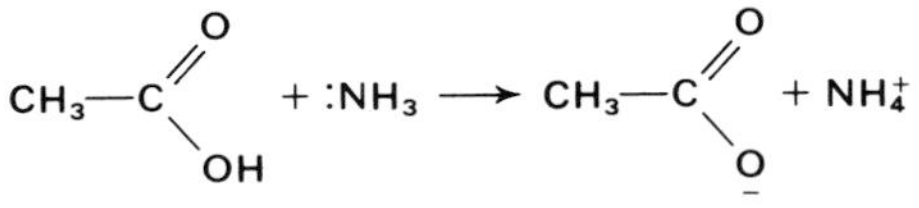

$$R-C\overset{\overset{O}{\|}}{\underset{OH}{}} + B: \longrightarrow R-C\overset{\overset{O}{\|}}{\underset{O^-}{}} + BH^+ \qquad CH_3-C\overset{\overset{O}{\|}}{\underset{OH}{}} + :NH_3 \longrightarrow CH_3-C\overset{\overset{O}{\|}}{\underset{O^-}{}} + NH_4^+$$

acetic acid   acetate ion

2. Acid chloride formation (Section 8.9)

$$R-\overset{\overset{\displaystyle O}{\|}}{C}-OH \ + \ SOCl_2 \ \longrightarrow \ R-\overset{\overset{\displaystyle O}{\|}}{C}-Cl \ + \ SO_2 \ + \ HCl$$

$$CH_3-CH_2-\overset{\overset{\displaystyle O}{\|}}{C}-OH \ + \ SOCl_2 \ \longrightarrow \ CH_3CH_2-\overset{\overset{\displaystyle O}{\|}}{C}-Cl \ + \ SO_2 \ + \ HCl$$

propanoic   thionyl   propanoyl
acid   chloride   chloride

SUBSTITUTION REACTIONS OF ACID CHLORIDES

1. Ester formation with alcohols (Section 8.16)

$$R-\overset{\overset{\displaystyle O}{\|}}{C}-Cl \ + \ R'OH \ \longrightarrow \ R-\overset{\overset{\displaystyle O}{\|}}{C}-O-R' \ + \ HCl$$

$$CH_3-\underset{\underset{\displaystyle CH_3}{|}}{CH}-\overset{\overset{\displaystyle O}{\|}}{C}-Cl \ + \ CH_3(CH_2)_3OH \ \longrightarrow \ CH_3-\underset{\underset{\displaystyle CH_3}{|}}{CH}-\overset{\overset{\displaystyle O}{\|}}{C}-O-(CH_2)_3CH_3 \ + \ HCl$$

2-methyl-   1-butanol   butyl
propanoyl   (butyl   2-methylpropanoate
chloride   alcohol)

2. Thioester formation with thiols (Section 8.19)

$$R-\overset{\overset{\displaystyle O}{\|}}{C}-Cl \ + \ R'-SH \ \longrightarrow \ R-\overset{\overset{\displaystyle O}{\|}}{C}-S-R' \ + \ HCl$$

$$CH_3-\overset{\overset{\displaystyle O}{\|}}{C}-Cl \ + \ CH_3-\underset{\underset{\displaystyle CH_3}{}}{\overset{\overset{\displaystyle CH_3}{|}}{CH}}-SH \ \longrightarrow \ CH_3-\overset{\overset{\displaystyle O}{\|}}{C}-S-\overset{\overset{\displaystyle CH_3}{|}}{CH}-CH_3 \ + \ HCl$$

acetyl   2-propanethiol   isopropyl
chloride   thioacetate

3. Amide formation with ammonia and amines (Section 8.18)

$$R-\overset{\overset{\displaystyle O}{\|}}{C}-Cl \ + \ 2\ R'-NH_2 \ \longrightarrow \ R-\overset{\overset{\displaystyle O}{\|}}{C}-NH-R' \ + \ R'NH_3^+Cl^-$$

$$CH_3-CH_2-CH_2-\overset{\overset{\displaystyle O}{\|}}{C}-Cl \ + \ 2\ NH_3 \ \longrightarrow \ CH_3-CH_2-CH_2-\overset{\overset{\displaystyle O}{\|}}{C}-NH_2 \ + \ NH_4^+Cl^-$$

butanoyl chloride   butanamide

$$CH_3CH_2CH_2\overset{\overset{\displaystyle O}{\|}}{C}-Cl \ + \ 2\ (CH_3)_3C-NH_2 \ \longrightarrow \ CH_3CH_2CH_2\overset{\overset{\displaystyle O}{\|}}{C}-NH-C(CH_3)_3 \ + \ (CH_3)_3C-NH_3^+Cl^-$$

butanoyl chloride   *tert*-butylamine   N-*tert*-butylbutanamide

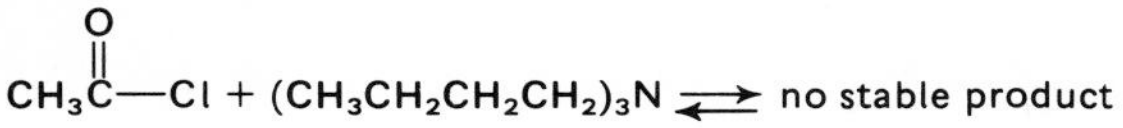

$$CH_3\overset{\overset{\displaystyle O}{\|}}{C}-Cl + (CH_3CH_2CH_2CH_2)_3N \rightleftharpoons \text{no stable product}$$

tributylamine

4. Anhydride formation with carboxylate ions (Section 8.19)

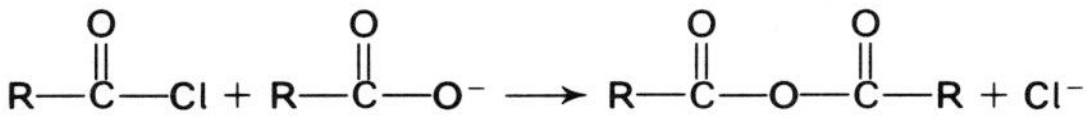

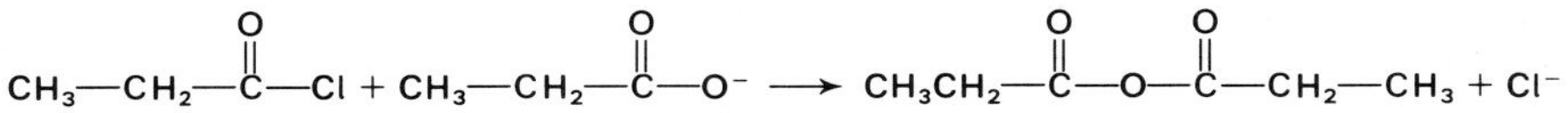

$$R-\overset{\overset{\displaystyle O}{\|}}{C}-Cl + R-\overset{\overset{\displaystyle O}{\|}}{C}-O^- \longrightarrow R-\overset{\overset{\displaystyle O}{\|}}{C}-O-\overset{\overset{\displaystyle O}{\|}}{C}-R + Cl^-$$

$$CH_3-CH_2-\overset{\overset{\displaystyle O}{\|}}{C}-Cl + CH_3-CH_2-\overset{\overset{\displaystyle O}{\|}}{C}-O^- \longrightarrow CH_3CH_2-\overset{\overset{\displaystyle O}{\|}}{C}-O-\overset{\overset{\displaystyle O}{\|}}{C}-CH_2-CH_3 + Cl^-$$

propanoyl chloride — propanoate ion — propanoic anhydride

5. Hydrolysis with water (Section 8.19)

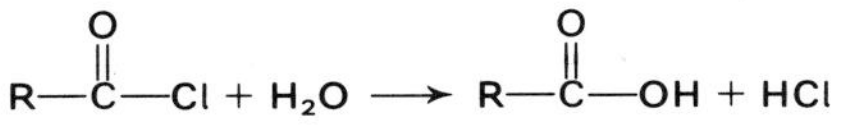

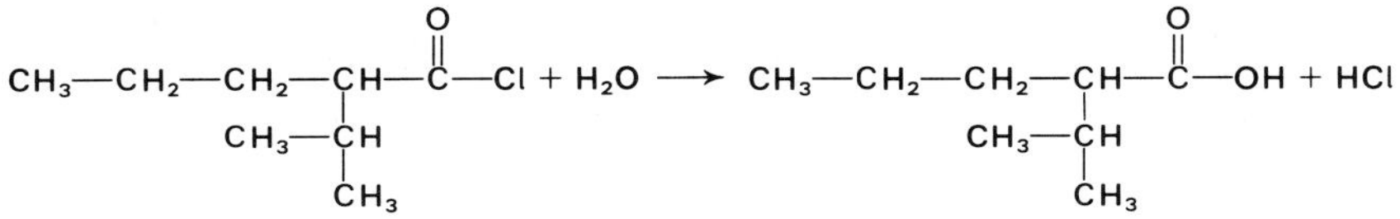

$$R-\overset{\overset{\displaystyle O}{\|}}{C}-Cl + H_2O \longrightarrow R-\overset{\overset{\displaystyle O}{\|}}{C}-OH + HCl$$

$$CH_3-CH_2-CH_2-\underset{\underset{\displaystyle CH_3-CH}{\underset{\displaystyle CH_3}{|}}}{CH}-\overset{\overset{\displaystyle O}{\|}}{C}-Cl + H_2O \longrightarrow CH_3-CH_2-CH_2-\underset{\underset{\displaystyle CH_3-CH}{\underset{\displaystyle CH_3}{|}}}{CH}-\overset{\overset{\displaystyle O}{\|}}{C}-OH + HCl$$

2-isopropylpentanoyl chloride — 2-isopropylpentanoic acid

DECOMPOSITION OF DICARBOXYLIC ACIDS AND β-KETOACIDS BY HEATING

1. Loss of H₂O—anhydride formation (Section 8.20)

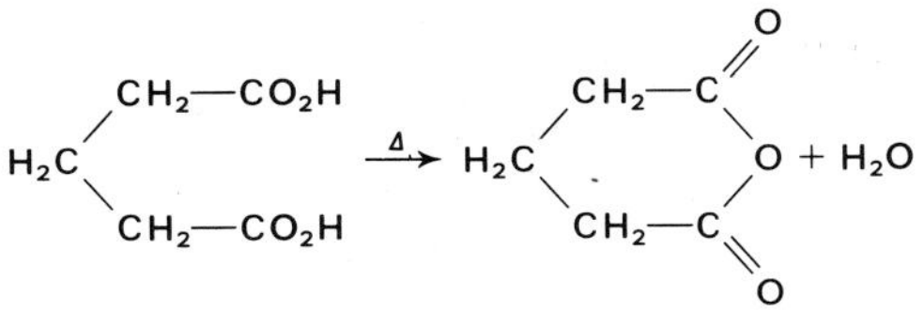

glutaric acid — glutaric anhydride 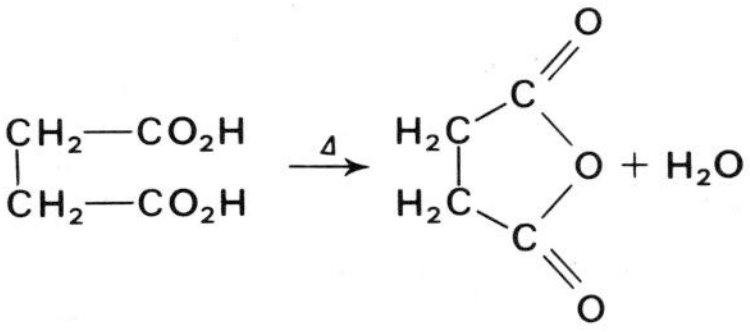 — succinic acid — succinic anhydride

2. Loss of CO₂—decarboxylation (Section 8.20)

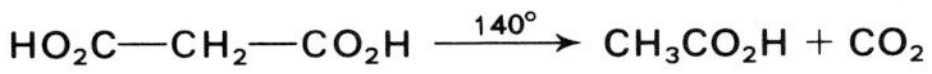

$$HO_2C-CH_2-CO_2H \xrightarrow{140°} CH_3CO_2H + CO_2$$

malonic acid — acetic acid

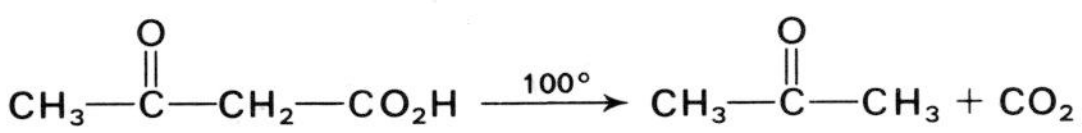

$$CH_3-\overset{\overset{\displaystyle O}{\|}}{C}-CH_2-CO_2H \xrightarrow{100°} CH_3-\overset{\overset{\displaystyle O}{\|}}{C}-CH_3 + CO_2$$

acetoacetic acid — acetone

REACTIONS OF HYDROXYACIDS

1. α,β-Unsaturated acids from β-hydroxyacids (Section 8.12)

$$R-\underset{\underset{OH}{|}}{CH}-CH_2CO_2H \xrightarrow{H^+} R-CH=CH-CO_2H + H_2O$$

$$CH_3\underset{\underset{OH}{|}}{CH}-CH_2CO_2H \xrightarrow{H^+} CH_3CH=CH-CO_2H + H_2O$$

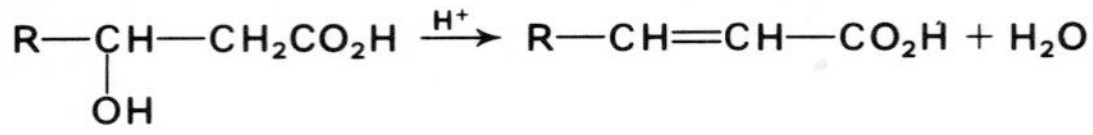

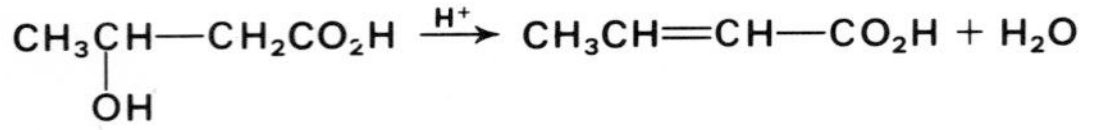

3-hydroxybutanoic acid   2-butenoic acid
(β-hydroxybutyric acid)

2. Lactones from hydroxyacids (Section 8.12)

$$\underset{\underset{OH}{|}}{CH_2}(CH_2)_xCO_2H \xrightarrow[trace]{H^+} CH_2-(CH_2)_x-C=O + H_2O \quad x = 2,3$$

$$\underset{\underset{OH}{|}}{CH_2}CH_2CH_2CH_2CO_2H \longrightarrow \text{(ring)}=O + H_2O$$

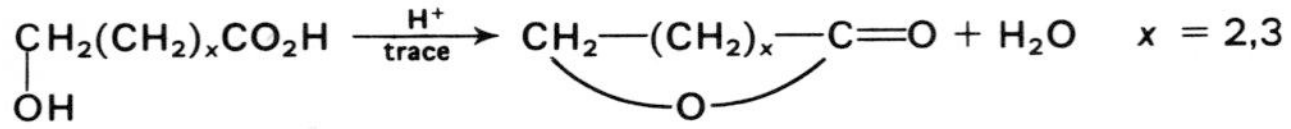

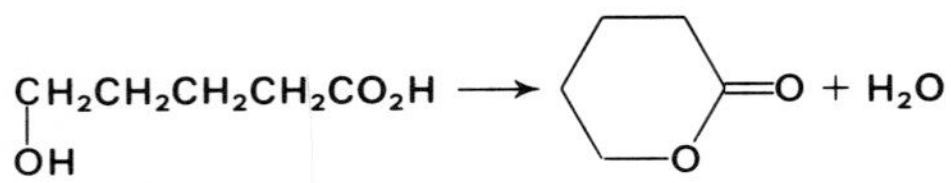

5-hydroxypentanoic
acid

Problems

8.20 Draw structures for the following compounds.

  a) heptanoyl chloride   **b)** *cis*-2-methoxy-3-pentenoyl chloride

  c) 2-hydroxyoctanedioic acid   **d)** malonic acid   **e)** succinic anhydride

  f) *cis*-2-hydroxycyclopentanecarboxylic acid   **g)** octadecanoic acid

  h) 2-butynoic acid   **i)** glutaric acid   **j)** phthalic anhydride

8.21 Name the following compounds by the IUPAC or common names.

a) 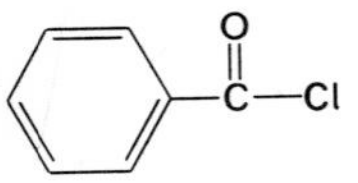   **b)** 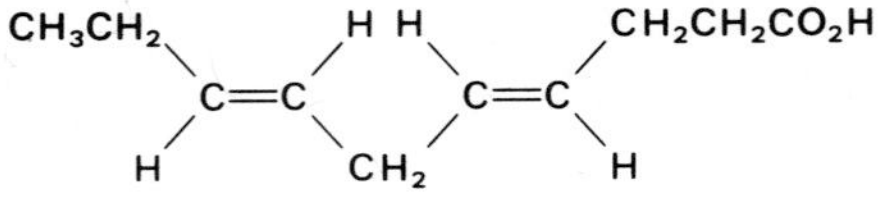   **c)**

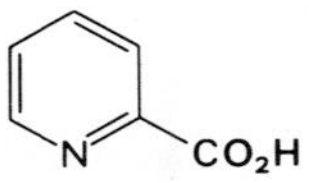

d) 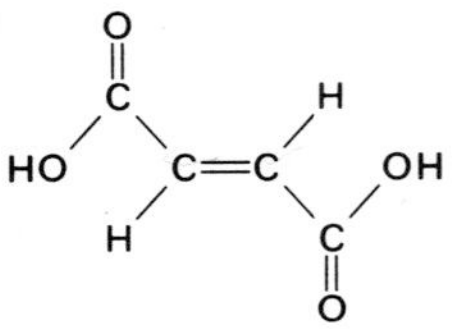   **e)** 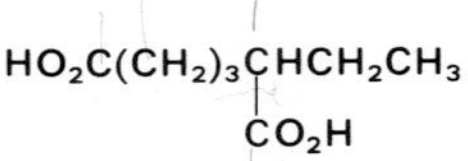 $HO_2C(CH_2)_3\underset{\underset{CO_2H}{|}}{C}HCH_2CH_3$ **f)**

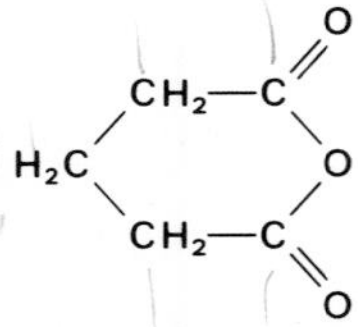

8.22 Write equations for the following reactions using structural formulas for all reactants and products.

a) benzoic acid + aqueous sodium hydroxide solution   **b)** 2-methylbutanoic acid + thionyl chloride

c) propanoic acid + phosphorus trichloride   **d)** maleic anhydride + methanol

e) benzoyl chloride + sodium benzoate   **f)** propenoyl chloride + aniline

g) acetyl chloride + water   **h)** 2-bromopropanoyl chloride + phenol

8.23 Give the structures of the starting materials and of compounds A through K.

a) acetic acid $\xrightarrow{\text{thionyl chloride}}$ A $\xrightarrow{\text{isopropylamine}}$ B *amide*

b) cyclohexanecarboxylic acid $\xrightarrow{\text{phosphorus trichloride}}$ C $\xrightarrow{\text{sec-butyl alcohol}}$ D *ester*

c) 2-ethylpentanoic acid $\xrightarrow{\text{thionyl chloride}}$ E $\xrightarrow{\text{methanethiol}}$ F *thioester*

d) disodium dimethylmalonate $\xrightarrow{\text{H}_3\text{O}^+}$ G $\xrightarrow{140°}$ H

e) succinic anhydride $\xrightarrow{\text{methanol}}$ I $\xrightarrow{\text{thionyl chloride}}$ J $\xrightarrow{\text{phenol}}$ K

8.24 The structures of phthalic, isophthalic, and terephthalic acids are given. Which of these do you guess will form cyclic anhydrides when heated to drive off water? Write the structures of anhydrides that might form from these three acids.

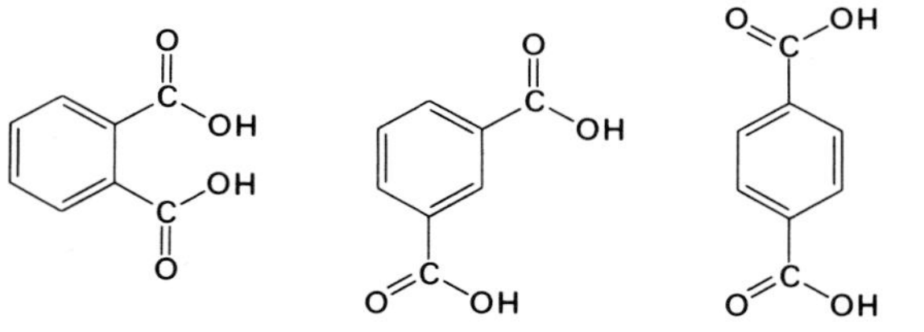

phthalic acid   isophthalic acid   terephthalic acid

8.25 Write the series of reactions needed to convert these reactants to the indicated products. Draw structures of the reactants and products for each step and give the necessary reagent over the arrow.

a) chloroacetic acid to methyl chloroacetate   **b)** 1-hexanol to 1-hexyl acetate

c) propenoic acid to ethyl propenethioate   **d)** phenylacetic acid to phenylacetamide

8.26 How many substituents on the ring, double bonds, and asymmetric carbons do prostaglandin E_2 and $F_{2\alpha}$ have?

8.27 To convert these starting materials to the products requires several reactions. Some of the products of the intermediate reactions are indicated but others are not. Give the reagents and formulas of reactants and products for each required reaction.

a) 1-hexanol to heptanonitrile to heptanoic acid

b) 1-hexanol to 2-chlorohexane to 2-methylhexanoic acid

CHAPTER 9

Esters and Fats

Carboxylic acid esters possess the unusual fea'ure of being derived structurally from two different kinds of hydroxyl-containing compounds, the carboxylic acids and the alcohols. Ester formation from a molecule of acid and a molecule of alcohol also produces a molecule of water. Esters are the product of the chemical merging of two polar functional groups, the carboxyl and the hydroxyl, to form the much less polar ester group. In the generation of an ester, two nonpolar hydrocarbon R groups, one formerly attached to the carboxyl and one to the hydroxyl, are now joined in the same ester molecule. Both of the original polar hydroxyl groups have been modified or used to form water.

$$R-C\overset{O}{\underset{O-H}{\Big\langle}} \quad + H-O-R' \longrightarrow R-C\overset{O}{\underset{O-R'}{\Big\langle}} \quad + H-O-H$$

Esters abound in the biological world. The most important esters associated with animals are **fats**, which belong to a larger group known as **lipids**. The term **lipophilic** means "fat loving," or in practice fat soluble or nonpolar. The term **hydrophilic** denotes "water loving," or more practically, water soluble or polar. Fats are esters of carboxylic acids having 12–24 carbons in a long, nonpolar, alkyl chain with one functional group, and are quite lipophilic.

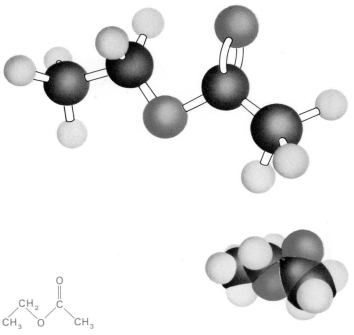

$$\underset{CH_3}{\overset{CH_2}{\diagdown}} O \underset{CH_3}{\overset{\overset{\displaystyle O}{\|}}{C}}$$

ethyl acetate

Although saturated hydrocarbons are largely resistant to bio-chemical degradation, the ester groups of the fats make their hydro-carbon groups subject to rapid biological degradation. Thus fats provide an evolutionary device by which compounds that are *mainly hydrocarbon* can be stored, retrieved, and used in a variety of ways. The ester group provides the key that allows the energy stored in the alkyl groups to be unlocked. The fats also compose much of the cell membranes that embrace and separate the small chemical factories of nature.

Smaller esters which are quite volatile perform more exotic functions in the natural world. Esters attract humming birds and bees to flowers. Esters enable male moths to find female moths to reproduce the species. Esters provide fruits and beverages with their delightful odors and flavors. Thus esters embellish, activate, preserve, store, fuel, segregate, and dissolve, and this is only a parital list of their functions.

The formula and models of ethyl acetate, one of the smallest esters, are shown on the preceding page.

9.1 SOURCES AND USES OF ESTERS

Esters of carboxylic acids are abundantly and widely distributed in nature. Of greatest familiarity are vegetable oils (liquids) and animal fats (solids), both of which are triesters of glycerol with long-chain carboxylic acids (C_{12} through C_{24}). As usually written these triesters resemble in shape the head of a pitch fork, with the three long acyl residues as the prongs.

glyceryl trioctadecanoate
(glyceryl tristearate)

pitch fork head

Interestingly, the fats and oils contain only the even-numbered carbon chain acid residues. The fatty acids are synthesized from and degraded to acetic acid, which accounts for the even number of carbons in their carbon chains. These esters are synthesized in the organism and are stored in fat reservoirs for use as fuel when food intake is insufficient to fulfill the needs of the organism.

Waxes like beeswax and carnauba are esters of C_{16} carboxylic acids and alcohols of C_{26} through C_{34} unbranched chains.

Other naturally occurring esters are volatile compounds with pleasant, somewhat sweet odors which are the characteristic fragrances of many flowers and fruits. The chief ester constituents present in some fruits and leaves are listed below.

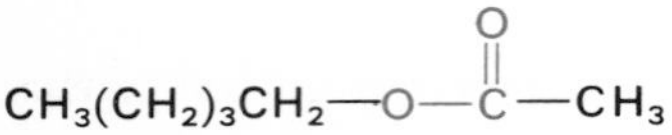

pentyl acetate
(banana)

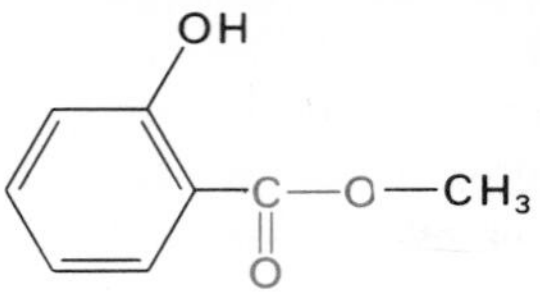

methyl salicylate
(oil of wintergreen)

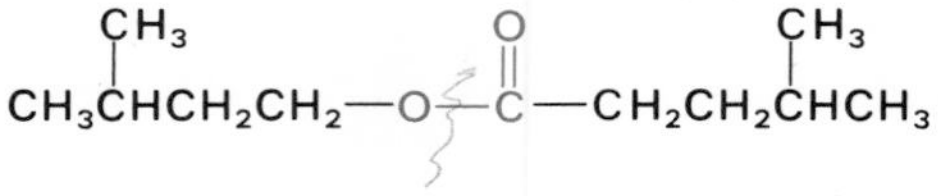

3-methyl-1-butyl 4-methylpentanoate
(apple)

"Polyesters" are large synthetic polymers containing many repeating units of ester groups arranged in long unbranched chains. The polyester "Dacron" is spun into a fiber for weaving or knitting. It is made from ethylene glycol (a diol) and 1,4-benzenedicarboxylic acid (terephthalic acid).

a dicarboxylic acid
(terephthalic acid)

ethylene glycol
(1,2-ethanediol)

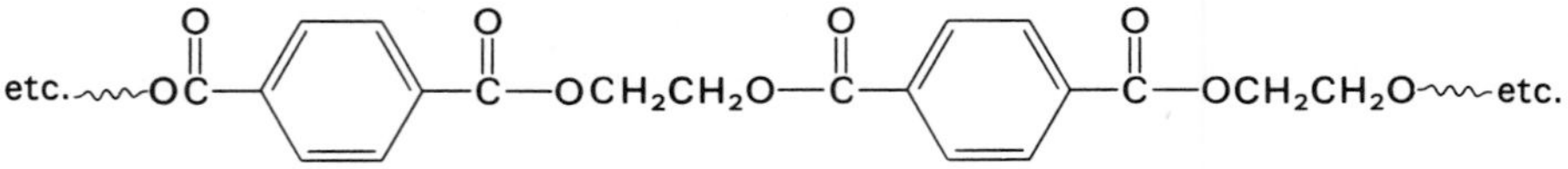

Dacron
(a polyester)

Most esters of small acids and alcohols are noncorrosive, nontoxic liquids with good properties for use as solvents. Their boiling points are low and they evaporate easily. Esters are used frequently as solvents for lacquers; for example, butyl acetate serves as the solvent for some nail polishes, and ethyl acetate as a solvent in purification of reaction products by crystallization and chromatography.

Esters are obtained from carboxylic acids by reaction of acid chlorides with alcohols (Section 8.16) and by the acid-catalyzed esterification of acids by alcohols (Section 9.11).

9.2 THE ESTER GROUP   The ester functional group contains the polar carbonyl group $C=O$ double bond and two $C-O$ single bonds to the same oxygen.

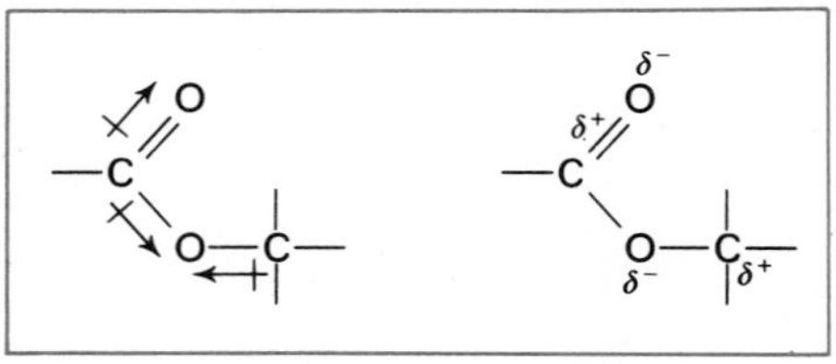

There are four atoms with a partial charge which are attractive to charged reagents. This carbonyl group has the same properties as the carbonyl group of the acid chloride. The C—O single bond to the acyl group is similar to the C—Cl bond in the sense that it is frequently broken in substitution reactions.

The O—C linkage of the alcohol portion of the ester is similar to that of alcohols in the sense that it is not easily broken. In esters the carbonyl group is the usual site of reaction because of its polar character. The reactions of esters therefore are expected and are observed to be more like those of acid chlorides than of alcohols or ethers.

9.3 NOMENCLATURE OF ESTERS

In nomenclature as in synthesis, esters are treated as derivatives of both carboxylic acids and alcohols. The specific name of the ester indicates its relationship to both classes of compounds. The ester name consists of two words, one denoting the alcohol and one the acid from which the ester was formed.

1. The alcohol portion is designated as the corresponding hydrocarbon substituent: 2-butyl or *sec*-butyl, ethyl, 1-octyl, *tert*-butyl, etc.
2. The carboxylate portion is named from the acid by substituting the ending "ate" for "ic acid."

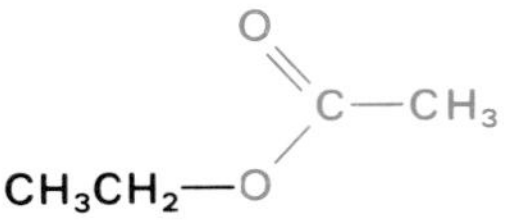

ethyl acetate 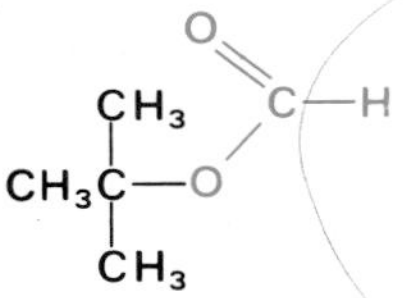   *tert*-butyl formate 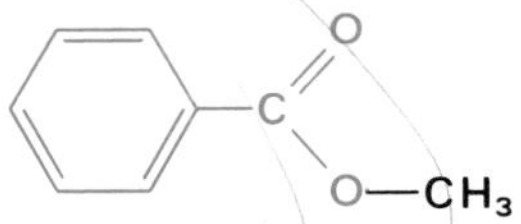   methyl benzoate

Cyclic esters in which both the alcohol and acid functions belong to the same carbon chain are called **lactones**.

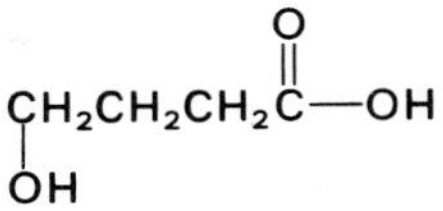

4-hydroxybutanoic acid 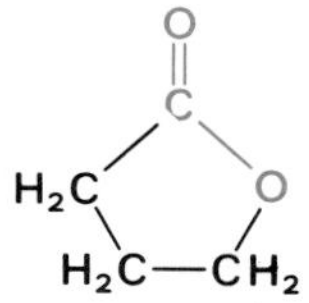   γ-butyrolactone
                             (tetrahydro-2-furanone)

Problems   **9.1** Name the following compounds.

a) 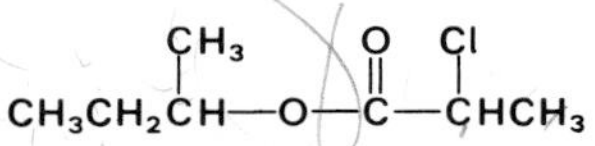 CH₃CH₂CH—O—C—CHCH₃
   with CH₃, O, Cl substituents

b) HC—O—CHCH₃
   with O, CH₃ substituents

c) CH₃(CH₂)₄—C—O—CH₂CHCH₃
   with O, CH₃ substituents

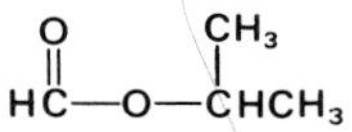

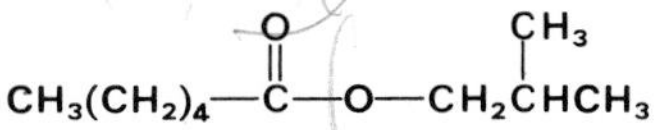

(handwritten margin note, top left)

$$CH_3\text{—}CH_2\text{—}O\text{—}\overset{\overset{\textstyle O}{\|}}{C}\text{—}\bigcirc$$

$$\begin{matrix} CH_3 \\ CH_2\text{—}O \\ CH_3 \end{matrix}$$

9.2 Write structures for the following compounds.

a) isobutyl benzoate   b) 2-octyl trifluoroacetate

c) phenyl 2-hydroxypentanoate   d) ethyl 2-phenylbutanoate

9.4 HYDROLYSIS OF ESTERS BY BASES

(handwritten margin note, left)

$$F\text{—}\overset{\overset{\textstyle O}{\|}}{\underset{\underset{\textstyle F}{\overset{\textstyle |}{C}}}{C}}\text{—}O$$

Esters react more slowly than acid chlorides, but undergo the same substitution reactions with the more reactive nucleophiles. Sometimes catalysts and heat are required for the reaction of esters.

The uncatalyzed reaction of an ester with water to give the acid and the alcohol components is exceedingly slow. Hydrolysis of an ester as usually performed in the laboratory involves either the use of hydroxide ion, a better nucleophile than water, or the use of a strong acid with the water to catalyze the reaction (Section 9.11).

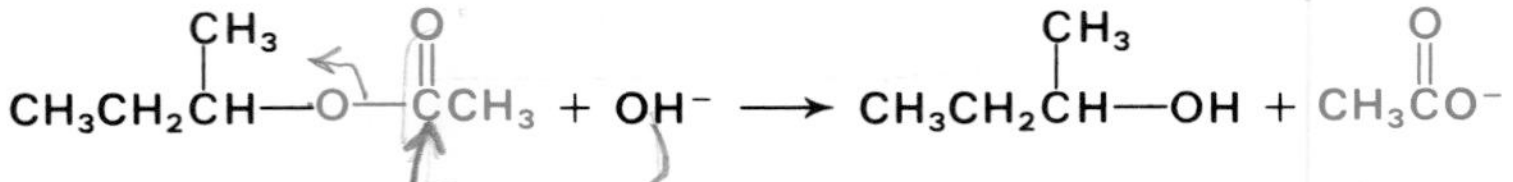

$$\begin{matrix} CH_3 \\ | \\ CH_3CH_2CH\text{—}O\text{—}\overset{\overset{\textstyle O}{\|}}{C}CH_3 + OH^- \end{matrix} \longrightarrow \begin{matrix} CH_3 \\ | \\ CH_3CH_2CH\text{—}OH \end{matrix} + CH_3\overset{\overset{\textstyle O}{\|}}{C}O^-$$

sec-butyl acetate   *sec*-butyl alcohol   acetate ion
                       (2-butanol)

$$CH_3\text{—}O\text{—}\overset{\overset{\textstyle O}{\|}}{C}(CH_2)_{16}CH_3 + OH^- \longrightarrow CH_3OH + CH_3(CH_2)_{16}CO_2^- \overset{H^+}{\longrightarrow} CH_3(CH_2)_{16}CO_2H$$

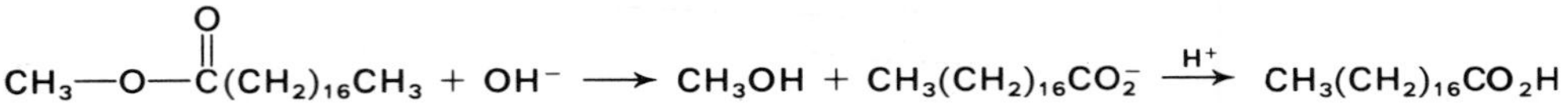

methyl octadecanoate   methanol   octadecanoate ion   octadecanoic acid

Sodium or potassium hydroxide is the usual reagent for hydrolysis. Esters are poor solvents for ionic substances and are not themselves very soluble in water. A mutual solvent for water and ester, such as ethanol or the higher-boiling ethylene glycol, is used to give a homogeneous (one-phase) reaction medium. Heat is also necessary for the reaction to proceed in a reasonable length of time (hours). Acidification of the final solution with hydrochloric acid forms the carboxylic acid.

Hydrolysis of an ester by hydroxide ion converts 100% of the ester and an equivalent molar quantity of hydroxide ion to the alcohol and the acid anion. The reaction can be used to measure the weight of material per ester unit. A known amount of base is used for the reaction with a known weight of ester. The excess base remaining at the end of the reaction is determined by titration with acid. The amount of base consumed by the hydrolysis indicates the amount of ester group present.

The mechanism of ester hydrolysis by a base is illustrated by the reaction of ethyl acetate with sodium hydroxide in the following steps.

Step 1 $$CH_3\text{—}\overset{\overset{\textstyle O}{\|}}{C}\text{—}O\text{—}CH_2CH_3 + {}^-OH \rightleftarrows CH_3\text{—}\overset{\overset{\textstyle O^-}{|}}{\underset{\underset{\textstyle OH}{|}}{C}}\text{—}OCH_2CH_3$$

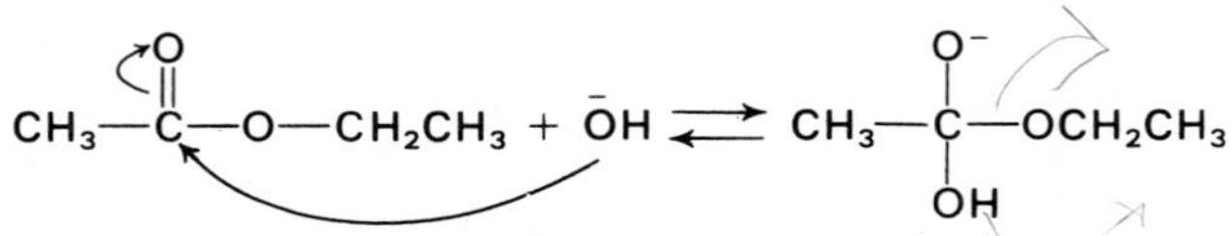

Step 2

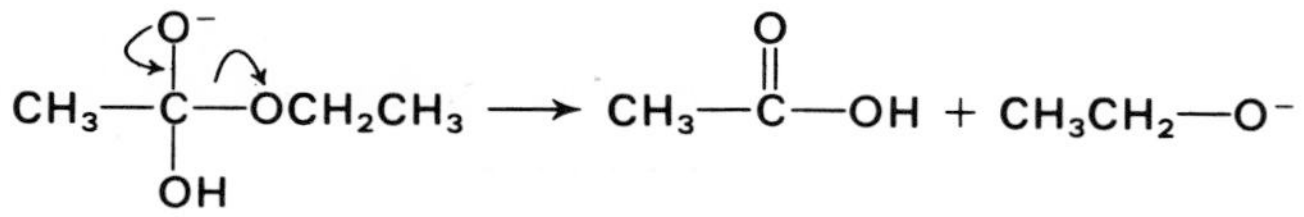

$$CH_3-\overset{\overset{\displaystyle O^-}{|}}{\underset{\underset{\displaystyle OH}{|}}{C}}-OCH_2CH_3 \longrightarrow CH_3-\overset{\overset{\displaystyle O}{||}}{C}-OH + CH_3CH_2-O^-$$

Step 3

$$CH_3-\overset{\overset{\displaystyle O}{||}}{C}-OH + CH_3CH_2-O^- \longrightarrow CH_3-\overset{\overset{\displaystyle O}{||}}{C}-O^- + CH_3CH_2-OH$$

The hydrolysis of esters is theoretically a reversible reaction. The reaction is driven to completion, however, by the formation of the carboxylate ion in the basic solution in the last step. The carboxylate ion is the weakest anion that can be formed in the reaction mixture. The reaction, although reversible in principle, is not reversible in practice in basic solution.

This mechanism of ester hydrolysis by hydroxide ion is well supported by experimental evidence. First, the rate of reaction depends upon the concentrations of both ester and hydroxide ion. Thus both species participate in the rate-limiting slow step.

Hydrolysis experiments with water and hydroxide ion labeled with the ^{18}O isotope gave alcohol which did not contain the ^{18}O label. The label appeared in the carboxylate ion. This experiment demonstrated that in the reaction the C—O bond that was broken was the acyl–oxygen bond.

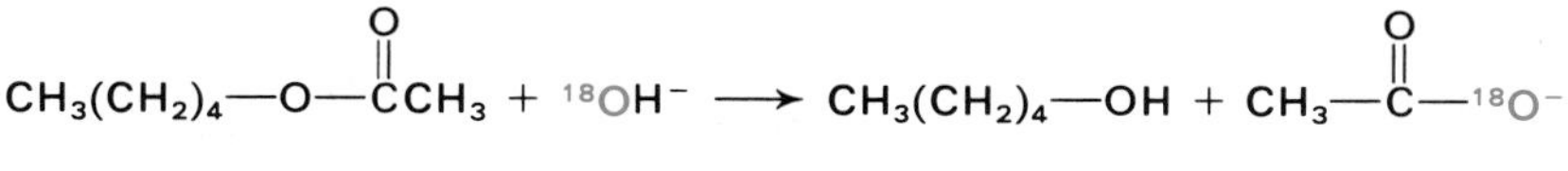

$$CH_3(CH_2)_4-O-\overset{\overset{\displaystyle O}{||}}{C}CH_3 + {}^{18}OH^- \longrightarrow CH_3(CH_2)_4-OH + CH_3-\overset{\overset{\displaystyle O}{||}}{C}-{}^{18}O^-$$

Another isotopic experiment indicated the reversible formation of an intermediate, and confirmed the two-step mechanism. In this experiment the ^{18}O was originally in the doubly bonded oxygen of the ester, which was treated with ordinary hydroxide ion.

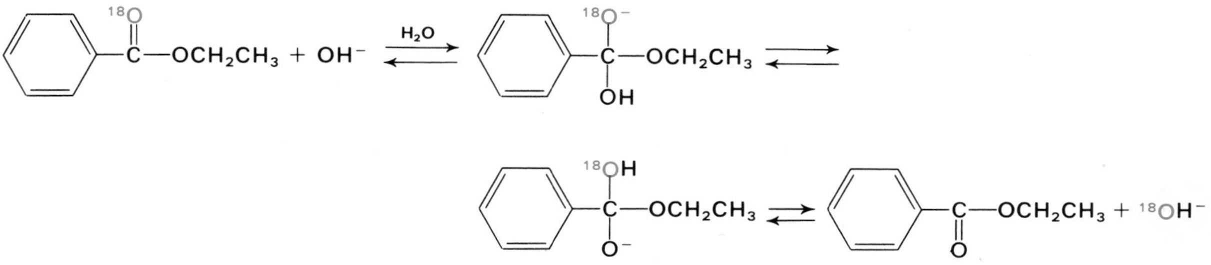

Examination of the unhydrolyzed ester recovered after partial hydrolysis showed that some of the ^{18}O was lost before the ester was hydrolyzed. The carbonyl ^{18}O was exchanged with ordinary ^{16}O of the water. This exchange could happen only if the intermediate was formed and reverted to the ester as well as going on to give acid.

9.3 Account for the fact that the reaction of hydroxide ion with ethyl 2,2-dimethylpropanoate is exceedingly slow by comparison with ethyl propanoate. (*Hint:* Note the similarity with the unreactivity of neopentyl halides—2,2-dimethyl-1-propyl halides—in nucleophilic substitutions.)

**9.5 TRANSESTERI-
FICATION**

Interchange of the alcohol group of an ester occurs when the ester is heated with a metal alkoxide. The interchange is called transesterification.

$$CH_3(CH_2)_5\!-\!O\!-\!\overset{\overset{\textstyle O}{\|}}{C}\!-\!C_6H_5 + CH_3\!-\!O^-Na^+ \rightleftharpoons CH_3\!-\!O\!-\!\overset{\overset{\textstyle O}{\|}}{C}\!-\!C_6H_5 + CH_3(CH_2)_5\!-\!O^-Na^+$$

hexyl benzoate   sodium methoxide   methyl benzoate   sodium hexoxide

Since transesterification products contain the same functional group as the starting material, why doesn't the reaction go both ways? In fact it does, and an equilibrium is reached with both esters present. What measures can be taken to shift the equilibrium in the desired direction?

Two equilibria exist, both involving methoxide and hexoxide ions. Besides the one written above, another is produced between alcohols and alkoxide ions.

$$CH_3\!-\!O\!-\!H + CH_3(CH_2)_5\!-\!O^- \rightleftharpoons CH_3\!-\!O^- + CH_3(CH_2)_5\!-\!O\!-\!H$$

methanol   1-hexoxide ion   methoxide ion   1-hexanol

Any shift in this equilibrium directly affects the transesterification equilibrium. For example, an increase in the concentration of methanol shifts the alcohol–alkoxide equilibrium to the right, *decreasing* the concentration of hexoxide ion and *increasing* the concentration of methoxide ion. An increase in methoxide ion shifts the transesterification equilibrium toward the right with formation of more methyl benzoate.

The production of methyl benzoate can be maximized by the use of methanol in large excess (usually as solvent), which shifts both equilibria to the right. The reverse reaction can be driven by distilling the methanol as it is formed. Methanol is the lowest-boiling material in the reaction mixture.

9.6 ESTERS TO AMIDES

Esters treated with an excess of ammonia or an amine are slowly converted to the amide when heated. The fact that the reaction goes in this direction supports our earlier experience that nitrogen makes a better nucleophile than does oxygen.

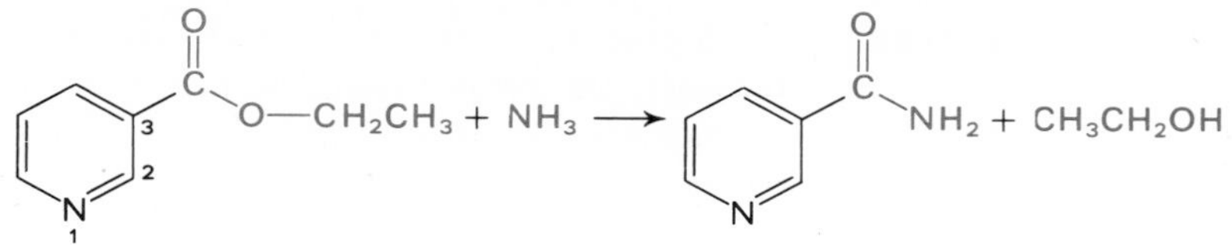

ethyl 3-pyridinecarboxylate (ethyl nicotinate) — 3-pyridinecarboxamide (nicotinamide)

Problems **9.4** Write equations for these reactions of esters.

a) *tert*-butyl 3-ethylpentanoate + diethylamine

b) 1-decyl acetate + sodium methoxide in methanol

c) methyl octadecanoate + sodium hydroxide in ethanol

9.5 Write the mechanism of the reaction of isopropyl hexanoate with propylamine.

9.7 ESTERS TO ALCOHOLS—REDUCTION BY LITHIUM ALUMINUM HYDRIDE

Lithium aluminum hydride, $LiAlH_4$, is a reagent which furnishes a nucleophilic hydrogen, the hydride ion H:⁻, to form a new C—H bond. The reaction of lithium aluminum hydride with an ester produces two alcohols, one from the acyl part and one from the alkyl part of the ester, as in this example using methyl pentanoate.

$$\underset{\text{methyl pentanoate}}{CH_3CH_2CH_2CH_2\overset{O}{\overset{\|}{C}}—OCH_3} \xrightarrow{LiAlH_4} \xrightarrow{H^+} \underset{\text{1-pentanol}}{CH_3CH_2CH_2CH_2CH_2—OH} + \underset{\text{methanol}}{CH_3—OH}$$

In the reaction the hydride ion adds to the carbonyl group of the ester and the alkoxide ion of the alkyl group becomes the leaving group. The compound formed from the acyl portion is an aldehyde whose carbonyl group then reacts with lithium aluminum hydride again. The final two products are two alkoxide ions which are converted to alcohols on treatment with dilute hydrochloric acid.

$$CH_3(CH_2)_3\overset{O}{\overset{\|}{C}}—OCH_3 \xrightarrow{LiAlH_4} CH_3(CH_2)_3\overset{O}{\overset{\|}{C}}—H + CH_3O^- \xrightarrow{LiAlH_4} \xrightarrow{H^+} CH_3(CH_2)_3CH_2—OH + CH_3—OH$$

Problem **9.6** Give the structures of the products of the reaction of lithium aluminum hydride with each of these esters, followed by acidification.

a) **b)**

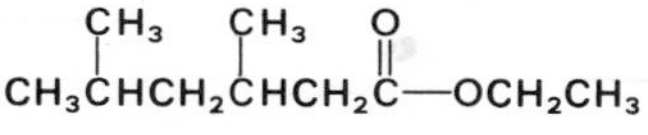

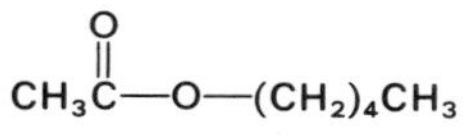

9.8 THIOESTERS In thioesters sulfur replaces the oxygen of ordinary esters as the link between the two carbon chains. Thioesters are very important compounds in the animal metabolic processes. Nearly all reactions of esters at the carboxyl group and elsewhere occur *more readily* with thioesters than with ordinary esters. The best-known thioesters in biological processes are those composed from fatty acids or amino acids and the thiol coenzyme A (CoA-SH).

$$HS-CoA = HS-CH_2CH_2-NH-\overset{\overset{\displaystyle O}{\|}}{C}CH_2CH_2-NH-\overset{\overset{\displaystyle O}{\|}}{C}-\overset{\overset{\displaystyle OH}{|}}{CH}-\overset{\overset{\displaystyle CH_3}{|}}{\underset{\underset{\displaystyle CH_3}{|}}{C}}CH_2-OP\bar{O}_3-OP\bar{O}_3-ribose-adenine$$

$\underbrace{\hspace{3cm}}$ thiol group $\underbrace{\hspace{5cm}}$ pantothenic acid (a vitamin) $\underbrace{\hspace{3cm}}$ adenosine

The remarkable thing about coenzyme A (and many other coenzymes) is the appendage of a large, complicated, and nonfunctioning group to the simple functioning SH group. The large substituent actually carries the functional group to the proper reaction site and probably aids in holding it in the right place.

Acetyl coenzyme A might well be the most important single reagent in biological reactions in mammals. It is a major compound in the synthesis and/or degradation of such substances as fatty acids, carbohydrates, cholesterol, and some amino acids.

$$CH_3-\overset{\overset{\displaystyle O}{\|}}{C}-S-(CoA)$$

acetyl coenzyme A

In the laboratory thioesters are little used. The reactivity of oxygen esters can be increased by higher temperature and strongly acidic or basic conditions, thus thioesters are not needed. Thioesters are also less desirable to handle and less readily available. Thiols, from which they are made, usually have an unpleasant, skunklike odor. Thioesters are prepared in the laboratory by the reaction of acid chlorides or anhydrides with thiols (Section 8.19).

9.9 REACTIONS OF THIOESTERS The thioesters undergo all the reactions of the oxygen esters with hydroxide and alkoxide ions, thiolate ions, ammonia, and amines. The reactions follow the same mechanistic scheme of other acid derivatives (Section 8.17).

The transesterification reaction of thioesters to give oxygen esters or other thioesters occurs faster in the laboratory when the anions of the alcohols and thiols are used than when the neutral compounds are employed.

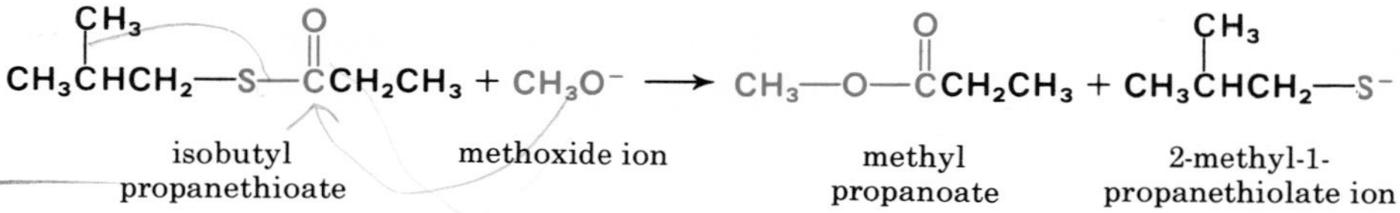

$$CH_3CHCH_2\!-\!S\!-\!\overset{\displaystyle O}{\overset{\|}{C}}CH_2CH_3 + CH_3O^- \longrightarrow CH_3\!-\!O\!-\!\overset{\displaystyle O}{\overset{\|}{C}}CH_2CH_3 + CH_3CHCH_2\!-\!S^-$$

isobutyl propanethioate methoxide ion methyl propanoate 2-methyl-1-propanethiolate ion

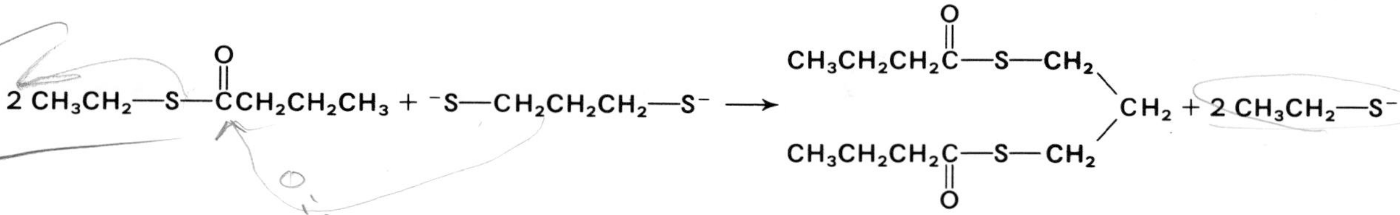

$$2\,CH_3CH_2\!-\!S\!-\!\overset{\displaystyle O}{\overset{\|}{C}}CH_2CH_2CH_3 + {}^-S\!-\!CH_2CH_2CH_2\!-\!S^- \longrightarrow$$

$$CH_3CH_2CH_2\overset{\displaystyle O}{\overset{\|}{C}}\!-\!S\!-\!CH_2\!\diagdown\!CH_2 + 2\,CH_3CH_2\!-\!S^-$$
$$CH_3CH_2CH_2\underset{\displaystyle O}{\overset{\|}{C}}\!-\!S\!-\!CH_2\!\diagup$$

An enzyme-catalyzed transesterification of an acyl coenzyme A to give an oxygen ester is an essential step in the biosynthesis of glyceryl esters (Section 9.16). The conversion of simple metabolic compounds to carbon dioxide via the citric acid cycle (Section 14.12) also involves transacylations and other reactions of coenzyme A derivatives of acids.

Amide formation by the reaction of ammonia or amines with thioesters takes place more readily than the analogous reactions with oxygen esters.

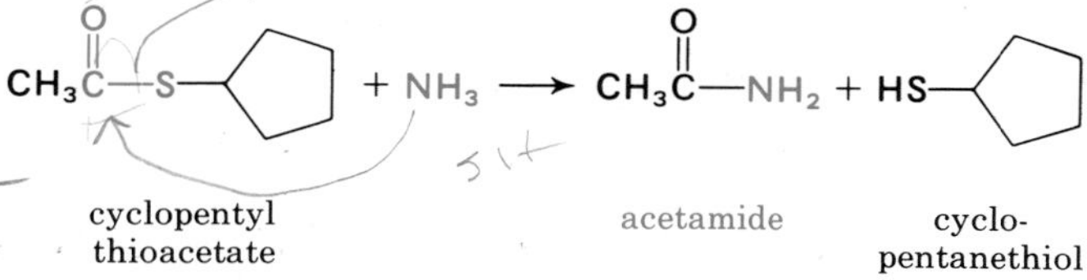

$$CH_3\overset{\displaystyle O}{\overset{\|}{C}}\!-\!S\!-\!\bigpentagon + NH_3 \longrightarrow CH_3\overset{\displaystyle O}{\overset{\|}{C}}\!-\!NH_2 + HS\!-\!\bigpentagon$$

cyclopentyl thioacetate acetamide cyclopentanethiol

The thioesters are hydrolyzed by treatment with hydroxide ion to the carboxylate ion and the thiolate ion. The thiolate ion is a weaker base than hydroxide ion.

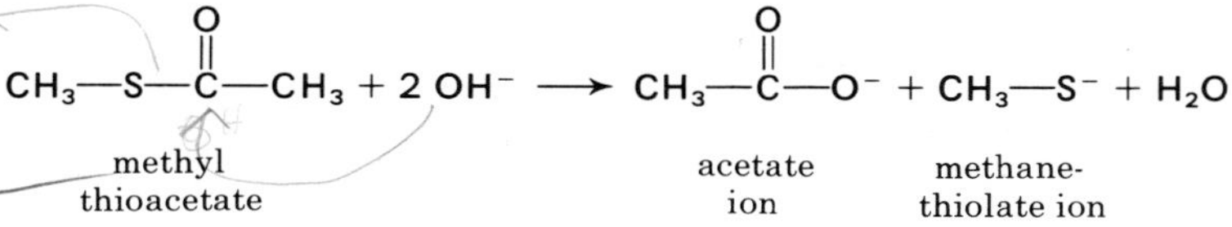

$$CH_3\!-\!S\!-\!\overset{\displaystyle O}{\overset{\|}{C}}\!-\!CH_3 + 2\,OH^- \longrightarrow CH_3\!-\!\overset{\displaystyle O}{\overset{\|}{C}}\!-\!O^- + CH_3\!-\!S^- + H_2O$$

methyl thioacetate acetate ion methanethiolate ion

Problem **9.7** Most of the cholesterol in the tissues of higher organisms exists as the ester of long-chain fatty acids. The liver contains an enzyme that catalyzes the formation of the ester by the reaction of cholesterol (formula, Section 5.13) with a fatty acid ester of coenzyme A such as hexadecanoyl coenzyme A. Write the equation for the reaction.

9.10 RELATIVE REACTIVITY OF CARBOXYL FUNCTIONAL GROUPS

The carboxyl functional groups can be put in order of decreasing reactivity.

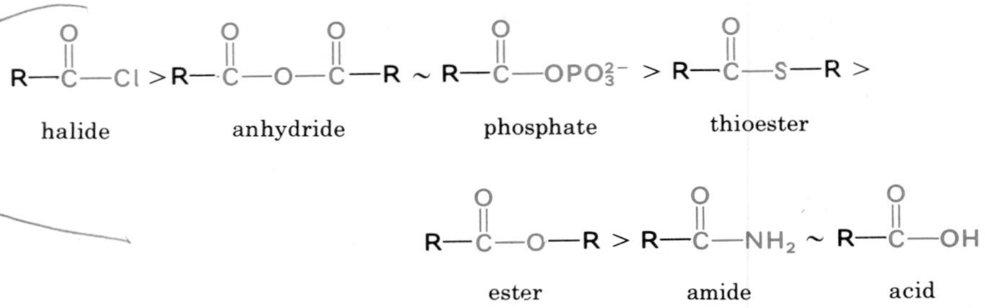

$$R-\underset{\underset{halide}{}}{\overset{O}{\underset{\|}{C}}}-Cl > R-\overset{O}{\underset{\|}{C}}-O-\underset{\underset{anhydride}{}}{\overset{O}{\underset{\|}{C}}}-R \sim R-\underset{\underset{phosphate}{}}{\overset{O}{\underset{\|}{C}}}-OPO_3^{2-} > R-\underset{\underset{thioester}{}}{\overset{O}{\underset{\|}{C}}}-S-R >$$

$$R-\underset{\underset{ester}{}}{\overset{O}{\underset{\|}{C}}}-O-R > R-\underset{\underset{amide}{}}{\overset{O}{\underset{\|}{C}}}-NH_2 \sim R-\underset{\underset{acid}{}}{\overset{O}{\underset{\|}{C}}}-OH$$

Reactions proceed from structures of high reactivity to structures of lower reactivity. The role of a reactive carboxyl function is to transfer the acyl group RC=O to an acceptor group; thus these reactions are called transacylations. In this example of a transacylation, the acid chloride is used to acylate the alcohol.

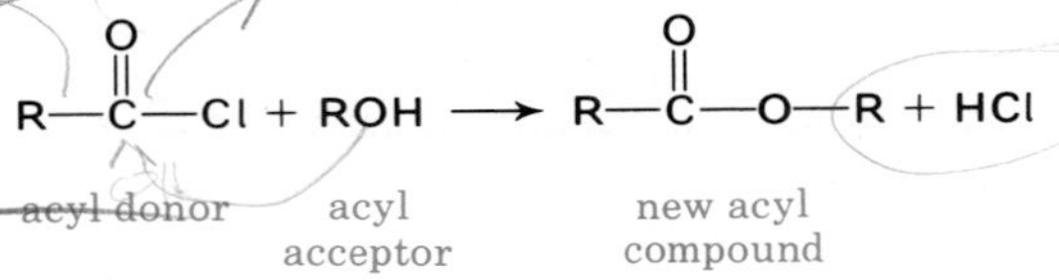

$$R-\overset{O}{\underset{\|}{C}}-Cl + ROH \longrightarrow R-\overset{O}{\underset{\|}{C}}-O-R + HCl$$

acyl donor · acyl acceptor · new acyl compound

The carboxylic acid group is very low on the scale. It is made more reactive when it is converted to its acid chloride. This change makes use of the great reactivity of the inorganic acid chloride $SOCl_2$ and the great stability of its products SO_2 and HCl.

Instead of acid chlorides, the biological acylating agent is the acyl phosphate (a mixed carboxylic-phosphoric anhydride) and an alkyl ester of an acyl phosphate (Section 9.15). Biological transacylation reactions are described in Section 9.16.

9.11 ACID-CATALYZED HYDROLYSIS OF ESTERS AND ESTERIFICATION OF ACIDS

In the presence of a strong acid (sulfuric acid is most often used) an ester is easily hydrolyzed by reaction with water to give the carboxylic acid and alcohol.

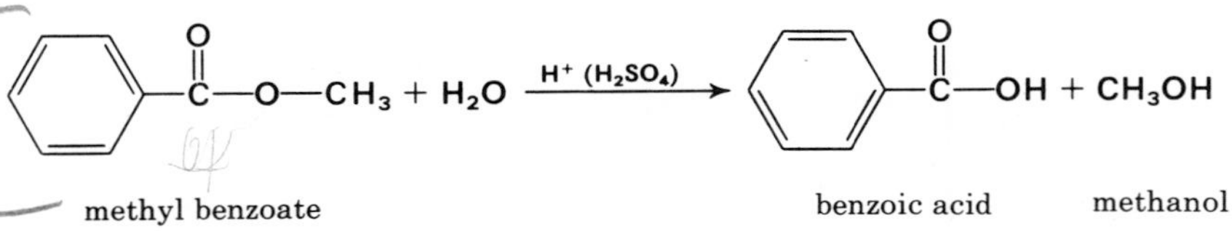

methyl benzoate · benzoic acid · methanol

The reaction proceeds to an equilibrium state from which both reactants and products can be isolated. Both the hydrolysis and esterification (the reverse reaction) are catalyzed by the strong acid, which is used in far less than a molar equivalent, in "catalytic amount." In the absence of a strong acid, the conversion takes place exceedingly slowly. Heat provided to the reaction mixture increases the rates of both the forward and reverse reactions.

Strong acids catalyze the reaction by protonating the doubly bonded oxygen of an ester or acid, which increases the electron deficiency of the carbonyl carbon and thus increases its reactivity toward nucleophilic reagents.

$$CH_3-\overset{\overset{\textstyle O}{\|}}{C}-O-CH_3 + H_3O^+ \longrightarrow$$

$$\left[CH_3-\overset{\overset{\textstyle +OH}{\|}}{C}-O-CH_3 \longleftrightarrow CH_3-\overset{\overset{\textstyle OH}{|}}{C}=\overset{+}{O}-CH_3 \longleftrightarrow CH_3-\overset{\overset{\textstyle OH}{|}}{\underset{+}{C}}-O-CH_3 \right] + H_2O$$

Since both the forward and reverse reactions are catalyzed, equilibration between ester–water and acid–alcohol is catalyzed.

Problem
9.8 How successful would you expect the reaction of an ester with 1-propanethiol to be under acidic conditions? With ammonia under acidic conditions? Justify your answers.

9.12 THE NATURE OF
CATALYSIS

A substance which, by its presence, increases the rate of a reaction is called a **catalyst**. The catalyst participates in at least one step (the rate-limiting or slow step) of the reaction and is regenerated in a further step to participate again. In the usual examples of catalysis, only a small amount of catalyst is needed, because each molecule can be utilized many times. A true catalyst does not affect the relative stabilities of starting and product states, but usually makes possible more stable intermediate states.

Catalyzed reactions are abundant both in laboratory syntheses and in natural processes. Laboratory catalysts include (a) acids, (b) bases, (c) electrophiles, such as metal cations, (d) nucleophiles, and (e) surfaces of metals and metal oxides.

The catalyst accelerates the reaction rate by providing a lower energy pathway that involves more stable transitional structures not available in the absence of the catalyst. The catalyst may work in a variety of ways, three of which are listed here.

1. It may increase the susceptibility of the original compound, the substrate, to attack.
2. It may enhance the ability of a leaving group to depart.
3. It may provide a better attacking group, such as a better nucleophile or electrophile.

The role of the catalyst is illustrated in the following examples of catalyzed reactions. In the acid-catalyzed esterification and hydrolysis of esters discussed in the previous section, acidification produces a reactant that is more readily subject to nucleophilic attack. The electron deficiency of the carbonyl carbon is increased by protonation of the carbonyl oxygen.

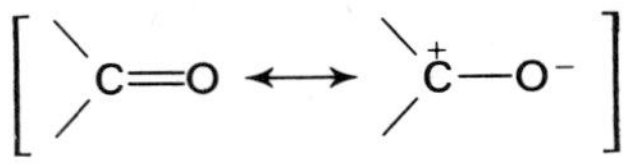

$$\left[\ \ce{>C=O} \longleftrightarrow \ce{>\overset{+}{C}-O^-}\ \right] \qquad \left[\ \ce{>C=\overset{+}{O}H} \longleftrightarrow \ce{>\overset{+}{C}-OH}\ \right]$$

carbonyl group 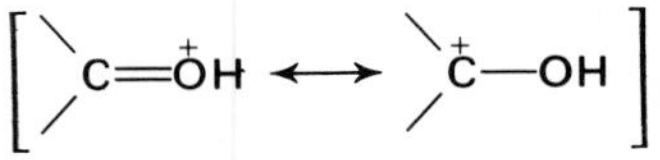   protonated carbonyl group

In the dehydration of alcohols by sulfuric acid, the acid catalysis enhances the ability of the leaving group to depart (Section 5.6).

Problem   9.9 Demonstrate with the use of the mechanism of dehydration of a tertiary alcohol that the acid is indeed acting as a catalyst.

Pyridine serves as a nucleophilic catalyst in reactions of acid chlorides with alcohols, thiols, amines and water. The mechanism of nucleophilic catalysis is shown in these steps.

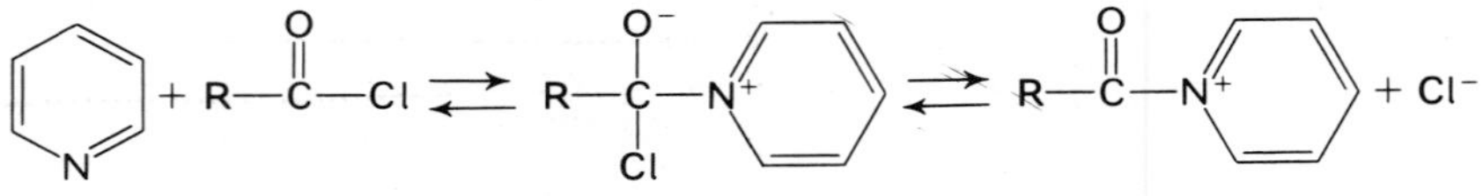

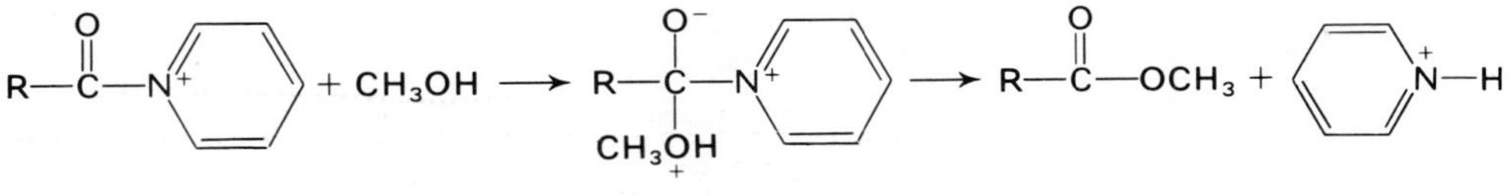

pyridine   acyl   acylpyridinium
chloride   ion

Pyridine reacts more rapidly with the acid chloride than does the alcohol. The substitution product with pyridine, the acylpyridinium ion, cannot lose a proton, and is unstable itself but forms a stable product by reacting with the alcohol. The catalysis operates in two ways. Pyridine is a better leaving group than Cl⁻ and a better nucleophile than the alcohol. The two processes together occur faster than the substitution by alcohol alone. In addition, pyridine neutralizes the HCl formed. In this reaction pyridine acts both as a catalyst and a reactant and therefore must be used in more than a stoichiometric amount.

Enzymes, nature's catalysts, are very large protein molecules. Side chains on the proteins have functional groups which may catalyze reactions in one or in several of the ways listed above. Compare the diagram of catalysis by an enzyme in Section 7.11. Our knowledge of the specific chemical steps that occur at the active sites of enzymes is limited. A few enzymes have been purified and studied extensively, and the number is increasing. The study of the mechanisms of enzyme catalysis is under active investigation in many laboratories. What is already known indicates that catalysis by enzymes combines in one overall reaction a procession of very fast, catalyzed steps. The reactants are absorbed into a cavity of the enzyme and oriented in a way advantageous for catalysis by the functional group bound to the enzyme. The combination of both acidic and basic groups in the enzyme that catalyzes the hydration of fumarate ion (the enzyme *fumarase*) was shown in Section 7.11. The structure and the catalytic role of another enzyme, *chymotrypsin*, which catalyzes transacylation reactions, is described in detail in Section 10.18.

Coenzymes are organic compounds whose presence is required for some enzyme-catalyzed reactions to occur. They are sometimes called "cofactors." In many cases the coenzyme is a reactant which produces a more reactive substrate with which the final reagent combines. The coenzyme is regenerated by the main reaction or by a subsequent reaction. An example is the formation of an acyl phosphate and of a thioester, acyl coenzyme A, in the conversion of fatty acids and alcohols to esters shown in Section 9.16.

The relationship between the water-soluble vitamins and the coenzymes is now reasonably well known, except for vitamin C. Each of the water-soluble vitamins is needed in the diet in trace amounts to act as coenzymes, or to be incorporated into the structure of a coenzyme: pantothenic acid in coenzyme A (Section 9.8); riboflavin (B_2) in FAD (Section 14.9); nicotinic acid or nicotinamide ("niacin") in NAD^+ (Section 11.13); thiamine (B_1), pyridoxal (B_6) (Section 11.14), lipoic acid, cobalamins (B_{12}), folic acid, and biotin as coenzymes themselves. No specific coenzyme function has been found for vitamin C or for the four fat-soluble vitamins, A, D, E, and K.

One of the current great challenges to the organic chemist is to try to synthesize compounds which can catalyze reactions in the laboratory in the manner of coenzymes.

9.13 FATS AND PHOSPHOLIPIDS

Lipids are water-insoluble components of cells that can be extracted from the tissue by nonpolar solvents. Two important classes of lipids are fats and phospholipids. You know already that fats are triesters of glycerol and long straight-chain fatty acids. Phospholipids are glyceryl esters of two fatty acids and one phosphate ester.

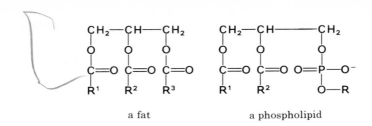

a fat   a phospholipid

The fatty acid residues (acyl residues, $RC=O$) in one molecule may possess the same or different structures. They usually contain an even number of carbon atoms, are from 12 to 24 carbon atoms long, and may be saturated or unsaturated.

The distribution of different acyl groups in fats varies with the source of the fat. Historical common names of acids often are derived from the source of the fat and may indicate the principal acid constituent in that organism—stearic in animal fat, palmitic in palm oil, oleic in olive oil, linoleic and linolenic in linseed oil. Structures of commonly occurring fatty acids are given in Fig. 9.1.

FIGURE 9.1   SOME NATURALLY OCCURRING FATTY ACIDS

Carbons	Formula	Name IUPAC	Common
		Saturated	
12	$CH_3(CH_2)_{10}CO_2H$	Dodecanoic	Lauric
14	$CH_3(CH_2)_{12}CO_2H$	Tetradecanoic	Myristic
16	$CH_3(CH_2)_{14}CO_2H$	Hexadecanoic	Palmitic
18	$CH_3(CH_2)_{16}CO_2H$	Octadecanoic	Stearic
20	$CH_3(CH_2)_{18}CO_2H$	Eicosanoic	Arachidic
		Unsaturated (all cis $CH=CH$)	
16	$CH_3(CH_2)_5CH=CH(CH_2)_7CO_2H$	9-Hexadecenoic	Palmitoleic
18	$CH_3(CH_2)_7CH=CH(CH_2)_7CO_2H$	9-Octadecenoic	Oleic
18	$CH_3(CH_2)_4CH=CHCH_2CH=CH(CH_2)_7CO_2H$	9,12-Octadecadienoic	Linoleic
18	$CH_3CH_2CH=CHCH_2CH=CHCH_2CH=CH(CH_2)_7CO_2H$	9,12,15-Octadecatrienoic	Linolenic

Glyceryl triesters are low-melting solids or liquids at room temperature. Olive oil solidifies in the refrigerator. The melting points are directly related to the degree of unsaturation of the acyl chains. The greater the number of double bonds present, the lower the melting point. The double bonds usually have a *cis* arrangement in the carbon chains as shown in Figs. 9.1 and 9.2.

FIGURE 9.2   MOLECULAR MODELS OF SOME FATTY ACIDS

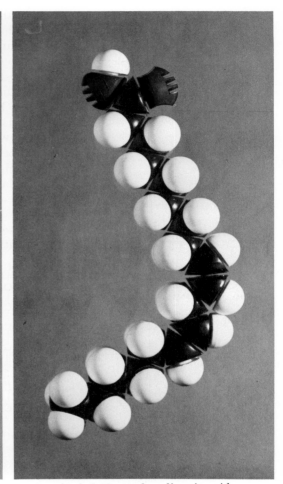

octadecanoic acid   *cis*-9-octadecenoic acid   *cis,cis*-9,12-octadecadienoic acid

	Acid mp, °C	Glyceryl triester mp, °C
Octadecanoic	77	53
cis-9-Octadecenoic	16	−5
trans-9-Octadecenoic	45	38

Solid margarine is prepared from liquid vegetable oils by the catalytic hydrogenation of some of the C=C double bonds of the unsaturated acyl groups. The label on the package says it contains "partially hydrogenated" corn oil or soybean oil.

Packing of fats in a crystalline matrix becomes more difficult with an increase in the number of *cis* double bonds. For the same reason, polyunsaturated fatty acids deposit as cholesteryl esters less easily in membranes than do saturated fats. Liquid vegetable oils have more unsaturation than solid animal fats. Not surprisingly, arctic mammals are richer in unsaturated fats than mammals of moderate climates.

Fats are one of the three main classes of food. Of the three, they provide the most heat per unit weight in metabolic processes, because of the large number of C—C and C—H bonds which are changed to C—O and H—O bonds. On total combustion, fats provide about 9.5 kcal/g, which is over twice the average value of 4.0 for proteins and carbohydrates which are already rich in C—O, O—H, and N—H bonds.

Lipids that occur in cellular membranes of brain, liver, and kidney are more complex. They differ from fats in that one of the fatty acid residues is replaced by a phosphate ester. The most abundant of these **phosphatides** (or phosphoglycerides) are **phosphatidyl choline** (lecithin) and **phosphatidyl ethanolamine** (cephalin from brain tissue). In these compounds, R^1 is usually a saturated chain, and R^2 is usually unsaturated. Phosphatides are white waxy solids.

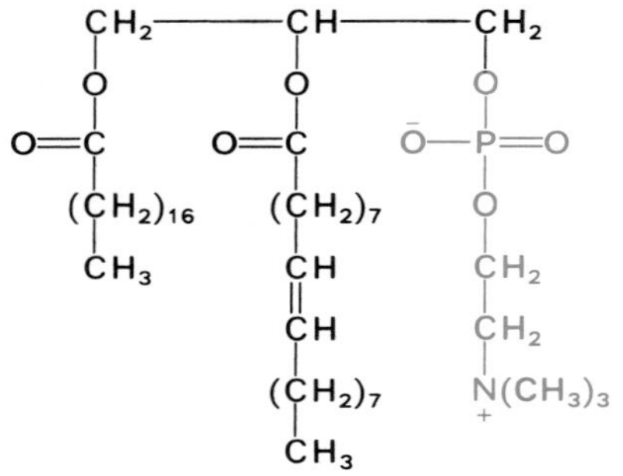

phosphatidyl choline
(lecithin)

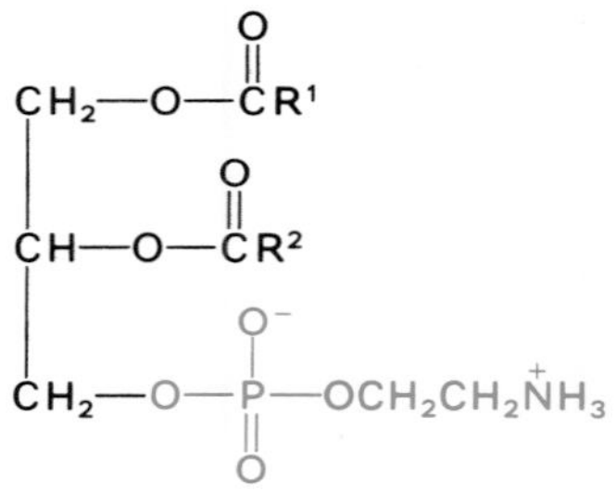

phosphatidyl ethanolamine
(cephalin)

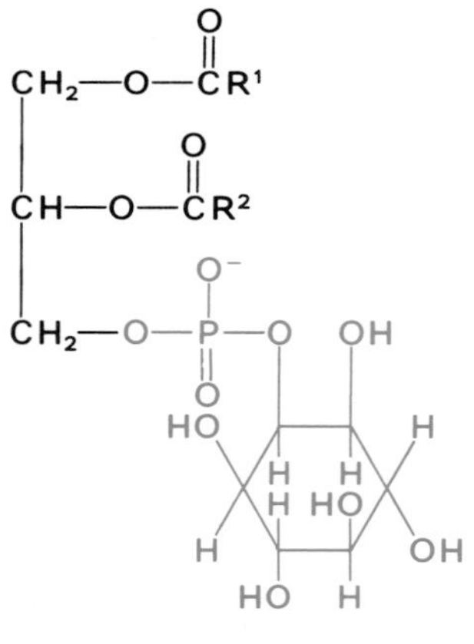

phosphatidyl inositol

In the laboratory, under weakly basic conditions, hydrolysis of phosphatides occurs at the two carboxylic ester units leaving the phosphate diester intact.

$$\underset{\substack{\text{CH}_2\text{—O—}\overset{\displaystyle\text{O}}{\overset{\|}{\text{C}}}(\text{CH}_2)_{16}\text{CH}_3 \\ \text{CH—O—}\overset{\displaystyle\text{O}}{\overset{\|}{\text{C}}}(\text{CH}_2)_{16}\text{CH}_3 \\ \text{CH}_2\text{—O—}\overset{\displaystyle\text{O}^-}{\underset{\displaystyle\text{O}}{\overset{\|}{\text{P}}}}\text{—OCH}_2\text{CH}_2\text{NH}_2}}{} \xrightarrow[\text{mild conditions}]{2\,\text{Na}^+\text{OH}^-} \underset{\substack{\text{CH}_2\text{OH} \\ \text{CHOH} \quad \text{O}^- \\ \text{CH}_2\text{—O}\overset{\|}{\underset{\text{O}}{\text{P}}}\text{OCH}_2\text{CH}_2\text{NH}_2}}{} + 2\,\text{CH}_3(\text{CH}_2)_{16}\text{CO}_2^-\,\text{Na}^+$$

$$\xrightarrow[\text{severe conditions}]{3\,\text{Na}^+\text{OH}^-} \underset{\substack{\text{CH}_2\text{OH} \\ \text{CHOH} \quad \text{O}^- \\ \text{CH}_2\text{—O—}\overset{\|}{\underset{\text{O}}{\text{P}}}\text{—O}^-}}{} + 2\,\text{CH}_3(\text{CH}_2)_{16}\text{CO}_2^-\,\text{Na}^+ + \text{HOCH}_2\text{CH}_2\text{NH}_2$$

Under more strenuous basic conditions, hydrolysis of one of the alkyl phosphate linkages also occurs, leaving the glyceryl phosphate. The two negative charges on the remaining alkyl phosphate repel attacking hydroxide ions and the glyceryl phosphate is relatively stable to alkaline hydrolysis. The phosphate ester can cleave under slightly acidic conditions by reacting with neutral water.

Problem **9.10** Although glycerol is optically inactive, carbon #2 becomes asymmetric whenever the ester substituents on carbons #1 and #3 are different. The natural phosphoglycerides and phosphatidic acids (without the second alkyl group on the phosphate) exist as the enantiomer shown. The carbons are numbered with the acyl group #1 and the phosphate group #3. Is this natural enantiomer R or S?

$$\underset{\substack{\text{CH}_2\text{—O—}\overset{\displaystyle\text{O}}{\overset{\|}{\text{C}}}\text{R}^1 \\ \text{R}^2\overset{\displaystyle\text{O}}{\overset{\|}{\text{C}}}\text{—O}\blacktriangleright\text{C}\blacktriangleleft\text{H} \quad \text{O} \\ \text{CH}_2\text{—O—}\overset{\|}{\underset{\text{OH}}{\text{P}}}\text{—O}^-}}{}$$

a phosphatidic acid

9.14 SOAPS, DETERGENTS, AND OTHER EMULSIFIERS

From earliest times, it has been known that animal fats heated with wood ash ("lye") produce a substance called "soap" which was used in washing. The modern process of soap making, or **saponification**, is the hydrolysis of the triester with sodium hydroxide to give glycerol and a mixture of sodium salts of the fatty acids. The salt produced is used as soap. "Ivory" soap is sodium stearate. "Palmolive" soap contains sodium palmitate.

$$(R = \text{long carbon chain})$$

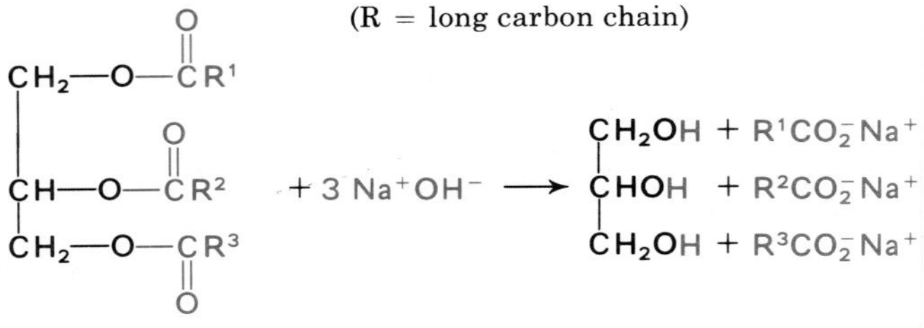

$$
\begin{array}{l}
CH_2-O-\overset{O}{\underset{}{C}}R^1 \\[6pt]
CH-O-\overset{O}{\underset{}{C}}R^2 \quad + \; 3\,Na^+OH^- \longrightarrow \\[6pt]
CH_2-O-\underset{O}{\overset{}{C}}R^3
\end{array}
\qquad
\begin{array}{l}
CH_2OH + R^1CO_2^-Na^+ \\[6pt]
CHOH \;\; + R^2CO_2^-Na^+ \\[6pt]
CH_2OH + R^3CO_2^-Na^+
\end{array}
$$

glyceryl triester   glycerol   soap

A soap is an emulsifier. It is capable of holding together in one phase oil and water, two mutually immiscible substances. The emulsifier is not wholly soluble in either oil or water, but one part of the structure is strongly polar (usually ionic) and one part is nonpolar.

Ordinary soaps, the sodium salts of large fatty acids, are excellent emulsifiers. They have an ionic end $-CO_2^-Na^+$ which is quite compatible with water, and a long nonpolar hydrocarbon chain which is incompatible with water, but compatible with oils. In a mixture of oil and water these ions become oriented with the ionic end well solvated by water molecules. The carbon chains, excluded from the water, dissolve in the oil droplets. The oily droplets containing the hydrocarbon portion have a coating of negatively charged carboxylate groups. The negative charges on the oil repel other droplets and prevent their coming together to form a single layer of oil. The emulsion can persist for a long time. These small oily pockets distributed throughout the water are called **micelles**.

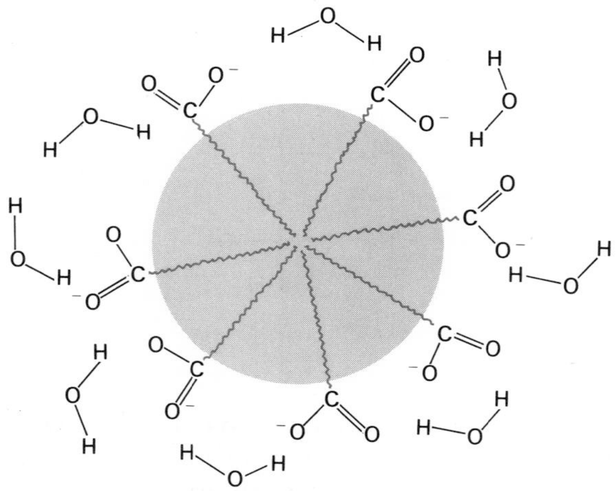

Calcium and other metal salts of fatty acids are not soluble in water. Addition of soap to "hard water" which contains calcium ions causes calcium fatty acid salts to precipitate. "Bathtub ring" is such a precipitate. "Softened water" is water in which the calcium ions have been replaced by sodium ions.

Detergents act in the same fashion as do soaps. They are sodium salts of alkyl-benzenesulfonic acids instead of carboxylic acids. They do not precipitate with calcium and other metal ions.

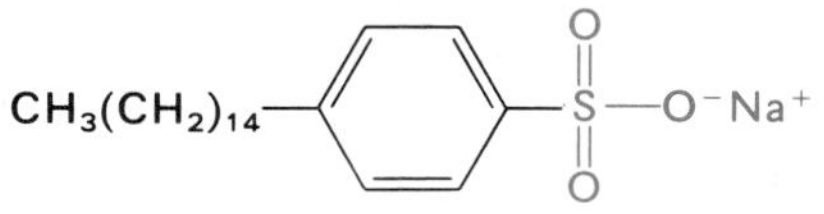

$$CH_3(CH_2)_{14} - \bigcirc - \overset{\overset{\displaystyle O}{\|}}{\underset{\underset{\displaystyle O}{\|}}{S}} - O^- Na^+$$

a detergent

The role of emulsifying agent is played by a number of biological compounds. Most of the cholesterol in mammals is converted to bile acids, which are steroids containing a carboxyl group in their side chains, such as **cholic acid**. The anion of cholic acid has three hydroxyls and the carboxylate ion group attached to a hydrophobic polycyclic nucleus. All of the hydroxyls are localized on one face of the molecule.

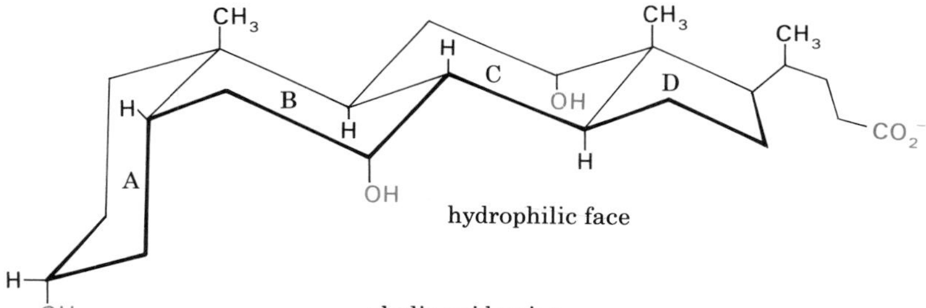

cholic acid anion

Micelles of cholic acid salts interact on the hydrophobic face with nonpolar lipids—triglycerides and cholesterol. The hydrophilic face interacts with an aqueous environment.

The bile acid anions aid in absorption of lipids through the walls of the small intestine by forming droplets of triacylglycerols stabilized by a small amount of protein.

Cholic acid also occurs as its amide with the amino group of the amino acid glycine (α-aminoacetic acid).

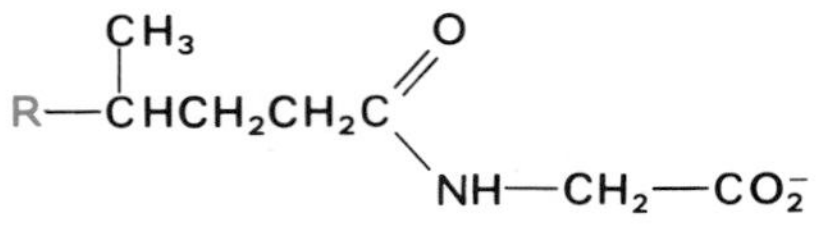

R = tetracyclic nucleus of cholic acid

cholylglycine

Problem **9.11** How many asymmetric carbons are there in cholic acid? Are the bonds joining ring B to ring A *cis* or *trans* on ring A? What about bonds joining ring B to C? Ring C to D?

The structural requirements for an emulsifier are also fulfilled by the phosphoglycerides, which contain strongly polar or ionic heads and long carbon chain tails. At pH 7 the phosphate group is an anion with one negative charge and the ammonium ions of ethanolamine and choline have positive charges.

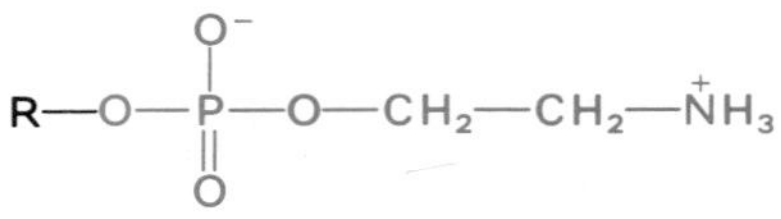

A model of phosphatidyl choline is shown in Fig. 9.3.

Phosphoglycerides dissolve only slightly in water, but are dispersed as micelles. Lecithin (phosphatidyl choline) is a constituent of hen eggs and makes an egg yolk a good emulsifying agent. Lecithin is used in some prepared foods as an emulsifying agent.

Phosphoglycerides also form a bilayer between two aqueous layers as shown in Fig. 9.4. Phospholipid bilayers of this sort have been studied extensively since their properties are very similar to those of natural cellular membranes. Because phospholipids readily form such bilayers spontaneously, it has been suggested that natural membranes may contain a lipid bilayer core. The study of transport of ions through such bilayers and through natural membranes is currently being vigorously pursued.

Variations in the size, shape, polarity, and charge of the polar heads of the phosphoglycerides play a significant role in the structures of micelles, of monolayers, and bilayers of lipids and of membranes.

**FIGURE 9.3
SPACE-FILLING MODEL OF
PHOSPHATIDYL CHOLINE
AT pH 7**

**FIGURE 9.4
MOLECULAR BILAYER OF
PHOSPHOLIPID BETWEEN
TWO AQUEOUS LAYERS**

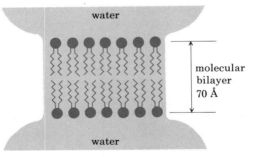

Bilayer is freely permeable to water,
but is impermeable to simple metal ions.

9.12 Would triacyl-, diacyl- or monoacylglycerols have properties appropriate for forming micelles?

9.15 MODELS OF BIOLOGICAL PHOSPHATE COMPOUNDS

In biological systems, none of the reactions previously discussed are appropriate for the conversion of a carboxylic acid to its ester or amide. Why is this true?

In the evolutionary chemistry of nature, an acylating agent emerged with high enough energy to react readily with a thiol or alcohol without being immediately hydrolyzed by the aqueous medium. The activated carboxyl derivative employed in biological conversion from acids to esters and amides is an **acyl phosphate** (a mixed anhydride of carboxylic and phosphoric acids) or an **acyl phosphate ester**.

$$CH_3-\overset{\overset{\displaystyle O}{\|}}{C}-O-\overset{\overset{\displaystyle O^-}{|}}{\underset{\underset{\displaystyle O}{\|}}{P}}-OH \qquad\qquad CH_3-\overset{\overset{\displaystyle O}{\|}}{C}-O-\overset{\overset{\displaystyle O^-}{|}}{\underset{\underset{\displaystyle O}{\|}}{P}}-O-CH_3$$

acetyl phosphate (an acyl phosphate)   methyl ester of acetyl phosphate

The mixed anhydride reacts readily with nucleophilic reagents. There are two points of electron deficiency and two good leaving groups. Attack on the carboxyl carbon by a nucleophile gives C—O cleavage, and the phosphate ion becomes a leaving group in a reaction which resembles the reaction of an acid chloride. Attack on phosphorus by a nucleophile gives P—O cleavage, and acetate ion is the leaving group.

$$CH_3-\overset{\overset{\displaystyle O}{\|}}{C}-O-\overset{\overset{\displaystyle O^-}{|}}{\underset{\underset{\displaystyle O}{\|}}{P}}-OH + Nu^-\Big\langle\begin{array}{l} CH_3-\overset{\overset{\displaystyle O}{\|}}{C}-Nu + HPO_4^{2-} \\[2ex] CH_3-\overset{\overset{\displaystyle O}{\|}}{C}-O^- + Nu-\overset{\overset{\displaystyle O^-}{|}}{\underset{\underset{\displaystyle O}{\|}}{P}}-OH \end{array}$$

The laboratory reactions of acetyl phosphate with alcohols and thiols (preferably the thiolate ion) yield the expected carboxylate esters and phosphate ion. Reaction with water at pH 7 is very slow, but hydroxide ion present at higher pH gives the two acid anions, acetate and phosphate. These reactions obviously proceed by nucleophilic attack on the carbonyl group of the acetyl phosphate.

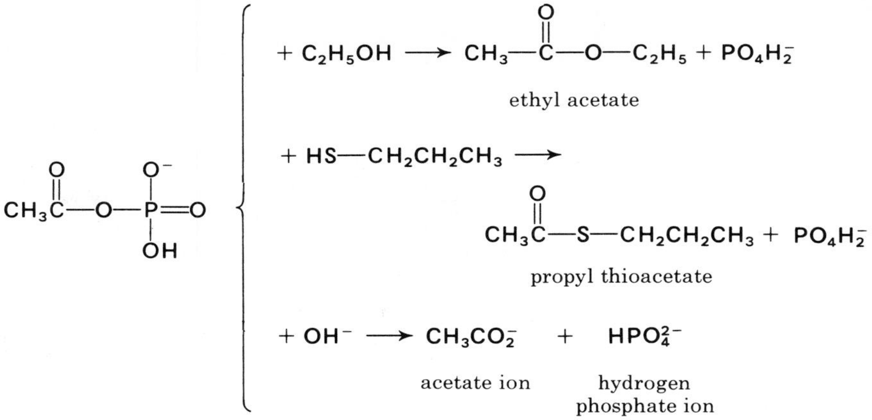

$$\text{CH}_3\overset{\overset{\displaystyle O}{\|}}{\text{C}}-\text{O}-\overset{\overset{\displaystyle O^-}{|}}{\underset{\underset{\displaystyle OH}{|}}{\text{P}}}=\text{O} \left\{ \begin{array}{l} \end{array} \right.$$

$$+ \text{ C}_2\text{H}_5\text{OH} \longrightarrow \text{CH}_3-\overset{\overset{\displaystyle O}{\|}}{\text{C}}-\text{O}-\text{C}_2\text{H}_5 + \text{PO}_4\text{H}_2^-$$

ethyl acetate

$$+ \text{ HS}-\text{CH}_2\text{CH}_2\text{CH}_3 \longrightarrow$$

$$\text{CH}_3\overset{\overset{\displaystyle O}{\|}}{\text{C}}-\text{S}-\text{CH}_2\text{CH}_2\text{CH}_3 + \text{PO}_4\text{H}_2^-$$

propyl thioacetate

$$+ \text{ OH}^- \longrightarrow \text{CH}_3\text{CO}_2^- \quad + \quad \text{HPO}_4^{2-}$$

acetate ion   hydrogen phosphate ion

Amines tend to attack the phosphorus atom and yield acetate ion and the amides of phosphoric acid.

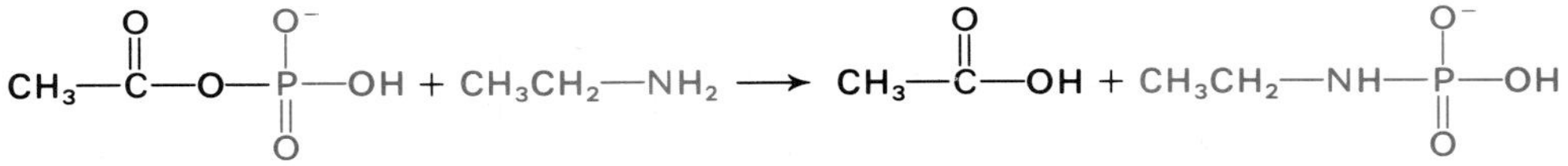

$$\text{CH}_3-\overset{\overset{\displaystyle O}{\|}}{\text{C}}-\text{O}-\overset{\overset{\displaystyle O^-}{|}}{\underset{\underset{\displaystyle O}{\|}}{\text{P}}}-\text{OH} + \text{CH}_3\text{CH}_2-\text{NH}_2 \longrightarrow \text{CH}_3-\overset{\overset{\displaystyle O}{\|}}{\text{C}}-\text{OH} + \text{CH}_3\text{CH}_2-\text{NH}-\overset{\overset{\displaystyle O^-}{|}}{\underset{\underset{\displaystyle O}{\|}}{\text{P}}}-\text{OH}$$

9.16 BIOLOGICAL ESTER FORMATION

The alkyl ester of an acyl phosphate is the usual form of the reactive carboxyl derivative employed in the biosynthesis of esters. The alkyl group in these compounds is **adenosine**, which you first encountered in S_N2 transmethylation reactions in Section 6.10. The adenosyl ester of butanoyl phosphate is shown.

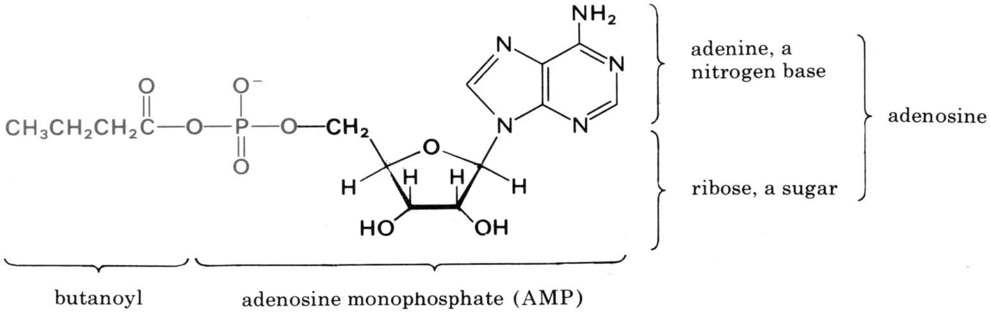

butanoyl   adenosine monophosphate (AMP)

butanoyl adenosyl monophosphate

The acyl adenosyl monophosphate is formed in the reaction of ATP (adenosine triphosphate) with the carboxylate ion acting as nucleophile and diphosphate ion as leaving group. This reaction is analogous to that of an acyl chloride with carboxylate ion which produces an anhydride (Section 8.19).

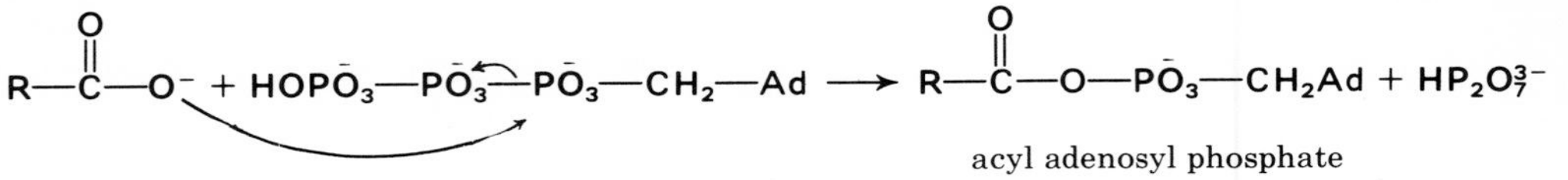

$$\text{R} - \overset{\overset{\displaystyle O}{\|}}{\text{C}} - \text{O}^- + \text{HOPO}_3 - \text{PO}_3^- - \text{PO}_3 - \text{CH}_2 - \text{Ad} \longrightarrow \text{R} - \overset{\overset{\displaystyle O}{\|}}{\text{C}} - \text{O} - \text{PO}_3 - \text{CH}_2\text{Ad} + \text{HP}_2\text{O}_7^{3-}$$

acyl adenosyl phosphate

The reaction of a nucleophile with the butanoyl adenosyl phosphate is analogous to that of acetyl phosphate in the previous section. Nucleophilic attack on the carbonyl carbon displaces the adenosyl phosphate group and produces a new carboxylic acid derivative.

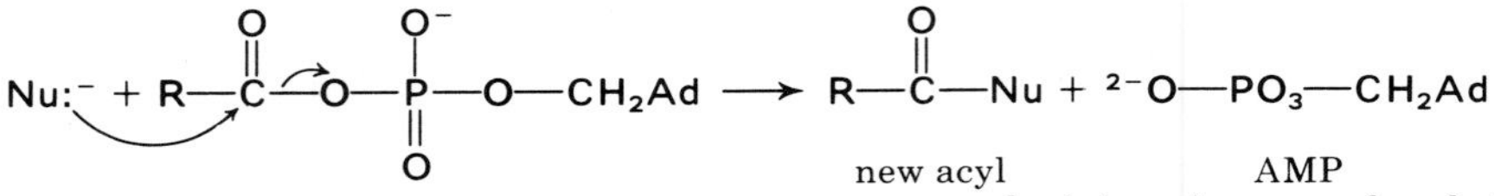

$$\text{Nu:}^- + \text{R} - \overset{\overset{\displaystyle O}{\|}}{\text{C}} - \text{O} - \overset{\overset{\displaystyle O^-}{|}}{\underset{\underset{\displaystyle O}{}}{\text{P}}} - \text{O} - \text{CH}_2\text{Ad} \longrightarrow \text{R} - \overset{\overset{\displaystyle O}{\|}}{\text{C}} - \text{Nu} + {}^{2-}\text{O} - \text{PO}_3 - \text{CH}_2\text{Ad}$$

new acyl compound   AMP (adenosine monophosphate)

A model sequence of reactions for the biosynthesis of an ester is given in Fig. 9.5.

**FIGURE 9.5
MODEL FOR BIOLOGICAL
ESTER FORMATION**

Reaction 1 Carboxylate ion nucleophile reacts with phosphorus atom at ester end of adenosine triphosphate, displacing the diphosphate ion as leaving group.

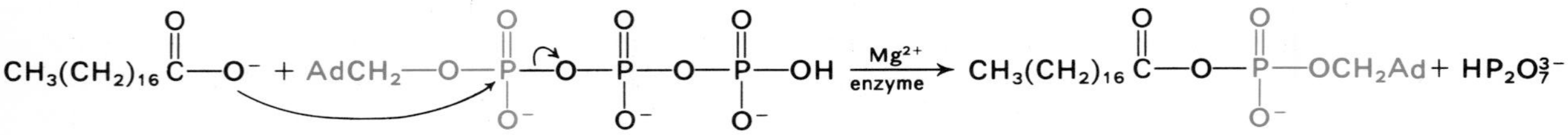

$$\text{CH}_3(\text{CH}_2)_{16}\overset{\overset{\displaystyle O}{\|}}{\text{C}} - \text{O}^- + \text{AdCH}_2 - \text{O} - \overset{\overset{\displaystyle O}{\|}}{\underset{\underset{\displaystyle O^-}{|}}{\text{P}}} - \text{O} - \overset{\overset{\displaystyle O}{\|}}{\underset{\underset{\displaystyle O^-}{|}}{\text{P}}} - \text{O} - \overset{\overset{\displaystyle O}{\|}}{\underset{\underset{\displaystyle O^-}{|}}{\text{P}}} - \text{OH} \xrightarrow[\text{enzyme}]{\text{Mg}^{2+}} \text{CH}_3(\text{CH}_2)_{16}\overset{\overset{\displaystyle O}{\|}}{\text{C}} - \text{O} - \overset{\overset{\displaystyle O}{\|}}{\underset{\underset{\displaystyle O^-}{|}}{\text{P}}} - \text{OCH}_2\text{Ad} + \text{HP}_2\text{O}_7^{3-}$$

octadecanoate ion   ATP   octadecanoyl adenosyl
(stearate ion)   adenosine triphosphate   phosphate

AdCH$_2$— = -sugar–nitrogen base = -ribose–adenine = -adenosine (formula above)

Reaction 2 The nucleophilic thiol, coenzyme A (formula, Section 9.8) displaces the phosphate ester as leaving group, forming the thioester.

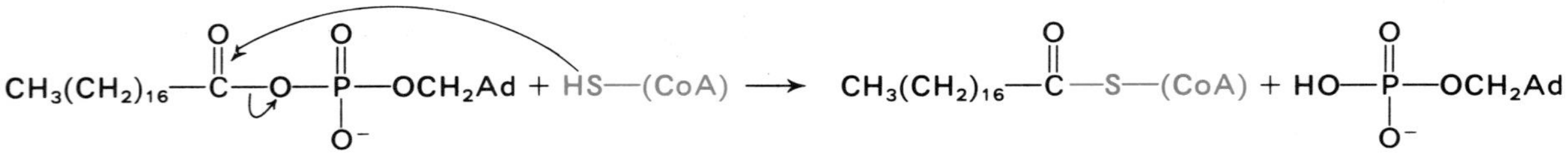

$$\text{CH}_3(\text{CH}_2)_{16} - \overset{\overset{\displaystyle O}{\|}}{\text{C}} - \text{O} - \overset{\overset{\displaystyle O}{\|}}{\underset{\underset{\displaystyle O^-}{|}}{\text{P}}} - \text{OCH}_2\text{Ad} + \text{HS} - (\text{CoA}) \longrightarrow \text{CH}_3(\text{CH}_2)_{16} - \overset{\overset{\displaystyle O}{\|}}{\text{C}} - \text{S} - (\text{CoA}) + \text{HO} - \overset{\overset{\displaystyle O}{\|}}{\underset{\underset{\displaystyle O^-}{|}}{\text{P}}} - \text{OCH}_2\text{Ad}$$

acyl phosphate ester   coenzyme A   acyl coenzyme A   adenosine
monophosphate

Reaction 3 The thiol, coenzyme A, is displaced by the nucleophile methanol, forming the methyl ester.

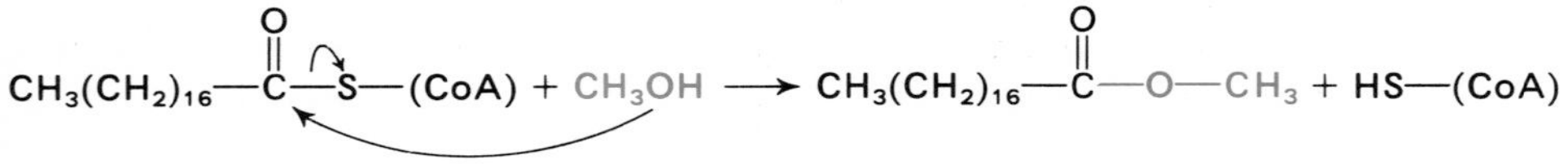

$$\text{CH}_3(\text{CH}_2)_{16} - \overset{\overset{\displaystyle O}{\|}}{\text{C}} - \text{S} - (\text{CoA}) + \text{CH}_3\text{OH} \longrightarrow \text{CH}_3(\text{CH}_2)_{16} - \overset{\overset{\displaystyle O}{\|}}{\text{C}} - \text{O} - \text{CH}_3 + \text{HS} - (\text{CoA})$$

octadecanoyl coenzyme A   methyl octadecanoate   coenzyme A

The biosynthesis of triacylglycerols from glycerol and fatty acids is known to require the action of the coenzymes ATP (adenosine triphosphate) and coenzyme A. The evidence available suggests that the chemical conversion goes through a series of nucleophilic substitutions on carboxylic acid derivatives, utilizing ATP and coenzyme A to form highly reactive intermediate compounds as depicted in Fig. 9.5.

Problem   **9.13** Give the structures of the products of these reactions.

a) ATP ($AdCH_2OPO_2$—OPO_2—OPO_3H^{3-}) + acetate ion $\xrightarrow{\text{enzyme}}$

b) ATP + CH_3—S—CH_2—CH_2—$\underset{\underset{\displaystyle ^+NH_3}{|}}{C}HCO_2^-$ $\xrightarrow[\text{($S_N$2, Section 6.10)}]{\text{enzyme}}$

(methionine)

c) acetyl phosphate anion + ethanethiol

d) acetyl phosphate anion + methylamine

e) Ac—O—$\overset{\displaystyle \overset{O}{\|}}{\underset{\displaystyle \underset{O^-}{|}}{P}}$—O—$CH_2Ad$ + coenzyme A $\xrightarrow{\text{enzyme}}$

acetyl adenosyl monophosphate

New Terms and Topics

Nomenclature of esters and thioesters (Section 9.3)
Polyesters (Section 9.1)
Types of catalysis—acid, base, nucleophilic, enzymic (Section 9.12)
Coenzymes (Sections 9.12 and 9.16)
Fatty acids, lipids, fats, phospholipids, triglycerides (Section 9.13)

Soaps, detergents, and emulsifiers (Section 9.14)
Saponification (Section 9.13)
Micelle, molecular bilayer membrane (Section 9.14)
Acyl phosphates, acyl alkyl phosphates (Section 9.15)
Biosynthesis of esters (Section 9.16)

Summary of Reactions

REACTIONS OF ESTERS

1. Amide formation (Section 9.6)

R—$\overset{\displaystyle \overset{O}{\|}}{C}$—OR + RNH_2 $\longrightarrow$ R—$\overset{\displaystyle \overset{O}{\|}}{C}$—NHR + ROH

$CH_3\underset{\displaystyle |}{\overset{\displaystyle |}{C}}H$—O—$\overset{\displaystyle \overset{O}{\|}}{C}$—$CH_2CH_2CH_3$ + $(CH_3CH_2)_2NH$ $\longrightarrow$ $CH_3CH_2CH_2$—$\overset{\displaystyle \overset{O}{\|}}{C}$—$N(CH_2CH_3)_2$ + CH_3—$\overset{\displaystyle \overset{OH}{|}}{C}H$—$CH_3$

isopropyl butanoate   diethylamine   N,N-diethylbutanamide   2-propanol

2. Transesterification (Section 9.5)

$$R \overset{\overset{O}{\parallel}}{-C} - OR^1 + R^2O^- \rightleftarrows R \overset{\overset{O}{\parallel}}{-C} - OR^2 + R^1O^-$$

$$CH_3(CH_2)_5\overset{\overset{CH_3}{|}}{CH} - O \overset{\overset{O}{\parallel}}{-C} CH_3 + CH_3O^- \rightleftarrows CH_3(CH_2)_5\overset{\overset{CH_3}{|}}{CH} - O^- + CH_3 - O \overset{\overset{O}{\parallel}}{-C} CH_3$$

| 2-octyl acetate | methoxide ion | 2-octoxide ion | methyl acetate |

3. Hydrolysis (Section 9.4)

$$R \overset{\overset{O}{\parallel}}{-C} - OR + OH^- \longrightarrow R \overset{\overset{O}{\parallel}}{-C} - O^- + ROH$$

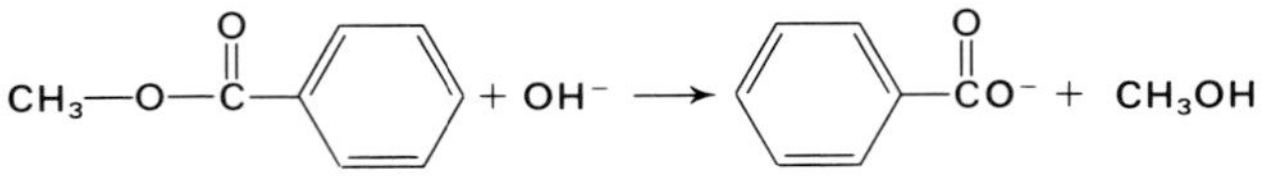

$$CH_3 - O \overset{\overset{O}{\parallel}}{-C} \langle\text{benzene ring}\rangle + OH^- \longrightarrow \langle\text{benzene ring}\rangle CO^- + CH_3OH$$

| methyl benzoate | benzoate ion |

4. Reduction by lithium aluminum hydride (Section 9.7)

$$R \overset{\overset{O}{\parallel}}{-C} - O - R' \xrightarrow{\text{LiAlH}_4} \xrightarrow{\text{H}^+} R - OH + R'OH$$

$$CH_3(CH_2)_3\overset{\overset{O}{\parallel}}{C} - O - CH_3 \xrightarrow{\text{LiAlH}_4} \xrightarrow{\text{H}^+} CH_3(CH_2)_3CH_2 - OH + CH_3 - OH$$

| methyl pentanoate | 1-pentanol | methanol |

REACTIONS OF THIOESTERS

1. Transesterification (Section 9.9)

$$R \overset{\overset{O}{\parallel}}{-C} - SR^1 + R^2O^- \longrightarrow R \overset{\overset{O}{\parallel}}{-C} - OR^2 + {}^-SR^1$$

$$CH_3(CH_2)_3 \overset{\overset{O}{\parallel}}{-C} - S - CH_2CH_2CH_3 + {}^-OCH_2CH_3 \longrightarrow CH_3(CH_2)_3 \overset{\overset{O}{\parallel}}{-C} - O - CH_2CH_3 + {}^-SCH_2CH_2CH_3$$

| propyl pentanethioate | ethoxide ion | ethyl pentanoate | propanethiolate ion |

2. Hydrolysis (Section 9.9)

$$R\overset{\overset{O}{\parallel}}{C} - SR^1 + 2 OH^- \longrightarrow R \overset{\overset{O}{\parallel}}{-C} - O^- + R^1S^- + H_2O$$

$$CH_3CH_2 \overset{\overset{O}{\parallel}}{-C} - S - CH_2CH_3 + 2 OH^- \longrightarrow CH_3CH_2 \overset{\overset{O}{\parallel}}{-C} - O^- + H_2O + {}^-SCH_2CH_3$$

| ethyl propanethioate | propanoate ion | ethanethiolate ion |

3. Amide formation (Section 9.9)

$$R\overset{\overset{\displaystyle O}{\|}}{C}\!-\!SR^1 + NH_3 \longrightarrow R\overset{\overset{\displaystyle O}{\|}}{C}\!-\!NH_2 + HSR^1$$

$$CH_3\overset{\overset{\displaystyle O}{\|}}{C}\!-\!S\!-\!\bigcirc + NH_3 \longrightarrow CH_3\overset{\overset{\displaystyle O}{\|}}{C}\!-\!NH_2 + HS\!-\!\bigcirc$$

| cyclopentyl | acetamide | cyclopentanethiol |
| thioacetate | | |

REACTIONS OF CARBOXYLIC ACIDS CATALYZED BY ACID

1. Acid-catalyzed esterification (an equilibrium) (Section 9.11)

$$RCO_2H + R^1OH \underset{\longleftarrow}{\overset{H^+(H_2SO_4)}{\longrightarrow}} RCO_2R^1 + H_2O$$

$$\bigcirc\!\!-\!\overset{\overset{\displaystyle O}{\|}}{C}\!-\!OH + CH_3OH \underset{\longleftarrow}{\overset{H^+(H_2SO_4)}{\longrightarrow}} \bigcirc\!\!-\!\overset{\overset{\displaystyle O}{\|}}{C}\!-\!OCH_3 + H_2O$$

benzoic acid   methyl benzoate

Problems

9.14 Write structures for the following compounds.

- **a)** butyl formate
- **b)** ethyl 3-methylhexanoate
- **c)** isobutyl chloroacetate
- **d)** isopropyl lactate
- **e)** diethyl malonate
- **f)** sodium tartrate
- **g)** acetyl coenzyme A (partial structure)
- **h)** *tert*-butyl butanethioate
- **i)** glyceryl tributanoate
- **j)** ethyl hexadecanoate
- **k)** methyl *cis,cis*-9,12-octadecadienoate (methyl linoleate)
- **l)** *sec*-butyl 2,2-dimethylpropanethioate
- **m)** ethyl 3-isopropylbenzoate
- **n)** phenyl 3-methylbutanoate

9.15 Name the following compounds.

a) $CH_3\overset{\overset{\displaystyle CH_3}{|}}{C}H\!-\!S\!-\!\overset{\overset{\displaystyle O}{\|}}{C}CH_2CH_3$

b) $CH_3CH_2\overset{\underset{\displaystyle CH_3}{|}}{C}H\!-\!O\!-\!CH\overset{\displaystyle O}{\underset{\|}{}}$

c) $CH_3\overset{\overset{\displaystyle CH_3}{|}}{C}HCH_2\!-\!O\!-\!\overset{\overset{\displaystyle O}{\|}}{C}CH_2\overset{\overset{\displaystyle CH_3}{|}}{C}HCH_3$

d) $Cl\!-\!\bigcirc\!\!-\!O\!-\!\overset{\overset{\displaystyle O}{\|}}{C}CH_3$

e) $(CH_3)_3C\!-\!S\!-\!\overset{\overset{\displaystyle O}{\|}}{C}\!-\!\bigcirc$

f) $C_2H_5\!-\!O\!-\!\underset{\underset{\displaystyle O}{\|}}{C}CH_2CH_2\underset{\underset{\displaystyle O}{\|}}{C}\!-\!O\!-\!C_2H_5$

g) $CH_3CH_2\overset{\underset{\displaystyle CH_3}{|}}{C}H\!-\!O\!-\!\underset{\underset{\displaystyle O}{\|}}{C}(CH_2)_7\underset{\underset{\displaystyle O}{\|}}{C}\!-\!OCH_3$

9.16 Write equations for these reactions.

    a) isopropyl acetate + sodium hydroxide solution

    b) isobutyl thiobenzoate + potassium hydroxide solution

    c) *tert*-butyl benzoate + sodium methoxide in methanol

    d) *sec*-butyl 2,3-dimethyl-3-ethylhexanoate + ammonia

    e) glyceryl tributanoate + potassium hydroxide solution

    f) dioctadecanoyl phosphotidyl choline + dilute sodium hydroxide

    g) ethyl butanethioate + propylamine

    h) 4-hydroxypentanoic acid + trace of sulfuric acid

9.17 Give the reagents needed to produce the compounds shown in each step.

    a) $CH_3CH_2CHCO_2H \xrightarrow{A} CH_3CH_2CHCOCl \xrightarrow{B} CH_3CH_2CHCO_2C_6H_5$
            $\underset{CH_3}{|}$ $\underset{CH_3}{|}$ $\underset{CH_3}{|}$

    b) $CH_3CH{=}CHCO_2H \xrightarrow{C} CH_3CH{=}CHCOCl \xrightarrow{D} CH_3CH{=}CHC{-}S{-}CH_2CH_3 \xrightarrow{E} CH_3CHCH_2C{-}S{-}CH_2CH_3$
                                                                 $\underset{O}{\|}$ $\underset{Br}{|}$ $\underset{O}{\|}$

    c) $CH_3CH{=}CHCO_2H \xrightarrow{F} CH_3CH{=}CHCO_2CH_3 \xrightarrow{G} CH_3\overset{OH}{\underset{|}{C}}HCH_2CO_2CH_3 \xrightarrow{H} CH_3\overset{O}{\overset{\|}{C}}O\overset{CH_3}{\underset{|}{C}}HCH_2CO_2CH_3$

9.18 The reactivity of acylating agents varies with the leaving group (L) in the following order of decreasing reactivity:

$$R{-}\overset{O}{\underset{\|}{C}}{-}Cl, \quad R{-}\overset{O}{\underset{\|}{C}}{-}O{-}\overset{O}{\underset{\|}{P}}(OH)_2, \quad R{-}\overset{O}{\underset{\|}{C}}{-}O{-}\overset{O}{\underset{\|}{C}}{-}R, \quad R{-}\overset{O}{\underset{\|}{C}}{-}S{-}Ar, \quad R{-}\overset{O}{\underset{\|}{C}}{-}O{-}Ar, \quad R{-}\overset{O}{\underset{\|}{C}}{-}S{-}R,$$

$$R{-}\overset{O}{\underset{\|}{C}}{-}O{-}R, \quad R{-}\overset{O}{\underset{\|}{C}}{-}NR_2$$

With what property of H—L does the above order correlate?

9.19 Explain the following facts.

    a) has a higher dipole moment than $H_3C{\diagdown}O{\diagdown}\overset{O}{\overset{\|}{C}}{\diagdown}CH_2{\diagup}CH_3$.

    b) boils at a higher temperature than $H_3C{\diagdown}O{\diagdown}\overset{O}{\overset{\|}{C}}{\diagdown}CH_2{\diagup}CH_3$.

    c) The fats of certain sea mammals that live in the polar regions are richer in *unsaturated* fatty acid groups than those that inhabit the equatorial regions of the earth.

d) Stearic acid is a compound which tends to arrange itself linearly both in crystals and in membranes, whereas linoleic acid cannot do this.

9.20 What common general structural feature and general physical property is shared by lecithin, cephalin, sodium salt of cholic acid, sodium stearate, and

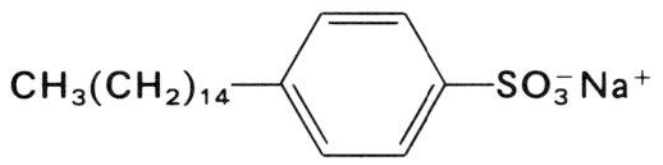

$$CH_3(CH_2)_{14}\text{—} \langle \text{benzene ring} \rangle \text{—}SO_3^-Na^+ \text{ ?}$$

9.21 Account for the following facts.

a) The hydrolysis of ethyl trifluoroacetate by sodium hydroxide is much faster than that of ethyl acetate.

b) The hydrolysis of ethyl 2,6-dimethylbenzoate by sodium hydroxide is exceedingly slow.

9.22 For biochemical experiments, ^{18}O-labeled ethyl benzoate was required. Unlabeled ethyl benzoate and $H_2^{18}O$ were available. Describe an experiment in which ^{18}O can be incorporated into the ester.

9.23 Sketch a series of reactions for the conversion of butanoic acid to its ethyl ester that would be a model for the biosynthesis of an ester using the coenzymes ATP and coenzyme A.

9.24 The biosynthesis of the amide of cholic acid and the amino group of glycine $(H_2N\text{—}CH_2\text{—}CO_2H)$, N-cholylglycine, requires the presence of coenzymes ATP and coenzyme A. Give a series of reactions to convert cholic acid to the amide.

CHAPTER 10

Amines, Amides, and Proteins

Ancient peoples probably became aware of amines earlier than most classes of organic compounds. The vile odors exuding from dead fish and decaying flesh acted as olfactory indicators of the age of potential food. The bitter taste of many plant seeds, leaves, twigs, barks, and roots is due to amines called alkaloids. Two of these alkaloids, quinine and morphine, are among the few ancient plant remedies that have survived the scrutiny of modern medicine.

Just as carboxylic acids are the acids of organic chemistry, so amines are the organic bases. A simple proton transfer between a carboxylic acid and an amine forms an ionic alkylammonium carboxylate salt. These groups individually and together help to maintain the proper pH of their environment.

Amino and carboxyl groups can also be combined by a covalent attachment to give amide groups. These amides are nearly neutral compounds of great laboratory stability, but are easily hydrolyzed biologically to ammonium and carboxylate ions again.

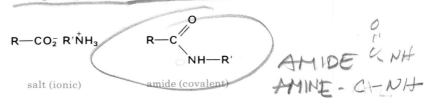

$$R\text{—}CO_2^- \; R'\overset{+}{N}H_3 \qquad\qquad R\text{—}C\overset{\displaystyle O}{\underset{\displaystyle NH\text{—}R'}{}}$$

salt (ionic)   amide (covalent)

While lipids are the nonpolar compounds of animals, amino acids are their most polar and water-soluble organic compounds. The combination of a carboxyl and an amino group in the same molecule exists largely as an intramolecular salt, produced by proton transfer from acid to amine base.

A covalent attachment of a large number of amino acids produces chains of amide linkages. In the process small, water-soluble, and mobile organic compounds are fashioned in the cell into large, immobile, and highly structured protein molecules. Proteins come closer than any other compounds to justifying the epithet "the compounds of life."

This chapter starts with the chemistry of amines and closes with proteins.

10.1 SOURCES AND USES OF AMINES

Small amines occur in living organisms as components of large molecules. Ethanolamine and choline cation in phosphoglycerides are examples (Section 9.13).

$$HO\text{—}CH_2\text{—}CH_2\text{—}NH_2 \qquad HO\text{—}CH_2\text{—}CH_2\text{—}\overset{+}{N}(CH_3)_3$$

ethanolamine   choline cation

Larger amines include the five nitrogen bases found in the nucleic acids of the genetic code, DNA and RNA, and in many coenzymes.

Adenine, one of the nucleic acid bases, also forms a part of ATP (adenosine triphosphate), coenzyme A, and the oxidation-reduction coenzymes NAD^+ and FAD (Section 13.9).

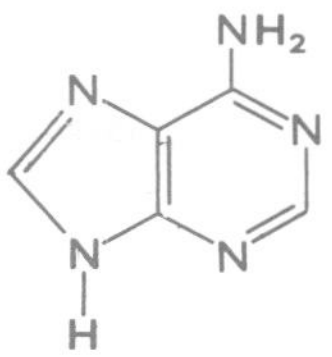

adenine
(nitrogen base in nucleic
acids, DNA and RNA)

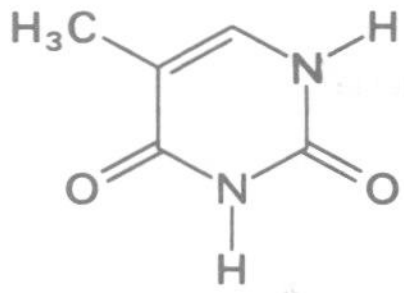

thymine
(base in DNA)

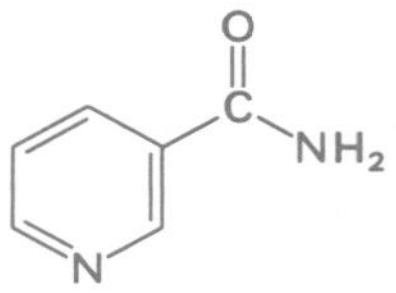

nicotinamide
(part of coenzyme NAD^+)

Larger multicyclic amines from plants, called alkaloids, include some that have marked physiological effects when taken orally: strychnine and brucine (poisonous), morphine and codeine (pain killers), lysergic acid amide (hallucinogen), quinine (medicinal). Note the variety of ring structures containing nitrogen atoms.

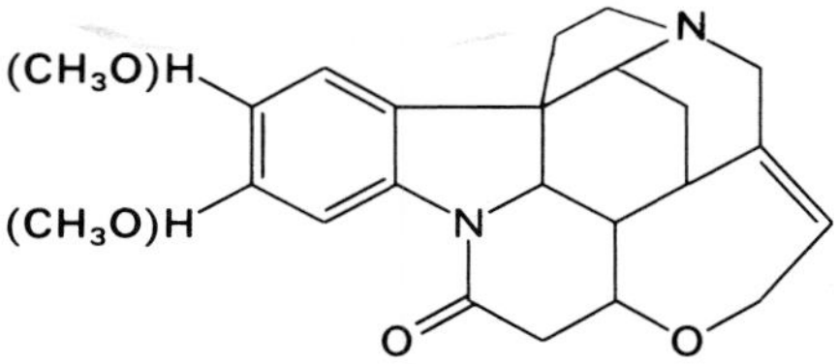

strychnine (H)
brucine (CH₃O)

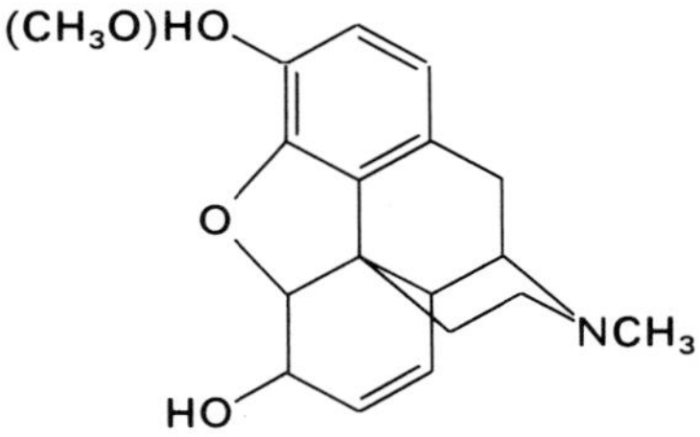

morphine (OH)
codeine (CH₃O)

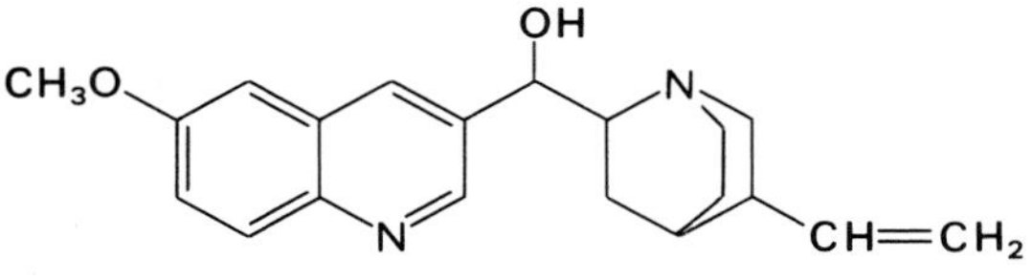

quinine

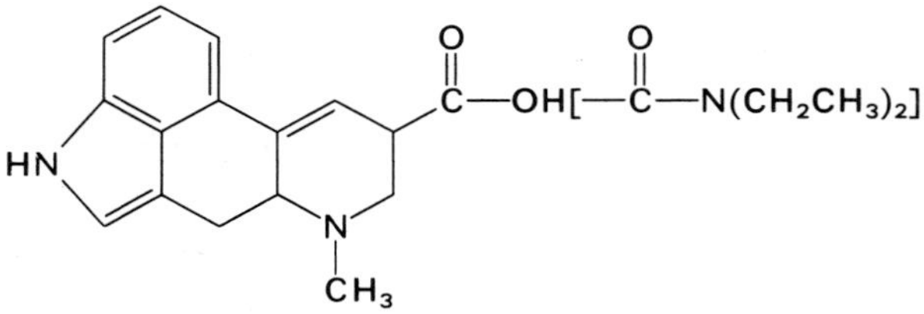

lysergic acid (C—OH)

LSD [—C—N(CH₂CH₃)₂]

The amino acids obtained from proteins are α-amino acids. Some also have other amino functional groups.

$$H_2N-CH_2CH_2CH_2CH_2\overset{\overset{\displaystyle +NH_3}{|}}{C}HCO_2^-$$

lysine

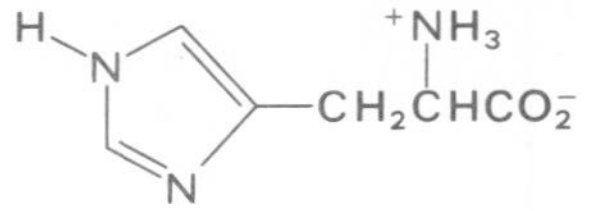

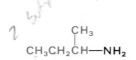

histidine

Problems

10.1 Adenine is one of the most useful and ubiquitous compounds found in animal metabolism. Its structure is shown above. Adenine is a bicyclic compound containing four nitrogens in a fused two-ring system.

a) What is the size and degree of unsaturation of each ring?

b) How are the fused rings joined?

c) How many of each kind of amino group (primary, secondary, tertiary) are there?

10.2 Strychnine is a multicyclic compound. How many rings are there and what size are they?

10.2 NOMENCLATURE OF AMINES

Recall that amines are classified as **primary, secondary, or tertiary,** according to the number of alkyl or aryl groups attached to the nitrogen. Primary amines have one group, secondary amines two groups, and tertiary amines have three groups attached to the nitrogen (Section 8.18).

$$CH_3-NH_2 \qquad (CH_3)_2NH \qquad (CH_3)_3N$$

methylamine   dimethylamine   trimethylamine
(primary)   (secondary)   (tertiary)

An ammonium salt having four carbon groups attached to the nitrogen is called a **quaternary salt.**

$$(CH_3)_4N^+Cl^-$$

tetramethylammonium chloride
(a quaternary ammonium salt)

1. In the IUPAC system primary amines are named by adding "amine" to (a) the name of the alkyl group or (b) the name of the parent hydrocarbon after deletion of the ending "e," whichever is simpler. For example, the compound having NH$_2$ on the #2 carbon of butane is usually called *sec*-butylamine, but the compound having the NH$_2$ on the #2 carbon of hexane is called 2-hexanamine or 2-hexylamine.

$$CH_3CH_2\overset{\overset{\displaystyle CH_3}{|}}{C}H-NH_2 \qquad CH_3CH_2CH_2CH_2-NH_2 \qquad \qquad CH_3(CH_2)_3\overset{\overset{\displaystyle CH_3}{|}}{C}H-NH_2$$

sec-butylamine   butylamine   ⬡—NH$_2$ 2-hexanamine
(2-butanamine)   (1-butanamine)   cyclohexylamine

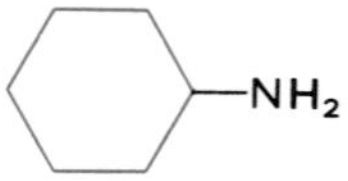

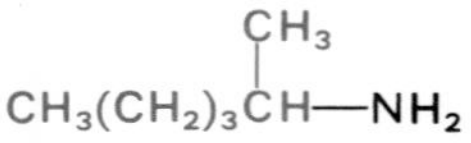

2. Symmetrical secondary and tertiary amines with simple hydrocarbon groups use the prefix "di-" or "tri-" to designate the number of groups, as in tripropylamine.

$(CH_3CH_2CH_2)_3N$

tripropylamine

Secondary and tertiary amines having a mixture of alkyl groups are named by utilizing the largest alkyl group as the base name. The other groups are designated by their alkyl group names, with N to indicate the position of attachment on nitrogen of the parent compound. The names of the smaller groups are placed as prefixes.

$CH_3CH_2CH_2CH_2CH_2$—$N(CH_3)_2$

N,N-dimethylpentylamine

3. Amines containing the NH_2 attached to the benzene ring are given the parent name **aniline**. The simplest compound is **aniline**.

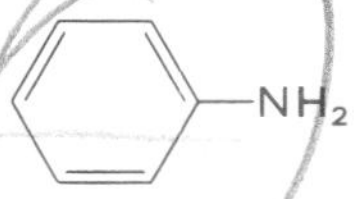

aniline

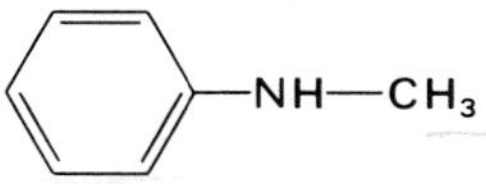

N-methylaniline

4-methylaniline

4. Cyclic amines, with N as a part of the ring, are usually called by their common names, as pyridine and piperidine.

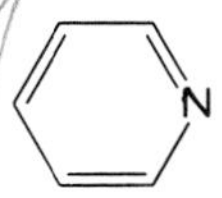

pyridine

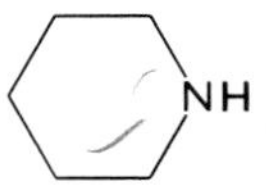

piperidine
(hexahydropyridine)

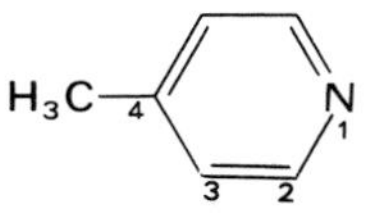

4-methylpyridine

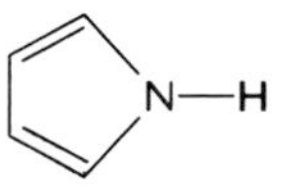

pyrrole

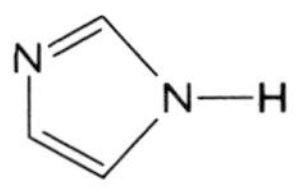

imidazole

5. Salts of amines are named by changing "amine" to "ammonium" for the cation and adding the name of the anion, as in N-methylethylammonium bromide. (The N may be dropped if no ambiguity results.)

CH_3—$\overset{+}{N}H_2$—CH_2CH_3 Br^-

N-methylethylammonium bromide

6. The names of salts of amines whose names do not end in "amine" are formed by replacing the "e" ending of the name with the ending "ium" and adding the name of the anion.

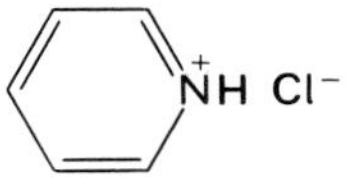

pyridinium chloride

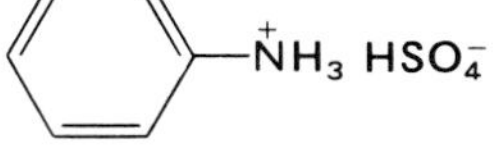

anilinium hydrogen sulfate

Problems **10.3** Name the following compounds.

a) 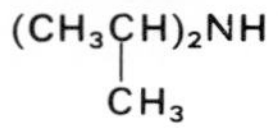 $(CH_3CH)_2NH$
   $\quad\quad\quad\;\; CH_3$

b) 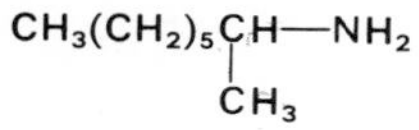 $CH_3(CH_2)_5CH—NH_2$
   $\quad\quad\quad\quad\quad\; CH_3$

c) 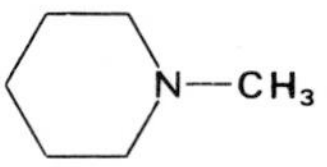 N—CH₃

d) 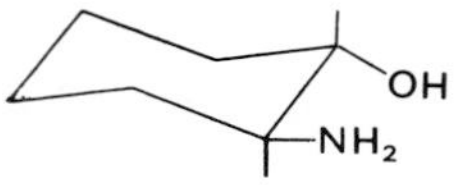 —OH, —NH₂

e) 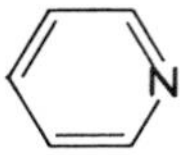 N

f) 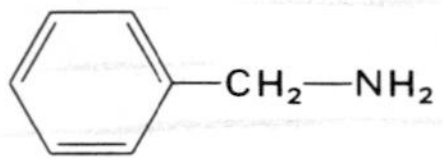 —CH₂—NH₂

10.4 Draw structures for these compounds.

    a) tributylamine   **b)** 2-methylpyridine

    c) diethyldimethylammonium ion

10.3 PHYSICAL PROPERTIES OF AMINES

The smallest amines, monomethyl-, dimethyl-, trimethylamine, and ethylamine, are gases. Most amines are liquids.

The gaseous and volatile liquid amines have the strong odor characteristic of dead fish. The common names of 1,4-butanediamine and 1,5-pentanediamine are "putrescine" and "cadaverine," terms which describe their odors better than any adjectives can. Larger amines have odors which are less unpleasant, though pyridine seems to be particularly obnoxious to many persons.

Pyridine and the smaller amines are water-soluble. The solubility in water decreases as the hydrocarbon parts of the molecules increase in size.

Primary and secondary amines are excellent hydrogen-bond donors and hydrogen-bond acceptors, whereas tertiary amines are only acceptors, since they have no hydrogen on nitrogen to donate. The boiling points of amines are higher than those of other compounds of comparable molecular weight except alcohols, acids, and amides, as shown in Fig. 10.1.

FIGURE 10.1 TRANSITION TEMPERATURES OF AMINES AND OTHER COMPOUNDS

Compound	Class	MW	mp, °C	bp, °C
$CH_3CH_2CH_2—NH_2$	Amine	59	−83	49
$CH_3CH_2CH_2—OH$	Alcohol	60	−127	97
CH_3CO_2H	Acid	60	17	118
CH_3CONH_2	Amide	59	82	221

10.4 AMINES AS ORGANIC BASES

Ammonia and amines are bases of moderate strength. One of the chief roles of an amine is in its action as a base.

$$CH_3CH_2CH_2{-}NH_2 + HCl \longrightarrow CH_3CH_2CH_2{-}\overset{+}{N}H_3\ Cl^-$$

propylamine (base)   propylammonium chloride (conjugate acid)

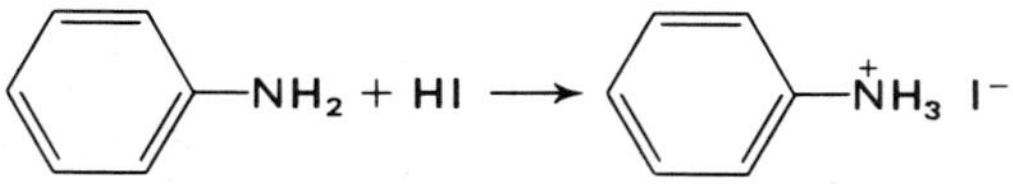

aniline + HI $\longrightarrow$ anilinium iodide

$-NH_2 + HI \longrightarrow -\overset{+}{N}H_3\ I^-$

aniline   anilinium iodide

The base strengths of amines depend on the structure of the carbon group. One alkyl group bonded to the nitrogen increases the basicity over that of ammonia by a factor of about 10 with a corresponding decrease in acid strength of the cation RNH_3^+. Further alkylation to secondary and tertiary amines has little effect on the basicity. In comparison with hydrogen, an alkyl group exerts a small electron-donating effect which increases the electron density and the basicity of attached atoms. (Compare the variation of carboxylic acids in Section 8.8.)

The strength of ammonium ions as acids is given in pK_a units in Fig. 10.2. Methylammonium ion has a pK_a of 10.6, which means that it is a weaker acid than acetic ($pK_a = 4.8$), but stronger than methanol ($pK_a \sim 16$). It also means that methylamine is a stronger base than acetate ion, but weaker than methoxide ion.

FIGURE 10.2
pK_a OF CONJUGATE ACIDS OF AMINES

Amine		Conjugate acid	pK_a* of RNH_3^+
Ammonia	NH_3	NH_4^+	9.2
Methylamine	CH_3NH_2	$CH_3NH_3^+$	10.6
Dimethylamine	$(CH_3)_2NH$	$(CH_3)_2NH_2^+$	10.7
Ethylamine	$CH_3CH_2NH_2$	$CH_3CH_2NH_3^+$	10.7
Cyclohexylamine	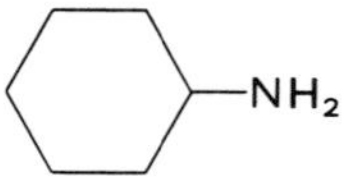 $-NH_2$	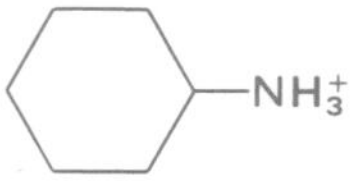 $-NH_3^+$	10.6
Aniline	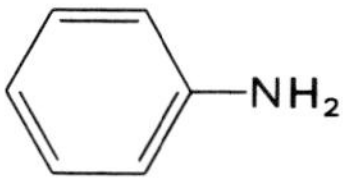 $-NH_2$	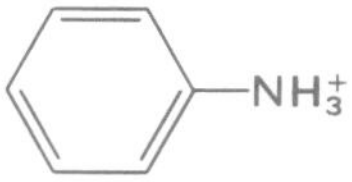 $-NH_3^+$	4.6
Pyridine	 N	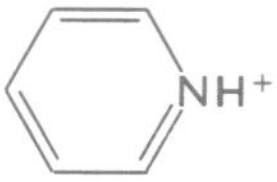 NH^+	5.2
Amide ion	NH_2^-	NH_3	34

* $pK_a = -\log_{10}K_a$
K_a = equilibrium constant for $RNH_3^+ + H_2O \rightleftharpoons RNH_2 + H_3O^+$

The introduction of unsaturation in amines in the form of $C=N-$ or $C=C-NH_2$ reduces substantially the base strength; pyridine and aniline are about 10^5 times weaker bases than alkylamines, and their cations are much stronger acids than RNH_3^+. Figure 10.2 gives the pK_a values of pyridinium and anilinium ions 5 and 6 units lower than the RNH_3^+ ions, showing these unsaturated cations to be 10^5 and 10^6 times stronger acids than the alkylammonium ions.

Most of the 10^6 difference in base strength between aniline and cyclohexylamine is accounted for as follows. In cyclohexylamine the unshared pair of electrons is completely localized on nitrogen. In aniline, however, the unshared pair of electrons on nitrogen is delocalized to some extent into the benzene ring, giving three resonance structures with a negative charge on a carbon of the ring and a positive charge on nitrogen. (Note the similarity of these resonance structures to those of phenol, Section 5.9.)

resonance structures of aniline

Donation of the unshared electrons of $-\dot{N}H_2$ to the aromatic ring of aniline reduces the electron density of the nitrogen and thus reduces the base strength. The lower energy and increased stability of the aniline molecule due to delocalization is lost when aniline accepts a proton, localizing the pair of electrons in the N—H bond.

On the other hand, the reduced basicity of pyridine arises from a different source. Protonation of pyridine does not destroy the stability of the molecule contributed by delocalization of electrons as it does with aniline, for the unshared pair on nitrogen is not a part of the delocalized system.

pyridine   pyridinium ion

Problem 10.5 Why can the pair of electrons on nitrogen in aniline delocalize into the ring while the pair on nitrogen of pyridine cannot? To answer this question draw the orbital picture of the pi electrons for both compounds and note the difference in the geometry of the N orbital containing the unshared pair of electrons. (Remember, in order to overlap side-to-side, orbitals must be parallel.)

The stronger acidity of pyridinium ion as compared to alkylammonium ions must be due to the difference in hybridization of bonds of the nitrogen atoms. In the pyridinium ion, nitrogen has sp^2 hybridization, and in a saturated ammonium ion it is sp^3. This difference is analogous to the increased acidity of acetylenic hydrogen ($-C\equiv C-H$) with sp hybridization over ethylenic hydrogen ($-CH=CH_2$), where the hybridization is sp^2 (Section 4.8).

Although the chief characteristic of amines is their basicity, they are also very weak proton donors ($pK_a \sim 33$). Like the more common stronger proton donors, water or alcohol, ammonia (liquid at $-33°$) reacts slowly with potassium metal to produce hydrogen and potassium amide. The amide ion is a very strong base.

$$2\ NH_3 + 2\ K^0 \longrightarrow H_2 + 2\ K^+NH_2^-$$

potassium amide

Amines are most conveniently converted to their lithium salts by reaction with the organometallic compound phenyllithium. The alkylamide ions produced are very strong bases.

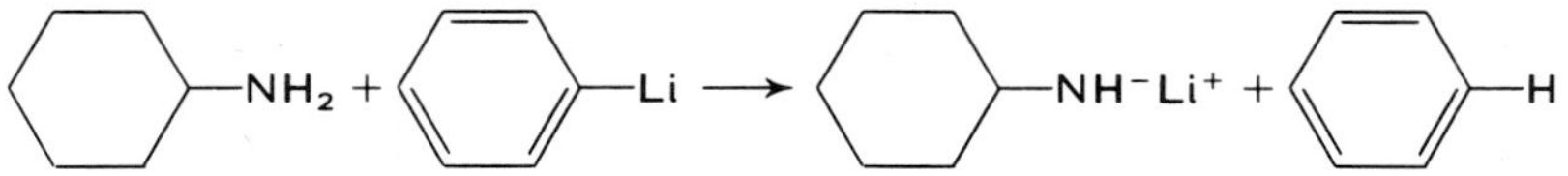

cyclohexylamine   phenyllithium   lithium   benzene
cyclohexylamide

Problem   **10.6** Arrange the compounds or ions in the following groups according to decreasing strength as bases.

a) $CH_3CH_2CH_2-NH_2$ CH_3-O^- $CH_3CO_2^-$ NH_2^-

b)

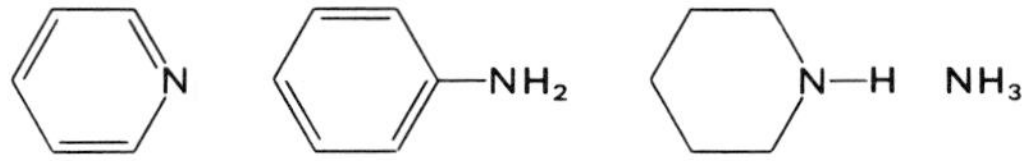

10.5 REACTIONS OF AMINES THAT FORM A NEW C—N BOND

Amines are better nucleophiles than are alcohols. We have seen that primary and secondary amines react with acyl chlorides to give N-substituted amides (Section 8.18). Tertiary amines, lacking a proton on the nitrogen, cannot form stable amides.

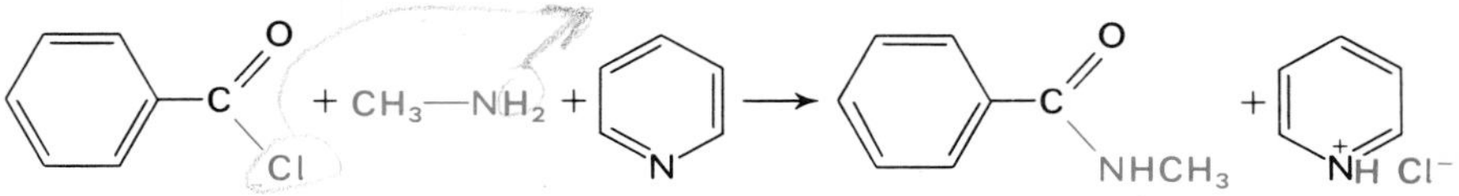

benzoyl   methyl-   pyridine   N-methyl-   pyridinium
chloride   amine   benzamide   chloride

Pyridine, a tertiary amine, absorbs the HCl produced in the above reaction, allowing all of the primary amine to react with the acyl chloride.

Similarly, amines react with alkyl halides in the usual substitution reaction. The products of these substitutions are secondary, tertiary, and quaternary ammonium salts. The ammonium ions which have a proton on nitrogen can be converted to free amine by treatment with sodium hydroxide.

$$CH_3-NH_2 + CH_3CH_2-Br \xrightarrow{-Br^-} CH_3-\overset{+}{N}H_2-CH_2CH_3 \xrightarrow{OH^-} CH_3-NH-CH_2CH_3 + H_2O$$

methylamine   bromoethane   methylethyl-   methylethylamine
                                    ammonium ion

This reaction is not a very useful one except in forming quaternary ammonium salts because a mixture of secondary, tertiary, and quaternary salts is formed. The newly formed amine also reacts with the alkyl halide.

The reaction of amines with phenyl isothiocyanate ($C_6H_5-N=C=S$) is used in the determination of the amino acid sequence of proteins (Section 10.16). The isocyanate group ($-N=C=O$) and the isothiocyanate group ($-N=C=S$) are very reactive to the neutral nucleophilic reagents, amines, alcohols, and water. The amine adds to the carbon–nitrogen double bond $-N=C=S$ to give a substituted thiodiamide. The reaction of pentylamine with phenyl isothiocyanate is shown here.

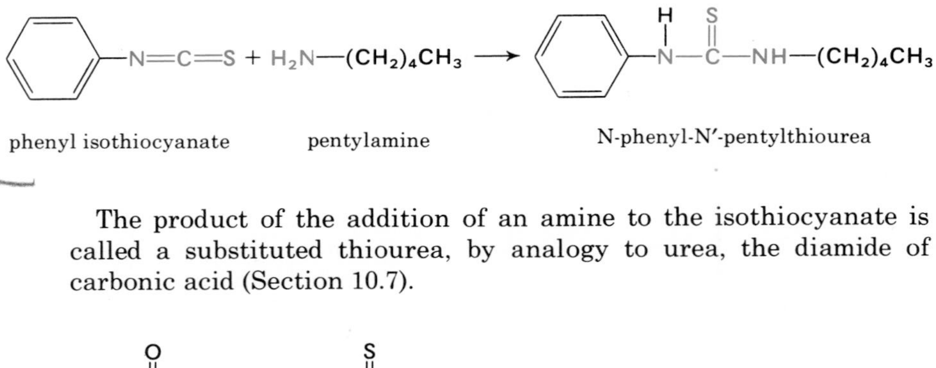

phenyl isothiocyanate   pentylamine   N-phenyl-N'-pentylthiourea

The product of the addition of an amine to the isothiocyanate is called a substituted thiourea, by analogy to urea, the diamide of carbonic acid (Section 10.7).

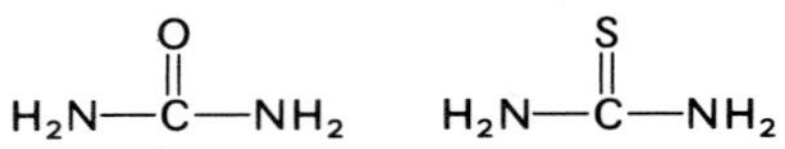

urea   thiourea

Amino groups are poor leaving groups and substitution reactions on amines for the purpose of displacing the amino group do not work except in unusual structures.

10.7 Write structures for the products of the reactions of isobutylamine with the following reagents.

a) 2-methylpropanoyl chloride   **b)** ethyl bromide

c) phenyl isothiocyanate

10.6 PREPARATION OF AMINES

In the laboratory amines are prepared by many methods. Primary amines may be obtained from alkyl halides on treatment with a large excess of ammonia. This reaction is an ordinary nucleophilic substitution on the alkyl halide with ammonia as the nucleophile. The excess of ammonia reduces the chance for the alkyl halide to react with the product amine.

$$CH_3CH_2CH_2\!-\!Br + NH_3 \xrightarrow{\quad} CH_3CH_2CH_2\!-\!\overset{+}{N}H_3\ Br^- \xrightarrow{\ NH_3\ } CH_3CH_2CH_2\!-\!NH_2 + NH_4^+\ Br^-$$

1-bromopropane   propylamine

Compounds which contain carbon–nitrogen double or triple bonds can be catalytically hydrogenated to give amines. Thus nitriles give primary amines and imines give secondary amines.

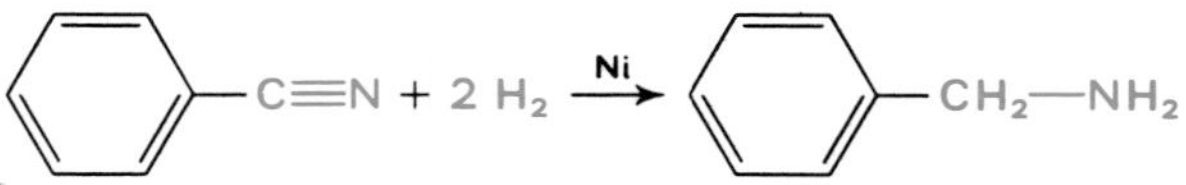

benzonitrile   benzylamine
(a primary amine)

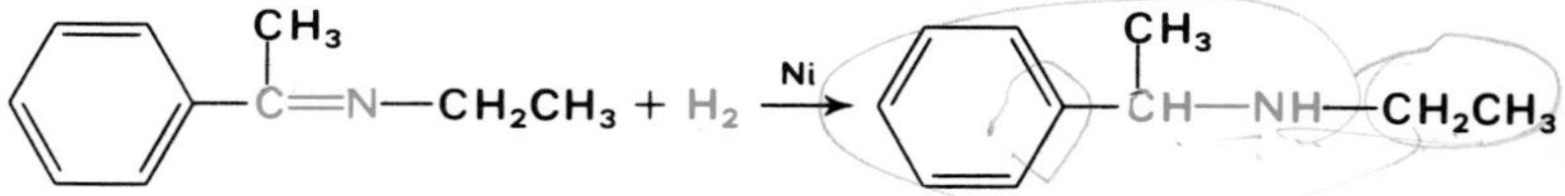

an imine
(prepared from a primary amine
and a ketone, Section 11.6)   N-ethyl-1-phenylethylamine
(a secondary amine)

Nitriles, imines, and amides are converted to amines by the very useful reagent which furnishes hydride ions, lithium aluminum hydride, LiAlH$_4$. The nucleophilic hydride adds to the polar C=N and C≡N groups of the imine and the nitrile.

$$CH_3\!-\!CH\!=\!N\!-\!CH_3 \xrightarrow[\quad]{LiAlH_4} \xrightarrow[\quad]{H_2O} CH_3\!-\!\overset{\overset{\displaystyle H}{|}}{CH}\!-\!NH\!-\!CH_3$$

methyl imine
of acetaldehyde   ethylmethylamine

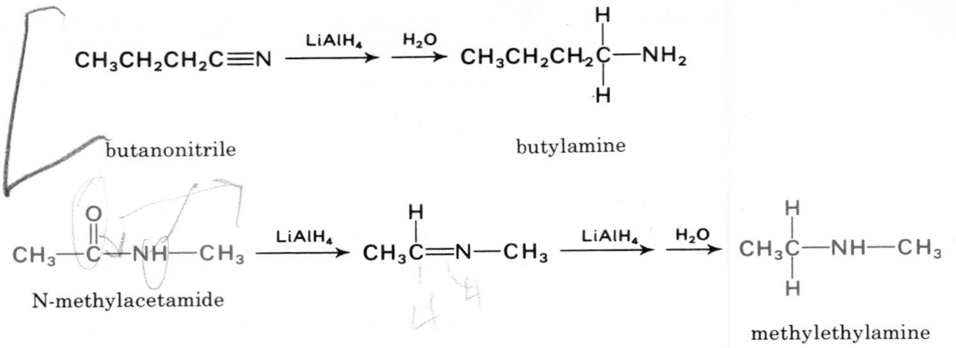

$$CH_3CH_2CH_2C\equiv N \xrightarrow{LiAlH_4} \xrightarrow{H_2O} CH_3CH_2CH_2\overset{\overset{\displaystyle H}{|}}{\underset{\underset{\displaystyle H}{|}}{C}}-NH_2$$

butanonitrile   butylamine

$$CH_3-\overset{\overset{\displaystyle O}{\|}}{C}-NH-CH_3 \xrightarrow{LiAlH_4} CH_3\overset{\overset{\displaystyle H}{|}}{C}=N-CH_3 \xrightarrow{LiAlH_4} \xrightarrow{H_2O} CH_3\overset{\overset{\displaystyle H}{|}}{\underset{\underset{\displaystyle H}{|}}{C}}-NH-CH_3$$

N-methylacetamide

methylethylamine

An imine is an intermediate in the reduction of the nitrile and of the amide. Since the imino group reacts so well with lithium aluminum hydride, all of the products from these three classes of nitrogen compounds are amines. Nitriles and unsubstituted amides yield primary amines on reduction by lithium aluminum hydride. Imines prepared from aldehydes and ketones (Section 11.6) give secondary amines. Substituted amides with lithium aluminum hydride yield secondary or tertiary amines, depending on the number of groups attached to N in the amide.

Reduction of an amide might possibly be expected to produce an alcohol (as from an ester, Section 9.7) instead of an amine. However, in the first addition of a hydride to the amide the negatively charged adduct must eliminate either the oxygen or the nitrogen group. We have seen earlier, in the reaction of esters with amines to give amides, that the tetrahedral adduct normally chooses to retain the nitrogen group and eliminate the better-leaving oxygen group. The same preference is found in the above reaction.

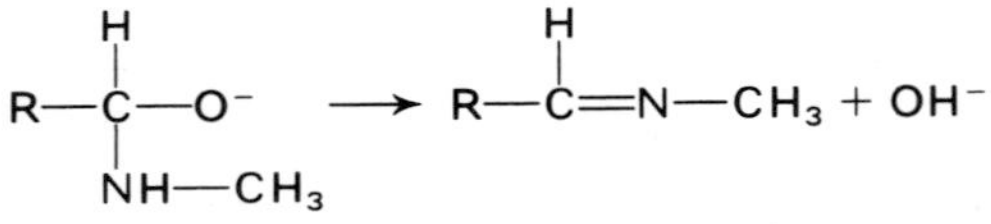

$$R-\overset{\overset{\displaystyle H}{|}}{\underset{\underset{\displaystyle NH-CH_3}{|}}{C}}-O^- \longrightarrow R-\overset{\overset{\displaystyle H}{|}}{C}=N-CH_3 + OH^-$$

Problem   **10.8** Write the structures of compounds A through F in these synthetic sequences.

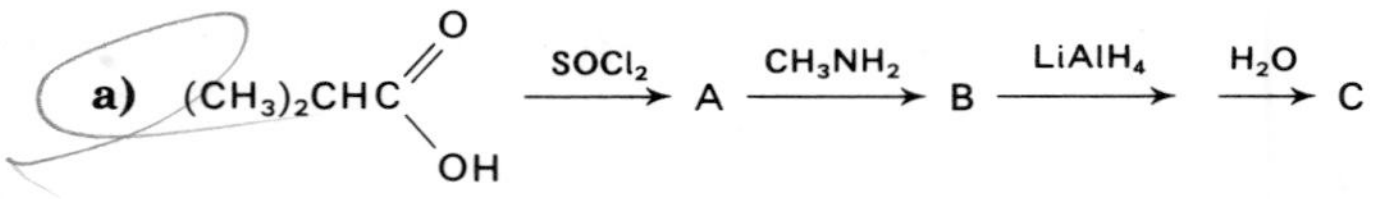

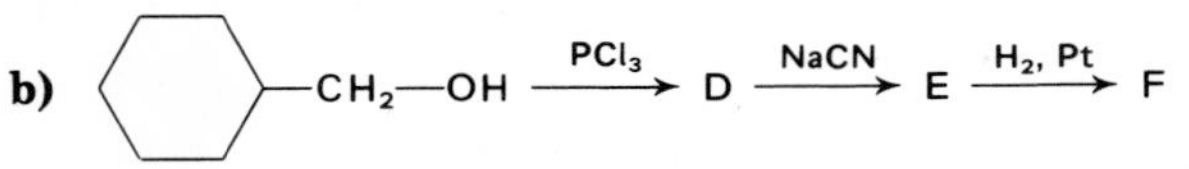

a) $(CH_3)_2CHC\overset{\overset{\displaystyle O}{\|}}{\underset{\underset{\displaystyle OH}{}}{}} \xrightarrow{SOCl_2} A \xrightarrow{CH_3NH_2} B \xrightarrow{LiAlH_4} \xrightarrow{H_2O} C$

b) ⬡—CH_2—OH $\xrightarrow{PCl_3}$ D $\xrightarrow{NaCN}$ E $\xrightarrow{H_2, Pt}$ F

10.7 SOURCES AND USES OF AMIDES

Besides amines, the most common nitrogen-containing compounds are the amides, the combination of carboxylic acids and amines. Amides are extremely stable compounds and find many uses as natural and synthetic products.

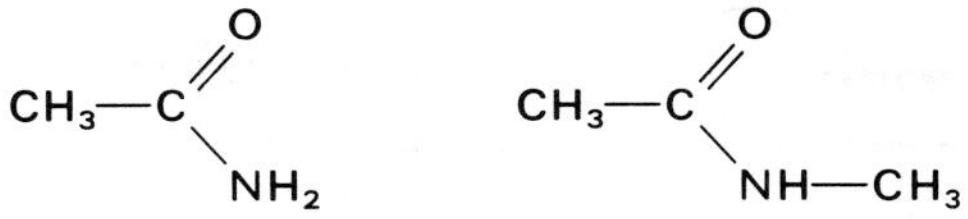

acetamide N-methylacetamide

Polyamides were the first synthetic fibers to be developed and have remained some of the most versatile products of modern technology. One nylon is made from a dicarboxylic acid and a diamine with a structure similar to the polyester made from a diacid and a diol (Section 9.1).

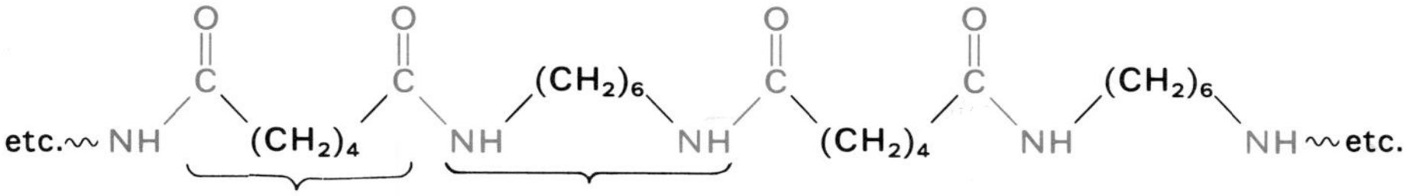

6-6 nylon—a polyamide

Urea is the diamide of carbonic acid, and its fully nitrogen counterpart is guanidine.

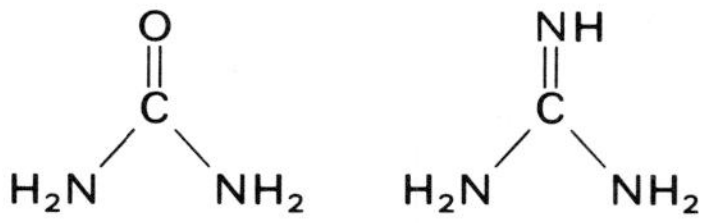

urea guanidine

The nitrogen analog of an anhydride is an **imide**.

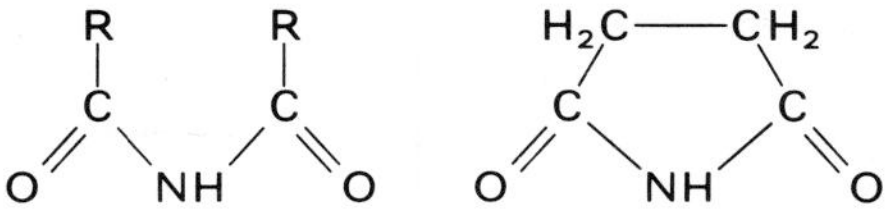

an imide succinimide

Cyclic compounds incorporating the imido, urea, and guanidino structures into rings are well-known natural and synthetic compounds.

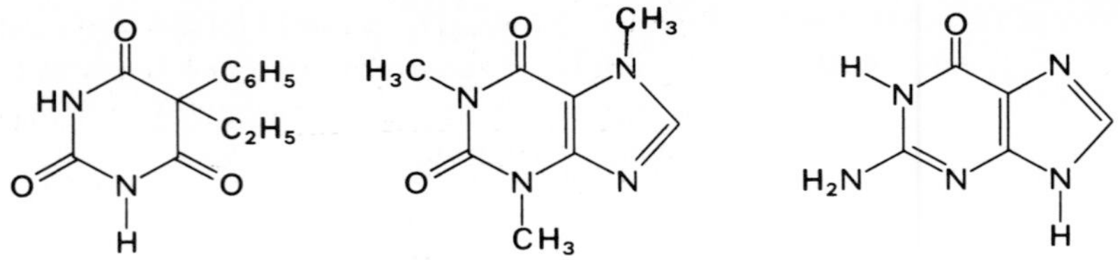

phenobarbital   caffeine   guanine (a base of DNA)

The amide linkage in natural products is found primarily in proteins, which are polymers of amino acids containing a hundred or more amide groups. The amide units of proteins are all between α-amino acids.

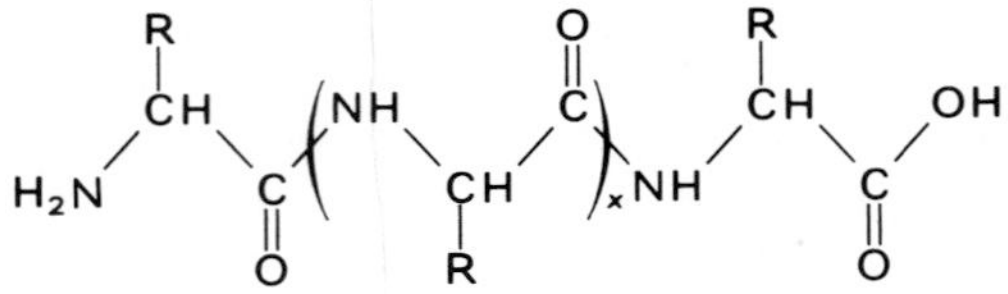

a polyamide of α-amino acid units

Problem   **10.9** In the structures of phenobarbital, caffeine, and guanine, identify atomic assemblies corresponding to those of the imido, urea, or guanidino groups. Do this by placing brackets on the outside of the formulas beside the appropriate atomic assemblies.

10.8 NOMENCLATURE OF AMIDES   The names of amides are based on the names of the corresponding carboxylic acids. The ending "ic acid" or "oic acid" is replaced by the ending "amide." Substituents on the nitrogen are designated by the prefix N; substituents on the carbon chain are numbered as for a carboxylic acid.

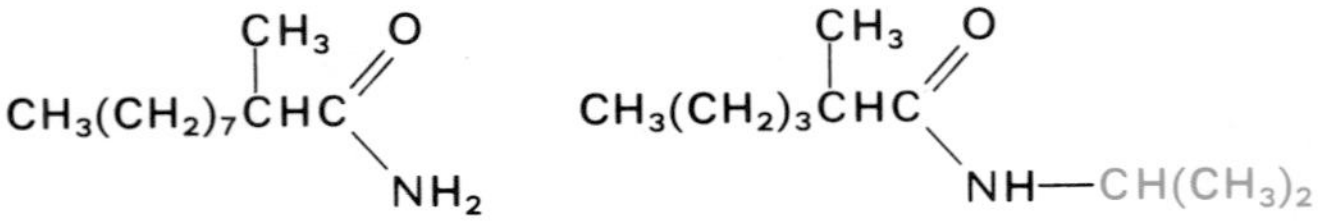

2-methyldecanamide   N-isopropyl-2-methylhexanamide

For acids which are known by their common names, the names of amides use these common names as the base name.

acetamide   benzamide

Problems   **10.10** Name the following compounds.

a) 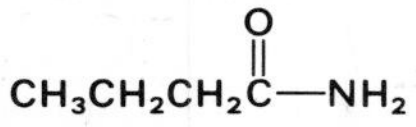 CH₃CH₂CH₂C(=O)—NH₂

b) 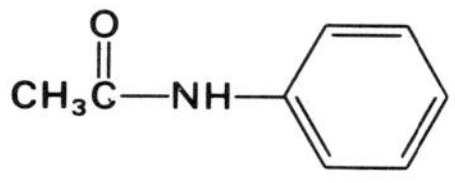 CH₃C(=O)—NH—C₆H₅

c) 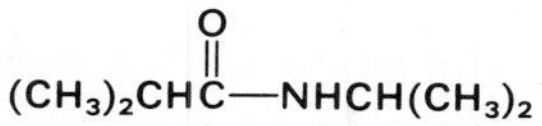 (CH₃)₂CHC(=O)—NHCH(CH₃)₂

d)

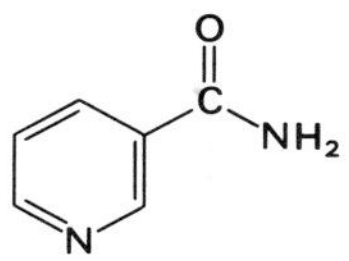

e)

H₃C, H on one carbon; CH₂C(=O)NH₂ and CH₂CH₃ on the other carbon, C=C

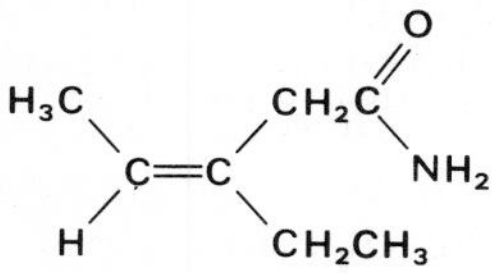

10.11 Write structures for the following compounds.

a) formamide
b) N,N-dimethylbenzamide
c) 2-chloro-N-methylpropanamide
d) 2-phenylacetamide
e) *trans*-2-butenamide

10.9 THE AMIDE FUNCTIONAL GROUP

The most stable of the carboxylic acid derivatives is the amide.

H₃C—C(=O)—N(H)(H)

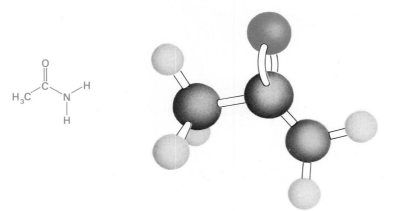

acetamide

The amide group is stabilized by resonance contributions from structures with separated charges.

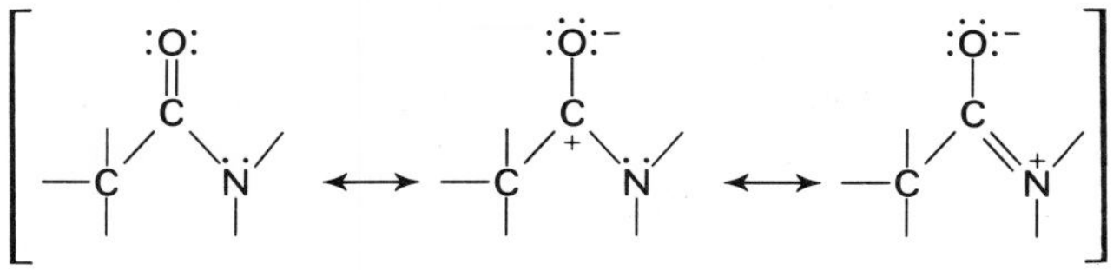

$$
\left[\quad \overset{\ddot{O}:}{\underset{|}{C}}-\overset{}{\underset{|}{C}}-\ddot{N}\diagup \quad \longleftrightarrow \quad \overset{:\ddot{O}:^-}{\underset{|}{\overset{+}{C}}}-\overset{}{\underset{|}{C}}-\ddot{N}\diagup \quad \longleftrightarrow \quad \overset{:\ddot{O}:^-}{\underset{|}{C}}-\overset{}{\underset{|}{C}}=\overset{+}{\underset{|}{N}}\diagup \quad \right]
$$

The resonance hybrid of the amide group is estimated to have only 60% double-bond character in the C=O bond and 40% double-bond character in the C—N bond. The oxygen has considerable electron density and negative charge and the nitrogen has a considerable positive charge.

The geometry of a group of atoms bound by partial double bonds tends to be the same as it would be if they were bound by full double bonds. Rotation of groups about the C—N bond is restricted. The doubly bonded atoms and the atoms attached to them must all lie in one plane. (Compare the structure of ethylene in Section 3.2). The bond angles of the carbonyl carbon and the nitrogen are close to 120°. As a result, the amide group is rigid and possesses a planar structure. The geometry of the amide group is particularly important to the structure of proteins (Section 10.19).

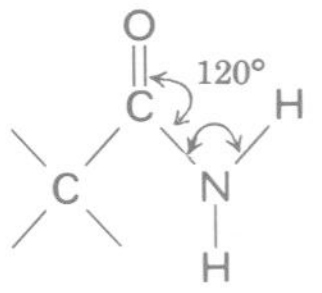

The separation of charge makes the amide group strongly polar. As a result, even the simple amides tend to be nonvolatile.

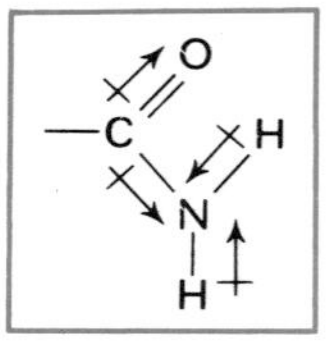

The substituted amide, N,N-dimethylformamide $HCON(CH_3)_2$, is an important polar solvent for ionic compounds.

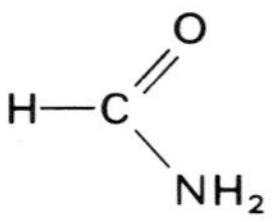

formamide 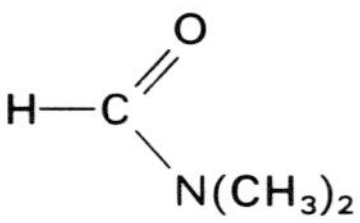 N,N-dimethylformamide

With the exception of formamide, a high-boiling liquid, most unsubstituted amides are crystalline solids at room temperature. The strong attractions between molecules of such small molecular weight are due to the number and strength of the —N—H···O= hydrogen bonds which bind the molecules together in crystals. Hydrogen bonding in amides is especially important in the structure of proteins (Section 10.18).

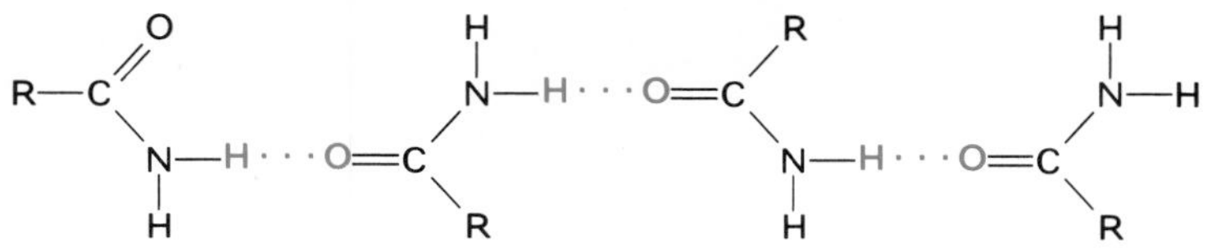

10.10 ACIDITY AND BASICITY OF AMIDES

Amides are both very weak bases and very weak acids. The basicity of the amide group is reduced considerably, in comparison with amines, by the polarity of the C=O bond and the electron delocalization of the group. The effect of the carbonyl group is to reduce the electron density on nitrogen and thus reduce its base strength.

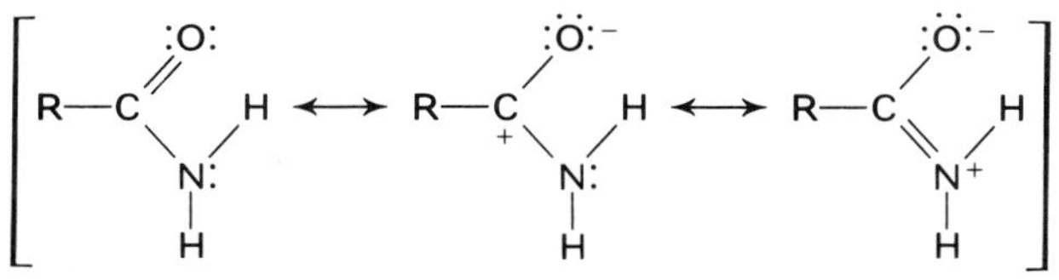

resonance structures of an amide

In the third contributing structure nitrogen carries a positive charge and has no unshared pair of electrons. In the second structure the carbonyl carbon, attached to the nitrogen, has a positive charge which has the effect of drawing electrons away from the nitrogen. In both of these structures, the electron density on nitrogen is reduced, and the base strength is correspondingly lower.

At the same time the C=O bond of the amide increases the acid strength of the N—H of the amide over an amine. The effect is similar to that of a C=O on the O—H of a carboxylic acid, which makes an acid more acidic than an alcohol. The amide N—H is still a very weak acid, but not so much as an amine.

10.11 HYDROLYSIS OF AMIDES

Nucleophilic substitution reactions of amides proceed by the same general mechanism as that found for the other acid derivatives. Amides are much less reactive than esters. Heated in solution with hydroxide ion, amides slowly form the carboxylate ion and ammonia or amine.

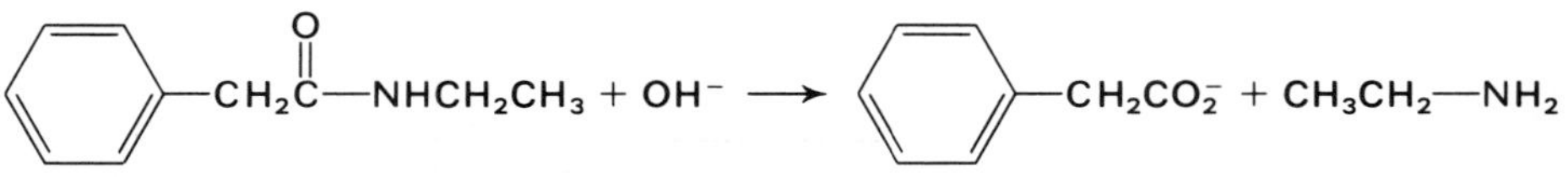

N-ethylphenylacetamide   phenylacetate ion   ethylamine

The reversibility of the substitution reaction is upset by the formation of the carboxylate ion, which does not react with amines. The reaction proceeds slowly to completion.

Amides are more easily hydrolyzed by acid-catalysts with hot sulfuric acid. The hydrolysis products in this reaction are ammonium ion and the carboxylic acid.

$$CH_3\overset{\underset{\|}{O}}{C}-NHCH_3 + H_3O^+ \longrightarrow CH_3CO_2H + CH_3NH_3^+$$

N-methylacetamide   acetic acid   methylammonium ion

10.12 PREPARATION OF AMIDES

One of the uses of acid chlorides as synthetic intermediates described in Section 8.14 is the conversion of a carboxylic acid to the corresponding amide. The ease of this reaction has been discussed in Sections 8.18 and 10.5. Acyl chlorides react with ammonia to give unsubstituted amides, and with primary and secondary amines to give N-substituted amides. Tertiary amines cannot form a stable amide.

$$CH_3CH_2C\overset{\overset{O}{\diagup}}{\diagdown}_{Cl} + 2\ NH_3 \longrightarrow CH_3CH_2C\overset{\overset{O}{\diagup}}{\diagdown}_{NH_2} + NH_4^+Cl^-$$

propanoyl chloride   propanamide

The intense interest in the synthesis of proteins has led to the development of methods for forming an amide linkage directly from a carboxylic acid and an amine under mild conditions. A reagent which promotes amide formation is dicyclohexylcarbodiimide.

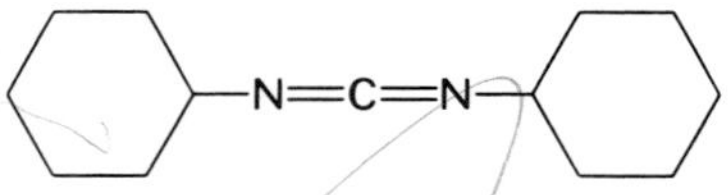

$$\bigcirc-N{=}C{=}N-\bigcirc$$

dicyclohexylcarbodiimide

$$C_6H_{11}-N{=}C{=}N-C_6H_{11} + CH_3C\overset{\overset{O}{\diagup}}{\diagdown}_{OH} + NH_2-CH_2CH_3 \longrightarrow$$

dicyclohexyl-carbodiimide   acetic acid   ethylamine

$$CH_3C\overset{\overset{O}{\diagup}}{\diagdown}_{NHCH_2CH_3} + C_6H_{11}N-\overset{\underset{\|}{O}}{C}-NC_6H_{11}$$
$$\qquad\qquad\qquad\qquad\qquad\qquad H\qquad\quad H$$

N-ethylacetamide   dicyclohexylurea

The diimide reacts with the carboxylic acid to form an acid derivative (a mixed anhydride) whose reactivity as an acylating agent is greatly enhanced over that of the original acid.

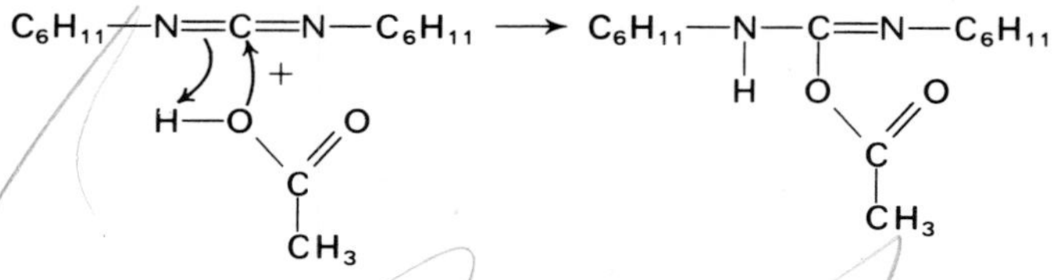

$$C_6H_{11}-N=C=N-C_6H_{11} \longrightarrow C_6H_{11}-N-C=N-C_6H_{11}$$

highly reactive acid derivative

The amine reacts with the activated intermediate to form the amide. The displaced group becomes a disubstituted urea.

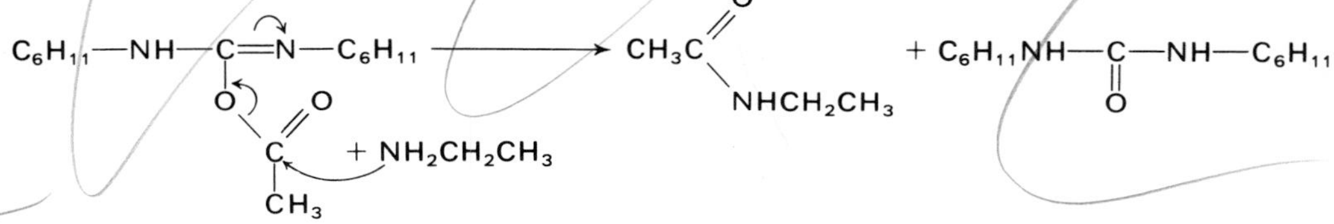

reactive acyl derivative   amine   amide   dicyclohexylurea

This reaction has many advantages over the acid chloride route. Both steps take place at room temperature. No strong acids or bases are required nor are any produced by the reaction. The diimide serves as the means of activating the carboxyl group as well as taking up the water produced by the amide formation. The use of this reagent in the synthesis of proteins is described in Section 10.17.

10.13 AMINO ACIDS   Amino acids incorporate both the amino and the carboxyl functional groups into the same molecule. Since one group is a moderately strong base and the other is a moderately strong acid, interaction between the two is expected.

The most important interactions occur in acids which have the amino group attached to the #2, #4, or #5 carbon. Amino acids with the amino group on the #4 or #5 carbon form internal amides or lactams, when heated.

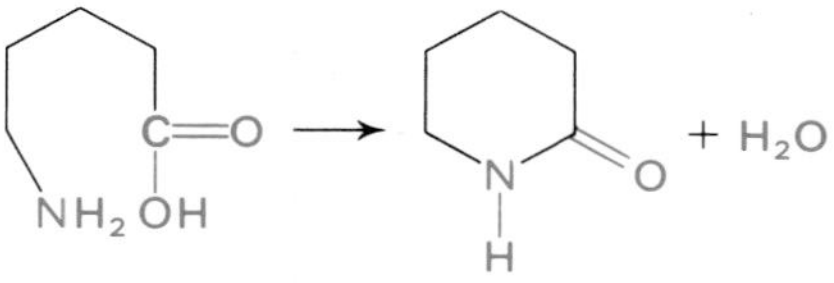

5-aminopentanoic acid   2-piperidone

This reaction resembles the formation of a cyclic ester, a lactone, from a 4- or 5-hydroxyacid (Section 8.12). The five- or six-membered rings are unstrained and are easily formed.

Crystalline 2-amino acids, better known as α-amino acids, exist in an ionic salt form, called a **zwitterion**.

$$H_3\overset{+}{N}-CH_2-CO_2^-$$

zwitterion of glycine (aminoacetic acid)

The physical properties of these α-amino acids confirm the ionic character of the compound. The melting points are very high, above 200°, indicating strong electrostatic attractions within the crystal lattice of a compound with such a small molecular weight. These acids are much more soluble in water than in less polar solvents.

10.14 α-AMINO ACIDS— PROTEIN BUILDING BLOCKS

The amino acids of most general biological importance are those which compose proteins, the macromolecular polyamides found in all living organisms.

$$\text{etc.}\sim NH-\underset{R^1}{CH}-\overset{O}{\overset{\|}{C}}-NH-\underset{R^2}{CH}-\overset{O}{\overset{\|}{C}}-NH-\underset{R^3}{CH}-\overset{O}{\overset{\|}{C}}-NH-\underset{R^4}{CH}-\overset{O}{\overset{\|}{C}}-NH-\underset{R^5}{CH}-\overset{O}{\overset{\|}{C}}-NH-\underset{R^6}{CH}-\overset{O}{\overset{\|}{C}}\sim\text{etc.}$$

Individual amino acids are obtained from proteins by acid hydrolysis of the amide linkages. The 20 amino acids that are found in proteins are *all* α-amino acids. Many have other functional groups as well, such as other amino groups, another carboxylic acid group, hydroxyl or thiol groups, etc. *All*, however, have an amino group on the #2 carbon.

The hydrolyzates from proteins yield a mixture of a number of these amino acids. The percentage and character of the components varies with the source and use of the protein. The structures of these amino acids given in Fig. 10.3 are grouped according to the nature of the R group: nonpolar, polar uncharged, negatively and positively charged at pH 6–7.

The acid–base qualities of these amino acids provide many of the properties of the simple compounds and of the proteins they form. For example, they are used as a means of separating and identifying the acids on ion-exchange columns. The zwitterion is amphoteric, since it can act either as an acid or as a base. The ammonium ion is the acidic group of the zwitterion and the carboxylate ion is the base.

$$H_3\overset{+}{N}CH_2CO_2H \underset{H^+}{\overset{OH^-}{\rightleftarrows}} H_3\overset{+}{N}CH_2CO_2^- \underset{H^+}{\overset{OH^-}{\rightleftarrows}} H_2NCH_2CO_2^-$$

protonated acid   zwitterion   amino acid anion

FIGURE 10.3 AMINO ACIDS OBTAINED FROM PROTEINS

L-amino acid

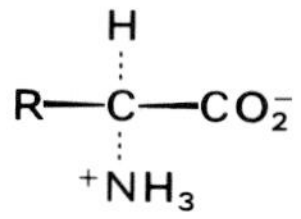

$$R\!-\!\overset{\overset{\displaystyle H}{|}}{\underset{\underset{\displaystyle {}^{+}NH_3}{|}}{C}}\!-\!CO_2^-$$

Nonpolar R-groups   R—

glycine (Gly)*   H—

alanine (Ala) CH_3—

valine (Val) CH_3CH—
                     CH_3

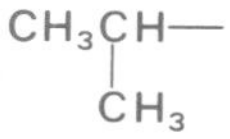

leucine (Leu) CH_3CHCH_2—
                       CH_3

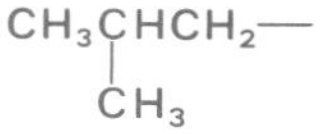

isoleucine (Ile) CH_3CH_2CH—
                         CH_3

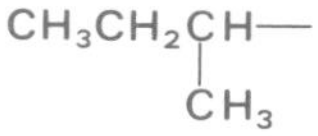

phenylalanine (Phe)   ⟨benzene ring⟩—CH_2—

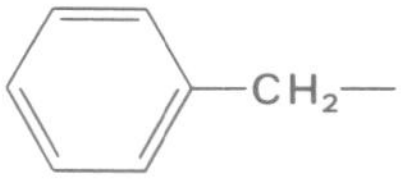

methionine (Met) $CH_3\!-\!S\!-\!CH_2CH_2$—

tryptophan (Trp)   ⟨indole ring with CH_2— and NH⟩

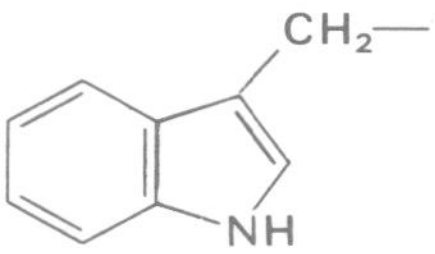

proline (Pro) (entire structure shown)   ⟨pyrrolidine ring with $CHCO_2^-$ and ${}^{+}NH_2$⟩

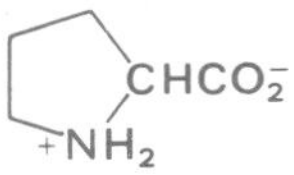

Polar uncharged R-groups

serine (Ser) $HOCH_2$—

threonine (Thr) CH_3CH—
                     OH

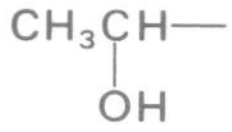

cysteine (Cys) $HSCH_2$—

tyrosine (Tyr)   HO—⟨benzene ring⟩—CH_2—

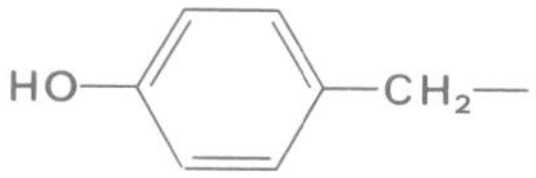

asparagine (Asn) $NH_2\!-\!\underset{\underset{\displaystyle O}{\|}}{C}CH_2$—

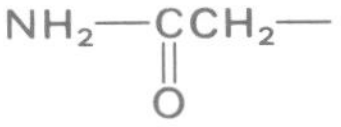

glutamine (Gln) $NH_2\!-\!\underset{\underset{\displaystyle O}{\|}}{C}CH_2CH_2$—

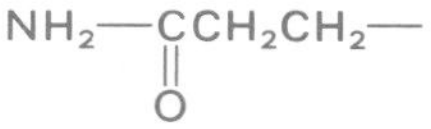

Charged R-groups (at pH 6–7)

  Negatively charged

  aspartic acid (Asp) $^{-}O_2CCH_2$—

  glutamic acid (Glu) $^{-}O_2CCH_2CH_2$—

  Positively charged

  lysine (Lys) $H_3\overset{+}{N}(CH_2)_4$—

  arginine (Arg) $H_2\overset{+}{N}\!=\!\underset{\underset{\displaystyle NH_2}{|}}{C}\!-\!NH\!-\!(CH_2)_3$—

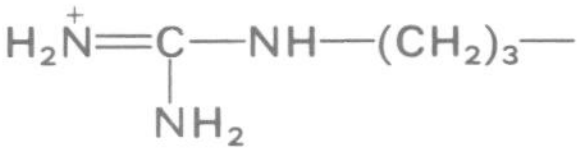

  histidine (His)   ⟨imidazole ring with NH, HN^{+}, and CH_2—⟩

* Gly is the shorthand symbol for glycine. Symbols for other amino acids are given beside the names.

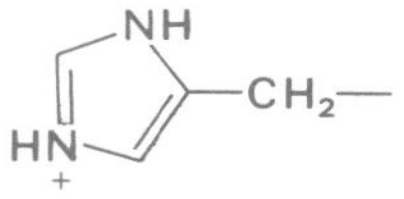

In its fully protonated form, the amino acid is capable of furnishing two protons to a base. The carboxylic acid group is a stronger acid than the ammonium ion. Alanine is characteristic of the amino acids with nonpolar groups. The lower pK_a (2.34) represents the reaction of the carboxylic acid, and the higher pK_a (9.69) represents the reaction of the ammonium ion. At a pH midway between the two pK_a's, no net electrical charge exists for alanine. This midway pH is called the isoelectric pH.

Problem

10.12 It will be of value to you to learn the names of the simplest amino acids and their structures. Draw the structures of these compounds.

a) alanine	**b)** protonated glycine	**c)** methionine anion
d) serine	**e)** cysteine	**f)** histidine

All of the amino acids obtained from the hydrolysis of proteins, with the exception of glycine where R = H, are optically active and *exist in only one* of the two possible enantiomeric structures. Equally significant is the fact that *all amino acids from the hydrolyzates of proteins of animals and higher plants have the same arrangement of R, H, NH_3^+, and CO_2^- groups around the asymmetric α-carbon.* The amino acid enantiomers having this same relative configuration (shown in Fig. 10.4) are identified as L-amino acids. Their enantiomers, which are not found normally in proteins, are called D-amino acids.

FIGURE 10.4
CONFIGURATIONS OF L-
AND D-AMINO ACIDS

L-(−)-serine   D-(+)-serine   L-(+)-alanine   D-(−)-alanine
(natural enantiomer)   (natural enantiomer)

Note that the L and D prefixes are not related to the direction of rotation given in parentheses.

The shapes of protein molecules are very dependent on the fact that all the α-amino acids which form the polypeptide chain have the same configuration. The configurations of the amino acids are of such significance that it has been convenient to identify their relationships in the family groups, the L-amino acids and the D-amino acids. The prefix L says nothing about the direction of optical rotation nor about the R or S notation, both of which depend on the actual structure of the R-group of the amino acid.

The choice of L and D prefixes to identify the enantiomers of these configurations is historical. It arises from the relationship of the configuration of these compounds to the configurations of L- and D-glyceraldehyde, the hydroxyaldehydes which were first used as the reference compounds for the carbohydrate family (see Section 13.3).

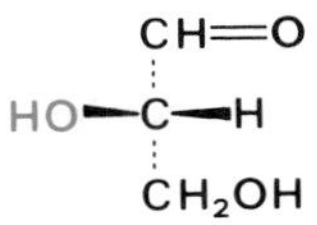

$$CH{=}O \qquad\qquad CH{=}O$$
$$HO{-}C{-}H \qquad\qquad H{-}C{-}OH$$
$$CH_2OH \qquad\qquad CH_2OH$$

L-(−)-glyceraldehyde   D-(+)-glyceraldehyde
(S)-(−)-glyceraldehyde   (R)-(+)-glyceraldehyde

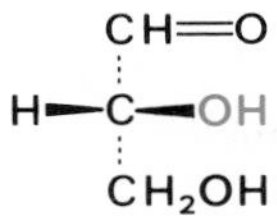

The relationships between the configurations of amino acids and those of the glyceraldehydes were worked out with tremendous labor by earlier chemists.

10.15 PROTEINS   Proteins are large natural polymers found in all living cells. The partial list of the uses of proteins in Fig. 10.5 is sufficient to indicate their immense importance to biological systems and to evolutionary chemistry.

**FIGURE 10.5
SOME TYPES OF PROTEINS
AND THEIR USES**

Type	Name	Use
Hormone	Insulin	Regulate glucose metabolism
Enzyme	Trypsin	Catalyze peptide hydrolysis
	Chymotrypsin	Catalyze peptide hydrolysis
Storage	Egg albumin	Nourish the young
Transport	Hemoglobin	Transport oxygen in blood
Contractile	Myosin	Muscles
Protective	Antibodies	Complex foreign proteins
Structure	α-keratins	Skin, hair, nail, hoof, wool
	Collagen	Tendon, cartilage
	Elastin	Ligament

Proteins are polymers of α-amino acids linked through amide or peptide groups to give polypeptide chains. A segment of the insulin chain is shown in Fig. 10.6.

Proteins range in molecular weight from 12,000 to one million. Some extremely large proteins composed of several polypeptide chains wound together have molecular weights in the millions.

Virtually all proteins are composed of chemically bound L-α-amino acids (Fig. 10.6). The sequence of amino acids is specific for each protein obtained from a different kind of cell. Amino acid sequences of proteins of 200–300 units have been determined.

Smaller proteins have been synthesized in the laboratory, notably insulin (51 units) and ribonuclease (124 units). The amino acid sequence

FIGURE 10.6 A SEGMENT OF THE INSULIN CHAIN

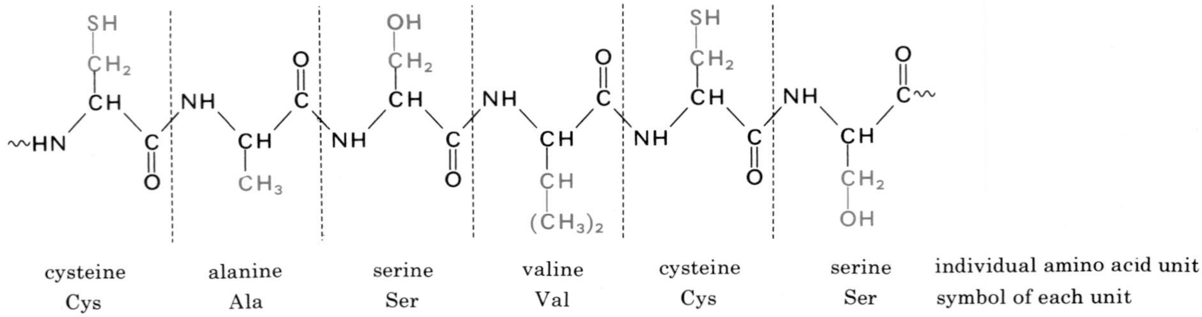

cysteine	alanine	serine	valine	cysteine	serine	individual amino acid unit
Cys	Ala	Ser	Val	Cys	Ser	symbol of each unit

of beef insulin and the positions of the —S—S— cross linkages of cystine are shown below.

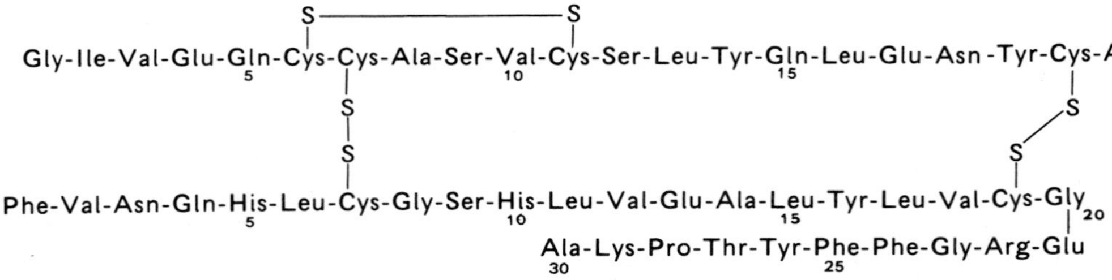

amino acid sequence of beef insulin

Not all proteins contain all 20 amino acids. Some have a high proportion of nonpolar R-groups and are insoluble in water. Some are predominantly basic, containing extra amino groups, while some, like pepsin in the stomach, have R-groups which are predominantly acidic, since they contain extra carboxylic acid groups.

Three known hormones are small polypeptides, not classed as proteins—oxytocin and vasopressin, from the pituitary gland, and bradykinin, from blood plasma, are nonapeptides.

A *tripeptide* has three amino acid units and two peptide (amide) linkages. The units are named beginning at the NH_2-terminal end.

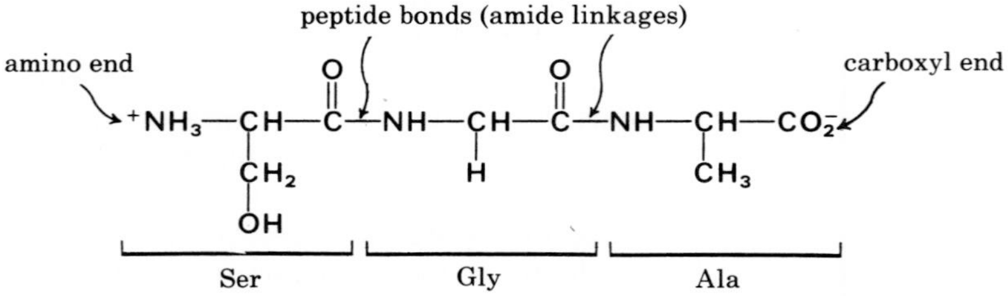

serylglycylalanine (Ser–Gly–Ala)

10.16 DETERMINATION OF AN AMINO ACID SEQUENCE OF PROTEIN

To determine the amino acid sequence in a given protein, the protein must first be isolated and purified. The amino acid content is then obtained after the total hydrolysis of the amide linkages, which is accomplished by treatment with hydrochloric acid at 100–120° for about 24 hours. The amino acids are separated and analyzed on an ion-exchange column that makes use of differences in their pK_a values. An automated "amino acid analyzer" performs all these steps and draws a graph indicating the nature and quantity of the acids present.

The amino and carboxyl terminal amino acid residues are labeled in the whole protein by special chemical reactions. After the total hydrolysis, the labeled acids are identified. Such labels must survive the hydrolysis reaction.

An early method of labeling the amino group of the terminal unit involves the reaction of the free amino group with 2,4-dinitrofluorobenzene—the Sanger reaction. (This reaction is discussed in detail in Section 12.12.) Use of the Sanger reagent to label the amino terminal unit of a dipeptide is shown.

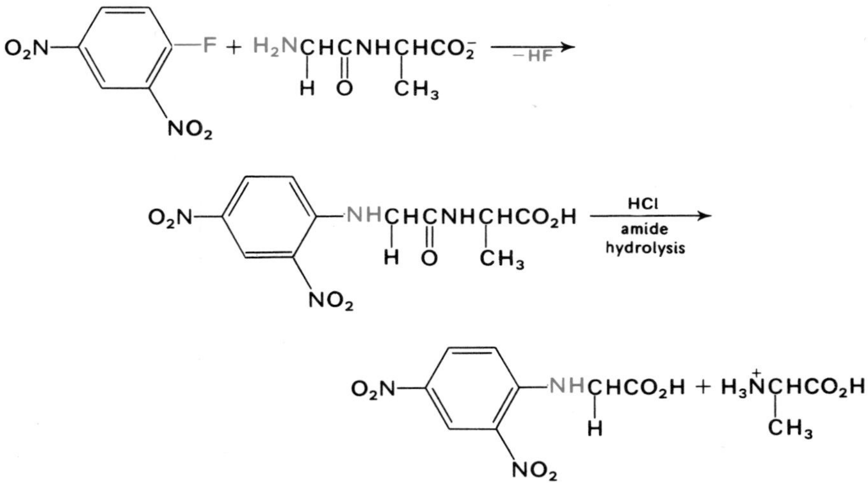

The product, an N-phenyl substituted amino acid, is stable to the conditions of acid hydrolysis of the amides. The amino acid labeled by a dinitrophenyl group is easily identified by ultraviolet spectroscopy (Chapter 15).

The carboxyl terminal acid is labeled by changing the carboxylic acid to an alcohol by reduction with lithium borohydride $LiBH_4$, a reagent similar to $LiAlH_4$ but one which doesn't reduce amides. After amide hydrolysis this end unit is a β-amino alcohol, $H_2N—CRH—CH_2OH$.

Free primary amino groups and free carboxyl groups in the R side chains of the protein also react with the labeling reagents, but this poses no problem in the determination.

A more sensitive and useful reaction for identifying the NH_2-terminal amino acid is the **Edman degradation**. This reaction has the attractive feature that the NH_2-terminal peptide link is cleaved while all the other peptide linkages remain intact.

In the Edman degradation the free α-amino group reacts with phenyl isothiocyanate in the presence of base to form a thiourea (Section 10.5). A subsequent mild treatment with hydrochloric acid breaks only that one peptide linkage adjacent to the thiourea group.

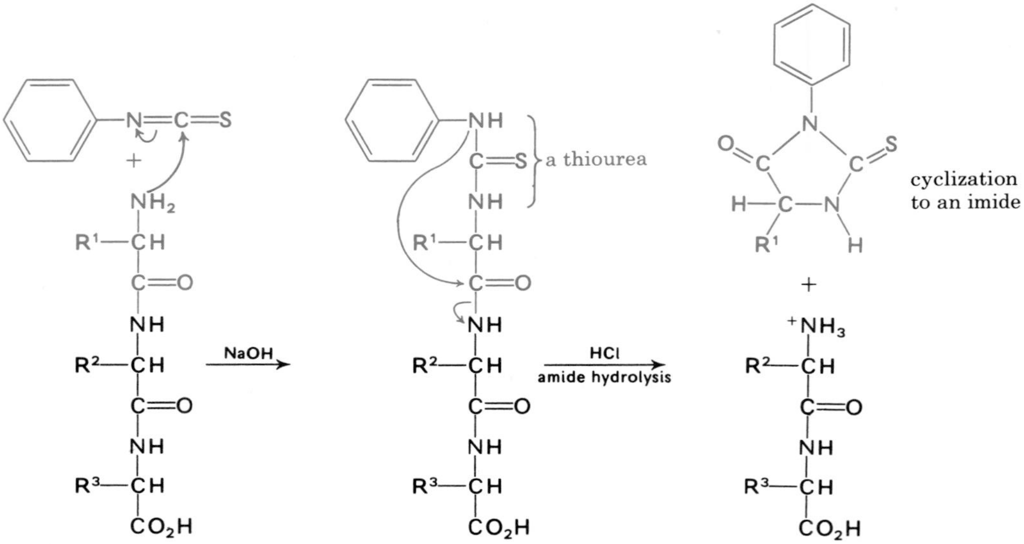

| tripeptide + phenyl isothiocyanate | labeled tripeptide | dipeptide + labeled NH_2-terminal unit |

Successive Edman degradations of a polypeptide break off the NH_2-terminal units one at a time. Most amino acid sequence determinations have proceeded by the degradation of small polypeptides obtained by the partial enzymatic hydrolysis of the much larger protein.

The enzyme trypsin, from the pancreas, catalyzes the hydrolysis of peptide bonds in which the carboxyl functional group belongs to lysine or arginine, the amino acids having additional nitrogen base functional groups. This catalytic selection is quite specific.

Lys-Ser-Ala-Arg-Gly-Trp-Ser-Lys-Ala-Gly $\xrightarrow{\text{trypsin}}$ Lys + Ser-Ala-Arg + Gly-Trp-Ser-Lys + Ala-Gly

The small fragments can be separated and analyzed for amino acid content, and the terminal units of each fragment can then be determined. Repeated Edman degradations are applied to tetrapeptides or larger fragments to determine the order of nonterminal amino acids. The procedure is shown schematically in Fig. 10.7.

FIGURE 10.7
AMINO ACID SEQUENCE
DETERMINATION

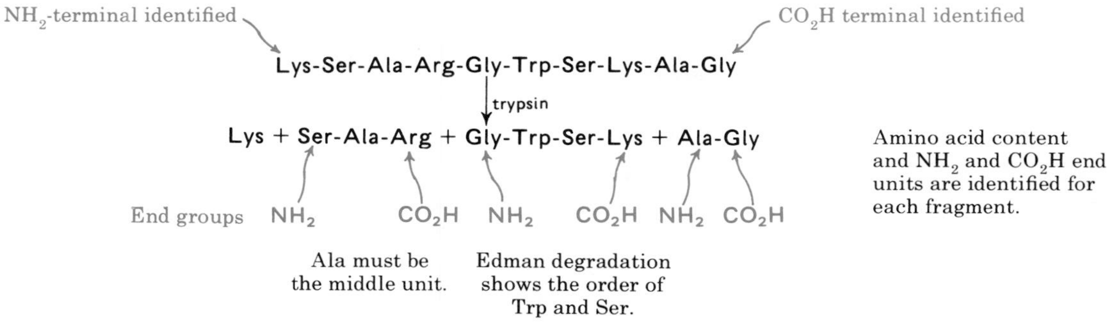

NH$_2$-terminal identified

CO$_2$H terminal identified

Lys-Ser-Ala-Arg-Gly-Trp-Ser-Lys-Ala-Gly

↓trypsin

Lys + Ser-Ala-Arg + Gly-Trp-Ser-Lys + Ala-Gly

Amino acid content and NH$_2$ and CO$_2$H end units are identified for each fragment.

End groups   NH$_2$ CO$_2$H   NH$_2$ CO$_2$H NH$_2$ CO$_2$H

Ala must be the middle unit.

Edman degradation shows the order of Trp and Ser.

What information is still missing for the complete determination of the order of amino acids? The order of the fragments in the original polypeptide is not known. In our simple example, the end fragments are obvious, since the end units of the whole polypeptide are known, but the order of the two larger fragments must be determined by an additional experiment. Partial hydrolysis of polypeptide chains also is accomplished through catalysis by the enzyme chymotrypsin. This enzyme is less specific for particular peptide bonds, but hydrolysis of linkages in which the carboxyl group has an aromatic group, Ar—CH$_2$CHCO—NH—, are catalyzed preferentially—phenylalanine, tyrosine, and tryptophan. (See Section 10.18 for the mechanism of chymotrypsin catalysis.) Hydrolysis of our polypeptide shown in Fig. 10.7 at the tryptophan carboxyl gives us two fragments of unequal length—six and four units, now with new end units Trp and Ser. It doesn't take a Sherlock Holmes now to see the original order of this small polypeptide.

Lys-Ser-Ala-Arg-Gly-Trp-Ser-Lys-Ala-Gly $\xrightarrow{\text{chymotrypsin}}$ Lys-Ser-Ala-Arg-Gly-Trp + Ser-Lys-Ala-Gly

Among the longest acid sequences which have been deduced are those of *beef trypsinogen* (229) and *beef chymotrypsinogen* (245), which are inactive precursors of the two enzymes which catalyze the hydrolysis of some peptide linkages. Perhaps the longest amino acid sequence that is known is *glyceraldehyde-3-phosphate-dehydrogenase* (333) of lobster muscle (!) which catalyzes a reaction step in the metabolism of glucose.

Problem **10.13** For an unknown tripeptide, the following reactions were used to determine that the order of the amino acids was Met–Ser–Val. What information did each reaction furnish in the determination? Give the structures of the products obtained from each set of reactions.

a) Met–Ser–Val + LiBH$_4$ followed by acid hydrolysis of all peptide bonds

b) Met–Ser–Val + C$_6$H$_5$—N=C=S with NaOH followed by mild hydrolysis with HCl

After the chemical activity of determining end groups and amino acid content for a polypeptide and for the smaller fragments obtained from partial hydrolysis, there still remains the puzzle of putting the amino acid units together in the proper order. If you like puzzles, try this one.

Problem **10.14** A polypeptide is composed of 12 amino acid units:

2 glycine	2 serine	1 arginine	1 tryptophane
3 alanine	1 histidine	1 lysine	1 phenylalanine

The NH$_2$-terminal unit is serine and the CO$_2$H-terminal unit is glycine. Trypsin hydrolysis gives three peptide fragments. Assume all yields are 100%.

	NH$_2$ end	CO$_2$H end	Also contains (in unknown order) one each of
A	Ser	Arg	Phe, Ala, His
B	Ala	Gly	Trp, Gly
C	Ser	Lys	Ala

Chymotrypsin hydrolysis gives three peptide fragments.

	NH$_2$ end	CO$_2$H end	Also contains
D	Ser	Phe	Ser, Lys, Ala, His
E	Ala	Trp	Ala, Gly, Arg
F		Gly	

What is the amino acid sequence?

10.17 CHEMICAL SYNTHESIS OF PROTEINS

The chemical synthesis of polypeptides provides a means of investigating the relationship between structure and biological activity. It can also provide much information for determining the effect of specific amino acids on the optimum conformation of the polypeptide chain.

In the laboratory syntheses of peptides, a reagent is employed to assist in the formation of the peptide bond between the amine and the carboxylic acid. Amide formation using dicyclohexylcarbodiimide for this purpose was described in Section 10.12.

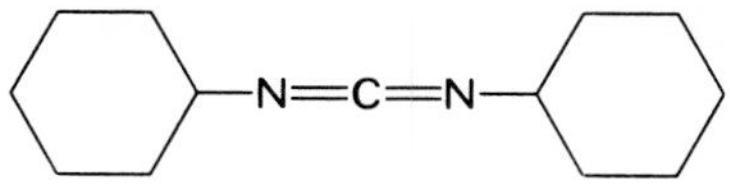

dicyclohexylcarbodiimide

$$C_6H_{11}-N=C=N-C_6H_{11} + RCO_2H + H_2N-R' \longrightarrow R\overset{O}{\overset{\|}{C}}NHR' + C_6H_{11}NH\overset{O}{\overset{\|}{C}}NHC_6H_{11}$$

dicyclohexylcarbodiimide   acid   amine   amide   dicyclohexylurea

When amino acids are used in peptide formation, the amino and carboxyl groups which are *not* to be linked must be protected to avoid reaction at these points also. Carboxyl groups may be blocked by formation of the methyl ester, which is removed at the end of the synthesis by mild hydrolysis with base.

$$H_3\overset{+}{N}-\underset{R}{CH}-CO_2H + CH_3-OH \xrightarrow{H_2SO_4} H_3\overset{+}{N}-\underset{R}{CH}-\underset{O}{\overset{\|}{C}}-O-CH_3 + H_2O$$

(Section 9.11)

A common amine-blocking agent involves making a benzyloxy-carbonyl derivative of the amine.

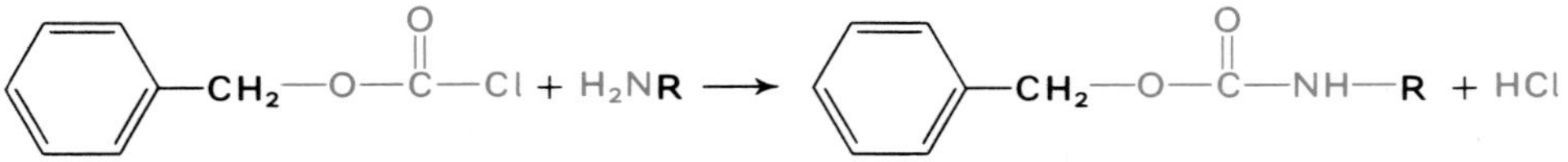

benzyloxycarbonyl   amine   benzyloxycarbonyl derivative
chloride

The benzyloxycarbonyl group may be readily removed on acidification with HBr in glacial acetic acid, which leaves the ammonium salt. The peptide synthesis starts at the CO_2H-terminal end. The course of the synthesis is shown diagramatically in Fig. 10.8.

FIGURE 10.8 SCHEMATIC OUTLINE OF SYNTHESIS OF A PEPTIDE LINKAGE

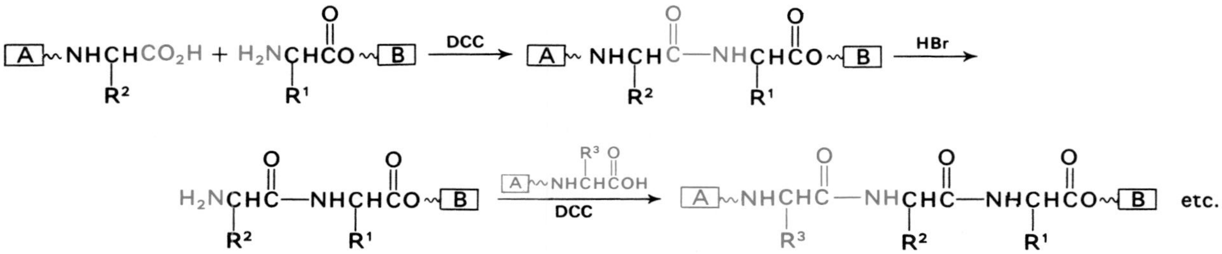

$\boxed{A}$ $\boxed{B}$ = protecting groups   DCC = dicyclohexylcarbodiimide

---

Problem **10.15** Give the sequence of reactions needed to synthesize Gly–Ala–Met using the methyl ester of methionine and other necessary compounds.

Another ingenious method for preparing large polypeptides, known as the Merrifield polypeptide synthesis, combined invention and science. First the carboxyl group of the amino acid which will be the carboxyl terminal unit is attached to polystyrene beads. This serves to protect the CO_2H group against reaction as well as to establish a large molecular weight difference between the growing polypeptide molecule and the other reagents, facilitating the purification procedure of the product after each reaction. The amino group of each new amino acid brought in for reaction is blocked by a *tert*-butoxycarbonyl group $—CO_2—C(CH_3)_3$. This easily removed blocking agent forms the gas isobutylene, which is another simplification of the purification procedure. The noteworthy invention in this polypeptide synthesis is that the series of reactions and purification procedures needed to add each new amino acid unit is *automated by computer control*. The automated synthesis of ribonuclease A, an enzyme of 124 units, by the Merrifield procedure was accomplished in 1969 in *21 days*. The natural process in a cell, however, is brought about in 20 minutes!

10.18 THE WORK OF AN ENZYME: CHYMOTRYPSIN CATALYSIS OF AMIDE HYDROLYSIS

Chymotrypsin is an enzyme which catalyzes the hydrolysis of protein peptide linkages, preferentially at carboxyl groups having aromatic R side chains. Chymotrypsin is obtainable in pure crystalline form. It has 241 amino acid units in three chain sections held together by five S—S linkages. The sequence of amino acids is known, and the shape of the molecule has been determined by X-ray crystallography (see Fig. 10.9).

Enzymes fold their polypeptide chains in such a way as to create cavities which the reacting molecules (substrates) enter and occupy temporarily. The portion of an enzyme which is directly involved in the chemical reaction it catalyzes is called the *active site* of the enzyme. The active site is located on the walls of the cavity.

The cavity serves many purposes.

1. It provides an environment of lower polarity than the water solution on the outside. Reacting molecules are stripped of their solvating water shell and become much more reactive in this less polar environment.
2. The cavity lining contains the parts of the enzyme which will participate in the reaction. The parts of the enzyme and substrate which must interact are brought together.
3. The shape and binding sites of the cavity may serve to select some substrates for reaction and make it more difficult for others. The selection is particularly marked between enantiomers.

FIGURE 10.9
THE SHAPE OF AN ENZYME
—CHYMOTRYPSIN

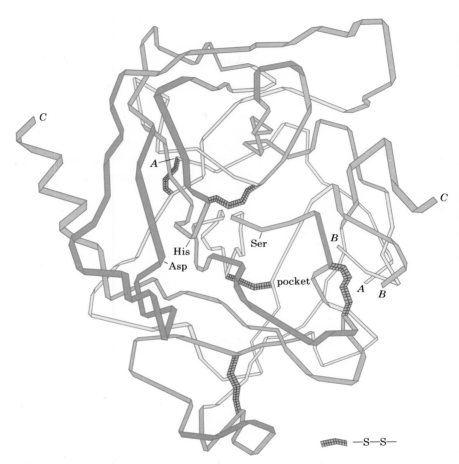

Chymotrypsin has three peptide chains held together by five —S—S— linkages. The cavity is shown by the three labeled amino acid units of the active site on the front side of the enzyme and the pocket lined with lipophilic side chains on the lower right.

    For chymotrypsin the planar aromatic rings of the R-groups of the carboxyl portion of the peptide must fit into a pocket, thereby holding the amide group in position more successfully than other R-groups do.

    Three amino acid units are known to be present at the active site of chymotrypsin: aspartic acid with a free carboxylate ion; histidine with the free cyclic diamine group; and serine with a free hydroxyl group.

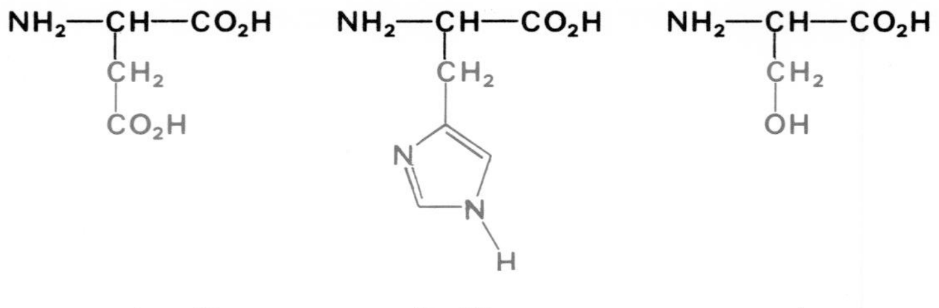

aspartic acid   histidine   serine

These three amino acids occur at positions #102, #57, and #195, respectively, in the order of the 241 units. Folding of the chain allows the three units to be held in close proximity at the active site as seen in Fig. 10.9. Hydrogen bonds between the functional parts of the R-groups of the amino acids hold them in proper alignment as shown in Fig. 10.10.

**FIGURE 10.10
AMINO ACIDS IN ACTIVE
SITE OF CHYMOTRYPSIN**

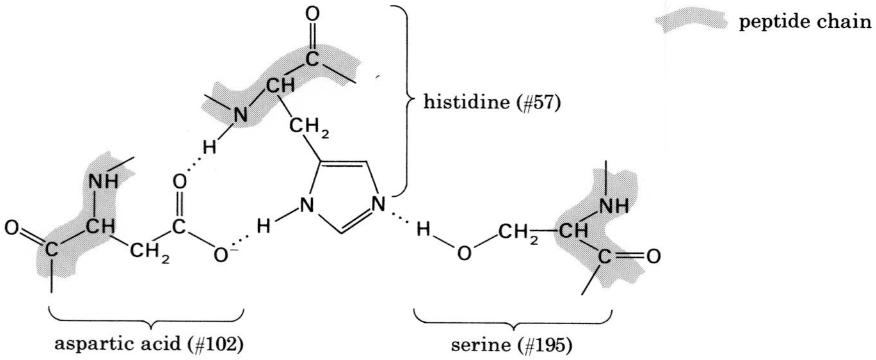

The reaction which the enzyme catalyzes is the hydrolysis of one amide linkage to the amine and the carboxylic acid. Catalysis of amide hydrolysis by enzyme is one billion times (10^9) faster than catalysis by ordinary organic molecules.

The proposed mechanism of catalysis of amide hydrolysis by chymotrypsin involves two transacylation steps.

1. An amide-to-ester interchange of peptide carboxyl from amine to alcohol on the enzyme (Fig. 10.11).

2. The hydrolysis of the acyl-enzyme ester formed in the first step (Fig. 10.12).

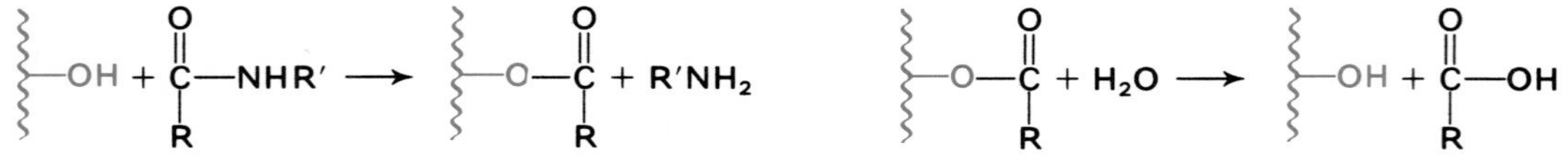

FIGURE 10.11 PROPOSED MECHANISM OF CHYMOTRYPSIN CATALYSIS—AMIDE-TO-ESTER, THE RATE-LIMITING STEP

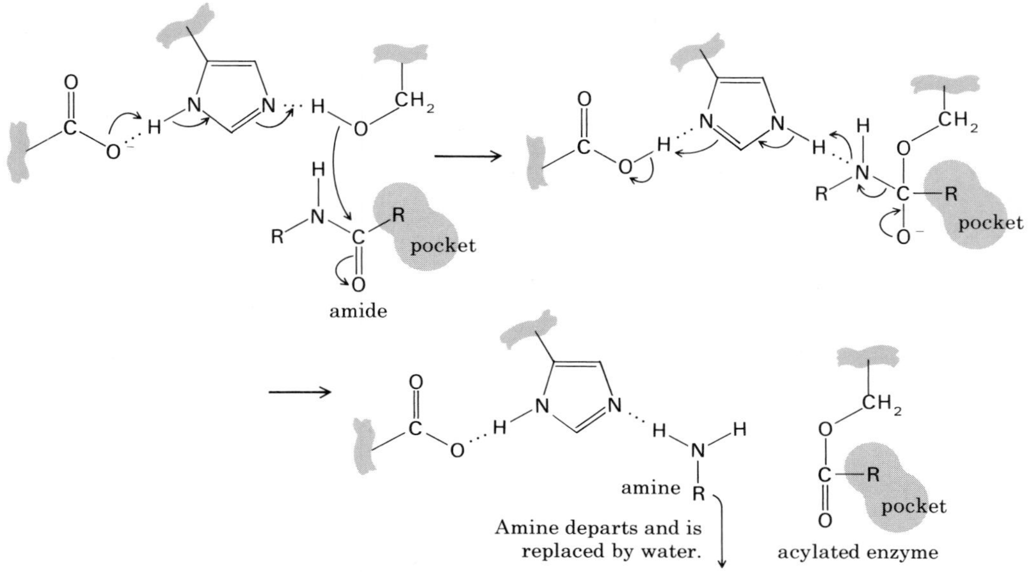
amide
pocket
amine
Amine departs and is replaced by water.
acylated enzyme

The hydroxyl hydrogen shifts from serine to the ring nitrogen of histidine and then to the liberated amine. The hydrogen which bonds the second ring nitrogen of histidine with the carboxylate oxygen of aspartic acid shifts from N to O and back to N in response to the action at the first ring nitrogen.

FIGURE 10.12 PROPOSED MECHANISM OF CHYMOTRYPSIN CATALYSIS—HYDROLYSIS OF ESTER, THE FAST STEP

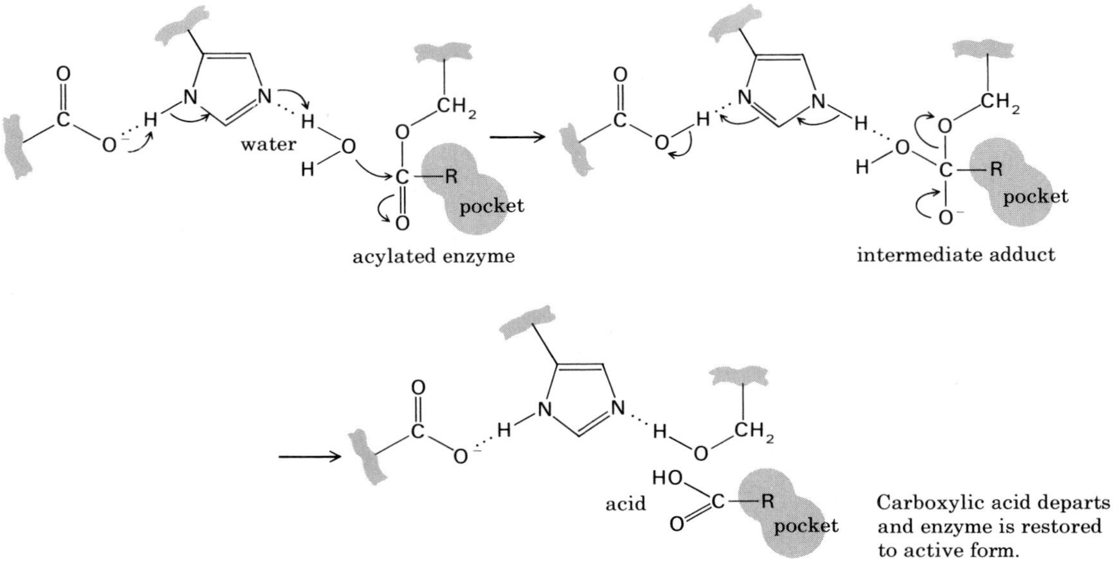
water
pocket
acylated enzyme
intermediate adduct
acid
Carboxylic acid departs and enzyme is restored to active form.

Hydrogen from the water passes first to the ring nitrogen and onto the hydroxyl group. The hydrogen between the second ring nitrogen and the carboxylate ion shifts from N to O and back to N.

The effectiveness of the sequence of reactions hinges on the ability of the imidazole ring of the histidine to be protonated and deprotonated successively on two ring nitrogens.

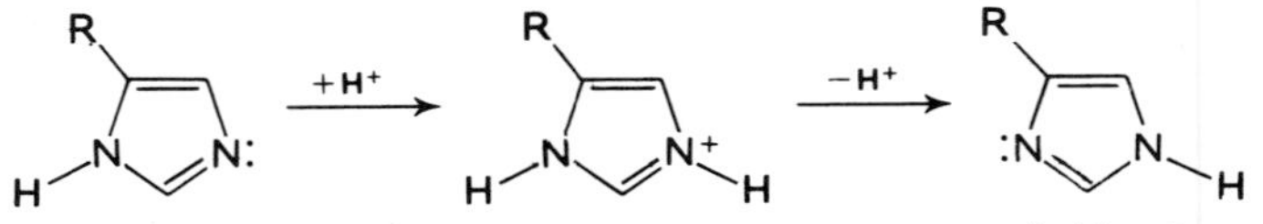

imidazole   imidazole

10.19 STRUCTURAL ORGANIZATION OF PROTEINS

The chain of the protein molecule is a sequence of peptide links. We have seen that groups may rotate freely about two of the three types of bonds composing the polypeptide chain, but that rotation around the third, the peptide bond, is restricted (Section 10.9). The amide group is stabilized by delocalization of pi electrons of the C=O bond into the C—N bond with the imposition of restricted rotation on the C—N bond.

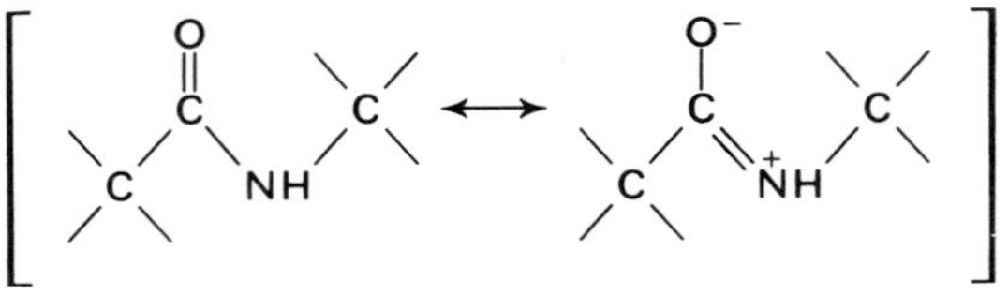

The C=N is estimated to have about 40% double-bond character. The C=N double-bond contribution to the amide structure requires that atoms bonded to this carbon and nitrogen all must lie in the same plane. The bond angles at both this C and N are approximately 120°, as compared with the ordinary bond angle at an amine N of about 107°.

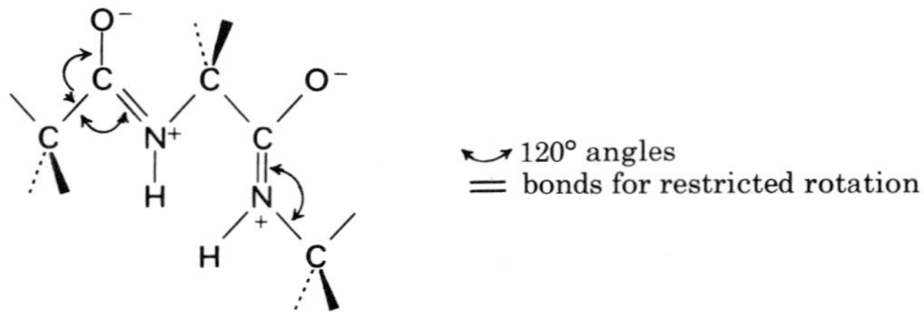

⤻ 120° angles
══ bonds for restricted rotation

The restriction of rotation about this carbonyl–nitrogen bond and the planar geometry of the six atoms shown gives the polypeptide chain rigid, planar sections alternating with flexible sections in which R-groups may rotate around the α-carbon.

Another restraint on the flexibility of the protein chain is provided by —S—S— bonds between cysteine units making cystine units (R—S—S—R). These disulfide links may be between cysteines in the same chain or in separate chains, as shown by insulin (Section 10.15).

The sequence of atoms determined by the covalent bonds comprising the peptide links and —S—S— bonds constitutes the **primary structure** of the protein.

It is now known that under normal biological conditions the polypeptide chain of a protein has only one conformation (or a few) which confers biological activity. This conformation is so stable that the protein can be isolated and retained in its native state. The major source of information about the conformation, or three-dimensional shape, of the protein is X-ray crystallography.

Proteins are classed as fibrous or globular. **Fibrous proteins** consist of polypeptide chains running parallel along a single axis to give long fibers or sheets. They are tough, insoluble in water or dilute salt solution, and are the structural elements in the connective tissue of higher animals. Examples are collagen of tendons and bone matrix, α-keratins of hair, wool, skin, nails, and feathers, and elastin of elastic connective tissue.

Globular proteins have the polypeptide chains tightly folded into a compact spherical or globular shape, and most are soluble in aqueous solutions. They usually have a dynamic function in the cell, including enzymal, hormonal, and transport actions.

Besides the covalent bonds, there are an enormous number of hydrogen bonds between peptide groups of the protein chains. The amide group can form hydrogen bonds at the carbonyl oxygen and the nitrogen at the same time. In water solution hydrogen bonds exist between amide groups and water. Protein chains in their usual conformations, however, fold back on themselves many times, with the result that a substantial portion of the molecule is isolated from contact with water. The hydrogen bonds of amide groups in these sections are satisfied by amide-to-amide-to-amide interactions.

$$\text{H}-\text{N}\backslash\ \ \diagdown\text{C}=\text{O}\cdots\text{H}-\text{N}\diagup\ \ \diagdown\text{C}=\text{O}\cdots\text{H}-\text{N}\diagup\ \ \diagdown\text{C}=\text{O}$$

The **secondary structure** of proteins is a regular, recurring arrangement in space of the polypeptide chains along one dimension. It is a result of the rigidity and planarity of the amide groups, the amide-to-amide hydrogen bonds, and the size and charge of the R-groups. Secondary structure is particularly evident in fibrous proteins and occurs in segments of globular proteins. Two general structural types are formed—helices and pleated sheets.

Analysis by X-ray of the fibrous α-keratins shows major portions of the chain arranged in a right-handed helical coil. A portion of the coil is pictured in Fig. 10.13.

An α-helix can form with either L- or D-amino acids, but not with a mixture of both enantiomers. The L-amino acids can form a right-handed or a left-handed helical coil, but the right-handed one is significantly more stable. Parallel chains of right-handed α-helical

FIGURE 10.13
POLYPEPTIDE CHAIN IN AN
α-HELIX

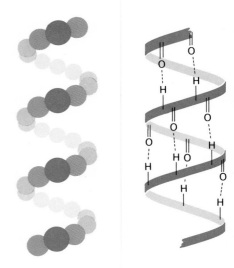

Each carbonyl oxygen is hydrogen bonded to the
nitrogen of the amide group several residues away in
order (below in the diagram). Hydrogen bonds are
nearly parallel to the long axis of the helix. R-groups
radiate outward from the coil. There are 3.6 residues
per turn.

coils running in the same direction may be held together by disulfide
linkages.

Wool is a good example of the properties of an α-helical structure. The
flexibility and elasticity and the lack of strength of the fiber are direct
results of the helical structure with few covalent bonds between chains.

Silk forms a pleated sheet structure in which chains are stretched to
their fullest length (called the β-conformation) and every amide group
is hydrogen-bonded to amide groups in the chains on both sides of it.
The pleated sheet is depicted in Fig. 10.14.

FIGURE 10.14
POLYPEPTIDE CHAINS IN A
PLEATED SHEET

Hydrogen bonds between chains are perpendicular
to the length of the sheet. R-groups lie above and
below the zigzagging planes of pleated sheet.
Adjacent chains run in opposite directions. Most
R-groups are small—H and CH_3. Bulky R-groups
cannot fit together in this structure.

hydrogen bonds

antiparallel pleated sheet

The **tertiary structure** of a protein refers to the way the polypeptide chain is bent or folded in three dimensions to form the compact, tightly folded structure of globular proteins. The globular proteins whose conformations are now known include chymotrypsin, hemoglobin, and myoglobin. Each has the polypeptide chain folded so compactly that there is room inside for only a few molecules of water. The polar R-groups are on the outside and are hydrated. The nonpolar R-groups are on the inside, hidden from the water medium. Chymotrypsin is roughly spherical. Myoglobin contains segments of α-helices whose rigid structures serve as walls around the iron-carrying heme portion. (See Section 12.10 for structure of heme.)

Some proteins are further organized by metal cations which coordinate with oxygen or nitrogen atoms. Calcium, zinc, and other cations are often required in the active enzyme.

Proteins lose their characteristic conformation with heat, on dissolution in a solvent of high dipole moment, and in other ways. The hardening of egg white on being heated is an example. The process of losing the characteristic conformation is called **denaturation**.

Problem **10.16** Not all polypeptide chains can form a stable α-helix. In a study of polypeptides in which all the amino acid units are identical, it was shown that polyalanine, whose R-groups are small and uncharged, spontaneously forms an α-helix in aqueous solution at pH 7.0. Polylysine at pH 7.0 exists in an irregular random form, but at pH 12 polylysine spontaneously forms an α-helix. Similarly, polyglutamic acid has a random form at pH 7.0 but an α-helical form at pH 2.0. Polyisoleucine fails to form an α-helix. Polyglycine does form an α-helix but seems to prefer the stretched conformation of a pleated sheet. Account for these findings on the basis of the nature of the R-groups involved. (Structures of the amino acids are given in Fig. 10.3, Section 10.14.)

New Terms and Topics

Nomenclature of amines (Section 10.2)

Effect of structure on basicity of amines, pK_a's of conjugate acids (Section 10.4)

Amines as nucleophiles (Section 10.5)

Geometry, polarity, and hydrogen bonding of the amide group (Section 10.9)

Nomenclature of amides (Section 10.8)

Acidity and basicity of amides (Section 10.10)

Lactams from 4- and 5-amino acids (Section 10.13)

Zwitterions (Section 10.13)

α-Amino acids derived from proteins—names and structures (Section 10.14)

L- and D-Amino acids—relative configuration of amino acids (Section 10.14)

Polypeptides and proteins (Section 10.15)

Systematic degradation of polypeptides, end-group labeling (Section 10.16)

Protective groups in synthesis of polypeptides (Section 10.17)

Catalysis by chymotrypsin (Section 10.18)

Cavity and active site of enzyme (Section 10.18)

Planarity of amide group (Section 10.19)

Hydrogen bonding of amide groups in protein (Section 10.19)

Primary, secondary, and tertiary structure of proteins (Section 10.19)

α-Helix and pleated sheet (Section 10.19)

Fibrous and globular proteins (Section 10.19)

Denaturation (Section 10.19)

Summary of Reactions

REACTIONS OF AMINES

1. Acid–base reactions (Section 10.4)

$$R—NH_2 + HA \longrightarrow R—NH_3^+ + A^-$$

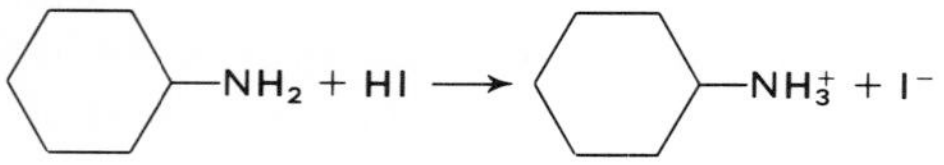

cyclohexyl-NH$_2$ + HI $\longrightarrow$ cyclohexyl-NH$_3^+$ + I$^-$

$$R—NH_2 + C_6H_5Li \longrightarrow R—NH^-Li^+ + C_6H_6$$

cyclohexyl-NH$_2$ + C$_6$H$_5$Li $\longrightarrow$ cyclohexyl-NH$^-$Li$^+$ + C$_6$H$_6$

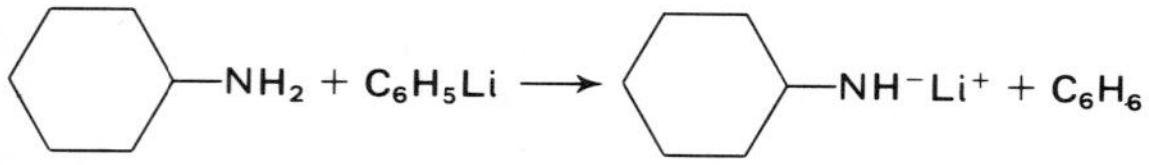

lithium cyclohexylamide

2. Amines as nucleophiles

a) With acyl chlorides (Section 10.5)

$$RCOCl + R_2'NH \xrightarrow{\text{pyridine}} RCONR_2' + HCl$$

$$CH_3COCl + CH_3NH_2 \xrightarrow{\text{pyridine}} CH_3CONHCH_3 + HCl$$

N-methylacetamide

b) With alkyl halides (Section 10.5)

$$RNH_2 + R—X \longrightarrow R_2NH_2^+ + X^-$$

$$CH_3NH_2 + CH_3CH_2—Br \longrightarrow CH_3CH_2—\overset{+}{N}H_2—CH_3\ Br^-$$

ethylmethylammonium bromide

c) With phenyl isothiocyanate (Section 10.5)

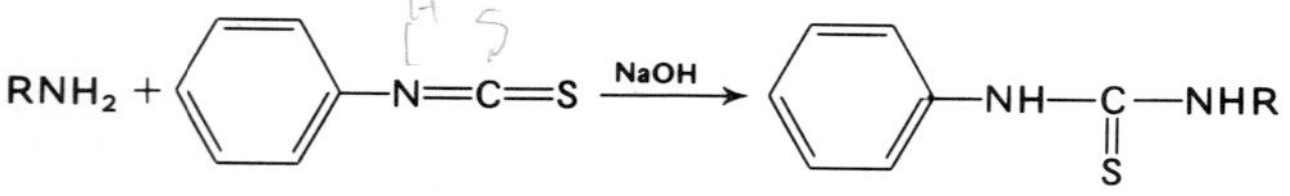

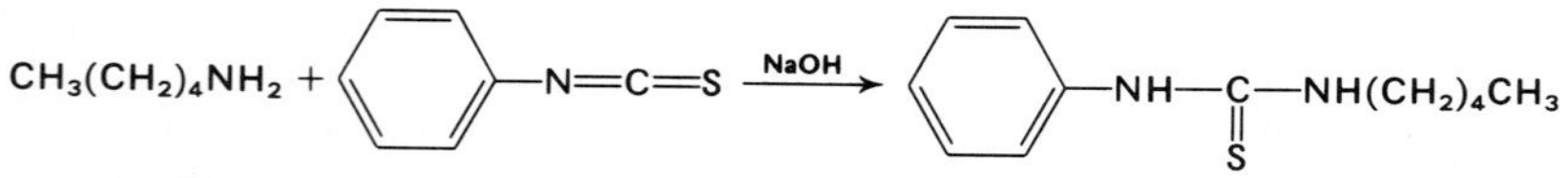

RNH$_2$ + —N=C=S $\xrightarrow{\text{NaOH}}$ —NH—C(=S)—NHR

CH$_3$(CH$_2$)$_4$NH$_2$ + —N=C=S $\xrightarrow{\text{NaOH}}$ —NH—C(=S)—NH(CH$_2$)$_4$CH$_3$

pentylamine   phenyl isothiocyanate   N-pentyl-N′-phenylthiourea

REACTIONS OF AMIDES

1. Hydrolysis of amides (Section 10.11)

a) With acid

$$RCONH_2 + H_3O^+ \longrightarrow RCO_2H + NH_4^+$$

$$CH_3CONHCH_3 + H_3O^+ \longrightarrow CH_3CO_2H + CH_3NH_3^+$$

b) With base

$$RCONR_2 + OH^- \longrightarrow RCO_2^- + R_2NH$$

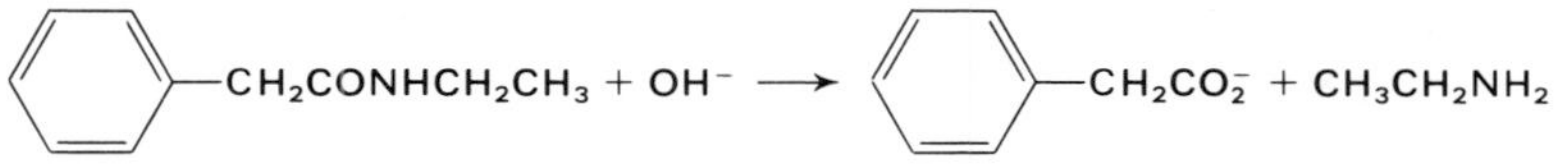

$$\text{(phenyl)}-CH_2CONHCH_2CH_3 + OH^- \longrightarrow \text{(phenyl)}-CH_2CO_2^- + CH_3CH_2NH_2$$

2. Reduction of amides by lithium aluminum hydride (Section 10.6)

$$RCONH_2 \xrightarrow{LiAlH_4} \xrightarrow{H_2O} RCH_2-NH_2$$

$$RCONHR' \xrightarrow{LiAlH_4} \xrightarrow{H_2O} RCH_2-NH-R'$$

$$RCONR'_2 \xrightarrow{LiAlH_4} \xrightarrow{H_2O} RCH_2-NR'_2$$

$$CH_3-\overset{\overset{\displaystyle O}{\|}}{C}-NH-CH_3 \xrightarrow{LiAlH_4} \xrightarrow{H_2O} CH_3CH_2-NH-CH_3$$

N-methylacetamide   methylethylamine

REDUCTIONS OF NITRILES

1. Catalytic hydrogenation (Section 10.6)

$$R-C{\equiv}N + 2\ H_2 \xrightarrow{Ni} R-CH_2-NH_2$$

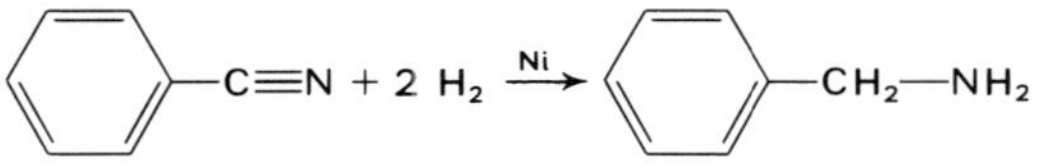

$$\text{(phenyl)}-C{\equiv}N + 2\ H_2 \xrightarrow{Ni} \text{(phenyl)}-CH_2-NH_2$$

benzonitrile   benzylamine

2. Reduction by lithium aluminum hydride (Section 10.6)

$$R-C{\equiv}N \xrightarrow{LiAlH_4} \xrightarrow{H_2O} R-CH_2-NH_2$$

$$CH_3CH_2CH_2-C{\equiv}N \xrightarrow{LiAlH_4} \xrightarrow{H_2O} CH_3CH_2CH_2-CH_2-NH_2$$

REACTIONS OF AMINO ACIDS

1. Lactams from 4- and 5-amino acids (Section 10.13)

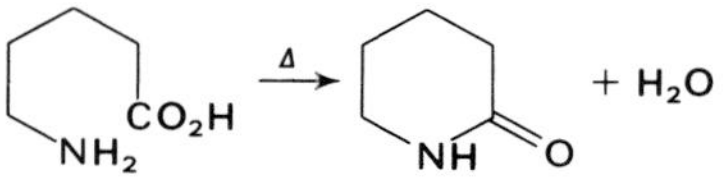

5-aminopentanoic acid

2. Acid–base reactions of zwitterions of α-amino acids (Section 10.13)

$$Cl^- H_3\overset{+}{N}CH_2CO_2H \underset{H^+}{\overset{OH^-}{\rightleftarrows}} H_3\overset{+}{N}CH_2CO_2^- \underset{H^+}{\overset{OH^-}{\rightleftarrows}} H_2NCH_2CO_2^- Na^+$$

protonated amino   zwitterion   amino acid
acid salt   carboxylate salt

Problems

10.17 Draw formulas for the following amino acid enantiomers.

    a) L-alanine   **b)** glycine   **c)** L-serine   **d)** L-histidine   **e)** L-cysteine

10.18 What is the R,S notation for the isomers in Problem 10.17?

10.19 Draw structures for the following compounds.

    a) triethylammonium chloride   **b)** N-methylaniline   **c)** N,N-diethyl-2-phenylbutanamide

    d) N-methylpyridinium bromide   **e)** alanylserine   **f)** glutamic acid

10.20 Name the following compounds by their common names.

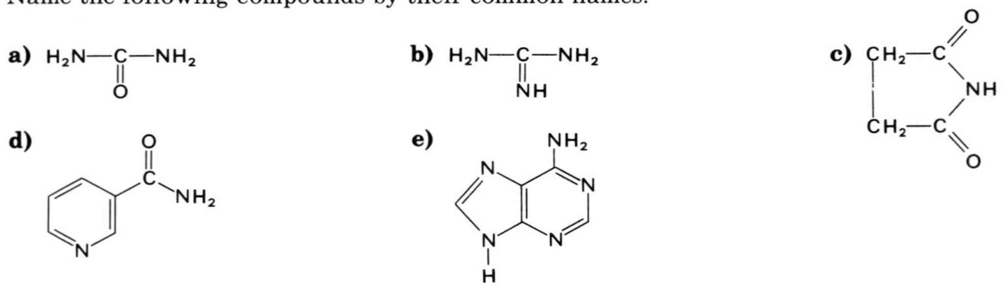

a) $H_2N-\overset{\overset{\displaystyle O}{\|}}{C}-NH_2$

b) $H_2N-\overset{\overset{\displaystyle NH}{\|}}{C}-NH_2$

c)

d)

e)

10.21 Nylon, the polyamide fiber, disintegrates when treated with a strong acid or strong base.

    a) Write the equations for the hydrolysis of 6-6 nylon with aqueous sulfuric acid (formula, in Section 10.7) and name the products.

    b) Write the equation for the hydrolysis of 6-6 nylon with aqueous sodium hydroxide and name the products.

10.22 The structure of strychnine (Formula, Section 10.1) has two nitrogens. Which of these nitrogen groups is the stronger base to be protonated first?

10.23 Write structures for the products of the following reactions.

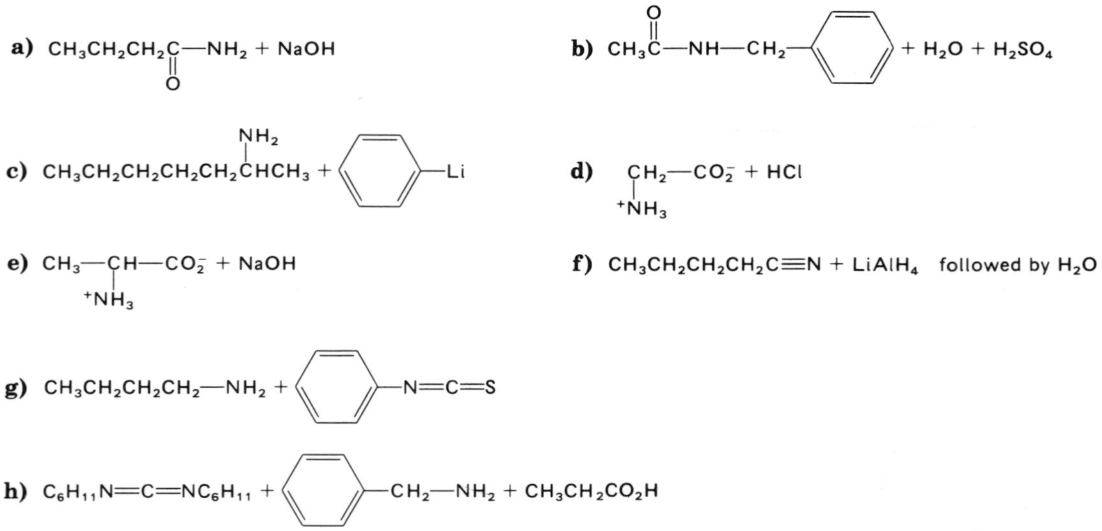

a) $CH_3CH_2CH_2\overset{\overset{\displaystyle O}{\|}}{C}-NH_2 + NaOH$

b) $CH_3\overset{\overset{\displaystyle O}{\|}}{C}-NH-CH_2- + H_2O + H_2SO_4$

c) $CH_3CH_2CH_2CH_2CH_2\overset{\overset{\displaystyle NH_2}{|}}{C}HCH_3 + -Li$

d) $\overset{}{\underset{\overset{|}{^+NH_3}}{CH_2}}-CO_2^- + HCl$

e) $CH_3-\overset{\overset{}{|}}{\underset{\overset{|}{^+NH_3}}{C}}H-CO_2^- + NaOH$

f) $CH_3CH_2CH_2CH_2C\equiv N + LiAlH_4$ followed by H_2O

g) $CH_3CH_2CH_2CH_2-NH_2 + -N=C=S$

h) $C_6H_{11}N=C=NC_6H_{11} + -CH_2-NH_2 + CH_3CH_2CO_2H$

i)

(pyridine ring) N + CH_3CH_2-Br

j) $CH_3CH_2CH_2CH_2\overset{\overset{\displaystyle O}{\|}}{C}-NH-CH_2CH_3$ + $LiAlH_4$ followed by H_2O

10.24 Give the reagents needed for the following series of reactions.

a) $CH_3CH_2\underset{\underset{\displaystyle CH_3}{|}}{C}H-CH_2OH$ $\xrightarrow{\text{A}}$ $CH_3CH_2\underset{\underset{\displaystyle CH_3}{|}}{C}HCH_2Cl$ $\xrightarrow{\text{B}}$ $CH_3CH_2\underset{\underset{\displaystyle CH_3}{|}}{C}HCH_2CN$ $\xrightarrow{\text{C}}$ $CH_3CH_2\underset{\underset{\displaystyle CH_3}{|}}{C}HCH_2CH_2-NH_2$

b) (phenyl)$-CH_2CO_2H$ $\xrightarrow{\text{D}}$ (phenyl)$-CH_2COCl$ $\xrightarrow{\text{E}}$

(phenyl)$-CH_2\overset{\overset{\displaystyle O}{\|}}{C}-NHCH_3$ $\xrightarrow{\text{F}}$ (phenyl)$-CH_2-CH_2-NHCH_3$

10.25 When subjected to moist heat, fibers of α-keratin, such as hair, can be stretched to twice the original length. In the stretched condition, their X-ray pattern resembles that of silk. The cooled fiber reverts to its original length and once again gives an X-ray pattern of an α-helix.

a) What happens to the protein structure when the fiber is heated and stretched?

b) Give a reasonable explanation for the spontaneous reversion of the protein to its original α-helix structure.

10.26 Assume a dipeptide valinylserine was prepared from optically pure L-valine and racemic serine. Write three-dimensional structures for the isomers produced. What are the configurational relationships between the isomers? Give each isomer a configurational symbol using R,S notation.

10.27 Explain why protonated glycine is such a much stronger acid than either glycine zwitterion or acetic acid.

CH_3CO_2H $H_3\overset{+}{N}-CH_2CO_2^-$ $H_3\overset{+}{N}-CH_2CO_2H$

acetic acid   glycine zwitterion   protonated glycine
$pK_a \sim 5$ $pK_a \sim 9.6$ $pK_a \sim 2$

The answers to Problems 10.28 and 10.29 are not found in the text, but they are reasonable extensions of the information found there.

10.28 When compound 1 was heated in a nonhydroxylic solvent in the presence of a stoichiometric amount of acid, and the reaction mixture was cooled and neutralized, compound 2 was produced. When compound 2 was heated in a nonhydroxylic solvent in the presence of a catalytic amount of base, and the reaction mixture was cooled and neutralized, compound 1 was produced. Explain.

$C_6H_5-\underset{\underset{\displaystyle OH}{|}}{C}H-\underset{\underset{\displaystyle NH-\underset{\overset{\|}{\displaystyle O}}{C}-C_6H_5}{|}}{C}H-CH_3$ $\underset{\underset{\displaystyle OH^-}{\longleftarrow}}{\overset{\overset{\displaystyle H^+}{\longrightarrow}}{}}$ $C_6H_5-\underset{\underset{\displaystyle C_6H_5-\underset{\overset{\|}{\displaystyle O}}{C}-O}{|}}{C}H-\underset{\underset{\displaystyle NH_2}{|}}{C}H-CH_3$

(1)   (2)

10.29 The N—H proton of succinimide is more acidic than the N—H proton of succinamide. Explain in terms of the resonance structures of the anions formed from each.

succinimide   succinamide

10.30 The imidazole group is found in the amino acid histidine (1). Imidazole itself (2) is amphoteric. Its conjugate acid (3) has a pK_a of about 7. Write the resonance structures for compounds 2 and 3, and for the conjugate base of imidazole (4).

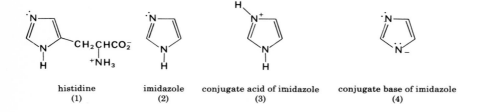

histidine	imidazole	conjugate acid of imidazole	conjugate base of imidazole
(1)	(2)	(3)	(4)

CHAPTER 11

Aldehydes and Ketones

Natural products whose molecular weights are moderate and which do not form hydrogen bonds are usually quite volatile. In flowering plants the major role for such compounds is to attract insects, bees, and butterflies to aid in pollination. Esters, ketones, and stable aldehydes (usually aromatic aldehydes) are often associated with this role.

In other examples the great reactivity of the carbonyl group of the aldehydes and ketones makes them excellent materials for building many needed structures. This role for all aldehydes and ketones occurs in both biosyntheses and degradations, and in laboratory sequences. The reactions of aldehydes and ketones are among the most interesting and varied that are found in organic chemistry.

11.1 SOURCES AND USES OF ALDEHYDES AND KETONES

Aldehydes and ketones are found widely distributed in the animal and plant worlds. Many carbonyl compounds are responsible for sweet, fragrant, or penetrating odors in flavors, fragrances, and wines.

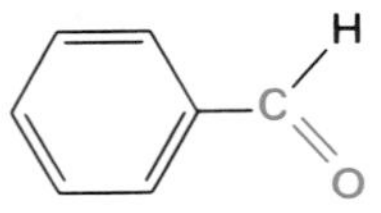

benzaldehyde
(oil of almonds)

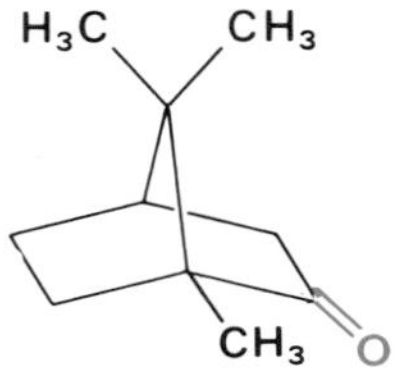

camphor

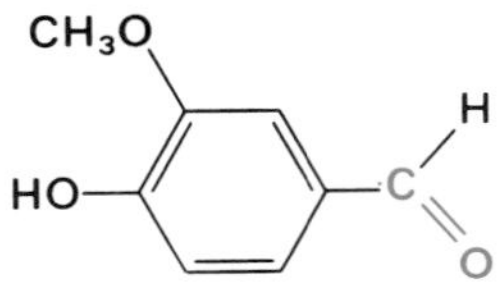

vanillin
(vanilla)

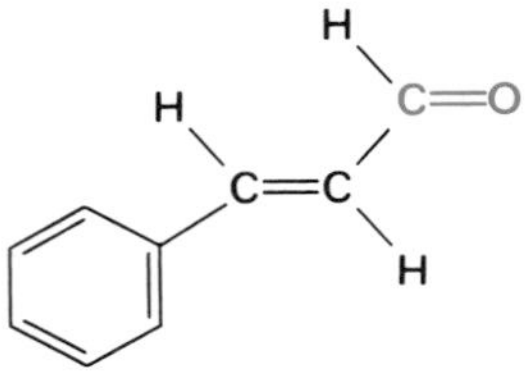

cinnamaldehyde
(cinnamon)

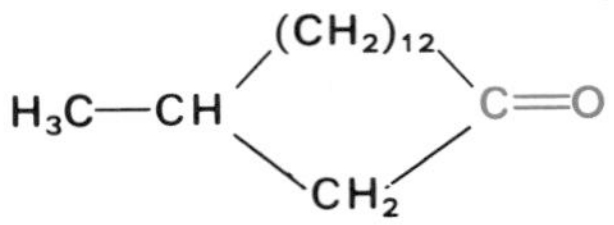

muscone
(extract from scent gland of
male musk deer, used in perfumes)

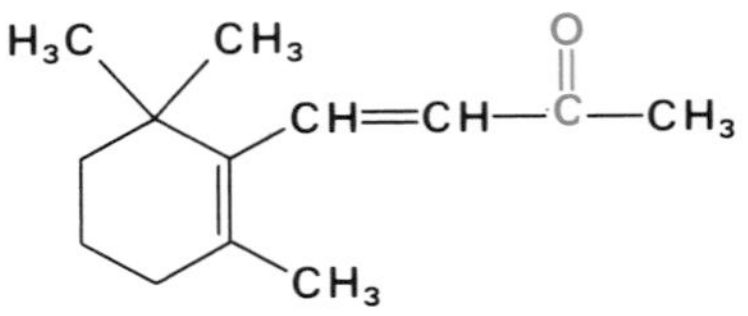

β-ionone
(fragrant scent of violets)

One of the busiest coenzymes known is pyridoxal, which participates in many of the metabolic reactions of amino acids.

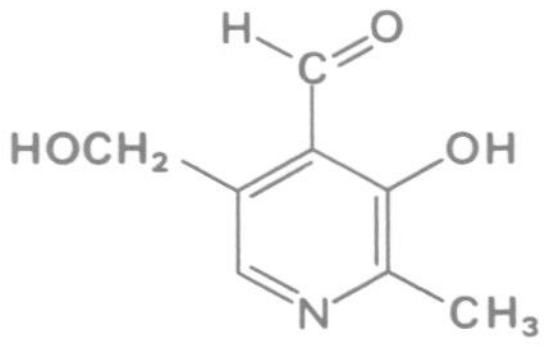

pyridoxal

Glyceraldehyde and pyruvic acid are fundamental components in cell metabolism of carbohydrates. Glucose and ribose, examples of simple sugars or carbohydrates, are polyhydroxyaldehydes. Ribose is a component of larger essential compounds such as coenzymes and nucleic acids, and glucose is the monomeric unit for the polymers starch and cellulose (Chapter 13).

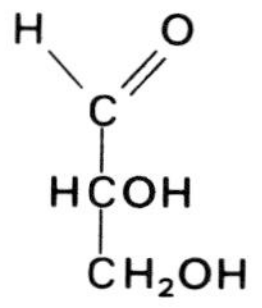

glyceraldehyde 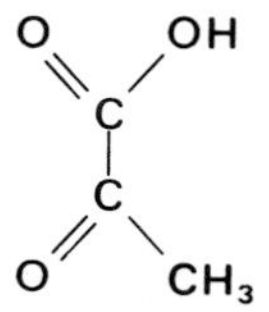   pyruvic acid 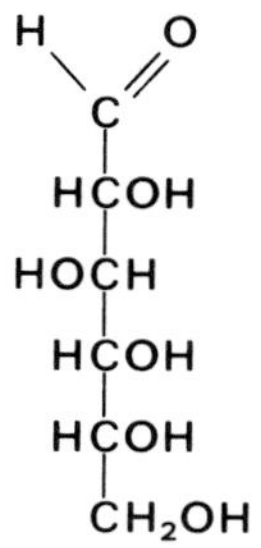   glucose 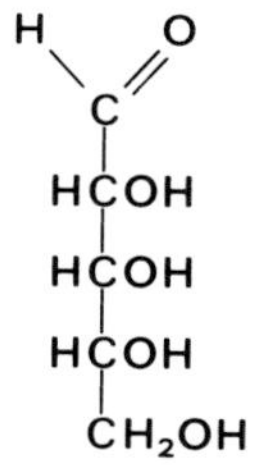   ribose

In the laboratory, under appropriate conditions, aldehydes and ketones become reactants in the syntheses of compounds having larger carbon skeletons. Ketones are not as sensitive to atmospheric oxygen, acids, or bases as are aldehydes, and therefore they find more direct use as compounds. For example, ketones are excellent solvents. The simplest one, acetone CH_3COCH_3, is miscible with both water and organic solvents and easily dissolves lacquers. It is one of the cheapest and most powerful of the commercial solvents.

Although carbonyl compounds accept hydrogen bonds to oxygen, they do not donate hydrogen to hydrogen bonds. Like esters they are considerably more volatile than acids or alcohols, which do form strong intermolecular hydrogen bonds.

Problem   **11.1** List the various functional groups and ring structures of these compounds, whose formulas are written above.

     a) camphor   **b)** vanillin   **c)** pyridoxal

     d) glyceraldehyde   **e)** glucose

11.2 THE CARBONYL FUNCTIONAL GROUP   The functional group of aldehydes and ketones is the carbonyl group, a carbon–oxygen double bond $C=O$.

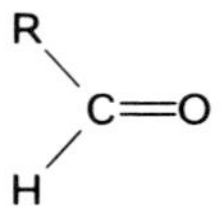

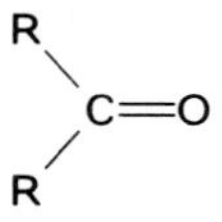

Aldehydes *always* have one hydrogen attached to C=O and the functional group is written −CH=O. The carbonyl group is also bonded to carbon, except in the case of formaldehyde, when it has a second hydrogen attached. Ketones *always* have two carbon groups attached to the carbonyl functional group. Models of acetaldehyde and acetone are depicted below.

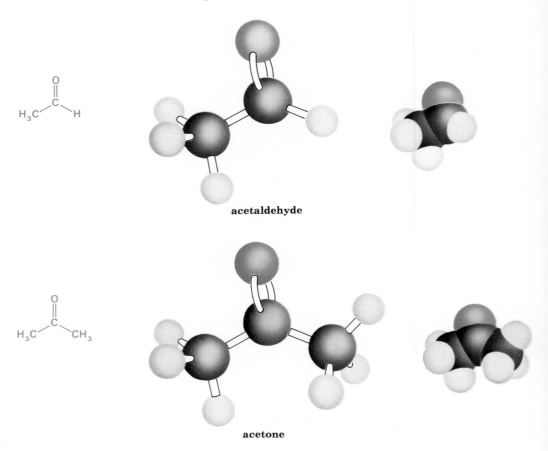

acetaldehyde

acetone

The carbon and oxygen atoms of the carbonyl groups and the two atoms bonded directly to the carbon lie in one plane. The bond angles at the carbonyl carbon are 120°. Figure 11.1 shows the atomic orbitals used to form the sigma and pi bonds of the carbonyl group.

As in the carboxylic acid family, the carbon–oxygen double bond is polarized to give the carbon a partial positive charge and the oxygen a partial negative charge. Alternatively, the carbonyl group may be considered a resonance hybrid, as shown.

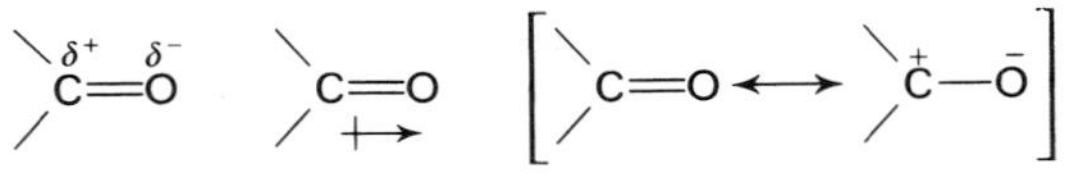

resonance structures

FIGURE 11.1
THE CARBONYL GROUP

top view   side view

Four atoms lie in one plane.

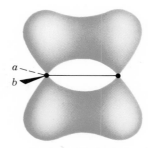

Carbon forms σ bonds with three sp^2 orbitals lying in one plane.

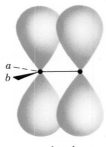

π bond   molecular orbital of π bond

Carbon and oxygen form a π bond composed of two parallel p orbitals perpendicular to the σ bonds.

The positive carbon of aldehydes and ketones thus attracts reagents with a high electron density on one atom—*nucleophilic reagents.*

11.3 NOMENCLATURE OF ALDEHYDES AND KETONES

1. In the IUPAC system the names of aldehydes are derived from the names of the corresponding hydrocarbons by replacing the "e" ending with "al."

2. The longest carbon chain containing the —CH=O carbon is used as the parent compound. Since the —CH=O group is always at the beginning of the chain, this carbon is always the #1 carbon and need not be specified.

$$CH_3{-}CH_2{-}CH_3 \qquad CH_3{-}CH_2{-}\overset{\displaystyle O}{\underset{\displaystyle H}{C}}$$

propane   propanal

3. Positions of carbon–carbon multiple bonds are designated by the lower number of the two unsaturated carbons.

4. Groups such as alkoxyl, amino, halo, hydroxyl, keto, and sulfhydryl are named as substituents.

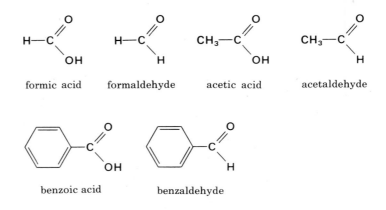

Z-3-chloro-2-pentenal   cis-2-butenal   3-hydroxybutanal

5. The name of an aldehyde which has the —CH=O group directly attached to a carbon of a ring system is formed by adding the suffix "carbaldehyde" to the name of the ring system. The carbon which bears the —CH=O group is #1.

cyclohexanecarbaldehyde

When the corresponding carboxylic acid is known by a common name, the common name of the aldehyde may be formed by changing "ic acid" or "oic acid" to "aldehyde."

formic acid   formaldehyde   acetic acid   acetaldehyde

benzoic acid   benzaldehyde

Ketones are named by dropping the "e" ending from the hydrocarbon name and adding the ending "one." The position of the =O is designated by the lowest number possible, using the longest carbon chain containing the carbonyl group.

4-hexen-2-one
(C=O given lowest
number possible)

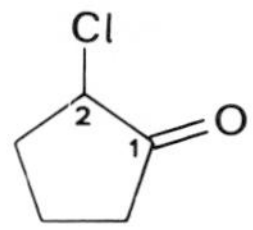

2-chlorocyclopentanone
(=O on #1 carbon)

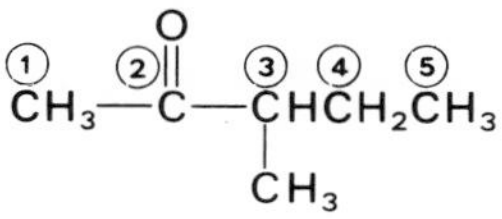

3-methyl-2-pentanone
(longest carbon chain
containing C=O)

Another method of naming simple ketones uses the carbon groups attached to C=O as substituents, for example, ethyl methyl ketone instead of 2-butanone. Some common names are also in use. The name "acetone" is as well established as it is poor, and illustrates the general rule that evolutionary nomenclature is confusing. "Acetophenone" is another common name in use.

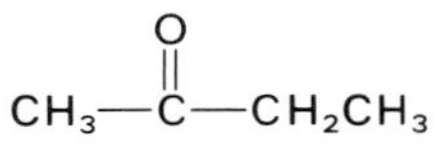

ethyl methyl ketone 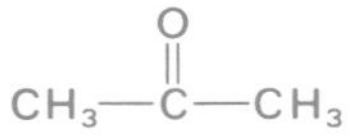   acetone 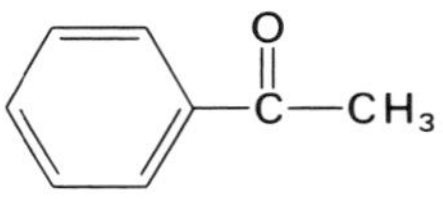   acetophenone

As a substituent the =O is given the name "oxo" (or "keto" in a common name).

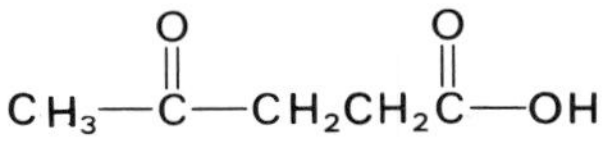

4-oxopentanoic acid 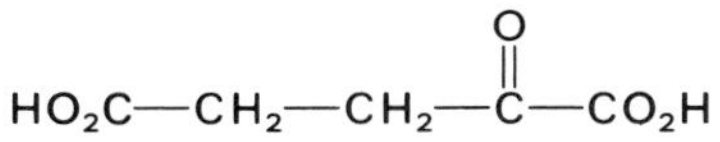   α-ketoglutaric acid

Problems   **11.2** Name the following structures by the IUPAC method.

a)

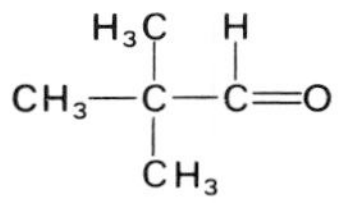

b) 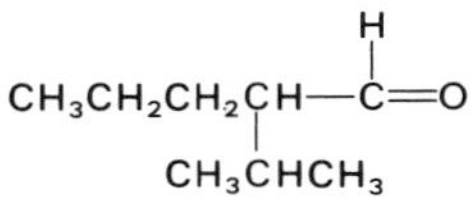 $CH_3CH_2CH_2CH-C=O$ with H on the carbonyl carbon and CH_3CHCH_3 substituent

c)

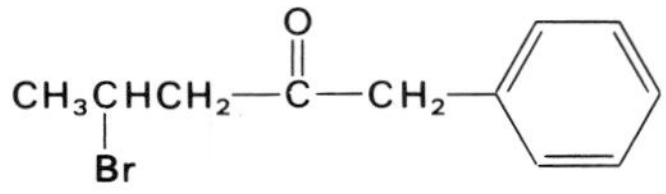

d)

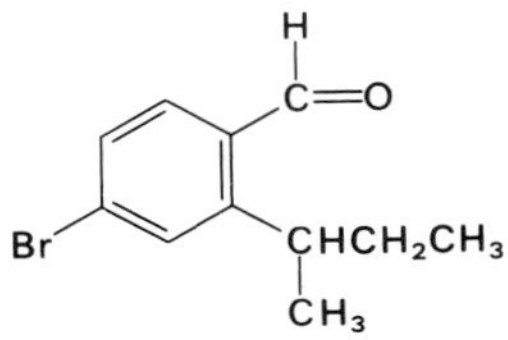

e) 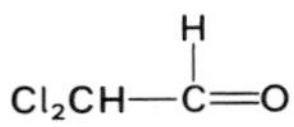 $Cl_2CH-C=O$ with H

f)

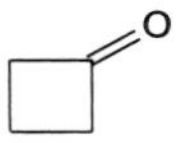

11.3 Write structures for the following compounds.

a) 1,1,3-trichloroacetone   **b)** phenylacetaldehyde   **c)** 2-butynal

d) 2-trifluoromethyl-4-isobutylbenzaldehyde

e) 1-cyclopropyl-2,4-pentanedione   **f)** 2-methyl-4-pyridinecarbaldehyde

11.4 MECHANISMS OF REACTIONS OF CARBONYL COMPOUNDS

The carbonyl group of aldehydes and ketones, like that of carboxylic acid derivatives, is open to attack on the electron-deficient carbon by nucleophilic reagents. With different reagents, however, the tetrahedral structure formed by the reaction of carbonyl with nucleophile proceeds in a variety of ways to form a stable product. One of these ways resembles the substitution reaction of the acid derivatives.

A carbon or hydrogen nucleophile adds to the C=O bond to give a stable saturated product.

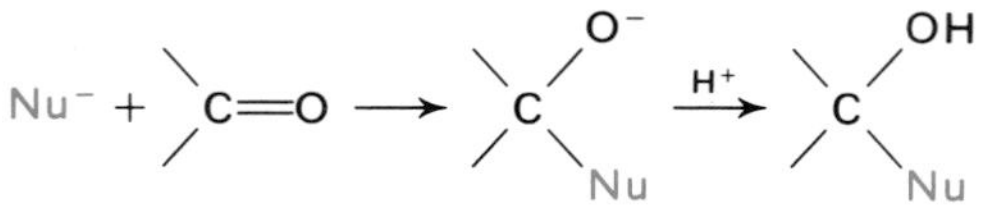

carbon or hydride nucleophile

With other nucleophiles, the tetrahedral reaction intermediate is unstable and loses water by one of two courses toward a stable product. This reaction intermediate can lose a leaving group H_2O in the same manner that an acid derivative intermediate does (Section 8.17). A stable unsaturated product is formed in this manner.

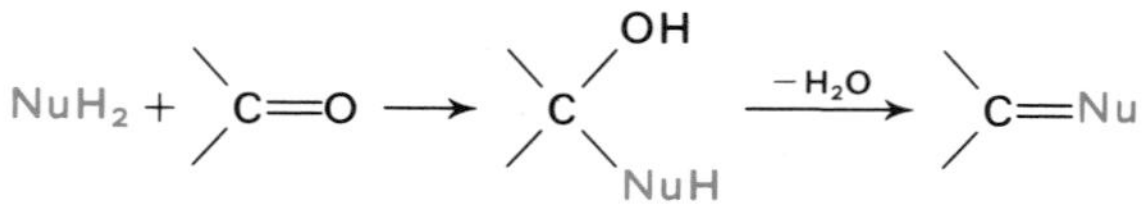

reaction intermediate

Under acid conditions this tetrahedral reaction intermediate can lose water giving a carbonium ion. The carbonium ion then reacts with a second nucleophile to give a saturated product as shown.

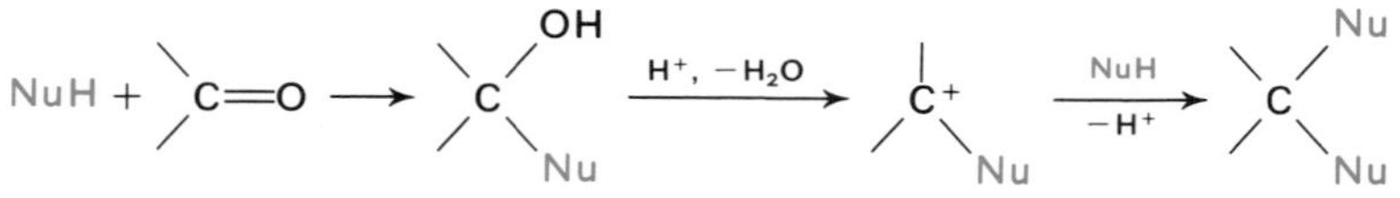

reaction intermediate

Examples of all of these types of reactions are given in the following sections.

The carbonyl groups of ketones are less reactive than those of aldehydes. The attached organic groups on ketones are more bulky than the hydrogen of the aldehyde and provide some steric hindrance to the formation of the tetrahedral adduct. The ketone carbonyl carbons are also less positive than those of aldehydes, because the alkyl and aryl groups of the ketone are electron-donating compared to hydrogen, and they partially counteract the positive charge on carbon.

11.5 ADDITION OF WATER AND ALCOHOL TO CARBONYL COMPOUNDS

Aldehydes and ketones add water reversibly to form dihydroxy compounds. These compounds have the two OH's on the same carbon and are called *gem*-diols. (The term *geminal* refers to two identical groups on the same carbon.)

$$H-\overset{\overset{\textstyle H}{|}}{C}=O + H_2O \rightleftarrows H-\overset{\overset{\textstyle H}{|}}{\underset{\underset{\textstyle OH}{|}}{C}}-OH$$

formaldehyde   formaldehyde hydrate

Such *gem*-diols are usually stable only in water solution. When removed from water, they revert to their parent carbonyl-containing compounds. Formaldehyde in the 40% aqueous solution (formalin) that is used to preserve biological specimens shows no detectable trace of C=O present. However, only formaldehyde can be isolated from the solution.

Structures having two functional groups, except two halogens, attached to the same carbon are ordinarily unstable relative to the carbonyl compound, if there is at least one hydrogen attached to one of the two heteroatoms $C(ZH)_2$ or RZ—C—ZH. The heteroatom Z is N, O, S, or Cl or any combination.

Only those carbonyls which have strong electron-withdrawing substituents form *gem*-diols which are stable enough to be isolated from water solution, such as "chloral hydrate" from trichloroacetaldehyde. Another *gem*-diol which is stable is ninhydrin, a compound which reacts with α-amino acids to form a dye used in identification of the individual amino acids.

trichloroacetaldehyde   "chloral hydrate"   ninhydrin

The addition of alcohols to carbonyl compounds, on the other hand, produces structures which sometimes have great importance. Under weakly acidic conditions, aldehydes and ketones readily add alcohols to form compounds having an alkoxy and a hydroxy group on the same carbon. These compounds are called **hemiacetals** and **hemiketals**.

$$CH_3\text{—}OH + CH_3CH_2\text{—}\overset{\displaystyle O}{\underset{\displaystyle H}{C}} \xrightarrow{H^+} CH_3CH_2\text{—}\overset{\displaystyle OH}{\underset{\displaystyle H}{C}}\text{—}O\text{—}CH_3$$

methanol   propanal   1-methoxy-1-propanol
(a hemiacetal)

Like the *gem*-diols, hemiacetals are unstable relative to the carbonyl compounds. The rapid and reversible reactions prevent the isolation of such adducts as pure substances except in special structures.

In the presence of strong acid and excess alcohol, the hemiacetal reacts with another molecule of alcohol to give an **acetal** RCH(OR′)$_2$, a 1,1-dialkoxy compound.

$$CH_3CH_2\text{—}\underset{\displaystyle OH}{CH}\text{—}OCH_3 + CH_3OH \underset{\text{dry HCl}}{\rightleftharpoons} CH_3CH_2\text{—}\underset{\displaystyle OCH_3}{CH}\text{—}OCH_3 + H_2O$$

1-methoxy-1-propanol   1,1-dimethoxypropane
(a hemiacetal)   (an acetal)

The steps involving the second molecule of methanol are acid catalyzed, and do not occur in the absence of acid. Note that this reaction is reminiscent of the two-step ionization mechanism for nucleophilic substitution of saturated compounds like alkyl halides (Section 6.4).

$$CH_3CH_2CH_2\text{—}\overset{\displaystyle H}{\underset{\displaystyle OH}{C}}\text{—}OCH_3 + H^+ \rightleftharpoons CH_3CH_2CH_2\text{—}\overset{\displaystyle H}{\underset{\displaystyle {}^+OH_2}{C}}\text{—}OCH_3 \rightleftharpoons \left[CH_3CH_2CH_2\text{—}\overset{\displaystyle H}{\underset{\displaystyle +}{C}}\text{—}OCH_3 \updownarrow CH_3CH_2CH_2\text{—}\overset{\displaystyle H}{C}{=}\overset{\displaystyle +}{O}CH_3 \right] + H_2O$$

Hemiacetal is protonated on hydroxyl.   Water is lost.   carbonium ion

$$CH_3CH_2CH_2\text{—}\overset{\displaystyle H}{\underset{\displaystyle +}{C}}\text{—}OCH_3 + CH_3OH \rightleftharpoons CH_3CH_2CH_2\text{—}\overset{\displaystyle H}{\underset{\displaystyle \underset{+}{HOCH_3}}{C}}\text{—}OCH_3 \rightleftharpoons CH_3CH_2CH_2\text{—}\overset{\displaystyle H}{\underset{\displaystyle OCH_3}{C}}\text{—}OCH_3 + H^+$$

Carbonium ion reacts rapidly with methanol.   Protonated acetal loses a proton, forms a neutral, stable product.

In the presence of acid, the products and reactants of the first and second reactions are all in equilibrium. With a large quantity of methanol, or with the removal of water, the equilibrium favors the acetal. With an excess of water and a catalytic amount of acid, the aldehyde is favored.

If the equilibrated system produced during acetal formation is neutralized or made basic, the acetal is made unreactive to water. The acetal, which is a stable compound, can then be isolated from the solution. Regeneration of the carbonyl compound from the acetal is easily accomplished by treatment of the acetal with dilute aqueous acid.

$$\text{CH}_3\text{CH}_2\text{CH}_2\text{CH}(\text{OCH}_3)_2 + \text{H}_2\text{O} \xrightarrow{\text{HCl}} \text{CH}_3\text{CH}_2\text{CH}_2\text{CH}{=}\text{O} + 2\ \text{CH}_3\text{OH}$$

    1,1-dimethoxybutane   butanal   methanol

In the same way that hydroxyacids easily form cyclic esters, a cyclic hemiacetal or cyclic acetal can be formed when the hydroxyl function is on the aldehyde molecule itself. A hydroxyl group on the #4 or #5 carbon reacts internally with the carbonyl to give the expected cyclic hemiacetal incorporated into a five- or six-membered ring. The cyclization of 4-hydroxybutanol is a good example.

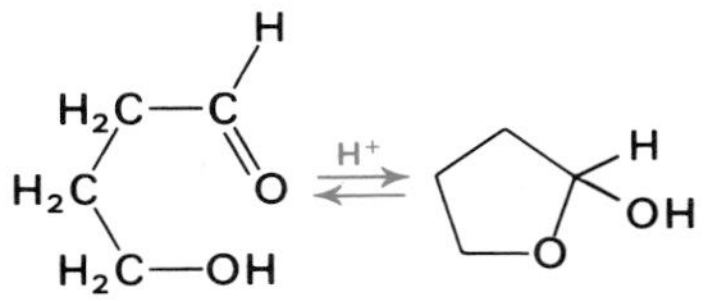

4-hydroxybutanal   cyclic hemiacetal

Further reaction of the cyclic hemiacetal with methanol in an acidic solution provides a cyclic acetal in which the two alkoxy groups came from different alcohols—one internal and one external.

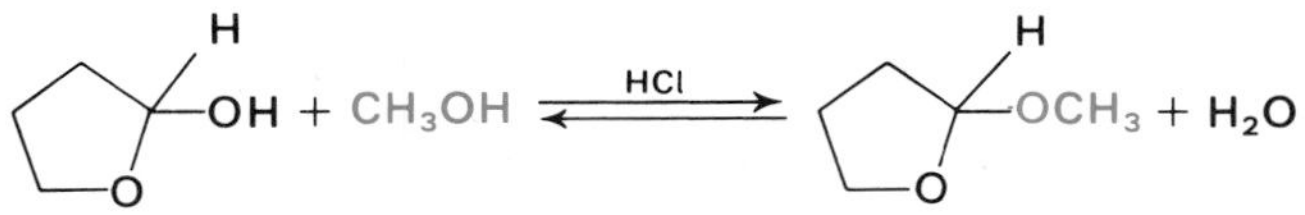

Ring formation to give cyclic hemiacetals and cyclic acetals is characteristic of carbohydrates, such as glucose, which is a hydroxyaldehyde.

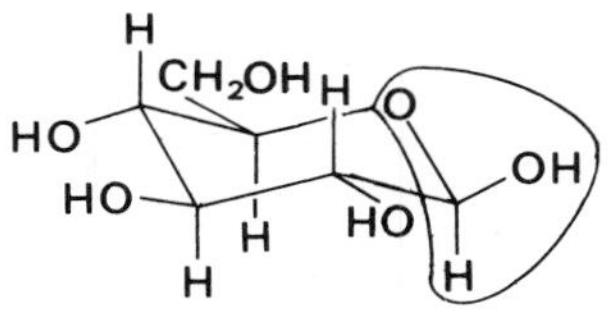

         glucose

11.4 Write equations for the following reactions.

  a) 3-ethylhexanal + ethanol + dry HCl

  b) 1,1-dimethoxypentane + H_2O + HCl

11.6 REACTION OF CARBONYL COMPOUNDS WITH AMMONIA AND AMINES

a) Formation of imines

Ammonia and amines react rapidly with aldehydes and ketones to form tetrahedral adducts which have a hydroxyl and an amino group attached to the same carbon. Since this tetrahedral intermediate has an H on both the OH and NH groups, it eliminates water to provide a new product, an imine, with an unsaturated functional group C=N—. Ammonia forms a simple imine (sometimes called a Schiff's base) with aldehydes and ketones, while primary amines give N-substituted imines.

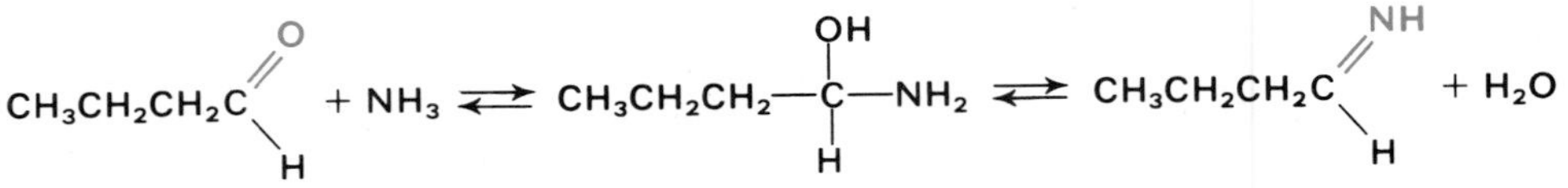

butanal   tetrahedral adduct   imine of butanal
                                                       (a Schiff's base)

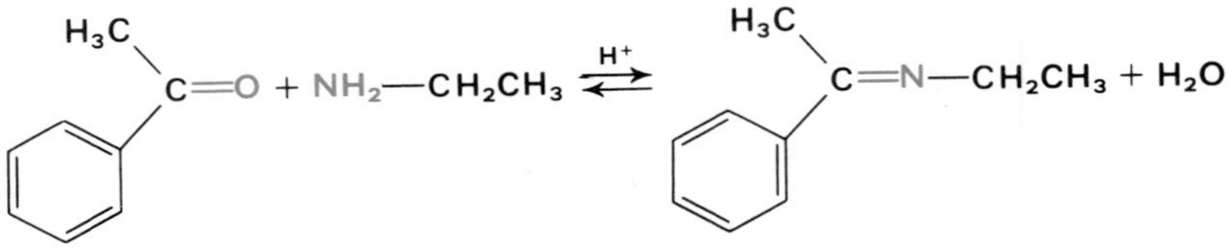

acetophenone

The reaction of a carbonyl compound and an amine to give an imine and water and the reverse reaction are catalyzed by moderately strong acids, such as acetic acid. The hydronium ion concentration must be small (10^{-3} or 10^{-4} M), but the concentration of acetic acid can be much higher in a carefully buffered solution. Too much strong acid will convert the nucleophilic amine to its ammonium ion, which is not a nucleophile and does not react.

Imines must be protected from water, for even atmospheric moisture made acidic by atmospheric CO_2 can hydrolyze the imine to the carbonyl compound and alkylammonium carbonate.

Imines in which either the carbonyl or the amino group is on an aromatic ring are sufficiently stable to be isolated when water is removed from the reaction mixture. Such an imine is formed in the equation above using acetophenone and in the following one with aniline and benzaldehyde.

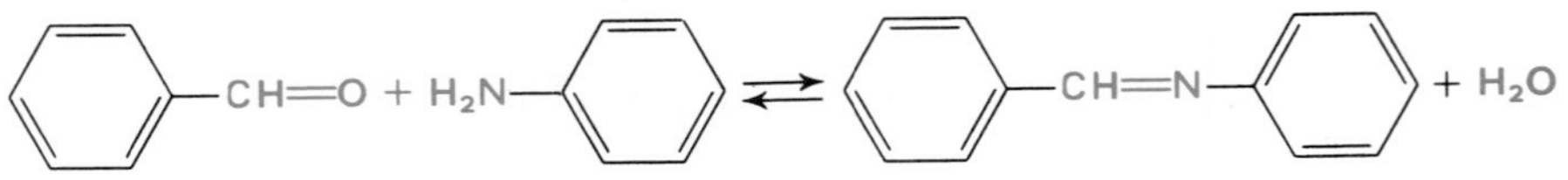

benzaldehyde   aniline   imine

Although imines are unimportant end products because of their instability, their formation as highly reactive, but not isolated, intermediates in many reactions is well established. The imine is more reactive than the corresponding carbonyl compound with nucleophilic reagents. In the first reaction of the synthesis of α-amino acids from aldehydes, ammonia, and hydrogen cyanide (the Strecker synthesis in next section), a rapidly formed imine reacts with the cyanide ion as the nucleophile.

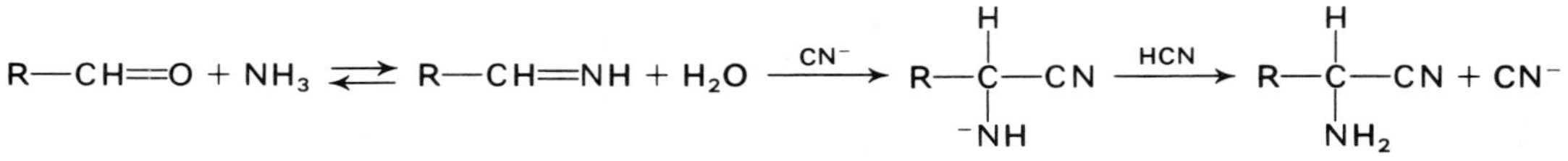

$$R-CH=O + NH_3 \rightleftarrows R-CH=NH + H_2O \xrightarrow{CN^-} R-\overset{\overset{\displaystyle H}{|}}{\underset{\underset{\displaystyle -NH}{|}}{C}}-CN \xrightarrow{HCN} R-\overset{\overset{\displaystyle H}{|}}{\underset{\underset{\displaystyle NH_2}{|}}{C}}-CN + CN^-$$

In biological sequences, the coenzyme pyridoxal reacts with amino acids to form imines. The stability of these imines is sufficient to allow study of pyridoxal reactions in the laboratory. The coenzyme's functions have thus been elucidated in enzyme model studies. Imine formation by pyridoxal and amino acid esters is shown in Fig. 11.2. Reactions of imines of pyridoxal and amino acid esters are described in Section 11.14.

FIGURE 11.2   IMINE FROM PYRIDOXAL AND ALANINE ESTER

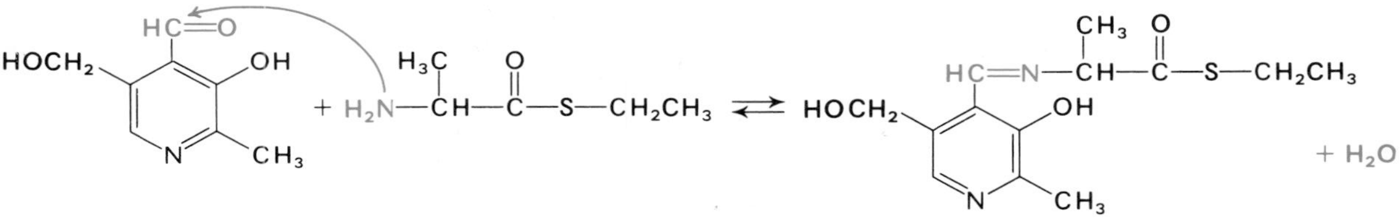

pyridoxal   ethyl thioester   imine
                   of alanine

b) Formation of nitrogen-base derivatives   Other nitrogen bases having the required Z—NH$_2$ group react with aldehydes and ketones to give products containing the C=N—Z group. **Phenylhydrazine** and **hydroxylamine** are representative of a number of similar bases having —NH—NH$_2$ or —O—NH$_2$ structures.

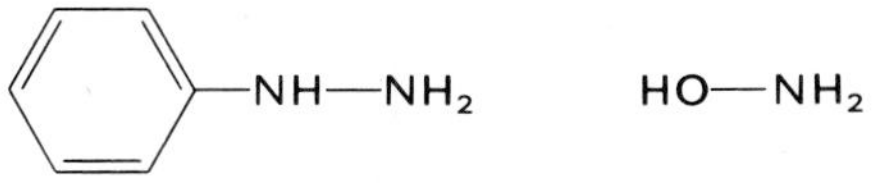

phenylhydrazine   hydroxylamine

Reactions of both reagents with aldehydes and ketones are catalyzed by acetic acid. The products are known as **phenylhydrazones** and **oximes**, respectively.

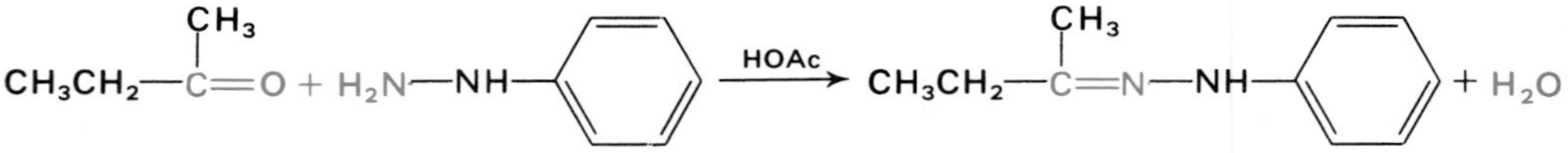

$$CH_3CH_2-\overset{\overset{\displaystyle CH_3}{|}}{C}=O + H_2N-NH-\bigcirc \xrightarrow{\text{HOAc}} CH_3CH_2-\overset{\overset{\displaystyle CH_3}{|}}{C}=N-NH-\bigcirc + H_2O$$

butanone   phenylhydrazine   butanone phenylhydrazone

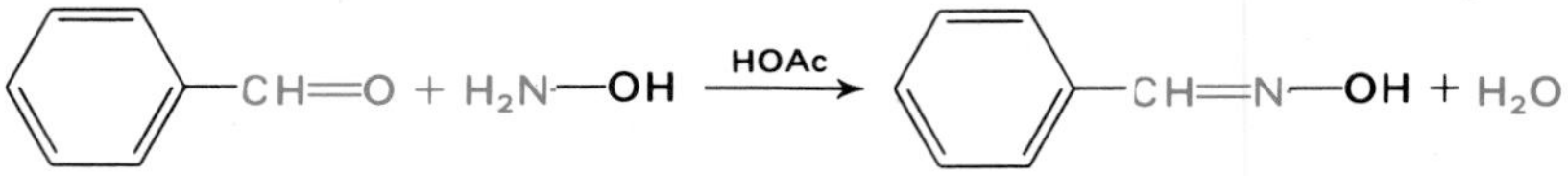

$$\bigcirc-CH=O + H_2N-OH \xrightarrow{\text{HOAc}} \bigcirc-CH=N-OH + H_2O$$

benzaldehyde   hydroxylamine   benzaldehyde oxime

The oxygen or nitrogen attached to the imine nitrogen greatly reduces the reactivity of the product toward water. Thus hydrolysis of phenylhydrazones and oximes is difficult, and carbonyl compounds are not easily recovered from these derivatives.

Phenylhydrazones frequently are insoluble in the aqueous alcohol solvent used for the reaction. They are easily crystallizable solids, which find use primarily as identifying solid derivatives of the parent carbonyl compounds. Melting points of these derivatives of hundreds of aldehydes and ketones are found tabulated in reference books. Identification of compounds by their solid derivatives was an important step in structure proof before the universal use of spectroscopic instruments.

Problem   **11.5** Write equations for the following reactions.

     a) phenylhydrazine + propanal + acetic acid

     b) 3-methylcyclopentanone + aniline + hydrogen chloride

     c) $CH_3CH_2CH=N-C_6H_5$ + water + hydrogen chloride

     d) 2-hydroxy-3-methylbutanal + *tert*-butylamine + hydrogen chloride

     e) cyclohexanone + hydroxylamine + acetic acid

11.7 ADDITION OF HYDROGEN CYANIDE— FORMATION OF A NEW CARBON–CARBON BOND

Remember that in the reactions of alkyl halides, the cyanide ion served as a nucleophilic carbon reagent. Aldehydes and simple ketones also react with the cyanide ion nucleophile. When the CN^- adds to the C=O group, the carbon chain is lengthened as a new carbon–carbon bond is formed. The product is an α-hydroxynitrile (or cyanohydrin).

$$CH_3CH_2-\overset{\overset{\displaystyle H}{|}}{C}=O + HC\equiv N \xrightarrow{OH^-} CH_3CH_2-\overset{\overset{\displaystyle H}{|}}{\underset{\underset{\displaystyle OH}{|}}{C}}-C\equiv N$$

<table>
<tr><td>propanal</td><td>2-hydroxybutanonitrile</td></tr>
</table>

$$CH_3-\overset{\overset{\displaystyle O}{||}}{C}-CH_3 + HC\equiv N \xrightarrow{OH^-} CH_3-\overset{\overset{\displaystyle OH}{|}}{\underset{\underset{\displaystyle CH_3}{|}}{C}}-C\equiv N$$

<table>
<tr><td>acetone</td><td>2-hydroxy-2-methylpropanonitrile</td></tr>
</table>

Note that the product of the addition of HCN to propanal has a carbon chain that is one carbon longer than that of the propanal, and that the functional group, the CN, is at the end of the chain. The addition product of CN^- with a ketone, like acetone, also has one more carbon and the chain now is branched.

Hydrogen cyanide is an acid and not in itself a nucleophilic reagent. The addition of HCN to carbonyl compounds is catalyzed by a base, usually hydroxide ion, which generates the nucleophilic reagent CN^- by reacting with the HCN. The cyanide ion then reacts with the carbonyl carbon.

$$OH^- + HC\equiv N \longrightarrow H_2O + {}^-C\equiv N$$

$$C_6H_5-\overset{\overset{\displaystyle H}{|}}{C}=O + {}^-C\equiv N \longrightarrow C_6H_5-\overset{\overset{\displaystyle H}{|}}{\underset{\underset{\displaystyle O^-}{|}}{C}}-C\equiv N \xrightarrow{HCN} C_6H_5-\overset{\overset{\displaystyle H}{|}}{\underset{\underset{\displaystyle OH}{|}}{C}}-C\equiv N + {}^-C\equiv N$$

<table>
<tr><td>benzaldehyde</td><td>anion adduct</td><td>2-hydroxy-2-phenyl-
acetonitrile</td></tr>
</table>

The addition step is reversible and the negatively charged adduct is unstable until it is protonated. The reaction of the adduct anion with more HCN forms the neutral hydroxynitrile and generates a new cyanide ion.

Problem **11.6** The addition of HCN to an aldehyde requires a catalytic amount of NaOH. What would you expect the outcome to be if each of the following situations existed instead?

  a) The gas HCN is bubbled into a solution containing an aldehyde but no NaOH.

  b) The salt NaCN is added to the aldehyde solution as a good source of CN^- (as in the reaction with alkyl halides), with no other reagent.

Acid-catalyzed hydrolysis converts the nitrile to the acid (Section 8.10).

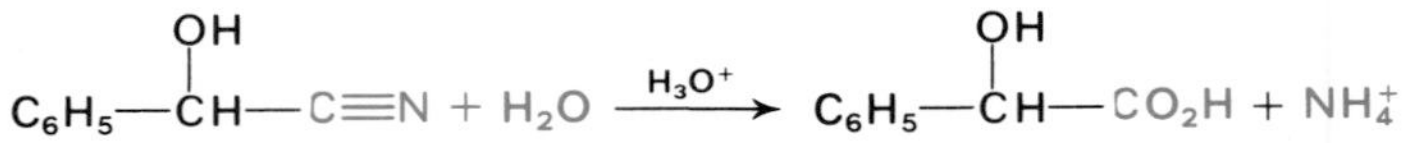

$$C_6H_5-\overset{\overset{\displaystyle OH}{|}}{C}H-C\equiv N + H_2O \xrightarrow{H_3O^+} C_6H_5-\overset{\overset{\displaystyle OH}{|}}{C}H-CO_2H + NH_4^+$$

hydroxyphenylacetonitrile   hydroxyphenylacetic acid
(mandelic acid)

A variation of this reaction may be used to synthesize α-amino acids instead of α-hydroxyacids. The key step to introduce the amino group may be accomplished if the hydrogen cyanide is used with the aldehyde in the presence of ammonia as well as the catalytic hydroxide ion. Ammonia itself is not a strong enough base to generate the cyanide ion in the reaction.

The carbonyl compound and ammonia form the imine in an equilibrium reaction. Cyanide ion adds to the imine, since the imine is more reactive than the aldehyde or ketone. The product is an α-aminonitrile. In a subsequent reaction, hydrolysis of the α-aminonitrile produces the α-amino acid. The overall sequence of reactions, known as the **Strecker synthesis of α-amino acids**, is shown in Fig. 11.3.

**FIGURE 11.3
STRECKER SYNTHESIS
OF α-AMINO ACIDS**

Reaction 1 Formation of the imine

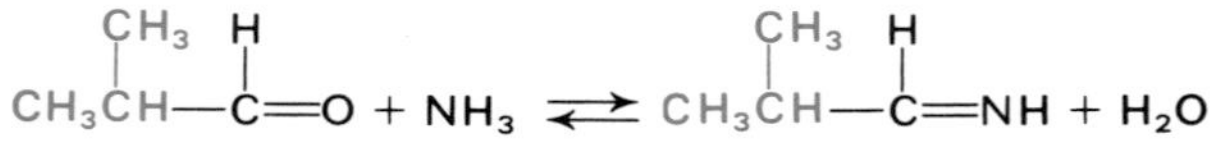

$$CH_3\overset{\overset{\displaystyle CH_3}{|}}{C}H-\overset{\overset{\displaystyle H}{|}}{C}=O + NH_3 \rightleftarrows CH_3\overset{\overset{\displaystyle CH_3}{|}}{C}H-\overset{\overset{\displaystyle H}{|}}{C}=NH + H_2O$$

2-methylpropanal   imine

Addition of hydrogen cyanide

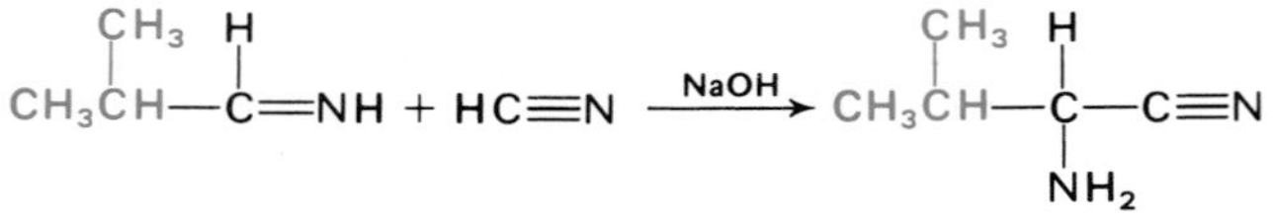

$$CH_3\overset{\overset{\displaystyle CH_3}{|}}{C}H-\overset{\overset{\displaystyle H}{|}}{C}=NH + HC\equiv N \xrightarrow{NaOH} CH_3\overset{\overset{\displaystyle CH_3}{|}}{C}H-\underset{\underset{\displaystyle NH_2}{|}}{\overset{\overset{\displaystyle H}{|}}{C}}-C\equiv N$$

imine   α-aminonitrile

Reaction 2 Hydrolysis of the nitrile

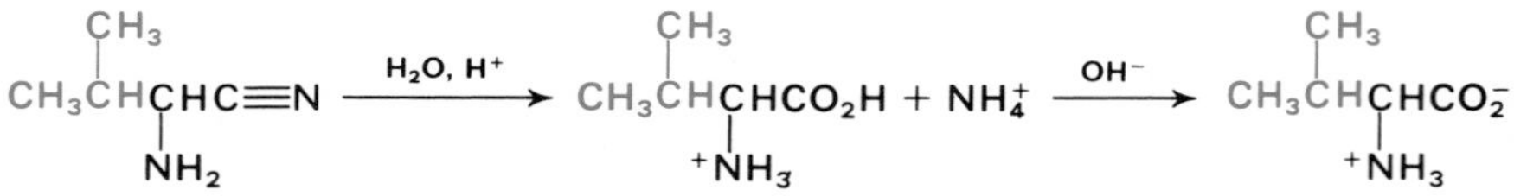

$$CH_3\overset{\overset{\displaystyle CH_3}{|}}{C}H\underset{\underset{\displaystyle NH_2}{|}}{C}HC\equiv N \xrightarrow{H_2O, H^+} CH_3\overset{\overset{\displaystyle CH_3}{|}}{C}H\underset{\underset{\displaystyle ^+NH_3}{|}}{C}HCO_2H + NH_4^+ \xrightarrow{OH^-} CH_3\overset{\overset{\displaystyle CH_3}{|}}{C}H\underset{\underset{\displaystyle ^+NH_3}{|}}{C}HCO_2^-$$

valine

Note that the synthesis produces the racemic form of the α-amino acid. Efficient methods to resolve the essential α-amino acids are extremely important, since only one of the two enantiomers is useful as a nutritional supplement in protein-poor diets. The unwanted enantiomer can be racemized and put through the resolving process again. In search of a practical stereospecific synthesis, some industrial methods for the production of α-amino acids now employ enzyme-catalyzed processes which are totally or partially stereospecific in yielding the desired enantiomer.

Problem **11.7** Draw the structure for the carbonyl compound which by reacting with HCN (and NH_3 if needed) would give each of the following acids after hydrolysis of the nitriles.

a) $CH_3CH_2CH_2\overset{\text{OH}}{\underset{|}{C}}HCO_2H$

b) $CH_3CH_2\overset{\text{CH}_3}{\underset{|}{C}}H\overset{}{\underset{|}{C}}HCO_2^-$ (isoleucine), with $\overset{+}{N}H_3$ below

c) $CH_3\overset{\text{CH}_3}{\underset{|}{C}}HCH_2\overset{+\text{NH}_3}{\underset{|}{C}}HCO_2^-$ (leucine)

d) $HO-\langle\bigcirc\rangle-CH_2\overset{+\text{NH}_3}{\underset{|}{C}}HCO_2^-$ (tyrosine)

e) $C_6H_5-CH=CH-CO_2H$

f) $^-O_2CCH_2CH_2\overset{+\text{NH}_3}{\underset{|}{C}}HCO_2^-$ (glutamic acid)

11.8 REACTIONS OF CARBONYL COMPOUNDS THAT INVOLVE THE HYDROGEN ON THE α-CARBON

Ordinary C—H bonds are difficult to break. The hydrogen alpha to a carbonyl group H—C—C=O, however, is readily removed by a strong base from an aldehyde and less readily from a ketone. A **carbanion**, a negatively charged ion with the charge on carbon, is produced as is shown by the reaction with propanal.

$$CH_3\overset{H}{\underset{|}{C}}H-\overset{H}{\underset{|}{C}}=O + OH^- \; \rightleftarrows \; \left[CH_3\overset{-}{\underset{}{C}}H-\overset{H}{\underset{|}{C}}=\ddot{O}: \; \longleftrightarrow \; CH_3CH=\overset{H}{\underset{|}{C}}-\ddot{O}:^- \right] + H_2O$$

resonance structures of α-carbanion

This is an acid–base reaction in which the proton is donated by a carbon atom rather than by the more usual electronegative atoms. The equilibrium lies very far in the direction of the aldehyde, which is a very weak acid. The α-carbanion formed is stabilized by the delocalization of electrons and negative charge. The resonance structure having the negative charge on oxygen is of lower energy (more stable) than the structure with the charge on carbon. A carbanion formed at an α-carbon and stabilized by charge and electron delocalization into a carbonyl group is considerably more stable than a carbanion derived from any other carbon of the aldehyde chain. Charge delocalization is possible only from the α-carbon. The delocalization of the α-carbanion makes the α-hydrogen a more acidic proton than hydrogens bonded to ordinary alkyl or alkenyl carbons.

The reactions of the following sections involve these α-carbanions as reaction intermediates. Before we consider the new reactions, it is very important that you develop a facility in recognizing and drawing the structures of the α-carbanions produced by treating aldehydes or ketones with base. Be sure to study and answer the following problem carefully.

Problem   **11.8** Draw resonance structures of all α-carbanions derived from the following compounds.

a) CH₃CHCH=O (with CH₃ substituent)

b) CH₃CCH₂CH₃ (with O)

c) γ CH₃ β CH=α CHCH=O
(γ-carbanion)

d) [ring]—CH₂CH=O

11.9 KETO–ENOL EQUILIBRIA   In the presence of a strong acid or base as catalyst, isomers of aldehydes and ketones, known as **enols**, exist in equilibrium with the carbonyl compounds. The enol structure is C=C—OH (alk*ene*–alcoh*ol*). For acetone, and for most simple aldehydes and ketones, the equilibrium lies far to the side of the carbonyl compound.

$$CH_3-\overset{O}{\overset{\|}{C}}-CH_3 \underset{}{\overset{H^+ \text{ or } OH^-}{\rightleftharpoons}} CH_3-\overset{OH}{\overset{|}{C}}=CH_2$$

acetone   1-propen-2-ol
(keto form)   (enol form)
> 99%   < 1%

The keto–enol isomers differ only in the positions of one proton and of one single and one double bond. This specific kind of isomerism is known as **tautomerism**, and the isomers are called **tautomers**. The conversion is fast and occurs in the presence of an acid or a base.

As a class enols, other than phenols, are very much less stable than the carbonyl compounds. Some stable enols are shown in Fig. 11.4.

It is difficult to isolate an enol and sometimes even to detect its presence when its equilibrium concentration is low. The stability of an enol is increased by hydrogen bonding of the OH group to a nearby =O when a six-membered ring is formed, and by the development of a conjugated double-bond system (as shown in Fig. 11.4).

Tautomerization by basic catalysis proceeds through the α-carbanion intermediate introduced in the previous section. In the first step an

FIGURE 11.4
SOME STABLE ENOL FORMS

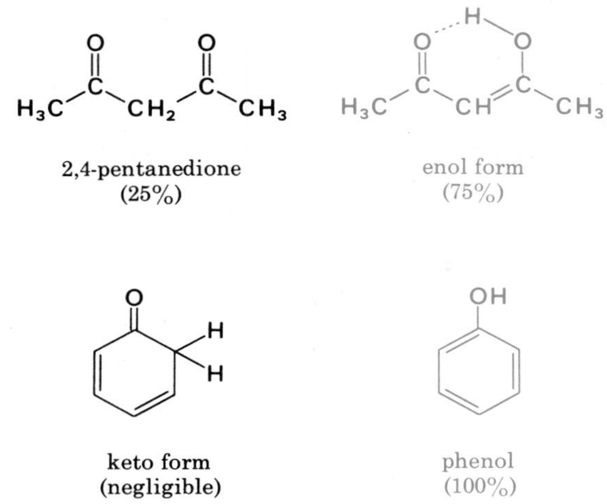

2,4-pentanedione
(25%)

enol form
(75%)

Enol is stabilized by
a) hydrogen bond in six-
membered ring,
b) conjugation of C=C
and C=O bonds.

keto form
(negligible)

phenol
(100%)

Keto form is destabilized
by losing aromatic
resonance stability.

α-hydrogen is removed from the carbonyl compound by a strong base to give an α-carbanion. In the second step the carbanion is protonated either on the carbonyl oxygen or on the α-carbon. Protonation of the oxygen produces the corresponding enol, while reprotonation on carbon merely reforms the original carbonyl compound. An equilibrium is established between the two isomers. The steps of the base-catalyzed tautomerization are illustrated by 3-pentanone.

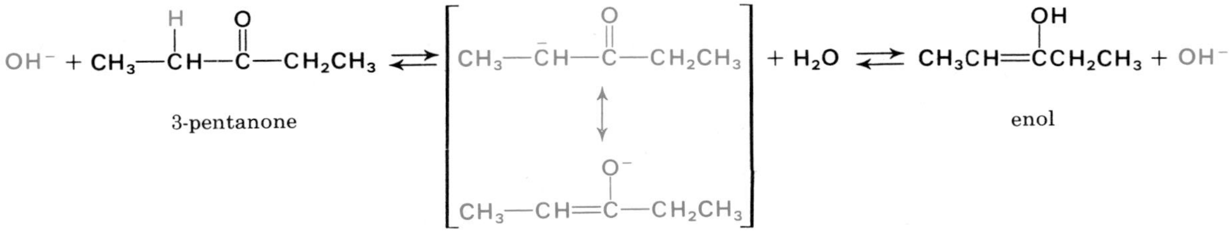

3-pentanone

resonance stabilized
carbanion

enol

Keto-enol tautomerization is also catalyzed by strong acid. In this case, the ketone is first protonated on the carbonyl oxygen to give a cation which is slightly stabilized by delocalization of charge and electrons. If a proton is then lost from the α-carbon instead of the oxygen, an enol is formed. The entire process is reversible, as shown with 3-pentanone.

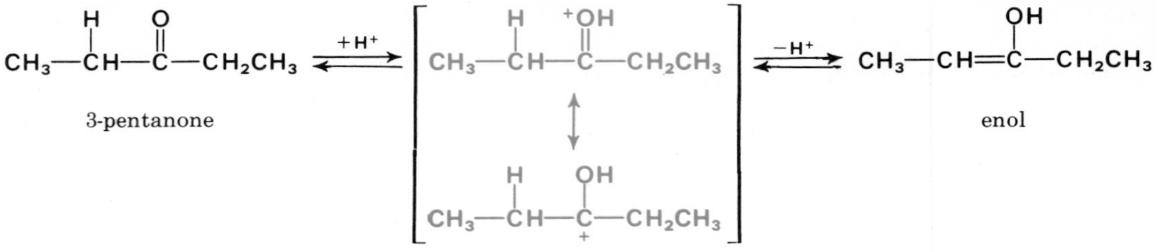

CH₃—CH—C—CH₂CH₃ $\underset{-H^+}{\overset{+H^+}{\rightleftarrows}}$ [CH₃—CH—C—CH₂CH₃ ⇅ CH₃—CH—C—CH₂CH₃] $\underset{}{\overset{-H^+}{\rightleftarrows}}$ CH₃—CH=C—CH₂CH₃

3-pentanone

resonance stabilized
cation

enol

Problem **11.9** The following compounds were warmed in D_2O and NaOD solution until all possible hydrogens had been replaced with deuterium through keto–enol equilibrium. Write the structures of the compounds produced.

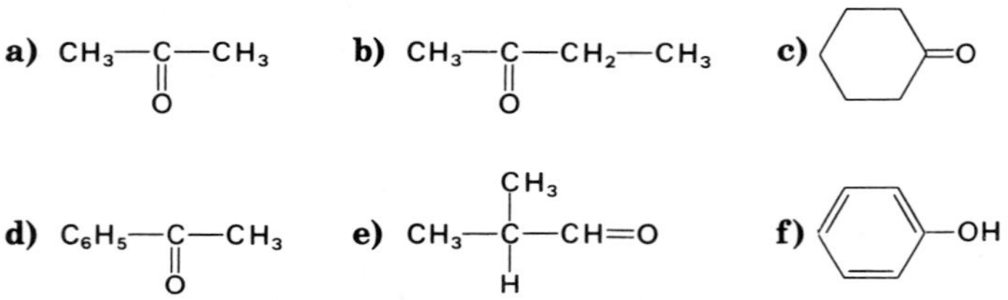

a) CH₃—C—CH₃
           ‖
           O

b) CH₃—C—CH₂—CH₃
           ‖
           O

c) ⬡=O

d) C₆H₅—C—CH₃
             ‖
             O

e) CH₃—C—CH=O
             |CH₃
             |
             H

f) ⬡—OH

11.10 THE ALDOL CONDENSATION REACTION   We might expect carbanions to be excellent nucleophiles, and as such, to add to carbonyl groups. And so they do, quite readily. In the presence of a strong base, one molecule of an aldehyde with a hydrogen on the α-carbon reacts with a second molecule of the same aldehyde to give a β-hydroxyaldehyde of twice the molecular weight of the starting compound. The initial product readily loses water to give an α,β-unsaturated aldehyde, as in this example of the reaction of propanal.

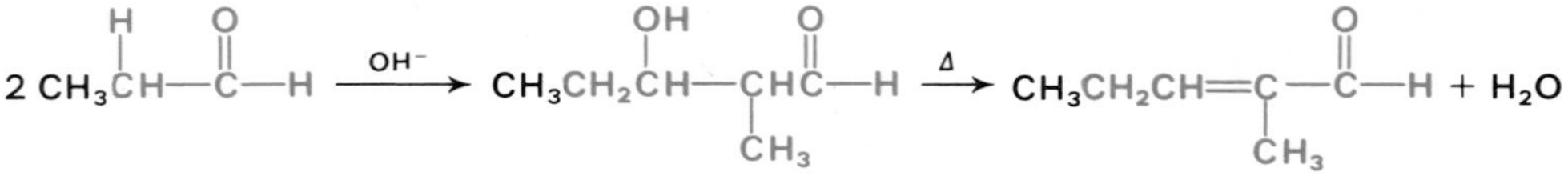

2 CH₃CH—C—H $\xrightarrow{\text{OH}^-}$ CH₃CH₂CH—CHC—H $\xrightarrow{\Delta}$ CH₃CH₂CH=C—C—H + H₂O

propanal

3-hydroxy-2-methyl-
pentanal
(a,β-hydroxyaldehyde)

2-methyl-2-pentenal
(an α,β-unsaturated aldehyde)

A reaction in which two molecules join together and split out a small molecule is a condensation reaction. This particular condensation reaction is called the **aldol condensation**. It is catalyzed by bases as strong as sodium hydroxide and proceeds through a series of four steps beginning with the formation of an α-carbanion of the aldehyde.

Step 1 Formation of the α-carbanion by base

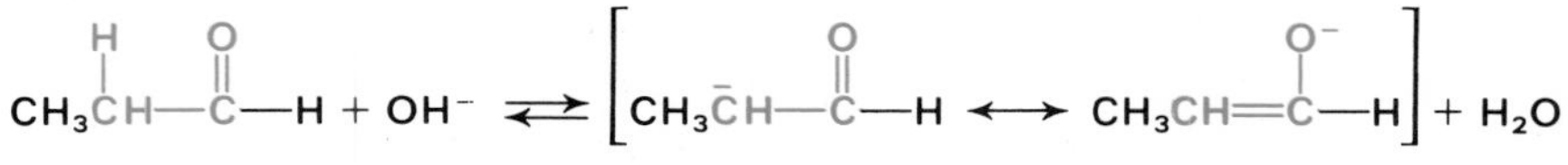

$$CH_3\overset{H}{\underset{}{C}}H-\overset{O}{\underset{}{C}}-H + OH^- \rightleftharpoons \left[CH_3\bar{C}H-\overset{O}{\underset{}{C}}-H \longleftrightarrow CH_3CH=\overset{O^-}{\underset{}{C}}-H \right] + H_2O$$

propanal   α-carbanion

Step 2 Nucleophilic addition of carbanion to carbonyl group

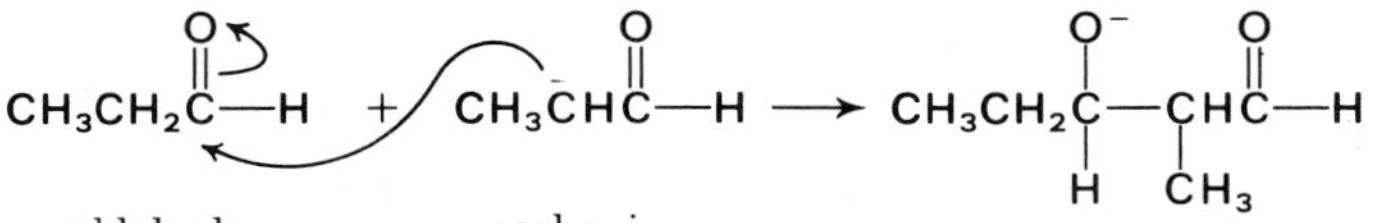

$$CH_3CH_2\overset{O}{\underset{}{C}}-H \ + \ CH_3\bar{C}H\overset{O}{\underset{}{C}}-H \longrightarrow CH_3CH_2\overset{O^-}{\underset{}{C}}-\overset{}{\underset{CH_3}{C}}H\overset{O}{\underset{}{C}}-H$$

aldehyde   carbanion

Step 3 Protonation of the adduct

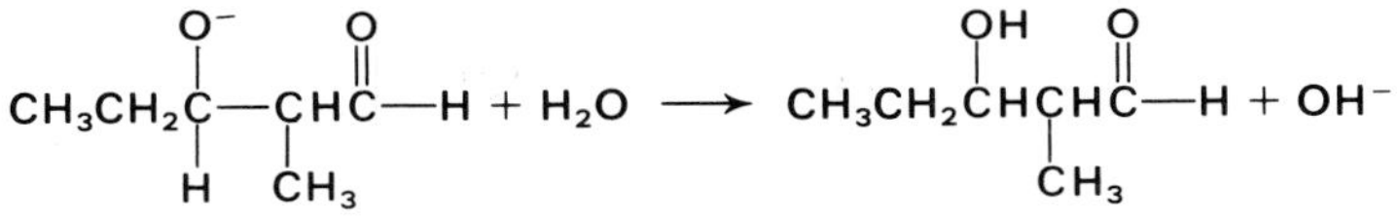

$$CH_3CH_2\overset{O^-}{\underset{H}{C}}-\overset{}{\underset{CH_3}{C}}H\overset{O}{\underset{}{C}}-H + H_2O \longrightarrow CH_3CH_2\overset{OH}{\underset{}{C}}H\overset{}{\underset{CH_3}{C}}H\overset{O}{\underset{}{C}}-H + OH^-$$

3-hydroxy-2-methylpentanal

Step 4 Loss of water

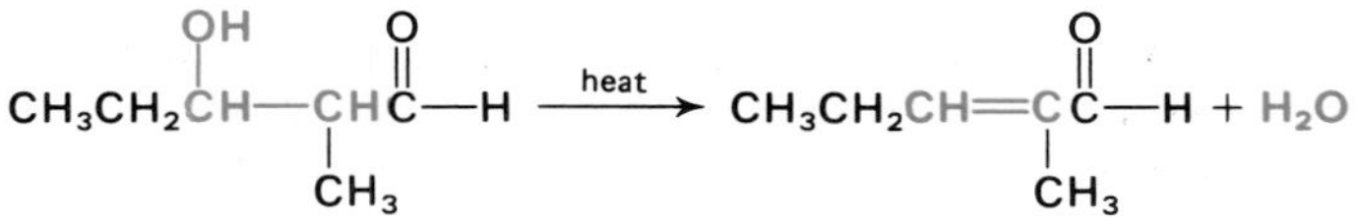

$$CH_3CH_2\overset{OH}{\underset{}{C}}H-\overset{}{\underset{CH_3}{C}}H\overset{O}{\underset{}{C}}-H \xrightarrow{heat} CH_3CH_2CH=\overset{O}{\underset{CH_3}{C}}\overset{}{C}-H + H_2O$$

2-methyl-2-pentenal

An α-carbanion is formed in very low concentration by the reaction of the aldehyde with hydroxide ion (step 1). The carbanion is a very reactive *nucleophile* and adds to the carbonyl group of another molecule of propanal (step 2). The alkoxide ion which is produced by the addition in step 2 receives a proton from water to give the β-hydroxyaldehyde and to regenerate the hydroxide ion catalyst (step 3). The β-hydroxy-aldehydes, where it is possible, lose water and form α,β-unsaturated aldehydes (step 4).

The aldol condensation is a very important example of carbon acting as a nucleophile. The nucleophilic reagent is generated in the reaction medium and is so reactive that it adds to the carbonyl group faster than any other reaction can occur in the same mixture. The aldol reaction provides a useful method for building larger molecules with longer carbon chains from smaller ones. The aldol and the related reactions discussed in Chapter 14 provide the most common means of forming new carbon–carbon bonds to build large molecules in biological systems. An example is the biosynthesis of the 6-carbon sugars, fructose and glucose, from two 3-carbon units (Section 13.6).

Mixed condensations that involve two different aldehydes (A and B) give all possible products (AA, AB, BA, and BB), and are therefore less useful.

Problems **11.10** Write the structures for the products formed by the aldol condensation of the following compounds and of the unsaturated compounds produced by subsequent loss of water.

    a) $CH_3CH{=}O$ **b)** $CH_3CH_2CH_2CH{=}O$

11.11 Write the structures for all final products (after water is lost) of the aldol condensation obtained from the following starting materials.

    a) $CH_3\overset{\overset{\textstyle H}{|}}{C}{=}O$ **b)** $C_6H_5{-}CH_2{-}\overset{\overset{\textstyle H}{|}}{C}{=}O$

    c) mixture of $C_6H_5{-}\overset{\overset{\textstyle H}{|}}{C}{=}O$ and $CH_3{-}\overset{\overset{\textstyle H}{|}}{C}{=}O$

    d) mixture of $CH_3{-}\overset{\overset{\textstyle H}{|}}{C}{=}O$ and $CH_3{-}CH_2{-}\overset{\overset{\textstyle H}{|}}{C}{=}O$

    e) mixture of $CH_3{-}\overset{\overset{\textstyle H}{|}}{C}{=}O$ and $H_2C{=}O$

11.11 REDUCTION AND OXIDATION OF CARBONYL COMPOUNDS

Aldehydes and ketones are the middlemen in the hierarchy of oxygen compounds of carbon. They can be reduced to alcohols readily. Aldehydes can also be extremely easily oxidized to carboxylic acids. Ketones are not oxidized except under severe conditions which break carbon–carbon bonds. The next few reactions illustrate the conversions of aldehydes and ketones to the other oxygen compounds.

It is more difficult to recognize the oxidation states of carbon compounds than those of inorganic compounds. In general the formation of a carbon–hydrogen bond is a reduction reaction, with the product at a lower oxidation state than the reactant. The replacement of a carbon–hydrogen bond by a carbon–oxygen bond is usually an oxidation reaction and gives a product at a higher oxidation state. Thus, for example, the conversion of an aldehyde to an alcohol is called a reduction, and the reverse conversion of an alcohol to an aldehyde is called an oxidation.

$$R{-}\overset{\overset{\textstyle H}{|}}{C}{=}O \quad \text{to} \quad R{-}\overset{\overset{\textstyle H}{|}}{\underset{\underset{\textstyle H}{|}}{C}}{-}OH \qquad\qquad R{-}\overset{\overset{\textstyle H}{|}}{\underset{\underset{\textstyle H}{|}}{C}}{-}OH \quad \text{to} \quad R{-}\overset{\overset{\textstyle H}{|}}{C}{=}O$$

        reduction   oxidation

a) Reduction of carbonyl compounds The carbonyl group of aldehydes and ketones is particularly reactive to reagents which furnish a hydride ion as a nucleophilic reagent. The addition of lithium aluminum hydride to an aldehyde or ketone gives a tetraalkoxyaluminate anion.

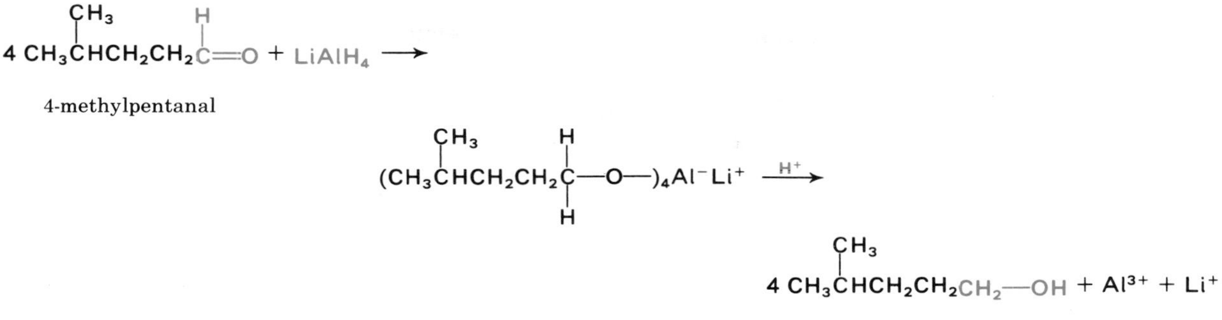

$$4\ CH_3\overset{\overset{\displaystyle CH_3}{|}}{C}HCH_2CH_2\overset{\overset{\displaystyle H}{|}}{C}=O + LiAlH_4 \longrightarrow$$

4-methylpentanal

$$(CH_3\overset{\overset{\displaystyle CH_3}{|}}{C}HCH_2CH_2\overset{\overset{\displaystyle H}{|}}{\underset{\underset{\displaystyle H}{|}}{C}}-O-)_4Al^-Li^+ \xrightarrow{\ H^+\ }$$

$$4\ CH_3\overset{\overset{\displaystyle CH_3}{|}}{C}HCH_2CH_2CH_2-OH + Al^{3+} + Li^+$$

4-methyl-1-pentanol

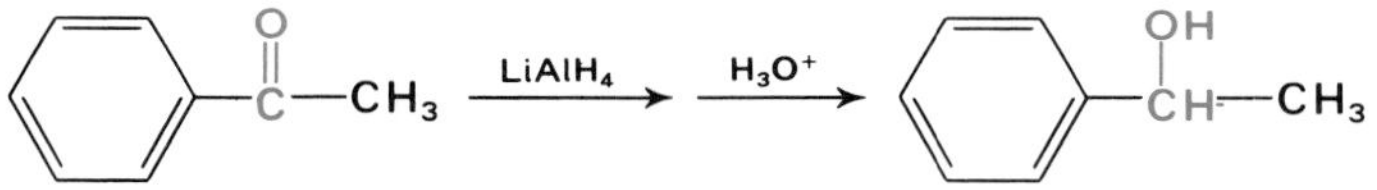

acetophenone   1-phenylethanol

Since only one hydride is required for each carbonyl group, one mole of LiAlH$_4$ reacts with four moles of aldehyde or ketone. Acidification of the tetraalkoxyaluminate ion formed by the addition yields the alcohol corresponding to the original carbonyl compound. An aldehyde adds a hydride to form a primary alcohol while a ketone and a hydride give a secondary alcohol.

A similar but milder reagent than LiAlH$_4$ which also furnishes a hydride is sodium borohydride NaBH$_4$. This hydride reagent has some specific advantages over the more reactive LiAlH$_4$. First, sodium borohydride reacts with the carbonyl group of *only* an aldehyde or a ketone. It does not react with esters, nitriles, amides, or carboxylate ions. The limited reactivity of NaBH$_4$ means that a compound which has both a ketone and an ester functional group is reduced at only the ketone function, and the ester group remains. Second, NaBH$_4$ reacts only slowly with hydroxyl groups in alcohols or water. Thus a hydroxyl group may be present in the compound to be reduced or an alcohol may be used as solvent.

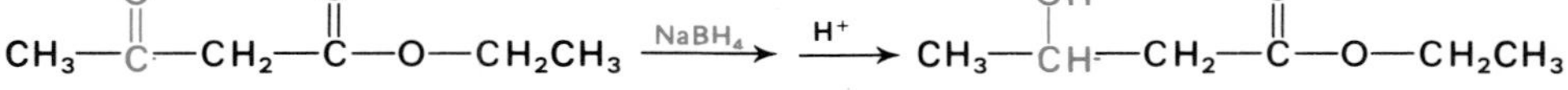

ethyl acetoacetate   ethyl 3-hydroxybutanoate
(ethyl 3-oxobutanoate)

b) Oxidation of aldehydes

Aldehydes are extremely easily oxidized to carboxylic acids by most of the usual inorganic oxidizing agents—atmospheric oxygen, bromine, nitric acid, silver oxide, or copper complex ions. The more powerful reagents such as chromic acid and potassium permanganate also are used. "Chromic acid" $H_2Cr_2O_7$ is a strong acid which cannot be isolated from solution. It is prepared in the reaction mixture from sodium dichromate with sulfuric acid. In the following reaction propanal is oxidized to propanoic acid.

$$3\ CH_3CH_2CH{=}O\ +\ Cr_2O_7^{2-}\ +\ 8\ H^+\ \longrightarrow\ 3\ CH_3CH_2CO_2H\ +\ 2\ Cr^{3+}\ +\ 4\ H_2O$$

propanal   propanoic acid

A simple test-tube diagnostic test to distinguish between aldehydes and ketones is based on the formation of an identifiable solid product in the oxidation of aldehydes by silver complex ions. Oxidation of an aldehyde results in the reduction of the silver complex ion $Ag(NH_3)_2^+$ to metallic silver, which is observed as a silver mirror or a fine black powder.

$$CH_3CH_2CH{=}O\ \xrightarrow{Ag(NH_3)_2^+OH^-}\ CH_3CH_2CO_2^-\ +\ Ag_{(s)}$$

11.12 METHODS OF PREPARATION OF ALDEHYDES AND KETONES

In synthetic sequences we usually find carbonyl compounds somewhere in the middle of the sequence. This occurs because they are readily prepared from other classes of compounds and participate in so many different kinds of reactions to form so many other classes of compounds. Two general routes are available. Carbonyl compounds may be prepared from alcohols, alkynes, and acids having the same carbon skeleton by reactions which merely alter the functional group. These reactions include hydration of alkynes and oxidation of alcohols. Carbonyl compounds may also be synthesized from smaller carbonyl compounds or esters by condensation reactions such as the aldol and Claisen reactions (Sections 11.10 and 14.7).

a) Acid-catalyzed hydration of alkynes

The acid-catalyzed hydration of alkynes is a method of synthesizing ketones directly from hydrocarbons having the same carbon skeleton. Remember that the direction of addition of water to an alkyne follows Markownikoff's rule, by which the oxygen becomes bonded to the more highly substituted carbon of the $-C{\equiv}C-$ group (Section 4.9).

$$CH_3-C{\equiv}C-H + H_2O \xrightarrow[HgSO_4]{H_2SO_4} CH_3-\overset{\displaystyle OH}{\underset{\displaystyle |}{C}}{=}CH_2 \longrightarrow CH_3-\overset{\displaystyle O}{\overset{\displaystyle \|}{C}}-CH_3$$

propyne   acetone enol   acetone

The unsaturated alcohol first formed by the addition of water to the alkyne is actually an enol. Under the strongly acidic conditions of the reaction, tautomerization of the enol to the ketone gives the ketone as the only isolable product.

Acetaldehyde is the only aldehyde which can be obtained by this procedure. Because of the direction of addition, hydration of any substituted acetylene leads to a ketone and not an aldehyde.

$$\text{H-C}\equiv\text{C-H} + \text{H}_2\text{O} \xrightarrow[\text{HgSO}_4]{\text{H}_2\text{SO}_4} \text{CH}_3\text{-CH}=\text{O}$$

acetylene   acetaldehyde

b) Oxidation of alcohols Ketones are readily obtained from the corresponding secondary alcohols by the reaction of the alcohols with a strong inorganic oxidizing agent, such as chromic acid formed from sodium dichromate and sulfuric acid.

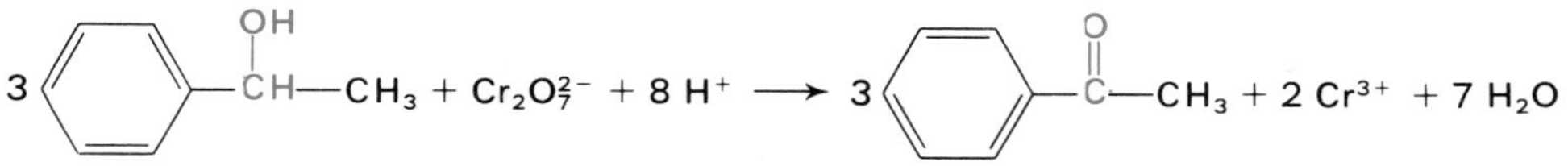

1-phenylethanol   acetophenone

Milder variations of the sodium dichromate–aqueous sulfuric acid reagent have been developed. Chromium trioxide added very slowly to pyridine forms a CrO_3—C_5H_5N complex which in a nonaqueous solution oxidizes alcohols to aldehydes, without further oxidation to acids. Unsaturated alcohols can also be oxidized this way.

$$\text{CH}_2=\text{CH-CH}_2\text{-OH} \xrightarrow{\text{CrO}_3\text{-pyridine}} \text{CH}_2=\text{CH-CH}=\text{O}$$

allyl alcohol   propenal

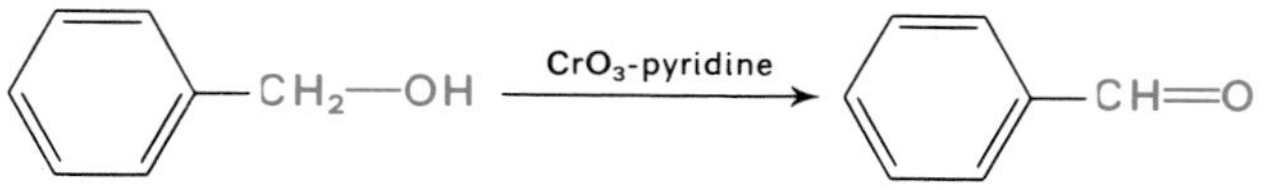

benzyl alcohol   benzaldehyde

Problem **11.12** Name a reagent or series of reagents which can be used to convert each of these compounds into a new compound containing a carbonyl group. Give the structure of the product formed.

a) CH_3—CH_2—$C\equiv CH$ **b)**    **c)** $CH_3CH_2CH_2OH$

11.13 A BIOLOGICAL OXIDATION-REDUCTION SYSTEM THAT INTERCONVERTS CARBONYL- AND HYDROXYL-CONTAINING COMPOUNDS

Biochemical reactions employ reagents that are structurally different from laboratory reagents. However, the two reagents perform the same kinds of reactions by somewhat similar mechanisms, as in the example NAD^+ -NADH. These compounds containing nicotinamide were the first coenzymes to be recognized (1904). They are synthesized biologically from nicotinamide ("niacin," a member of the vitamin B complex group) or nicotinic acid. The coenzyme system NAD^+-NADH consists of hydride donor-acceptor reagents which are not unlike those of the laboratory. These coenzymes are required in most enzyme-catalyzed hydrogenations of ketones and dehydrogenations of alcohols.

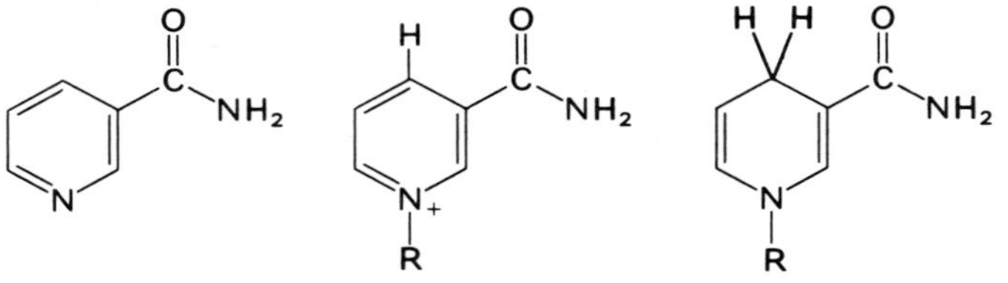

nicotinamide — NAD^+ (oxidized form) — NADH (reduced form)

Among the many enzyme-catalyzed oxidation reactions involving the coenzyme NAD^+ is the oxidation of ethanol to acetaldehyde by fermenting yeast.

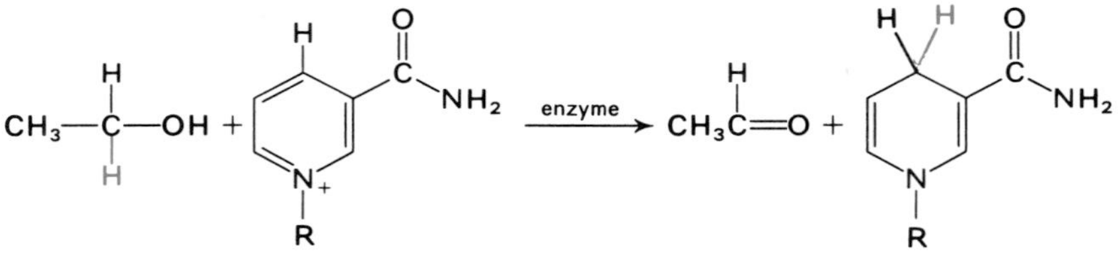

ethanol — NAD^+ — acetaldehyde — NADH

There is a direct transfer of the hydrogen as the hydride ion H^- from the *carbon* bearing the hydroxyl group to the carbon atom in the #4 position of the positively charged pyridine ring. The proton attached to the oxygen is subsequently lost to a nearby basic group on the enzyme.

$$CH_3-\overset{H}{C}=\underset{+}{O}H + B^- \longrightarrow CH_3\overset{H}{C}=O + HB$$

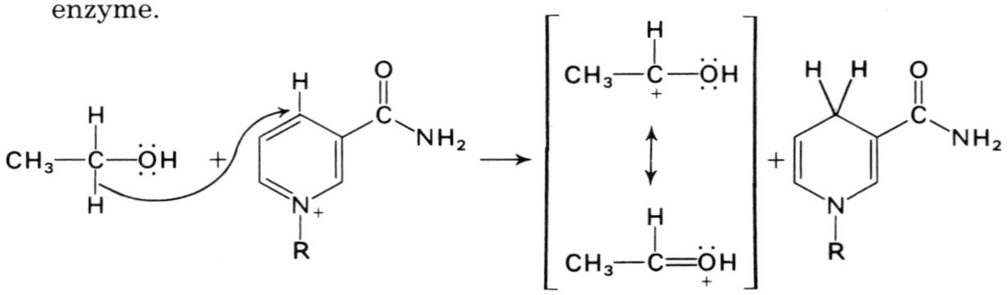

The structure of the R-group of NAD^+ (nicotinamide adenine dinucleotide) is shown here and considered in detail in Chapter 13.

$$R = \text{-ribose-O}—\bar{P}O_2—O—\bar{P}O_2—\text{O-ribose-adenine}$$

$$\underbrace{\qquad}_{\text{(a sugar)}} \qquad\qquad \underbrace{\qquad\qquad}_{\text{(adenosine)}}$$

The two-way nature of the hydride transfer system of NAD^+-NADH is illustrated in the synthesis and in the degradation of fatty acids. The oxidation of fatty acids (metabolic degradation) involves the conversion of β-hydroxythioesters into β-ketothioesters by NAD^+. The thiol part of the thioester is coenzyme A (Section 9.8).

$$\underset{\text{OH}}{\text{CH}_3}\overset{\text{OH}}{\underset{|}{\text{C}}}\text{HCH}_2\overset{\text{O}}{\underset{\|}{\text{C}}}—\text{S}—\text{(CoA)} + NAD^+ \longrightarrow \text{CH}_3\overset{\text{O}}{\underset{\|}{\text{C}}}\text{CH}_2\overset{\text{O}}{\underset{\|}{\text{C}}}—\text{S}—\text{(CoA)} + NADH + H^+$$

On the other hand, the *synthesis* of fatty acids requires the reduction of β-ketothioesters to β-hydroxythioesters by NADH. The thiol used in the thioester in the synthesis sequence has an SH group attached to a protein, an "acyl-carrier protein" (ACP). This thiol is abbreviated HS-(ACP).

$$NADH + \text{CH}_3\overset{\text{O}}{\underset{\|}{\text{C}}}\text{CH}_2\overset{\text{O}}{\underset{\|}{\text{C}}}—\text{S}—\text{(ACP)} + H^+ \longrightarrow NAD^+ + \text{CH}_3\overset{\text{OH}}{\underset{|}{\text{C}}}\text{HCH}_2\overset{\text{O}}{\underset{\|}{\text{C}}}—\text{S}—\text{(ACP)}$$

The thioesters utilized in syntheses involve a different thiol than the coenzyme A used in the oxidative degradation of the fatty acids. The completely dissimilar thiol structures keep the two routes of synthesis and degradation from becoming involved with each other.

Problem **11.13** Draw the structures for the reactants and products of the conversion of the coenzyme A thioester of β-hydroxydecanoic acid to the β-ketodecanoyl coenzyme A with NAD^+. For your structures of NAD^+ and NADH use the symbol R for the side chain which does not enter into the reaction. Is this reaction an oxidation or a reduction of the thioester?

11.14 REACTIONS OF IMINES OF PYRIDOXAL AND AMINO ACIDS Up to now we haven't said much about imines and their reactions. They are such highly reactive compounds that they appear primarily as intermediate compounds in the reactions of either carbonyl compounds or primary amines.

$$R_2C{=}O + H_2N—R \rightleftarrows R_2C{=}N—R + H_2O \qquad\qquad \text{(Section 11.6)}$$

We have seen imines as very reactive intermediates in the addition of HCN and NH_3 to aldehydes to give α-aminonitriles (Section 11.7). Reactions catalyzed by enzymes employ imine intermediates in a variety of reactions. Some of the noteworthy examples of biosyntheses and degradations which go through imines are those for which pyridoxal serves as a coenzyme. We will consider some of these interesting reactions now.

Recall that the $C=N$ bond is similar in polarity to the $C=O$ bond, and that imines react in a manner analogous to carbonyl compounds (Section 11.6). Imines undergo addition and substitution reactions more readily than do carbonyl compounds. Also, like the $C=O$ group, the $C=N$ group increases the acidity of an α-hydrogen. With imines, however, there are two kinds of α-hydrogen, and two types of isomerization involving α-hydrogen transfer are possible. The first type is a tautomerization similar to keto–enol tautomerization, where a hydrogen shifts between a carbon and a nitrogen. The hydrogen attached to the α-carbon on the carbon side of the $C=N$ double bond ($CH-C=N$) becomes involved in a tautomeric equilibrium between the imine and the enamine (*ene + amine*). At equilibrium the imine is strongly favored, just as ketones are usually favored in keto–enol equilibria.

imine   enamine

A second type of tautomerization of imines involves the α-hydrogen on the nitrogen side of the double bond ($C=N-CH$). In this isomerization, which is catalyzed by a strong base, the proton shifts from carbon to carbon as imine 1 changes to imine 2. A delocalized carbanion is an intermediate.

imine 1   delocalized carbanion   imine 2

Isomerizations of this latter kind between two imines are extremely important reactions in the biosynthesis and degradation of amino acids. The position of the equilibrium between these two imines is determined by the structures of the carbon groups making up the imines. Hydrolysis of the two imines gives different carbonyl compounds and different amines. The overall reaction—formation of imine 1, isomerization of imine 1 to imine 2, and hydrolysis of imine 2—is called **transamination**, because it involves the transfer of an amino group from one compound to another.

ketone 1   amine 1   imine 1   imine 2   **amine 2**   ketone 2

Problem **11.14 a)** Write the steps involved in the isomerization of starting imine A to product imine B catalyzed by potassium *tert*-butoxide in *tert*-butyl alcohol.

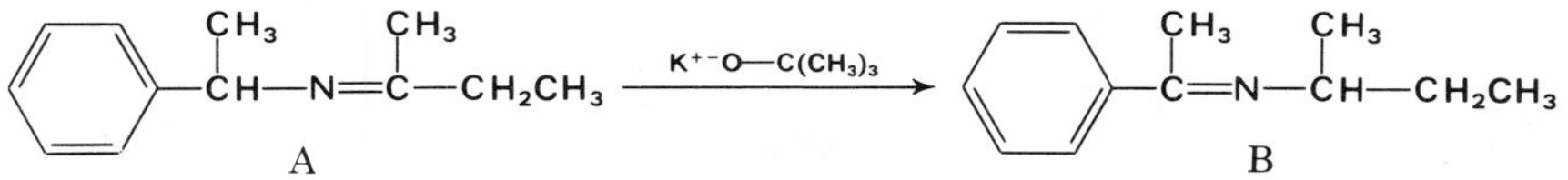

b) Give the structures of the hydrolysis products of imines A and B.

In the laboratory moderate bases like amines bring about the isomerization of imines containing pyridine groups. This finding is interesting because amines are potentially present in the active site of enzymes and pyridoxal is a coenzyme for many reactions of amino acids. Imines of pyridoxal and amino acid have been shown to be intermediates in these reactions.

One type of reaction of amino acids for which pyridoxal phosphate acts as a coenzyme is a transamination producing a transfer of an NH_2 group from an α-amino acid to an α-ketoacid as shown here.

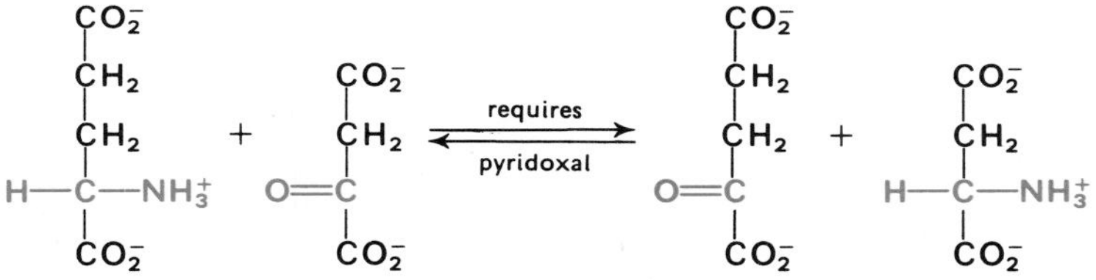

glutamate ion   oxaloacetate ion α-ketoglutarate ion   aspartate ion
(amino acid 1)   (ketoacid 2)   (ketoacid 1)   (amino acid 2)

The observed transamination shown by this equation is actually two successive transamination reactions with pyridoxal and pyridoxamine as the other reactants (see formulas below).

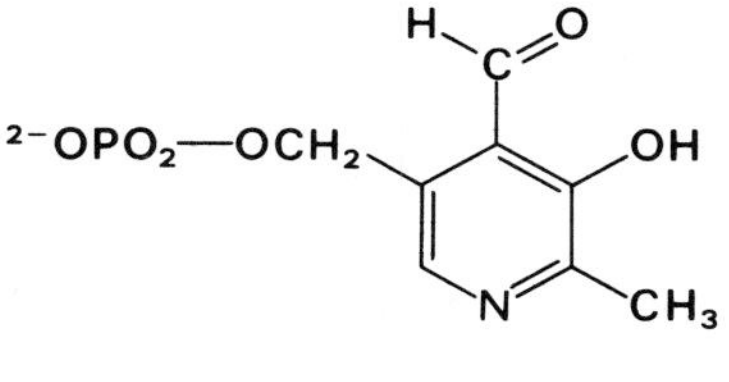

pyridoxal phosphate   pyridoxamine phosphate
(vitamin B_6)

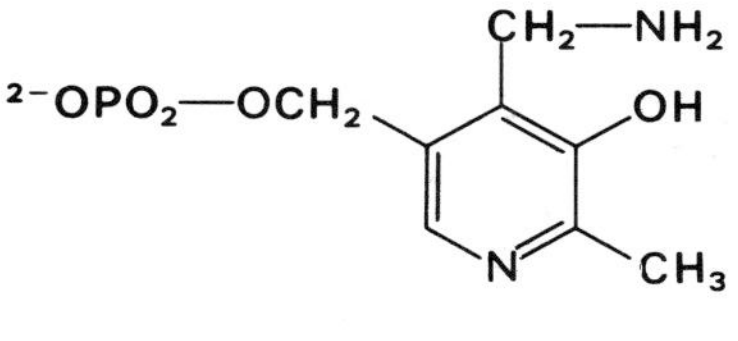

The first reaction proceeds through an imine of pyridoxal and the amino acid ion, glutamate ion. The ease of the transamination depends upon the combined acidifying effects on the α-hydrogen of the pyridine group, the C=N, and the C=O groups. The products of the first reaction are pyridoxamine and a ketoacid ion, the NH_2 having been transferred from the amino acid to the pyridoxal.

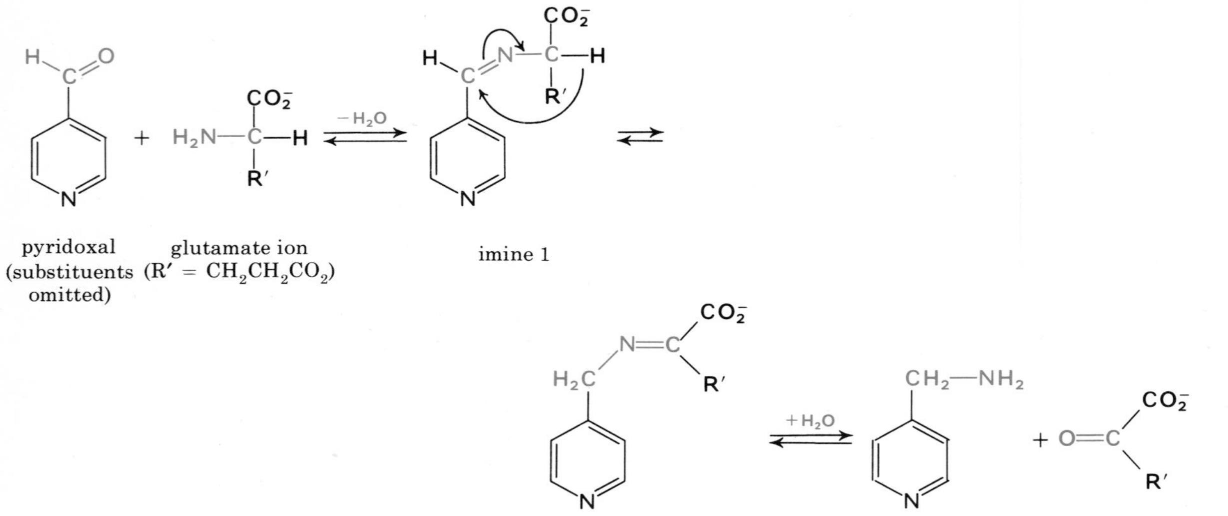

pyridoxal
(substituents (R′ = $CH_2CH_2CO_2$)
omitted)

glutamate ion

imine 1

imine 2

pyridoxamine

α-ketoglutarate ion
(R′ = $CH_2CH_2CO_2^-$)

In the second transamination, this sequence of reactions is reversed, starting with pyridoxamine and a new α-ketoacid, oxaloacetate ion. The imine formed undergoes isomerization to a new imine. Hydrolysis of the new imine gives pyridoxal again and a second amino acid ion.

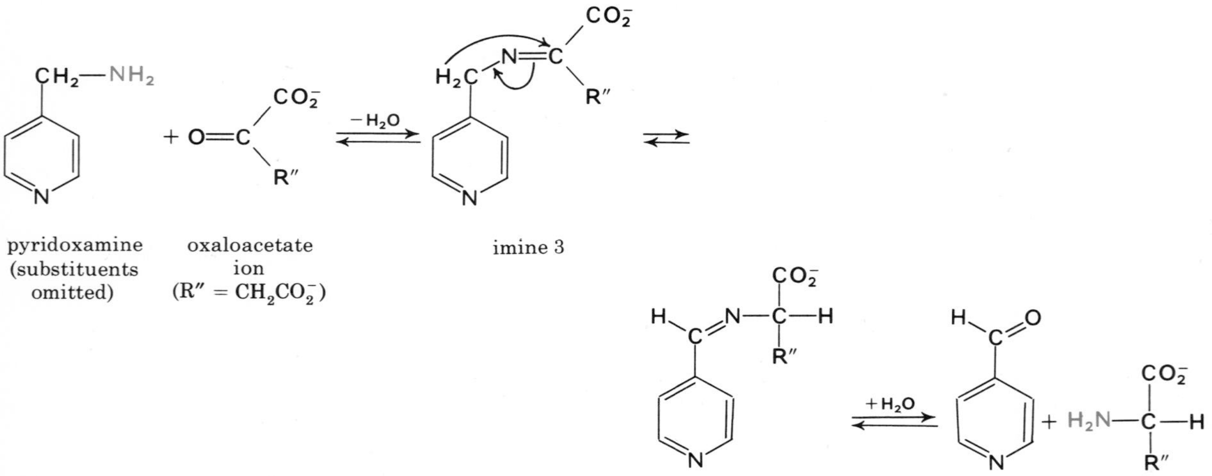

pyridoxamine
(substituents
omitted)

oxaloacetate
ion
(R″ = $CH_2CO_2^-$)

imine 3

imine 4

pyridoxal

aspartate ion
(R″ = $CH_2CO_2^-$)

The transaminations depicted above are important in the biosyntheses of the amino acids that can be synthesized in the body. The ketoacid is synthesized and then receives the NH_2 group from the glutamate ion. Interestingly, in animals and many plants, the NH_2 group for glutamate can come from NH_3 itself. From the external source of nitrogen, the NH_2 group is incorporated into glutamic acid and is then transferred to various ketoacids to form some of the needed amino acids for protein synthesis.

New Terms and Topics

Addition reactions of aldehydes and ketones (Section 11.4)

Acetals, hemiacetals, ketals, hemiketals (Section 11.5)

Substitution reactions (Section 11.6)

Imines, hydrazones, oximes (Section 11.6)

α-Carbanions; electron delocalization of carbanions (Section 11.8)

Nucleophilic carbon reagents; aldol condensation (Section 11.10)

Keto–enol equilibria; tautomerization (Section 11.9)

Coenzyme NAD^+-NADH system (Section 11.13)

Imine–imine isomerization (Section 11.14)

Transamination (Section 11.14)

Role of pyridoxal phosphate in transamination (Section 11.14)

Summary of Reactions

REACTIONS OF ALDEHYDES AND KETONES

1. With water (Section 11.5)

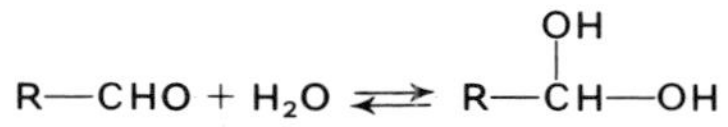

$$R—CHO + H_2O \rightleftharpoons R—\overset{\overset{\displaystyle OH}{|}}{C}H—OH \quad \text{usually unstable except in water}$$

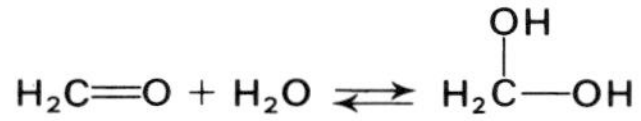

$$H_2C{=}O + H_2O \rightleftharpoons H_2\overset{\overset{\displaystyle OH}{|}}{C}—OH$$

formaldehyde

2. With ammonia (Section 11.6)

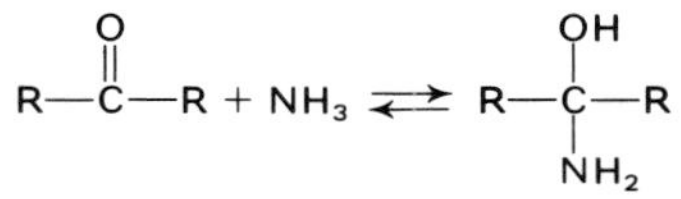

$$R—\overset{\overset{\displaystyle O}{||}}{C}—R + NH_3 \rightleftharpoons R—\underset{\underset{\displaystyle NH_2}{|}}{\overset{\overset{\displaystyle OH}{|}}{C}}—R \quad \text{unstable except in solution}$$

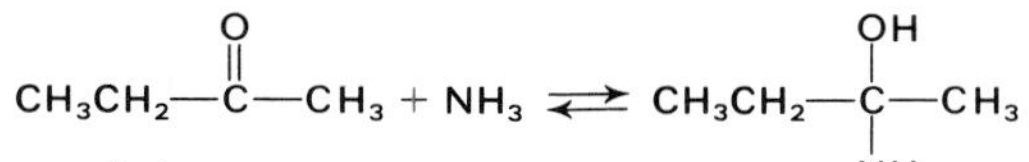

$$CH_3CH_2—\overset{\overset{\displaystyle O}{||}}{C}—CH_3 + NH_3 \rightleftharpoons CH_3CH_2—\underset{\underset{\displaystyle NH_2}{|}}{\overset{\overset{\displaystyle OH}{|}}{C}}—CH_3$$

butanone

3. With alcohols (Section 11.5)

$$R-CH=O + R'OH \xrightarrow{H^+} R-\underset{\underset{OH}{|}}{C}H-OR' \xrightarrow{\text{dry HCl, R'OH}} R\underset{\underset{OR'}{|}}{C}H-OR' + H_2O$$

hemiacetal   acetal

$$CH_3CH_2CH_2CH=O + CH_3OH \xrightarrow{H^+} CH_3CH_2CH_2\underset{\underset{OH}{|}}{C}H-OCH_3 \xrightarrow{\text{dry HCl, CH}_3\text{OH}} CH_3CH_2CH_2\underset{\underset{OCH_3}{|}}{C}H-OCH_3 + H_2O$$

butanal   a hemiacetal   1,1-dimethoxybutane
(butanal dimethylacetal)

$$\underset{\underset{OH}{|}}{C}H_2CH_2CH_2CH=O \xrightarrow{H^+} \quad \xrightarrow{\text{CH}_3\text{OH}}_{\text{dry HCl}} \quad + H_2O$$

4-hydroxybutanal   2-hydroxy-
tetrahydrofuran
(a cyclic hemiacetal)   2-methoxy-
tetrahydrofuran
(a cyclic acetal)

4. With hydrogen cyanide (aldehydes and methyl ketones only) (Section 11.7)

$$R-CH=O + Na^+CN^- \xrightarrow{\text{HCl}} R-\underset{\underset{OH}{|}}{C}H-C\equiv N \xrightarrow[\Delta]{\text{H}_3\text{O}^+} R-\underset{\underset{OH}{|}}{C}H-CO_2H + NH_4^+$$

$$\text{benzaldehyde} + HC\equiv N \xrightarrow{\text{NaOH}} \quad \xrightarrow[\text{H}_2\text{O}]{\text{H}^+} \quad + NH_4^+$$

benzaldehyde   2-phenyl-2-hydroxy-
acetonitrile   2-phenyl-2-
hydroxyacetic acid
(mandelic acid)

5. With hydrogen cyanide and ammonia (Section 11.7)

$$R-CH=O + HC\equiv N + NH_3 \xrightarrow{\text{OH}^-} R\underset{\underset{NH_2}{|}}{C}H-C\equiv N \xrightarrow{H^+} R\underset{\underset{NH_3^+}{|}}{C}H-CO_2H + NH_4^+$$

$$CH_3\underset{\underset{CH_3}{|}}{C}H-CH=O + HC\equiv N + NH_3 \xrightarrow{\text{OH}^-} CH_3\underset{\underset{CH_3}{|}}{C}H-\underset{\underset{NH_2}{|}}{C}H-C\equiv N \xrightarrow{H^+} CH_3-\underset{\underset{CH_3}{|}}{C}H\underset{\underset{+NH_3}{|}}{C}HCO_2H + NH_4^+$$

2-methylpropanal   3-methyl-2-amino-
butanonitrile   3-methyl-2-amino-
butanoic acid (valine)

6. With amines—carbonyl–imine interconversions (Section 11.6)

$$R-\underset{\underset{R}{|}}{C}=O + R'NH_2 \underset{\text{dry HCl}}{\rightleftarrows} R-\overset{\overset{OH}{|}}{\underset{\underset{R}{|}}{C}}-NHR' \rightleftarrows R-\underset{\underset{R}{|}}{C}=NR' + H_2O$$

$$CH_3CH_2CH_2CH_2-\overset{O}{\underset{}{C}}H + \text{aniline} \underset{\text{dry HCl}}{\rightleftarrows} CH_3CH_2CH_2CH_2-\overset{H}{\underset{\underset{OH}{|}}{C}}-NH \rightleftarrows CH_3CH_2CH_2CH_2\overset{H}{\underset{\underset{N}{||}}{C}} + H_2O$$

pentanal   aniline

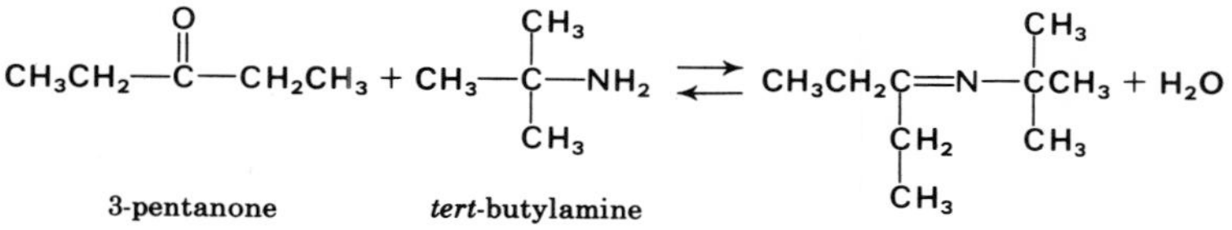

3-pentanone   *tert*-butylamine

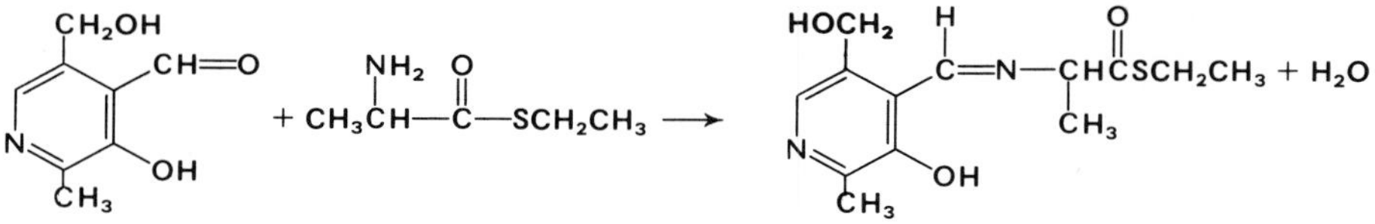

pyridoxal

ethyl 2-amino-
propanethioate

7. Carbonyl hydrazone formation (Section 11.6)

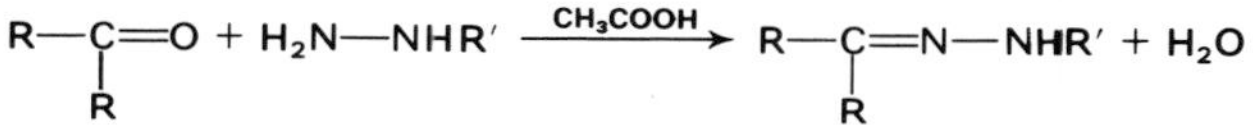

$$R—\underset{R}{\overset{}{C}}=O + H_2N—NHR' \xrightarrow{CH_3COOH} R—\underset{R}{\overset{}{C}}=N—NHR' + H_2O$$

a hydrazone

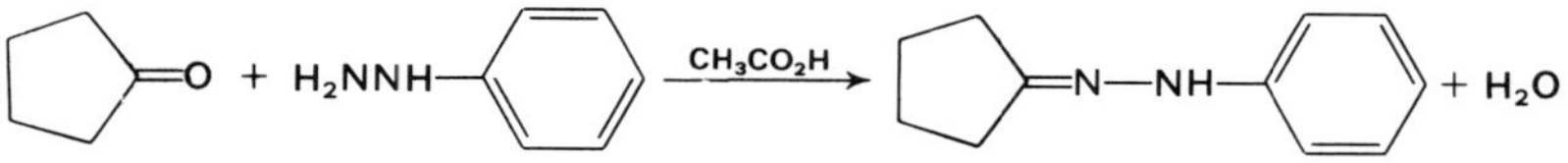

cyclopentanone   phenylhydrazine

cyclopentanone
phenylhydrazone

8. Keto–enol equilibria (Section 11.9)

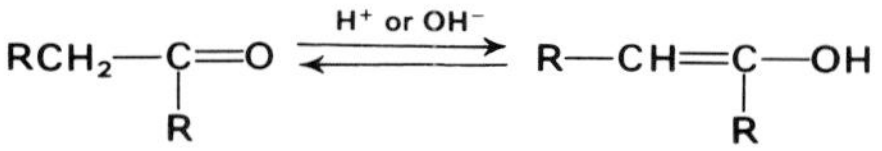

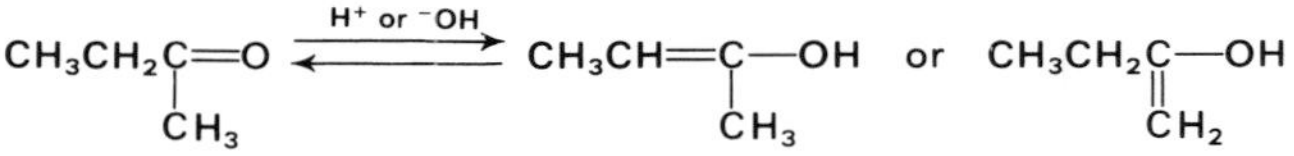

$$RCH_2—\underset{R}{\overset{}{C}}=O \underset{\longleftarrow}{\overset{H^+ \text{ or } OH^-}{\rightleftharpoons}} R—CH=\underset{R}{\overset{}{C}}—OH$$

$$CH_3CH_2\underset{CH_3}{\overset{}{C}}=O \underset{\longleftarrow}{\overset{H^+ \text{ or } ^-OH}{\rightleftharpoons}} CH_3CH=\underset{CH_3}{\overset{}{C}}—OH \quad \text{or} \quad CH_3CH_2\underset{CH_2}{\overset{}{C}}—OH$$

butanone   butanone enol   butanone enol

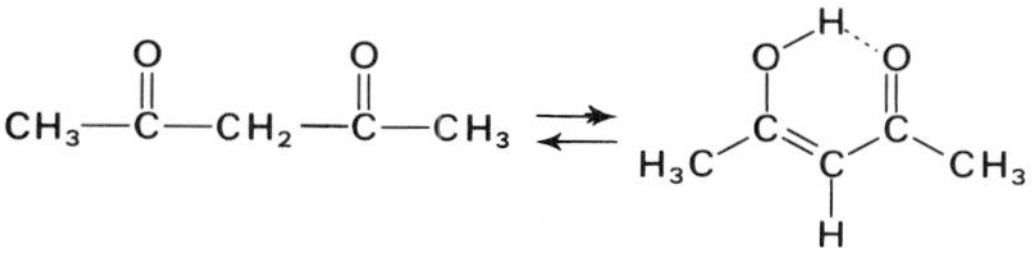

2,4-pentanedione   2,4-pentanedione enol

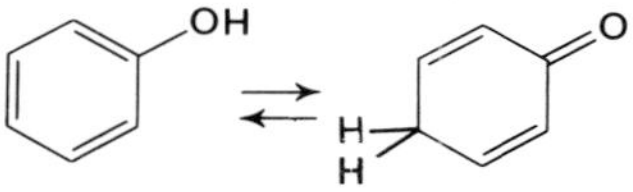

phenol   keto form

9. Aldol condensation (Section 11.10)

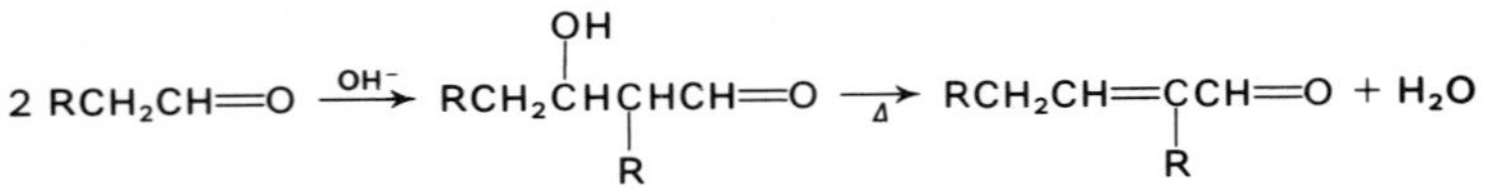

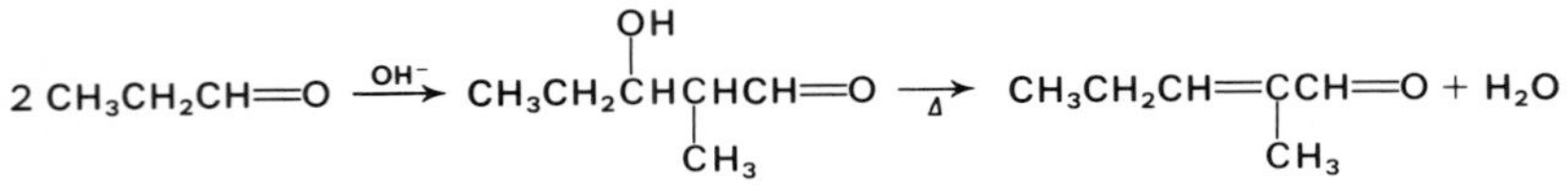

$$2 \ RCH_2CH{=}O \xrightarrow{\ OH^-\ } RCH_2\overset{\displaystyle OH}{\underset{\displaystyle R}{C}HCHCH{=}O} \xrightarrow[\Delta]{} RCH_2CH{=}\underset{\displaystyle R}{C}CH{=}O + H_2O$$

$$2 \ CH_3CH_2CH{=}O \xrightarrow{\ OH^-\ } CH_3CH_2\overset{\displaystyle OH}{\underset{\displaystyle CH_3}{C}HCHCH{=}O} \xrightarrow[\Delta]{} CH_3CH_2CH{=}\underset{\displaystyle CH_3}{C}CH{=}O + H_2O$$

propanal	2-methyl-3-hydroxy-pentanal	2-methyl-2-pentenal

10. With metal hydrides (Section 11.11)

$$R_2C{=}O \xrightarrow{LiAlH_4} \xrightarrow{H^+} R_2CHOH$$

$$RCH{=}O \xrightarrow{NaBH_4} \xrightarrow{H^+} RCH_2{-}OH$$

$$CH_3CH{=}O \xrightarrow{LiAlH_4} \xrightarrow{H^+} CH_3CH_2{-}OH$$

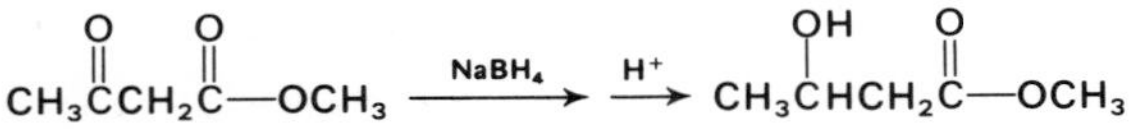

$$CH_3\overset{\displaystyle O}{\overset{\|}{C}}CH_2\overset{\displaystyle O}{\overset{\|}{C}}{-}OCH_3 \xrightarrow{NaBH_4} \xrightarrow{H^+} CH_3\overset{\displaystyle OH}{C}HCH_2\overset{\displaystyle O}{\overset{\|}{C}}{-}OCH_3$$

methyl acetoacetate	methyl 3-hydroxybutanoate

11. With oxidizing agents (Section 11.11)

$$RCH{=}O \xrightarrow{Na_2Cr_2O_7,\ H_2SO_4} RCO_2H + Cr^{3+}$$

$$CH_3CH_2CH{=}O \xrightarrow{Na_2Cr_2O_7,\ H_2SO_4} CH_3CH_2CO_2H + Cr^{3+}$$

$$RCH{=}O \xrightarrow{Ag(NH_3)_2^+OH^-} RCO_2^- NH_4^+ + Ag_{(s)}$$

$$HO{-}CH_2CH_2CH{=}O \xrightarrow{Ag(NH_3)_2^+OH^-} HO{-}CH_2CH_2CO_2^- NH_4^+ + Ag_{(s)}$$

Problems

11.15 Name the following compounds by common names or IUPAC system.

a)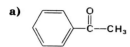

b)

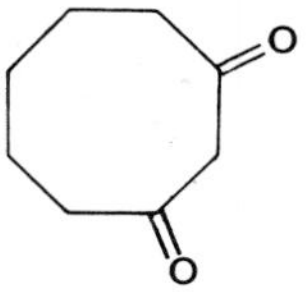

c) 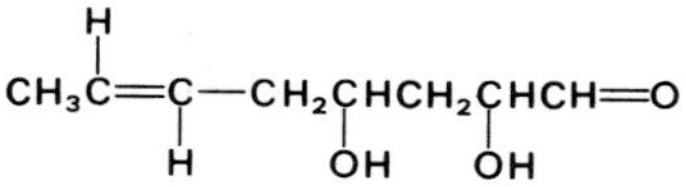 CH₃C=C—CH₂CHCH₂CHCH=O

d) 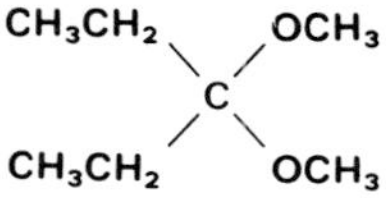 CH₃CH₂   OCH₃
            C
      CH₃CH₂   OCH₃

e) Cl₃CCH=O

f) 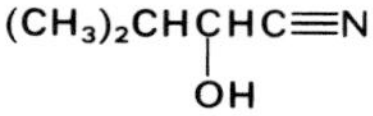 (CH₃)₂CHCHC≡N
                OH

g) 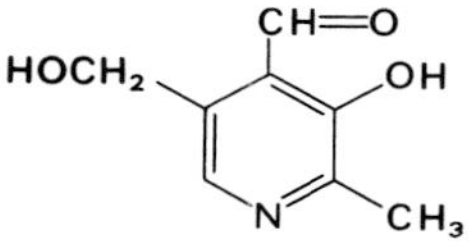 HOCH₂ CH=O
              OH
              N   CH₃

h) 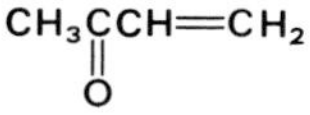 CH₃CCH=CH₂
            O

11.16 Write structures for the following compounds.

a) hydrate of trichloroacetaldehyde
c) aniline imine of 2-methylbutanal
e) *tert*-butyl imine of pyridoxal
g) enol of 3,5-heptanedione

b) oxime of phenyl propyl ketone
d) acetone 1,3-propanediol ketal
f) phenylhydrazone of 2-hexanone

11.17 Write structures for the products of the following reactions.

a) CH₃CH=CHCH₂CH₂CH=O + CH₃OH —HCl dry→

b) 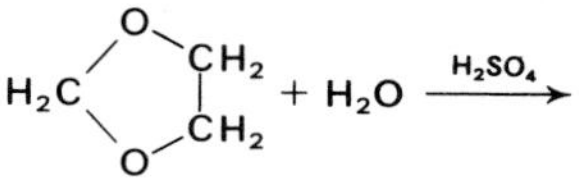 + H₂O —H₂SO₄→

c) C₆H₅CH=O + H₂NCH₂CH₂OH —high boiling solvent / to distill out water→ a cyclic compound

d) CH₃CH=O + H₂NCH₂CH₂SH —high boiling solvent / to distill out water→ a cyclic compound

e) C₆H₅CH₂CH=O + HCN + NH₃ —NaOH→

f) 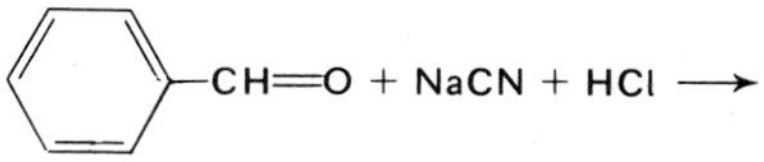 —CH=O + NaCN + HCl →

g) =O —NaBH₄→

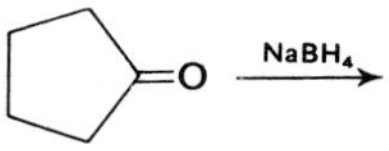

h) CH₃CH₂CH₂CH₂CH=O —NaOH→

11.18 Starting materials which do not contain asymmetric centers often give products which do. Under special circumstances, such as enzyme catalysis, one enantiomer is produced exclusively or preferentially; however, under most conditions a racemic mixture of the product is obtained. Which products of the following reactions would be racemic forms? Draw the structures of racemic products.

a) $CH_3CHCH_2CH{=}O + HCN \xrightarrow{NaOH}$
      |
     CH_3

b) $CH_3CH_2CH{=}O + CH_3CH_2CH_2{-}OH \xrightarrow{HCl}$

c) $CH_3CH_2CCH_2CH_3 + NH_3 + HCN \xrightarrow{NaOH}$
        ‖
        O

d) $HO{-}CH_2CH_2CH_2CH_2CH{=}O \xrightarrow{HCl}$

11.19 Write the final products (after water is lost) of the aldol condensation obtained from the following starting materials.

a) $CH_3CH_2CH{=}O$ **b)** $C_6H_5CH_2CH{=}O$

11.20 Testosterone contains $-C{=}C-C{=}O$ an α,β-unsaturated ketone group. Write the principal resonance forms of this linkage. Why should both the carbonyl carbon and the β-carbon be susceptible to attack by nucleophiles?

11.21 With structural formulas, write equations for reactions needed to convert the specified starting materials and inorganic reagents into the products required.

a) $C_6H_5CHCO_2^-$ from benzaldehyde
     |
    $^+NH_3$

b) $C_6H_5CHCO_2H$ from benzaldehyde
     |
    OH

11.22 Write structures for an enol form of the following compounds. Where intramolecular hydrogen bonds are possible, write the structure in that manner.

a) $C_6H_5{-}C{-}CH_2{-}C{-}C_6H_5$
       ‖   ‖
       O   O

b) $CH_3{-}C{-}CH_2{-}C{-}O{-}CH_3$
       ‖   ‖
       O   O

c) $C_6H_5{-}C{-}CH_2{-}C{\equiv}N$
       ‖
       O

d) $CH_3{-}C{-}CH_2{-}C_6H_5$
       ‖
       O

e) $CH_3{-}C{-}C{-}CH_3$
       ‖ ‖
       O O

f)

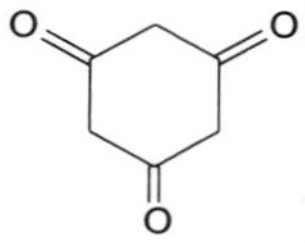

11.23 Write the series of equations for the enzyme-catalyzed transamination starting with each of the following.

    a) glutamate ion and pyridoxal

    b) pyridoxamine and pyruvate ion $(CH_3{-}\underset{\overset{\|}{O}}{C}{-}CO_2^-)$

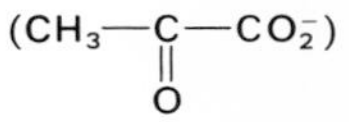

11.24 The following questions concern the structure of streptomycin, an antibiotic.

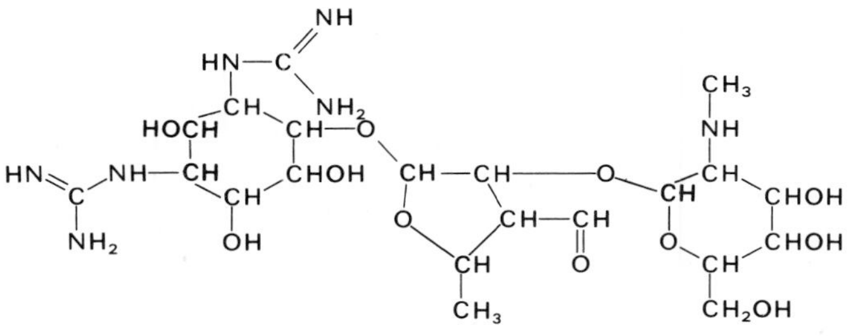

ring 1   ring 2   ring 3

streptomycin

a) What is the molecular formula of this compound?

b) Which of the functional groups listed are found attached to ring 1, ring 2, or ring 3—acid, amide, ester, hydroxyl, amine, ether, guanidine, urea, aldehyde, ketone, acetal, ketal?

c) How many asymmetric carbons are in each ring? How many stereoisomers are possible?

d) Draw the R configuration for any two carbons in the middle ring. (A difficult question because priority assignments require tedious attention to detail.)

e) Treatment of the antibiotic with hydrochloric acid could protonate three base positions. Identify them and write the partial structures for the conjugate acids.

f) Prolonged treatment of the antibiotic with hydrochloric acid could cleave the molecule into three separate units. What reaction would be involved? Write the structures for the three products.

g) Could the hydrochloric acid treatment cause any of the asymmetric carbons to lose the specific configuration? Why? Which ones? Write the structures of intermediates produced, which account for the change in configuration.

CHAPTER 12

Reactions of
Aromatic Rings

The greatest distinction between laboratory and biological syntheses lies in the way compounds containing aromatic rings are formed. Biosyntheses of compounds containing benzene rings or heterocyclic structures, like adenine, usually build the cyclic structure in place. The formation of carbon rings from carbonyl precursors or of nitrogen heterocyclic rings from amino acids is followed by elimination sequences which develop aromatic unsaturation.

In the laboratory naturally occurring benzene and many alkyl-benzenes are among the most readily available and inexpensive starting materials. Reactions converting these materials to other useful products containing the aromatic ring have long been known. Changes in which new groups are attached directly to the aromatic carbon usually occur only in the presence of powerful catalysts and have no direct counterparts in living systems. These reactions are the subject of this chapter.

The evolution of our knowledge of organic chemistry has passed through whimsical periods, full of paradoxes and controversies. Of the varieties of chemical properties that have generated the most puzzlement, that of aromaticity stands near the summit. About the time that the structural theory of organic chemistry brought cohesion and order to the science, an anomaly appeared. Although benzene possessed three double bonds, the bonds could equally well be located in two different but equivalent positions. Few problems have stimulated as many vigorous arguments and as many clever experiments. The good investigator is the most resourceful when challenged, and this problem challenged theoreticians and experimentalists alike. The first section gives the story.

12.1 THE BENZENE STORY—AROMATICITY

Benzene C_6H_6 is the simplest and best-known example of the **aromatic compounds**, or **arenes**. The structure of this simple compound represents the longest, the most controversial, and the most important single problem in the development of the structural theory of organic compounds.

The early sources of benzene and the substituted benzenes, such as chlorobenzene, nitrobenzene, toluene, benzaldehyde, and benzyl alcohol, had fragrant aromas. The term "aromatic compounds" was invented and correlated the pleasant odors and the common structural features of these compounds. Shortly afterwards, evil-smelling aromatic compounds such as aniline and benzenethiol were discovered, but as so often happens, the term persisted.

Benzene was first isolated in 1825. Its empirical formula, CH, and later its molecular formula, C_6H_6, showed an unexpectedly high C:H ratio and presumably a high degree of unsaturation. In 1865 Kekulé proposed the cyclic 1,3,5-cyclohexatriene formula for benzene.

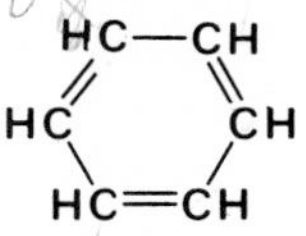

Yet the chemists of the time could not account for the chemical properties of benzene, which proved to be entirely different from those of the ordinary unsaturated hydrocarbons. For example:

1. Benzene and other aromatic hydrocarbons proved unusually stable (unreactive) toward reagents, such as bromine and hydrogen bromide, which normally reacted quickly with alkenes at room temperature.
2. The number of isomeric disubstituted benzene compounds that were identified was fewer than expected from a 1,3,5-cyclohexatriene parent compound with alternating double and single bonds. For example, only one 1,2-dibromobenzene was ever identified.

Kekulé then suggested that "oscillation" between two structures of benzene might explain the unusual properties. He wrote two formulas for benzene and two for the single compound 1,2-dibromobenzene.

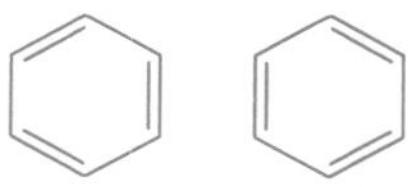

benzene
(two Kekulé structures)

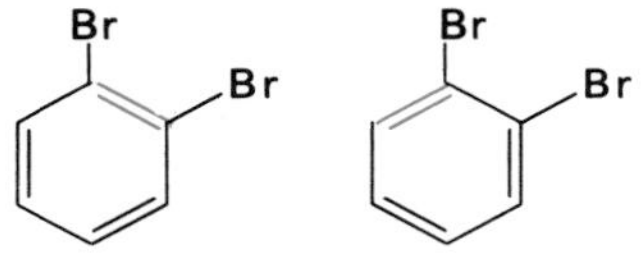

1,2-dibromobenzene
(two Kekulé structures)

The theory that benzene possessed a structure which was a resonance hybrid of simple valence-bond structures was developed in the 1930's. The hybrid exists because it has a lower energy than either of the classical Kekulé valence-bond formulas that can be written for the compound. These classical formulas which contribute to the "actual" structure are called resonance structures.

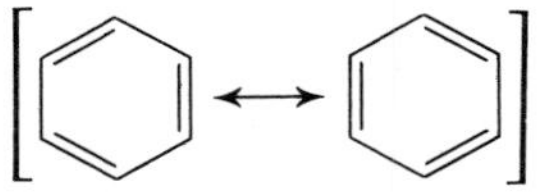

The subsequent measurement of bond lengths by X-ray crystallography showed equal bond lengths, 1.39 Å, for all carbon–carbon bonds in benzene, with a value between that of the normal single C—C bond (1.54 Å) and that of the normal double C=C bond (1.33 Å). Thus experimentally the compound possesses bond lengths expected of a hybrid and not of a classical structure. Since all six carbon and all six hydrogen atoms lie in the same plane, and since all six carbon–carbon bonds are of the same length, the carbon skeleton forms a perfect hexagon.

The lowered energy and increased stability of benzene are now attributed to delocalization of the six pi electrons. This delocalization of benzene accounts for its unreactivity with reagents which normally add to olefinic linkages. The stability of the ring structure also accounts for the fact that when benzene does react under more strenuous conditions, it undergoes substitution rather than addition. In the substitution reaction much of the delocalization of the pi electrons of the ring is maintained. Thus under appropriate conditions benzene reacts with bromine to give bromobenzene and HBr, products of the substitution of a bromine for a hydrogen on the ring (see Section 12.6 for more details).

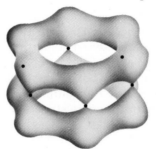

benzene   bromobenzene

The term aromatic and the quality of aromaticity very early became synonymous with a six-membered ring containing unsaturation equivalent to three double bonds—the "aromatic ring." Hydrocarbons containing one or more such rings were designated as aromatic hydrocarbons, as distinct from alkanes, alkenes, and alkynes—the aliphatic hydrocarbons.

The best description of the bonding of an aromatic ring is by molecular orbitals which encompass all six carbons of the ring. (See Section 3.4.)

a molecular orbital of benzene

Was there something unique about a six-membered ring? Could aromaticity be associated with rings of other sizes in which double and single bonds alternated, such as 1,3-cyclobutadiene (4 carbons) and 1,3,5,7-cyclooctatetraene (8 carbons)? The first part of the answer to these questions came with a laborious synthesis of cyclooctatetraene.

cyclooctatetraene

Interestingly, cyclooctatetraene in its reactions resembled the conjugated and noncyclic polyenes. The substance did not exhibit the unusual stability associated with benzene and its derivatives. Cyclooctatetraene, with its four *cis*-double bonds, has been found to be in a tub conformation. None of the possible conformations can be planar without considerable angle strain.

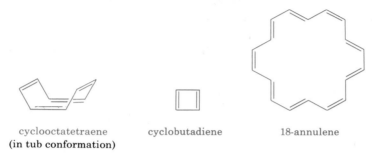

cyclooctatetraene
(in tub conformation)

cyclobutadiene

18-annulene

Cyclobutadiene is so reactive that it can be observed only at very low temperatures. Not only does it not have aromatic stability, but it is much less stable than butadiene, its open-chain counterpart.

Quantum mechanical calculations of the molecular orbitals provided the generalization that significant pi-bond delocalization and aromatic properties should be associated with rings having $4n + 2$ pi electrons, $n = $ small digit, 1–6. Neutral compounds with 6, 10, 14, 18, 22, and 26 pi electrons and atoms in the ring were expected to show aromatic properties, while rings of 4, 8, 12, and 16 would not.

Most of these cyclic polyenes have been synthesized. The rings that contain 10 and 14 carbons cannot become planar due to nonbonded repulsion between their attached hydrogen atoms. Thus pi orbitals of some of the bonded carbons could not overlap, and the systems were not stabilized by electron delocalization. The 18-carbon ring with 9 double bonds (18-annulene) is planar, has plenty of room for its hydrogens, and does show aromatic stability, but not nearly so much as benzene. For all practical purposes, the magic of the six-membered ring persists.

Problems   **12.1** Given their chemical properties, how many different carbon–carbon bond lengths do you expect to find in each of the following?

    a) cyclobutadiene   **b)** cyclooctatetraene

    c) benzene   **d)** 18-annulene

12.2 How many inward and how many outward turned H's does 18-annulene have? How many *cis* and how many *trans* double bonds?

12.3 Draw the carbon skeleton of 18-annulene. Fill in the hole of the structure with hexagons by drawing the appropriate bonds between vertices. Put in additional double bonds until you have six fused benzene rings forming a circle. The compound you have drawn is known, and is called coronene. It is planar, aromatic, and beautifully symmetrical.

Aromatic compounds are distinguished by the presence of at least one six-membered ring containing unsaturation equivalent to three double bonds. All carbons in the ring have sp^2 hybridized orbitals and form bonds with three other atoms. The six carbons and six other atoms bonded to them all lie in the same plane. The benzene ring may have attached to it as many as six functional groups or hydrocarbon chains. All types of functional groups and hydrocarbon chains are found in aromatic compounds.

Hydrocarbon substituents attached to a benzene ring may be either aliphatic (saturated or unsaturated) or aromatic groups. Many of the simple benzene derivatives are known by their common names: **toluene** (methylbenzene), three isomeric **xylenes** (dimethylbenzenes), **styrene** (vinylbenzene), **cumene** (isopropylbenzene), and **biphenyl** (phenylbenzene).

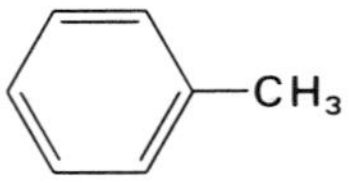

—CH$_3$	—CH=CH$_2$	—CH(CH$_3$)$_2$	
toluene (methylbenzene)	styrene (vinylbenzene)	cumene (isopropylbenzene)	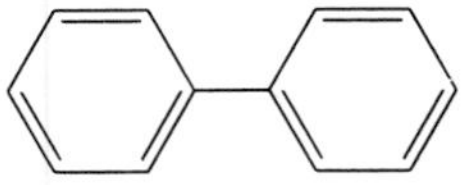 biphenyl

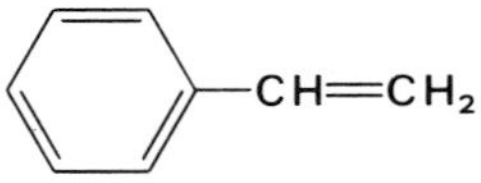

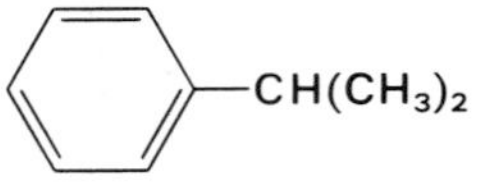

para-xylene (1,4-dimethyl-benzene)   *meta*-xylene (1,3-dimethyl-benzene)   *ortho*-xylene (1,2-dimethyl-benzene)

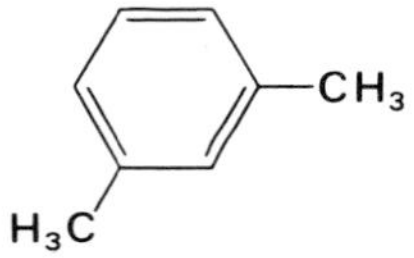

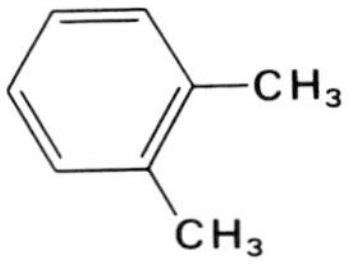

Two substituents attached to a benzene ring are correctly designated by numbers, e.g., 1,3-dimethylbenzene. However, the prefixes *ortho* (abbreviated *o*), *meta* (*m*), and *para* (*p*) are often used in place of the numbers. *Ortho* means on adjacent carbons, or 1,2-; alternate positions are *meta* or 1,3-; positions on opposite sides of the ring are *para* or 1,4-.

Aromatic hydrocarbons were historically obtained by the distillation of coal tar. Now they are prepared in vast quantities by catalytic cycloaromatization (dehydrogenation) of petroleum fractions. Simple alkylbenzenes are used as solvents or as basic starting materials for many industrial products.

The benzene ring occurs in the structures of many natural products and drugs. Possibly the prevalence of the ring is due to its planarity and rigidity coupled with its lack of reactivity. Substituents are held rigidly in place at specific distances from one another. The inherent stability of benzene rings may inhibit undesirable metabolic reactions. Whatever the reasons for the presence, benzene rings are found in large numbers of compounds, many of which have been shown in earlier chapters. A few are illustrated here.

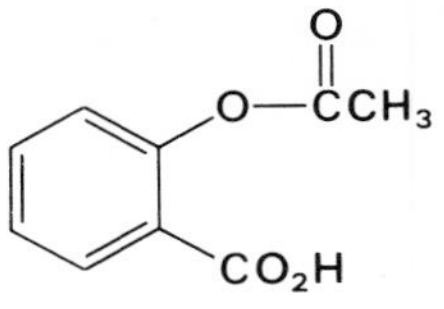

aspirin

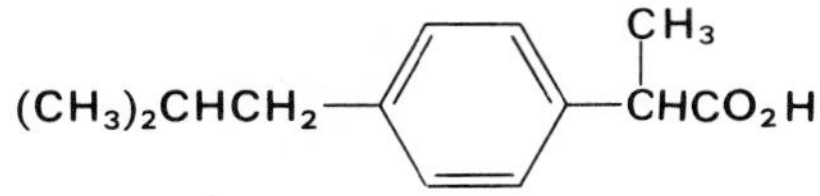

moltrin
(an aspirin substitute)

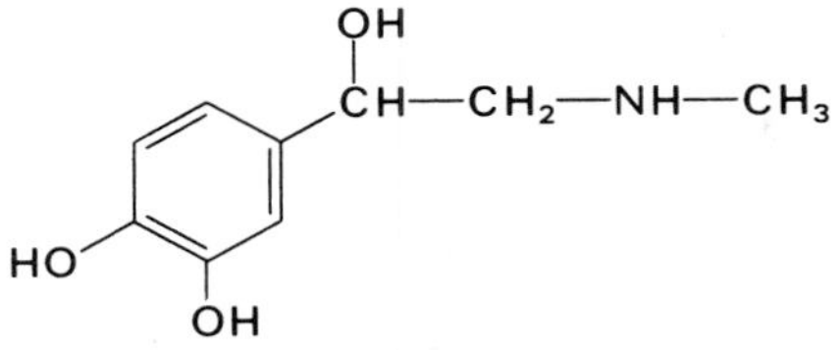

epinephrine
(adrenaline)

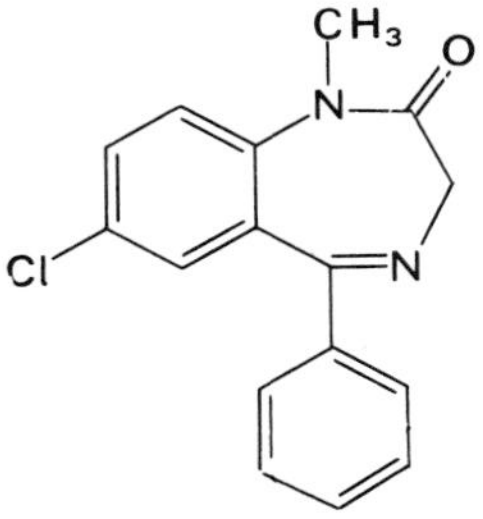

valium

Many important compounds have aromatic rings in which two of the carbons are common to two rings. These structures are known as fused aromatic ring systems. The simplest examples of fused rings are those of **naphthalene, anthracene,** and **phenanthrene.** In some of the resonance structures for fused rings, all of the rings do not contain three double bonds. In the resonance structure shown here for anthracene, the ring on the right does not have three double bonds. In general, the greater the number of fused benzenes, the lower is the stability per double bond due to pi electron delocalization.

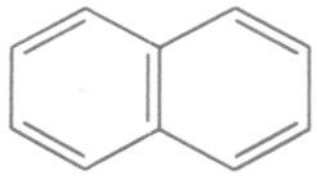

naphthalene
$C_{10}H_8$

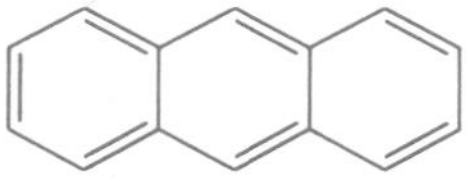

anthracene
$C_{14}H_{10}$

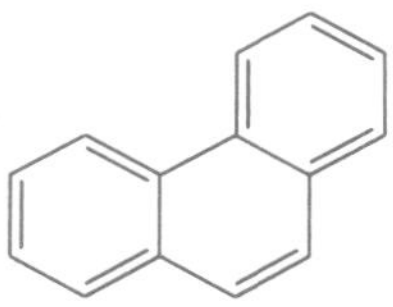

phenanthrene
$C_{14}H_{10}$

Bicyclic (two rings) and tricyclic (three rings) aromatic hydrocarbons are white solids that crystallize as platelets. They sublime at low temperatures. The use of naphthalene as mothballs is based on its volatility and its moth-repellent properties.

Problem 12.4 a) Draw the resonance structures for naphthalene, anthracene, and phenanthrene.

b) Unlike benzene or naphthalene, phenanthrene adds one mole of bromine. Guess the structure of the dibromide formed, and explain why the reaction occurs in terms of the resonance structures for phenanthrene.

Larger multicyclic aromatic compounds are produced by the incomplete combustion of coal and cigarettes. Some of them have been demonstrated to be carcinogenic, cancer-causing, such as this pentacyclic one.

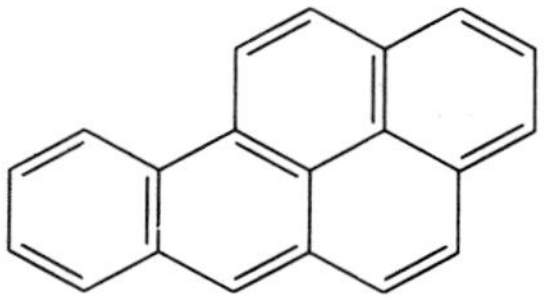

benzopyrene

Some compounds with highly unsaturated heterocyclic rings exhibit aromatic properties, though to a lesser extent than benzene. Rings containing nitrogen are frequently encountered in natural products. A few of the parent compounds are shown here.

pyridine   pyrimidine   purine   pyrrole   furan   thiophen

12.3 DELOCALIZATION ENERGY OF BENZENE

The stabilization of benzene and other aromatic compounds provided by delocalization of electrons has been described. A measurement of this delocalization energy for benzene can be made by comparing the heat of hydrogenation of benzene with a value calculated for the heat of hydrogenation of a ficticious 1,3,5-cyclohexatriene whose bonds are localized.

Benzene can be hydrogenated to cyclohexane under more drastic conditions than are required for hydrogenation of cyclohexene.

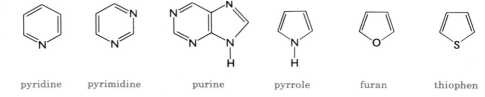

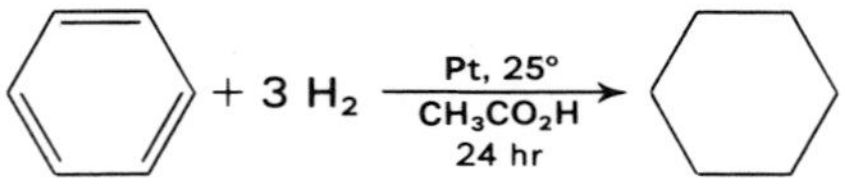

benzene   cyclohexane

Hydrogenation of cyclohexene gives $\Delta H = -28.6$ kcal/mole. Multiplication of this value by three gives $\Delta H = -85.8$ kcal/mole, the value expected for a hypothetical 1,3,5-cyclohexatriene with three localized double bonds.

Benzene evolves on hydrogenation 36 kcal/mole less heat than that calculated for localized 1,3,5-cyclohexatriene. This 36 kcal/mole difference is the pi electron delocalization energy which stabilizes benzene relative to a hypothetical Kekulé structure. Thus we say that benzene is stabilized by 36 kcal of delocalization energy.

12.4 GENERAL MECHANISM OF ELECTROPHILIC AROMATIC SUBSTITUTION

The pi bonds of the benzene ring produce a region of high electron density on either face of the ring. Electron-deficient reagents, electrophiles, are attracted to the ring. With very strong electrophiles benzene undergoes a substitution reaction in which the electrophile replaces a hydrogen on the ring.

The general mechanism for **electrophilic aromatic substitution** is shown in two steps. The electrophile adds to a double bond of the benzene ring forming a tetrahedral carbonium ion stabilized by delocalization of the positive charge and the other pi electrons in the ring. The second step is the loss of the proton on the tetrahedral carbon to give the product with a regenerated aromatic ring.

Step 1 Addition of the electrophile

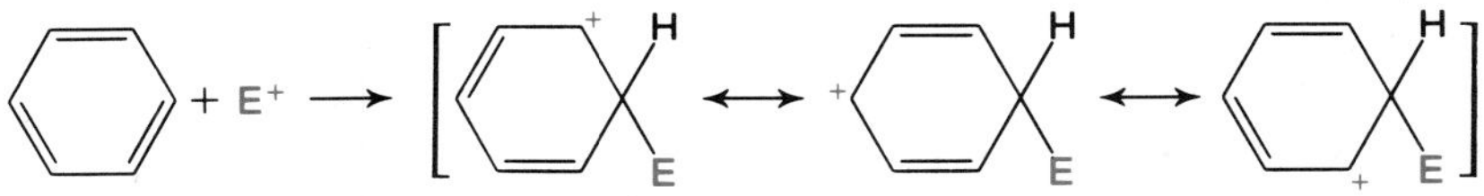

Step 2 Elimination of a proton

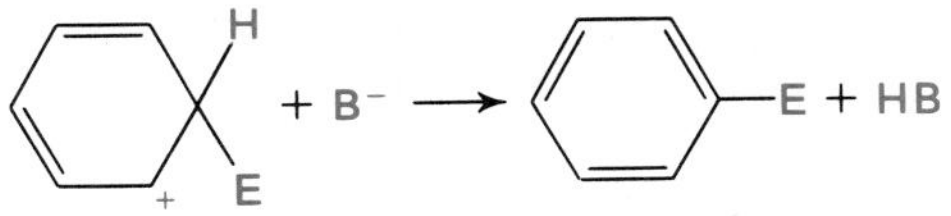

Substitution reactions of benzene with four types of reagents are described. In each of the examples which follow, special conditions are employed to generate an unusually strong electrophilic reagent, since benzene does not react with the usual ones.

Problem

12.5 Draw the resonance structures for the intermediate carbonium ion formed in the reaction of electrophile E^+ with each of the following substances.

a) benzene

b) toluene with reaction at the position *ortho* to CH_3

c) toluene with reaction at the *meta* position

d) toluene with reaction at the *para* position

12.5 NITRATION

In the nitration of benzene, hydrogen is replaced by a nitro group, $—NO_2$, with the formation of a C—N bond. The product is called **nitrobenzene**.

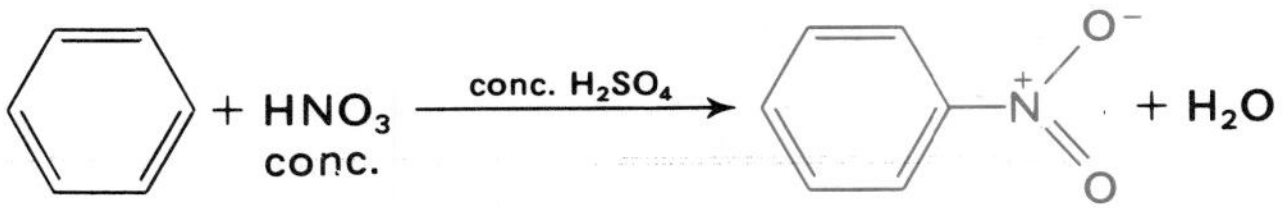

nitrobenzene

The nitronium ion, $O=\overset{+}{N}=O$, a powerful electrophile, is produced by the reaction of concentrated nitric acid with concentrated sulfuric acid. The reaction goes to completion.

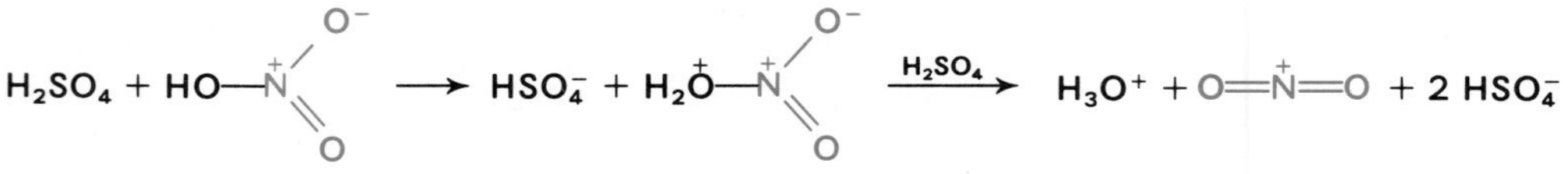

nitric acid　　　　　　　　　　　　　　　　　　nitronium ion

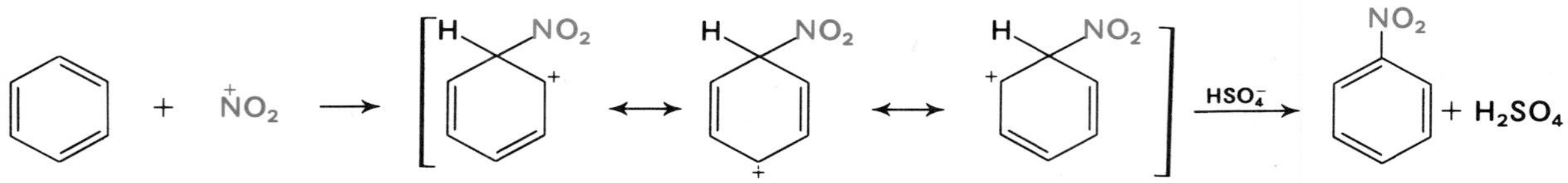

benzene　　nitronium ion　　　　　　　　　　　　　　　　　　nitrobenzene

Nitrobenzene is a yellow, oily liquid with a sweet odor. It is a polar compound due to the large dipole moment of the nitro group itself. Nitrobenzene is very unreactive and is an excellent solvent due to its polarity and unreactivity.

Under strongly forcing conditions trinitro-substituted aromatic compounds can be produced.

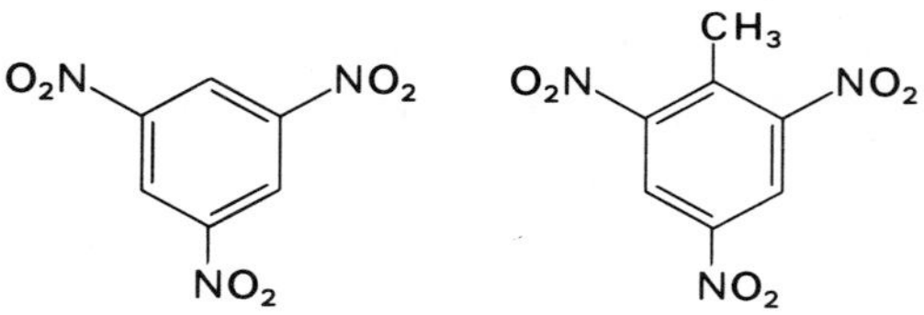

trinitrobenzene　　　　trinitrotoluene (TNT)

Both trinitrobenzene and trinitrotoluene (TNT) are powerful explosives when activated by a detonator. The reactions produce, from solids, gases such as CO_2, CO, H_2O, and N_2, plus large amounts of energy. In effect, a substance of low volume goes to substances that occupy large volumes.

Nitration of aromatic compounds provides the first step in the synthesis of a large number of substances. For example, a nitro compound is readily reduced by many reagents to a primary amine. Nitration of benzene and reduction of the nitrobenzene to aniline is the most direct synthesis of aniline from readily available starting materials.

The nitro group is converted smoothly to an amino group by iron and hydrochloric acid or by catalytic hydrogenation. Note the ready reduction of the nitro group without change in the benzene ring.

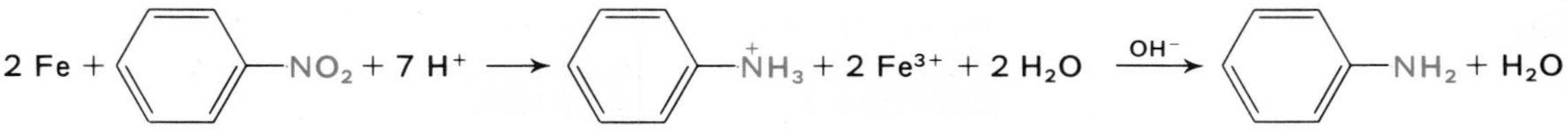

$$2\ Fe + C_6H_5{-}NO_2 + 7\ H^+ \longrightarrow C_6H_5{-}\overset{+}{N}H_3 + 2\ Fe^{3+} + 2\ H_2O \xrightarrow{\ OH^-\ } C_6H_5{-}NH_2 + H_2O$$

nitrobenzene   anilinium ion   aniline

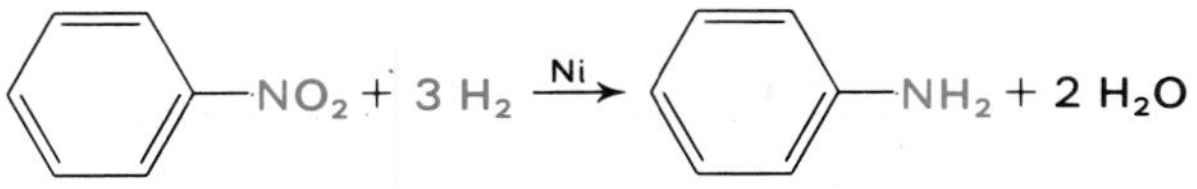

$$C_6H_5{-}NO_2 + 3\ H_2 \xrightarrow{\ Ni\ } C_6H_5{-}NH_2 + 2\ H_2O$$

nitrobenzene   aniline

Problems   **12.6** Draw the structures for the intermediate carbonium ion and for the product for the reaction of each of the following substances with concentrated nitric and sulfuric acids.

     a) benzene   **b)** toluene in the *para* position   **c)** 1,4-dimethylbenzene

12.7 Write the formulas for the starting material and the product of the reaction of each substance with iron and hydrochloric acid.

     a) nitrobenzene   **b)** *p*-nitrotoluene   **c)** 1,4-dimethyl-2-nitrobenzene

12.6 BROMINATION   Under the condition for the addition of bromine to alkenes, benzene is unreactive. However, the substitution of a hydrogen by a bromine is achieved with catalysis by ferric bromide.

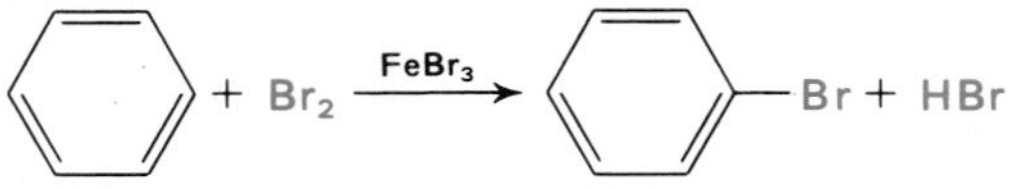

benzene   bromobenzene

     In actual practice an iron nail added to the solution of benzene containing some bromine produces a trace of ferric bromide catalyst. The ferric bromide then reacts with more bromine to form two ions, the positive bromonium ion Br^+ and $FeBr_4^-$.

$$2\ Fe + 3\ Br_2 \longrightarrow 2\ FeBr_3; \qquad FeBr_3 + Br_2 \longrightarrow Br^+ + FeBr_4^-$$

     The bromonium ion, a very strong electrophile, adds to the benzene ring, giving a delocalized carbonium ion. The carbonium ion loses the proton to the base $FeBr_4^-$ to give HBr and regenerated catalyst $FeBr_3$.

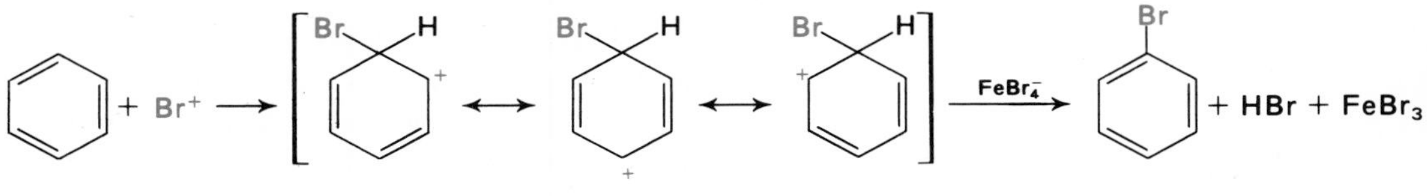

Chlorine also reacts with benzene in the presence of $FeCl_3$. The gas is less convenient to handle than liquid bromine. Both halogens are extremely corrosive, and are handled in hoods.

12.7 FRIEDEL-CRAFTS ALKYLATION AND ACYLATION

A reaction in which an alkyl group is substituted for hydrogen on a benzene ring involves an alkyl halide and anhydrous aluminum chloride as catalyst. The reaction, known as the **Friedel-Crafts alkylation**, produces an alkylbenzene and HCl.

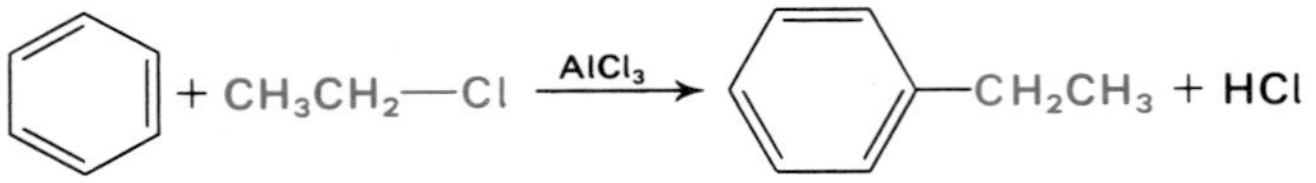

ethyl chloride   ethylbenzene

The alkylating agent, an alkyl cation, is produced by the reaction of ethyl chloride with anhydrous aluminum chloride.

$$CH_3CH_2-Cl + AlCl_3 \longrightarrow CH_3CH_2^+ \; AlCl_4^-$$

ethyl chloride   ethyl cation

The ethyl cation adds to benzene by the usual mechanism for aromatic substitution. A proton is then lost, and the aromatic structure is reformed. Hydrogen chloride gas is evolved and the $AlCl_3$ catalyst is regenerated.

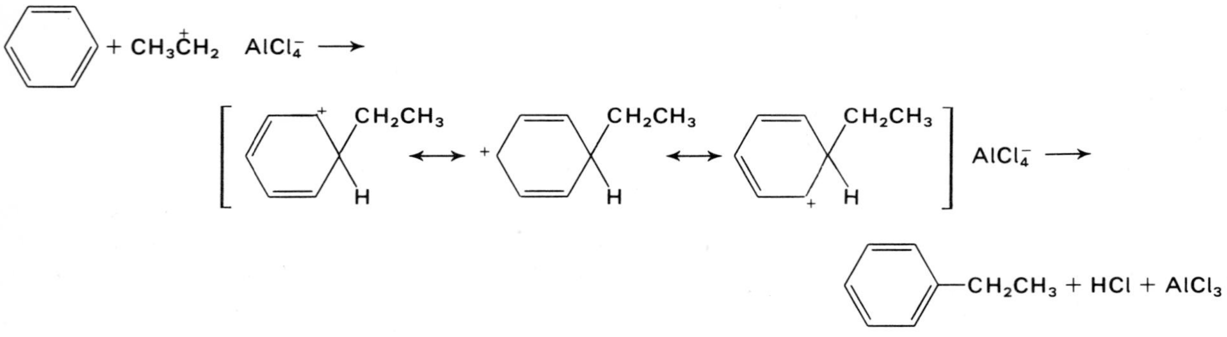

ethylbenzene

Because ethylbenzene is more reactive than benzene toward electrophilic reagents (see Section 12.9), *para*-diethylbenzene is also formed during the Friedel-Crafts alkylation.

Alkylation of aromatic compounds is usually limited to methyl, ethyl, isopropyl, and *tert*-butyl groups. When primary or secondary halides containing three or more carbons are used for alkylation, a rearrangement of the positive charge and the carbon structure of the carbonium ion occurs which prevents the attachment of an alkyl group at the #1 carbon. For example, 1-chlorobutane and benzene with aluminum chloride give only *sec*-butyl and *tert*-butylbenzene, but no butylbenzene.

Problem **12.8** Write the structures for the carbonium ion intermediate and for the product of the reaction of benzene with each of these reagents and anhydrous aluminum chloride.

    a) methyl iodide   **b)** isopropyl bromide   **c)** *tert*-butyl chloride

Anhydrous aluminum chloride also catalyzes a substitution reaction on benzene by an acid chloride. The product of this **Friedel-Crafts acylation** is an alkyl phenyl ketone.

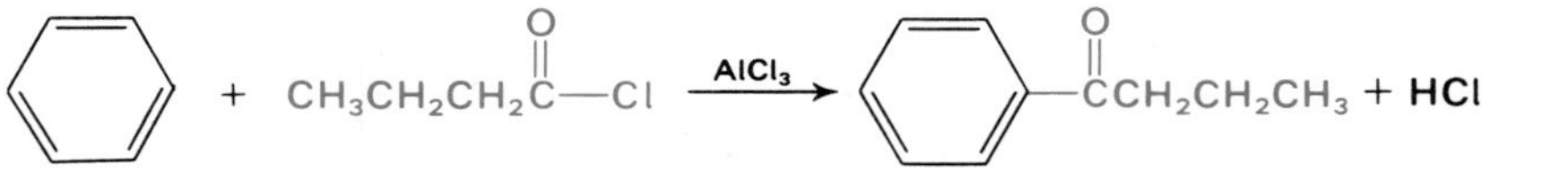

           butanoyl chloride   **phenyl propyl ketone**
                                         (1-phenyl-1-butanone)

The reaction resembles alkylation in mechanism. The $AlCl_3$ reacts with butanoyl chloride to form the butanoyl cation, $CH_3CH_2CH_2C^+{=}O$, and $AlCl_4^-$. Addition of the butanoyl cation to benzene followed by loss of a proton produces the phenyl ketone and HCl, and regenerates the $AlCl_3$ catalyst.

$$AlCl_3 + CH_3CH_2CH_2COCl \longrightarrow CH_3CH_2CH_2\overset{+}{C}{=}O + AlCl_4^-$$

           butanoyl chloride   **butanoyl cation**

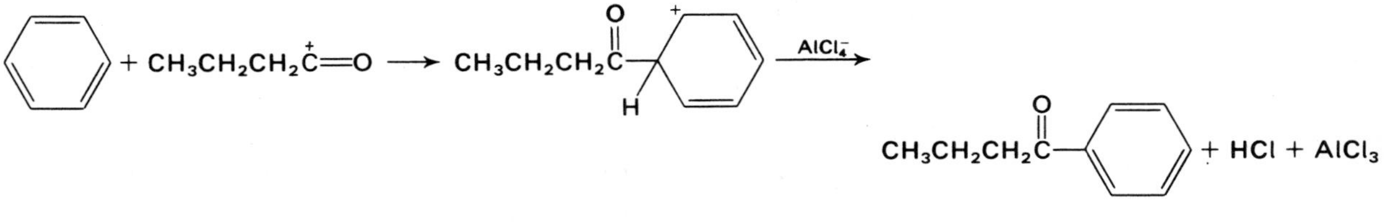

                          1-phenyl-1-butanone

The reaction is free of disubstituted products since no further reaction of the ketone with the acylating agent occurs. Acylation can also be carried out using an acid anhydride like acetic or succinic anhydride, instead of the acid chloride.

For the introduction of straight-chain alkyl groups onto the benzene ring, the best route involves the use of an acyl chloride, which does not give rearranged products. The ketone initially produced is reduced directly to the hydrocarbon by the Wolff-Kishner reaction, with hydrazine and potassium hydroxide.

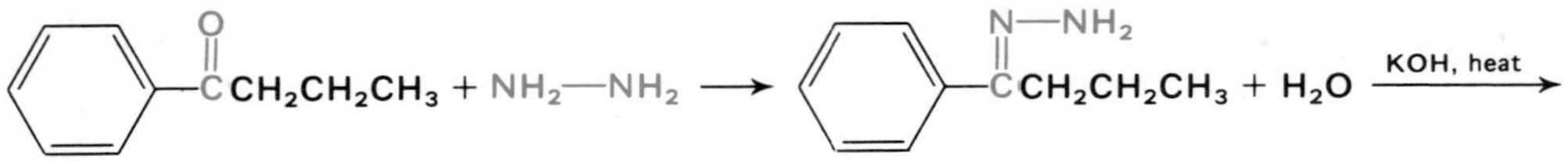

1-phenyl-1-butanone   hydrazine   a hydrazone

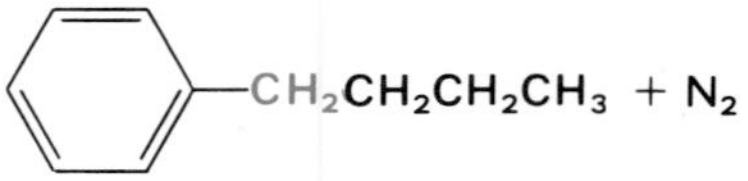

butylbenzene

Problems   **12.9**   Write equations for the following reactions.

a) benzoyl chloride + benzene + anhydrous aluminum chloride

b) acetyl chloride + chlorobenzene (in the *ortho* position) + $AlCl_3$

c) acetic anhydride + biphenyl (in the *para* position) + $AlCl_3$

12.10 Give the structures for the reactants and products of these reactions.

a) 1,4-dichloro-2-acetylbenzene + hydrazine + potassium hydroxide heated strongly

b) 1-butanoyl-2-methoxybenzene + sodium borohydride

12.11 Write a series of reactions that would convert benzene into propylbenzene, using any other starting materials you need.

12.8 SULFONATION   The production of aromatic sulfonic acids, **sulfonation**, is one of the most important substitution reactions. The reagent used to substitute the aromatic ring is fuming (100%) sulfuric acid. The reactive electrophile is either SO_3 or $^+SO_3H$.

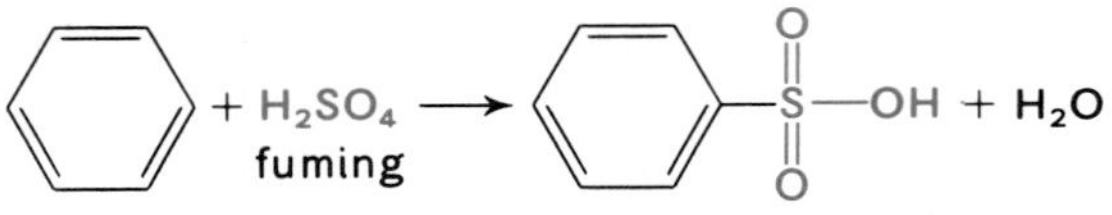

benzenesulfonic acid

Sulfonation is reversible. Sulfonic acid groups are removed if the sulfonic acids are heated with 50% aqueous sulfuric acid. Sulfonic acid groups are often used as blocking groups while other substitution reactions are taking place and later are removed.

Aromatic sulfonic acids are strong acids, comparable to the strong inorganic acids. They are cheap, are more soluble than inorganic acids in organic solvents, and find considerable use as acid catalysts.

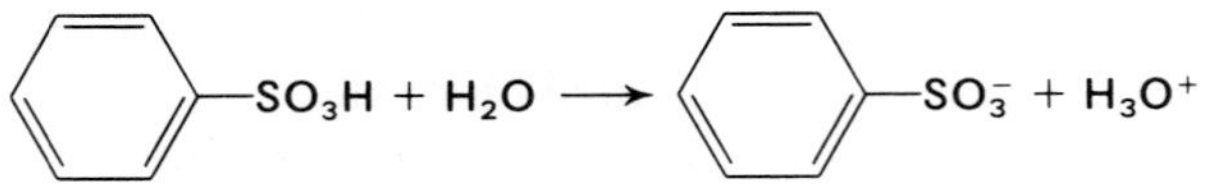

benzenesulfonic acid   benzenesulfonate ion

The presence of a $-SO_3H$ group or its salt $-SO_3^- Na^+$ increases the water-solubility of aromatic compounds. An important use is in dyes and detergents, which are frequently water-soluble salts of sulfonic acids. (See the structure of the dye Congo red in Section 12.11.)

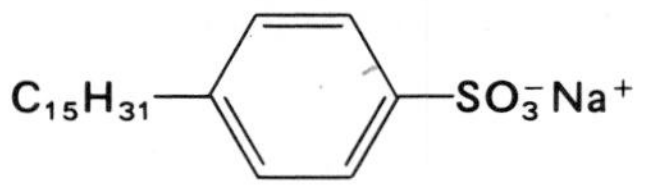

a detergent

Problem **12.12** Write equations for these reactions.

a) butylbenzene + fuming sulfuric acid (in the *para* position)

b) *p*-dichlorobenzene + fuming sulfuric acid

c) 3,5-dichloro-4-methyl-1-benzenesulfonic acid heated with 50% aqueous sulfuric acid

A summary of the introduction of side chains onto a benzene ring and subsequent transformations to other side chains is given in the diagram in Fig. 12.1.

FIGURE 12.1
REACTIONS OF BENZENE
AND BENZENE DERIVATIVES

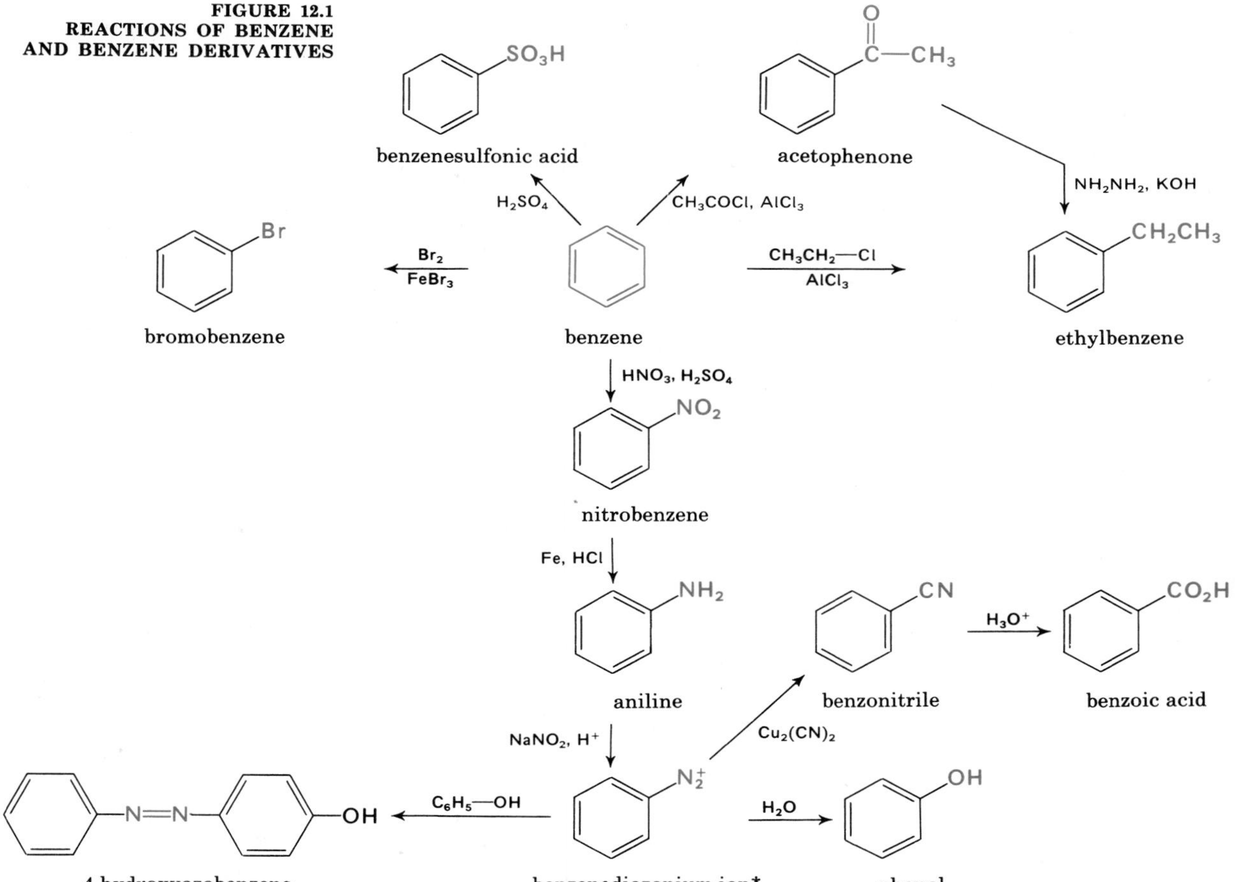

* Preparation and reactions of the diazonium ion are described in Section 12.10.

Aromatic compounds with hydrocarbon or functional groups attached to benzene rings undergo the same substitutions we have just discussed for benzene itself. Two questions immediately come to mind. With one substituent already present on the ring, the positions are no longer equivalent. Where does the second reagent enter in relation to the original group? Will the second substitution proceed as easily as the first?

The answers to these questions lie in the nature of the original substituent, which has a strong influence on further reactions of the ring itself. Substituent effects can be accounted for by the resonance structures that can be drawn for the intermediate carbonium ions for reaction at each of the ring positions.

The hydroxyl group of phenol has about the greatest effect on the course of further substitution of any group. Substitution reactions of phenol are *very much faster* than benzene, and the products formed are the *ortho* and *para* isomers only. Remember that the resonance structures for phenol include these in which the oxygen has a positive charge (Section 5.9).

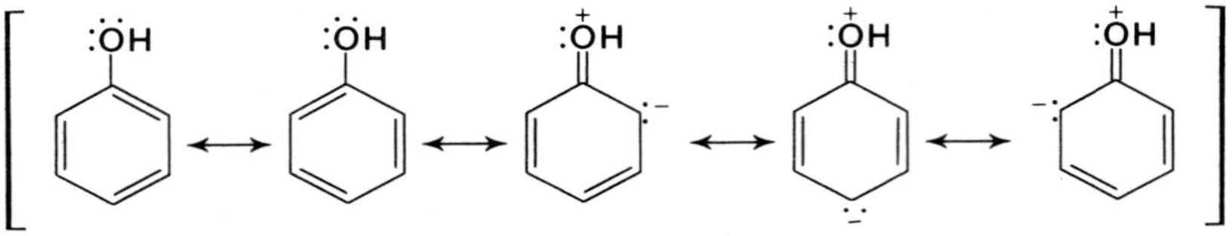

phenol

In the resonance structures for the intermediate carbonium ion formed when an electrophile adds to phenol, delocalization of a pair of electrons of the oxygen plays an important role. The carbonium ions produced by the addition of E^+ to phenol at each of the possible positions on the ring—*ortho*, *meta*, or *para* to the hydroxyl group—are formulated below.

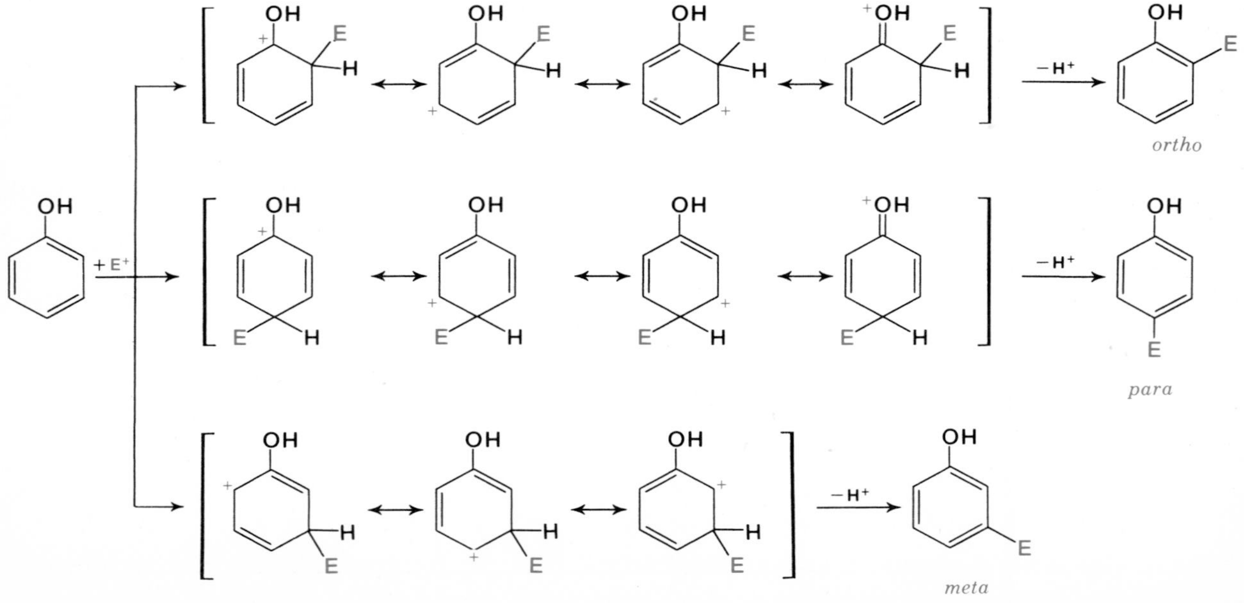

ortho

para

meta

Substitution is expected and found to predominate at that position for which the most stable reaction intermediate is generated. An examination of the intermediates written above shows that the one for the electrophile adding to the position *meta* to the resident hydroxyl group has one less resonance structure, the important one involving the hydroxyl group participation. Thus the *meta* carbonium ion is *much less* stable than those of the *ortho* or *para* positions. *Ortho*- and *para*-substituted phenols are produced in almost total exclusion of the *meta*-substituted phenol.

From the formulas drawn above, we cannot distinguish between the preference for *ortho* and *para* products. Usually a mixture of the two products is formed and must be separated. The hydroxyl group directs an incoming reagent to the *ortho* and *para* positions.

The tremendous activating effect of the hydroxyl group on the *ortho* and *para* positions is attributed to the strong stabilization afforded by that contributing resonance structure containing the positive oxonium ion.

Substituent groups such as —ÖH, —ÖR, —ÖCOR, and —ṄHCOR have unshared pairs of electrons available to distribute the positive charge of the intermediate carbonium ions in aromatic substitution. All of the groups are strong electron-donating functions in the resonance structures. They are all strong *activators* for the reactions and are *ortho-para directors* for further substitution.

Note that it is the strong electron donation through resonance, and not the electron-withdrawing inductive effect, which governs the activation and orientation by these groups.

Toluene reacts faster than benzene, and the methyl group is an *ortho-para* director. The effect is not as pronounced, however, as with the substituents which have unshared electrons that can participate.

Problems　**12.13** Write the resonance structures for methyl phenyl ether. Write the structures for the expected product(s) of its reaction with bromine and ferric bromide.

12.14 Write equations for the following reactions, giving the predominant products.

　　a) N-phenylacetamide + acetyl chloride + aluminum chloride

　　b) ethylbenzene + fuming sulfuric acid

　　c) methyl phenyl ether + propyl chloride + aluminum chloride

　　d) phenyl acetate + bromine + ferric bromide

　　e) toluene + conc. nitric acid + conc. sulfuric acid

　　f) 1-phenyl-1-propanone + hydrazine + potassium hydroxide heated

　　g) *o*-nitrotoluene + iron + hydrochloric acid

Substituents which are electron-withdrawing groups *deactivate* aromatic rings toward further substitution. Deactivation by reduction of electron density of the ring is greatest at the *ortho* and *para* positions. The *ortho* and *para* intermediate carbonium ions are less stable than the *meta* carbonium ion, as shown by these resonance structures for nitrobenzene. The nitro group is one of the strongest deactivating groups.

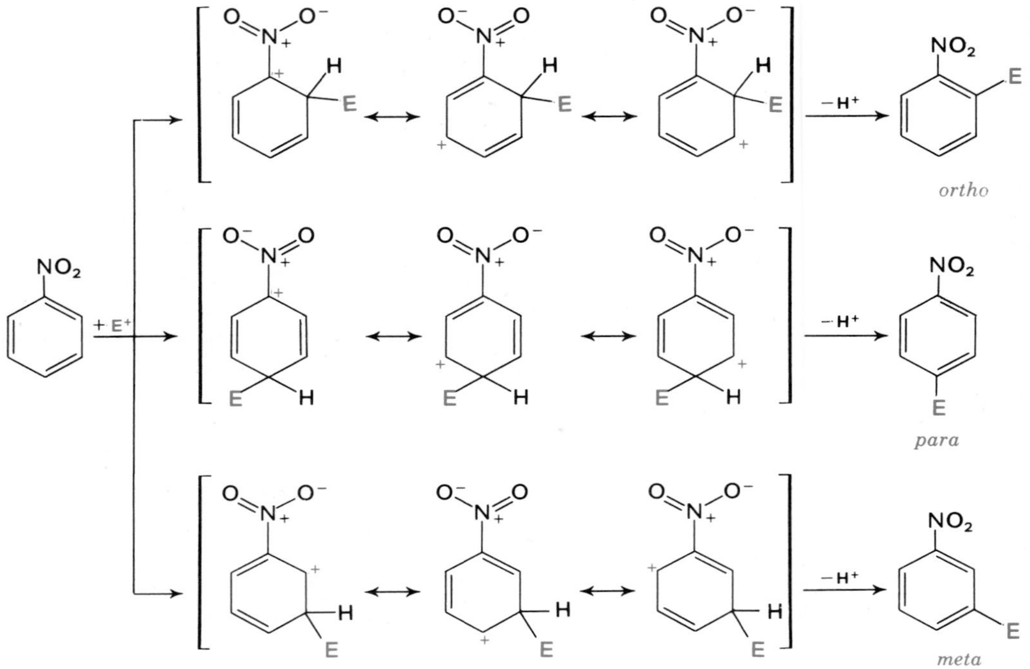

The structures written first for *ortho* and *para* carbonium ions have positive charges on adjacent atoms, which reduces the stability of these intermediates relative to the *meta* carbonium ion. The nitro group is both deactivating and *meta*-directing.

Other deactivating *meta*-directing groups are —CN, —CO$_2$H, —CH=O, —COR, —SO$_3$H, —NH$_3^+$. These compounds react with bromine, nitric acid, and sulfuric acid. None of the compounds having *meta*-directing, deactivating groups are sufficiently reactive to undergo alkylation or acylation.

Chloro-, bromo-, and iodobenzene are less reactive than benzene, but give *ortho-para* substitution predominantly. Fluorobenzene orients *ortho-para* and is about equal in reactivity to benzene. The curious effects are due to the opposing forces of electron-withdrawing inductive and *weak* electron-donating resonance effects. Halogens are *ortho-para* directing, but deactivating, substituents.

The resonance structures for the intermediate carbonium ion for electrophilic attack at the *para* position of chlorobenzene are shown.

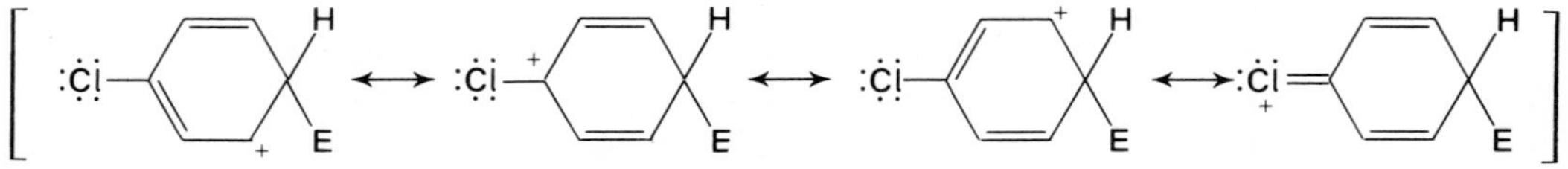

p-chloro carbonium ion intermediate

Note that in electrophilic substitution of the aromatic ring, it is a hydrogen which is replaced. Other substituents are unaffected in the reactions discussed above. In the case of the replacement of the sulfonate group by hydrogen in 50% sulfuric acid, the reaction is merely the reverse of the sulfonation reaction.

Problems **12.15** Give the structures of the predominant product(s) in each reaction.

    a) bromobenzene + methyl iodide + aluminum chloride

    b) benzenesulfonic acid + conc. nitric acid + conc. sulfuric acid

    c) methyl benzoate + bromine + ferric bromide

    d) chlorobenzene + fuming sulfuric acid

    e) N-phenylacetamide + bromine + ferric bromide

12.16 By a change in the order of introduction of substituents, it is possible to prepare different isomers of disubstituted benzenes. Write the structures for compounds A through J.

a) benzene $\xrightarrow{\text{HNO}_3,\ \text{H}_2\text{SO}_4}$ A $\xrightarrow{\text{Br}_2,\ \text{FeBr}_3}$ B $\xrightarrow{\text{Fe, HCl}}$ C

b) benzene $\xrightarrow{\text{Br}_2,\ \text{FeBr}_3}$ D $\xrightarrow{\text{HNO}_3,\ \text{H}_2\text{SO}_4}$ E $\xrightarrow{\text{Fe, HCl}}$ F

c) benzene $\xrightarrow{\text{HNO}_3,\ \text{H}_2\text{SO}_4}$ A $\xrightarrow{\text{Fe, HCl}}$ G $\xrightarrow{\text{CH}_3\text{CO}-\text{Cl}}$

H $\xrightarrow{\text{Br}_2,\ \text{FeBr}_3}$ I $\xrightarrow[-\text{CH}_3\text{CO}_2\text{H}]{\text{H}_3\text{O}^+}$ J

12.10 DIAZONIUM IONS

Nucleophilic displacement of the usual functional groups directly attached to the aromatic ring does not occur under ordinary conditions as in nucleophilic substitutions of alcohols and alkyl halides. There is, however, a particular substituted benzene which is readily made and whose functional group is very easily replaced by a wide variety of nucleophiles, including water, halide ions, and cyanide ions. This important substance is a diazonium salt $C_6H_5-{}^+N\equiv N: X^-$, which is stable only in aqueous solution at about $0-10°$.

Diazonium salts are prepared by the reaction of nitrous acid with the corresponding anilines and are used immediately to prepare various types of products. Nitrous acid HNO_2 is prepared in solution by combining sodium nitrite and a strong mineral acid, e.g., hydrochloric acid.

$$-NH_2 + Na^+NO_2^- + 2\ HCl \longrightarrow -\overset{+}{N}\equiv N\quad Cl^- + Na^+Cl^- + 2\ H_2O$$

aniline   sodium nitrite   benzenediazonium chloride

Aromatic diazonium ions undergo loss of N_2 when allowed to warm to room temperature. The very short-lived aryl cation $C_6H_5^+$ reacts rapidly with water to form phenol. This is a good method to introduce an OH group onto an aromatic ring.

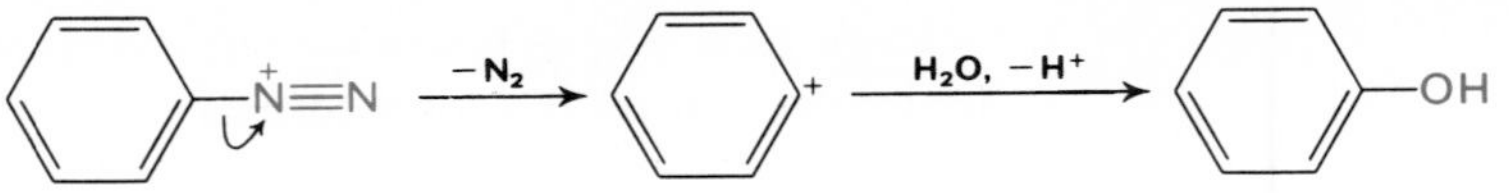

benzenediazonium ion   phenol

The introduction of another nucleophile is usually catalyzed by its copper (I) salt. The reaction leads to the attachment of the anion to the aryl group in the position vacated by the nitrogen. The reaction, known as the **Sandmeyer reaction**, is useful for introducing chloride, bromide, iodide, and cyanide ions with the corresponding copper (I) salt. The strong acid used with the reaction corresponds to the halide ion reacting. In the case of cyanide ion, sulfuric acid is used. Examples with three substituted anilines are shown.

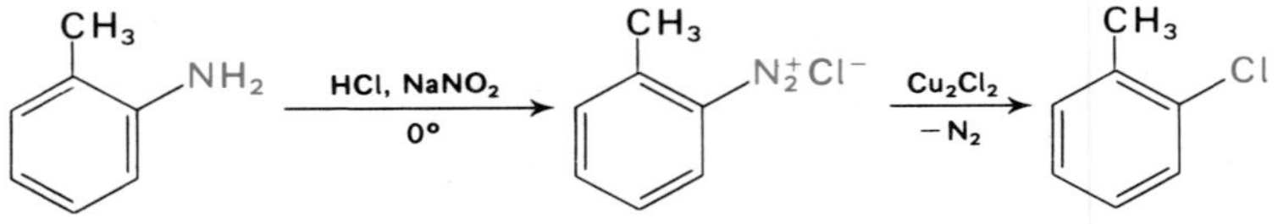

o-methylaniline   o-chlorotoluene

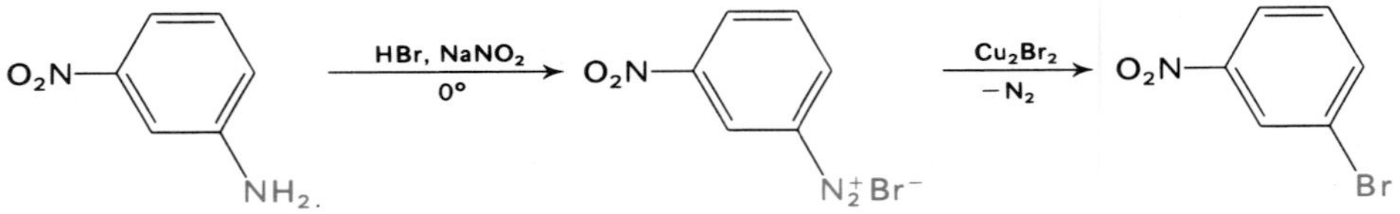

m-nitroaniline   m-nitrobromobenzene

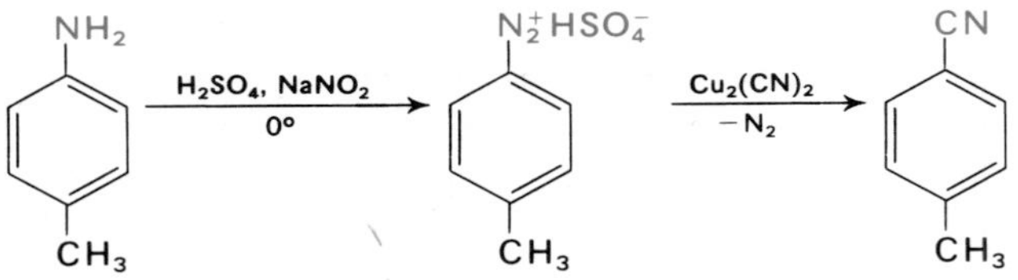

p-methylaniline   p-methylbenzonitrile

With the exception of —CH=O, which would be oxidized by the nitrous acid, almost any other substituent may be present in any ring position. The Sandmeyer reaction often allows synthetic chemists to prepare compounds that would be denied to them were they dependent only on substituent-orienting effects in substitution reactions. It is often possible to locate two or more substituents in benzene rings in desirable positions relative to one another only through the Sandmeyer reaction.

Problem   **12.17** Give the formulas of compounds A through L.

a) benzene $\xrightarrow[]{\text{HNO}_3,\ \text{H}_2\text{SO}_4}$ A $\xrightarrow{\text{Fe, HCl}}$ B $\xrightarrow[0°]{\text{NaNO}_2,\ \text{H}_2\text{SO}_4}$ C $\xrightarrow{\text{H}_2\text{O, heat}}$ D

b) nitrobenzene $\xrightarrow{\text{Br}_2, \text{FeBr}_3}$ E $\xrightarrow{\text{Fe, HCl}}$ F $\xrightarrow[0°]{\text{NaNO}_2, \text{HBr}}$ G $\xrightarrow{\text{Cu}_2\text{Br}_2}$ H

c) acetanilide (N-phenylacetamide) $\xrightarrow{100\% \text{ H}_2\text{SO}_4}$ I $\xrightarrow{\text{HCl, H}_2\text{O}}$

J $\xrightarrow[0°]{\text{NaNO}_2, \text{H}_2\text{SO}_4}$ K $\xrightarrow{\text{Cu}_2(\text{CN})_2}$ L

Aryldiazonium ions are electrophilic reagents in their own right which attack the highly reactive aromatic rings of phenol and aniline. The result is a diazonium coupling of two aromatic rings through the two nitrogens, N=N (the azo group), without the loss of N_2.

benzenediazonium ion   N,N-dimethylaniline   4-dimethylaminoazobenzene

A large number of colored compounds used as dyes, pigments, and indicators are prepared by this reaction. The colors of the compounds can be controlled by the design of their structures, and dyes of all hues have been prepared. Some examples of these colored compounds are described in the next section.

12.11 DYES AND PIGMENTS

Colored organic compounds have been used as dyes and pigments for centuries to impart various tints and hues to fibers and paints. In general pigments are insoluble in water and oil, and are used as a dispersion. Dyes for fibers must contain some polar groups for attachment to the polymer molecule and usually have some degree of water-solubility. Other than in the attached groups, which impart differences in solubility properties, compounds which serve as dyes and pigments can have similar structural units.

Color in organic compounds is based on the delocalization of electron pairs or positive charge through a long chain of conjugated double bonds (Section 15.2). The length of the conjugated system and the extent of electron or charge delocalization correlate with the hue of the dye.

A good example of a colored compound is β-carotene, the tetraterpene having a series of 11 conjugated C=C bonds (Section 4.10).

β-carotene (yellow)

The presence of aromatic rings in compounds, particularly naphthalene and the nitrogen heterocycles, increases the possibilities of electron or charge delocalization without increasing the size of the molecule. The substitution of amino, hydroxyl, and nitro groups for hydrogen on aromatic rings involves the electrons of these groups in the bond delocalization. Many compounds, both those extracted from plants by the ancients and those synthesized by the modern chemist, derive their color properties from aromatic structures like those shown below.

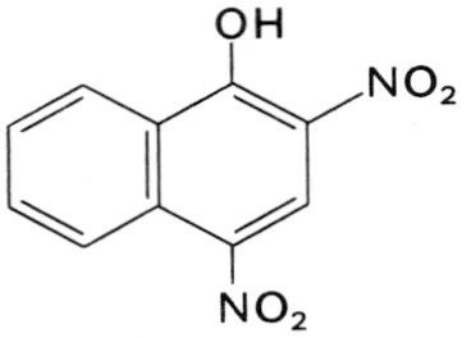

Martius yellow
(for silk and wool)

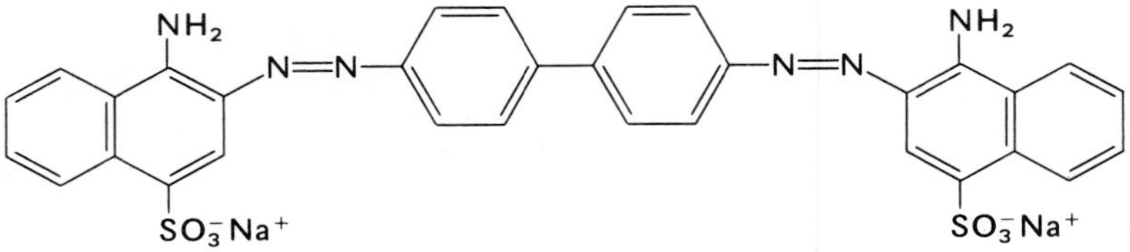

Congo red
(for cotton)

synthetic dyes

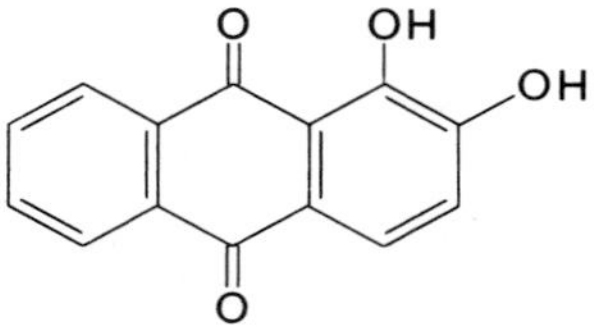

alizarin
("Turkey Red")

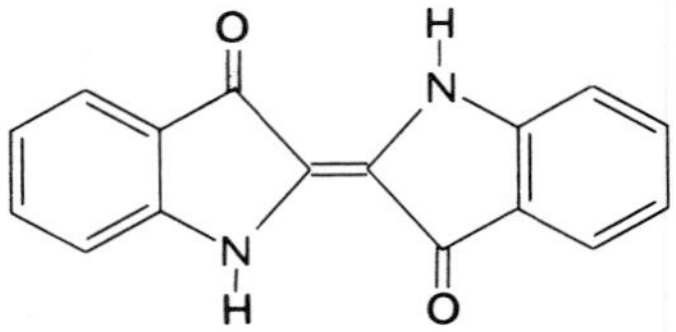

indigo

ancient plant dyes

Problems   **12.18** List all of the functional groups and ring structures which make up the formula for Congo red.

12.19 Draw resonance structures for 4-dimethylaminoazobenzene (formula in Section 12.10).

A group of plant pigments have structures in which four pyrrole rings are assembled to form a macrocyclic ring, called the **porphyrin** ring. They also contain a metal ion which just fits in the central hole and is bonded to the nitrogens. Examples of porphyrins are chlorophyll, hemin, and vitamin B_{12}.

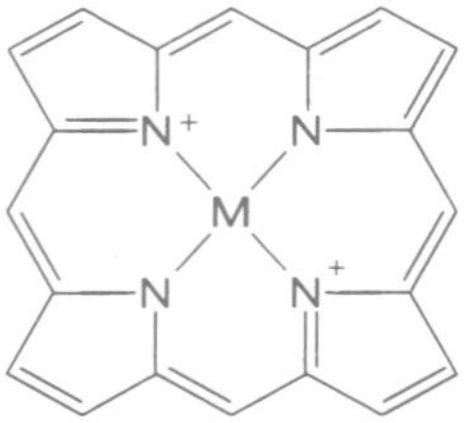

porphyrin structure 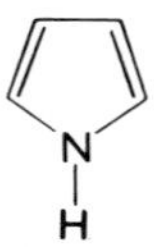   pyrrole 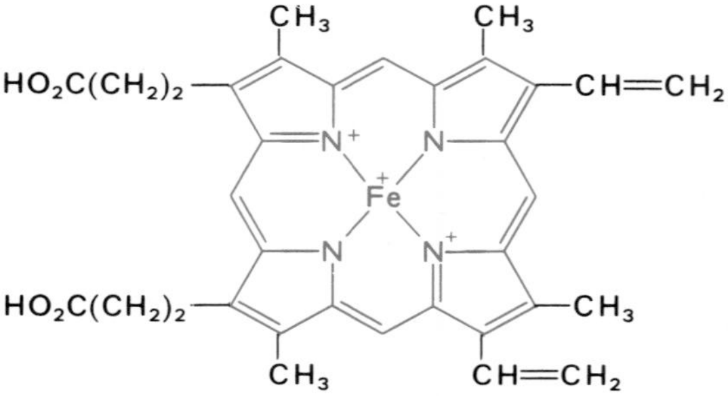   hemin

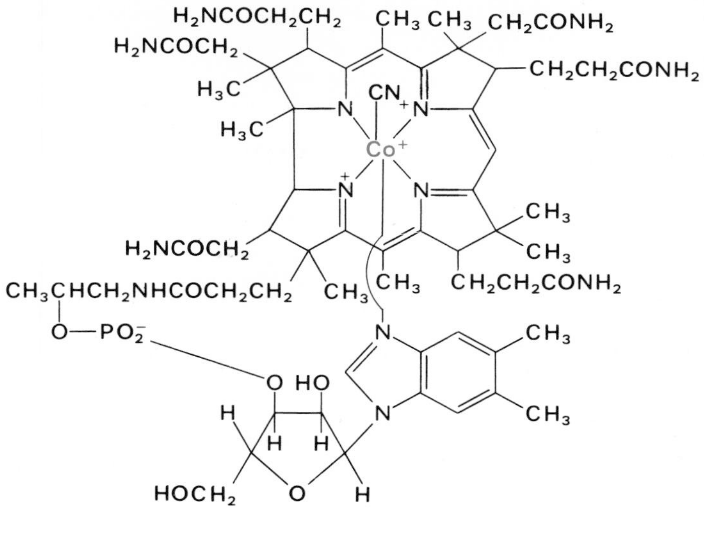

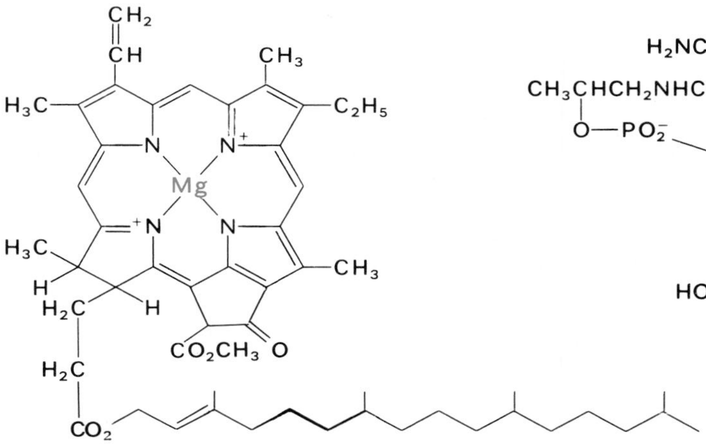

chlorophyll *a*

vitamin B$_{12}$

Vitamin B$_{12}$ is the most complex organic compound to have been synthesized in the laboratory from simple starting materials. How many side chains are attached to the porphyrin ring?

12.12 ARYL HALIDES—NUCLEOPHILIC AROMATIC SUBSTITUTION

Aryl halides are normally unreactive except with magnesium, forming Grignard reagents (Section 14.2). Electron-withdrawing groups located *ortho* or *para* to the halogen greatly facilitate substitution of the halogen by nucleophilic reagents. Nitro groups are particularly effective. Picryl chloride (2,4,6-trinitrochlorobenzene) has a chlorine that is as labile as an acyl chloride. Water as a nucleophile readily displaces the chlorine to give picric acid (2,4,6-trinitrophenol), a strong acid, pK_a = 0.75.

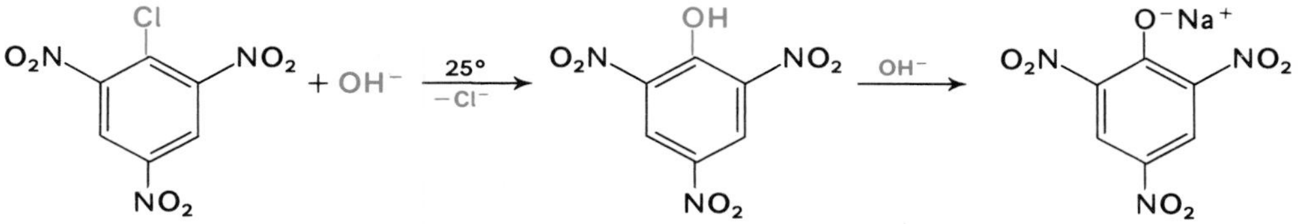

picryl chloride

picric acid
(2,4,6-trinitrophenol)
pK_a = 0.75

Other strong nucleophiles, such as sodium methoxide or an amine, react readily with aryl halides having a nitro group in the *ortho* or *para* position.

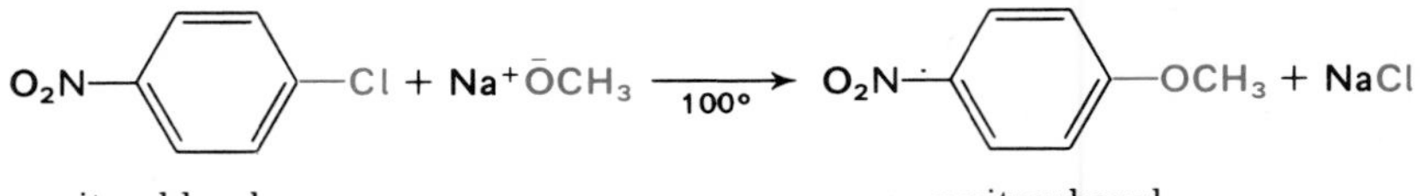

p-nitrochlorobenzene

p-nitrophenyl methyl ether

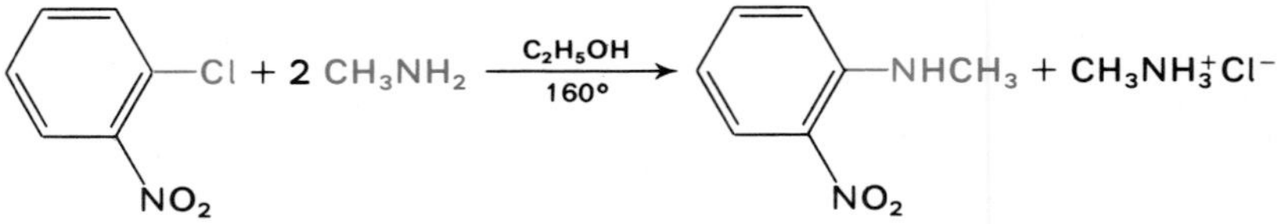

o-nitrochlorobenzene

N-methyl-*o*-nitroaniline

The order of reactivity of halides as leaving groups in aromatic nucleophilic substitution is usually $F \gg Cl > Br > I$. The superiority of fluoride as a leaving group led to the use of 2,4-dinitrofluorobenzene (Sanger's reagent) as a specific means for marking terminal amino groups of proteins and polypeptides before hydrolysis of the peptide units (Section 10.16). Identification of the amino acid attached to the dinitrophenyl group specifies the amino terminal acid present in the polypeptide or protein.

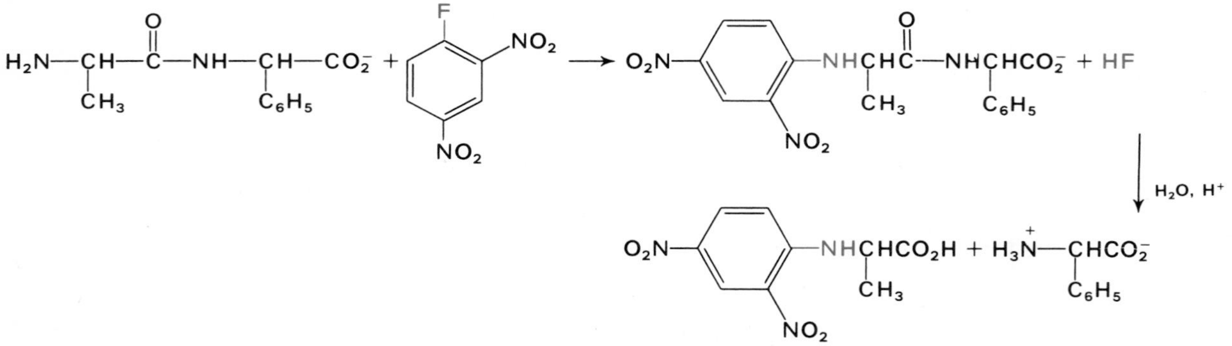

Problem 12.20 Write the structure for the product of each reaction.

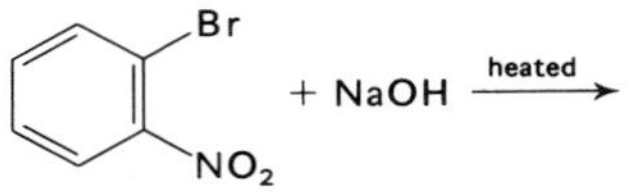

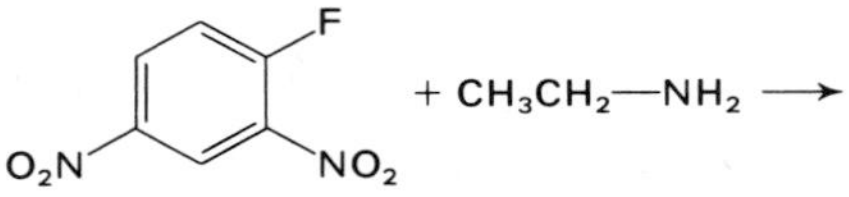

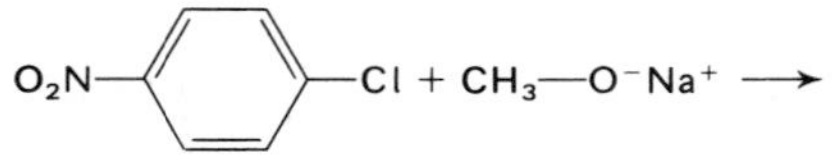

New Terms and Topics

Aromaticity; $4n + 2$ rule (Section 12.1)

Systematic and common names for aromatic compounds (Section 12.2)

Fused-ring aromatic hydrocarbons; heterocyclic aromatics—pyridine, pyrimidine, purine, pyrrole (Section 12.2)

Measurement of delocalization energy (Section 12.3)

Electrophilic aromatic substitution (Section 12.4)

Resonance structures of intermediate cation of substitution (Section 12.4)

Nitration and the nitro group (Section 12.5)

Bromination and sulfonation (Sections 12.6 and 12.8)

Friedel-Crafts alkylation and acylation (Section 12.7)

Orientation of entering second group by first substituent (Section 12.9)

Ortho-para directors; *meta*-directors; activating and deactivating groups (Section 12.9)

Diazonium salts (Section 12.10)

Substitution reaction of diazonium ions (Section 12.10)

Diazonium coupling (Section 12.10)

Dyes; color through delocalization of electrons and charge (Section 12.11)

Nucleophilic aromatic substitution; activation of halide by nitro group (Section 12.12)

Summary of Reactions

ELECTROPHILIC SUBSTITUTION REACTIONS OF BENZENE

1. Nitration (Section 12.5)

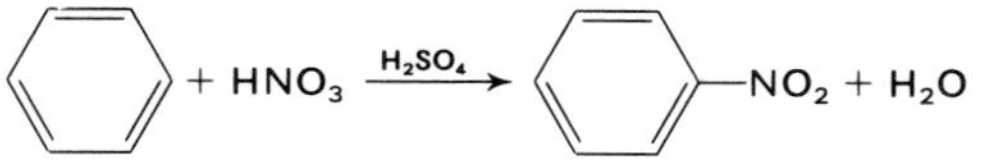

$$\text{benzene} + HNO_3 \xrightarrow{H_2SO_4} \text{C}_6\text{H}_5-NO_2 + H_2O$$

nitrobenzene

2. Bromination (Section 12.6)

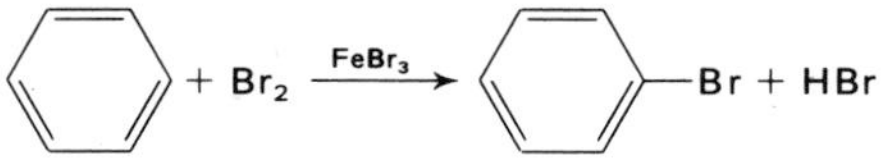

$$\text{benzene} + Br_2 \xrightarrow{FeBr_3} \text{C}_6\text{H}_5-Br + HBr$$

bromobenzene

3. Alkylation—Friedel-Crafts reaction (Section 12.7)

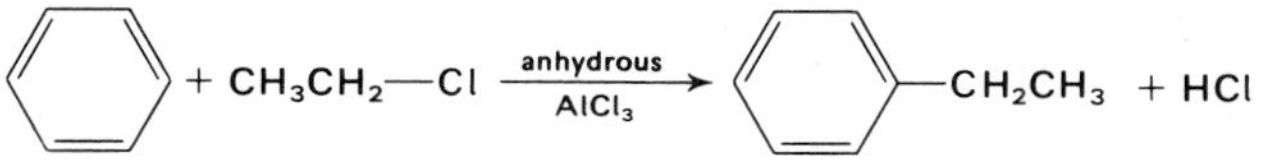

$$\text{benzene} + CH_3CH_2-Cl \xrightarrow[\text{AlCl}_3]{\text{anhydrous}} \text{C}_6\text{H}_5-CH_2CH_3 + HCl$$

ethylbenzene

4. Acylation (Section 12.7)

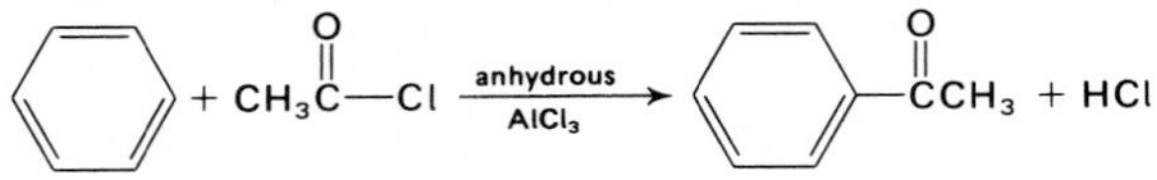

methyl phenyl ketone
(acetophenone)

5. Sulfonation (Section 12.8)

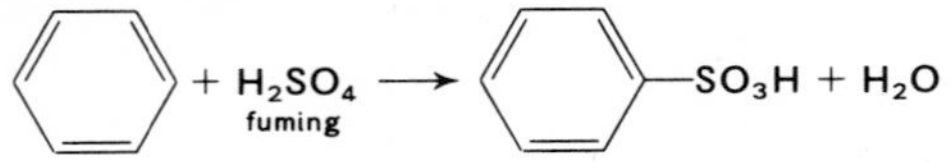

ORIENTATION OF ENTERING GROUP IN SUBSTITUTED BENZENES

1. Oxygen, nitrogen, and halogen-substituted benzenes—*ortho-para* directors (Section 12.9)

a)

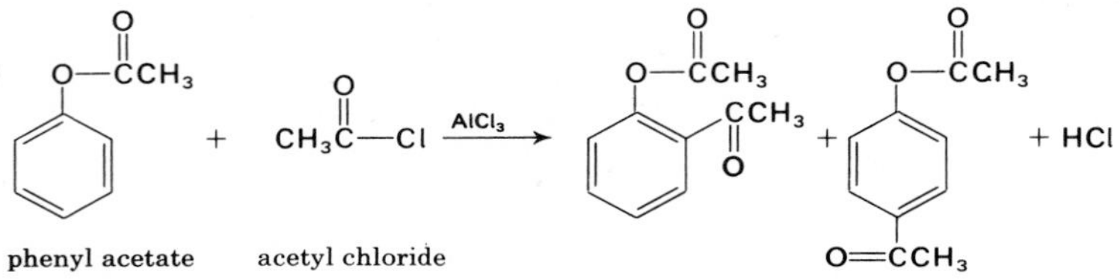

phenyl acetate   acetyl chloride

o-acetylphenyl
acetate

p-acetylphenyl
acetate

b)

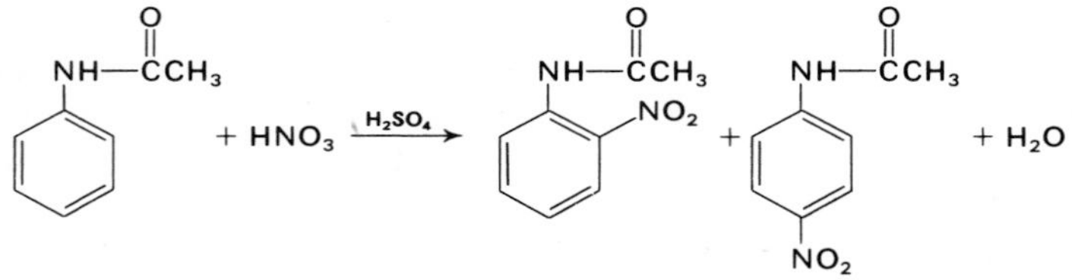

N-phenylacetamide
(acetanilide)

o-nitroacet-
anilide

p-nitroacet-
anilide

c)

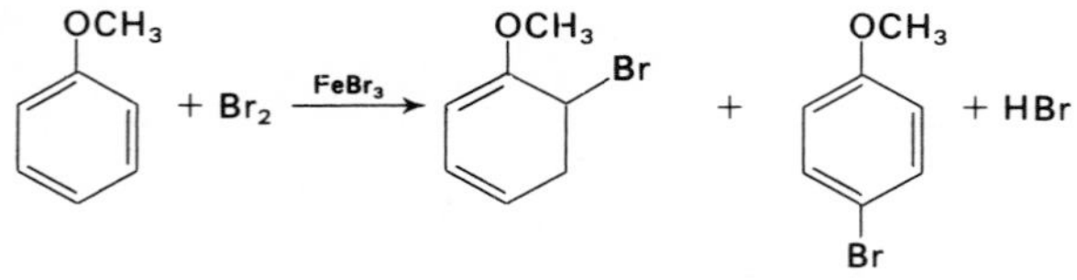

methyl phenyl
ether

o-bromophenyl
methyl ether

p-bromophenyl
methyl ether

d)

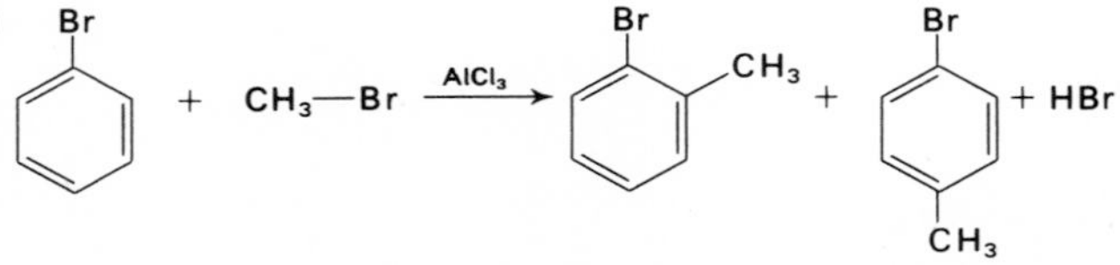

bromobenzene   methyl bromide

o-bromotoluene

p-bromotoluene

2. Alkylbenzenes—*ortho-para* directors (Section 12.9)

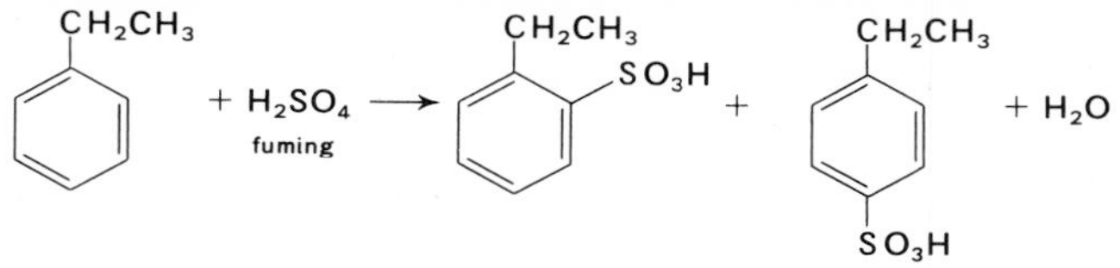

ethylbenzene

o-ethylbenzene-
sulfonic acid

p-ethylbenzene-
sulfonic acid

3. Phenol (Section 12.9)

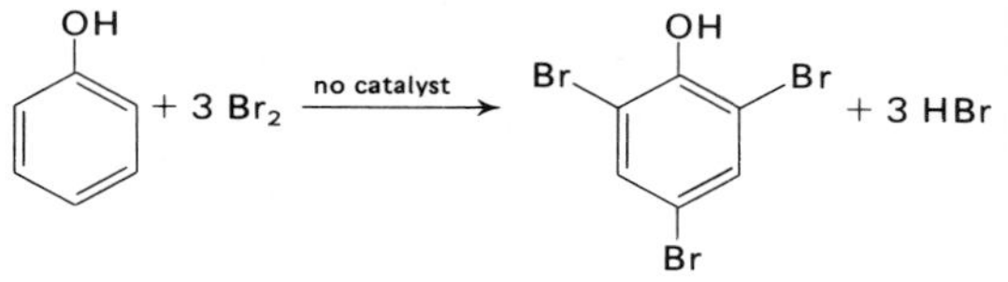

2,4,6-tribromophenol

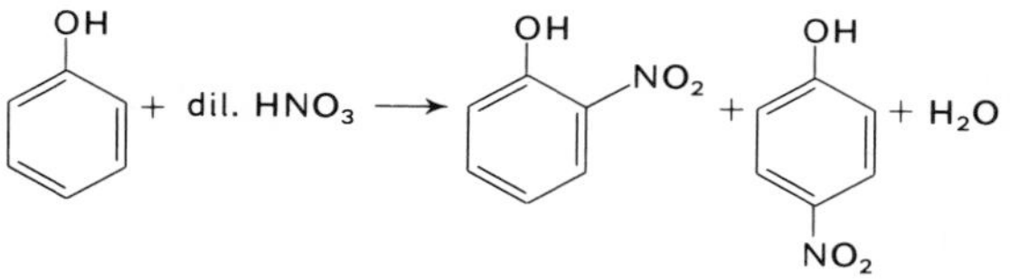

o-nitrophenol   *p*-nitrophenol

4. Nitrobenzene, aryl carbonyl, carboxyl and sulfonyl compounds—*meta* directors (Section 12.9)

a)

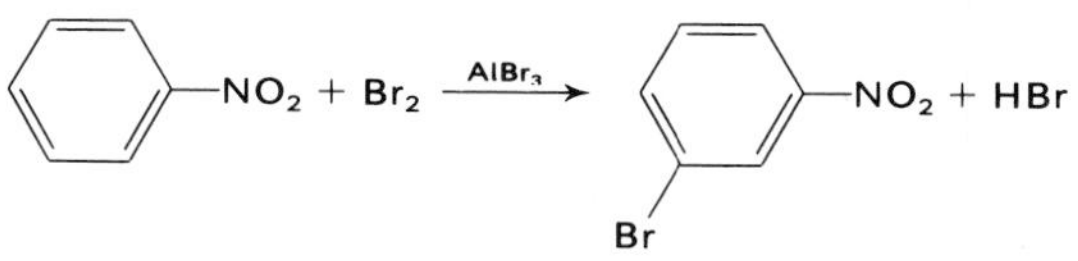

nitrobenzene

m-nitrobromobenzene

b)

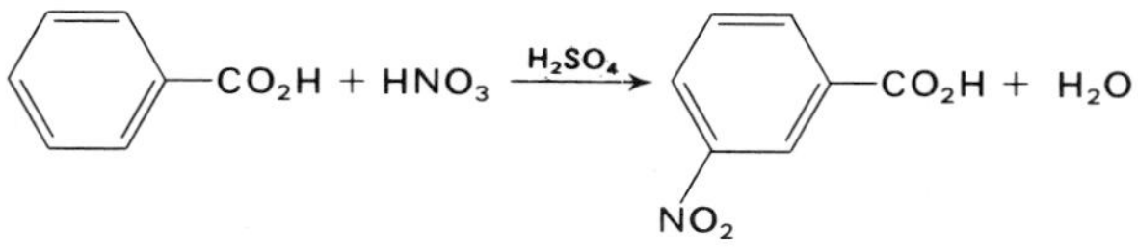

benzoic acid

m-nitrobenzoic acid

c)

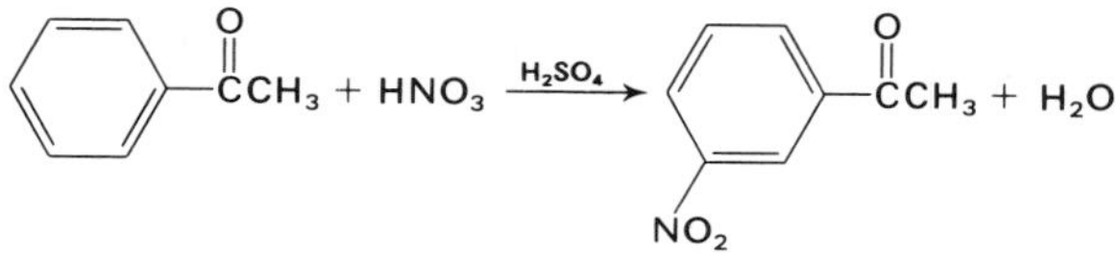

acetophenone

m-nitroacetophenone

d)

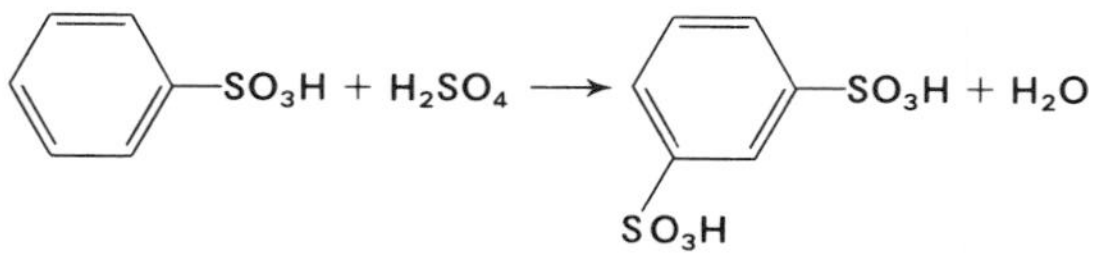

benzenesulfonic acid

m-benzenedisulfonic acid

None of these *meta*-directing compounds react under reasonable conditions with Friedel-Crafts alkylation or acylation reagents.

REDUCTION OF AROMATIC COMPOUNDS (Section 12.3)

1. Reduction of the aromatic ring of hydrocarbons and other derivatives

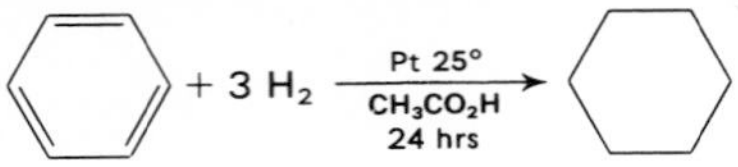

benzene + 3 H$_2$ $\xrightarrow[\substack{CH_3CO_2H \\ 24\ hrs}]{Pt\ 25°}$ cyclohexane

benzene   cyclohexane

2. Reduction of side chains (Section 12.7)

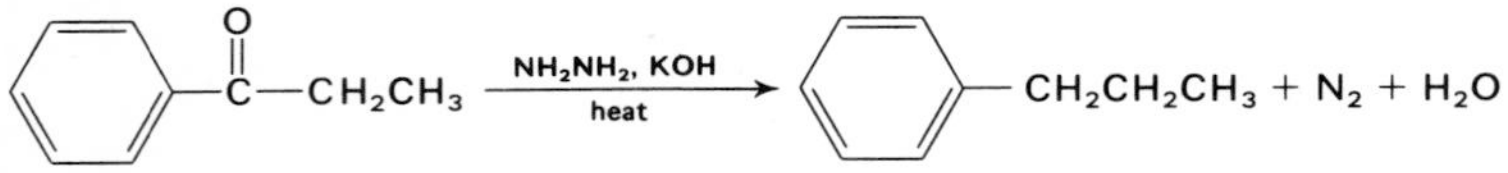

1-phenyl-1-propanone $\xrightarrow[heat]{NH_2NH_2,\ KOH}$ CH$_2$CH$_2$CH$_3$ + N$_2$ + H$_2$O

1-phenyl-1-propanone   propylbenzene

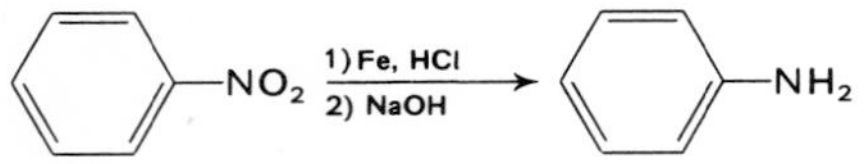

nitrobenzene —NO$_2$ $\xrightarrow[\substack{2)\ NaOH}]{1)\ Fe,\ HCl}$ —NH$_2$ aniline

nitrobenzene   aniline

DIAZOTIZATION OF ANILINES (Section 12.10)

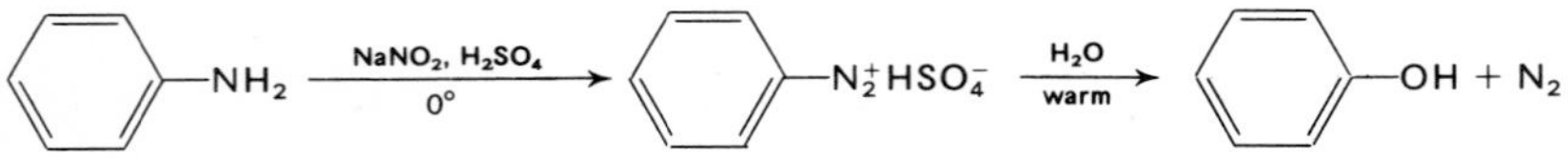

—NH$_2$ $\xrightarrow[0°]{NaNO_2,\ H_2SO_4}$ —N$_2^+$HSO$_4^-$ $\xrightarrow[warm]{H_2O}$ —OH + N$_2$

aniline   benzenediazonium   phenol
hydrogen sulfate

REACTIONS OF DIAZONIUM SALTS (Section 12.10)

1. Displacement—Sandmeyer reaction

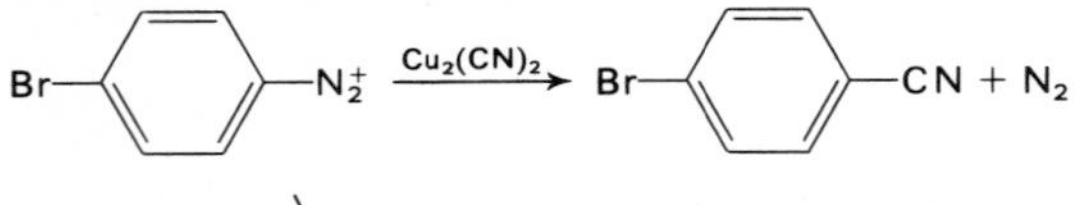

Br—⟨ ⟩—N$_2^+$ $\xrightarrow{Cu_2(CN)_2}$ Br—⟨ ⟩—CN + N$_2$

p-bromobenzonitrile

$\xrightarrow{Cu_2Br_2}$ Br—⟨ ⟩—Br + N$_2$

p-dibromobenzene

2. Coupling with reactive rings (Section 12.10)

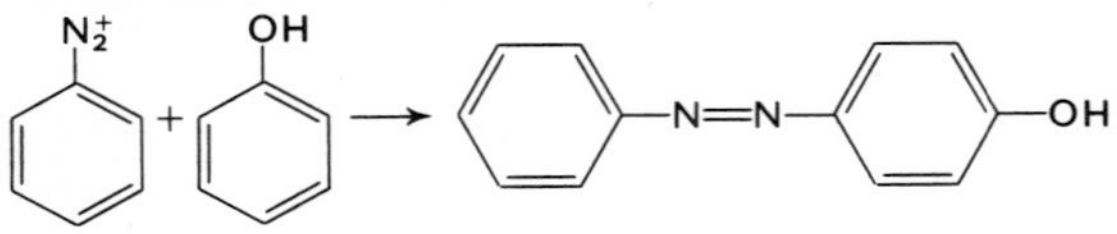

N$_2^+$ + OH $\longrightarrow$ —N=N—⟨ ⟩—OH

4-hydroxyazobenzene

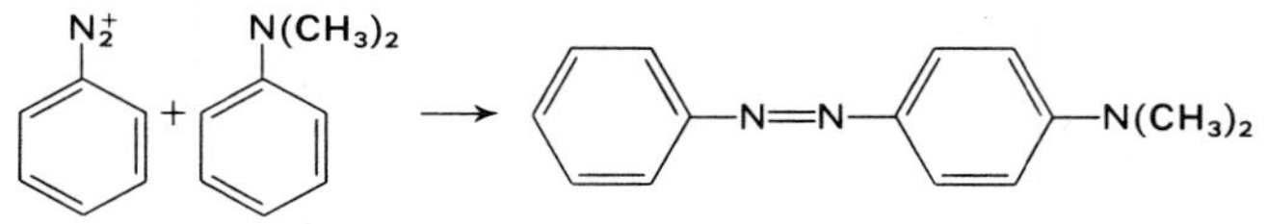

4-dimethylamino-
azobenzene

NUCLEOPHILIC SUBSTITUTION OF AROMATIC HALIDES ACTIVATED BY NITRO GROUP (Section 12.12)

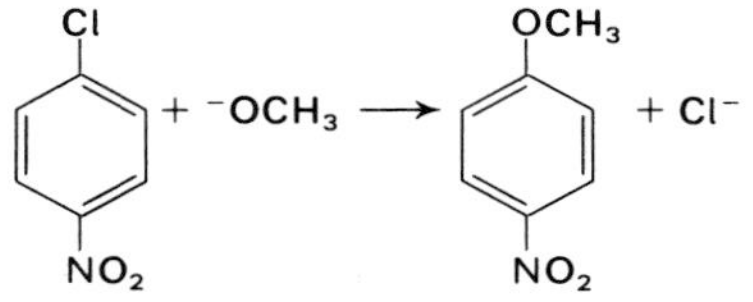

| p-nitrochlorobenzene | p-nitrophenyl methyl ether | 2,4-dinitro-fluorobenzene | 2,4-dinitro-N-methylaniline |

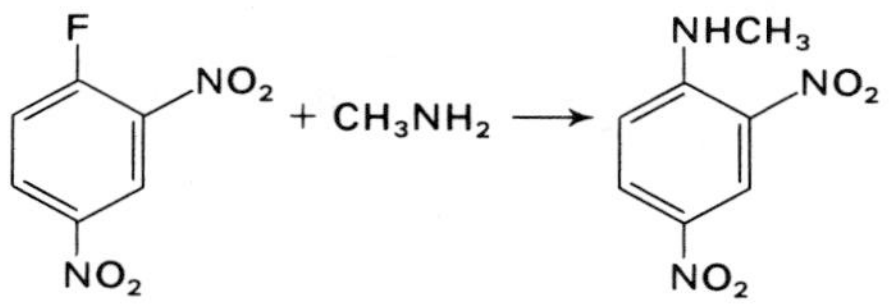

(Useful for labeling terminal amino group in polypeptides.)

Problems

12.21 Write formulas for the following compounds.

a) *m*-nitrophenol　　　　　**b)** *o*-fluorophenyl acetate　　　**c)** 2-methoxybenzoic acid

d) *p*-aminobenzenesulfonic acid　**e)** biphenyl　　　　　　　**f)** *o*-methylaniline

g) 2,4-dinitro-1-fluorobenzene　**h)** *p*-dichlorobenzene　　　**i)** 3-ethyl-N-methylaniline

12.22 In the early days of organic chemistry, determination of structure of compounds was a tedious process requiring careful separations of products. A chemist had three dibromobenzenes. They were identified by the number of tribromobenzenes each formed on bromination. What number of tribromo-benzenes would *ortho-*, *meta-*, and *para*-dibromobenzene each form?

12.23 Compounds i) and ii) ($C_7H_7O_3N$) were isomers obtained from the treatment of benzyl alcohol with nitric acid and sulfuric acid. Each isomer was put through the following series of reactions. Give the structures of i) and ii) and of compounds A through J.

i) $\xrightarrow{\text{Fe, HCl}}$ A $\xrightarrow{\text{NaNO}_2,\ \text{H}_2\text{SO}_4}$ B $\xrightarrow{\text{Cu}_2(\text{CN})_2}$ C $\xrightarrow{\text{H}_3\text{O}^+}$ D　　ii) $\xrightarrow{\text{Fe, HCl}}$ F $\xrightarrow{\text{NaNO}_2,\ \text{H}_2\text{SO}_4}$ G $\xrightarrow{\text{Cu}_2(\text{CN})_2}$ H $\xrightarrow{\text{H}_3\text{O}^+}$ J

12.24 Give the formulas for the product(s) of each reaction.

a)

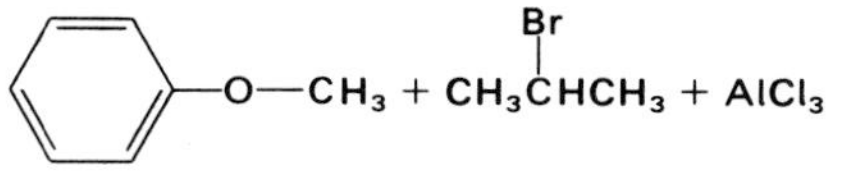

b)

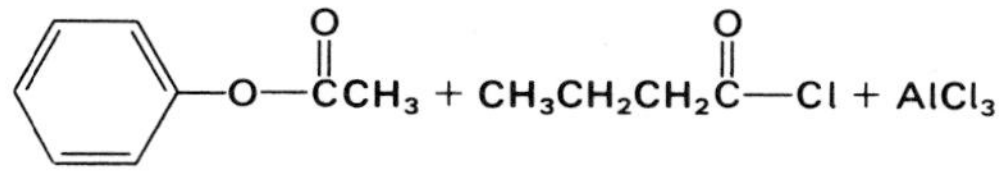

c)

d)

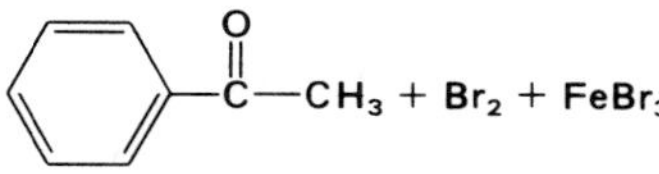

e)

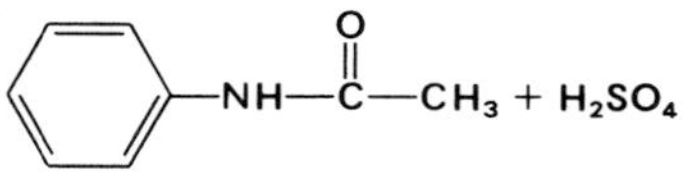

f)

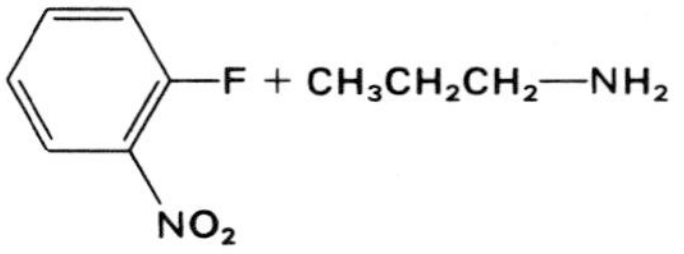

12.25 Draw formulas for compounds A through L.

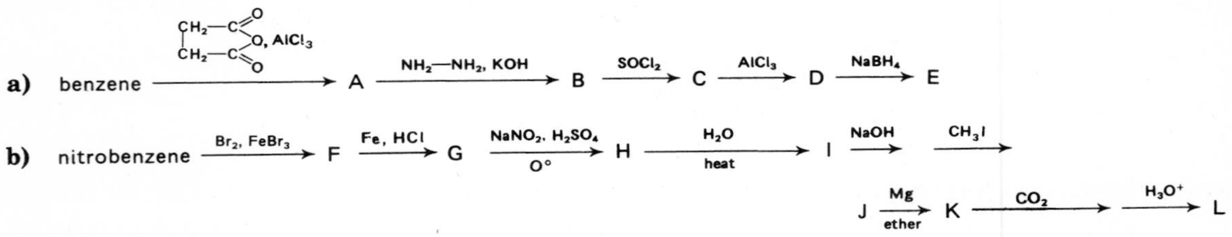

a) benzene $\xrightarrow{\substack{CH_2-C\diagup O \\ CH_2-C\diagdown O}, AlCl_3}$ A $\xrightarrow{NH_2-NH_2, KOH}$ B $\xrightarrow{SOCl_2}$ C $\xrightarrow{AlCl_3}$ D $\xrightarrow{NaBH_4}$ E

b) nitrobenzene $\xrightarrow{Br_2, FeBr_3}$ F $\xrightarrow{Fe, HCl}$ G $\xrightarrow[0°]{NaNO_2, H_2SO_4}$ H $\xrightarrow[heat]{H_2O}$ I $\xrightarrow{NaOH}$ $\xrightarrow{CH_3I}$

J $\xrightarrow[ether]{Mg}$ K $\xrightarrow{CO_2}$ $\xrightarrow{H_3O^+}$ L

12.26 How many functional groups and how many asymmetric centers does the porphyrin portion of vitamin B_{12} have?

12.27 Give formulas for these nitrogen heterocyclic compounds.

a) pyridine   **b)** pyrimidine   **c)** purine   **d)** pyrrole   **e)** imidazole

12.28 For each of the heterocyclic structures named in Problem 12.27, name one important biological compound in which it forms part of the structure.

12.29 Write resonance structures for the following (formulas in Section 12.11).

a) Martius yellow   **b)** Congo red

12.30 Provide explanations for the following experimental observations.

a) When benzene was treated with isobutylene and sulfuric acid only tert-butylbenzene was produced.

b) When treated with bromine at low temperatures, compound 1 gave compound 2 in good yield.

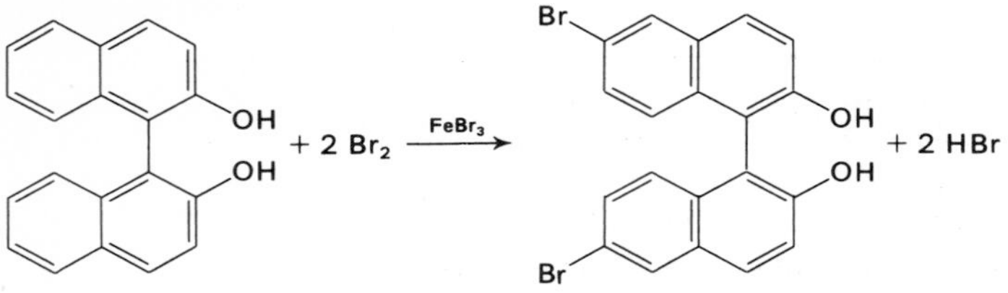

(1)   (2)

c) Nitration of N,N,N-trimethylanilinium nitrate gave m-nitrated product.

d) When aniline is treated with dilute nitric acid at elevated temperature, a mixture of o- and p-nitroanilines is produced. With more concentrated nitric acid, the m-isomer dominates.

e) When benzene is treated with one mole of methyl chloride and aluminum chloride, a mixture of benzene, toluene, and xylenes is obtained. When benzene is treated with acetyl chloride and aluminum chloride, acetophenone (methyl phenyl ketone) is the only product.

f) Bromination of tert-butylbenzene gives no o-substituted, but largely p- and a little m-substituted product.

CHAPTER 13

Carbohydrates and Sugar-Containing Compounds

Every day most of us use two rather pure chemical compounds. You are already extremely familiar with common table salt, which is nearly pure sodium chloride. You may not realize that another common substance you eat is actually one of the purest chemicals available. The white crystalline table sugar is highly refined sucrose, whose molecular formula is $C_{12}H_{22}O_{11}$. The compound contains a large number of hydroxyl groups and therefore the very high solubility of sucrose in water is not surprising. Fudge is sucrose that has been crystallized from water in the presence of delicious, occluded impurities.

The sweet taste of sucrose is characteristic of many other compounds which have similar structures. Lactose, an isomer of sucrose, is the sweetener found in milk. Fructose $C_6H_{12}O_6$, which makes up half of the sucrose structure, is somewhat sweeter to the taste than sucrose. Sugars belong to the family of natural compounds called **carbohydrates**. Because of the sweet taste of some of the compounds, all of the carbohydrates are referred to as sugars or **saccharides**.

This chapter treats first the carbohydrates whose structures are principally composed only of saccharide units. Many biological compounds which perform essential functions contain saccharides bonded to other types of structures. Some of these substances are already familiar to you and others are new. The later sections of the chapter are devoted to the structures of these sugar-containing compounds and some of their reactions and biological functions.

Many of the molecules we shall examine are large and multifunctional. In studying them, keep your eye on the portion of the molecule which undergoes reaction. Often, in the reactions of carbohydrates, only one or two of the many functional groups are involved.

13.1 CARBOHYDRATES The family of carbohydrates contains a large number of naturally occurring compounds whose principal functional group is an aldehyde or its derivative, such as an acetal, which is readily hydrolyzed to a parent aldehyde. A few examples are ketones or ketone derivatives. Carbohydrates are polyhydroxy aldehydes, or ketones, in which the distinguishing feature is the very high ratio of oxygen to carbon atoms, 1:1 or very nearly. The term carbohydrate is derived from the fact that, by chance, some of the earliest studied compounds possessed the general formula $C_n(H_2O)_n$—in effect, a carbon hydrate. Many carbohydrates are now known that do not possess this general formula.

The presence of the many hydroxyl groups along with the carbonyl group determines most of the chemical and physical properties of this family. All but the large polymers are water-soluble or strongly hydrophilic (water-loving).

Polymeric carbohydrates, such as **starch** and **cellulose**, furnish a substantial portion of the food supply for animals. Starch occurs principally in roots and seeds, for example in potatoes and grain, where it serves as nutrient for young plants. Cellulose provides the structural material for grass and for plants such as trees, whose woody parts are more than 50% cellulose. Cotton and linen fibers are almost pure cellulose.

Starch is easily hydrolyzed to smaller digestible carbohydrates by most animals. On the other hand, the hydrolysis of cellulose requires a special enzyme which bacteria in ruminants, such as the cow, possess.

Carbohydrates are divided into three groups: **monosaccharides**, which contain only one carbonyl function; **disaccharides** and **oligosaccharides**, which consist of two or several monosaccharides connected by acetal linkages; and **polysaccharides**, which have many units connected by acetal linkages.

13.2 MONOSACCHARIDES

Monosaccharides are straight-chain aldehydes or ketones with hydroxyl groups on all, or nearly all, the other carbons. The structures have many asymmetric carbons. Because of the many diastereoisomers possible, each diastereoisomeric monosaccharide is known by a common name, which usually ends in "ose." The monosaccharides are also classified according to the type of carbonyl group, aldo or keto, and the number of carbons in the chain, tri-, tetra-, pent-, or hexose. Thus a five-carbon sugar with an aldehyde group is an **aldopentose**, and a six-carbon ketone sugar is a **ketohexose**.

The common monosaccharides found in nature have three to eight carbons. The smallest one is **glyceraldehyde**, an aldotriose. Other important ones in our food supply and metabolism are **glucose** ("dextrose" or "dextrin") and **fructose** ("levulose"). Glucose is an aldohexose, and fructose, a ketohexose. **Ribose**, an aldopentose, is a constituent of many important coenzymes (see Section 13.9) and of nucleic acids which contain genetic information and direct the biosynthesis of proteins (see Sections 13.10 and 13.11).

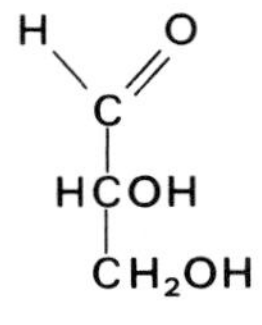

glyceraldehyde
(aldotriose)

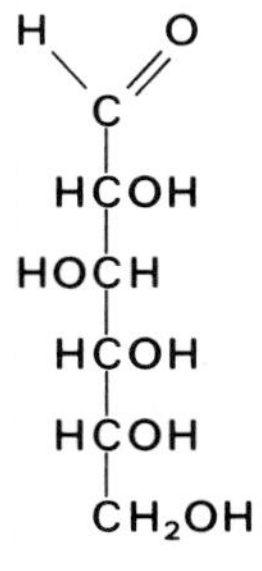

glucose
(aldohexose)

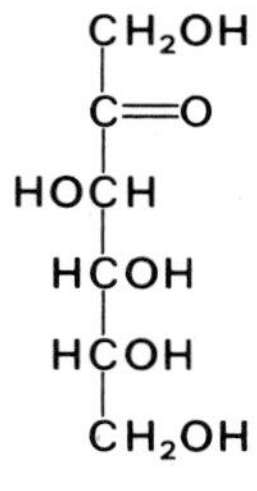

fructose
(ketohexose)

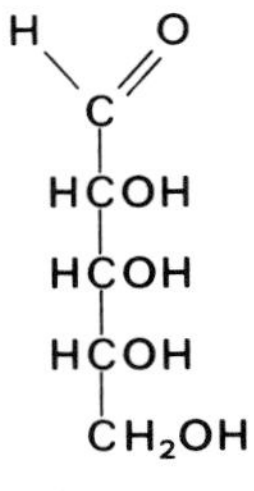

ribose
(aldopentose)

Carbohydrate molecules have one or more asymmetric carbons, and only one enantiomer is commonly found in nature.

Problem **13.1** Star the asymmetric carbon atoms of glucose. Predict from their number the total number of stereoisomers of glucose. Include glucose in your number.

13.3 DIASTEREOISOMERS OF MONOSACCHARIDES

The configurations of the asymmetric carbons of the monosaccharides were extensively investigated early in the study of carbohydrates. In the late 19th century, by laborious purifications and ingenious chemical interconversions, Emil Fischer determined the configurations of the asymmetric carbons relative to each other in each compound. He also related the structure of each diastereoisomer to the structures of the glyceraldehyde enantiomers.

The compounds which are configurationally related to D-glyceraldehyde were historically grouped as a family, the D-sugars. The enantiomers of the D-compounds, related to L-glyceraldehyde, were given the prefix L. As we observed with amino acids, the designation of D or L names a configurational family, and does not indicate the direction of rotation of polarized light. The prefix D given to a carbohydrate indicates that the *highest* numbered asymmetric carbon contains the same configuration as the single asymmetric center of D-glyceraldehyde. The prefix L indicates that the highest numbered asymmetric center possesses the same configuration as L-glyceraldehyde.

$$
\begin{array}{ccc}
CH{=}O & R & CH{=}O \\
 & & (CHOH)_2 \\
H{-}C{-}OH & H{-}C{-}OH & H{-}C{-}OH \\
CH_2OH & CH_2OH & CH_2OH
\end{array}
$$

D-glyceraldehyde   compound in   a D-aldopentose
                     D-family

A simple method of depicting the configurations of all the asymmetric carbons in the formula of a sugar is a **Fischer projection** formula. The three-dimensional formula is projected in a certain conventional way into two dimensions. In the convention, a three-dimensional formula can be translated into a two-dimensional drawing, and vice versa. The structure of the sugar is first written with the terminal hydroxyl group at the bottom and the carbonyl carbon at the top. The bonds of the carbon chain thus run vertically, and bonds to the hydrogens and hydroxyl groups attached to asymmetric carbons are shown horizontally, one on one side and the other on the other side of the carbon. By convention the horizontal bonds of the Fischer projection are imagined to extend *above* the plane of the paper and the vertical bonds to extend *below* the plane of the paper. In this way a two-dimensional drawing can be translated into a three-dimensional structure. The three-dimensional structures and Fischer projections of glyceraldehyde enantiomers illustrate the convention.

$$\begin{array}{cccc}
\text{CH}{=}\text{O} & \text{CH}{=}\text{O} & \text{CH}{=}\text{O} & \text{CH}{=}\text{O} \\
\text{H}{-}\text{C}{-}\text{OH} & \text{H}{-}\text{C}{-}\text{OH} & \text{HO}{-}\text{C}{-}\text{H} & \text{HO}{-}\text{C}{-}\text{H} \\
\text{CH}_2\text{OH} & \text{CH}_2\text{OH} & \text{CH}_2\text{OH} & \text{CH}_2\text{OH} \\
\text{three-} & \text{Fischer} & \text{three-} & \text{Fischer} \\
\text{dimensional} & \text{projection} & \text{dimensional} & \text{projection}
\end{array}$$

D-(+)-glyceraldehyde   L-(−)-glyceraldehyde

Fischer projection formulas cannot be rearranged in the way a three-dimensional formula can. To compare formulas of enantiomers, you may rotate a Fischer projection *only* in the plane of the paper and *only through 180°.*

Problems

13.2 a) Draw Fischer projections of the two aldotetroses that belong to the D-family.

b) Draw Fischer projections of the two L-aldotetroses, the enantiomers of the compounds drawn in (a).

13.3 Some aldopentoses are shown as Fischer projections. Identify each as belonging to the D or L family. Identify a pair of enantiomers.

$$\begin{array}{cccc}
\text{CH}{=}\text{O} & \text{CH}{=}\text{O} & \text{CH}{=}\text{O} & \text{CH}{=}\text{O} \\
\text{H}{-}\text{C}{-}\text{OH} & \text{H}{-}\text{C}{-}\text{OH} & \text{HO}{-}\text{C}{-}\text{H} & \text{HO}{-}\text{C}{-}\text{H} \\
\text{H}{-}\text{C}{-}\text{OH} & \text{H}{-}\text{C}{-}\text{OH} & \text{HO}{-}\text{C}{-}\text{H} & \text{H}{-}\text{C}{-}\text{OH} \\
\text{HO}{-}\text{C}{-}\text{H} & \text{H}{-}\text{C}{-}\text{OH} & \text{H}{-}\text{C}{-}\text{OH} & \text{HO}{-}\text{C}{-}\text{H} \\
\text{CH}_2\text{OH} & \text{CH}_2\text{OH} & \text{CH}_2\text{OH} & \text{CH}_2\text{OH} \\
\text{A} & \text{B} & \text{C} & \text{D}
\end{array}$$

The configurations of the asymmetric carbons of the aldotetroses were established by syntheses using separate enantiomers of glyceraldehyde as starting materials. The method of synthesis leaves untouched the configuration of the original asymmetric carbon of glyceraldehyde, which is the highest numbered asymmetric carbon in the compound. The configuration of this particular carbon in all larger saccharides produced from D-glyceraldehyde is the same as the configuration of the D-glyceraldehyde carbon. Thus all diastereoisomeric sugars and related compounds formed in the synthesis given below from D-glyceraldehyde belong to the D-family.

In the synthesis, two nitriles are formed from one enantiomer of glyceraldehyde by the addition of HCN to the carbonyl group. A second asymmetric center is created, and two diastereoisomeric nitriles are produced. Hydrolysis of the nitriles gives diastereoisomeric acids which form lactones. A reduction of the lactones by sodium amalgam produces the two D-aldotetroses, D-threose and D-erythrose. The synthesis is shown with Fischer projection formulas.

*C = asymmetric carbon

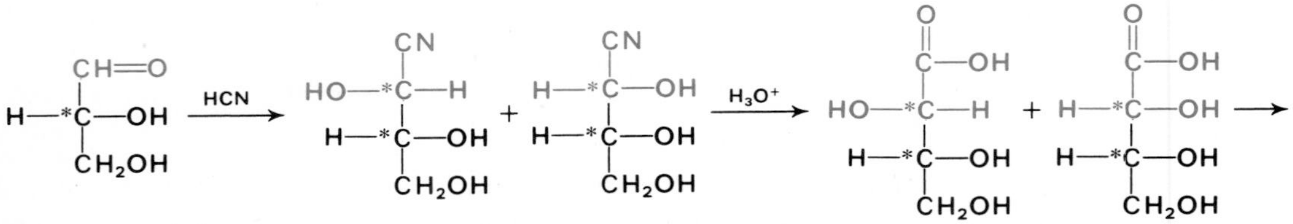

D-glyceraldehyde   diastereoisomeric D-nitriles   diastereoisomeric D-acids

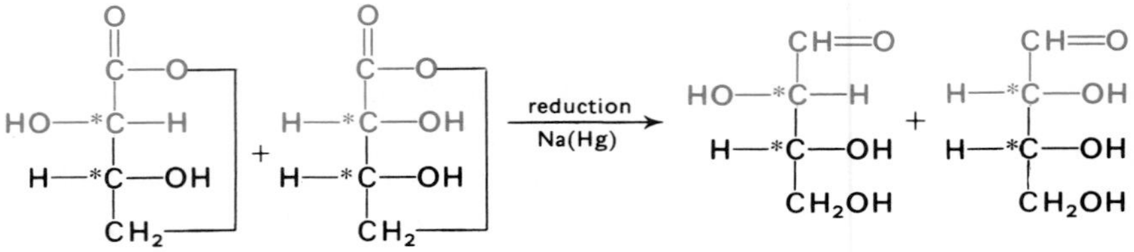

diastereoisomeric D-lactones   D-(−)-threose   D-(−)-erythrose diastereoisomeric aldotetroses

To establish the configurations of asymmetric centers in the two tetroses, the diastereoisomers are separated and each is oxidized separately by nitric acid to the corresponding tartaric acid. The configurational assignments for the newly created asymmetric centers are established by the optical activity of the tartaric acid diastereoisomer produced. Tetrose A is oxidized to an optically active tartaric acid. Therefore tetrose A must have the formula in which the hydroxyl groups in the Fischer projection are on opposite sides of the carbon chain (i.e., threose). Tetrose B on oxidation gives *meso*-tartaric acid. Tetrose B must have the two hydroxyls on the same side of the carbon chain, a formula which has a potential plane of symmetry (i.e., erythrose).

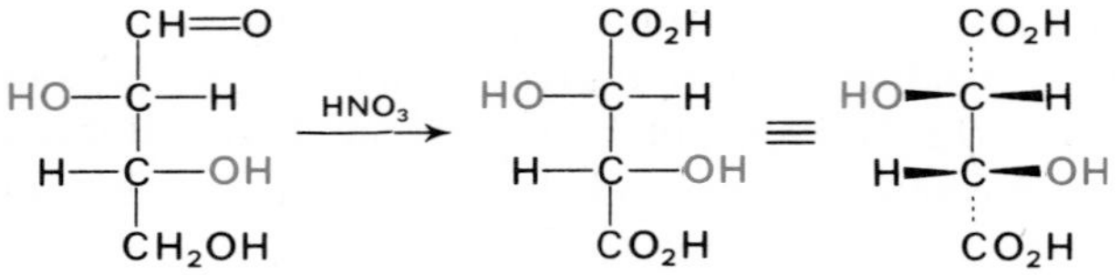

D-(−)-threose   optically active tartaric acid, D-(−) 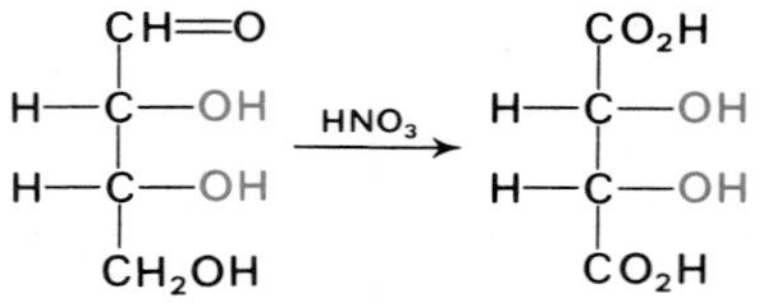   D-(−)-erythrose   *meso*-tartaric acid

Similarly, from L-(−)-glyceraldehyde, L-(+)-threose and L-(+)-erythrose were produced.

13.4 Write out the synthesis and proof of structure for the L-aldotetroses from L-glyceraldehyde.

Monosaccharides that contain from three to eight carbons are found in nature. Of these, D-ribose and D-glucose occur in quantity as major constituents of larger molecules, and D-glyceraldehyde and D-fructose appear as intermediates in metabolism.

The structures of the *aldoses* D-ribose (an aldopentose) and D-glucose (an aldohexose) were related to D-glyceraldehyde by successive elongation, hydrolysis, reduction to pentoses, and oxidation of the pentoses in sequences similar to those shown above. The configurations of the #4 carbon in ribose and #5 carbon in glucose provide the configurational family name (D or L).

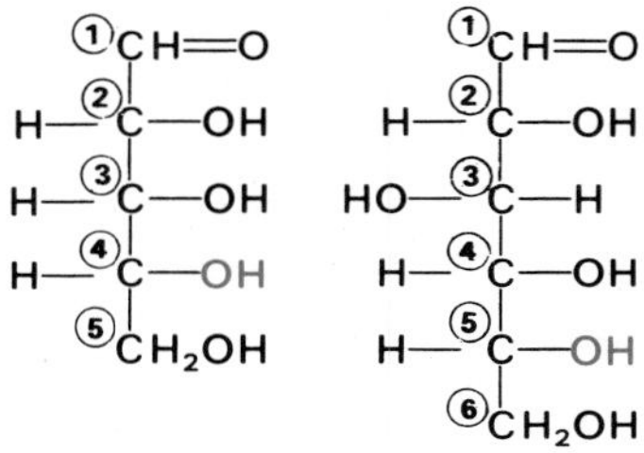

D-(−)-ribose   D-(+)-glucose

Problems **13.5** Examine the structures of ribose, glucose, and the two D-tetroses. Which tetrose would you use to start the synthesis of D-ribose? Of D-glucose?

13.6 a) Using D-erythrose, outline the steps required to confirm the configuration of D-ribose.

b) Continue the outline for steps to confirm the configuration of D-glucose.

Two biologically important ketoses are D-fructose and dihydroxyacetone. Note that the three asymmetric carbons of D-fructose are identical in configuration to the three corresponding carbons in D-glucose. The interconversion of these two hexoses, glucose and fructose, plays an important role in carbohydrate metabolism (see Sections 13.5b and 13.6).

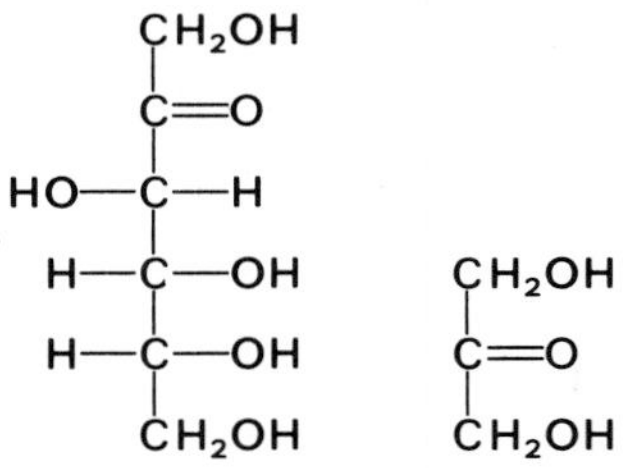

D-(−)-fructose   dihydroxyacetone

Problem **13.7** How many stereoisomers of aldopentoses are there? Draw them with Fischer formulas. Label the enantiomeric pairs.

13.4 HEMIACETALS AND ACETALS OF GLUCOSE— GLUCOSIDES

The reactions of monosaccharides in the laboratory are those expected of alcohols and carbonyl compounds. The interactions of hydroxyl and carbonyl groups in the same molecule are particularly important and provide carbohydrates with many of their unique properties.

In hot methanol with dry hydrogen chloride, glucose gives an acetal which contains only one methoxyl group. The acetal is formed from the internal hydroxyl on the #5 carbon of the glucose and from one external methanol molecule.

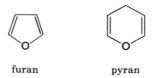

D-glucose   Fischer projection   Haworth projection

D-glucopyranose
(intramolecular hemiacetal)

In converting from a Fischer to a Haworth projection, all H and OH on the left in the Fischer are up in the Haworth; those on the right are down. The ring O is placed on the backside. For a D-sugar the CH$_2$OH group is up.

methyl D-glucopyranoside
(an acetal)

Cyclic hemiacetals (like lactones) form easily with hydroxyl groups four or five carbons away from the carbonyl group, giving five- or six-membered ether rings. Glucose normally uses the OH of carbon #5 to form a six-membered cyclic acetal, but the five-membered is also known. The two structures are named as **furan**-derived for the five-membered ether ring, or as **pyran**-derived for the six-membered ether ring.

furan   pyran

The name methyl D-gluco*pyranoside* indicates the presence of the six-membered ring in the acetal structure, and D-gluco*furanoside* indicates a five-membered ring.

With the formation of the hemiacetal or acetal, a new asymmetric center is created at carbon #1, and two methyl D-glucopyranosides are obtained. Diastereoisomers that differ in configuration only at the acetal or hemiacetal center are known as **anomers**. The formation of the two diastereoisomers of the acetal is depicted below. Formulas in the chair form of a six-membered ring are a better representation of the cyclic structure than either a Fischer or a Haworth projection is. The groups written up and down are the same as a Haworth projection.

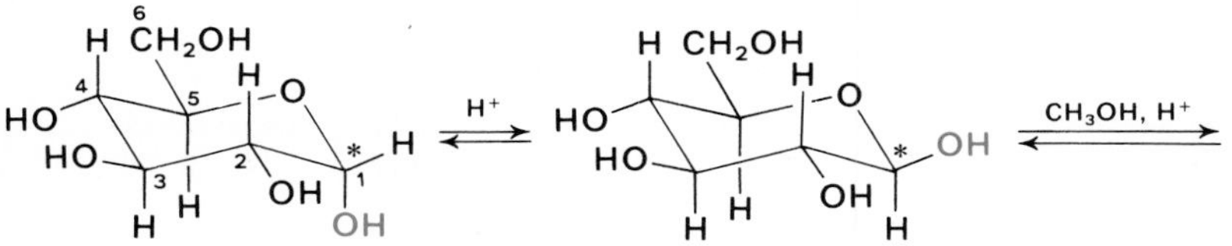

α-D-glucopyranose
(an α-hemiacetal—
*—OH down)

β-D-glucopyranose
(a β-hemiacetal—
*—OH up)

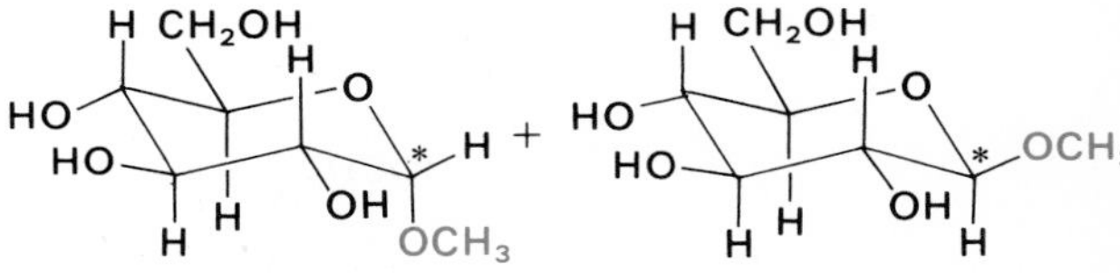

methyl α-D-glucopyranoside
(an α-acetal —OCH₃ down)

methyl β-D-glucopyranoside
(a β-acetal —OCH₃ up)

In the structures of the cyclic hemiacetal and acetal of D-glucose, the hydrogens of carbons #2, #3, #4, and #5 are all axial in the cyclohexane chair conformation. The OH groups and more bulky CH_2OH are all in the more favorable equatorial positions. The β-acetal has the OCH_3 of carbon #1 also in the equatorial position. The β-D-glucopyranose molecule in the conformation with all its bulky groups equatorial is probably the most stable single structure that can be written for the hemiacetals of the D-aldohexopyranoses. In that form D-glucose is the most stable diastereoisomer of the D-aldohexoses.

In a Fischer projection, the hemiacetal hydroxyl of the α-anomer is on the same side as the oxygen of C-4 or C-5 which forms the ether link to the carbonyl group, for D-glucose, D-fructose and D-ribose on the right. In the cyclic formula the α-hydroxyl is down.

Glucose crystallizes in either the α-hemiacetal structure or the β-hemiacetal structure. When dissolved in water, the crystalline hemiacetal equilibrates with trace amounts of the open-chain aldehyde and further with the other hemiacetal. This equilibration involves the aldehyde as an intermediate, and is called **mutarotation**. The equilibration reaction can be followed by the change in the optical rotation of a solution made from either anomer. The mutarotation is both acid- and base-catalyzed.

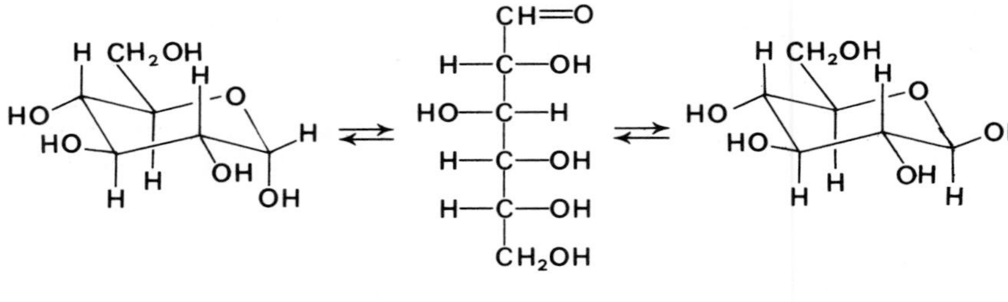

α-D-glucopyranose
$[\alpha]_D^{20} = +112°$ (in water)

D-glucose

β-D-glucopyranose
$[\alpha]_D^{20} = +19°$ (in water)

equilibrium mixture $[\alpha]_D^{20} = +52°$ (in water)

The cyclic acetals of carbohydrates are given the general class name **glycosides**. Derivatives of specific sugars have the following specific names: glucose, **glucoside**; fructose, **fructoside**; ribose, **riboside**.

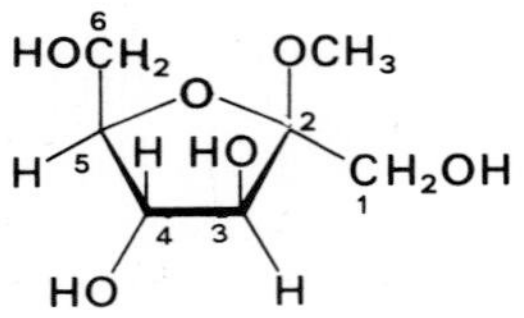

methyl β-D-fructofuranoside

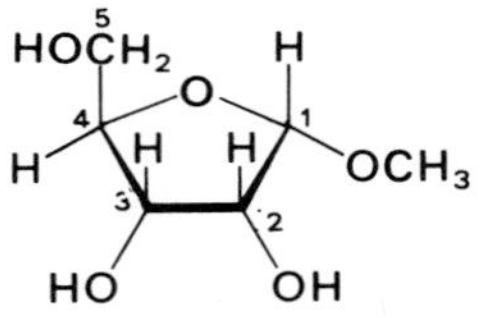

methyl α-D-ribofuranoside

It is interesting that the primary hydroxyl group does not enter into the cyclic acetal in most natural products, but is left available for other use.

Problems

13.8 The rotations of separate compounds in the same solution are additive. Calculate the percent of each anomer present in the equilibrium established for glucopyranose in water as shown above.

13.9 Like the equilibrium for glucose, an equilibrium exists between α- and β-ribofuranose and the aldehyde form of ribose. Draw structures for the two cyclic hemiacetals (Haworth projections) and for the open-chain aldehyde of the D-compounds.

13.10 As with an ordinary alcohol, all the free hydroxyl groups of methyl β-D-glucopyranoside react with acetic anhydride to form esters. How many acetic anhydrides per glucoside are needed? Write the equation for the reaction, using chair formulas for the cyclic acetals.

13.5 OTHER REACTIONS OF MONOSACCHARIDES

a) Oxidation-reduction

Although most monosaccharides are in hemiacetal form, an equilibrium exists between the hemiacetal and the open-chain aldehyde. With the small concentration of aldehyde present, sugars having this equilibrium give the usual aldehyde group reactions. Oxidation of the aldehyde is brought about by mild oxidizing agents, such as bromine, silver, or copper (II) complex ions.

$$HO-CH_2(CHOH)_4-CH=O \xrightarrow{Ag(NH_3)_2^+ \; OH^-} HO-CH_2(CHOH)_4-CO_2^- + Ag_{(s)}$$

Sugars which have acetal instead of hemiacetal groups are *not* oxidized by these reagents without prior hydrolysis of the acetal group. Such sugars are called nonreducing sugars, in contrast to the hemiacetals, which do react with oxidizing agents and are reducing sugars.

Nitric acid oxidizes both the carbonyl and the primary alcohol groups to carboxylic acids, but the yields are low because carbon–carbon bonds are also cleaved in competing reactions.

$$HO-CH_2(CHOH)_4CH=O \xrightarrow{HNO_3} HO_2C(CHOH)_4CO_2H$$

The reactions which convert both the end functional groups of a sugar to the same function, that is, oxidation to carboxylic acids or reduction to alcohols, have played a large role in the determination of configuration of the asymmetric centers. With matching ends, the possibility of *meso* structures gives considerable information about the original compound (see Section 13.3).

The aldehyde and ketone groups can be reduced to hydroxyl by sodium borohydride. The reduction of ketones is nonstereospecific, which means that a mixture of diastereoisomers is obtained.

$$HO-CH_2(CHOH)_4CH=O \xrightarrow{NaBH_4} HO-CH_2(CHOH)_4-CH_2OH$$

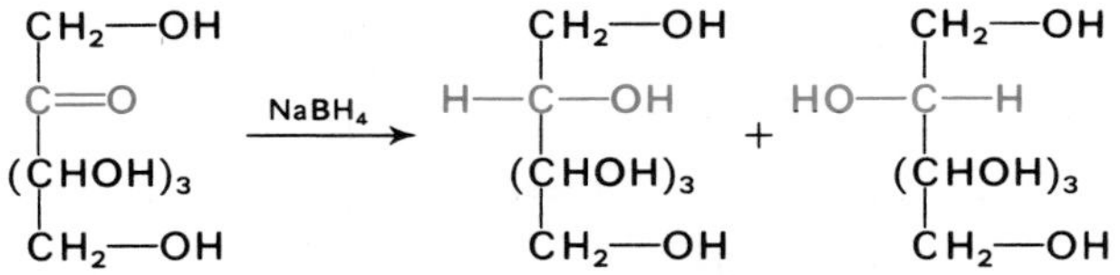

diastereoisomers

b) Enolization of saccharides

As with ordinary aldehydes and ketones, enolization occurs with saccharides. In the presence of a base the open-chain carbonyl group is in equilibrium with its enol.

When glucose is treated with a base, it forms an equilibrium mixture of glucose, fructose, and mannose, an aldohexose. The enol of D-glucose may be converted to any one of the three carbonyl compounds, D-glucose, D-fructose, and D-mannose.

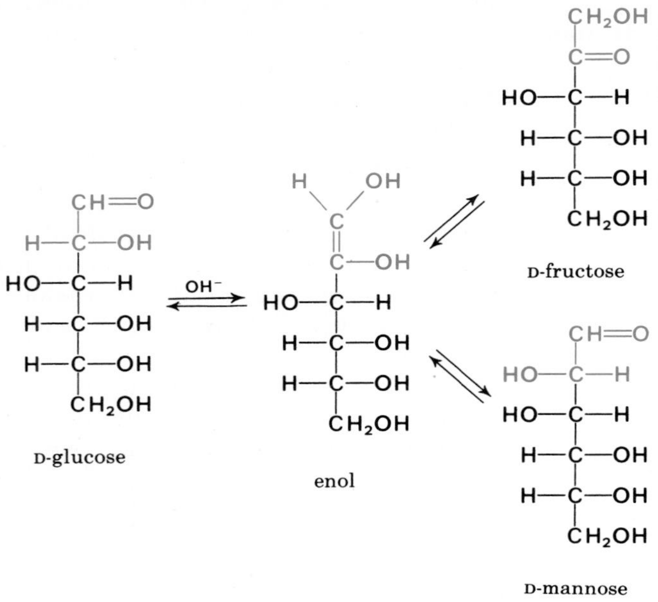

```
  CH=O              H    OH              CH2OH
H—C—OH              C                    C=O
HO—C—H       OH⁻    C—OH         HO—C—H
H—C—OH   ⇌   HO—C—H              H—C—OH
H—C—OH             H—C—OH              H—C—OH
  CH2OH            H—C—OH                CH2OH
                     CH2OH
 D-glucose                              D-fructose
                     enol
                                          CH=O
                                        HO—C—H
                                        HO—C—H
                                        H—C—OH
                                        H—C—OH
                                          CH2OH

                                        D-mannose
```

The fact that this interconversion occurs tells us something about the configurations of the asymmetric carbons of D-mannose. The carbons of all three D-saccharides not involved in the enolization must have identical configurations. Therefore D-mannose must differ from D-glucose only in the configuration of the #2 carbon.

13.6 BIOSYNTHESIS OF GLUCOSE BY AN ALDOL CONDENSATION OF TRIOSES

The biosynthesis of a hexose is accomplished by a simple aldol condensation of two trioses, an aldose and a ketose. The analogous aldol reaction between the simpler compounds, propanal and acetone, occurs in the laboratory, where it is in competition with the condensation of propanal with itself (Section 11.10).

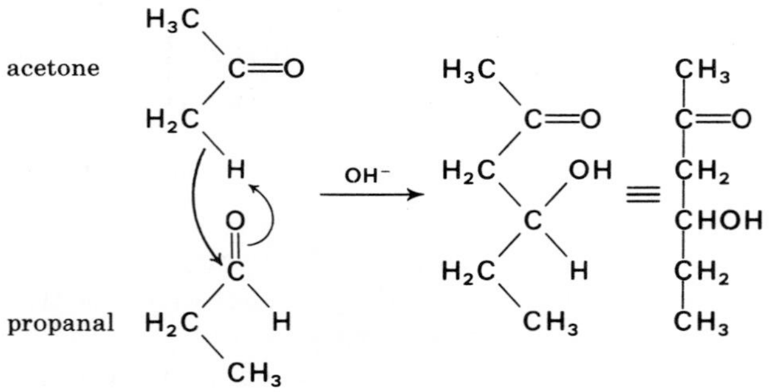

```
          H3C
acetone       C=O
          H2C                H3C                  CH3
              H       OH⁻        C=O               C=O
              |     ⟶      H2C       OH           CH2
              O                    C          ≡    CHOH
              ‖                                    CH2
              C              H2C    H              CH3
propanal  H2C    H                 CH3
              CH3
```

In the synthesis of D-fructose 1,6-diphosphate, the carbon chain is increased when the carbanion at one of the α-carbons of dihydroxyacetone phosphate adds as a nucleophile to the carbonyl carbon of D-glyceraldehyde 3-phosphate. The presence of the extra hydroxyl and phosphate groups in the two starting materials complicates the writing of the reaction, but should not obscure its simplicity.

dihydroxyacetone phosphate

D-glyceraldehyde 3-phosphate

D-fructose 1,6-diphosphate (ketone form)

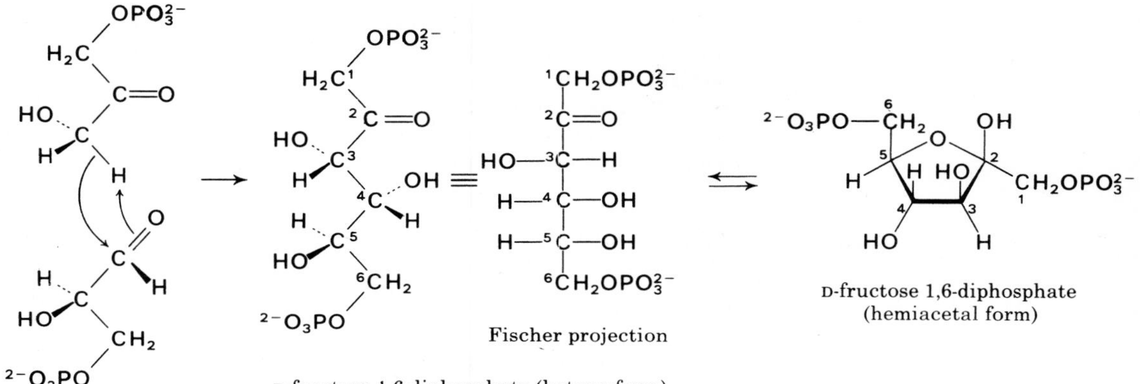

Fischer projection

D-fructose 1,6-diphosphate (hemiacetal form)

In the enzyme-catalyzed reaction, the starting trioses possess only one asymmetric carbon between them; the other two asymmetric centers in fructose are created by the reaction which forms the new carbon–carbon bond. The configurations of the two new centers are, of course, determined by the enzyme which orients the two molecules and removes the appropriate α-hydrogen from dihydroxyacetone phosphate. As in other enzyme-catalyzed reactions the proton removal is stereospecific; only one of the two protons on the α-carbon is removed.

After loss of the phosphate group on carbon #1 in the reaction shown above, D-fructose 6-phosphate is subsequently isomerized through the common enol to glucose 6-phosphate. Remember that the remaining three asymmetric centers are identical in fructose and glucose (Section 13.5).

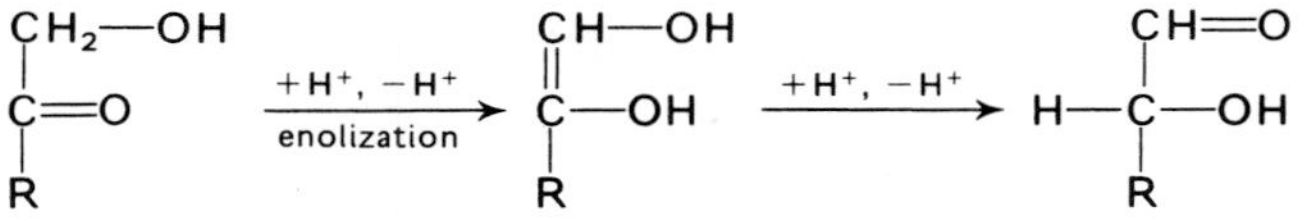

fructose   enol   glucose

By these reactions excess trioses from the metabolism of various foodstuffs are converted to glucose. Glucose in turn is polymerized to glycogen (a polyglucose, see Section 13.8) and stored in the liver. Glycogen serves as the reserve fuel for mammals.

The aldol condensation of trioses to hexose is reversible. The route is reversed when quick energy is required and is supplied by the degradation of blood glucose. Glucose is isomerized to fructose, which is degraded by a reverse aldol to dihydroxyacetone and glyceraldehyde (Section 14.6). The trioses are also converted to pyruvate ion, which ultimately goes to acetate and CO_2.

Problem   **13.11** A 2-ketoheptose, sedoheptulose 1,7-diphosphate, is synthesized biologically by aldol condensation of erythrose 4-phosphate and dihydroxyacetone phosphate. Draw the formulas of the starting compounds (for erythrose see Section 13.3). Indicate the course of the aldol condensation. Give the formula for the product of the condensation without designating the configurations of the two new asymmetric centers.

13.7 DISACCHARIDES   A disaccharide consists of two simple sugar units bound together by a glycosidic (acetal) linkage. Of the naturally occurring disaccharides, sucrose and lactose are most important because of their food value to animals.

    Lactose is found in the milk of mammals. It consists of a D-glucose unit whose 4-hydroxyl is bound to the aldehyde carbon of a D-galactose unit (a galactopyranoside). Galactose is a diastereoisomer of glucose in which C-4 has a configuration opposite that of glucose. Lactose is a β-galactoside.

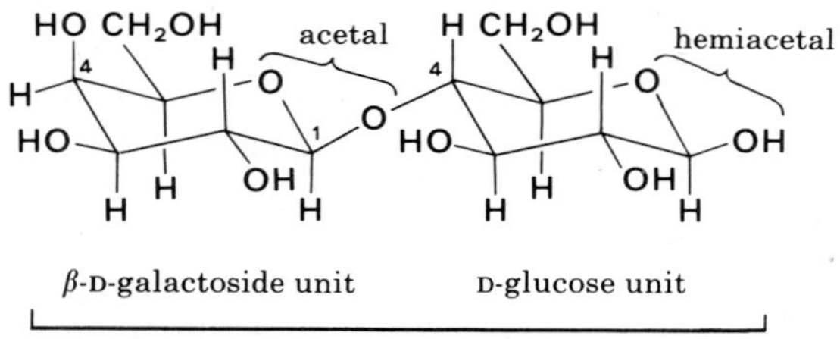

β-D-galactoside unit   D-glucose unit

lactose

    Table sugar is pure **sucrose**, a material found in sugar beets and sugar cane. Sucrose is composed of D-glucose and D-fructose joined through the carbonyl carbons of both hexoses. There is no hemiacetal or hemiketal group in sucrose, only an acetal connecting link.

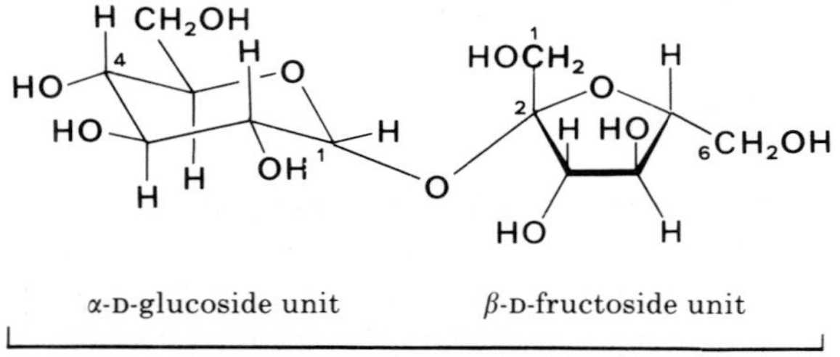

α-D-glucoside unit   β-D-fructoside unit

sucrose

Maltose and **cellobiose** are disaccharides obtained from the controlled hydrolysis of starch and cellulose respectively. Both contain two glucose units, but they differ in the configuration of the acetal linkage. Maltose has the α-glucosidic linkage, while cellobiose has the β-glucosidic linkage.

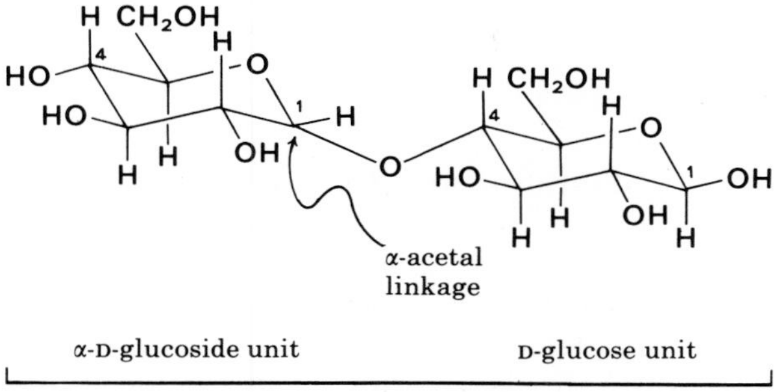

α-acetal linkage

α-D-glucoside unit   D-glucose unit

maltose

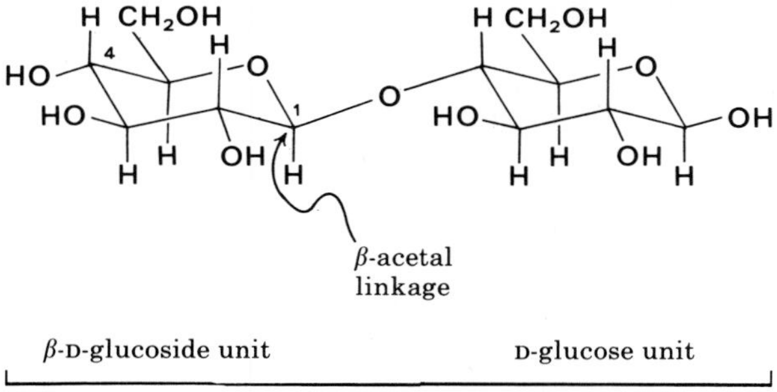

β-acetal linkage

β-D-glucoside unit   D-glucose unit

cellobiose

Disaccharides are hydrolyzed to monosaccharides by the action of acids or enzymes.

Problem   **13.12** Using Fischer projection formulas, write the structures for the products of the reactions for the hydrolysis by HCl of lactose, sucrose, and cellobiose to their respective monosaccharides.

13.8 POLYSACCHARIDES   Starch is a polymer of glucose. The glucose units are joined by acetal linkages in which the hydroxyl of C-4 is bound to the C-1 of the preceding glucose with an α-configuration.

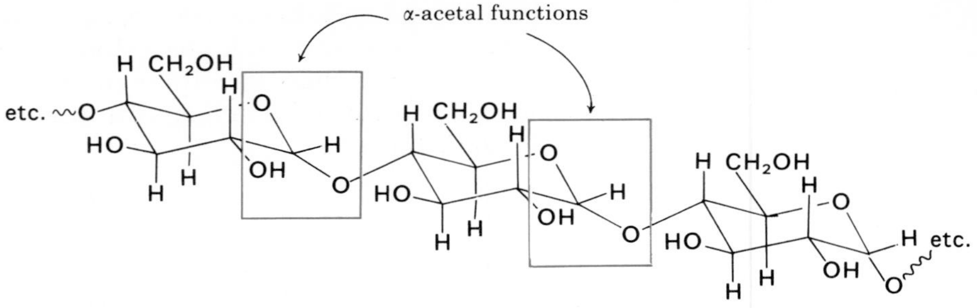

α-acetal functions

starch

Starch is found in roots, seeds, and fruits of plants (potato, corn, wheat, rice), where it serves as a nutrient source for new plants. Starch also serves as a nutrient for animals, where it is hydrolyzed to glucose units by enzymes which are able to act only on α-glucosidic linkages.

Animals use glucose as fuel or store it in another polymeric form, **glycogen**, the animal food reservoir. **Glycogen** is a polymer of glucose with α-linkages mainly between carbons #1 and #4. Glycogen has fewer units than starch and also contains many branches that involve 1,6-bonds between glucose units.

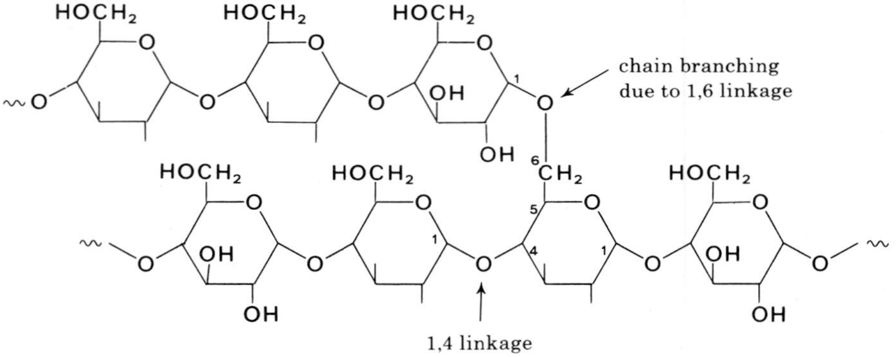

chain branching
due to 1,6 linkage

1,4 linkage

glycogen

Cellulose is the most abundant organic substance in living matter. It is found in woody and green parts of all plants and provides their structural material. About 50% of wood and 90% of cotton fiber are cellulose.

Cellulose is a polyglucose in which the glucosidic linkages are between the 4-hydroxyl of one glucose and the carbonyl carbon of another. The acetal linkages, however, are β-linkages, which make cellulose quite different from starch with its α-linkages. Grass- and leaf-eating animals have bacteria which produce an enzyme capable of hydrolyzing the β-glucosidic bonds to glucose.

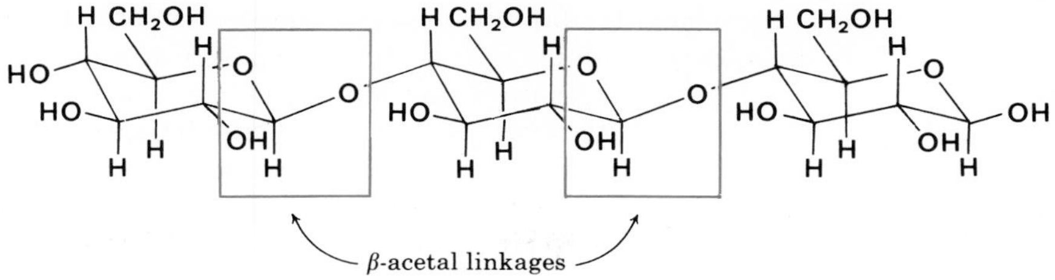

β-acetal linkages

cellulose

13.9 N-GLYCOSIDES OF D-RIBOSE IN THE STRUCTURES OF COENZYMES

Amines, like alcohols, are good nucleophiles. Suppose we have an amine instead of an alcohol in a glycoside structure of an aldose. The resulting compound is called an N-glycoside. In the laboratory such N-glycosides are usually unstable and are a curiosity.

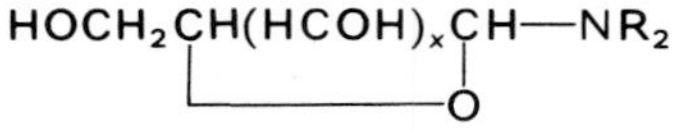

$$HOCH_2CH(HCOH)_xCH\!-\!NR_2$$

N-glycosidic linkage

In biological systems, however, certain N-glycosides of β-D-ribose form a part of an astonishing number of large molecules which play principal roles in genetic and metabolic processes. The sugar in such molecules seems to be largely restricted to D-ribose, just as the energy storage operation is restricted to D-glucose polymers.

The five-membered ring of β-D-ribofuranosides is basically rigid and flat. The ring oxygen reduces the number of eclipsed bonds on the ring.

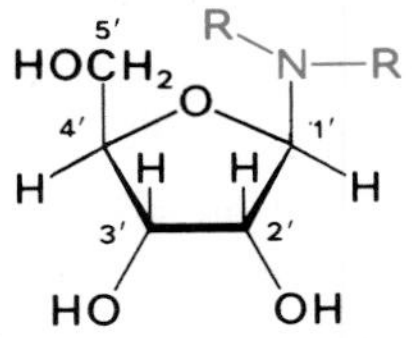

an N-β-D-ribofuranoside

The ribofuranoside has two adjacent hydroxyl groups on the lower side of the ring and the primary hydroxyl group on the upper side of the ring. The glycoside linkage is also on the upper side.

While the exact contribution of the ribose to the special performance of the large molecule may be obscure, the rigidity and position of functional groups make ribose distinct from glucose. Reexamine the α-D-glucopyranose molecule in this regard (Section 13.4), remembering the conformation and the flexibility of a six-membered ring. In the riboside molecule, the rigidity of the ring, the hydrogen-bonding potentialities of the 2′ and 3′-hydroxyls, the *cis*-geometry of 2′ and 3′-hydroxyls, and the positions of possible derivatives at 1′ and 5′-positions all must contribute to its evolutionary usefulness.

a) Adenosine triphosphate

By far the most frequently encountered N-glycoside in biochemical compounds is that composed of adenine and D-ribose, known by its common name adenosine. In the numbering of these important N-glycosides, the ordinary numbers are reserved for the attached heterocyclic amine, and primed numbers are used for the ribose portion. For the ribose itself, only the carbons are numbered and 1' is the potential aldehyde carbon.

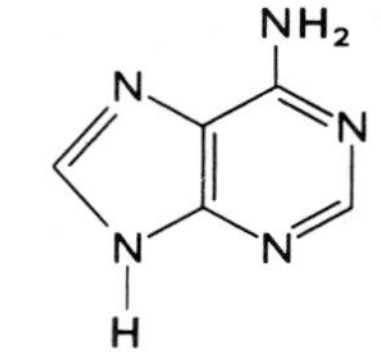

adenine

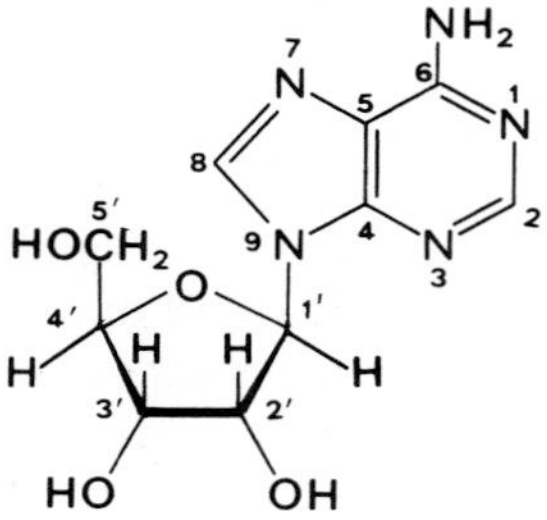

adenosine
(adenyl β-D-riboside)

The coenzyme, **adenosine triphosphate**, ATP, has a triphosphate group attached to the primary hydroxyl on the 5'-carbon of ribose.

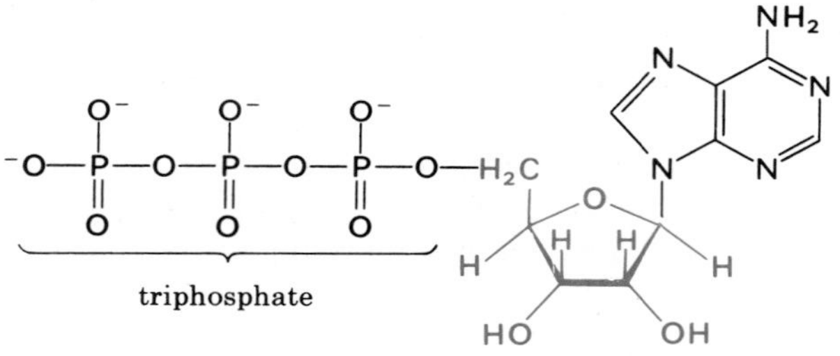

adenosine triphosphate (ATP)

The primary role of ATP as a coenzyme is to furnish the energy needed to bring about biochemical reactions which require energy to make them proceed. These reactions have products which are at a higher energy level than the reactants. We have seen an example in the production of high-energy acyl phosphate esters from carboxylate ions and ATP (Section 9.16). The acyl adenosine monophosphate is a transient high-energy intermediate in the conversion of a carboxylate ion and the thiol of coenzyme A to a thioester acyl-CoA and adenosine 5'-monophosphate (AMP). The reaction sequence is shown in Fig. 13.1.

Metabolic oxidation reactions give off energy. Accompanying many of these oxidations is a side reaction in which AMP or ADP is phosphorylated back to adenosine triphosphate. In this way the energy given off by the metabolic reaction is stored in the formation of the triphosphate linkage to be used as needed for reactions of synthesis that require energy. Adenosine triphosphate is the dollar in the currency of biochemical energy transfer and storage.

In other examples of reactions of adenosine triphosphate, one phosphate may be lost, or all three phosphates may be replaced. An

FIGURE 13.1   ROLE OF ATP IN THE FORMATION OF AN ESTER OF COENZYME A

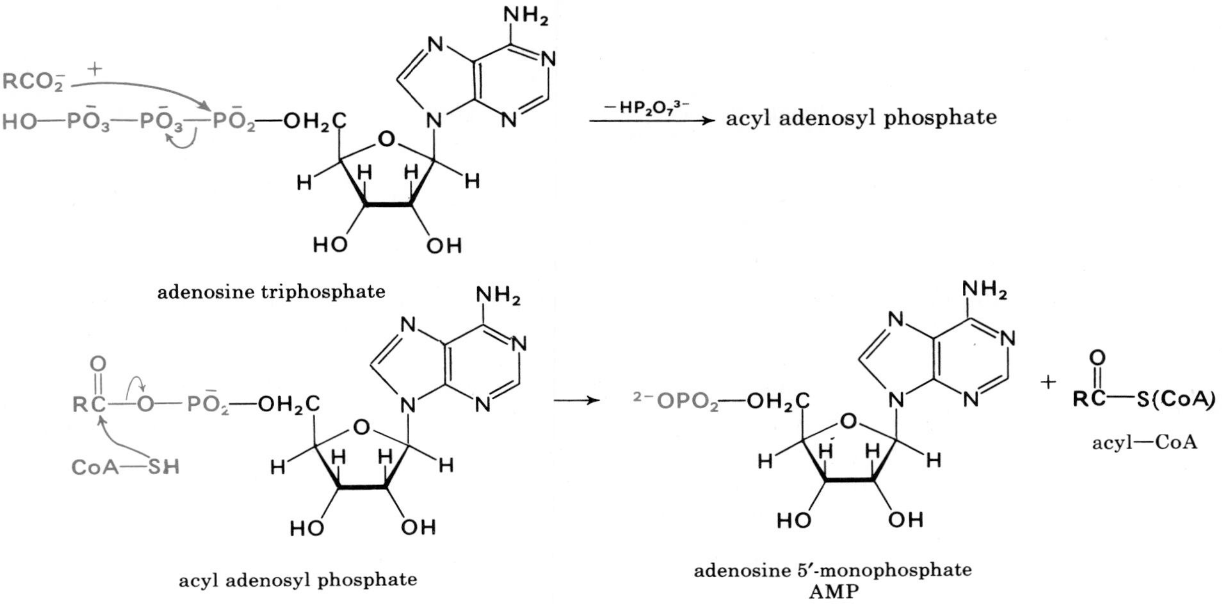

adenosine triphosphate

acyl adenosyl phosphate

adenosine 5'-monophosphate
AMP

example of the displacement of all three phosphate units is the biological transmethylation sequence (Section 6.10).

The acyl-transfer agent, coenzyme A, also incorporates adenosine diphosphate at the opposite end of the molecule from the thiol group. The exact role played by the nonreacting part of the large molecule is not known, but probably it binds the coenzyme to the enzyme in the right orientation using noncovalent forces.

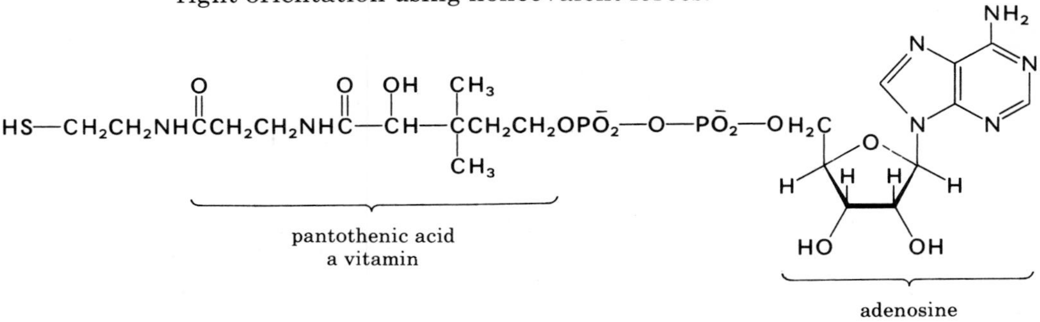

pantothenic acid
a vitamin

adenosine

coenzyme A

Adenosine and other N-glycosides are given the general class name nucleosides. These compounds of ribose are commonly known as **ribonucleosides**. The 5'-monophosphorylated riboside is called a ribonucleo**tide**. The formula for the ribonucleotide adenosine 5'-monophosphate is shown here.

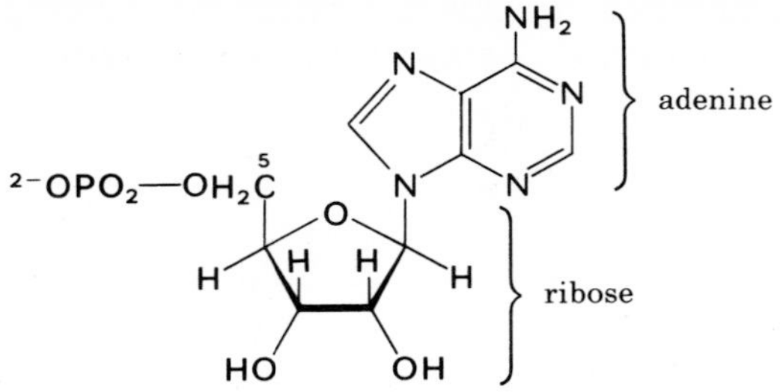

adenosine 5'-monophosphate AMP
(a ribonucleotide)

Problem **13.13** Hormones arriving via the bloodstream stimulate in the cell membrane the production from ATP of the nucleotide adenosine 3',5'-cyclic-monophosphate. This cyclic AMP transmits and amplifies the chemical signal delivered by the hormone. Draw the structure of the cyclic AMP.

b) Dinucleotides When two nucleotides are joined through the phosphate groups, making a diphosphate linkage, the resulting molecule is called a **dinucleotide**. The term dinucleotide seems to be used somewhat loosely, but most often it indicates two *ribo*nucleotides.

nitrogen base-ribose-$OP\bar{O}_3P\bar{O}_3$-ribose-nitrogen base

a dinucleotide

We have already used several dinucleotides in our study thus far. Can you recall them by this description?

We have used nicotinamide adenine dinucleotide (NAD$^+$) as an oxidizing agent for alcohols (Section 11.13). Can you construct the whole molecule from this name?

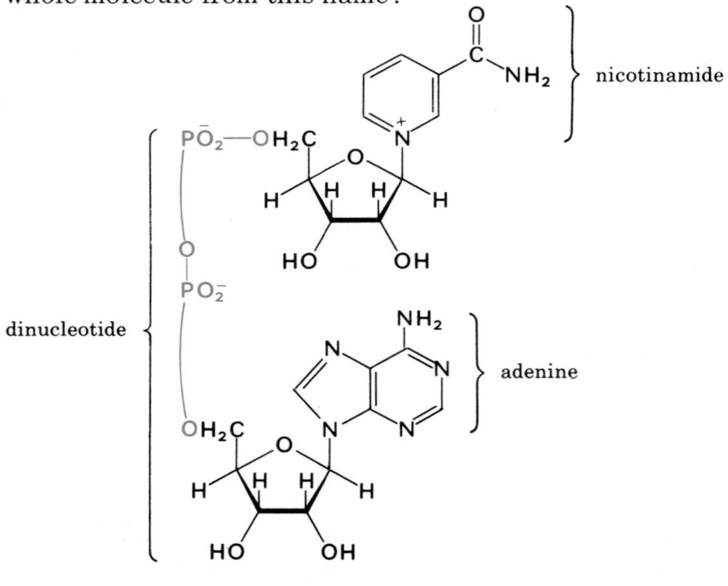

nicotinamide adenine dinucleotide (NAD$^+$)

13.10 POLYNUCLEOTIDES—THE NUCLEIC ACIDS, DNA AND RNA

One of the most exciting chapters in the unfolding story of the reduction of biology to organic chemistry involves the elucidation of the structures and functions of the nucleic acids. Nucleic acids, so called because they were first isolated from the nuclei of cells, are polymers with repeating units of nucleotides.

The polynucleotide chain has a backbone of alternating sugar and phosphate groups in which two hydroxyls of each sugar are bound to two different phosphates by ester linkages, and each phosphate is bound to two different sugar units to form an unsymmetrical diphosphate. A nitrogen base is attached to each sugar portion as an N-glycoside.

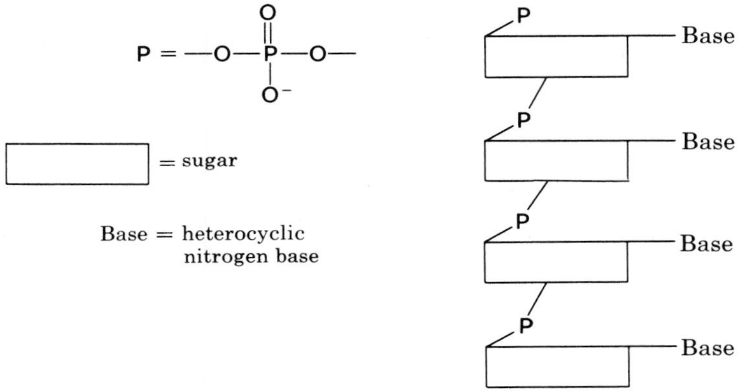

The phosphoric acid groups of the polynucleotides (nucleic acids) exist as anions above pH 4.

Two kinds of nucleic acids exist which differ in the structure of the sugar portion of the nucleotide. In **ribonucleic acids**, RNAs, the sugar is ribose throughout the polymer chain. For **deoxyribonucleic acids**, DNAs, the sugar is deoxyribose, a modified ribose in which the hydroxyl group on the 2′-carbon is missing. The prefix "deoxy" means "without oxygen."

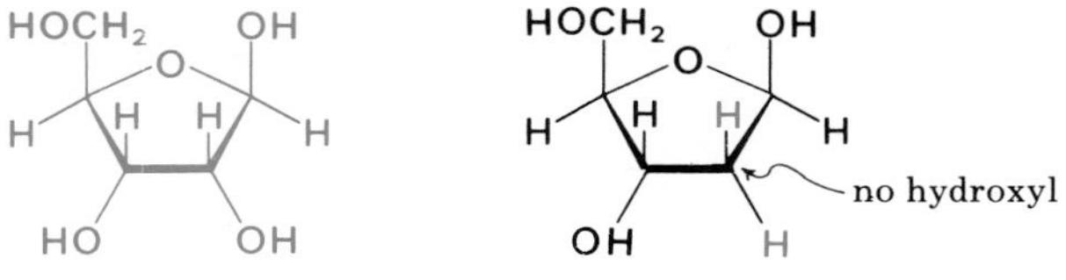

ribose   deoxyribose

In the nucleic acids the ribose (or deoxyribose) units are linked through one phosphate group which forms an ester with the 5′-hydroxyl of one ribose and with the 3′-hydroxyl of the next ribose.

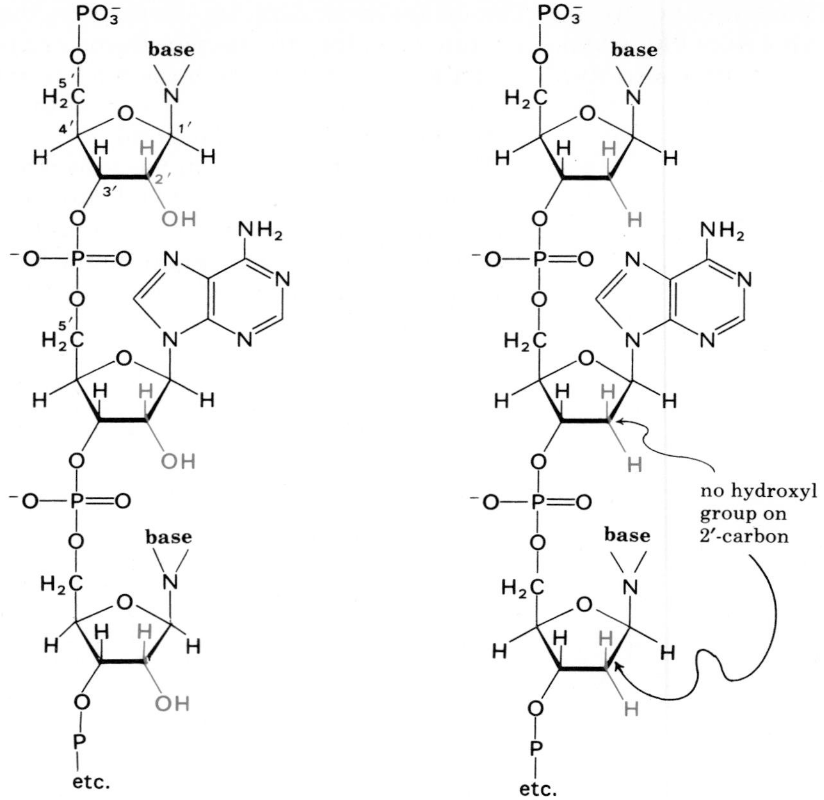

RNA model   DNA model

The same four bases occur in all the thousands of DNAs isolated from living cells. These bases are **adenine**, **guanine**, **thymine**, and **cytosine**, designated by the letters A, G, T, and C.

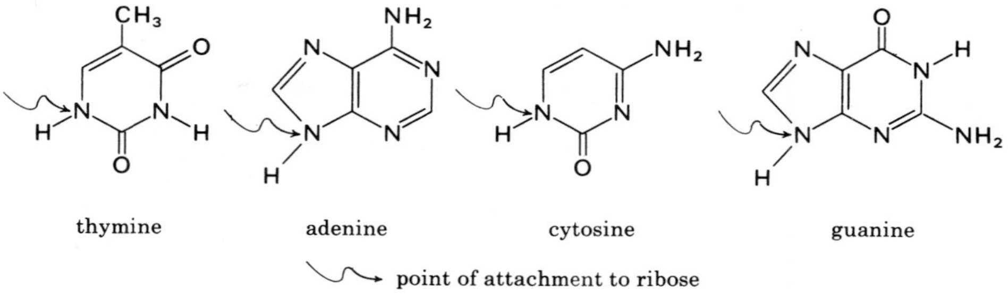

thymine   adenine   cytosine   guanine

point of attachment to ribose

Only the DNAs of some bacterial viruses are small enough to be extracted intact from cells, and these have molecular weights as high as 30 million.

13.14 Adenine, as shown above, is in a fully aromatic structure. The other three bases as written are not. Draw tautomeric structures for thymine, cytosine, and guanine in which the rings are fully aromatic.

The DNA in a cell is the chemical equivalent to the biological "gene." The gene determines the order of amino acids to be assembled into proteins, which ultimately control the biochemistry of the organism. This genetic information is stored in the structure of the DNA molecule. From there the information is retrieved in ordering protein synthesis. When cells divide, the gene (the DNA) is carried into the daughter cells, where it similarly orders protein synthesis.

The ribonucleic acids (RNAs) also contain only four major bases. Three—adenine, guanine, and cytosine—are identical to the bases in DNAs. In RNA uracil replaces thymine as the fourth major base.

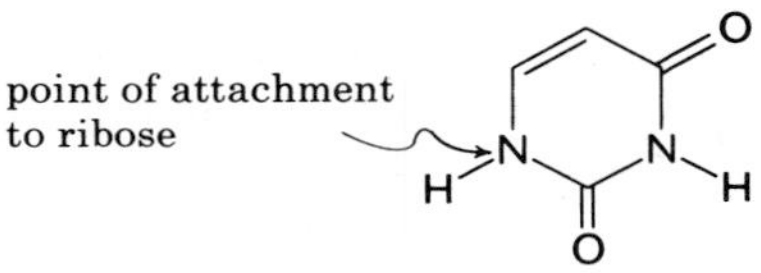

point of attachment to ribose

uracil

There are three distinctly different types of RNA, each of which performs a different function. The ribonucleic acids participate directly in the biosynthesis of proteins, as described in Section 13.11. Messenger RNA, *m*RNA, obtains the code for protein synthesis from DNA and serves as the template for peptide formation. Transfer RNA, *t*RNA, forms a bond with the amino acid to activate the acyl group, and delivers it to the growing peptide chain. As befits the transfer duties, *t*RNAs are relatively small polynucleotides with 75–90 units and 23–30,000 molecular weight. Each of the twenty different amino acids has its own *t*RNA. Ribosomal RNA, *r*RNA, provides the "active sites" on which the messenger RNA and acyl transfer RNA come together for polypeptide synthesis to occur. Ribosomal RNAs are large polynucleotides with molecular weights of up to one million.

13.11 TRANSFER RNA AS A REACTANT IN THE BIOSYNTHESIS OF A POLYPEPTIDE

The steps required to form a peptide (amide) linkage between two amino acids are very similar to the biological conversions of other carboxylic acids to esters or amides. Recall that the carboxyl group of the acid must be activated by the formation of a carboxylic acid derivative which is higher in the reactivity order than the carboxyl group itself (see Section 9.10). Recall also that in the biosynthesis of fatty acid esters (Section 9.16) the reactive carboxyl derivative is the acyl adenosyl monophosphate, RCO_2PO_3-adenosine, an acyl phosphoric anhydride. The mixed anhydride reacts with a thiol, coenzyme A, which displaces adenosine monophosphate and forms a thioester. The thioester is converted to an oxygen ester when an alcohol displaces the thiol.

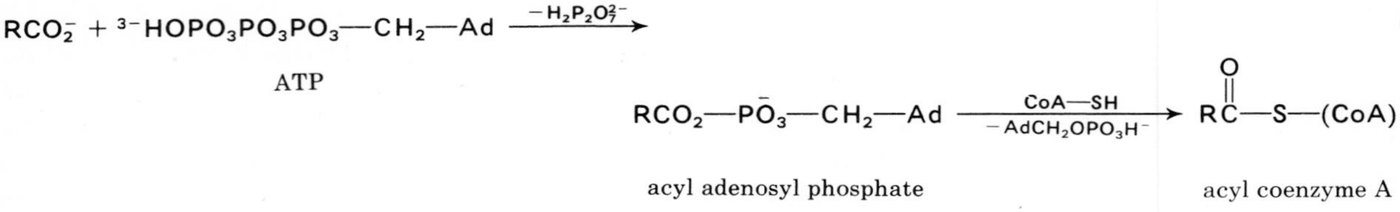

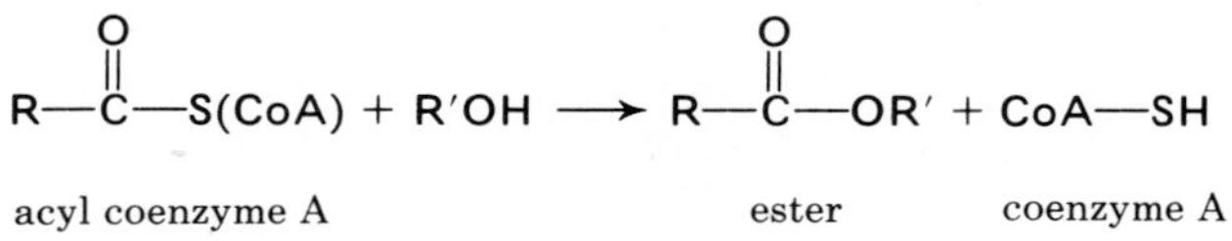

$$RCO_2^- + {}^{3-}HOPO_3PO_3PO_3{-}CH_2{-}Ad \xrightarrow{-H_2P_2O_7^{2-}}$$

ATP

$$RCO_2{-}P\bar{O}_3{-}CH_2{-}Ad \xrightarrow[-AdCH_2OPO_3H^-]{CoA{-}SH} \underset{\text{acyl coenzyme A}}{R\overset{O}{\overset{\|}{C}}{-}S{-}(CoA)}$$

acyl adenosyl phosphate

$$\underset{\text{acyl coenzyme A}}{R{-}\overset{O}{\overset{\|}{C}}{-}S(CoA)} + R'OH \longrightarrow \underset{\text{ester}}{R{-}\overset{O}{\overset{\|}{C}}{-}OR'} + \underset{\text{coenzyme A}}{CoA{-}SH}$$

Up to this point the formation of a peptide linkage follows closely the route just outlined. For example, the carboxyl group of alanine reacts with adenosine triphosphate to give alanyl adenosyl monophosphate and a diphosphate ion. It is probable that a sulfhydryl group on the enzyme displaces AMP (adenosine monophosphate) to give an enzyme-bound amino acid thioester, since all enzymes which catalyze the conversion of an amino acid to acyl-*t*RNA have one or two free sulfhydryl groups (SH) in the active site.

The thioester is converted to an amino ester of *t*RNA. In this ester the acyl group is bonded to the 2'-hydroxyl group of the ribose. The heterocyclic base of the nucleotide unit at this end of the *t*RNA molecule is adenine. The reaction using alanine is formulated below.

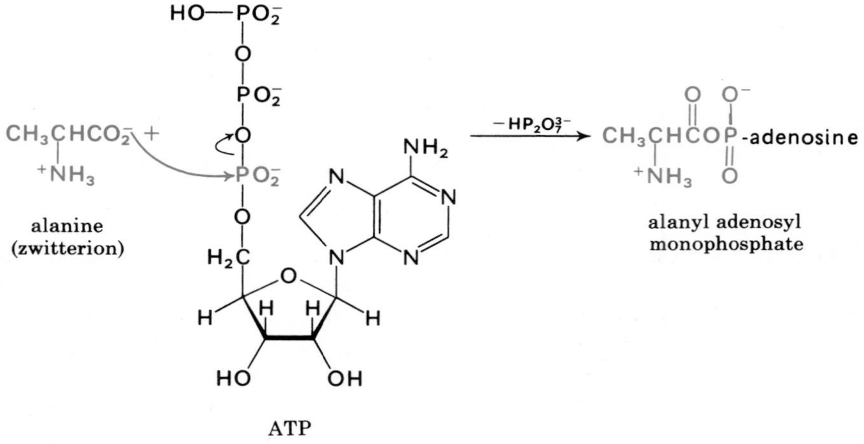

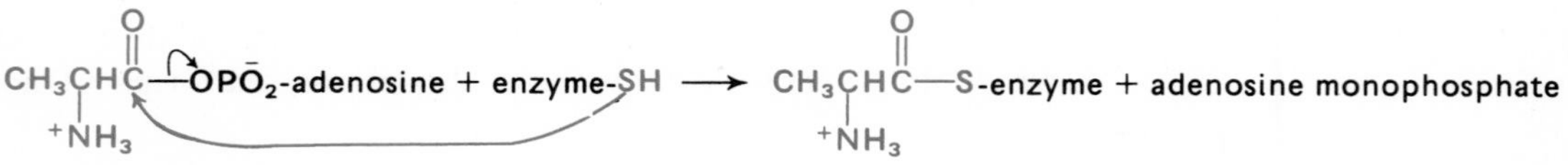

ATP   alanyl adenosyl monophosphate

alanine (zwitterion)

$$\underset{\overset{|}{{}^+NH_3}}{CH_3CH\overset{O}{\overset{\|}{C}}{-}\bar{O}PO_2\text{-adenosine}} + \text{enzyme-SH} \longrightarrow \underset{\overset{|}{{}^+NH_3}}{CH_3CH\overset{O}{\overset{\|}{C}}{-}S\text{-enzyme}} + \text{adenosine monophosphate}$$

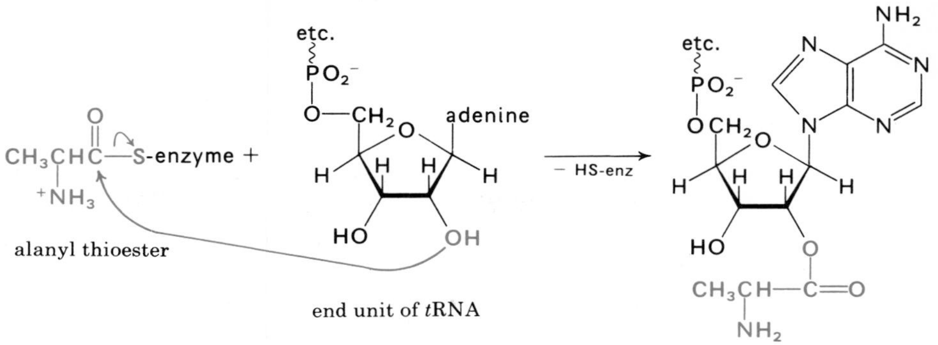

CH₃CHC—S-enzyme + ... end unit of tRNA → (− HS-enz) → alanyl tRNA (ester at 2′-hydroxyl group of ribose)

alanyl thioester

end unit of *t*RNA

alanyl *t*RNA
(ester at 2′-hydroxyl group
of ribose)

Problem **13.15** It is believed that the first amino acid in each protein synthesis is methionine. Write the series of reactions starting with N-formylmethionine, coenzyme A, ATP, and *t*RNA to form N-formylmethionyl transfer RNA.

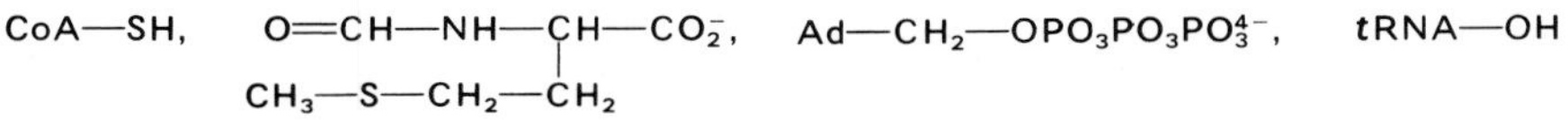

$$CoA-SH, \qquad O{=}CH-NH-CH-CO_2^-, \qquad Ad-CH_2-OPO_3PO_3PO_3^{4-}, \qquad tRNA-OH$$
$$CH_3-S-CH_2-CH_2$$

N-formylmethionine   ATP

In the polypeptide synthesis the amino group of the first amino acid is blocked as an N-formamide (O=CH—NH—). The new peptide bond is formed between one amino ester *t*RNA molecule (e.g., alanyl *t*RNA) and an amino ester *t*RNA whose amino group has been made into a formamide group to protect it from reaction. Thus only one of the two amino ester reactants possesses a free amino group to participate in the transacylation with the second reactant. The reaction is shown adding an alanine unit to an N-formylated methionine unit.

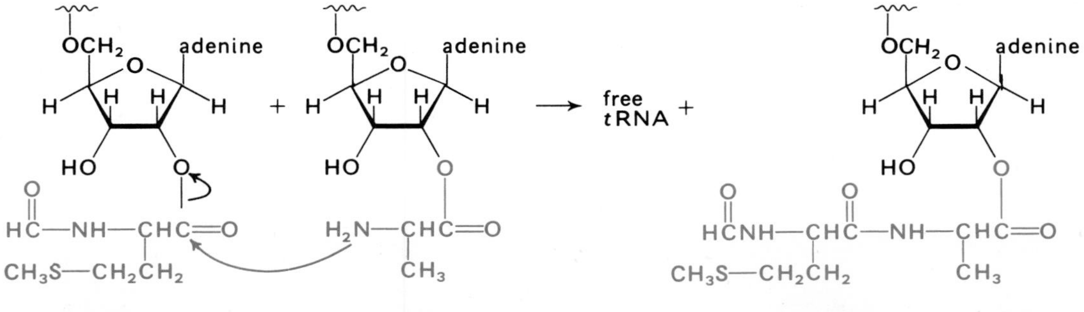

N-formylmethionyl *t*RNA   alanyl *t*RNA   N-formylmethionylalanyl *t*RNA

As the peptide chain grows, it is passed from one *t*RNA to another. At the end of the peptide synthesis both the formyl group and the last *t*RNA group are removed to produce the polypeptide or protein.

Note that the biosynthesis of polypeptides starts at the amino terminal end of the chain. This is in contrast to the Merrifield synthesis of polypeptides (Section 10.17), which starts at the carboxyl end.

13.12 THE DOUBLE HELIX OF DNA

The Watson–Crick model of DNA consists of two strands of DNA held together along the entire length of their chains by hydrogen bonding. The strands are antiparallel, that is they run in opposite directions.

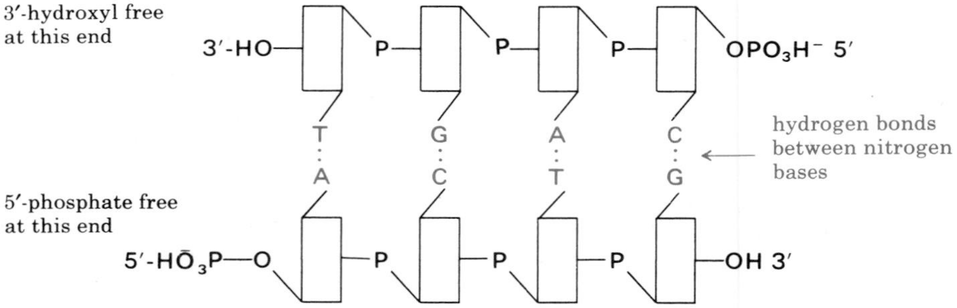

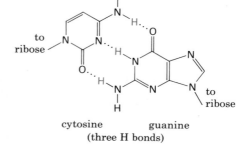

3'-hydroxyl free at this end

3'-HO— —P— —P— —P— OPO$_3$H$^-$ 5'

T⋮A G⋮C A⋮T C⋮G

hydrogen bonds between nitrogen bases

5'-phosphate free at this end

5'-HO$_3$P—O— —P— —P— —P— —OH 3'

The strands are not identical, but are complementary to each other since they are held together by hydrogen bonding between specific pairs of nitrogen bases. Adenine always hydrogen-bonds to its complementary base thymine, and guanine always hydrogen-bonds to cytosine. The complementary relationship between the two members of a bonded pair arises out of the number and placement of the H donors and H acceptors of the hydrogen bonds. The hydrogen-bonded pairs are shown below.

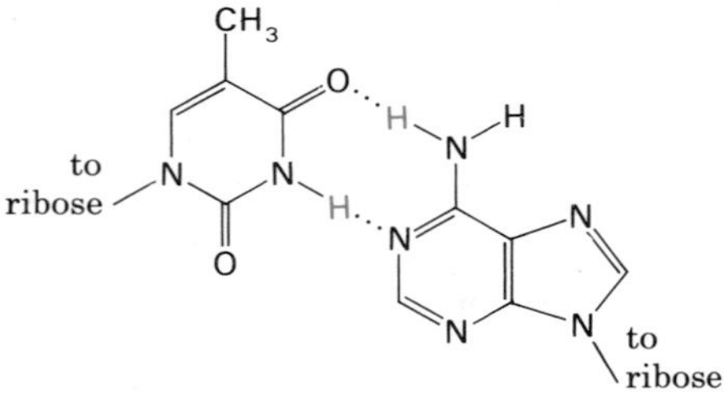

thymine   adenine
(two H bonds)

cytosine   guanine
(three H bonds)

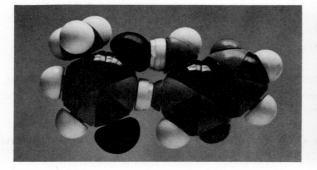

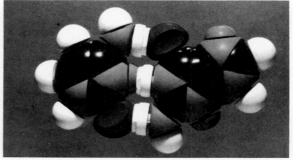

The bases, bonded end-to-end, lie in the same plane. The deoxyribose-phosphate chains lie on opposite sides of the base pairs running perpendicular to the planes of the bases (perpendicular to the plane of the paper).

The two strands of DNA form a double helix as the two chains coil around one central axis. The double helix resembles in shape a suspended spiral staircase whose treads are the hydrogen-bonded base pairs and whose railings are the deoxyribose-phosphate backbones. The hydrophilic (water-liking) sugar and phosphate groups are on the outside, while the hydrophobic (water-hating) bases are on the inside of the double helix.

axis

the double helix of
complementary strands of DNA

The double helix provides for the maximum number of hydrogen bonds between the bases and the geometry that maximizes their stability. No bases are included that are not paired. This arrangement beautifully combines great diversity with a high degree of order.

The process by which new and identical DNA double strands are made for new cells is called **replication**. It occurs by separation of the two strands of the old helix and the formation of a new complementary strand for each original strand. Thus each daughter DNA contains one new and one old strand.

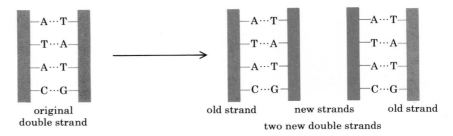

original
double strand

old strand   new strands   old strand

two new double strands

An analogy may illuminate the replication process. Imagine a couple dancing (the double strand). The couple separates. For each of the members of the couple, a new partner is found identical to the old partner. Two couples are now dancing, each containing one old member and one new member.

In a manner similar to the above replication process, a single strand of messenger RNA is formed with its series of bases complementary to the series of DNA. The RNA strand corresponds to only a small portion of the total DNA. In this way the genetic code contained in the order of the bases in the DNA is transcribed to a messenger RNA, which serves as a template for building the protein.

The coding units, or **codons**, are composed of three adjacent nitrogen bases taken from the four bases in RNA—adenine, uracil, cytosine, and guanine, A, U, C, and G. The order of the three bases in *m*RNA denotes a single amino acid. On the other hand, a single amino acid has several codons. A few codons are shown here.

Codon	Amino acid
AAA, AAG (2)	Lysine
GGA, GGU, GGC, GGG (4)	Glycine
UCG, UCU, UCA, UCC, AGU, AGC (6)	Serine

Of the 64 possible codons, 61 are used for individual amino acids. The remaining three are polypeptide-terminating codons, UAA, UGA, and UAG. The polypeptide-initiating codon seems to be the one for methionine AUG, for most if not all proteins.

The codons in *m*RNA must be matched one after another by the anticodon (the three complementary bases to the codon) in the transfer RNA which brings the amino acid to the reaction site.

Problem   13.16 Draw a segment of messenger RNA schematically from the 5′-phosphate end on the left to the 3′-OH end on the right, showing the sequence of bases uracil, cytosine, and adenine (one codon for serine).

New Terms and Topics

Carbohydrates (Section 13.1)

Monosaccharides—names, structures; aldohexose, etc. (Section 13.12)

Configurations of asymmetric carbons of monosaccharides (Section 13.3)

Relative configurations of saccharides; D- and L-families (Section 13.3)

Fischer and Haworth projection formulas (Sections 13.3 and 13.4)

Glycosides (Section 13.4)

Glucosides, acetals of glucose; α- and β-linkages (Section 13.4)

Mutarotation (Section 13.4)

Ribosides—acetals of ribose (Section 13.4)

Intramolecular hemiacetals—pyranose and furanose structures (Section 13.4)

Disaccharides—sucrose, lactose (Section 13.7)

Polysaccharides—starch, glycogen, cellulose (Section 13.8)

N-glycosides of D-ribose—N-ribosides (Section 13.9)

Structures of coenzymes containing N-ribosides: ATP, coenzyme A, NAD$^+$ (Section 13.9)

Nucleoside, nucleotide, dinucleotide, polynucleotide (Section 13.10)

Nucleic acids—DNA and RNA, structure and use (Section 13.10)

Biosynthesis of polypeptides; transfer RNA (Section 13.11)

Summary of Reactions

1. Aldehyde–hemiacetal equilibrium (Section 13.4)

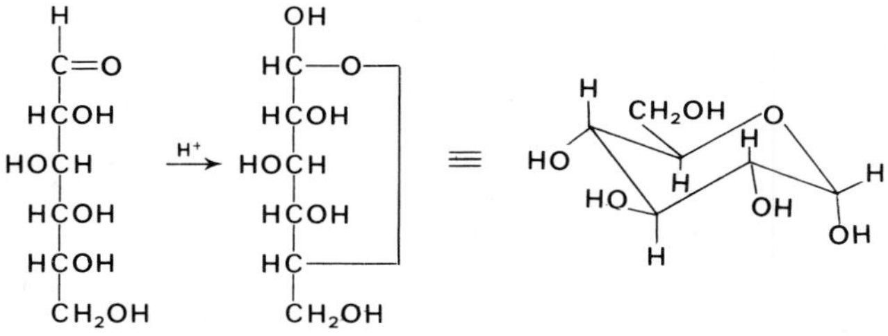

D-glucose   α-glucopyranose

2. Acetal formation (Section 13.4)

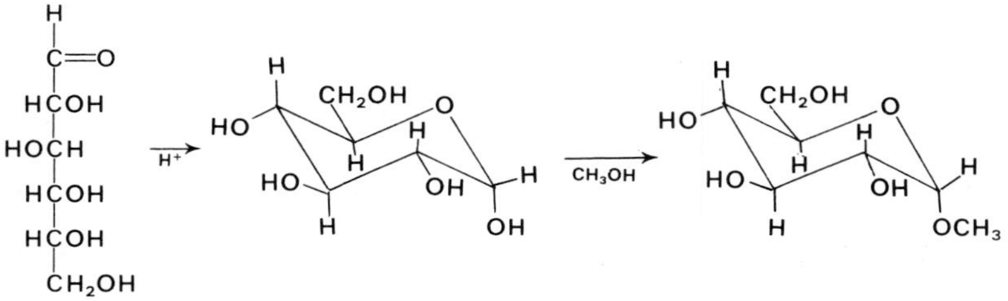

α-glucopyranose   methyl α-glucopyranoside

3. Oxidation (Section 13.5)

$$
\begin{array}{c}
CH{=}O \\
(H\overset{|}{C}OH)_n \\
CH_2OH
\end{array}
+ 2\ Ag(NH_3)_2^+ + 3\ OH^- \longrightarrow
\begin{array}{c}
CO_2^- \\
(H\overset{|}{C}OH)_n \\
CH_2OH
\end{array}
+ 4\ NH_3 + 2\ Ag_{(s)} + 2\ H_2O
$$

an aldose

$$
\begin{array}{c}
CH{=}O \\
(H\overset{|}{C}OH)_n \\
CH_2OH
\end{array}
\xrightarrow{\ HNO_3\ }
\begin{array}{c}
CO_2H \\
(H\overset{|}{C}OH)_n \\
CO_2H
\end{array}
$$

Problems

13.17 Write the products for the complete hydrolysis by treatment with acid for the following compounds.

    a) sucrose   **b)** methyl α-D-glucopyranoside   **c)** adenosine triphosphate

13.18 Write the structures for the products of the following reactions.

    a) L-ribose with $Ag(NH_3)_2^+\ OH^-$ **b)** D-fructose with $NaBH_4$ **c)** D-ribose with HNO_3

13.19 Write the structure of each of the following.

    a) an aldohexose which gives the same product that D-glucose gives with $NaBH_4$

    b) an aldopentose which gives the same product that D-ribose gives with $NaBH_4$

13.20 With Fischer projection formulas, write structures for the following compounds.

    a) the stereoisomers of $CH_3\overset{\underset{|}{OH}}{C}HCO_2H$

    b) the stereoisomers of deoxyribose (in aldehyde form)

    c) $CH_3OCH_2\overset{\underset{|}{OH}}{C}HCHO$ of the D-configuration

    d) *S*-configuration of $HOCH_2\overset{\underset{|}{SH}}{C}HCO_2H$

    e) the stereoisomers of $HOCH_2{-}\overset{\underset{|}{OH}}{C}H{-}\overset{\underset{|}{OH}}{C}H{-}\overset{\underset{|}{OH}}{C}H{-}CH_2OH$

    f) 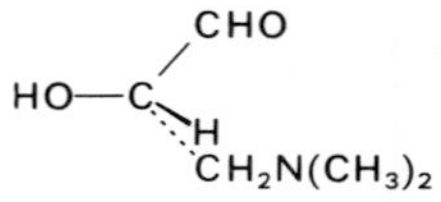 $HO{-}\overset{\overset{\textstyle CHO}{\diagup}}{\underset{\underset{\textstyle CH_2N(CH_3)_2}{\cdot\cdot H}}{C}}$ stereoisomers

13.21 In mono- and disaccharides the primary and secondary hydroxyl groups that are not tied up in acetal or ketal linkages are easily acetylated with acetic anhydride and pyridine. Write structures for fully acetylated derivatives of the following compounds. Use the symbol Ac for $CH_3\overset{\overset{\textstyle O}{\|}}{C}{-}$

    a) sucrose   **b)** methyl β-D-ribofuranoside   **c)** methyl α-D-deoxyribofuranoside

13.22 A triribonucleotide is known to contain one each of the three bases adenine (A), thymine (T), and guanine (G). Indicate all possible orders in which the three nucleotides might be assembled. Remember that the trimer in a sense has a head and a tail. Use an arrow between the letters to read from head to tail.

13.23 Deoxyribose can be synthesized by an aldol condensation between D-glyceraldehyde 3-phosphate and acetaldehyde. Give the structures for reactants and product (in aldehyde form).

13.24 In the laboratory the isomerization of glucose to an equilibrium mixture of glucose and fructose can be accomplished by treatment with sodium hydroxide. In the bioconversion of glucose to trioses, one step involves the isomerization of glucose 6-phosphate to fructose 6-phosphate. Assume that this reaction occurs through open-chain forms of both hexoses and that the phosphates play only a minor mechanistic role in the biological reaction. Draw a mechanistic scheme for the isomerization, assuming that the reaction is both acid- and base-catalyzed at some stages of the conversion. Use models for glucose and fructose as shown.

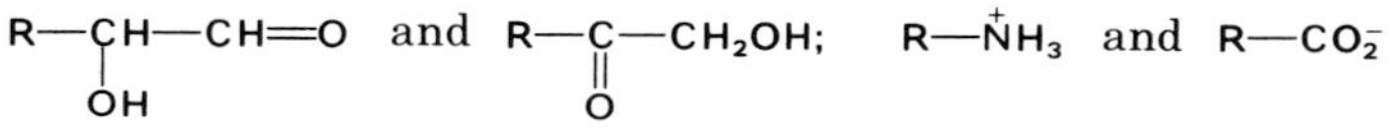

$$R-\underset{\underset{\displaystyle OH}{|}}{C}H-CH=O \quad \text{and} \quad R-\underset{\underset{\displaystyle O}{\|}}{C}-CH_2OH; \quad R-\overset{+}{N}H_3 \quad \text{and} \quad R-CO_2^-$$

glucose   fructose   acid   base

13.25 Vitamin C (L-ascorbic acid) is synthesized enzymatically from D-glucose. Examine the structure of L-ascorbic acid and list the changes necessary to convert D-glucose. Note that a 180° rotation of D-glucose in the plane of the paper gives a formula which bears a strong relationship to L-ascorbic acid.

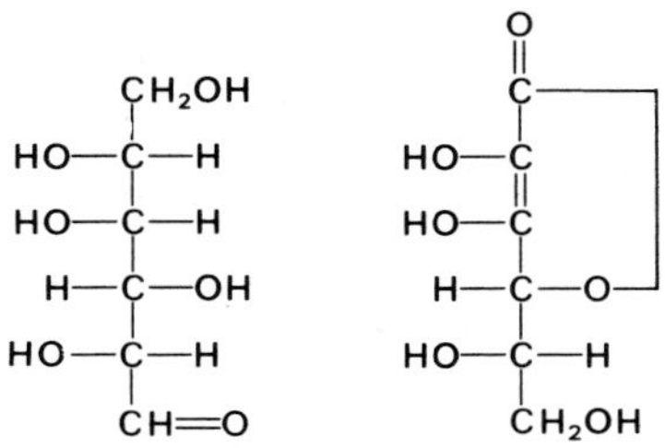

D-glucose (rotated 180°)   L-ascorbic acid

CHAPTER 14

Carbon Chain-
Building Reactions

Perhaps the most fascinating feature of organic chemistry is the opportunity it provides the scientist for designing and synthesizing new compounds from known starting materials. The structural possibilities are boundless for the synthetic organic chemist with a knowledge of organic reactions, a command of structural theory, and a creative imagination. The synthetic routes used to prepare naturally occurring compounds in the laboratory are sequences that sometimes utilize from ten to thirty different reactions. The thought processes involved in designing such syntheses are not dissimilar to those employed in the game of chess. Each move must be made with a large number of following alternative moves in mind. In the game of synthesizing new compounds, many more pieces must be moved over a vastly larger board and with less formal rules. The opponent is the general intractability of nature, which when misunderstood resists manipulation.

In the former chapters, most reactions involved the making and breaking of bonds between carbon and hydrogen, or between carbon and oxygen, nitrogen, or sulfur. Such reactions are used in the more obvious steps in most syntheses. Less obvious and not discussed until now are the reactions which extend and elaborate carbon chains. This chapter is concerned with those reactions that allow carbon chains to be extended. In other words, this chapter describes those reactions in which new carbon–carbon bonds are made. Since some of these reactions are reversible, the chapter also treats a few reactions in which carbon–carbon bonds are broken, and *compounds are degraded*.

Most reactions that increase the carbon chain length are based on combining a negative carbon with a positive carbon. Compounds with a positive carbon have been encountered in all of the functional groups that contain C—O, C—Cl, C=O, or C=N groups. Compounds that contain a negative carbon will be stressed in this chapter.

14.1 ORGANOMETALLIC COMPOUNDS

The most stable classes of compounds, other than hydrocarbons, have functional groups with carbon bonded to a strongly electronegative nonmetal—oxygen, nitrogen, sulfur, or chlorine. Very reactive compounds with completely different properties can be prepared in which carbon is bonded to electropositive metals. Compounds with carbon-metal bonds are called organometallics. Of the many organometallics known, we will limit our study to the extremely reactive compounds of magnesium or sodium.

The organomagnesium compounds have solubility properties characteristic of covalent substances. For example, butylmagnesium chloride is prepared and used in an ether solution. The bond to magnesium is strongly polarized to give carbon a partial negative charge and the metal a partial positive charge.

$$CH_3CH_2CH_2\overset{\delta-}{CH_2}\!\!-\!\!\overset{\delta+}{Mg}\!\!-\!\!\overset{\delta-}{Cl}$$

butylmagnesium chloride

Ionic compounds containing sodium ions and carbanions are possible when carbon has attached substituents that stabilize the negative charge. An example is the sodium salt of diethyl malonate (a sodium enolate salt).

$$[C_2H_5O-\overset{\overset{\displaystyle :O:}{\|}}{C}-\overset{Na^+}{\underset{\cdot\cdot}{\overset{}{C}}}H-\overset{\overset{\displaystyle :O:}{\|}}{C}-OC_2H_5 \longleftrightarrow C_2H_5O-\overset{\overset{\displaystyle :O:}{\|}}{C}-\overset{Na^+}{C}H=\overset{\overset{\displaystyle :\overset{-}{O}:}{|}}{C}-OC_2H_5 \longleftrightarrow C_2H_5O-\overset{\overset{\displaystyle :\overset{-}{O}:}{|}}{C}=\overset{Na^+}{C}H-\overset{\overset{\displaystyle :O:}{\|}}{C}-OC_2H_5]$$

sodium enolate salt of diethyl malonate

Other examples of these organometallics are listed below.

Organomagnesium halides (Grignard reagents) $R_3C-Mg-X$:

$$CH_3-Mg-I \qquad (CH_3)_3C-Mg-Cl \qquad C_6H_5-Mg-Br$$

methylmagnesium iodide   *tert*-butylmagnesium chloride   phenylmagnesium bromide

Sodium enolates of aldehydes or ketones (α-carbanions):

$$\left[R-\overset{\overset{\displaystyle :O:}{\|}}{C}-\overset{}{\underset{\cdot\cdot}{C}}H-\overset{\overset{\displaystyle :O:}{\|}}{C}-R \longleftrightarrow R-\overset{\overset{\displaystyle :O:}{\|}}{C}-CH=\overset{\overset{\displaystyle :\overset{-}{O}:}{|}}{C}-R \longleftrightarrow R-\overset{\overset{\displaystyle :\overset{-}{O}:}{|}}{C}=CH-\overset{\overset{\displaystyle :O:}{\|}}{C}-R \right] Na^+$$

$$\left[CH_3-\overset{\overset{\displaystyle :O:}{\|}}{C}-\overset{}{\underset{\cdot\cdot}{C}}H-\overset{\overset{\displaystyle :O:}{\|}}{C}-CH_3 \longleftrightarrow CH_3-\overset{\overset{\displaystyle :O:}{\|}}{C}-CH=\overset{\overset{\displaystyle :\overset{-}{O}:}{|}}{C}-CH_3 \longleftrightarrow CH_3-\overset{\overset{\displaystyle :\overset{-}{O}:}{|}}{C}=CH-\overset{\overset{\displaystyle :O:}{\|}}{C}-CH_3 \right] Na^+$$

sodium enolate of 2,4-pentanedione

Organometallics add to aldehydes, ketones, and esters to make compounds that have carbon skeletons composed of the carbons of the two starting materials. The preparation and the addition reactions of organomagnesium compounds are treated in the following sections (Sections 14.2 through 14.4). Reactions of sodium enolate salts are considered in later sections (Sections 14.5 through 14.10).

Note that the examples of the two types of organometallic reagents given above differ in the structure of the carbon parts. In the organomagnesium compound the partial charge on carbon is localized. These reagents act as extremely strong bases. In the enolate ion the delocalization of charge onto one or more oxygens provides a stabilized ion, making it a moderately strong base. The conjugate acids of these reagents (known as carbon acids) differ widely in the acidity ranges of the two groups. The pK_a values for the conjugate acids are listed in Fig. 14.1 along with those of familiar compounds that serve as calibration points.

FIGURE 14.1
ACIDITY VALUES (pK_a)
FOR CARBON–HYDROGEN
ACIDS AND OTHER
REPRESENTATIVE COM-
POUNDS (AGAINST WATER)

Acid		Base	pK_a
Methane	CH_4	CH_3^-	>45
Ethylene	$H_2C\!\!=\!\!CH_2$	$H_2C\!\!=\!\!\bar{C}H$	~ 45
Benzene	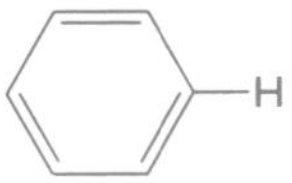 ⟨benzene⟩—H	 ⟨benzene⟩$^-$	43
Ammonia	NH_3	NH_2^-	34
Acetylene	$HC\!\!\equiv\!\!CH$	$HC\!\!\equiv\!\!C^-$	25
Ethyl propanoate	$CH_3CH_2CO_2C_2H_5$	$CH_3\bar{C}HCO_2C_2H_5$	25
Acetaldehyde	$CH_3CH\!\!=\!\!O$	$^-CH_2CH\!\!=\!\!O$	20
tert-Butyl alcohol	$(CH_3)_3COH$	$(CH_3)_3C\!\!-\!\!O^-$	19
Ethanol	CH_3CH_2OH	$CH_3CH_2\!\!-\!\!O^-$	18
Methanol	CH_3OH	$CH_3\!\!-\!\!O^-$	16
Water	H_2O	OH^-	16
Diethyl malonate	$CH_2(CO_2C_2H_5)_2$	$^-CH(CO_2C_2H_5)_2$	13
2,4-Pentanedione	$CH_3\overset{\displaystyle \|}{\underset{\displaystyle O}{C}}\!\!-\!\!CH_2\!\!-\!\!\overset{\displaystyle \|}{\underset{\displaystyle O}{C}}\!\!-\!\!CH_3$	$CH_3\overset{\displaystyle \|}{\underset{\displaystyle O}{C}}\!\!-\!\!\bar{C}H\!\!-\!\!\overset{\displaystyle \|}{\underset{\displaystyle O}{C}}\!\!-\!\!CH_3$	9
Hydrogen cyanide	HCN	^-CN	9.3
Phenol	$C_6H_5\!\!-\!\!OH$	$C_6H_5\!\!-\!\!O^-$	9.9
Ammonium ion	NH_4^+	NH_3	9.2
Acetic acid	CH_3CO_2H	$CH_3CO_2^-$	4.8

The range of pK_a values given in Fig. 14.1 is very wide and yet the strongest acid listed is acetic acid (pK_a = 4.8), which is only moderately strong. The degree of strength of an acid varies enormously. Acetic acid is a very much stronger acid than methanol (16), which in turn is very much stronger than ethyl acetate (25) or acetylene (25). Way up the scale, methane is an *exceedingly weak* acid, which fact makes its conjugate base, the methide ion (:CH_3^-), an *exceedingly strong* base.

14.2 GRIGNARD REAGENTS FROM ALKYL AND ARYL HALIDES

An alkyl or aryl halide in ether solution reacts with metallic magnesium to generate an organomagnesium halide, called a **Grignard reagent**. Isopropyl bromide and magnesium form isopropylmagnesium bromide. The carbon–magnesium bond replaces the carbon–bromine bond.

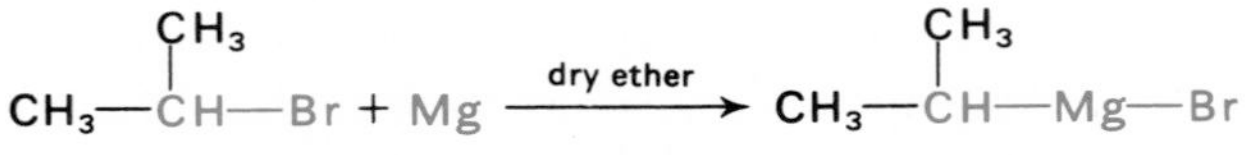

$$CH_3\!\!-\!\!\overset{\displaystyle CH_3}{\underset{\displaystyle |}{CH}}\!\!-\!\!Br + Mg \xrightarrow{\text{dry ether}} CH_3\!\!-\!\!\overset{\displaystyle CH_3}{\underset{\displaystyle |}{CH}}\!\!-\!\!Mg\!\!-\!\!Br$$

isopropyl   metallic   isopropylmagnesium bromide
bromide   magnesium   (isopropyl Grignard reagent)

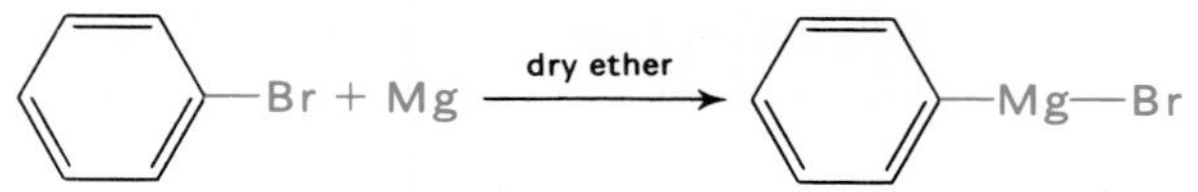

bromobenzene → phenylmagnesium bromide

The structures of the halides that form Grignard reagents vary widely. Primary, secondary, tertiary alkyl, allyl, benzyl, and phenyl halides all form the reagent. The halogen may be iodide, bromide, or chloride, but fluorides do not react.

The reaction of magnesium with alkyl halides, which occurs on the surface of the metal, is very highly exothermic and requires external cooling to retain the diethyl ether solvent (bp 34°) in the reaction flask.

Grignard reagents are used immediately, because they are very reactive and tend to decompose on long standing. Great care must be taken in the preparation to exclude contaminants which destroy the reagents, such as water, carbon dioxide, and oxygen. The stress on dry solvent and dry reactants arises from the fact that the organometallic compounds react rapidly and vigorously with water. The hydrocarbon of the reagent and the metal salt are formed.

$$CH_3CH_2CH_2\!\!-\!\!Mg\!\!-\!\!Br + H_2O \longrightarrow CH_3CH_2CH_3 + Mg(OH)Br$$

The reagents also react with all other compounds containing acidic hydrogens—acids, amines, alcohols—to give hydrocarbons by proton transfer. Grignard reagents cannot be prepared from halides having any of these groups present in the same molecule.

$$CH_3\!\!-\!\!Mg\!\!-\!\!I + (CH_3)_2NH \longrightarrow CH_4 + (CH_3)_2N\!\!-\!\!Mg\!\!-\!\!I$$

Grignard reagents react with virtually all compounds that contain C=O, C=N, or C≡N (Section 14.4). This means that the reagents cannot be prepared from halides that contain R_2C=O, RCH=O, RCO_2H, RCO_2R, RC≡N, RCONHR, RCH=NR groups.

Even with all these limitations, the Grignard reagent is one of the most useful compound classes we have for laboratory synthesis of larger compounds with more complex, branched carbon skeletons.

Oh, Grignard, the Beautiful

The carbonyl is polarized,
The carbon end is plus,
A nucleophile will thus attack
The carbon nucleus.
A Grignard makes an alcohol,
Of types there are but three,
It makes a bond to correspond
From C to shining C.

Professor FRANK H. WESTHEIMER

Organomagnesium halides R—Mg—X (Grignard reagents) react rapidly with aldehydes and ketones to produce alcohols having carbon chains composed of the combination of the two reactants. The reaction is an addition of the Grignard reagent to the C=O of the carbonyl compound with the negative carbon portion bonding to the carbonyl carbon and the positive magnesium bonding to the oxygen.

$$CH_3CH_2-\overset{}{Mg}-Cl + CH_3-\overset{\overset{O}{\|}}{C}-H \xrightarrow{\text{dry ether}} CH_3-\overset{\overset{O-Mg-Cl}{|}}{\underset{\underset{CH_2CH_3}{|}}{C}}-H \xrightarrow{H_3O^+} CH_3-\overset{\overset{OH}{|}}{\underset{\underset{CH_2CH_3}{|}}{C}}-H + Mg^{2+} + Cl^-$$

ethylmagnesium   acetaldehyde   2-butanol
chloride

The Grignard reagent acts as a nucleophilic carbon reagent. The initial product of the addition of the Grignard to the carbonyl group is the chloromagnesium alkoxide, which is then converted to the alcohol on treatment with mineral acid.

The carbonyl compound in ether solution is added slowly to a solution of the organometallic compound. The reaction is rapid and vigorous and evolves much heat. The ether solvent is retained in the flask by the use of an ice bath.

The combination of a Grignard reagent with a carbonyl compound to form an alcohol is a principal method for creating larger compounds with more complex carbon skeletons than the starting materials. There are almost no restrictions on the hydrocarbon part of the structure of a halide that can be used to generate an organomagnesium halide. The carbonyl compounds which add the Grignard reagent likewise have few restrictions. Only ketones with large, branched groups which sterically hinder the addition fail to react with Grignard reagents in this way.

Formaldehyde and Grignard reagents give primary alcohols, other aldehydes produce secondary alcohols, while ketones form tertiary alcohols.

$$\underset{\text{isobutyl-}}{\overset{\overset{CH_3}{|}}{CH_3\overset{}{C}HCH_2}-MgCl} + H-\overset{\overset{O}{\|}}{C}-H \longrightarrow CH_3\overset{\overset{CH_3}{|}}{\overset{}{C}}HCH_2-\overset{\overset{O-MgCl}{|}}{C}H_2 \xrightarrow{H_3O^+} CH_3\overset{\overset{CH_3}{|}}{\overset{}{C}}HCH_2-CH_2-OH$$

isobutyl-   formaldehyde   3-methyl-1-butanol
magnesium chloride

benzaldehyde   methyl Grignard   1-phenylethanol

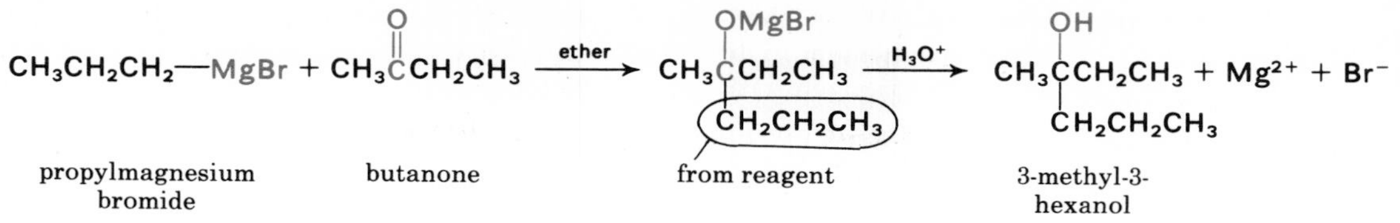

$$\text{CH}_3\text{CH}_2\text{CH}_2\text{—MgBr} + \underset{\displaystyle \overset{\text{O}}{\overset{\|}{\text{CH}_3\text{C}}}\text{CH}_2\text{CH}_3}{} \xrightarrow{\text{ether}} \underset{\displaystyle \overset{\text{OMgBr}}{\underset{\text{CH}_2\text{CH}_2\text{CH}_3}{|}}}{\text{CH}_3\text{C}\text{CH}_2\text{CH}_3} \xrightarrow{\text{H}_3\text{O}^+} \underset{\displaystyle \overset{\text{OH}}{\underset{\text{CH}_2\text{CH}_2\text{CH}_3}{|}}}{\text{CH}_3\text{C}\text{CH}_2\text{CH}_3} + \text{Mg}^{2+} + \text{Br}^-$$

propylmagnesium bromide	butanone	from reagent	3-methyl-3-hexanol

Almost any combination of Grignard reagent and carbonyl compound can be used to synthesize the desired compound. The 3-methyl-3-hexanol prepared above would also be the product of the reaction of methyl Grignard and 3-hexanone, or the product of the reaction of ethyl Grignard with 2-pentanone.

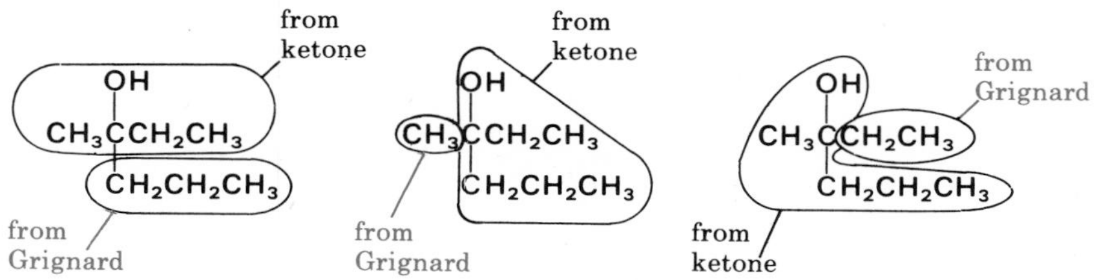

Problem   **14.1** Write equations for the following reactions in dry ether followed by acidification.

  a) phenylmagnesium bromide + 2-methylcyclohexanone

  b) isobutylmagnesium chloride + 2-pentanone

  c) benzaldehyde + propylmagnesium bromide

  d) cyclopropylmagnesium bromide + formaldehyde

To form carboxylic acids from alkyl halides, a solution of the Grignard reagent in ether is added to crushed dry ice CO_2. This reaction was first mentioned as a method of preparing carboxylic acids from alkyl and aryl halides (Section 8.10). The Grignard adds to one C=O bond, forming the magnesium salt of a carboxylic acid having one carbon more than the starting halide. Acidification of the magnesium carboxylate leads to the carboxylic acid.

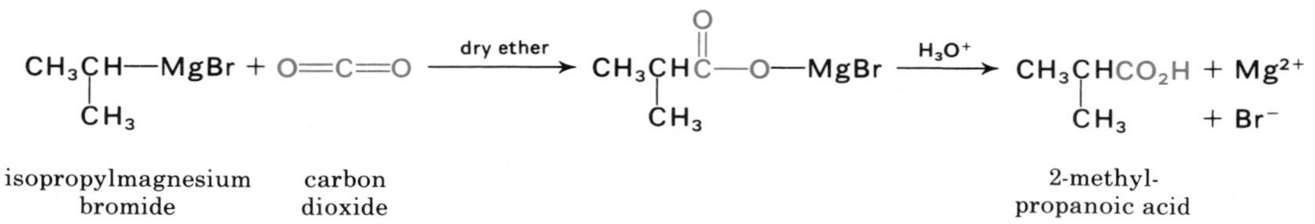

$$\underset{\displaystyle \underset{\text{CH}_3}{|}}{\text{CH}_3\text{CH—MgBr}} + \text{O}{=}\text{C}{=}\text{O} \xrightarrow{\text{dry ether}} \underset{\displaystyle \underset{\text{CH}_3}{|}}{\text{CH}_3\text{CH}\overset{\text{O}}{\overset{\|}{\text{C}}}\text{—O—MgBr}} \xrightarrow{\text{H}_3\text{O}^+} \underset{\displaystyle \underset{\text{CH}_3}{|}}{\text{CH}_3\text{CHCO}_2\text{H}} + \text{Mg}^{2+} + \text{Br}^-$$

isopropylmagnesium bromide	carbon dioxide		2-methyl-propanoic acid

14.2 Write the equation for the reaction of 2-methylbromobenzene with magnesium in dry ether. Write the equations for the subsequent reactions of this reagent with each of the following.

a) deuterium oxide   **b)** carbon dioxide followed by acidification

14.4 REACTIONS OF GRIGNARD REAGENTS WITH ESTERS

Many reagents which add to carbonyl groups of aldehydes and ketones also react with esters as well. This is especially true of organo-magnesium halides. Two moles of Grignard reagent add to one mole of ester to form magnesium derivatives of two alcohols. Acidification of the two alkoxides produces two alcohols.

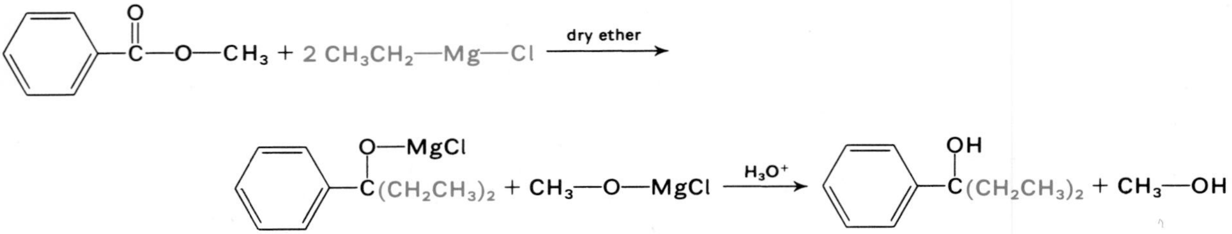

$$C_6H_5-\overset{O}{\underset{\|}{C}}-O-CH_3 + 2\ CH_3CH_2-Mg-Cl \xrightarrow{\text{dry ether}}$$

$$C_6H_5-\overset{O-MgCl}{\underset{|}{C}}(CH_2CH_3)_2 + CH_3-O-MgCl \xrightarrow{H_3O^+} C_6H_5-\overset{OH}{\underset{|}{C}}(CH_2CH_3)_2 + CH_3-OH$$

chloromagnesium 3-phenyl-3-pentoxide   chloromagnesium methoxide   3-phenyl-3-pentanol   methanol

The first addition of the Grignard reagent to the carbonyl group of an ester produces an intermediate magnesium alkoxide of a hemiacetal which immediately decomposes to a ketone. The carbonyl group of the ketone, however, is as reactive as that of the ester, and the addition of a Grignard to C=O occurs a second time.

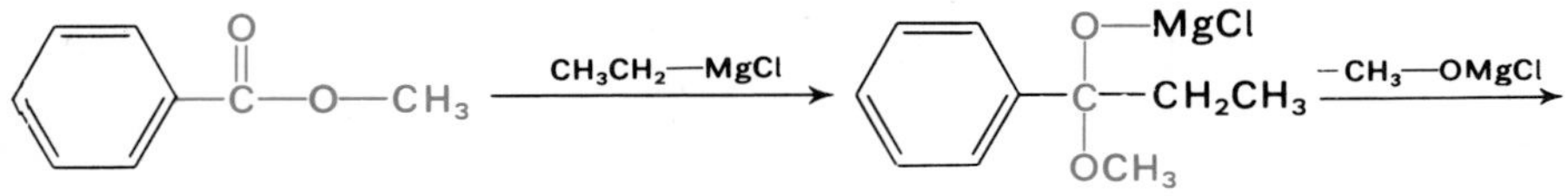

unstable hemiacetal salt intermediate

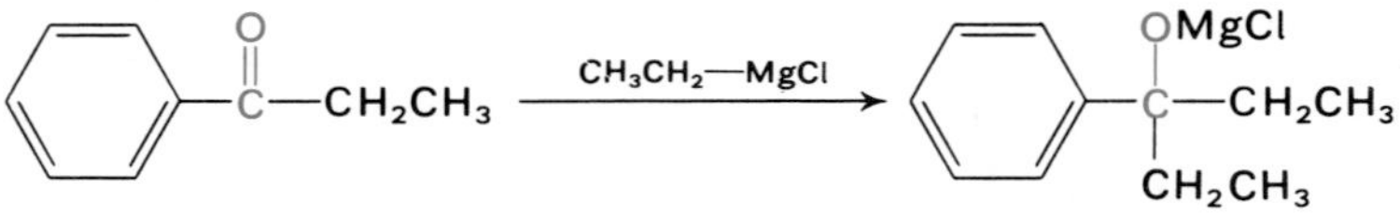

ketone intermediate   chloromagnesium alkoxide

The products of the addition of an organomagnesium halide to an ester are the magnesium salts of a tertiary alcohol and of the alcohol which originally formed the ester. Careful acidification of the reaction mixture produces the two alcohols.

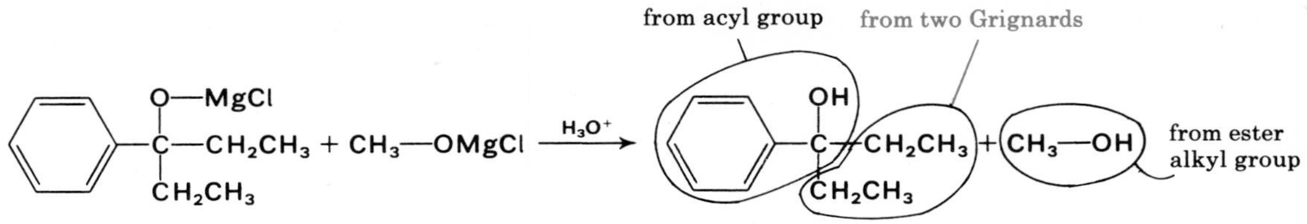

The tertiary alcohol produced by this reaction has one structural specification: two of the alkyl or aryl groups attached to the carbon bearing the hydroxyl are *identical* and come from the Grignard reagent, while the third is the aryl or alkyl portion originally attached to the carboxyl carbon of the ester. The alcohol portion of the ester is expelled to produce, after acidification, the corresponding alcohol as the second organic product. The Grignard reaction with esters is a simple reaction which produces a larger and more branched structure from three smaller components.

Aryl halides are less reactive in forming Grignard reagents than alkyl halides, but the bromides work well under the usual conditions in diethyl ether.

bromobenzene phenylmagnesium bromide

ethyl acetate

1,1-diphenylethanol

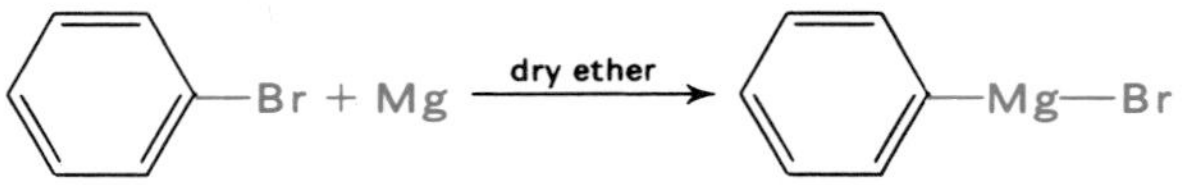

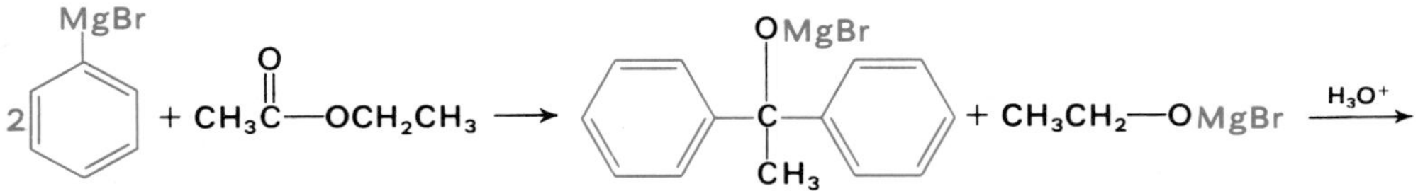

Grignard reagents can be prepared from alkyl or aryl halides that contain alkene, dialkylalkyne, ether, acetal, ketal, or tertiary amino groups. Grignard reagents cannot be formed from any halides that additionally contain any functional group with which Grignard reagents react, such as an amino, hydroxyl, thiol, carbonyl, carboxyl, ester, amido, cyano, anhydride, or terminal alkyne group.

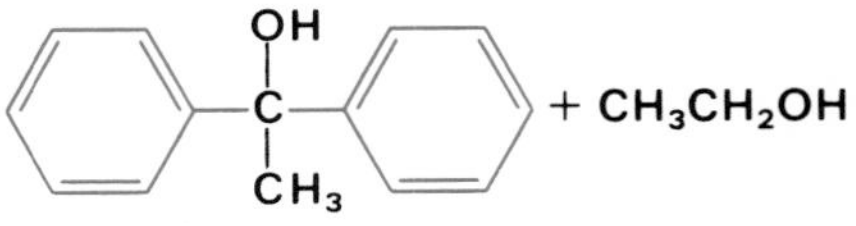

Problem **14.3** Write the structures for the final organic products (after acidification) of these reactions.

a) ethyl acetate + isobutylmagnesium chloride in dry ether

b) methyl benzoate + butylmagnesium chloride in dry ether

14.5 SODIUM ENOLATE SALTS

In the study of aldehydes and ketones, we learned that C=O and C=N groups increase the acidity of hydrogens on adjacent carbons (α-hydrogens) over the usual C—H (Section 11.8). The pK_a values listed in Fig. 14.1 show that aldehydes, ketones, and esters are vastly stronger acids than hydrocarbons. Loss of the α-proton to a strong base gives an α-carbanion (an enolate ion), whose negative charge is delocalized onto oxygen.

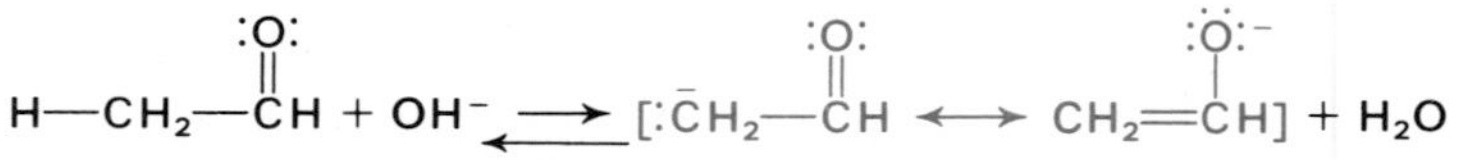

acetaldehyde · · · enolate ion of acetaldehyde

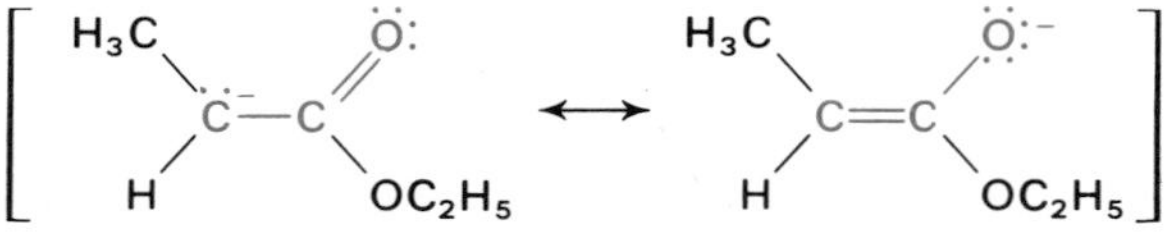

enolate ion of ethyl propanoate

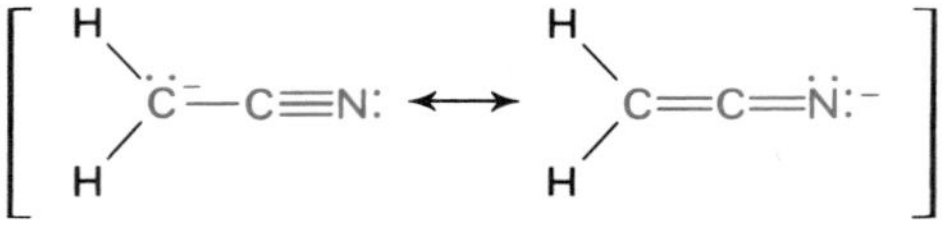

carbanion of acetonitrile

The carbanion has most of its negative charge on N or O, but a small portion of the charge is on the α-carbon. Remember that only removal of an α-hydrogen produces a charge-delocalized carbanion. Hydrogens in the β-and γ-positions are not as acidic because the charges of the carbanions are *not stabilized* by delocalization.

In Fig. 14.1 the pK_a's of water (16) and an aldehyde (20) differ by 4–5 units. This means that in the reaction of hydroxide ion with an aldehyde the equilibrium lies far on the side of the reactants. Only a trace of α-carbanion exists at any one time. This small amount, however, is enough to produce a reaction with another molecule.

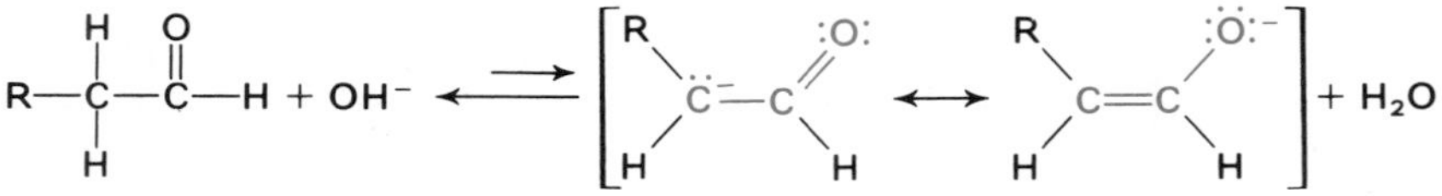

The formation of an α-carbanion from an ester (p$K_a \sim 25$) requires a base stronger than hydroxide ion. Ethoxide ion produces a sufficient concentration of the carbanion for it to react with other species.

When the C—H is between two carbonyl or cyano groups, the acidity of the proton is greatly enhanced, as is shown by the pK_a values in Fig. 14.1 for a β-diketone and a malonic ester. The charge of the anion is highly delocalized, and is distributed between both oxygens or both nitrogens.

$$
\left[
\begin{array}{c}
\underset{\displaystyle\underset{\text{H}}{|}}{\overset{\displaystyle\overset{\text{O}\quad\quad\text{O}}{}}{\text{R}\diagdown\text{C}\diagup\text{C}\diagdown\text{C}\diagup\text{R}}}
\end{array}
\longleftrightarrow
\cdots
\longleftrightarrow
\cdots
\right]
$$

enolate ion

The reaction equilibrium of a β-dicarbonyl compound ($pK_a \sim 9\text{–}13$) with ethoxide ion (ethanol, $pK_a = 18$) lies very far on the side of the enolate ion and the alcohol. Thus a reasonable concentration of the enolate salt can be obtained in solution.

$$
\text{C}_2\text{H}_5\text{O}\overset{\text{O}}{\overset{\|}{\text{C}}}\!\!-\!\text{CH}_2\!-\!\overset{\text{O}}{\overset{\|}{\text{C}}}\text{OC}_2\text{H}_5 + \text{C}_2\text{H}_5\text{O}^-\text{Na}^+ \longrightarrow \text{C}_2\text{H}_5\text{O}\overset{\text{O}}{\overset{\|}{\text{C}}}\!\!-\!\overset{-}{\text{C}}\text{H}\!-\!\overset{\text{O}}{\overset{\|}{\text{C}}}\text{OC}_2\text{H}_5 + \text{C}_2\text{H}_5\text{OH}
$$

diethyl malonate   ethoxide ion   enolate ion of   ethanol
($pK_a = 13$)   diethyl malonate   ($pK_a = 18$)

These concepts of acidity and delocalized α-carbanions have been reviewed in detail, because α-carbanion reactions are the most useful and important ones available in the laboratory and in nature for building larger compounds with longer carbon chains from smaller ones.

The basic principles of the addition of carbanions to aldehydes and ketones were described in connection with the aldol condensation (Section 11.10), which is reviewed in the next section. A similar condensation reaction of esters, known as the Claisen condensation, involves the carbonyl group of one ester molecule and the α-carbanion of another molecule (Section 14.7). An immense variety of related reactions have been and are being invented to accomplish special purposes in laboratory syntheses. The development of these reactions generates some of the great excitement in research at the present time. Their prototypes, the aldol and Claisen condensation reactions, are treated here in detail, particularly because of the importance of similar reactions in biological chemistry. The selective catalysis by enzymes of several varieties of carbanion additions to carbonyl and carboxyl functional groups has been developed by nature to a highly refined state. The mechanisms of these reactions have been elucidated in detail. You will see the similarity of nature's and the chemist's laboratory reactions in the remaining sections of this chapter.

14.6 THE ALDOL CONDENSATION A reaction of aldehydes which was discussed previously was one in which two aldehyde molecules react with each other to form an aldehyde having a molecular weight equal to the sum of the molecular weights of the two reactants. You will remember that this reaction, known as the **aldol condensation**, is catalyzed by a strong base and occurs if one of the aldehydes has a hydrogen on the α-carbon (Section 11.10).

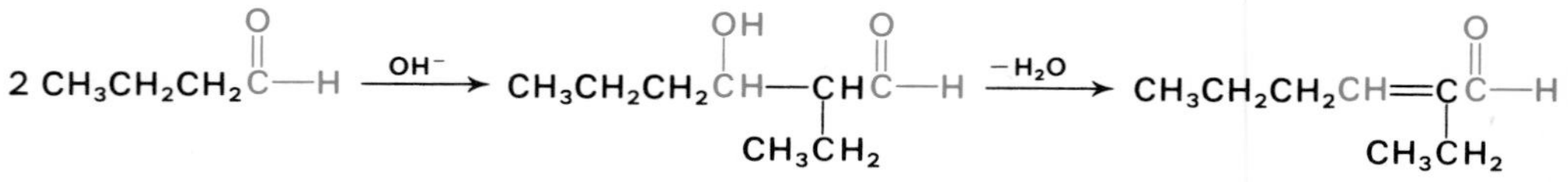

$$2\ CH_3CH_2CH_2\overset{\overset{\displaystyle O}{\|}}{C}-H \xrightarrow{\ OH^-\ } CH_3CH_2CH_2\overset{\overset{\displaystyle OH}{|}}{C}H-\underset{\underset{\displaystyle CH_3CH_2}{|}}{C}H\overset{\overset{\displaystyle O}{\|}}{C}-H \xrightarrow{\ -H_2O\ } CH_3CH_2CH_2CH=\underset{\underset{\displaystyle CH_3CH_2}{|}}{C}\overset{\overset{\displaystyle O}{\|}}{C}-H$$

butanal   2-ethyl-3-hydroxyhexanal   2-ethyl-2-hexenal

The basic catalyst removes an α-hydrogen from the aldehyde and generates an α-carbanion. The carbanion is a typical nucleophilic reagent which adds to the aldehyde carbonyl group in an ordinary addition reaction.

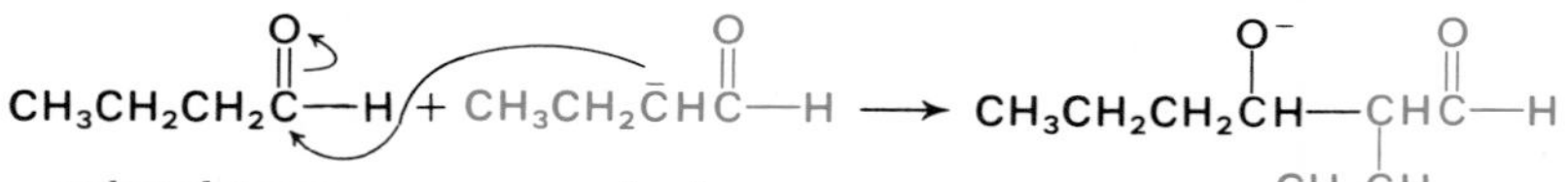

$$CH_3CH_2CH_2\overset{\overset{\displaystyle O}{\|}}{C}-H + CH_3CH_2\bar{C}H\overset{\overset{\displaystyle O}{\|}}{C}-H \longrightarrow CH_3CH_2CH_2\overset{\overset{\displaystyle O^-}{|}}{C}H-\underset{\underset{\displaystyle CH_3CH_2}{|}}{C}H\overset{\overset{\displaystyle O}{\|}}{C}-H$$

carbonyl group   α-carbanion

Unless very mild conditions are used, the hydroxyaldehyde ordinarily loses water to give the final product an α,β-unsaturated aldehyde.

We saw the example of an enzyme-catalyzed aldol condensation between phosphate derivatives of an aldehyde, glyceraldehyde, and a ketone, dihydroxyacetone, in the biosynthesis of fructose and glucose (Section 13.6).

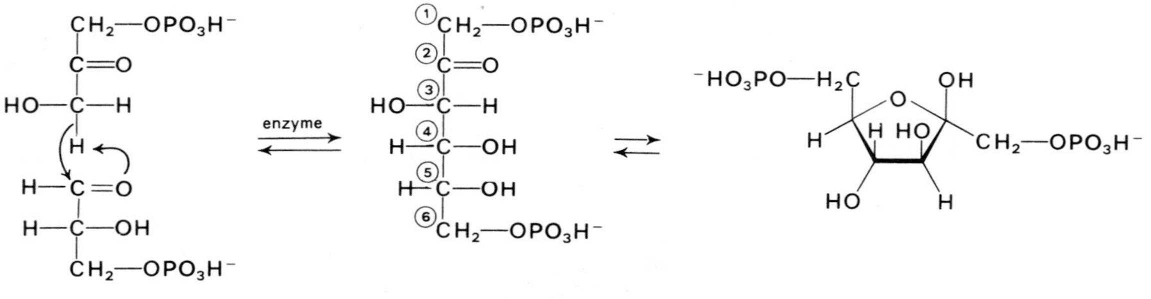

glyceraldehyde 3-phosphate   ketone form fructose 1,6-diphosphate   hemiketal form

14.4 Give the structure of the most probable product of the reaction of acetone with the terpene geranial catalyzed by NaOH.

$$CH_3C \overset{\overset{\displaystyle CH_3}{|}}{=} CHCH_2CH_2C \overset{\overset{\displaystyle CH_3}{|}}{=} CH—CH \overset{\displaystyle}{=} O$$

geranial

14.7 THE CLAISEN ESTER CONDENSATION AND THE REVERSE REACTION

A strong base catalyzes a self-condensation reaction of esters, analogous to the aldol condensation of aldehydes. The α-carbanion of one molecule of ester adds to the carbonyl carbon of a second molecule of ester in a nucleophilic addition reaction. The product formed is a β-ketoester. The reaction is known as the **Claisen ester condensation**.

$$2\ CH_3CH_2 \overset{\overset{\displaystyle O}{||}}{C}—OC_2H_5 \xrightarrow[\text{C}_2\text{H}_5\text{OH}]{\text{Na}^+ \text{-OC}_2\text{H}_5} CH_3CH_2 \overset{\overset{\displaystyle O}{||}}{C}—\underset{\underset{\displaystyle CH_3}{|}}{C}H \overset{\overset{\displaystyle O}{||}}{C}—OC_2H_5 + C_2H_5OH$$

ethyl propanoate

ethyl 2-methyl-3-oxopentanoate
(a β-ketoester)

Esters provide the most useful compounds for this kind of condensation reaction. Note from the pK_a's in Fig. 14.1 that esters are weaker acids than aldehydes or ketones. It is therefore an advantage to use a stronger base than hydroxide ion to produce the carbanion and to effect the condensation. It is particularly convenient for both the ester and the alkoxide ion catalyst to be derived from the same alcohol. In this situation, transesterification (Section 9.5) can occur without any effect on the condensation reaction.

The ester condensation proceeds through several steps. The α-carbanion, or enolate ion, of one ester molecule is formed by removal of the α-hydrogen by the ethoxide ion catalyst (step 1). The carbanion adds as a nucleophile to the carbonyl group of another ester molecule, giving the usual tetrahedral intermediate of nucleophilic substitution reactions of esters (step 2). (Compare reactions of other nucleophiles in Sections 9.4 through 9.7.) Loss of the ethoxide ion of the original ester group as leaving group from the tetrahedral intermediate reforms a carbonyl group, now as a ketone in a β-ketoester (step 3). The β-ketoester is a stronger acid ($pK_a = 11$) than ethanol and transfers its α-hydrogen to the ethoxide ion in the final step of the reaction (step 4). The mechanism is formulated here with ethyl acetate as the ester.

Step 1 Formation of the α-carbanion of the ester

$$CH_3 \overset{\overset{\displaystyle O}{||}}{C}—OC_2H_5 + C_2H_5O^- \rightleftharpoons \ \bar{C}H_2 \overset{\overset{\displaystyle O}{||}}{C}—OC_2H_5 + C_2H_5—OH$$

ethyl acetate   ethoxide ion α-carbanion

Steps 2 and 3 Addition of the carbanion to the carbonyl group and loss of ethoxide ion

$$CH_3\overset{O}{\overset{\|}{C}}-OC_2H_5 + \bar{C}H_2\overset{O}{\overset{\|}{C}}-OC_2H_5 \rightleftharpoons CH_3\underset{OC_2H_5}{\overset{O^-}{\overset{|}{\underset{|}{C}}}}-CH_2\overset{O}{\overset{\|}{C}}-OC_2H_5 \rightleftharpoons CH_3\overset{O}{\overset{\|}{C}}CH_2\overset{O}{\overset{\|}{C}}-OC_2H_5 + C_2H_5O^-$$

ethyl acetate   α-carbanion   tetrahedral intermediate   ethyl acetoacetate

Step 4 Formation of the enolate ion of the β-ketoester

$$CH_3-\overset{O}{\overset{\|}{C}}-CH_2-\overset{O}{\overset{\|}{C}}-OC_2H_5 + C_2H_5O^- \rightleftharpoons CH_3-\overset{O}{\overset{\|}{C}}-\bar{C}H-\overset{O}{\overset{\|}{C}}-OC_2H_5 + C_2H_5OH$$

(pK_a = 11)   enolate ion of   (pK_a = 18)
                                                β-ketoester

The first three steps of the condensation are reversible reactions whose equilibria favor the reactants and are unfavorable toward product formation. The equilibrium for the reaction of the ethyl acetate ($pK_a = 25$) with ethoxide ion lies far toward the ester and alkoxide ion in step 1. The position of equilibrium for the addition of the carbanion to the ester carbonyl also favors the reactants and not the products in step 2. The final step has an acid–base equilibrium between ethoxide ion and the β-ketoester ($pK_a = 11$) which strongly favors the product and determines the position of equilibrium for the overall reaction of ethyl acetate to enolate ion of ethyl acetoacetate. Acidification of the reaction mixture then gives the ethyl acetoacetate.

The mechanism explains why an ester like ethyl 2-methylpropanoate, which possesses only one α-hydrogen, does not undergo condensation. The ketoester which would be produced by the condensation does not have an α-hydrogen between the two carbonyl groups and cannot give a carbanion whose formation drives the reaction toward condensation product.

$$(CH_3)_2CHCO_2C_2H_5 \overset{\times}{\longrightarrow} (CH_3)_2CH\overset{O}{\overset{\|}{C}}-\underset{CH_3}{\overset{CH_3}{\overset{|}{\underset{|}{C}}}}-CO_2C_2H_5$$

ethyl 2-methylpropanoate   no α-hydrogen

Problems **14.5** Offer an explanation for the fact that ethyl esters and sodium ethoxide are preferred over methyl esters and sodium methoxide in condensation reactions.

14.6 Offer an explanation for the fact that a good yield of the 1,3-diketone can be obtained in this reaction of ethyl acetate with acetone to form 2,4-pentanedione.

$$CH_3\overset{O}{\overset{\|}{C}}OC_2H_5 + CH_3\overset{O}{\overset{\|}{C}}CH_3 \xrightarrow{Na^+\,{}^-OC_2H_5} CH_3\overset{O}{\overset{\|}{C}}CH_2\overset{O}{\overset{\|}{C}}CH_3$$

2,4-pentanedione

14.7 Write the structure of the most probable product of each reaction after final acidification.

a) ethyl butanoate + sodium ethoxide

b) diethyl hexanedioate + sodium ethoxide (cyclic product)

Even in laboratory reaction sequences, it is sometimes desirable to shorten carbon chains. All of the ester condensations are potentially reversible reactions. The reverse reaction *dominates* when concentrated sodium hydroxide is used as a catalyst instead of ethoxide ion. Cleavage of a β-ketoester results in the formation of two carboxylate salts, the most stable possible products that can be derived from the system. The cleavage shortens the carbon chain by two carbons.

$$CH_3\overset{O}{\overset{\|}{C}}CH_2\overset{O}{\overset{\|}{C}}OC_2H_5 + 2\,Na^+OH^- \longrightarrow 2\,CH_3\overset{O}{\overset{\|}{C}}O^-Na^+ + C_2H_5OH$$

In this degradation reaction, the hydroxide ion attacks the more reactive carbonyl group, the ketone. The tetrahedral intermediate formed breaks at the C—C bond to give one molecule of carboxylic acid and one of ester. The acid is neutralized, and the ester is hydrolyzed in subsequent reaction with the base.

$$CH_3\overset{O}{\overset{\|}{C}}CH_2\overset{O}{\overset{\|}{C}}O\!-\!C_2H_5 + OH^- \longrightarrow CH_3\overset{O^-}{\underset{OH}{\overset{|}{C}}}CH_2\overset{O}{\overset{\|}{C}}O\!-\!C_2H_5 \longrightarrow$$

tetrahedral intermediate

$$CH_3\overset{O}{\overset{\|}{C}}\!-\!OH + {}^-CH_2\overset{O}{\overset{\|}{C}}\!-\!OC_2H_5 \longrightarrow CH_3CO_2^- + CH_3\overset{O}{\overset{\|}{C}}\!-\!OC_2H_5 \xrightarrow{OH^-} C_2H_5OH + 2\,CH_3CO_2^-$$

14.8 ENZYME-CATALYZED CLAISEN CONDENSATIONS IN BIOSYNTHESIS

Examples of enzyme-catalyzed condensations of carboxylic acid derivatives abound in the biosyntheses of such compounds as fatty acids, cholesterol, and related steroids and terpenes. The enzyme systems have refined this one type of reaction and made use of it in widely different systems. This fact is not surprising because the reaction fundamentally requires only bases and acids as catalysts, and many bases and acids are readily available in the active sites of most enzymes in the form of amines or carboxylate ions and of ammonium ions or carboxylic acids.

In many of these reactions, the carboxylic acid derivative is a thioester, usually of the thiol of coenzyme A, (CoA)—SH. Two examples of the use of enzyme-catalyzed ester condensations are given in the following paragraphs and one reverse condensation for degradation of fatty acids in Section 14.9.

a) Initial reactions in the biosynthesis of cholesterol

The biosynthesis of cholesterol includes two simple illustrations of the Claisen ester condensation using the ester acetyl coenzyme A. The synthesis starts with the condensations of three molecules of acetyl coenzyme A to give one molecule of a six-carbon diester. The first reaction is a simple Claisen ester self-condensation of acetyl coenzyme A, which is analogous to the condensation of ethyl acetate in the previous section.

Reaction 1

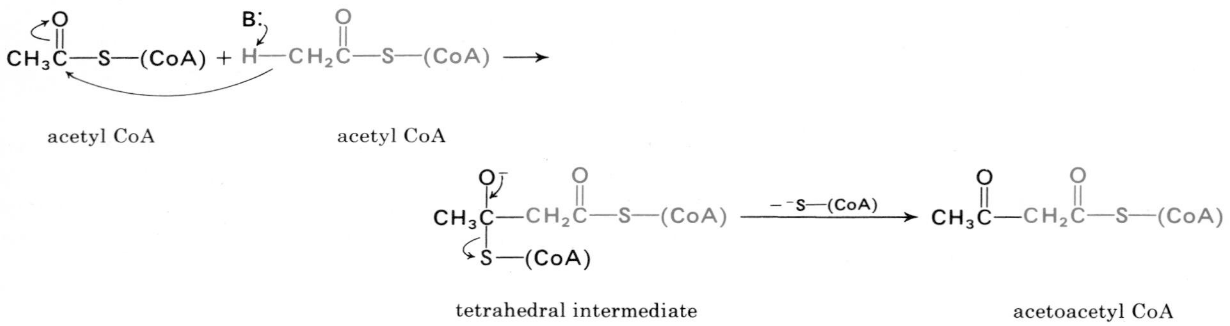

acetyl CoA   acetyl CoA

tetrahedral intermediate   acetoacetyl CoA

Reaction 2

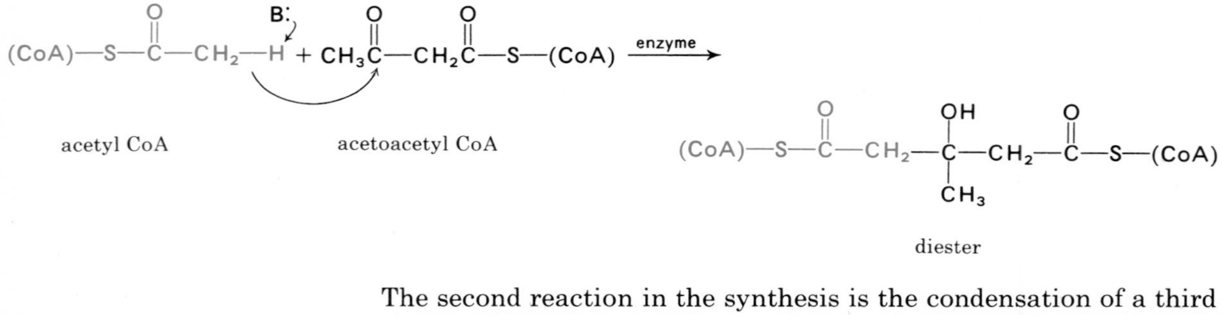

acetyl CoA   acetoacetyl CoA

diester

The second reaction in the synthesis is the condensation of a third molecule of acetyl CoA with the product of the first condensation, the acetoacetyl coenzyme A. Note that the enzyme acts selectively—it removes the proton from the acetate to give the acetate α-carbanion instead of removing the much more acidic proton between C=O groups of acetoacetyl CoA. The latter process would not lead to the required six-carbon diester.

The subsequent reduction of one ester group to the alcohol and hydrolysis of the second ester group gives mevalonic acid, an important intermediate in the synthesis of terpenes and steroids.

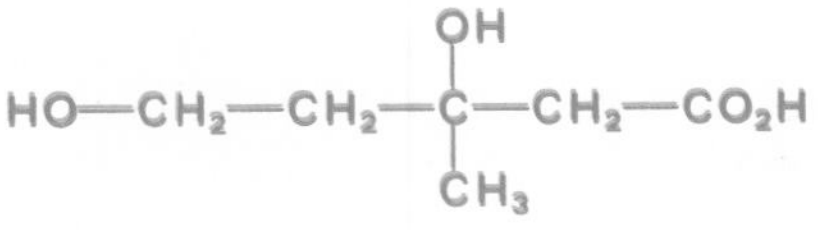

$$HO-CH_2-CH_2-\overset{\overset{\displaystyle OH}{|}}{\underset{\underset{\displaystyle CH_3}{|}}{C}}-CH_2-CO_2H$$

mevalonic acid

b) First reaction of the citric acid cycle— acetyl coenzyme A to citric acid

The biological sequence of reactions called the citric acid cycle is the route for the conversion of one molecule of acetyl coenzyme A to two molecules of CO_2. (See Section 14.12 for the entire cycle.) The cycle of reactions is started by the ester condensation which combines acetyl coenzyme A with oxaloacetic acid; this reaction is followed by hydrolysis of the thioester to give citric acid. The carbanion of acetyl CoA reacts with the keto group of the oxaloacetate.

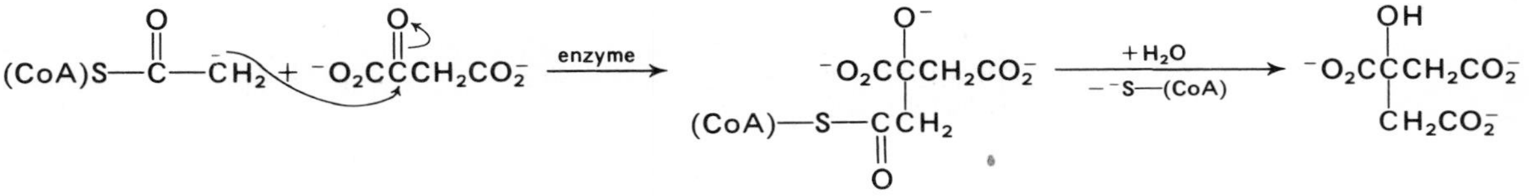

acetyl CoA   oxaloacetate ion   citrate ion

14.9 DEGRADATION OF FATTY ACIDS BY CLEAVAGE OF A β-KETOESTER

The cleavage of β-ketothioesters (reverse Claisen condensations) is employed by nature as the method of degrading the long fatty acids to acetyl coenzyme A. A sequence of the degradation is illustrated here.

The final reaction in a four-reaction sequence in the stepwise degradation of fatty acids involves a carbon–carbon bond cleavage of a β-ketoester of coenzyme A. In the reaction the ketothioester becomes two thioesters—one, acetyl coenzyme A and the other, an ester having two less carbons than the original one.

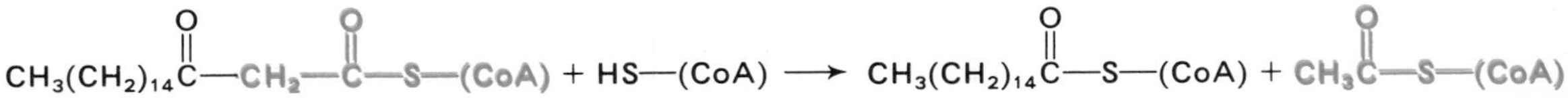

octadecanoyl CoA   CoA   hexadecanoyl CoA   acetyl CoA

The four-reaction sequence for the degradation of fatty acids starts with the formation of an α,β-unsaturated ester from the saturated ester, by the action of a coenzyme, FAD.

$$R-CH_2-CH_2-\overset{\overset{\displaystyle O}{\|}}{C}-S-(CoA) + FAD \longrightarrow R-CH=CH-\overset{\overset{\displaystyle O}{\|}}{C}-S-(CoA) + FADH_2$$

| acyl coenzyme A | coenzyme | unsaturated acyl CoA | hydrogenated coenzyme |

This biological dehydrogenation has no direct counterpart in the laboratory. The coenzyme FAD responsible for the removal of two hydrogens is made from **riboflavin** (vitamin B$_2$). The flavin portion of FAD (flavin adenine dinucleotide) contains a group of atoms including two unsaturated nitrogens capable of reacting with two hydrogens. The nitrogens are in 1,4-position to each other, and each becomes bonded to one hydrogen.

Single-pronged arrowhead indicates the movement of a single electron

This group of atoms is found in a tricyclic ring system with four nitrogens called **flavin**.

a flavin (oxidized form)   a flavin-H$_2$ (reduced form)

For riboflavin, R $= -CH_2-CHOH-CHOH-CHOH-CH_2-OH$.
The dehydrogenation coenzyme is composed of four attached parts, riboflavin-diphosphate-ribose-adenine.

$$\text{flavin-CH}_2-CHOH-CHOH-CHOH-CH_2-O-P\bar{O}_3-P\bar{O}_3\text{-ribose-adenine}$$

riboflavin   adenosine

FAD (flavin adenine dinucleotide)

All of the steps in the sequence for the oxidation and cleavage of fatty acids have been discussed. They are dehydrogenation of the saturated ester, addition of water, oxidation of the hydroxyl group to a keto group, and cleavage by reverse Claisen condensation.

Problem   **14.8** To illustrate the degradation of a fatty acid, write the structures for the products A through H for this series of reactions starting with hexanoyl coenzyme A.

a) $CH_3(CH_2)_4\overset{\overset{\displaystyle O}{\|}}{C}$—S(CoA) $\xrightarrow[\text{(Section 8.12)}]{\overset{\text{(dehydrogenation)}}{\text{FAD}}}$ A $\xrightarrow{\overset{\text{(addition)}}{\underset{}{H_2O, \text{ enzyme}}}}$

B $\xrightarrow[\text{(Section 11.13)}]{\overset{\text{(oxidation)}}{NAD^+, \text{ enzyme}}}$ C $\xrightarrow{\overset{\text{(cleavage)}}{HS—CoA}}$ D + E

b) D $\xrightarrow{\text{FAD}}$ F $\xrightarrow{H_2O}$ G $\xrightarrow{NAD^+}$ H $\xrightarrow{HS—CoA}$ E + E

14.10 MALONIC ESTER REACTIONS AND DECARBOXYLATIONS

Two carbonyl groups attached to the same carbon greatly increase the acidity of the bonded hydrogens, as the pK_a's in Fig. 14.1 indicate. We have seen that the carbanion is stabilized by delocalization of electrons and charge onto both carbonyl groups.

$$\left[\overset{\overset{\displaystyle O}{\|}}{\underset{}{C}}\underset{\overset{\displaystyle -}{CH}}{}\overset{\overset{\displaystyle O}{\|}}{\underset{}{C}} \longleftrightarrow \overset{\overset{\displaystyle O^-}{|}}{\underset{}{C}}\underset{CH}{}\overset{\overset{\displaystyle O}{\|}}{\underset{}{C}} \longleftrightarrow \overset{\overset{\displaystyle O}{\|}}{\underset{}{C}}\underset{CH}{}\overset{\overset{\displaystyle O^-}{|}}{\underset{}{C}} \right]$$

enolate ion

One of the results of this increased acidity is evident in the stabilization of the β-ketoester carbanion formed as the product in the ester condensation reaction (Section 14.8).

With a pK_a of 13, diethyl malonate is a much stronger acid than ethanol ($pK_a = 18$). The equilibrium established by the reaction of diethyl malonate with sodium ethoxide in ethanol strongly favors the products.

$C_2H_5O\overset{\overset{\displaystyle O}{\|}}{C}\!-\!CH_2\!-\!\overset{\overset{\displaystyle O}{\|}}{C}\!-\!OC_2H_5 + C_2H_5O^- \;\rightleftharpoons\; C_2H_5O\overset{\overset{\displaystyle O^-}{|}}{C}\!=\!CH\!-\!\overset{\overset{\displaystyle O}{\|}}{C}\!-\!OC_2H_5 + C_2H_5OH$

diethyl malonate

enolate ion of
malonic ester

By this reaction virtually all of the ester can be converted to enolate ion. This carbanion reacts with both saturated (e.g., alkyl halides) and unsaturated compounds (carbonyls) containing carbons which carry partial positive charges.

Of particular interest is the substitution reaction of the enolate ion of diethyl malonate with primary halides. This reaction is the key reaction in a synthetic sequence known as the malonic ester synthesis of carboxylic acids. By this method a carboxylic acid is prepared having two more carbons in the chain than the original alkyl halide.

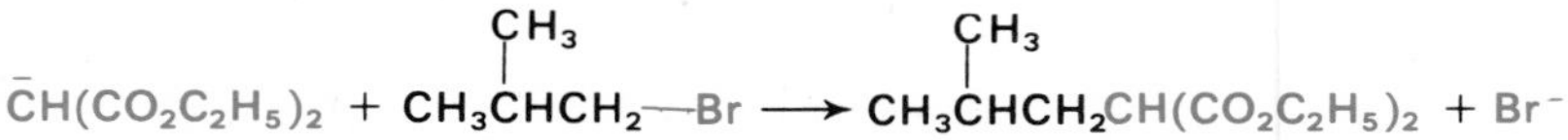

$$\overset{-}{C}H(CO_2C_2H_5)_2 + CH_3\overset{\underset{|}{CH_3}}{CH}CH_2\text{—}Br \longrightarrow CH_3\overset{\underset{|}{CH_3}}{CH}CH_2CH(CO_2C_2H_5)_2 + Br^-$$

enolate ion of diethyl malonate	isobutyl bromide	diethyl isobutylmalonate

The sequence of reactions for the malonic ester synthesis is shown as follows.

Reaction 1 Formation of the enolate ion of diethyl malonate

$$(C_2H_5O\overset{\overset{O}{||}}{C})_2CH_2 + C_2H_5O^-Na^+ \longrightarrow (C_2H_5O\overset{\overset{O}{||}}{C})_2CH^-Na^+ + C_2H_5OH$$

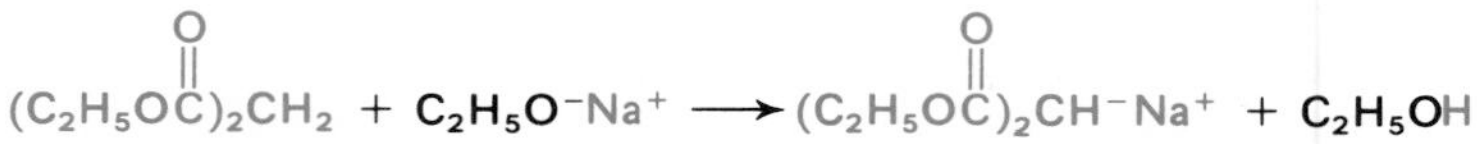

diethyl malonate   sodium enolate

Reaction 2 Substitution on the primary halide

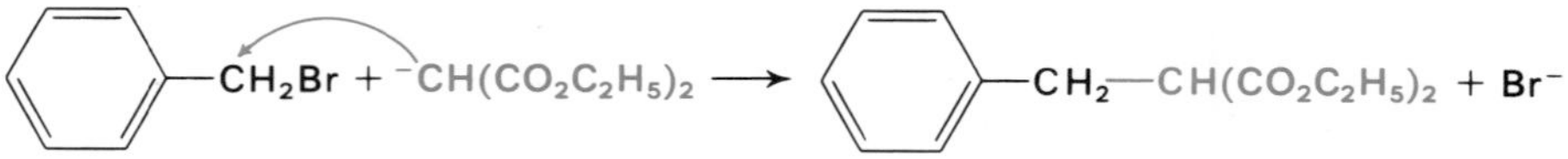

$$\langle\text{ph}\rangle\text{—}CH_2Br + {}^-CH(CO_2C_2H_5)_2 \longrightarrow \langle\text{ph}\rangle\text{—}CH_2\text{—}CH(CO_2C_2H_5)_2 + Br^-$$

benzyl bromide   diethyl benzylmalonate

Reaction 3 Hydrolysis of the diester and acidification of the salt

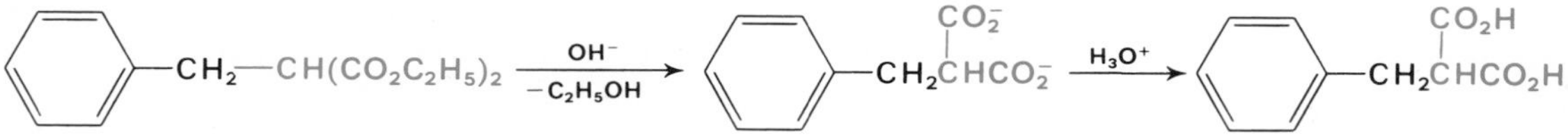

$$\langle\text{ph}\rangle\text{—}CH_2\text{—}CH(CO_2C_2H_5)_2 \xrightarrow[-C_2H_5OH]{OH^-} \langle\text{ph}\rangle\text{—}CH_2\overset{\underset{|}{CO_2^-}}{C}HCO_2^- \xrightarrow{H_3O^+} \langle\text{ph}\rangle\text{—}CH_2\overset{\underset{|}{CO_2H}}{C}HCO_2H$$

benzylmalonic acid

Reaction 4 Decarboxylation of substituted malonic acid

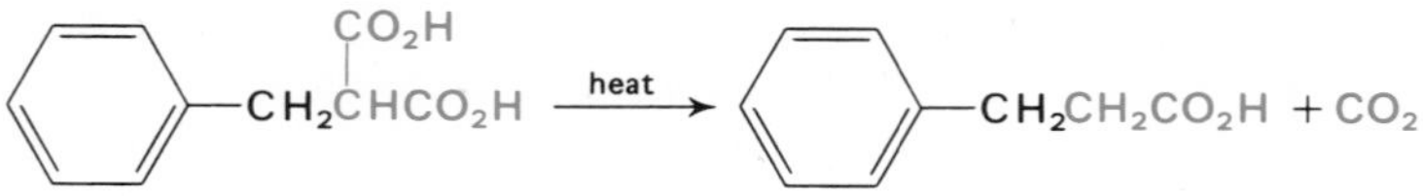

$$\langle\text{ph}\rangle\text{—}CH_2\overset{\underset{|}{CO_2H}}{C}HCO_2H \xrightarrow{heat} \langle\text{ph}\rangle\text{—}CH_2CH_2CO_2H + CO_2$$

3-phenylpropanoic acid

Remember that the first fact you learned about malonic acid (Section 8.20) was that when heated it loses CO_2 to give acetic acid. Mono- and disubstituted malonic acids also lose CO_2 readily as in reaction 4 above. By the use of the malonic ester, carboxylic acids with longer and branched carbon chains can be prepared from alkyl halides. Through the malonic ester synthesis, primary alkyl halides can be converted to carboxylic acids that contain two more carbon atoms in the chain.

The decarboxylations of 1,3-dicarboxylic acids and 3-keto acids are important in both syntheses and degradations. Loss of carbon dioxide from 3-keto acids occurs even more readily than from malonic acids. Ketones are produced, probably, through enol intermediates.

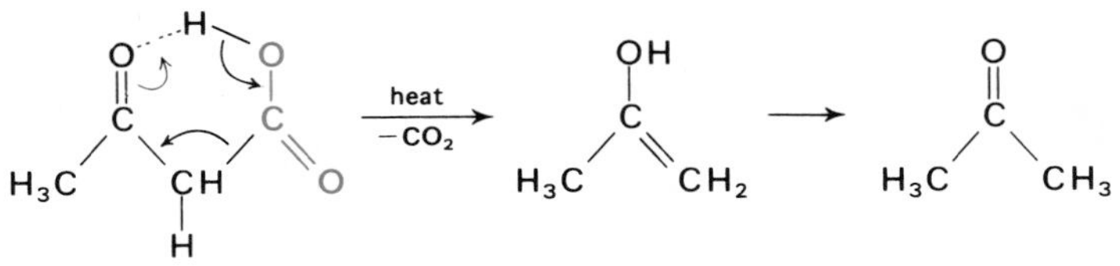

| acetoacetic acid | 2-propenol (enol) | acetone |

Problem **14.9** Write structures for products of the following series of reactions.

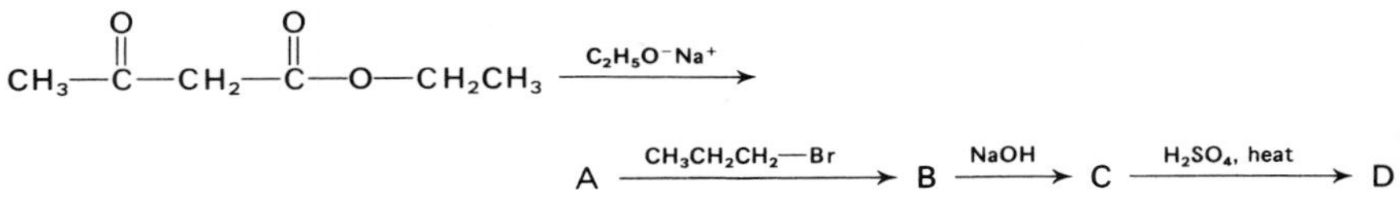

14.11 MALONIC ACID IN BIOSYNTHESIS

Biological systems employ many devices to direct the traffic necessary to the maintenance of life. Starting materials must get to the right places at the right times and in the proper concentrations without interfering with one another. One chemical traffic control system that has evolved resembles the use of two one-way streets whose vehicles go in opposite directions to and from two places. In a previous section (Section 14.9), we traced the degradation of fatty acid thioesters of coenzyme A to acetyl coenzyme A. The synthesis of fatty acids is, in principle, the reverse of the degradative sequence. In practice, two changes keep the two routes from interfering with each other, much as two one-way streets separate vehicles bound in opposite directions.

One of these differences involves the thioester group. Instead of coenzyme A, the synthesis employs a different thiol, HS—(ACP), to make the thioester. ACP are letters which stand for acyl-carrier protein, a small protein with a free sulfhydryl group, —SH. The condensations to construct the fatty acid carbon chain start with one acetyl—S—enzyme, but all other ester groups are malonyl—S—ACP. The condensation reaction is shown here between acetyl enzyme and malonyl ACP. The malonic acid formed loses CO_2 immediately.

The condensation reaction and decarboxylation are the first in the sequence of reactions which include reduction, dehydration, and hydrogenation.

Reaction 1 Condensation and decarboxylation reactions

$$
\underset{\text{acetyl Enz}}{CH_3\overset{\displaystyle O}{\overset{\|}{C}}-S-Enz} + \underset{\text{malonyl ACP}}{H-\underset{\underset{CO_2^-}{|}}{C}H\overset{\displaystyle O}{\overset{\|}{C}}-S(ACP)} \xrightarrow{-HS-Enz}
$$

$$
CH_3\overset{\displaystyle O}{\overset{\|}{C}}-\underset{\underset{CO_2^-}{|}}{C}H\overset{\displaystyle O}{\overset{\|}{C}}-S(ACP) \xrightarrow{-CO_2} \underset{\text{acetoacetyl ACP}}{CH_3\overset{\displaystyle O}{\overset{\|}{C}}CH_2\overset{\displaystyle O}{\overset{\|}{C}}-S(ACP)}
$$

Reactions 2, 3, and 4 Reduction, dehydration, and hydrogenation (Sections 11.13, 14.9)

$$
CH_3\overset{\displaystyle O}{\overset{\|}{C}}CH_2\overset{\displaystyle O}{\overset{\|}{C}}-S(ACP) \xrightarrow{NADH} CH_3\underset{\underset{OH}{|}}{C}HCH_2\overset{\displaystyle O}{\overset{\|}{C}}-S(ACP) \xrightarrow{-H_2O}
$$

$$
CH_3\overset{\displaystyle H}{\underset{\displaystyle H}{C}}=\overset{\displaystyle O}{\overset{\|}{C}}-S(ACP) \xrightarrow{FADH_2} \underset{\text{butanoyl ACP}}{CH_3CH_2CH_2\overset{\displaystyle O}{\overset{\|}{C}}-S(ACP)}
$$

Reaction 5 Condensation with malonyl ACP and decarboxylation

$$
CH_3CH_2CH_2\overset{\displaystyle O}{\overset{\|}{C}}-S(ACP) + \underset{\underset{CO_2^-}{|}}{C}H_2\overset{\displaystyle O}{\overset{\|}{C}}-S(ACP) \longrightarrow
$$

$$
CH_3CH_2CH_2\overset{\displaystyle O}{\overset{\|}{C}}\underset{\underset{CO_2^-}{|}}{C}H\overset{\displaystyle O}{\overset{\|}{C}}-S(ACP) \xrightarrow{-CO_2} CH_3CH_2CH_2\overset{\displaystyle O}{\overset{\|}{C}}CH_2\overset{\displaystyle O}{\overset{\|}{C}}-S(ACP)
$$

14.12 THE CITRIC ACID CYCLE The citric acid cycle is a sequence of reactions in which the acetyl group of acetyl coenzyme A is converted to CO_2 and H_2O.

All of the reactions of the citric acid cycle are summarized in the circular diagram given in Fig. 14.2. All but one of the reactions are familiar to you already. The carrier compound oxaloacetate picks up acetyl coenzyme A to form citric acid, a six-carbon compound (Section 14.8b). In succeeding reactions the substrate loses two carbon dioxide molecules, leaving a four-carbon oxaloacetate to begin the cycle again.

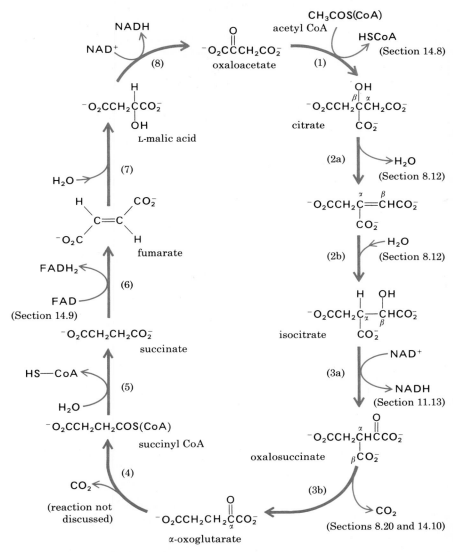

FIGURE 14.2
THE CITRIC ACID CYCLE

In this cyclic diagram, the biochemists' convention has been used to condense the equations for the reactions. Thus A $\xrightarrow[\quad]{NAD^+ \quad NADH}$ B indicates A was converted to B by reaction with NAD^+, and NADH was formed.

The reactions of the cycle in Fig. 14.2 (numbers in parentheses next to arrows) are described in the following list.

(1) Ester condensation to citric acid (Section 14.8)
(2) Dehydration of β-hydroxyacid; hydration of α,β-unsaturated acid (Section 8.12)
(3) Oxidation of alcohol and decarboxylation of β-ketoacid (Sections 11.13, 8.20, 14.10)
(4) Decarboxylation of α-ketoacid and formation of thioester (not discussed)
(5) Hydrolysis of thioester (Section 9.9)
(6) Dehydrogenation of acid to α,β-unsaturated acid (Section 14.9)
(7) Hydration to hydroxyacid
(8) Oxidation to ketoacid

14.13 VINYL POLYMERIZATION

Of all the methods for forming carbon–carbon bonds, the various reactions of vinyl polymerization produce the most spectacular results with the least labor. From these processes come most of our plastics—polystyrene, polyvinyl chloride, lucite (polymethyl methacrylate), and the acrylics (polymers of acrylonitrile CH_2=CH—CN).

Polystyrene is a very versatile material, tough, stable, and weather-resistant. It liquefies at a temperature of about 100°. The molten material can be poured into a mold for shaping or made into a fragile, spongy substance, "Styrofoam."

Polystyrene is obtained from monomeric styrene by three standard methods for forming polymers. It serves as a good example to illustrate vinyl addition polymerization.

The polymerization of styrene is an addition reaction in which a highly reactive intermediate formed from one molecule of styrene adds to an olefinic carbon of another styrene to produce a new, highly reactive intermediate of the same kind. This type of process is called a chain reaction.

$$R* + CH_2\!\!=\!\!\underset{\underset{C_6H_5}{|}}{CH} \longrightarrow R\!-\!CH_2\!-\!\underset{\underset{C_6H_5}{|}}{CH*} \xrightarrow{CH_2=CHC_6H_5} R\!-\!CH_2\!-\!\underset{\underset{C_6H_5}{|}}{CH}\!-\!CH_2\!-\!\underset{\underset{C_6H_5}{|}}{CH*} \text{ etc.}$$

reactive intermediate

The reactive intermediate designated by the asterisk (*) above may be a cation ($+$), an anion ($-$), or a free radical, a neutral fragment with an unpaired electron ($\cdot$); all of these are stabilized by electron or charge delocalization.

In each case an initiator starts a chain reaction which continues as each new molecule of styrene becomes involved.

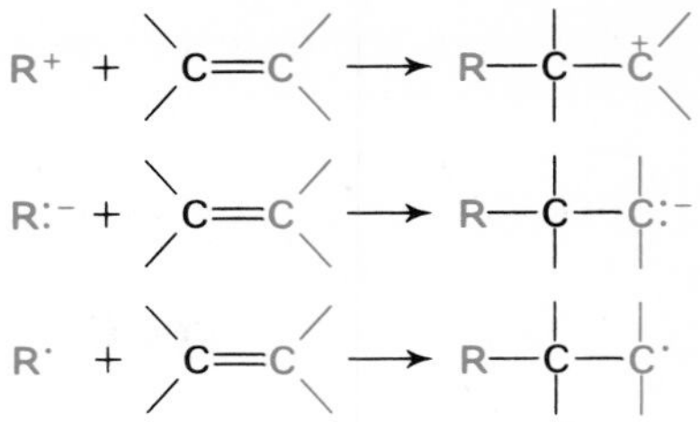

$$R^+ + ^{\backslash}_{/}C{=}C^{/}_{\backslash} \longrightarrow R{-}\overset{|}{\underset{|}{C}}{-}\overset{+}{\underset{}{C}}{\big\backslash}^{/}$$

$$R{:}^- + ^{\backslash}_{/}C{=}C^{/}_{\backslash} \longrightarrow R{-}\overset{|}{\underset{|}{C}}{-}\overset{}{\underset{|}{C}}{:}^-$$

$$R^{\cdot} + ^{\backslash}_{/}C{=}C^{/}_{\backslash} \longrightarrow R{-}\overset{|}{\underset{|}{C}}{-}\overset{}{\underset{|}{C}}{\cdot}$$

a) Cationic polymerization Polymerization of styrene can be initiated by sulfuric acid or by a nonprotic acid like boron trifluoride with a trace of water ($BF_3 + H_2O \rightarrow HOBF_3^- H^+$).

Chain initiation

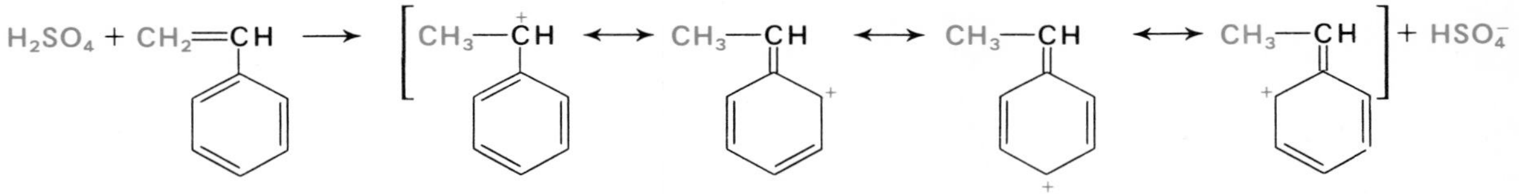

Chain propagation

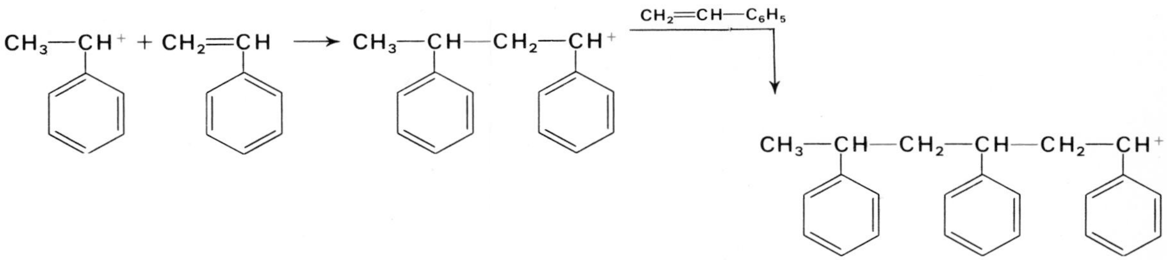

In each addition of the positive ion, the carbonium ion produced is a secondary carbonium ion stabilized by delocalization of charge into the benzene ring.

Chain termination

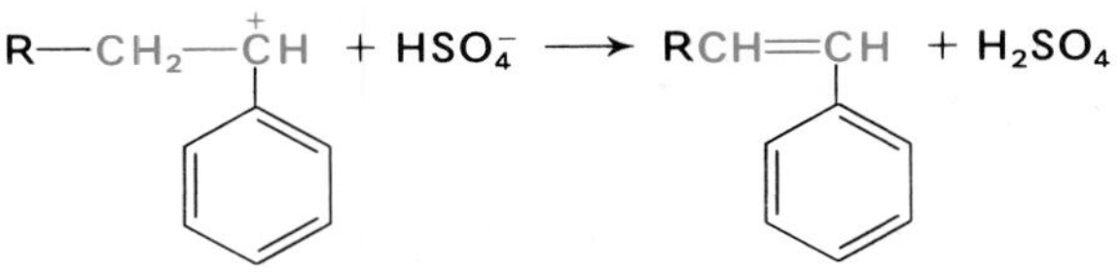

$$R{-}CH_2{-}\overset{+}{C}H + HSO_4^- \longrightarrow RCH{=}CH + H_2SO_4$$

Chain termination usually is the result of a transfer of a proton from the carbonium ion to some base, forming a terminal olefinic linkage.

b) Free-radical polymerization More important than cationic polymerization is the initiation and development of polymers through free radical intermediates. If a compound cleaves by breaking one bond in which one electron of the pair stays with one atom and the other electron goes with the second bonded atom, two neutral species are formed, each having one unpaired electron. The odd electron species are called free radicals.

$$A\!:\!B \longrightarrow A\cdot + B\cdot$$

Compounds having one weak bond (e.g., a peroxide, R—O—O—R) frequently decompose, when heated, to form two free radicals, RO·. Benzoyl peroxide decomposes slowly at about 70–80° to give benzoyl free radicals.

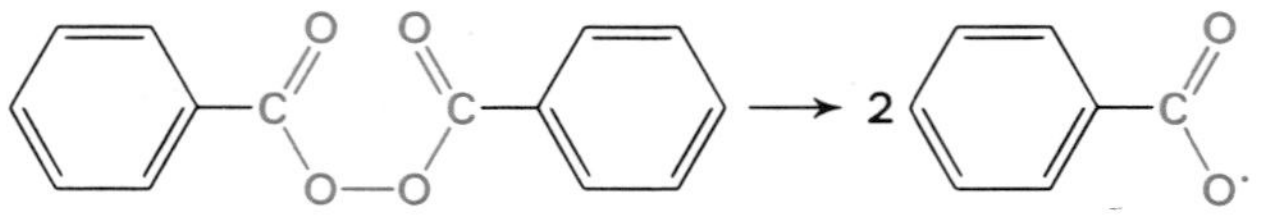

benzoyl peroxide   benzoyl radicals

Benzoyl radical then adds very rapidly to the olefinic carbon of styrene in a process which initiates the chain reaction for polymerization.

Chain initiation

$$C_6H_5-CO_2\cdot + CH_2{=}CH-C_6H_5 \longrightarrow C_6H_5-CO_2-CH_2-\underset{\underset{\displaystyle C_6H_5}{|}}{C}H\cdot$$

Chain propagation

$$C_6H_5CO_2-CH_2-\underset{\underset{\displaystyle C_6H_5}{|}}{C}H\cdot + CH_2{=}\underset{\underset{\displaystyle C_6H_5}{|}}{C}H \longrightarrow C_6H_5CO_2-CH_2-\underset{\underset{\displaystyle C_6H_5}{|}}{C}H-CH_2-\underset{\underset{\displaystyle C_6H_5}{|}}{C}H\cdot \quad \text{etc.}$$

Each free radical adds to the styrene to give the more stable new free radical, the one in which the unpaired electron is on the carbon adjacent to the aromatic ring. Delocalization of the single electron into the ring stabilizes this free radical exactly as the benzylic cation above is stabilized.

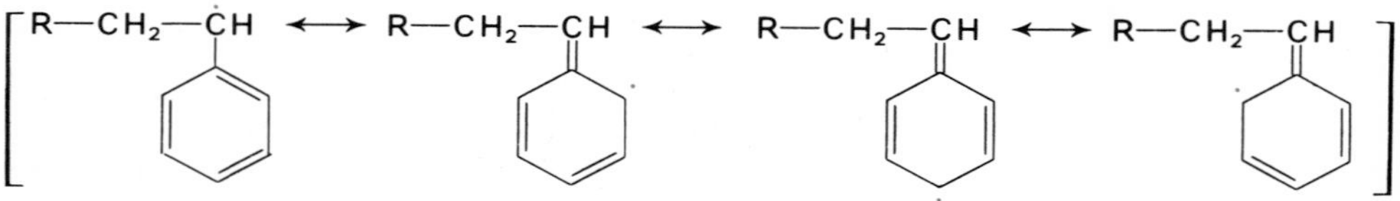

The chain is terminated by combination of two free radicals, forming a new C—C bond, or by disproportionation, in which a hydrogen with one electron H· is transferred from one free radical to a second.

Chain termination by combination of two free radicals

$$2 \ R—CH_2—\overset{\cdot}{C}H \longrightarrow R—CH_2—\underset{R}{C}H—\underset{R}{C}H—CH_2—R$$

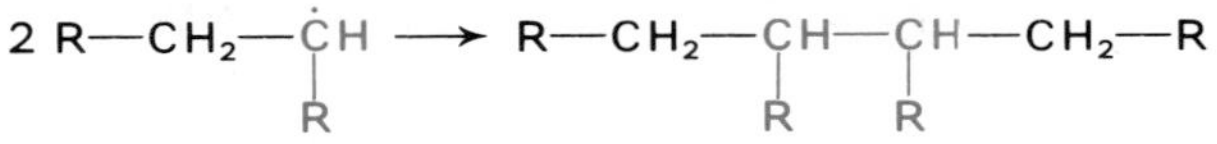

Chain termination by disproportionation—transfer of H·

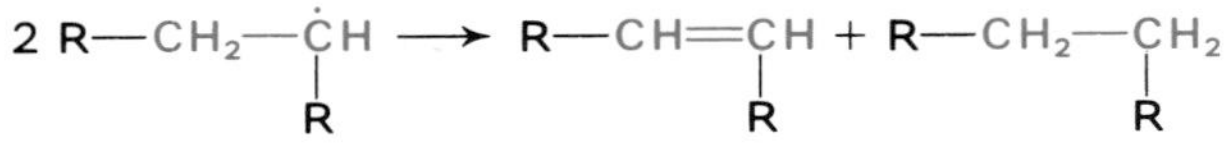

$$2 \ R—CH_2—\overset{\cdot}{C}H \longrightarrow R—CH=\underset{R}{C}H + R—CH_2—\underset{R}{C}H_2$$

c) Anionic polymerization

Anions for initiators are very strong bases like the amide ion NH_2^-. The same process of chain initiation, propagation, and termination occurs.

With styrene the anion adds to the olefinic linkage to give a new anion with negative charge on the carbon adjacent to the aromatic ring. This anion is stabilized by delocalization of the charge and electrons into the ring.

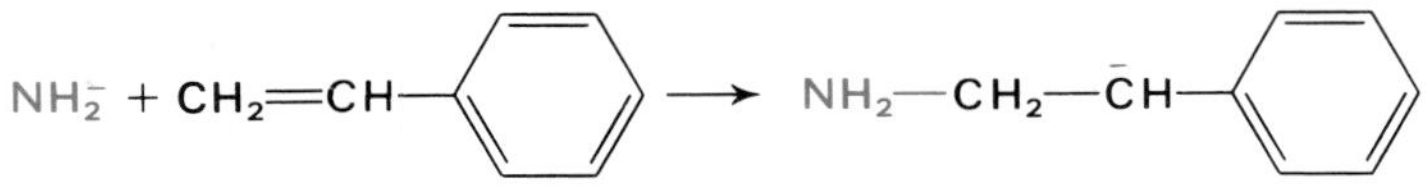

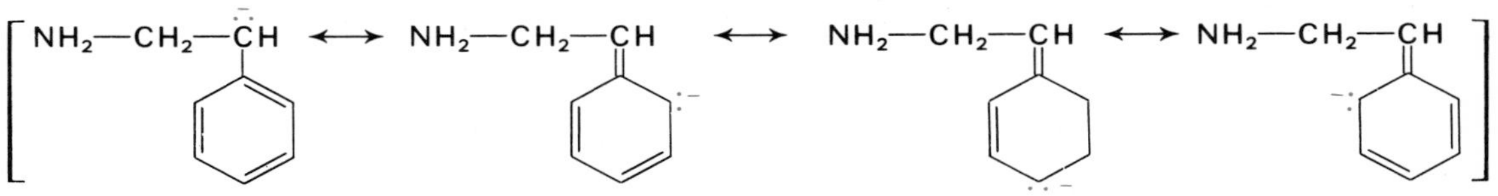

In all of these polymerization techniques, it is possible to add divinylbenzene to provide a high degree of cross linking between growing chains. Cross linking produces a rigid, tough polymer.

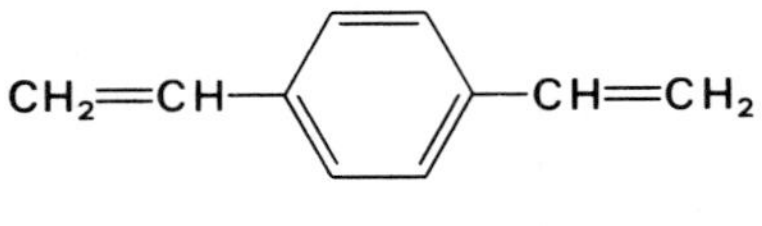

divinylbenzene

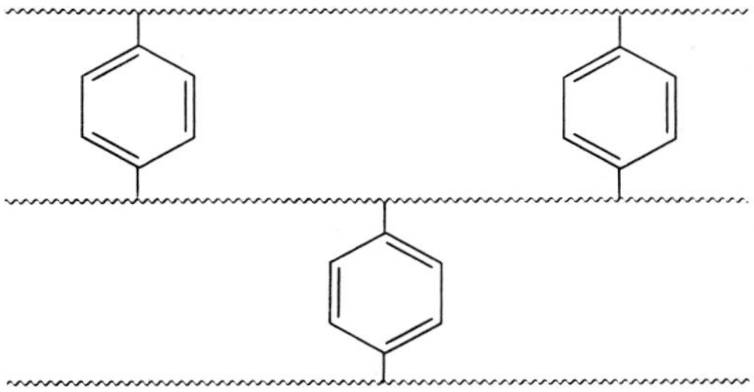

cross-linked polystyrene

Polymers which have elastic properties similar to natural rubber are prepared by the copolymerization by free radicals of butadiene and styrene. For such a growing copolymer, the free radical with styrene at the end reacts faster with the butadiene, and the free radical with butadiene at the end reacts faster with styrene. Thus the polymer is composed of alternating styrene and butadiene units.

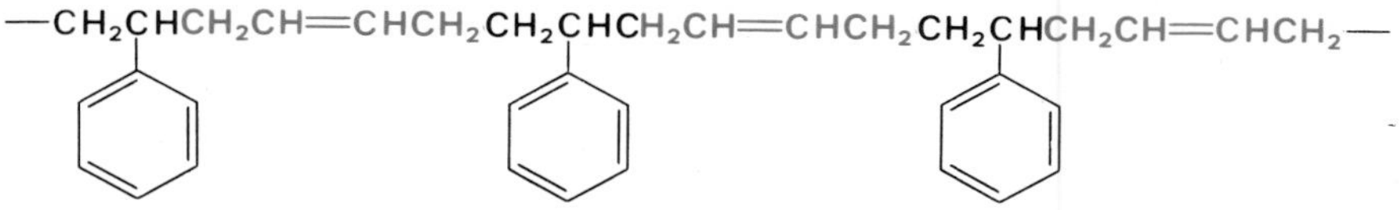

To have the properties of an elastomer, a copolymer must have double bonds in the chain, and they must have the *cis* configuration.

Problem **14.10** Styrene forms a copolymer with methyl methacrylate in which the monomer units alternate. The copolymer is initiated by free radicals. Write one series of addition reactions in the formation of the copolymer.

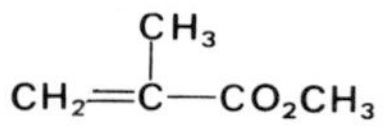

$$CH_2=\overset{\overset{\displaystyle CH_3}{|}}{C}-CO_2CH_3$$

methyl methacrylate

New Terms and Topics

Acidity range of very weak acids—α-protons of carbonyl and carboxyl compounds, hydrocarbons (Section 14.1)

Organometallic compounds of magnesium, lithium, and sodium (Section 14.1)

Grignard reagents as nucleophiles for aldehydes, ketones, and esters (Sections 14.3, 14.4)

Condensation reactions; aldol (aldehydes and ketones); Claisen (esters) (Sections 14.6, 14.7)

Reverse Claisen condensation for breaking carbon–carbon bonds (Section 14.7)

Enolate ions of malonic ester and β-ketoesters (Section 14.10)

Enolate ions as nucleophiles (Sections 14.6, 14.7, 14.10)

Decarboxylation of malonic acids and ketoacids (Section 14.10)

Biosyntheses and degradations using condensation and reverse reactions (Sections 14.8, 14.9)

The citric acid cycle (Section 14.12)

Polystyrene (Section 14.13)

Vinyl polymerization; cationic, anionic and free radical (Section 14.13)

Chain reaction; initiation, propagation, termination (Section 14.13)

Vinyl copolymer (Section 14.13)

Summary of Reactions

PREPARATION OF GRIGNARD REAGENTS

Alkyl and aryl halides with metallic magnesium (Section 14.2)

$$R-Cl + Mg \xrightarrow{\text{dry ether}} R-Mg-Cl$$

$$(CH_3)_2CH-Cl + Mg \xrightarrow{\text{dry ether}} (CH_3)_2CH-Mg-Cl$$

isopropyl   metallic   isopropylmagnesium
chloride   magnesium   chloride

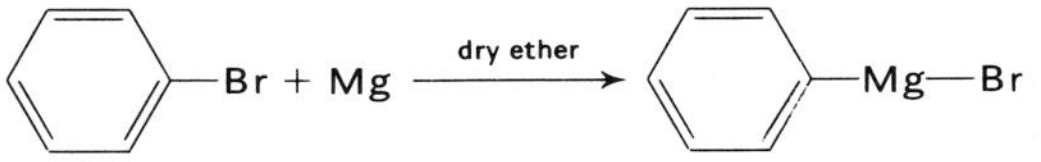

bromobenzene   phenylmagnesium bromide

REACTIONS OF GRIGNARD REAGENTS

1. With aldehydes and ketones (Section 14.3)

$$R-Mg-X + R_2C{=}O \xrightarrow{\text{dry ether}} R_3C-O-Mg-X \xrightarrow{H_3O^+} R_3C-OH$$

$$(CH_3)_2CH-Mg-Cl + CH_3CH_2\overset{O}{\overset{\|}{C}}CH_3 \xrightarrow{\text{dry ether}} CH_3CH_2\overset{O-MgCl}{\underset{CH_3}{C}}CH(CH_3)_2 \xrightarrow{H_3O^+} CH_3CH_2\overset{OH}{\underset{CH_3}{C}}CH(CH_3)_2$$

isopropyl-   2-butanone   2,3-dimethyl-3-
magnesium chloride   pentanol

2. With esters (Section 14.4)

$$2\ RMgX + RCO_2R' \xrightarrow{\text{dry ether}} R_3COMgX + R'OMgX \xrightarrow{2\ H^+} R_3COH + R'OH$$

$$2\ CH_3CH_2MgCl + C_6H_5CO_2CH_3 \longrightarrow C_6H_5\overset{OMgCl}{C}(CH_2CH_3)_2 + CH_3OMgCl \xrightarrow{2\ H^+} C_6H_5\overset{OH}{C}(CH_2CH_3)_2 + CH_3OH$$

ethylmagnesium   methyl benzoate   3-phenyl-3-pentanol   methanol
chloride

3. With carbon dioxide (Section 14.3)

$$R-Mg-X + CO_2 \longrightarrow R-CO_2-Mg-X \xrightarrow{H^+} R-CO_2H$$

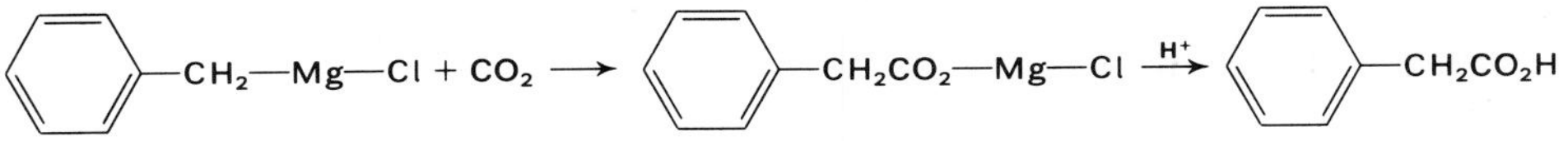

$$C_6H_5-CH_2-Mg-Cl + CO_2 \longrightarrow C_6H_5-CH_2CO_2-Mg-Cl \xrightarrow{H^+} C_6H_5-CH_2CO_2H$$

benzylmagnesium   phenylacetic acid
chloride

CONDENSATION REACTIONS

1. Aldehydes—aldol condensation (Section 14.6)

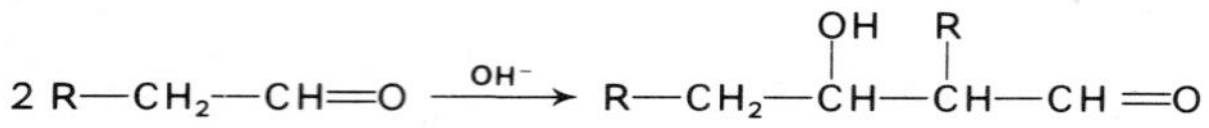

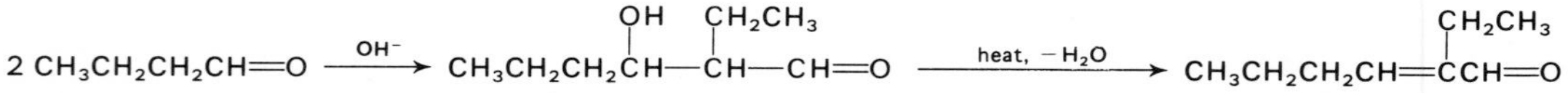

$$2\ R\!-\!CH_2\!-\!CH\!=\!O \xrightarrow{\ OH^-\ } R\!-\!CH_2\!-\!\underset{}{\overset{OH}{CH}}\!-\!\underset{}{\overset{R}{CH}}\!-\!CH\!=\!O$$

$$2\ CH_3CH_2CH_2CH\!=\!O \xrightarrow{\ OH^-\ } CH_3CH_2CH_2\underset{}{\overset{OH}{CH}}\!-\!\underset{}{\overset{CH_2CH_3}{CH}}\!-\!CH\!=\!O \xrightarrow{\ heat,\ -H_2O\ } CH_3CH_2CH_2CH\!=\!\underset{}{\overset{CH_2CH_3}{C}}CH\!=\!O$$

butanal   2-ethyl-3-hydroxyhexanal   2-ethyl-2-hexenal

2. Esters—Claisen condensation (Section 14.7)

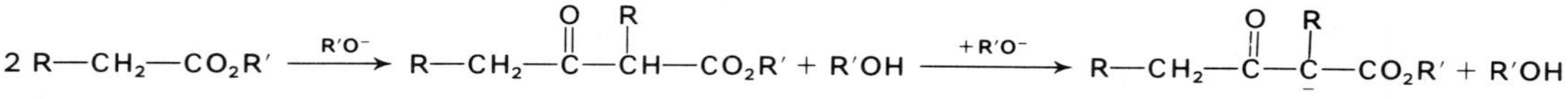

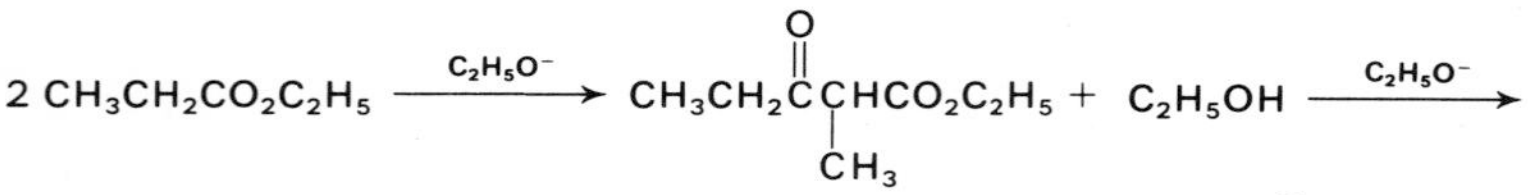

$$2\ R\!-\!CH_2\!-\!CO_2R' \xrightarrow{\ R'O^-\ } R\!-\!CH_2\!-\!\overset{O}{\overset{\|}{C}}\!-\!\underset{}{\overset{R}{CH}}\!-\!CO_2R' + R'OH \xrightarrow{\ +R'O^-\ } R\!-\!CH_2\!-\!\overset{O}{\overset{\|}{C}}\!-\!\underset{}{\overset{R}{\underset{-}{C}}}\!-\!CO_2R' + R'OH$$

$$2\ CH_3CH_2CO_2C_2H_5 \xrightarrow{\ C_2H_5O^-\ } CH_3CH_2\overset{O}{\overset{\|}{C}}CHCO_2C_2H_5 + C_2H_5OH \xrightarrow{\ C_2H_5O^-\ }$$
$$\underset{CH_3}{|}$$

ethyl propanoate

$$CH_3CH_2\overset{O}{\overset{\|}{C}}\underset{CH_3}{\underset{|}{C}}CO_2C_2H_5 + C_2H_5OH \xrightarrow{\ H^+\ } CH_3CH_2\overset{O}{\overset{\|}{C}}\underset{CH_3}{\underset{|}{CH}}CO_2C_2H_5$$

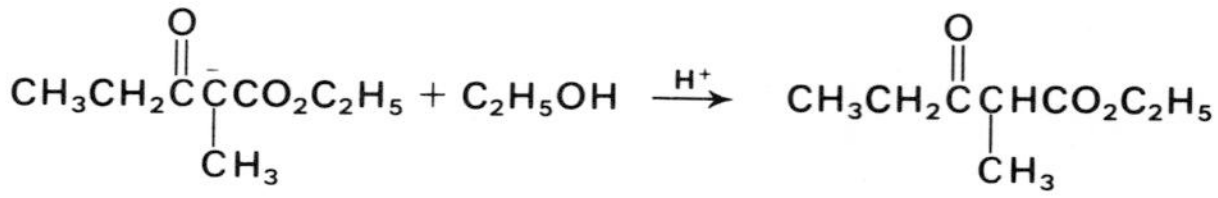

ethyl 2-methyl-3-oxopentanoate

REVERSE CONDENSATIONS WITH STRONG CONCENTRATED BASE (Section 14.7)

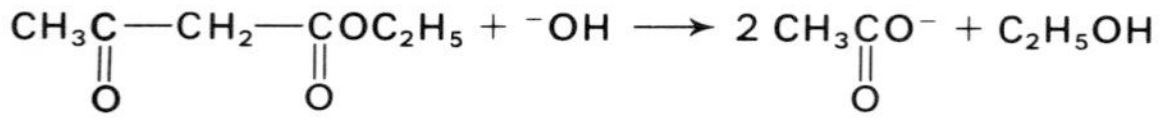

$$CH_3\underset{O}{\overset{}{\underset{\|}{C}}}\!-\!CH_2\!-\!\underset{O}{\overset{}{\underset{\|}{C}}}OC_2H_5 + {}^-OH \longrightarrow 2\ CH_3\underset{O}{\overset{}{\underset{\|}{C}}}O^- + C_2H_5OH$$

DECARBOXYLATIONS OF β-CARBONYL ACIDS (Section 14.10)

Malonic acid

$$RCH(CO_2H)_2 \xrightarrow{\ \Delta\ } RCH_2CO_2H + CO_2$$

β-Ketoacid

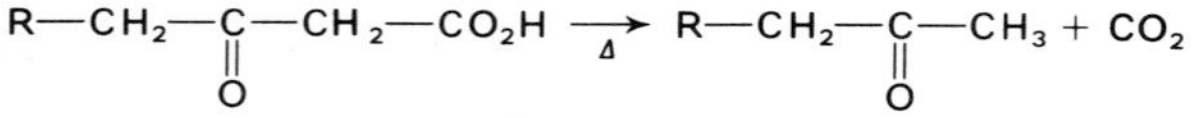

$$R\!-\!CH_2\!-\!\underset{O}{\overset{}{\underset{\|}{C}}}\!-\!CH_2\!-\!CO_2H \xrightarrow{\ \Delta\ } R\!-\!CH_2\!-\!\underset{O}{\overset{}{\underset{\|}{C}}}\!-\!CH_3 + CO_2$$

MALONIC ESTER SYNTHESIS (Section 14.10)

$$R\!-\!Cl + CH_2(CO_2C_2H_5)_2 \xrightarrow{\ RO^-\ } R\!-\!CH(CO_2C_2H_5)_2 + Cl^-$$

$$R\!-\!CH(CO_2C_2H_5)_2 \xrightarrow[\text{2) } H^+]{\text{1) } OH^-} R\!-\!CH_2CO_2H + CO_2 + 2\ C_2H_5OH$$

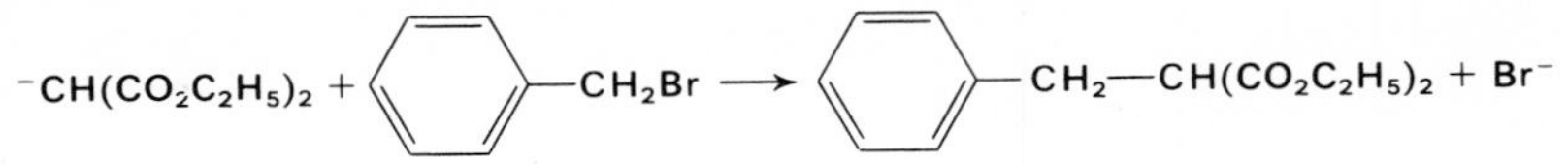

$^-CH(CO_2C_2H_5)_2$ + [benzene]—CH_2Br → [benzene]—CH_2—$CH(CO_2C_2H_5)_2$ + Br^-

malonic ester anion   benzyl bromide   diethyl benzylmalonate

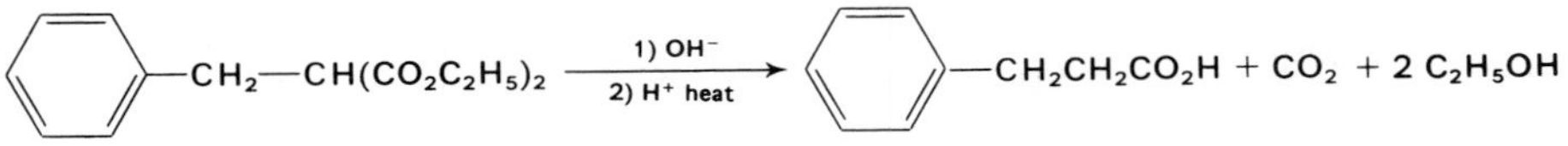

[benzene]—CH_2—$CH(CO_2C_2H_5)_2$ $\xrightarrow[\text{2) H}^+\text{ heat}]{\text{1) OH}^-}$ [benzene]—$CH_2CH_2CO_2H$ + CO_2 + 2 C_2H_5OH

3-phenylpropanoic acid

VINYL POLYMERIZATION OF STYRENE (Section 14.13)

1. Cationic polymerization

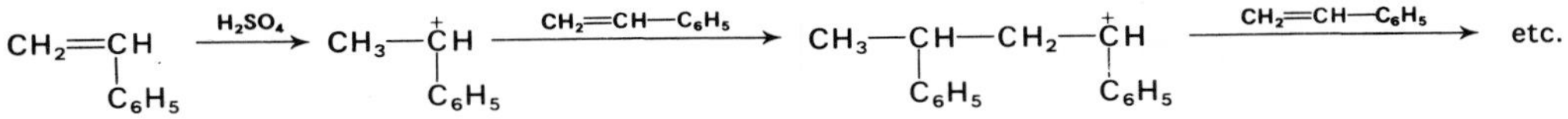

$CH_2{=}CH$ (C_6H_5) $\xrightarrow{H_2SO_4}$ $CH_3{-}\overset{+}{C}H$ (C_6H_5) $\xrightarrow{CH_2=CH-C_6H_5}$ $CH_3{-}CH{-}CH_2{-}\overset{+}{C}H$ (C_6H_5) (C_6H_5) $\xrightarrow{CH_2=CH-C_6H_5}$ etc.

2. Free radical polymerization

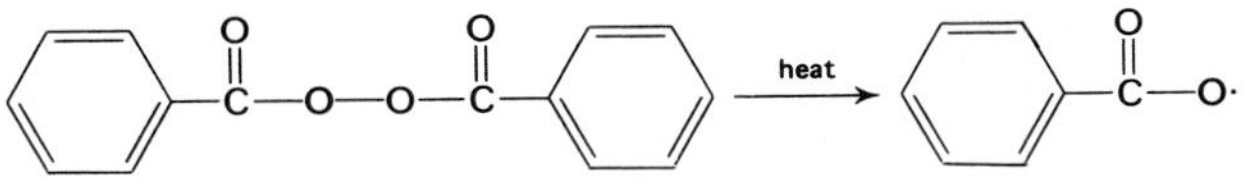

[benzoyl peroxide structure] $\xrightarrow{\text{heat}}$ [benzoyl free radical structure]

benzoyl peroxide   benzoyl free radical

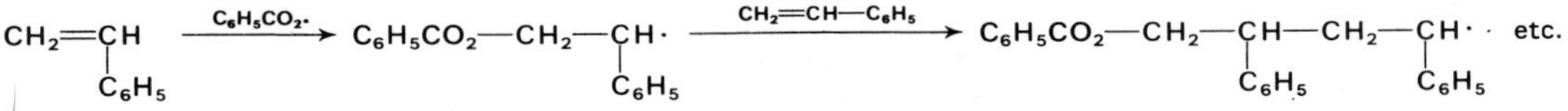

$CH_2{=}CH$ (C_6H_5) $\xrightarrow{C_6H_5CO_2\cdot}$ $C_6H_5CO_2{-}CH_2{-}\overset{.}{C}H$ (C_6H_5) $\xrightarrow{CH_2=CH-C_6H_5}$ $C_6H_5CO_2{-}CH_2{-}CH{-}CH_2{-}CH\cdot$ (C_6H_5) (C_6H_5) etc.

3. Anionic polymerization

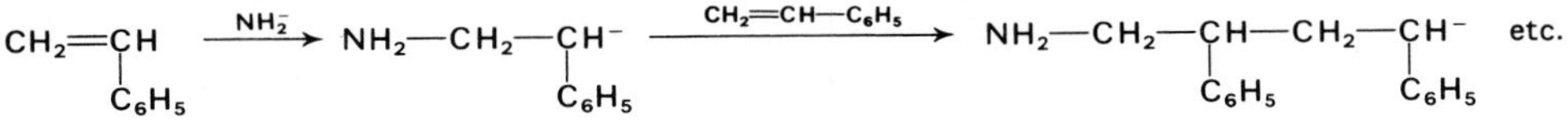

$CH_2{=}CH$ (C_6H_5) $\xrightarrow{NH_2^-}$ $NH_2{-}CH_2{-}\overset{-}{C}H$ (C_6H_5) $\xrightarrow{CH_2=CH-C_6H_5}$ $NH_2{-}CH_2{-}CH{-}CH_2{-}CH^-$ (C_6H_5) (C_6H_5) etc.

Problems

14.11 Write the principal resonance structures of the most stable carbanions derivable from the following compounds.

a) $CH_3CH_2CH_2CH{=}O$

b) 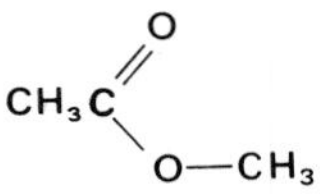 $CH_3C(=O){-}O{-}CH_3$

c) 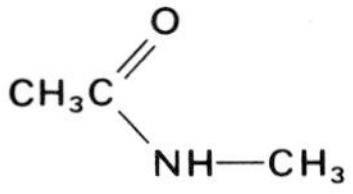 $CH_3C(=O){-}NH{-}CH_3$

d) $CH_3CH_2C{\equiv}N$

e) $CH_2(CO_2CH_3)_2$

f) 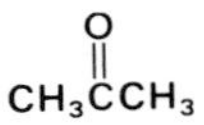 $CH_3\overset{O}{\overset{\|}{C}}CH_3$

g) 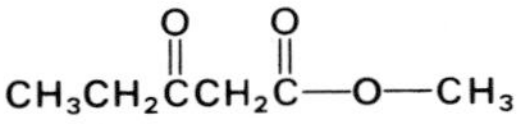 $CH_3CH_2\overset{O}{\overset{\|}{C}}CH_2\overset{O}{\overset{\|}{C}}{-}O{-}CH_3$

h) 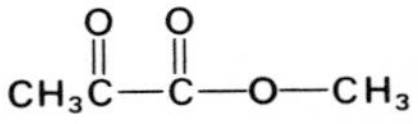 $CH_3\overset{O}{\overset{\|}{C}}{-}\overset{O}{\overset{\|}{C}}{-}O{-}CH_3$

i) 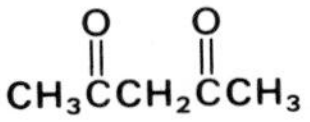 $CH_3\overset{O}{\overset{\|}{C}}CH_2\overset{O}{\overset{\|}{C}}CH_3$

14.12 Write equations for the following reactions.

    a) isopropylmagnesium chloride + pentanal in dry ether followed by acid

    b) ethyl butanoate + sodium ethoxide followed by acid

    c) 3-methylbutanal + sodium hydroxide heated

    d) methylmagnesium iodide + ethyl 2-propylhexanoate in dry ether followed by acid

    e) diethyl malonate + sodium ethoxide followed by benzyl chloride

14.13 Write structures for the products of the following reactions.

    a) $CH_2{=}CH{-}CH_2{-}MgCl + CO_2 \xrightarrow[\text{2) HCl}]{\text{1) ether}}$

    b) $C_6H_5{-}MgBr + CH_3CH_2CH_2CH_2CH{=}O \xrightarrow[\text{2) HCl}]{\text{1) ether}}$

    c)

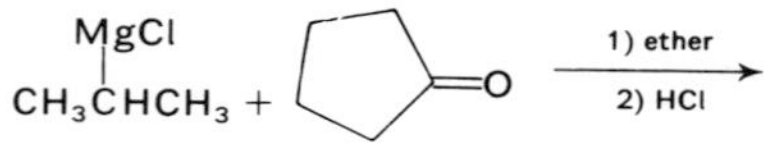

$\xrightarrow[\text{2) HCl}]{\text{1) ether}}$

    d) $CH_3O{-}\overset{O}{\overset{\|}{C}}CH_2CH_2\overset{O}{\overset{\|}{C}}{-}OCH_3 + CH_3CH_2MgCl \xrightarrow[\text{2) HCl}]{\text{1) ether}}$

    e) $CH_3CH_2CH_2CH_2CH{=}O \xrightarrow{\text{NaOH}}$

    f)

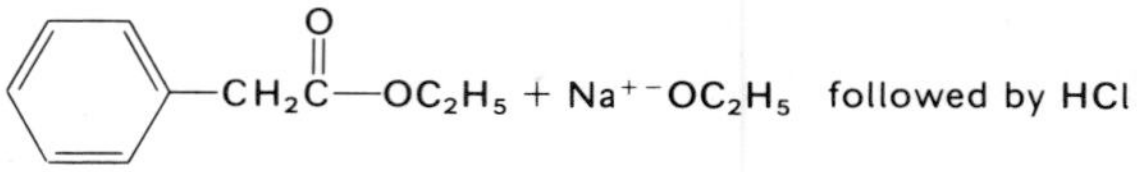

$+ Na^{+-}OC_2H_5$ followed by HCl

    g) $CH_3CH_2CH(CO_2C_2H_5)_2 \xrightarrow[\text{2) CH}_3\text{I}]{\text{1) Na}^{+-}\text{OC}_2\text{H}_5}$

    h)

$\xrightarrow[\text{2) HCl}]{\text{1) Na}^{+-}\text{OC}_2\text{H}_5}$ (cyclic product)

    i)

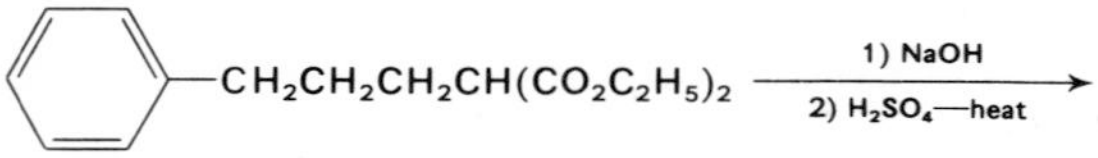

$\xrightarrow[\text{2) H}_2\text{SO}_4\text{---heat}]{\text{1) NaOH}}$

14.14 Write structures for appropriate combinations of Grignard reagent and reactant which could be used to produce the following compounds.

    a)

    b) $CH_3\overset{}{\underset{CH_3}{C}}HCH_2OH$

    c)

    d)

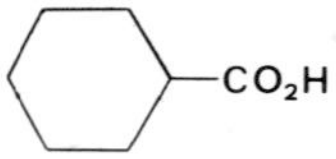

    e) $CH_3CH_2\overset{}{\underset{CH_2CH_2CH_3}{C}}H{-}OH$

14.15 Give the structures of products A through S.

    a) $CH_3CH_2CH{=}O \xrightarrow{\text{NaOH}} A \xrightarrow{\text{heat}} B \xrightarrow{\text{H}_2,\ \text{Pt}} C \xrightarrow{\text{Na}_2\text{Cr}_2\text{O}_7,\ \text{H}_2\text{SO}_4} D$

    b) $CH_3CH_2CH_2CH{=}O \xrightarrow{\text{NaBH}_4} E \xrightarrow{\text{hot H}_2\text{SO}_4} F \xrightarrow{\text{HBr}} G \xrightarrow{\text{Mg, ether}}$

$H \xrightarrow[\text{2) HCl}]{\text{1) CH}_3\text{CH}_2\text{CH}_2\text{CH}{=}\text{O}} I \xrightarrow{\text{SOCl}_2} J \xrightarrow{\text{Mg, ether}} K \xrightarrow[\text{2) HCl}]{\text{1) CO}_2} L$

    c) $CH_3CH_2CH{=}O \xrightarrow{\text{NaBH}_4} M \xrightarrow{\text{PCl}_3} N \xrightarrow{\text{Mg, ether}} O \xrightarrow[\text{2) HCl}]{\text{1) CH}_3\text{CO}_2\text{C}_2\text{H}_5} P$

    d)

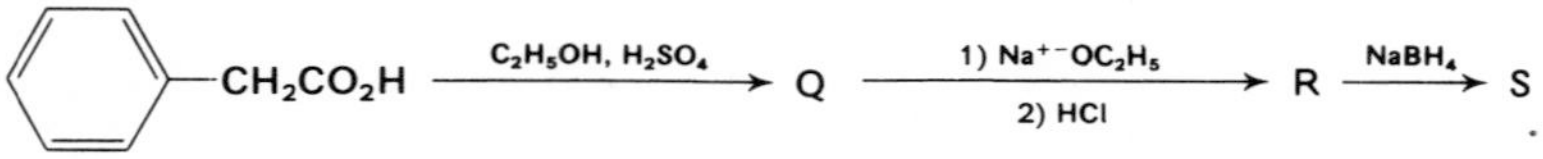

$\xrightarrow{\text{C}_2\text{H}_5\text{OH, H}_2\text{SO}_4} Q \xrightarrow[\text{2) HCl}]{\text{1) Na}^{+-}\text{OC}_2\text{H}_5} R \xrightarrow{\text{NaBH}_4} S$

14.16 By deuterium labeling of the CH_2—OH group of dihydroxyacetone phosphate, investigators have found that several enzyme systems cause the compound to condense stereospecifically with D-glyceraldehyde 3-phosphate to give D-fructose 1,6-diphosphate. In all cases the new C—C bond replaced the old C—D bond, in the compound shown below, with retention of configuration.

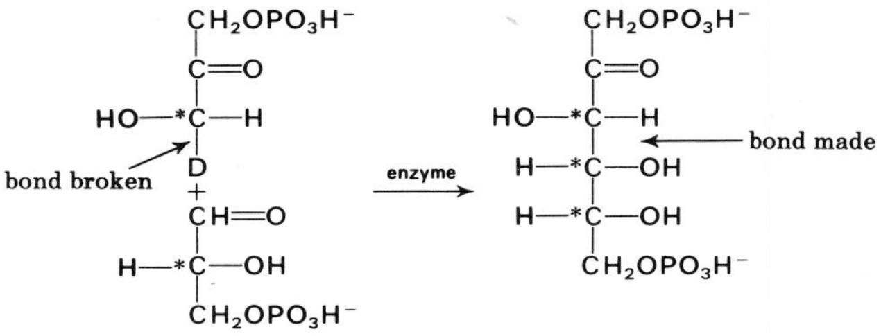

fructose 1,6-diphosphate
(Fischer projection)

a) Write the open-chain structure of the D-fructose 1,6-diphosphate that would have been obtained had the aldol reaction occurred with inversion of configuration.

b) How do you account for the stereospecificity (only the phenomenon, not the direction) of this process, as well as that exhibited in the generation of the new asymmetric carbon at C-4?

c) Which of the two enantiomers below would you expect is *more likely* to be substitutable for the dihydroxyacetone phosphate in the enzyme catalyzed reaction? Why?

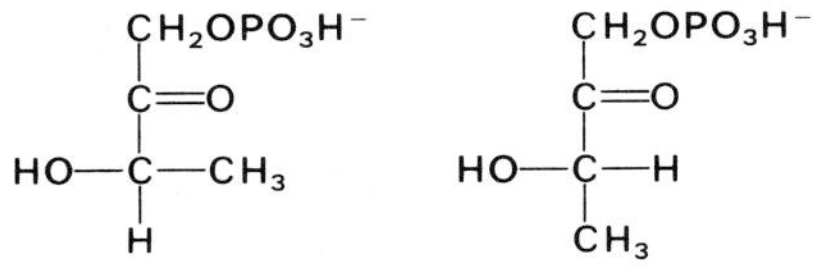

14.17 Trace the course of one round of the citric acid cycle using acetyl coenzyme A labeled with ^{14}C in the carboxyl group ($CH_3{}^{14}CO$—SCoA) and oxaloacetate ion. At what point will the $^{14}CO_2$ be given off?

CHAPTER 15

Spectroscopy of
Organic Compounds

Scientific and technological capabilities change with the decades. The methods of structure determination and the reactions which are routine processes for one generation of chemists become outdated in the next as a result of new discoveries and inventions. In the late 19th and early 20th centuries the information leading to the elucidation of the structures of compounds was accumulated bit by bit through elemental analysis, molecular weight determinations, and the study of many chemical reactions. The structures of common organic compounds were accepted when all reasonable alternatives had been eliminated.

During the course of the last fifty years the standard methods of determining structures of new compounds have changed from reaction vessels to refined instruments which yield more information in a fraction of the time. The structures of most new compounds prepared or discovered today are determined largely by spectroscopy, the study of the interaction of light with the molecules, along with further corroboration by their syntheses. Spectroscopy can be likened to a collection of knotholes in the fences around the construction site of a big building. The sidewalk observer can see a small piece of the structure through each knothole. With a collection of knotholes, it is often possible to see enough pieces to assemble an image of the entire building. The many types of spectroscopy are a series of small windows, each of which allows a glimpse of a different portion of the molecule's structure. With enough glimpses and with enough practice and skill, we can assemble a reasonable model of the real structure.

The many types of interactions of compounds with light are physical properties which depend upon the atomic sequences of the compounds. We will consider only the spectroscopic properties of very simple molecules, where similar atomic groups show similar spectroscopic features. More complex molecules have more complicated spectra whose interpretations require a greater degree of refinement and skill than we can develop here.

15.1 SPECTROSCOPY AND MOLECULAR STRUCTURE

The colors of plants—green leaves, red, blue, yellow flowers, and autumn leaves—provide visible evidence of the interaction of light with organic compounds. The compounds absorb some of the wavelengths of visible light. The remaining light, which is transmitted, appears colored to the eye. Suntan lotions screen out, by absorption, some of the sun's ultraviolet rays which are harmful to the skin. Particularly important for our health is the absorption of much ultraviolet light from the sun by oxygen and ozone in the atmosphere.

What is the nature of the interaction or absorption of light waves by compounds? How does the absorption correlate with structure? When a light beam strikes an object the radiation is either absorbed or transmitted. The beam of a single wavelength will be transmitted through the object unless the radiation has exactly the energy which the molecules of the substance can absorb. The energy absorbed by the particular molecules depends upon the structure of the molecules. Absorption of a light ray means the transference of the energy of the wave to the molecule.

The energy of the radiation depends upon the frequency of the wave; the greater the frequency, the greater the energy. The frequency of the wave is inversely related to the wavelength of the light. The shorter the wavelength, the greater is the frequency and the energy of the light, as shown by the following equation.

$v = c/\lambda$ v = frequency, in Hz (hertz, or cycles/sec)
c = velocity of light, in cm/sec
λ = wavelength, in cm

The rays of the electromagnetic spectrum vary from the long radio waves, with wavelengths measured in meters, to the much shorter infrared, visible, and ultraviolet waves, measured in 10^{-6} and 10^{-9} meters. X-rays and gamma rays are much shorter still. The waves move at a constant velocity of 3×10^{10} cm/sec.

Figure 15.1 gives the wavelength ranges associated with the various sections of the electromagnetic spectrum.

Instruments have been designed to produce a beam of light consisting of a very narrow range of frequencies. They also measure how much of the radiation is actually absorbed when the beam contacts solutions of organic compounds in nonabsorbing solvents. The information obtained from spectrometers in the form of a graph or spectrum

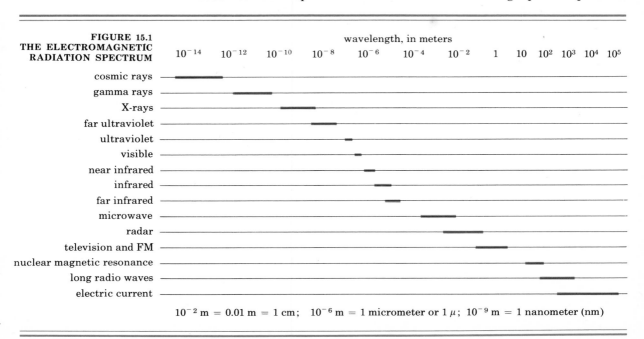

FIGURE 15.1
THE ELECTROMAGNETIC
RADIATION SPECTRUM

10^{-2} m = 0.01 m = 1 cm; 10^{-6} m = 1 micrometer or 1 μ; 10^{-9} m = 1 nanometer (nm)

shows that different compounds with similar groups of atoms absorb light of very nearly the same wavelengths or frequencies. The absorption is characteristic of different molecular parts and is used to detect their presence.

Absorption of ultraviolet and visible light is associated with multiple bonds (double and triple). Both single and multiple bonds are activated by infrared absorption. Absorption of radio waves by individual nuclei of compounds held in a strong magnetic field produces what is termed a **nuclear magnetic resonance spectrum**. Patterns of NMR spectra provide more detailed and useful information about molecular structure than any other single measurement of a physical property.

Spectral measurements are made rapidly and give the chemist much information about the structure of a compound in a short time. The present-day determination of molecular structure rests heavily on the known correlations of absorption and structure. In the sections which follow, we will examine some of the information that spectral measurements give us about the structure of compounds.

15.2 THE ULTRAVIOLET-VISIBLE SPECTRUM

Each of the major kinds of spectrometry—ultraviolet (UV), infrared (IR), and nuclear magnetic resonance (NMR)—provides a special type of information about compounds. The frequency and thus the energy of the radiation in the UV, IR, and NMR spectrometers vary greatly, so that the response of the compound to the radiation in each is quite different.

The energy of light in the ultraviolet-visible region is high enough to raise electrons in the molecule from the lowest energy orbitals they normally occupy (the **ground state**) to higher levels (an **excited state**). The absorption is known as electronic excitation.

The electrons of single bonds (sigma electrons) and of isolated multiple bonds (pi electrons) are tightly held and require a great deal of energy to be raised to a higher electronic level. The energy for these electronic transitions corresponds to the far ultraviolet region, with wavelengths below 200 nanometers (1 nm = 1×10^{-9} m), which are out of the range of the ordinary UV spectrometer.

Compounds which contain two or more double or triple bonds, however, may respond to ordinary ultraviolet light. These bonds may be C=C, C≡C, C=O, or C=N. If the double or triple bonds are separated by *only one single bond*, as in 1,3-pentadiene, less energy is required to raise the pi electrons to a higher electronic level. The lower energy requirement is a result of the interaction between the two double bonds. The double bonds in this arrangement are said to be **conjugated**.

$$CH_2=CH-CH=CH-CH_3 \qquad CH_2=CH-CH_2-CH=CH_2$$

1,3-pentadiene
(conjugated double
bonds)

1,4-pentadiene
(nonconjugated or
isolated double bonds)

Double bonds which are separated by at least two single bonds, as in 1,4-pentadiene, act as isolated multiple bonds, mentioned above. They absorb only in the far ultraviolet.

A UV absorption spectrum has axes of wavelength (λ) in nanometers and of intensity (ε). The value of ε is a constant for the compound at the wavelength measured and equals the observed absorbance for a solution in a 1-cm cell divided by the molarity of the solution. The UV spectrum for 1,3-butadiene consists of a curve or band with the greatest absorption ($\varepsilon = 21{,}000$) at a wavelength of 217 nm. The wavelength of greatest absorption is called λ_{max} and the intensity of absorption at that wavelength is ε_{max}, as shown in Fig. 15.2 for a diene and an α,β-unsaturated ketone.

FIGURE 15.2   ULTRAVIOLET ABSORPTION SPECTRA

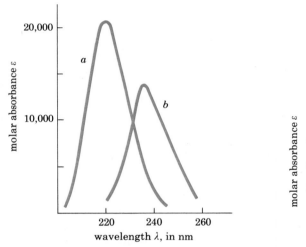
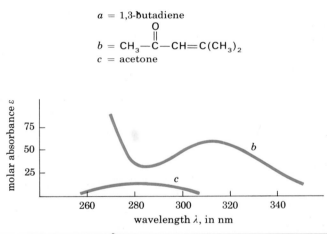

Carbonyl compounds have a very small second absorption band at about 280 nm, with the strong first band below 200 nm. When the C=O is conjugated with a C=C, the two bands move to longer wavelengths and often increase in intensity, as shown in Fig. 15.2 for acetone and an unsaturated ketone.

The addition of each new conjugated double bond moves the absorption band to longer wavelengths and often increases the intensity. With three conjugated double bonds, 1,3,5-hexatriene has a λ_{max} at 258 nm with a greater intensity than for butadiene ($\lambda_{max} = 217$ nm). The absorption of a compound with a long enough conjugated system occurs in the visible light range (400–700 nm), and the compound is colored to the eye. An example is β-carotene, the yellow-orange pigment in carrots and other plants, which has a system of 11 conjugated C=C bonds (formula given in Section 4.10). The β-carotene absorbs some of the blue light ($\lambda_{max} = 450$ nm, $\varepsilon = 140,000$) with a strong intensity and transmits only the yellow-orange light which the eye detects.

The UV spectra of benzene and pyridine are more complicated than those of the open-chain compounds. They have wide absorption bands in the ultraviolet: benzene ($\lambda_{max} = 204$ and 255 nm, $\varepsilon = 7900$ and 215); pyridine ($\lambda_{max} = 257$ and 270 nm, $\varepsilon = 2750$ and 450). Compounds containing these rings absorb in the ultraviolet. Active ingredients in suntan lotions are usually substituted anilines.

Problems **15.1** Compounds A and B are open-chain isomers with the formula C_3H_6O. The UV spectrum of A shows a weak band with $\lambda_{max} = 280$ nm and $\varepsilon = 15$; the spectrum of B shows no absorption above 210 nm. Write possible structures for the two compounds.

15.2 Compounds C and D are isomeric cyclohexadienes. Write structures for the compounds based on the UV spectra: C has a band with $\lambda_{max} = 256$, $\varepsilon = 8000$, and D shows no absorption above 210 nm.

Ultraviolet-visible absorption spectra are used to identify the length of the conjugated system of functional groups. The greatest use, however, is in analytical work. A spectrum can be obtained on a 10^{-5} to 10^{-2} M solution, depending on the intensity of the band. In dilute solution the observed absorbance is usually directly proportional to the concentration of the substance, and the absorbance at the wavelength is additive for all species present. The concentrations of dilute solutions and of individual compounds in mixtures can be determined with great accuracy by ultraviolet-visible spectra.

15.3 THE INFRARED SPECTRUM The energy of infrared radiation is lower than that of ultraviolet. The energy of infrared radiation corresponds to the energy needed to raise the compound to higher vibrational levels. Excitation may be observed in stretching vibrations of the bonds or in bending vibrations of a group of atoms, as depicted.

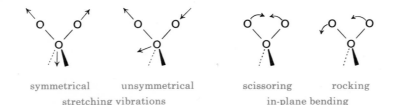

symmetrical   unsymmetrical   scissoring   rocking   wagging   twisting

stretching vibrations   in-plane bending   out-of-plane bending

With the infrared spectrum, we have a means of detecting certain groups of atoms, particularly functional groups, present in the compound. Certain groups of atoms give rise to bands at or near the same frequency regardless of the structure of the rest of the molecule. These characteristic absorptions give the chemist valuable structural information about the compound.

Infrared spectra are recorded as plots of either wavenumbers, in reciprocal centimeters cm^{-1} (frequency), or wavelengths, in microns (μ, 10^{-4} cm), versus % transmittance of light. The IR spectrum is recorded between 4000 and 660 cm^{-1} (2.5–16 μ). Absorption bands due to stretching absorptions are most easily identified, because there aren't many kinds of bonds and the absorption occurs in a part of the spectrum in which only a few bands appear, the region of 4000–1400 cm^{-1}. The ranges of absorption for some types of bonds and groups of atoms are depicted on this chart.

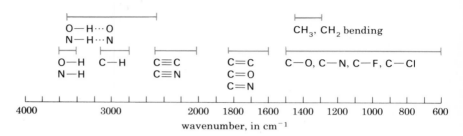

The IR spectrum of 1-butanol is shown in Fig. 15.3.

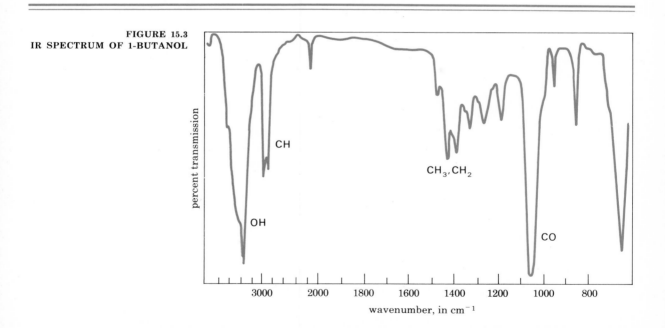

FIGURE 15.3
IR SPECTRUM OF 1-BUTANOL

The particular absorption characteristics of the groups in each class of compound are discussed in the following sections, and a table of characteristic absorptions is given in the chapter summary. Absorption bands of 1400–600 cm^{-1} are mostly associated with bending vibrations. Bending vibrations are not often as dependable in their positions as are stretching vibrations and they cannot be as clearly assigned to individual groups. The most reliable bending vibration absorptions are those of CH_3 and CH_2. Each compound, however, has its characteristic pattern of bands in this region, and this part of the spectrum is therefore called the "fingerprint region."

15.4 THE NUCLEAR MAGNETIC RESONANCE SPECTRUM

When held in a strong magnetic field, some kinds of nuclei behave like bar magnets and tend to align themselves parallel to the applied magnetic field, just as the magnetic needle of a compass aligns itself parallel to the earth's magnetic field. Absorption by these nuclei of radiation energy of radio waves causes the nuclear magnet to be flipped from a parallel to an antiparallel alignment, a state of higher energy. The spectrum produced from such absorption is called a nuclear magnetic resonance spectrum.

The energy required to flip a particular nucleus depends on the kind of atom. Only a few isotopes of certain elements show this absorption. The ^1H, ^{13}C, ^{15}N, ^{19}F, and ^{31}P isotopes have been particularly important in the determination of the sequence of atoms in compounds. We shall consider here only the *proton* magnetic resonance absorption, since it is the most commonly used. Neither deuterium ^2H nor the abundant isotope of carbon ^{12}C absorbs. The low natural abundance of ^{13}C in ordinary compounds reduces the general use of its spectra.

The proton magnetic resonance spectrum is the richest single source of information the chemist has concerning the structure of compounds. In a finely resolved spectrum, every peak can indicate something. For large molecules, however, the overlap of peaks becomes extensive and the spectra are difficult to interpret completely.

Absorption by hydrogen nuclei of a compound held in a magnetic field of 14,000 gauss occurs at a frequency of 60 megahertz (MHz or million cycles/sec). This radiation is in the radio frequency range. The frequency of the radio wave needed increases with the strength of the applied magnetic field, and finer spectra are obtained with larger magnets and 100 or 200 MHz radiation.

The aspect of proton magnetic resonance which makes it such a valuable tool for the organic chemist is that protons in different atomic groups respond to slightly different magnet-frequency conditions. The spectrum differentiates among protons on aromatic, alkene, or alkane carbons, and protons in aldehyde or carboxylic acid groups.

What is the basis for the separation of proton signals? The environment of a particular proton determines how much of the applied magnetic field the nucleus actually feels. Protons are shielded in the magnetic field by the electrons of atoms around them, with the result that the effective strength of the magnetic field is lowered. If excitation is to occur, the frequency of the radio wave must match the *effective* field strength. Instrumentally, greater accuracy is achieved with a constant frequency and a variable magnetic strength than the reverse. *At a constant frequency*, all protons absorb at the same effective field strength, but different *applied* field strengths are required to produce the needed effective field strength. In order to overcome the shielding effect, the applied field strength is increased. The required increase in field strength has a different value for excitation of protons in different molecular environments, and the spectrum has a signal or band for each kind of proton.

The spectrum furnishes a wealth of information about the structure of compounds with respect to the sequence of bonds as well as the spatial relationships of atoms. In the sections which follow, let us examine the type of information which can be obtained.

15.5 NUMBER OF SIGNALS IN THE NMR SPECTRUM

What protons give the same signal? We can answer this question by looking at some examples. Chemically equivalent protons give equivalent NMR signals. Compounds having all protons equivalent give one signal, e.g., methane, chloromethane, ethane, *tert*-butyl bromide, formaldehyde.

$$CH_4 \qquad CH_3\!-\!Cl \qquad CH_3\!-\!CH_3 \qquad (CH_3)_3C\!-\!Br \qquad \underset{H}{\overset{H}{>}}C\!=\!O$$

methane   chloromethane   ethane   *tert*-butyl bromide   formaldehyde

Problem   **15.3** What is a possible structure of each of the following compounds with one NMR signal and the specified molecular formula?

a) C_2H_6O **b)** $C_3H_6Cl_2$ **c)** C_3H_6O (strong ir band at 1715 cm^{-1})

The number of signals in the NMR spectrum represents the number of different chemical environments for protons in the molecule. The protons of a methyl group $CH_3\!-\!C\!-$ have a different signal from the protons of a methylene group $-C\!-\!CH_2\!-\!C\!-$. The protons of a methylene bonded to a bromine and a carbon $-C\!-\!CH_2\!-\!Br$ are different from those of a methylene bonded to two carbons $-C\!-\!CH_2\!-\!C\!-$. The number of signals in each of the following compounds is indicated by the letters above the formulas.

$$\overset{a}{C}H_3\!-\!\overset{b}{C}H_2\!-\!Br \qquad \overset{a}{C}H_3\!-\!\overset{b}{C}H_2\!-\!\overset{c}{C}H_2\!-\!Br \qquad \overset{a}{C}H_3\!-\!\underset{Br}{\overset{b}{C}H}\!-\!\overset{a}{C}H_3$$

ethyl bromide   1-bromopropane   2-bromopropane   chlorocyclopropane
(2 NMR signals)   (3 NMR signals)   (2 NMR signals)   (3 NMR signals)

15.4 How many signals would each of these compounds have?

a) $CH_3-O-CH_2-CH_3$

methyl ethyl ether

b)

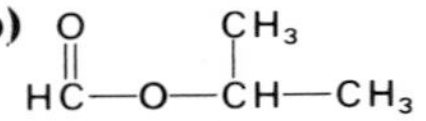

$$\begin{array}{ccc} O & & CH_3 \\ \parallel & & | \\ HC-O- & CH- & CH_3 \end{array}$$

isopropyl formate

c) 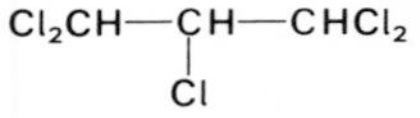 $Cl_2CH-CH-CHCl_2$

with Cl below the central CH

1,1,2,3,3-pentachloropropane

15.6 THE POSITION OF NMR SIGNALS: THE CHEMICAL SHIFT

Protons in similar molecular environments in different compounds give signals at approximately the same frequency. The actual position of the signal on the spectrum gives an indication of the type of environment the proton has.

Before we examine the positions of certain signals, we need to learn something about the frequency scale used in the spectrum. A reference compound, **tetramethylsilane** (TMS), is added to the solution of the compound to be studied.

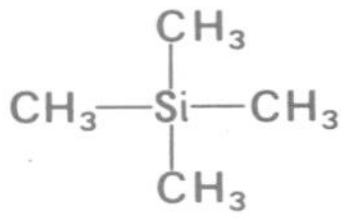

$$\begin{array}{c} CH_3 \\ | \\ CH_3-Si-CH_3 \\ | \\ CH_3 \end{array}$$

tetramethylsilane (TMS)

The position of a proton signal is defined in relation to the position of the signal of TMS. The difference in the position of the signal of a given proton and that of TMS is called the **chemical shift** of the proton. Almost all the protons attached to carbon are shielded less than those of TMS, and their absorptions appear to the left of the TMS signal. The chemical shift is stated in terms of a fraction: (the difference in frequency in hertz between the signal and that of TMS) over (the frequency of the radio wave in million hertz), that is, chemical shift is stated in parts per million. A difference in position of 300 Hz is recorded as a chemical shift of 5.00 ppm.

$$\text{chemical shift} = \frac{300 \text{ Hz}}{60.0 \text{ million Hz}} = 5.00 \text{ ppm}$$

The frequency scale is set with TMS at 0 ppm, with an increase in parts per million going downfield to the left. The chemical shift scale is called the δ-scale and is numbered from 10 to 0 ppm. With the perversity of chemists, it is not surprising that others prefer to give the downfield direction decreasing numbers and give TMS (upfield) a large number. In this case TMS is set at 10.0 ppm and the scale is labeled τ (tau). If you use NMR spectra you will quickly become accustomed to both scales. In the meantime we will use just the δ-scale with TMS at 0 ppm.

FIGURE 15.4
NUCLEAR MAGNETIC
RESONANCE SPECTRUM
OF METHYL ACETATE

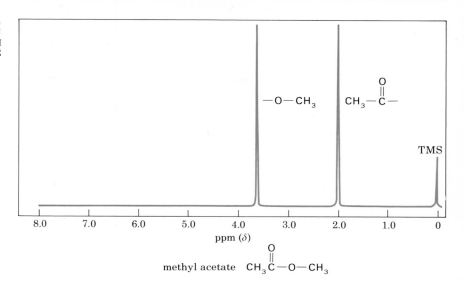

FIGURE 15.4 NUCLEAR MAGNETIC RESONANCE SPECTRUM OF METHYL ACETATE

In the spectrum of methyl acetate in Fig. 15.4, we see that the chemical shift for the methyl hydrogens CH_3—R varies with the structure of R.

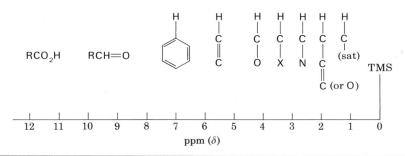

A listing of chemical shifts is given in the chapter summary and in the descriptions of NMR spectra for each class of compound in Sections 15.9 through 15.12. The ranges for the hydrogen signals are shown in Fig. 15.5.

FIGURE 15.5
DISTRIBUTION OF CHEMICAL
SHIFTS FOR PROTONS

The intensity of a signal is measured by the *area* under the curve forming that signal. When the area of each of the proton signals for the compound is measured, the ratios of the areas indicates the ratios of the number of hydrogens giving each signal. The process of obtaining the area of the signal is known as integrating the signal. The ratios of the integrals can be calculated from the values of the integrals of peak areas. This information is sometimes furnished by the instrument as a step graph and sometimes as a readout.

The signal of one kind of proton is slightly but definitely affected by the presence on adjacent carbons of protons having a different chemical shift. The signal of the methylene CH_2 protons of $Cl_2CH—CH_2Cl$ is affected by the presence of the tertiary hydrogen CH. The tertiary proton will at any instant be aligned with the magnetic field or against it. When it is *with* the field the effective field strength which the CH_2 protons feel is increased by a small amount. Conversely, when it is *against* the field, the effective field the CH_2 protons feel is slightly decreased by the same amount.

with   against

The signal for each methylene proton is split by the presence of the CH proton into two peaks, and each peak is shifted from the original position by the same small extent. The signal for CH_2 is now a **doublet**, with each peak the same size. The effect is called **spin-spin coupling**.

At the same time, the CH proton in $Cl_2CH—CH_2Cl$ will have the field it feels increased, unchanged, or decreased by the same coupling with the CH_2 protons. The CH proton signal will be unchanged when the two CH_2 protons are aligned in opposite directions and their effects cancel each other. The situation is described by the arrows which denote the direction of alignment of the CH_2 protons and indicate the relative sizes of the peaks of the split signal.

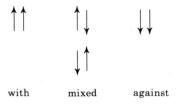

with   mixed   against

Half of the time the two methylene protons will be aligned opposite to each other and have no effect. The other half of the time they will both be with the field or both be against the field and will therefore cause the CH signal to be slightly increased or decreased. The proton signal for the CH group adjacent to the CH_2 is thus split into three peaks, a **triplet**, in which the center peak has twice the area of each of the two side peaks (1:2:1). The signal patterns are shown below as drawn by the instrument (left) and in a schematic fashion.

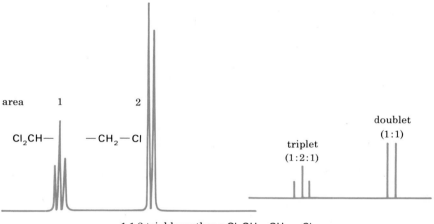

1,1,2-trichloroethane Cl_2CH-CH_2-Cl

Note that the spacing between peaks of the signal of the CH hydrogen shown above is identical, measured in Hz, to the spacing between peaks of the signal for the CH_2 protons with which the CH signal is coupled. Note also that in the real spectrum the split signal is slightly unsymmetrical, with peaks on the inside toward the coupling partner higher than peaks on the outside away from the coupling partner.

Splitting caused by three protons of a methyl group CH_3 produces a four-peak signal, a **quartet**, in which the two center peaks are three times the size of the two side peaks (1:3:3:1). A signal for the CH proton next to the CH_3 group in Cl_2CH-CH_3 will appear as a quartet. The methyl signal is a doublet due to coupling with the CH proton.

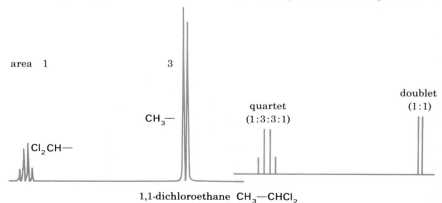

1,1-dichloroethane CH_3-CHCl_2

Problem **15.5** Work out the pattern for splitting by the three protons of a methyl group using arrows in the manner shown for splitting by two protons of the CH_2 group. The result should justify both the number of peaks and the relative sizes.

The number of peaks of a signal indicates the number of neighboring protons which have a different chemical environment and a different chemical shift from the one under consideration. The rule is that N equivalent neighboring protons produce $N + 1$ peaks in the signal.

When does coupling occur? Coupling occurs primarily between protons on adjacent carbons. Coupling does not occur between protons having the same chemical shift nor between protons separated by three or more atoms.

Problem **15.6** What signal patterns would you expect for the protons in each of these compounds?

a) CH_3—O—CH_2—CH_3 b) H—$\overset{\overset{O}{\|}}{C}$—O—$\overset{\overset{CH_3}{|}}{CH}$—$CH_3$ c) Cl_2CH—$\overset{\overset{Cl}{|}}{CH}$—$CHCl_2$

methyl
ethyl ether

isopropyl
formate

1,1,2,3,3-penta-
chloropropane

In summary, the four types of information available from NMR spectra are as follows.

1. The number of signals indicates the number of different kinds of proton environments.
2. The positions of the signals (chemical shifts) correlate with the degree of shielding of the proton, so that similar structural units absorb at similar positions.
3. The relative areas of the signals indicate the relative numbers of protons responsible for the signals.
4. The splitting of the signal into two or more peaks indicates the interaction of a proton with its neighboring protons on adjacent carbons. The number of these peaks indicates the number of neighboring protons.

15.9 SPECTRA OF HYDROCARBONS

The spectral characteristics of carbon-hydrogen groups appear both for hydrocarbons and for the remote hydrocarbon portions of other classes of compounds. Thus the parts of hydrocarbon spectra will be found in the spectra of all of the following classes as well.

a) IR spectra

Carbon–hydrogen bond vibrations give IR absorption bands which are characteristic of the orbital hybridization of the carbon atom. The stretching vibrations of the C—H bond which occur in the range 3300–2850 cm^{-1} can be separated into three distinct, narrow ranges.

The range for hydrogen attached to tetrahedral carbons (alkanes) is the widest, 3000–2850 cm^{-1}. We find hydrogens on trigonal carbons (=CH, olefinic and aromatic) absorbing at 3100–3000 and acetylenic hydrogens (≡CH) at 3300 cm^{-1}. The greater the carbon unsaturation, the higher is the frequency of the CH absorption. The major group bands are usually sufficiently separated to allow recognition of the unsaturated structures or, particularly, the complete absence of any group.

Carbon–hydrogen groups of three or four atoms (—CH$_2$— or —CH$_3$) have characteristic absorptions for the bending vibrations which are apparent in most spectra. These bands are found at 1375 for —CH$_3$ and in the range 1470–1430 cm^{-1} for —CH$_2$— and —CH$_2$—H.

Absorption occurs at lower frequencies for carbon–carbon bonds than for carbon–hydrogen bonds. Carbon–carbon double bond stretching (C=C) gives bands for alkenes at 1650 and for aromatics at 1600 and 1500 cm^{-1}. The aromatic bands are quite variable in intensity and may disappear completely. The C≡C stretch has a band at higher frequency at 2100 cm^{-1}, which is quite distinctive.

Note the bands for =CH, —CH, and C=C stretching and for CH$_3$ and CH$_2$ bending in the spectrum for 1-hexene (Fig. 15.6).

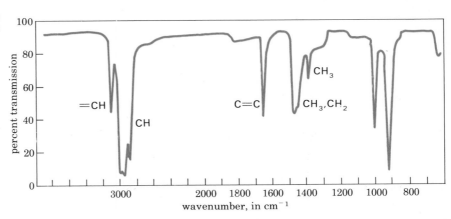

**FIGURE 15.6
IR SPECTRUM OF 1-HEXENE**

b) **NMR spectra**   The chemical shift for the hydrogens of a methyl group in a saturated hydrocarbon or separated by several carbons from any functional group in other classes of compounds has the lowest value of ordinary hydrogens, δ 0.85–0.95. The methylene group —C—CH$_2$—C— in a carbon chain has a chemical shift slightly higher than the methyl, at δ 1.15–1.25.

The presence of an unsaturated bond in the immediate vicinity shifts the absorption to higher δ values. There are three major absorptions of protons related to multiple carbon–carbon bonds: the terminal olefinic hydrogens, the internal olefinic hydrogens, and the hydrogens on the carbon adjacent to olefinic carbons, $CH_2=CH—CH_2—$. In alkenes the methylene group adjacent to the $C=C$ is less strongly affected by the unsaturated bond and has a chemical shift nearest to that of the alkane, δ 1.60–1.95. Absorption for the hydrogens directly attached to the olefinic carbons are much further downfield; the chemical shift for the terminal olefinic hydrogen is at δ 4.65 and for the internal olefinic hydrogen at δ 5.05–5.55.

The effect of delocalization of electrons in the ring on the CH signal for protons attached to aromatic rings is very marked and produces a chemical shift far downfield in the range of δ 6.5 to δ 8.0. The signal for all the aromatic hydrogens on the same ring may be a singlet or it may consist of several signals with many peaks in the same general region. In general the aromatic protons are treated as one signal. The spectrum of toluene has two singlets with relative areas of 5 and 3.

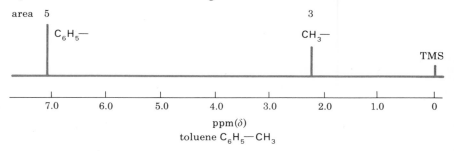

toluene $C_6H_5—CH_3$

Problems

15.7 How many signals in the NMR spectrum would 2-methyl-1-butene have? Sketch a reasonable NMR spectrum for this alkene showing the pattern and relative placement of each signal.

15.8 Give a structure for each compound that is consistent with the IR spectral bands.

a) C_5H_{10}:   3100, 2900, 1650, 1470, 1375 cm^{-1}

b) C_5H_8:   3300, 2900, 2100, 1470, 1375 cm^{-1}

15.10 SPECTRA OF ALCOHOLS, ETHERS, ALKYL HALIDES, AND AMINES

The spectra of alcohols, ethers, halides, and amines are surveyed together since these compounds consist of hydrocarbon groups plus a functional group of an electronegative atom attached to a tetrahedral carbon. The effects of these electronegative atoms on protons in NMR and on C—X bonds in IR spectra produce similar results.

a) IR spectra

The most obvious band in the IR spectra of alcohols, phenols, and amines is the OH or NH stretching vibration. The OH and NH bands absorb at the highest frequency for any bonds, about 3600 cm^{-1}. However, the usual IR spectrum of an alcohol or phenol shows hydrogen-bonded OH groups. The strong and very broad absorption band in the range 3600–3300 cm^{-1} is due to OH stretching of an —O—H···O— group. Amines show NH stretching bands in roughly the same region, about 3500–3200 cm^{-1}, but they are usually less broad and less strong than OH.

The stretching frequencies for bonds of carbons to atoms other than carbon and hydrogen are much lower than for C—H. The C—O stretching band is usually strong and broad and occurs at about 1200–1000 cm^{-1}. Alcohols and ethers both have this absorption band, but the OH band is absent from ethers, naturally. Phenols show C—O stretch at slightly higher wavenumbers, about 1230 cm^{-1}.

The C—X stretching vibrations of alkyl halides occur as strong bands in the fingerprint region of the IR spectrum: C—F at 1400–1000 cm^{-1} and C—Cl at 800–600 cm^{-1}. Since carbon tetrachloride and chloroform are often used as solvents in IR spectra, the strong C—Cl absorption blocks any compound absorption in the 850–750 region. The C—N stretching band occurs in the region 1350–1000 cm^{-1}, but this band is not clear enough in the midst of other bands to be useful for identification.

The spectra of 1-butanol (Section 15.3) and *tert*-butylamine (Fig. 15.7) illustrate the OH and NH bands particularly.

**FIGURE 15.7
IR SPECTRUM OF
TERT-BUTYLAMINE**

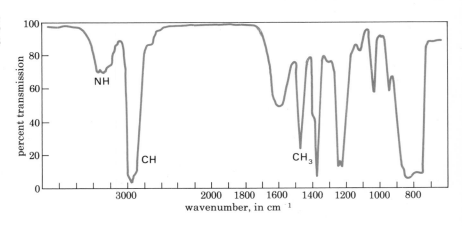

b) NMR spectra　Oxygen, nitrogen, and halogen atoms exert strong influences on protons attached to the same carbon and to the next carbon. The chemical shifts of hydrogens on these carbons move to higher δ values than CH signals of hydrocarbons. The —CH$_2$—O— chemical shift is in the range δ 3.4–4.0 in alcohols and ethers (and esters, Section 15.11). The —CH$_2$—Cl chemical shift is about δ 3.4 and the —CH$_2$—N— is about δ 2.25. The strong shift due to these electronegative atoms on the same carbon with the hydrogen gives valuable information for the determination of the compound structure.

Chemical shifts for protons on carbons next to the carbon bearing the functional group —CH_2—C—X (where X = O, Cl, N) show much less influence of the electronegative atom. These absorptions are at δ 1.20–1.55, only slightly higher than ordinary hydrocarbon proton signals.

The chemical shift of H attached to an O or N varies considerably for alcohols and amines, in the range δ 1 to δ 5. A very significant point in the absorption of these hydrogens is that they exchange between oxygens or nitrogens of different molecules so fast that they do not stay attached to any one oxygen or nitrogen long enough to be affected by the neighboring protons. Thus the OH or NH proton signal is not split by coupling with its neighbor protons nor does it split their signals.

In the spectrum of ethanol (Fig. 15.8) note the chemical shifts of the CH_3—C and C—CH_2—O— protons as well as the splitting pattern for each signal. The quartet-triplet combination is a familiar pattern of the ethyl group, CH_3—CH_2—.

FIGURE 15.8
NMR SPECTRUM OF ETHYL
ALCOHOL

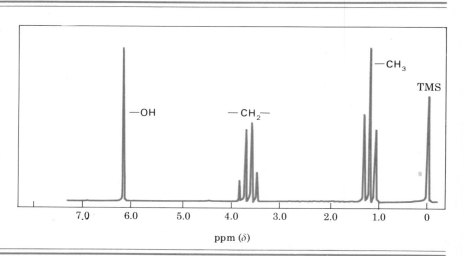

ppm (δ)

Problem   **15.9** The NMR spectrum for a compound $C_4H_{10}O$ shows three signals: δ 4.1, septet (1 H); δ 3.1, singlet (3 H); δ 1.55, doublet (6 H). The IR spectrum shows only one band above 2000, at 2950 cm^{-1}. Give a reasonable structure for the compound.

15.11 SPECTRA OF THE CARBOXYLIC ACID FAMILY AND OF ALDEHYDES AND KETONES   The spectra of all of these classes of compounds are dominated by the presence of the carbonyl group in the IR and NMR spectra. While the common C=O is most important, certain differences exist in the different classes that are recognizable and are explored in these paragraphs.

a) IR spectra   The IR spectra of all of these compounds containing the C=O group are dominated by the C=O stretching absorption. This C=O band is probably the most easily identified band in the entire infrared spectrum. It is a very strong spike whose frequency lies in a region which is free from most other absorptions, 1750–1640 cm^{-1}. Besides being a distinctive band, its position within this range is indicative of the particular carbonyl functional group present in the molecule. The C=O for each type of compound—acid, ester, amide, aldehyde, or ketone—has a very narrow and characteristic range of absorption frequency, shown in this list.

$$\begin{array}{c} O \\ \parallel \\ R-C-Y \end{array}$$

Class	Y	Saturated	Unsaturated
Acid	—OH	1725–1700 cm^{-1}	1700–1680 cm^{-1} (α,β-unsaturated)
Ester	—OR	1740–1725	
Amide	—NH$_2$, —NHR, —NR$_2$	1690–1640	
Aldehyde	—H	1725	1700 (Ar—CH=O)
Ketone	—R	1715	1690 (ArCOR)

The OH stretching absorption for the CO$_2$H group in the range 3000–2500 cm^{-1} is at lower wavenumbers than for alcohols. It is a strong, very broad band for the hydrogen-bonded OH in OH$\cdots$O=. The combination of this band with the C=O absorption band is usually unmistakable. The NH stretch for amides with —NH$_2$ and —NHR is similar to that for amines at 3500–3000 cm^{-1}, but in amides it is accompanied by the distinctive C=O band.

The two C—O stretching absorptions for an ester (C—O—C) group come in the 1300–1050 cm^{-1} region. These bands are characteristic for esters and are missing in spectra of aldehydes or ketones.

Note the strong C=O stretching absorption band in the IR spectrum of cyclohexanone (Fig. 15.9).

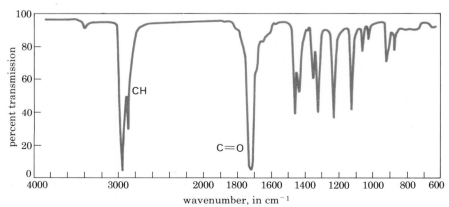

FIGURE 15.9
IR SPECTRUM OF
CYCLOHEXANONE

b) NMR spectra   The functional groups of acids CO_2H and of aldehydes $CH{=}O$ contain unusual protons, those closely associated with the $C{=}O$. In NMR spectra the absorptions by these protons are far downfield from any others and are quite distinctive. The chemical shift of the CO_2H of a carboxylic acid has a value of δ 10.5 to δ 12, and for the $CH{=}O$ of an aldehyde in the range of δ 9 to δ 10. The signal for the aldehydic hydrogen couples with that for an α-hydrogen. For amides the CO—NH chemical shift is in the range δ 5 to δ 8, a lower value than for acids, but higher than for amines.

The chemical shift for protons on α-carbons for acids, esters, amides, aldehydes, and ketones (protons not directly attached to the $C{=}O$) are in a range of δ values only slightly higher than for hydrocarbon protons, δ 2.0–2.2.

The protons of the alcohol portion of an ester resemble those for similar protons of an alcohol or ether. The chemical shift for CO_2—CH— is about δ 3.7–4.0.

Note the chemical shifts for benzyl propanoate.

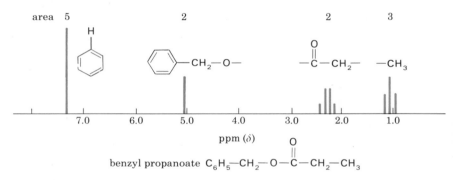

benzyl propanoate $C_6H_5{-}CH_2{-}O{-}\overset{\overset{O}{\|}}{C}{-}CH_2{-}CH_3$

Problem   **15.10** What similarities in IR spectra would the compounds below show? How would the pattern of NMR signals distinguish between them?

$HO{-}CH_2{-}CH{=}O$ $CH_3{-}CO_2H$

hydroxyacetaldehyde   acetic acid

New Terms and Topics

Spectrum, wavelength, frequency (Section 15.1)

Ultraviolet, visible, infrared, radio regions (Section 15.1)

Spectroscopy, absorption (Section 15.1)

Electronic excitation (Section 15.2)

Bond vibration, stretching and bending (Section 15.3)

Wavenumber (Section 15.3)

Nuclear magnetic resonance (Section 15.4)

Chemical shift (Section 15.6)

Spin-spin splitting; triplet, quartet patterns (Section 15.8)

Summary of Spectral Data

Bond	Class	Range, cm^{-1}
O—H	Alcohol	3600; hydrogen-bonded 3600–3300 strong
O—H	Acid	hydrogen-bonded 3000–2500
N—H	Amine, amide	3600; hydrogen-bonded 3500–3200
≡C—H	Alkyne	3300
=C—H	Alkene, aromatic	3100–3000
C—H	Alkane	3000–2850
C≡C	Alkyne	2200 weak
C=C	Alkene	1650
C=C	Aromatic	1600, 1500 often weak
C=O	Acid	1760–1720
C=O	Ester	1740–1720
C=O	Aldehyde	1725 (Ar—CH=O 1700)
C=O	Ketone	1715 (ArCOR 1690)
C=O	Amide	1690–1640
C—O	Ether, alcohol, ester	1200–1000
CH$_2$	Bending	1470–1430
CH$_3$	Bending	1470, 1375

Proton in	Chemical shift in ppm (δ)	Proton in	Chemical shift in ppm (δ)
CH$_3$—R	0.85–0.95	—CH$_2$—⬡	2.3
R—CH$_2$—R	1.15–1.25	—CH$_2$—C=O	2.0–2.2
R—CH$_2$—C≡C	1.60–1.95	—CH=C	5.0–5.5
—CH$_2$—O—	3.4–4.0	CH$_2$=C	4.65
—CH$_2$—O—C=O	3.4–4.0	HC≡C—	2.35
—CH$_2$—N—	2.5	—CH=O	9.25–10
—CH$_2$—N—C=O	3.2	—CO$_2$H	10.5–12
—CH$_2$—Cl	3.4	⬡—H	6.90–7.50
—CH$_2$—C—O—	1.20–1.55		

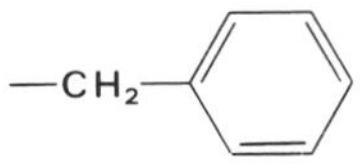

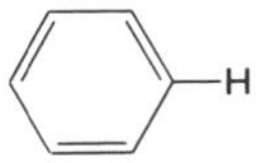

Problems

15.11 The IR spectrum for a compound with formula $C_6H_{12}O$ has only two bands higher than 1500 cm^{-1}: 2950 and 3350 (both strong bands). To what class does this compound belong? What features of the structure, besides the functional group, are probable? Give a reasonable structure for the compound.

15.12 A compound with molecular formula C_2H_4O has the following spectral characteristics: NMR—one signal, a singlet; IR—strong absorption band at 2950 cm^{-1}, no other band above 1470 cm^{-1}; UV—no absorption above 210 nm. What isomeric structures are ruled out by the spectra? What is a possible structure for the compound?

15.13 Compound A has an absorption band in the ultraviolet with λ_{max} = 242 nm and ε_{max} = 10,000, while compound B has a λ_{max} of 320 nm and ε_{max} of 8000, with no absorbance at 242 nm. A solution of a mixture of the two compounds shows an observed absorbance of 3000 at 242 nm and of 1600 at 320 nm. What is the relative number of moles of the two compounds in the mixture?

15.14 A deuteron in a compound in place of a proton does not absorb in the NMR spectrum nor does it affect the signal pattern of its neighboring protons in a significant way. Deuterium incorporation into a compound is often used to simplify the NMR spectrum by reducing the number of signals or the number of peaks in the signals. Describe the NMR spectrum you expect for diisopropyl ketone and the changes which you expect to take place in the spectrum if the compound is treated with NaOD in D_2O.

15.15 The significant feature of the IR spectrum of a compound with formula $C_4H_8O_2$ is a strong band at 1730 cm^{-1}. The NMR spectrum has three signals: δ 3.6, singlet (3H); δ 2.3, quartet (2H); δ 1.15, triplet (3H). What is the compound?

15.16 The NMR spectrum for the cyclic hydrocarbon called 18-annulene, $C_{18}H_{18}$ (Section 12.1), with 9 conjugated double bonds, has two broad bands at δ 8.9 and δ −1.9, with the ratio of hydrogens 2:1 respectively. Identify the hydrogens for each band.

15.17 Novocaine has the molecular formula $C_{13}H_{20}N_2O_2$; however, it is stored and used as the chloride salt $C_{13}H_{21}N_2O_2Cl$. Novocaine itself is hydrolyzed by a base to give *p*-aminobenzoic acid $C_7H_7NO_2$ and a second fragment. Deduce as many parts of the structure of novocaine as you can from the following spectral data.

$C_{13}H_{20}N_2O_2$: IR—a strong peak about 1700 cm^{-1}

NMR—δ 1.05 triplet (6H); δ 2.6 quartet (4H); δ 2.75 triplet (2H); δ 4.2 singlet (2H); δ 4.35 triplet (2H); δ 6.5–7.8 multiplet (4H). When the compound was shaken with D_2O, the singlet at δ 4.2 disappeared.

Answers to
Problems

Brief answers to all problems are given on these pages. More detailed answers, as well as explanations and solutions to many problems, are given under separate cover in the Study Guide.

Chapter 1

1.1 a) H—B̈r:　**b)** H—F̈:　**c)** H—S̈:　**d)** H—P̈—H　**e)** :C̈l—C—C̈l:

with H below in c) and d); and in e) there is :C̈l: above and :C̈l: below the central C.

1.2 a)

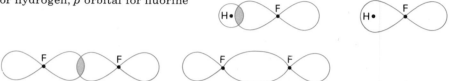

The three-dimensional formula may be turned in any of these directions.

1.3 a) HF—*s* orbital for hydrogen, *p* orbital for fluorine

b) F₂—*p* orbitals

1.4 Values of bond angles around N and O approach those for tetrahedral C and suggest that these atoms also use hybridized *sp³* orbitals for bonding. The unshared electron pairs would also occupy *sp³* orbitals.

a) C　N
　　sp³　*sp³*

b) C　N

1.5 a) 109°　**b)** 109°　**c)** 109°　**d)** 107°　**e)** 107°　**f)** 105°

1.6

H—C̈—Cl　H—C̈—N̈—H　H—C̈—O—H

1.7

Cl—C—C—O—H　Cl—C—C—O—H

with δ⁻ |δ⁺ |δ⁺ δ⁻ δ⁺ labels on the second structure.

1.8 a) An atomic orbital is the three-dimensional space around the nucleus of an atom in which one or a pair of electrons is most probably to be found. An *s* orbital is equidistant from the nucleus in all directions, that is, it is spherical in shape.

b) A covalent bond is the attraction which holds two atoms together at a prescribed distance between them. The bond consists of the attraction of one pair of electrons by the nuclei of both atoms simultaneously.

c) An *sp³* hybridized orbital is an atomic orbital which has a blend of the characteristics of *s* and *p* orbitals. An atom has four *sp³* orbitals, each with 1 part *s* and 3 parts *p* character.

d) A molecular bonding orbital is the three-dimensional space which is occupied by the pair of bonding electrons and which encompasses both nuclei of the atoms bonded together.

e) A sigma bond is a covalent bond whose molecular orbital (or overlapping atomic orbitals) is symmetrical about the axis connecting the two nuclei.

f) A bond angle is the angle formed by two bonds originating at the same central atom.

g) The bond length is the distance between the nuclei of two bonded atoms.

h) A tetrahedral carbon is bonded to four other atoms with the bonds radiating three-dimensionally from the carbon in the center toward the four atoms located at the vertices of a tetrahedron. The bond angles at the tetrahedral carbon are approximately 109°.

i) A hydrocarbon is a compound containing only carbon and hydrogen atoms.

j) The methyl group is a group of atoms with three hydrogens bonded to one carbon, CH_3-.

k) A functional group is a group of atoms containing atoms other than carbon and hydrogen which is attached to a hydrocarbon group.

l) A polar covalent bond is a bond in which the pair of electrons is shared between two atoms of greatly differing electronegatives. One atom attracts the electrons more strongly than the other, producing a separation of partial positive and negative charges.

m) An attraction between a strongly electronegative atom and a hydrogen attached to another such atom is called a hydrogen bond.

1.9 a) 109°

b) H—O—H and H—N—H slightly smaller than H—C—H, 105° and 107° respectively

c) same as H—C—H in methane, 109.5°

1.10 a) CH_3 **b)** OH **c)** NH_2 **d)** SH

1.11 a) CH_3—OH methanol **b)** CH_3—Br bromomethane

c) CH_3—NH_2 methylamine **d)** CH_3—SH methanethiol

1.12 a) alcohol **b)** alkyl halide **c)** amine **d)** thiol

1.13 a) CH_3—O—CH_3 **b)** CH_3—NH—CH_3

1.14 a) (i)   (ii)   (iii)

b) Yes. Since the Cl—C—Cl bond angle at C is about 109°, the bond moments for the two Cl—C bonds do not completely cancel each other. The same is true for the C—O—C angle in (iii) and for the dipole moment of (iii).

1.15 CH_3—O—H · · · O—CH_3 CH_3—O—H · · · O—H

Chapter 2

2.1 C_2H_6, C_3H_8, C_4H_{10}, C_5H_{12}; by CH_2

CH_3—CH_3 CH_3—CH_2—CH_3 CH_3—CH_2—CH_2—CH_3 CH_3—CH_2—CH_2—CH_2—CH_3

2.2 C_6H_{14}, C_8H_{18}, $C_{10}H_{22}$; $CH_3CH_2CH_2CH_2CH_2CH_3$, $CH_3CH_2CH_2CH_2CH_2CH_2CH_2CH_3$, $CH_3CH_2CH_2CH_2CH_2CH_2CH_2CH_2CH_2CH_3$

2.3 a) CH_3—CH_2—CH_2—CH_2—CH_2—CH_3 (2 methyls) **b)** CH_3—CH_2—CH—CH_2—CH_3 (3 methyls)
with CH_3 branch

c) CH_3—CH—CH_2—CH_2—CH_3 (3 methyls) **d)** CH_3—C—CH_2—CH_3 (4 methyls) **e)** CH_3—CH—CH—CH_3 (4 methyls)
with CH_3 branches

2.4 a) CH_3—C—CH_2—CH_2—CH_3 **b)** CH_3—CH—CH_2—CH—CH_2—CH_3
with CH_3 branches

c) CH_3—CH_2—CH—CH—CH_2—CH_2—CH_2—CH_3
with CH_2—CH_3 branches

2.5 a) hexane   **b)** 3-methylpentane   **c)** 2-methylpentane   **d)** 2,2-dimethylbutane   **e)** 2,3-dimethylbutane

2.6 a) Conformation **a** is preferred. **c > b > d**   **2.7 a)** $2\ C_4H_{10} + 13\ O_2 \longrightarrow 8\ CO_2 + 10\ H_2O$

                                                                  b) 634 kcal/mol butane (see Study Guide)

b) 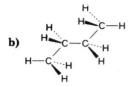   Preferred conformation **a** has no eclipsed bonds

2.8 butyl—H from terminal carbon of butane; *sec*-butyl—H from internal C of butane; isobutyl—H from terminal (or methyl) C of isobutane; *tert*-butyl—H from internal C of isobutane

2.9 a) $CH_3-CH_2-CH_2-\overset{\displaystyle |}{\underset{\displaystyle CH_3-\underset{\underset{\textstyle CH_3}{|}}{\overset{\overset{\textstyle |}{}}{C}}-CH_3}{CH}}-CH_2-CH_2-CH_2-CH_3$ **b)** $CH_3-CH_2-\overset{\displaystyle |}{\underset{\displaystyle CH_3-CH-CH_3}{CH}}-CH_2-CH_2-CH_2-CH_3$

c) $CH_3-\overset{\overset{\textstyle CH_3}{|}}{\underset{\underset{\textstyle CH_3}{|}}{C}}-CH_2-CH_2-CH_3$

2.10 a) 3-methyl-5-*tert*-butyloctane   **b)** 4-isopropylheptane   **c)** 2,5-dimethylhexane

2.11 $CH_3-CH_2-CH_2-CH_2-OH$ $CH_3-CH_2-\underset{\underset{\textstyle OH}{|}}{CH}-CH_3$ $CH_3-\underset{\underset{\textstyle CH_3}{|}}{CH}-CH_2-OH$ $CH_3-\overset{\overset{\textstyle OH}{|}}{\underset{\underset{\textstyle CH_3}{|}}{C}}-CH_3$

2.12 primary: 1-propanol, 1-butanol, and isobutyl alcohol; secondary: isopropyl and *sec*-butyl alcohols; tertiary: *tert*-butyl alcohol

2.13 a) $CH_3-O-CH_2CH_3$ **b)** $CH_3-O-CH_2-CH_2-CH_3$ $CH_3-O-\underset{\underset{\textstyle CH_3}{|}}{CH}-CH_3$ $CH_3-CH_2-O-CH_2-CH_3$

2.14 a) 1,1-diethylcyclopropane   **b)** 1-methyl-3-propylcyclopentane   **c)** 1-butyl-1,4-dimethylcyclohexane

2.15 a)

        H⋯⎡⎤⋯H   H⋯⎡⎤⋯CH₂CH₃

$CH_3CH_2CH_2$⎿⎦CH_2CH_3 $CH_3CH_2CH_2$⎿⎦H

b) five dimethylcyclobutanes and three dimethylcyclopropanes (see Study Guide for structures of these compounds)

2.16 a)   **b)**   **c)**

2.17 a) 2-methylbutane   **b)** ethanol   **c)** 2,3-dimethylhexane

d) 4-ethyl-4-methyloctane   **e)** isopropyl alcohol (2-propanol)   **f)** 1-butanol

g) 4-isopropyl-3-methylheptane   **h)** *sec*-butylcycloheptane   **i)** *trans*-1-ethyl-4-methylcyclohexane

2.18 a)   **b)** $CH_3-\underset{\underset{\textstyle CH_3}{|}}{CH}-CH_2-\underset{\underset{\textstyle CH_2-CH_3}{|}}{CH}-CH_2-CH_3$ **c)** $CH_3-\underset{\underset{\textstyle OH}{|}}{CH}-CH_2-CH_2-CH_3$

d) $CH_3-\underset{\underset{\textstyle OH}{|}}{CH}-CH_2-CH_3$ **e)** $CH_3-(CH_2)_6-CH_3$ **f)** $\begin{matrix} CH_3-CH_2 \\ | \\ CH_3-CH_2-C-CH_2-CH_3 \\ | \\ CH_2-CH_3 \end{matrix}$

g) $CH_3-\underset{\underset{\textstyle CH_3}{|}}{CH}-CH_2-Cl$ **h)** $CH_3-\overset{\overset{\textstyle CH_3}{|}}{\underset{\underset{\textstyle CH_3}{|}}{C}}-OH$ **i)**

                                                                                 H⋯△⋯H

                                                            CH_3CH_2⎓⎓$CH(CH_3)_2$

j) [cyclobutane/diamond shape]

k) $CH_3CH_2CH_2CHCH_2CH_2CH_3$
  $\quad\quad\quad\quad\quad\ |$
  $\quad\quad\quad\quad\ CH_2CH_2CH_3$

l) $CH_3\!-\!O\!-\!CH_3$

  or [cyclohexane with H, Br substituents]

m) $CH_3\!-\!CH_2\!-\!NH_2$

n) [cyclohexane chair with Br, Br]

2.19 a) $\quad\quad\quad\quad CH_2\!-\!OH$
  $\quad\quad\quad\quad\quad\ |$
  $CH_3\!-\!CH_2\!-\!CH\!-\!CH_3$

b) $\quad\quad OH\quad CH_3$
  $\quad\quad\ |\quad\quad\ |$
  $CH_3\!-\!CH\!-\!CH\!-\!CH_3$

c) $\quad\quad\quad\quad\quad CH_3$
  $\quad\quad\quad\quad\quad\ |$
  $CH_3\!-\!CH_2\!-\!C\!-\!CH_3$
  $\quad\quad\quad\quad\quad\ |$
  $\quad\quad\quad\quad\quad OH$

d) [cyclopentane]—OH

e) [cyclobutane]$\begin{smallmatrix}CH_3\\OH\end{smallmatrix}$

f) [cyclobutane]$-CH_2\!-\!OH$ (other examples are possible)

2.20 $CH_3CH_2CH_2CH_2CH_2\!-\!OH\quad\quad$ $CH_3CH_2CH_2\overset{\overset{\displaystyle OH}{|}}{C}HCH_3\quad\quad$ $CH_3CH_2\overset{\overset{\displaystyle OH}{|}}{C}HCH_2CH_3\quad\quad$ $CH_3CH_2\overset{\overset{\displaystyle CH_3}{|}}{C}HCH_2\!-\!OH$

$CH_3CH_2\overset{\overset{\displaystyle CH_3}{|}}{\underset{\underset{\displaystyle OH}{|}}{C}}CH_3\quad\quad$ $CH_3\overset{\overset{\displaystyle CH_3}{|}}{\underset{\underset{\displaystyle OH}{|}}{C}}HCHCH_3\quad\quad$ $HO\!-\!CH_2CH_2\overset{\overset{\displaystyle CH_3}{|}}{C}HCH_3\quad\quad$ $CH_3\overset{\overset{\displaystyle CH_3}{|}}{\underset{\underset{\displaystyle CH_3}{|}}{C}}CH_2\!-\!OH$

(See Study Guide for steps to write all formulas.)

$CH_3\!-\!O\!-\!CH_2CH_2CH_2CH_3\quad\quad$ $CH_3\!-\!O\!-\!\overset{\overset{\displaystyle CH_3}{|}}{C}HCH_2CH_3\quad\quad$ $CH_3\!-\!O\!-\!\overset{\overset{\displaystyle CH_3}{|}}{\underset{\underset{\displaystyle CH_3}{|}}{C}}CH_3\quad\quad$ $CH_3\!-\!O\!-\!CH_2\overset{\overset{\displaystyle CH_3}{|}}{C}HCH_3$

$CH_3CH_2\!-\!O\!-\!CH_2CH_2CH_3\quad\quad$ $CH_3CH_2\!-\!O\!-\!\overset{\overset{\displaystyle CH_3}{|}}{C}HCH_3$

2.21 primary: **a, d, e**; secondary: **b, c**; tertiary: **f**.

2.22 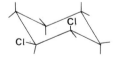 [six cyclohexane structures with Cl substituents]

2.23 *identical structures:* Set 2, **e** and **g**; Set 3, **h** and **i**; Set 4, **l** and **m**

cis-trans isomers: Set 2, **e** and **f**; Set 4, **l** and **o**

structural isomers: Set 1, **a** and **b**; Set 2, **d** and **e**, **d** and **f**; Set 3, **h, j,** and **k**;
Set 4, **l, n,** and **p**; **n, o,** and **p**.

2.24 a) [four cyclohexane chair conformations with Cl substituents]

$\quad\quad$ *cis*-1,2-dichlorocyclohexane $\quad\quad\quad\quad\quad$ *trans*-1,2-dichlorocyclohexane

b) The two *cis*-conformations are equal in that each has one axial and one equatorial chlorine. The *trans*-conformation with both chlorines in equatorial positions (left) is more stable than the one with both chlorines in axial positions (right).

2.25 a) $C_{10}H_{20}O$ menthol; $C_{10}H_{18}O$ borneol $\quad\quad$ **b)** no

c) 6-carbon cyclohexane ring, methyl, isopropyl substituents, hydroxyl functional group; 5-methyl and 1-hydroxyl groups are both *trans* to 2-isopropyl group.

d) same number of carbons, six-membered ring with hydroxyl functional group and methyl substituent—3-carbon isopropyl substituent in menthol, but 3-carbon group in borneol attached to two ring positions making a new ring.

e)

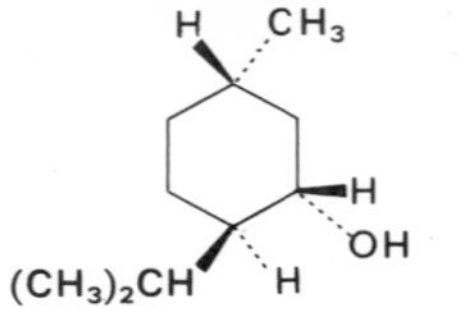

Formula shown is *one cis-trans* isomer of menthol. A change in the geometry (forward or backward) of any one of the three groups attached to the ring will produce a *cis-trans* isomer.

f)

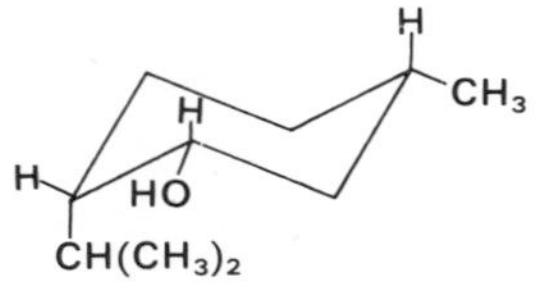

g)

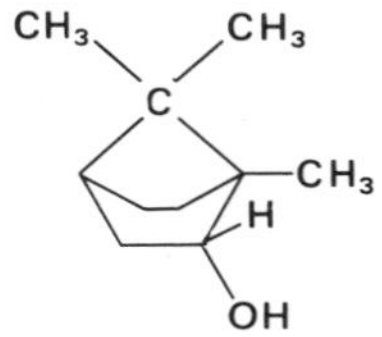

2.26 a)

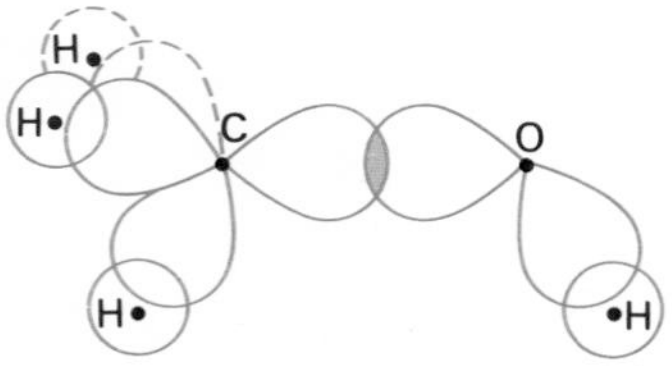

b)

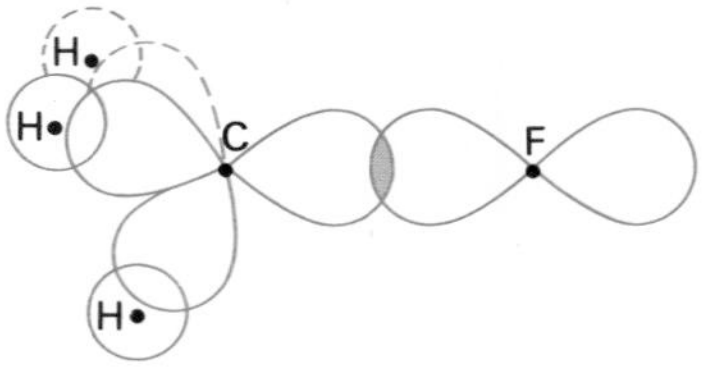

2.27 seven;

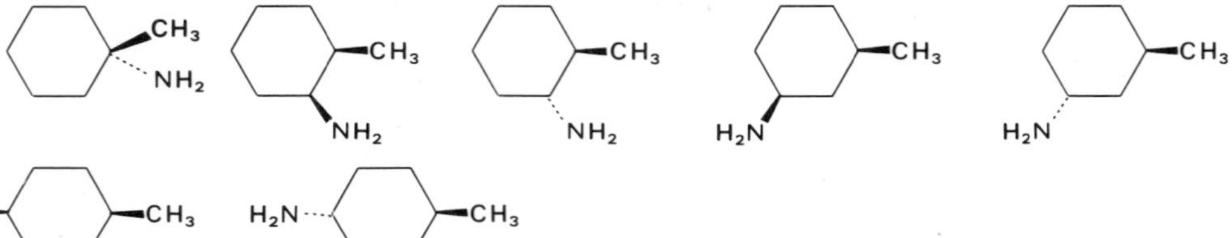

2.28 eight;

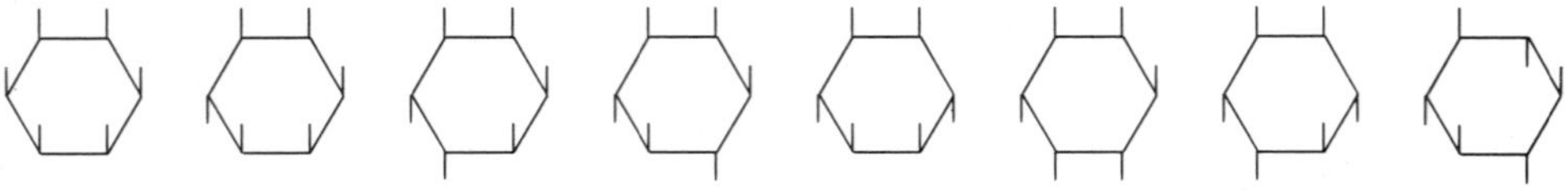

Chapter 3

3.1 a) $CH_3CH=CH_2$ **b)** $CH_3CH_2CH=CH_2$ **c)** $CH_3CH_2CH=CHCH_3$ **3.2 a)** 2,5,8-decatriene **b)** 2-butene

d)

e) $CH_3CH=CHCH=CH_2$ **f)**

c) propadiene **d)** cyclooctene

3.3 b, c, e and f:

b)

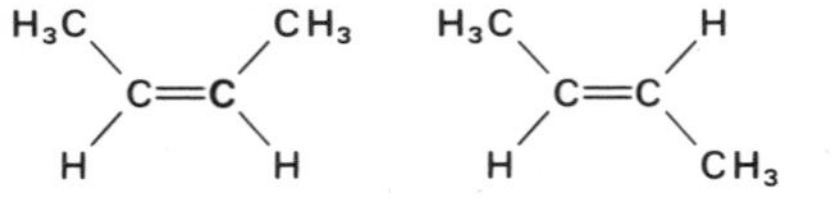

cis *trans*

c)

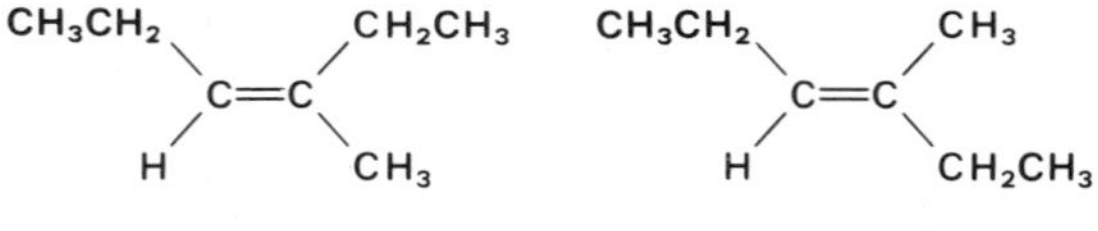

cis *trans*

e)

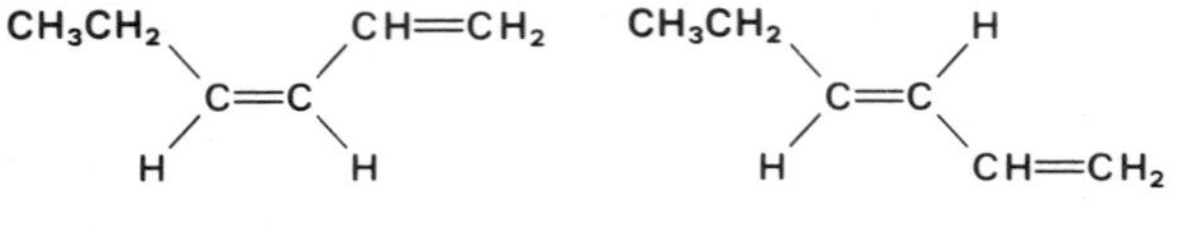

cis *trans*

f)

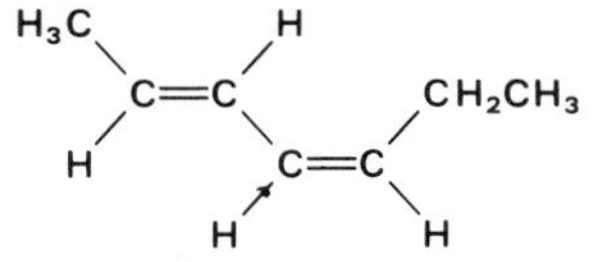

trans-cis 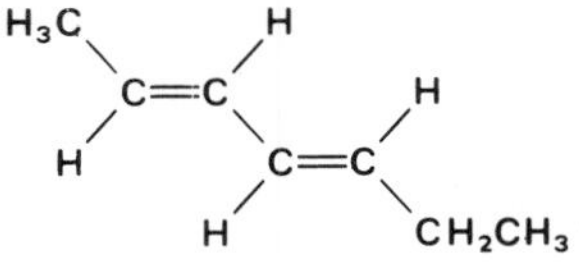 *trans-trans* 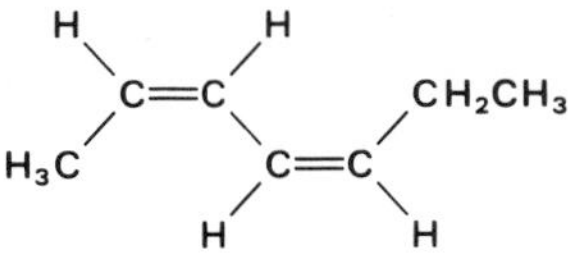 *cis-cis*

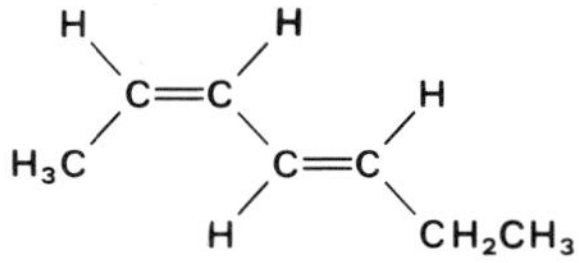

cis-trans

3.4 The four carbons shown by C in the formula must lie in one plane for the double bond to form. For a ring to be formed joining the two carbons which occupy *trans* positions on the double bond and still maintaining reasonably normal bond angles, at least four more carbon atoms and five bonds are required to span the distance.

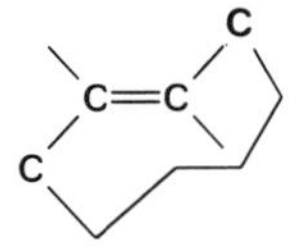

3.5 The value should be close to the dipole moment of vinyl chloride. Since two of the C—Cl bond moments cancel each other as in *trans*-dichloroethene, the dipole moment is due to only one C—Cl bond.

3.6

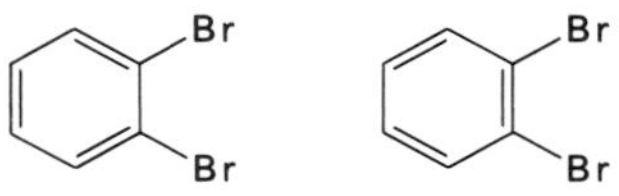

Instead of separate stable isomers, these structures represent resonance structures which contribute to the actual single structure of 1,2-dibromobenzene. In the actual structure the six carbon–carbon bond lengths are equal and each bond is about 50% double in character. The most stable molecular orbital encompasses all six carbons of the ring.

3.7 a)

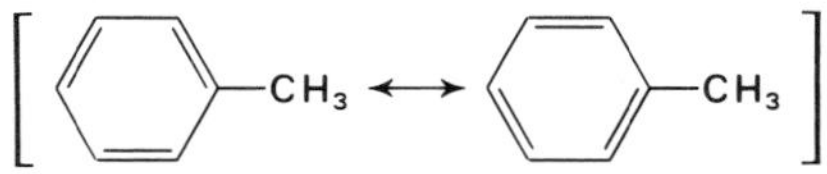

b)

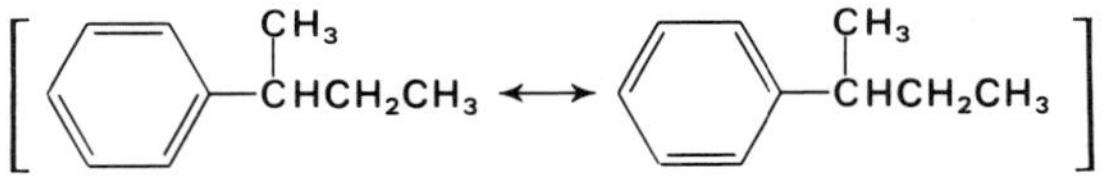

c)

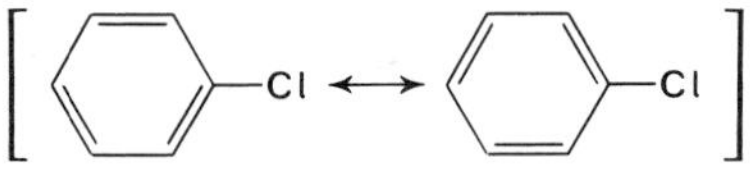

d)

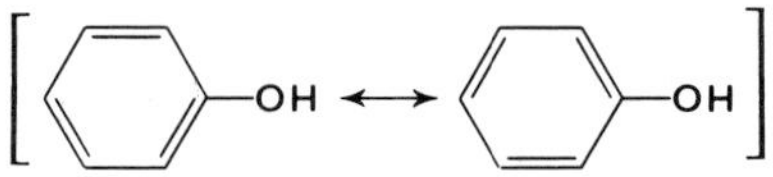

e)

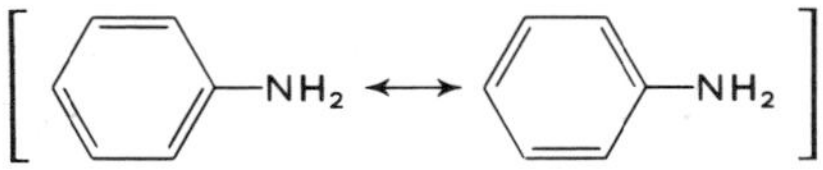

3.8 Some of the possible aromatic structures are as follows.

a) C_8H_{12}—no aromatic structures possible

b) C_8H_{10}

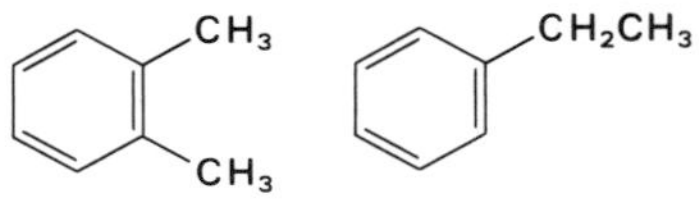

c) $C_{10}H_{10}$

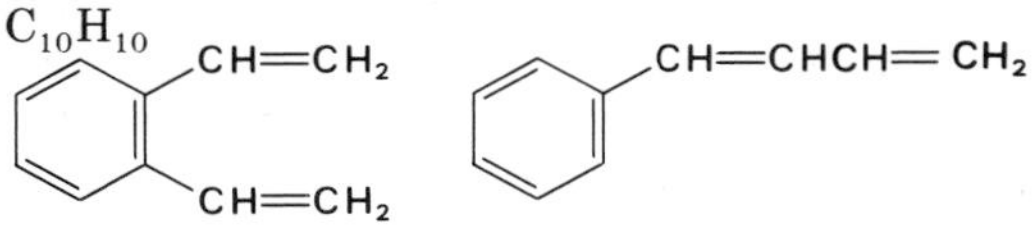

3.9 a)

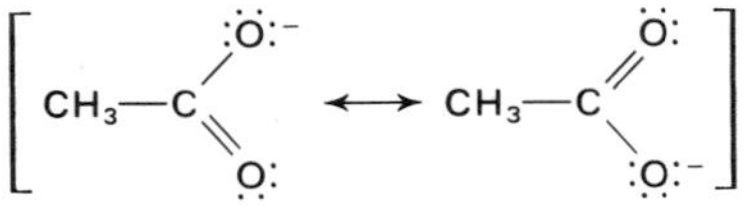

b)

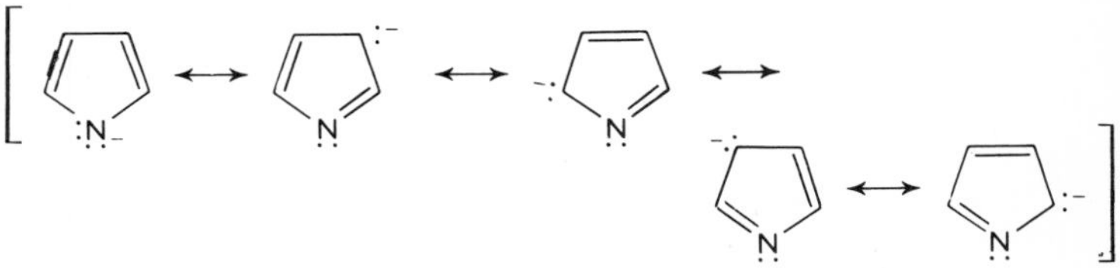

c) $\left[\text{CH}_3\text{—CH=CH—}\overset{+}{\text{C}}\text{H}_2 \longleftrightarrow \text{CH}_3\text{—}\overset{+}{\text{C}}\text{H—CH=CH}_2 \right]$

d) $\left[\text{H}_2\text{N—}\underset{\overset{|}{\overset{+}{\text{N}}\text{H}_2}}{\overset{\text{||}}{\text{C}}}\text{—}\overset{\cdot\cdot}{\text{N}}\text{H—CH}_2\text{—CH}_2\text{—OH} \longleftrightarrow \text{H}_2\overset{+}{\text{N}}\text{=}\underset{\overset{|}{:\text{NH}_2}}{\text{C}}\text{—}\overset{\cdot\cdot}{\text{N}}\text{H—CH}_2\text{—CH}_2\text{—OH} \longleftrightarrow \right.$

$\left. \text{H}_2\overset{\cdot\cdot}{\text{N}}\text{—}\underset{\overset{|}{:\text{NH}_2}}{\text{C}}\text{=}\overset{+}{\text{N}}\text{H—CH}_2\text{—CH}_2\text{—OH} \right]$

e) (resonance structures of cyclopentadienyl anion)

(See Study Guide for suggestions on writing resonance structures.)

3.10 five carbons needed; $\text{CH}_3\text{—}\overset{\overset{\text{O}}{\text{||}}}{\text{C}}\text{—CH}_2\text{CH}_2\text{CH}_3$ $\text{CH}_3\text{CH}_2\text{—}\overset{\overset{\text{O}}{\text{||}}}{\text{C}}\text{—CH}_2\text{CH}_3$ $\text{CH}_3\text{—}\overset{\overset{\text{O}}{\text{||}}}{\text{C}}\text{—}\overset{\overset{\text{CH}_3}{|}}{\text{C}}\text{HCH}_3$

3.11 $\text{CH}_3\text{CH}_2\text{CH}_2\text{CH}_2\overset{\overset{\text{O}}{\text{||}}}{\text{C}}\text{—H}$ $\text{CH}_3\text{CH}_2\overset{\overset{\text{O}}{\text{||}}}{\underset{\overset{|}{\text{CH}_3}}{\text{C}}}\text{HC—H}$ $\text{CH}_3\overset{\overset{\text{CH}_3}{|}}{\text{C}}\text{HCH}_2\overset{\overset{\text{O}}{\text{||}}}{\text{C}}\text{—H}$ $\text{CH}_3\overset{\overset{\text{H}_3\text{C}}{|}}{\underset{\overset{|}{\text{CH}_3}}{\text{C}}}\text{—}\overset{\overset{\text{O}}{\text{||}}}{\text{C}}\text{—H}$

3.12 a) aldehyde   **b)** carboxylic acid   **c)** amide   **d)** ketone   **e)** ester

3.13 hexane $\text{CH}_3(\text{CH}_2)_4\text{CH}_3$;   hexanal $\text{CH}_3(\text{CH}_2)_4\overset{\overset{\text{}}{}}{\underset{\overset{\text{||}}{\text{O}}}{\text{C}}}\text{—H}$;   2-hexanone $\text{CH}_3(\text{CH}_2)_3\overset{}{\underset{\overset{\text{||}}{\text{O}}}{\text{C}}}\text{—CH}_3$;

hexanoic acid $\text{CH}_3(\text{CH}_2)_4\overset{}{\underset{\overset{\text{||}}{\text{O}}}{\text{C}}}\text{—OH}$;   sodium hexanoate $\text{CH}_3(\text{CH}_2)_4\overset{}{\underset{\overset{\text{||}}{\text{O}}}{\text{C}}}\text{—O}^-\text{Na}^+$;

methyl hexanoate $\text{CH}_3(\text{CH}_2)_4\overset{}{\underset{\overset{\text{||}}{\text{O}}}{\text{C}}}\text{—O—CH}_3$;   hexanamide $\text{CH}_3(\text{CH}_2)_4\overset{}{\underset{\overset{\text{||}}{\text{O}}}{\text{C}}}\text{—NH}_2$

3.14 a) H_3O^+ **b)** NH_4^+ **c)** H_2O **d)** H_3O^+ **e)** H_2S

3.15 a) $\text{H}\overset{}{\underset{\overset{\text{||}}{\text{O}}}{\text{C}}}\text{—OH}$ **b)** $\text{CH}_3\overset{\overset{\text{OH}}{|}}{\text{C}}\text{HCH}_3$ **c)** H—O—H **d)** $\text{CH}_3\text{—NH—CH}_3$ **e)** $\text{CH}_3\text{—}\overset{+}{\text{N}}\text{H}_3$

f) (benzene ring)—OH   **g)** $\text{CH}_3\text{CH}_2\text{CH}_2\overset{\overset{\text{CH}_3}{|}}{\underset{\overset{|}{\text{OH}}}{\text{C}}}\text{CH}_3$ **h)** $\text{HO—}\overset{\overset{\text{O}}{\text{||}}}{\underset{\overset{\text{||}}{\text{O}}}{\text{S}}}\text{—OH}$ **i)** $\text{CH}_3\text{CH}_2\overset{\overset{\text{O}}{\text{||}}}{\text{C}}\text{—OH}$ **j)** $\text{CH}_3\text{CH}_2\text{CH}_2\text{—NH}_2$

See Study Guide for formulas of conjugate bases.

3.16 a) $\text{C}_{22}\text{H}_{23}\text{O}_8\text{N}_2\text{Cl}$; 478

b) four rings; six carbons in each ring; A, aromatic (three C=C); B, saturated; C, one C=C; D, one C=C

c) ten functional groups, five different kinds   **d)** eight H_2

e) 1—carbonyl C=O, 2—amide CONH$_2$, 3—hydroxyl OH, 4—amino N(CH$_3$)$_2$, 8—hydroxyl OH, 10—chloro Cl, 13—hydroxyl OH, 15—carbonyl C=O, 17—hydroxyl OH, 18—hydroxyl OH

f) No; #8 and #18 hydroxyls are attached to tertiary carbons; #3 and #17 hydroxyls are attached to unsaturated carbons; #13 hydroxyl is attached to aromatic ring.

3.17 a) $\text{HC≡CCH}_2\text{CH}_2\text{CH}_3$ $\text{HC≡CCH(CH}_3)_2$ $\text{CH}_3\text{C≡CCH}_2\text{CH}_3$

b) $\text{CH}_2\text{=C=CHCH}_2\text{CH}_3$ $\text{CH}_2\text{=CHCH=CHCH}_3$ $\text{CH}_2\text{=CHCH}_2\text{CH=CH}_2$ $\text{CH}_3\text{CH=C=CHCH}_3$

$\text{CH}_2\text{=C=}\overset{}{\underset{\overset{|}{\text{CH}_3}}{\text{C}}}\text{CH}_3$ $\text{CH}_2\text{=CH}\overset{}{\underset{\overset{|}{\text{CH}_3}}{\text{C}}}\text{=CH}_2$

3.18 a) $\overset{③}{\text{CH}}_2\text{=}\overset{②}{\text{CH}}\text{—}\overset{①}{\text{CH}}_2\text{OH}$ **b)** $\overset{⑤}{\text{CH}}_3\text{—}\overset{④}{\text{CH}}\text{=}\overset{③}{\text{CH}}\text{—}\overset{②}{\text{C}}≡\overset{①}{\text{C}}\text{—H}$

carbon #1 sp^3, #2 and #3 sp^2 and p #1 and #2 sp and p, #3 and #4 sp^2 and p, #5 sp^3

c) ring carbons sp^2 and p, methyl sp^3

3.19 a) aldehyde **b)** ester **c)** ether **d)** ketone **e)** ketone **f)** ether **g)** ester **h)** amide **i)** amine

3.20 structures not isomeric with others: **a**, one on right; **b**, middle one; **c**, middle one

3.21 b, c, f:

b) 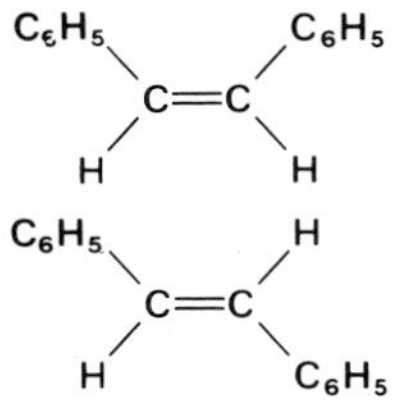 C_6H_5, C_6H_5, $C=C$, H, H; C_6H_5, H, $C=C$, H, C_6H_5

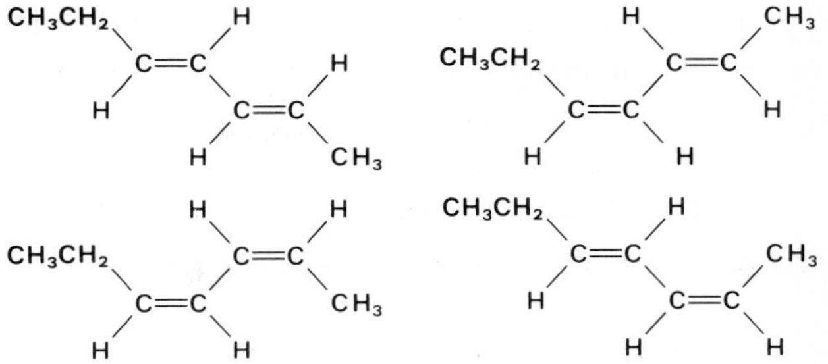

c) CH_3CH_2, H, $C=C$, H; H, $C=C$, H, CH_3; CH_3CH_2, H, $C=C$, CH_3, H; CH_3CH_2, H, $C=C$, H, CH_3; CH_3CH_2, H, $C=C$, H; H, $C=C$, CH_3, H

f) 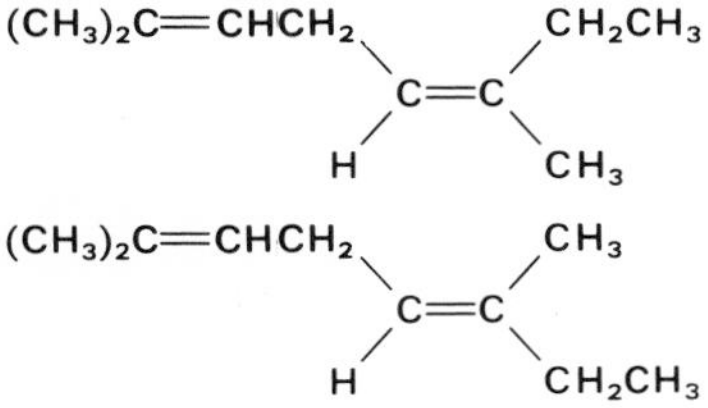 $(CH_3)_2C=CHCH_2$, CH_2CH_3, $C=C$, H, CH_3; $(CH_3)_2C=CHCH_2$, CH_3, $C=C$, H, CH_2CH_3

3.22 a) 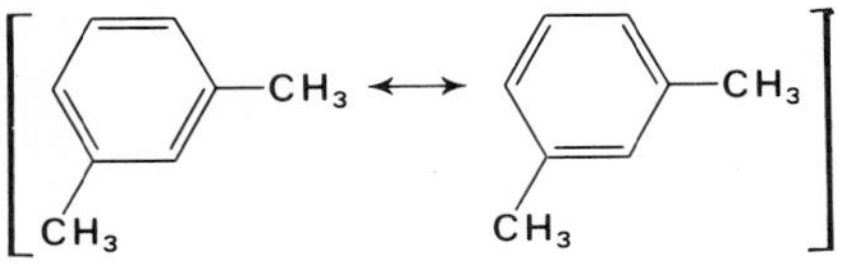 [resonance structures: dimethylbenzene CH_3, CH_3 ↔ CH_3, CH_3]

b) 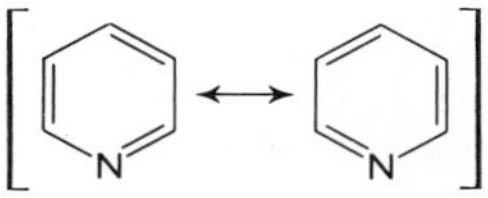 [pyridine ↔ pyridine]

c) 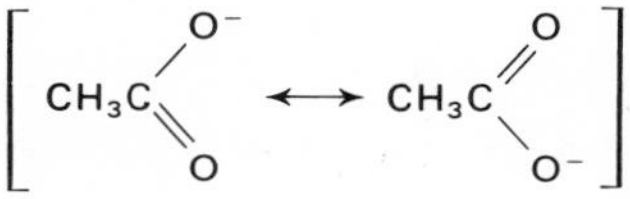 [CH_3C with O^- and O ↔ CH_3C with O and O^-]

d) 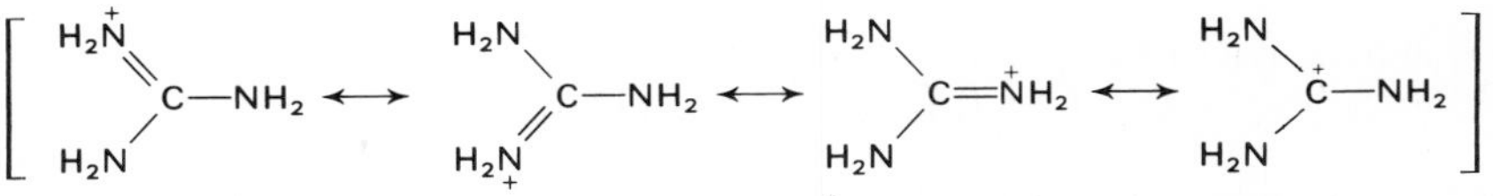 [$H_2\overset{+}{N}$, $C-NH_2$, H_2N ↔ H_2N, $C-NH_2$, $H_2\overset{+}{N}$ ↔ H_2N, $C=\overset{+}{N}H_2$, H_2N ↔ H_2N, $C-NH_2$, H_2N]

3.23 a) $CH_3CH_2CH_2-SH > CH_3CH_2CH_2-OH > CH_3CH_2CH_2-NH_2$

b) 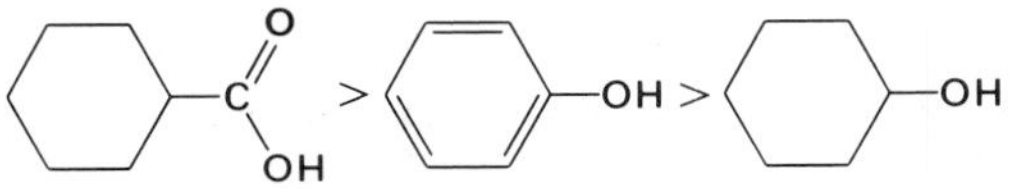 cyclohexyl-C(=O)OH > phenyl-OH > cyclohexyl-OH

c) $CH_3CH_2CH_2-\overset{+}{O}H_2 > CH_3CH_2CH_2-\overset{+}{N}H_3 > CH_3CH_2CH_2-OH$

3.24 a) 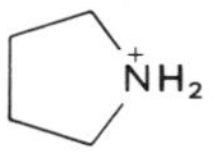 pyrrolidine $\overset{+}{N}H_2$ **b)** 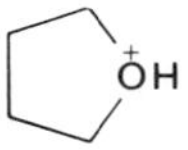 tetrahydrofuran $\overset{+}{O}H$ **c)** 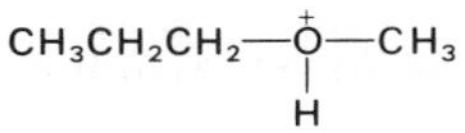 $CH_3CH_2CH_2-\overset{+}{\underset{H}{O}}-CH_3$ **d)** 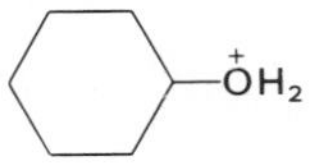 cyclohexyl-$\overset{+}{O}H_2$

e) 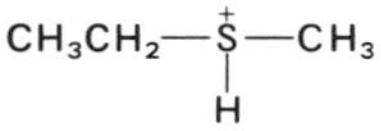 $CH_3CH_2-\overset{+}{\underset{H}{S}}-CH_3$ **f)** 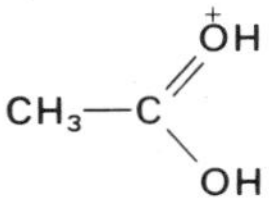 CH_3-C with $\overset{+}{O}H$ and OH

3.25 a) Glucose has hydroxyl and aldehyde groups. **b)** Fat has ester groups.
c) Tripeptide has amide, amino, hydroxyl, sulfhydryl (thiol), and carboxyl groups.

3.26 $CH_3CH_2\overset{O}{\overset{||}{C}}CH_2CH_3$ ketone, $CH_3CH_2CH_2CH_2\overset{O}{\overset{||}{C}}-H$ aldehyde, $CH_3CH=CH-O-CH_2CH_3$ unsaturated ether,

$CH_3CH=CHCH_2CH_2-OH$ unsaturated alcohol

Chapter 4

4.1 a) *E*-3-hexene   **b)** *Z*-1-chloro-1-butene   **c)** 2,3-dimethyl-2-pentene

    d) *Z*-3-methyl-2-heptene   **e)** *E*-3-methyl-1,3-pentadiene   **f)** 6-methyl-*E*-2-*Z*-5-nonadiene

4.2 a) *Z*-2,4-dimethyl-3-phenyl-3-hexene   **b)** *Z*-1-cyclopentyl-1-cyclopropyl-2-phenyl-1,3-butadiene

    c) *E*-2-phenyl-2-butene   **d)** *E*-1,4-hexadiene

    e) 2-chloro-1,3-cyclopentadiene   **f)** 1,3,5,7-cyclooctatetraene

    g) *E*-1-chloro-1,2-dimethylcyclopropane

4.3 a) $CH_3CH_2CH\!=\!CHCH_2CH_3 + HCl \longrightarrow CH_3CH_2\underset{\underset{H}{|}}{C}H\!-\!\underset{\underset{Cl}{|}}{C}HCH_2CH_3$

b) $+ H_2 \xrightarrow{Pt}$ **c)** $+ H_2O \xrightarrow{H_2SO_4}$

4.4 $CH_2\!=\!CH_2 + H_2SO_4 \longrightarrow CH_3\!-\!\overset{+}{C}H_2 + HSO_4^-;\ \ CH_3\overset{+}{C}H_2 + HO\!-\!CH_2CH_3 \longrightarrow CH_3CH_2\!-\!\underset{\underset{H}{|}}{\overset{+}{O}}\!-\!CH_2CH_3;$

    $CH_3CH_2\!-\!\underset{\underset{H}{|}}{\overset{+}{O}}\!-\!CH_2CH_3 + HSO_4^- \longrightarrow CH_3CH_2\!-\!O\!-\!CH_2CH_3 + H_2SO_4$

4.5 a) $\underset{}{\overset{\overset{H}{|}}{C}}H_2\!-\!\overset{+}{C}H\!-\!CH_2CH(CH_3)_2$ secondary, more stable than $\overset{\overset{H}{|}}{C}H_2\!-\!\overset{+}{C}HCH_2CH(CH_3)_2$ primary

b) tertiary, more stable than secondary

c) $CH_3\overset{+}{C}H\!-\!\underset{\underset{H}{|}}{C}HCH_2CH_2CH_3$ secondary, about equal to $CH_3\underset{}{C}H\!-\!\overset{}{\underset{\underset{H}{|}}{\overset{+}{C}}}HCH_2CH_2CH_3$ secondary

4.6 a) $CH_2\!=\!CHCH_2CH_2CH_2CH_3 + H_2O \xrightarrow{H_2SO_4}\ \underset{\text{major}}{CH_3\overset{\overset{OH}{|}}{C}HCH_2CH_2CH_2CH_3} + \underset{\text{minor}}{CH_2\overset{\overset{OH}{|}}{C}H_2CH_2CH_2CH_2CH_3}$

b) $CH_3\overset{\overset{C_6H_5}{|}}{C}\!=\!CHCH_3 + HCl \longrightarrow \underset{\text{major}}{CH_3\overset{\overset{C_6H_5}{|}}{\underset{\underset{Cl}{|}}{C}}CH_2CH_3} + \underset{\text{very minor}}{CH_3\overset{\overset{C_6H_5}{|}}{\underset{\underset{Cl}{|}}{C}}HCHCH_3}$

4.7 a) $\underset{}{\overset{}{C}}H_2\!-\!\underset{\underset{OH}{|}}{C}HCH_2CH_2\underset{\underset{OH}{|}}{C}H\!-\!\underset{\underset{OH}{|}}{C}H_2$ **b)**   **4.8 a)**   **b)**

4.9 a) $CH_3CH\!=\!O + O\!=\!CHCH_2CH_2CH\!=\!O + O\!=\!\overset{\overset{CH_3}{|}}{C}CH_2CH_3$ **b)** $CH_3\overset{\overset{O}{||}}{C}CH_2CH_2CH_2CH\!=\!O$

4.10 a) $CH_3CH_2\overset{\overset{CH_3}{|}}{C}\!=\!CHCH_2CH_2CH\!=\!\overset{\overset{CH_3}{|}}{C}CH_2CH_3$ **b)** $(CH_3)_2C\!=\!$$-CH_3$ (one of 3 possible answers)

4.11 a) $CH_3CH_2C\!\equiv\!CCH_2CH_3 + H_2O \xrightarrow[HgSO_4]{H_2SO_4} CH_3CH_2\overset{\overset{O}{||}}{C}CH_2CH_2CH_3$

b) $HC\equiv CCH_2CH_2C\equiv CH + 4\ HCl \longrightarrow$

$$CH_3\underset{\underset{Cl}{|}}{\overset{\overset{Cl}{|}}{C}}-CH_2CH_2\underset{\underset{Cl}{|}}{\overset{\overset{Cl}{|}}{C}}-CH_3$$

4.12 a) six-membered ring, one double bond, three methyls and one long unsaturated carbon chain

b) eleven in carotene, five in vitamin A; all conjugated (alternating double and single bonds)

tail-to-tail

c)

4.13 a) $CH_3CH=CHCH_2CH_3$

b) $CH_2=\underset{\underset{CH_2CH_3}{|}}{C}-CH=CH_2$

c) $HC\equiv C-CH=CH-CH_3$

d) $HC\equiv C-CH_2-C\equiv C-CH_3$

e)

$$\underset{C_6H_5}{\overset{H}{}}C=C\underset{C_6H_5}{\overset{H}{}}$$

f) $CH_2=CH-\underset{\underset{C_6H_5}{|}}{CH}-CH_3$

g)

$$\underset{C_6H_5}{\overset{H}{}}C=C\underset{H}{\overset{CH_3}{}}$$

h) $CH_2=\underset{\underset{CH_3}{|}}{C}-CH=CH_2$

i) $CH_2=CH-Cl$

j) $\sim(CH_2-\underset{\underset{Cl}{|}}{CH}-CH_2-\underset{\underset{Cl}{|}}{CH}-CH_2-\underset{\underset{Cl}{|}}{CH})_x\sim$

k) $CH_2=CH-CH_2-Cl$

l)

$$\underset{HO_2C}{\overset{H}{}}C=C\underset{H}{\overset{CO_2H}{}}$$

m)

n) $\underset{H}{\overset{H_3C}{}}C=C\underset{CH_2CH_3}{\overset{C(CH_3)_3}{}}$

o)

p)

$$\underset{H_3C}{\overset{Cl}{}}C=C\underset{H}{\overset{H}{}} \quad \underset{H}{}C=C\underset{CH_3}{\overset{H}{}}$$

4.14 a) 1,4-cyclohexadiene **b)** 3,3-dimethyl-1,4-pentadiyne **c)** 3-bromo-4-methyl-1-hexene

d) 5-chloro-2-heptyne **e)** 3,4-heptadiene

4.15 a) $CH_3CH_2CH_2\underset{\underset{OH}{|}}{C}(CH_3)_2$

b) $C_6H_5-\underset{\underset{Br}{|}}{CH}-\underset{\underset{Br}{|}}{CH}-CH_2-\underset{\underset{Br}{|}}{CH}-\underset{\underset{Br}{|}}{CH}-C_6H_5$

c) $CH_3CH_2CH_2\underset{\underset{Br}{|}}{CH}-C_6H_5$

d) $CH_3CH_2CH_2C\equiv C^-Na^+ + NH_3$

e)

$$\underset{H}{\overset{H_3C}{}}C=C\underset{H}{\overset{CH_3}{}} \quad CH_3CH_2CH_2CH_3$$

f)

g) $CH_3CCl_2CH_2CH_2CH_3$

h) $CH_3CH_2\underset{\underset{HO}{|}}{CH}\underset{\underset{OH}{|}}{CH}CH_3$

i) $CH_3CH_2CH-CHCH_2CH_3 + CH_3CO_2H$
$\underset{O}{\diagdown\diagup}$

j)

$$CH_3CH_2\overset{\overset{O}{\|}}{C}CH_3$$

4.16 a) $CH_3CH_2\underset{\underset{OH}{|}}{CH}-\underset{\underset{Cl}{|}}{CH_2}$

b) $CH_3CH_2\overset{+}{C}HCH_2-Cl$

4.17 a) 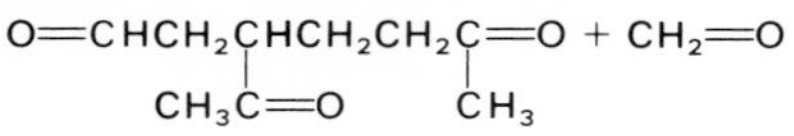 O=CHCH$_2$CHCH$_2$CH$_2$C=O + CH$_2$=O   **b)** 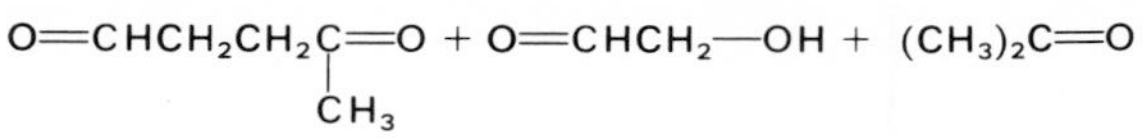 O=CHCH$_2$CH$_2$C=O + O=CHCH$_2$—OH + (CH$_3$)$_2$C=O

      CH$_3$C=O   CH$_3$ CH$_3$

4.18 A *cis* double bond puts carbons needed to bond together near each other. With *trans* C=C and C≡C, the carbons cannot get close enough together to form a bond in a six-membered ring.

4.19 The two CH$_2$ planes are perpendicular to each other and do not rotate.

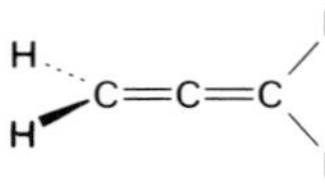

(See Study Guide for orbital representation.)

4.20 Bromonium cation reacts with the nucleophilic CH$_3$OH to give the ether instead of reacting with Br⁻ to give the expected dibromo compound.

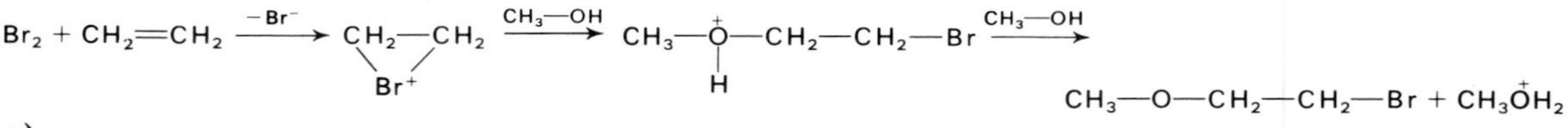

4.21 a)

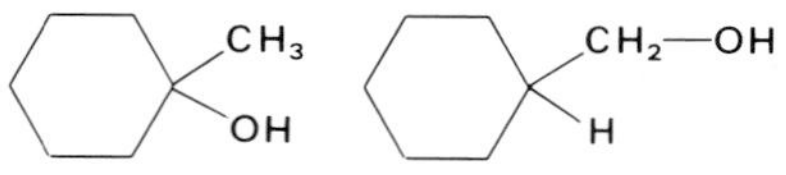

major   very minor

    Major product comes from tertiary carbonium ion, minor product from a primary carbonium ion.

b) 2-methyl-1-butene and 1-methylcyclopentene because they give tertiary carbonium ions while the other compounds form only secondary cations.

4.22 a) The unconjugated all-*cis* structure bends back on itself in a kind of spiral arrangement which allows more room for all carbons. When forced to stretch, it can straighten out by rotating around the single C—C bonds, but comes back to the original conformation when the pull is released. The all-*trans* structure normally is better accommodated by a stretched conformation and has no possibility of stretching further.

b) CH$_3$C≡C—C≡CCH$_3$ All carbons lie on the same line, allowing the rod-shaped molecules to pack easily in a crystal. Molecules with a CH$_2$ between C≡C bonds bend in the middle and pack less easily in a crystal. Less energy is required to separate the bent molecules, and the crystal melts at a lower temperature than for the straight molecules.

4.23 a)

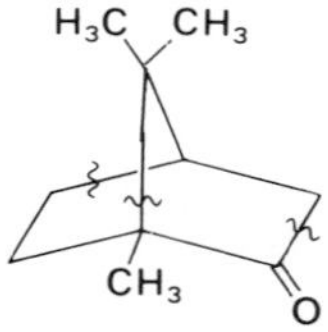

camphor

b)

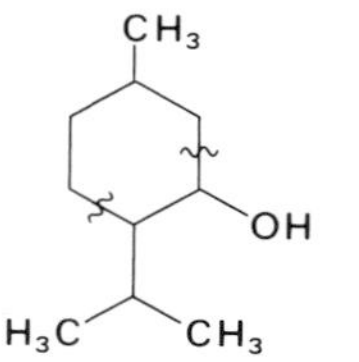

menthol

c)

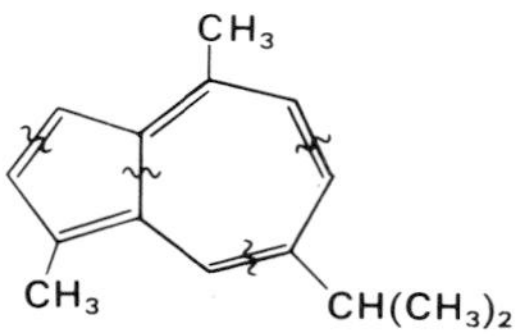

guaiazulene

d)

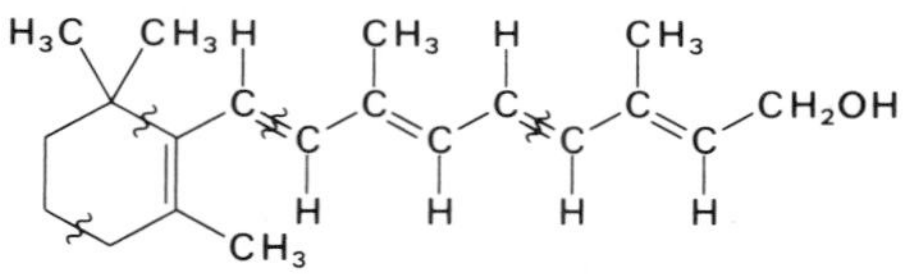

vitamin A

e)

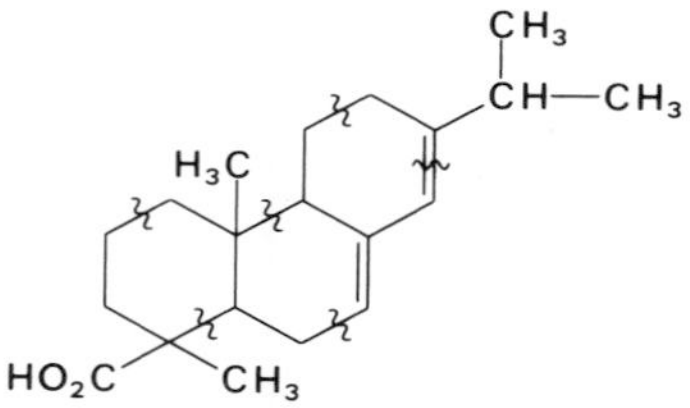

abietic acid

4.24

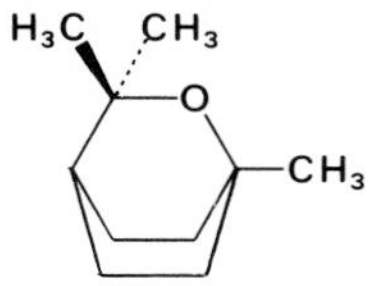

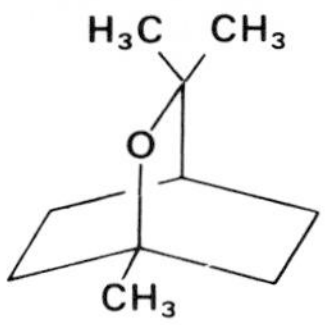

4.25 a)

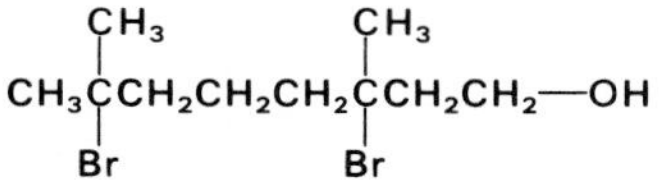

b) 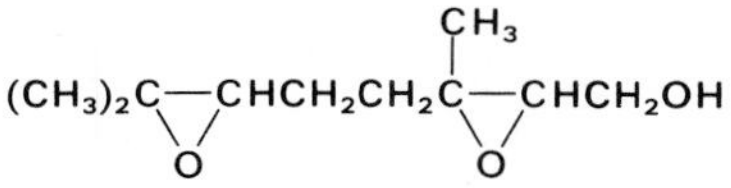 $(CH_3)_2C$—$CHCH_2CH_2C$—$CHCH_2OH$ (with epoxide oxygens and CH₃)

Chapter 5

5.1 a) 2-butanol (*sec*-butyl alcohol)

    b) 2,4-pentanediol

    c) 3-butyn-2-ol

    d) cyclobutanol

    e) cyclopropylmethanol

    f) *trans*-2-chlorocyclohexanol

    g) *cis*-1,3-cyclohexanediol

    h) 3-ethyl-1-pentanol

    i) *trans*-5-iodo-2-phenyl-3-penten-2-ol

    j) 6,7-dichloro-4,4-dimethyl-3-heptanol

    k) 2-isopropyl-5-methylcyclohexanol (menthol)

5.2 a) 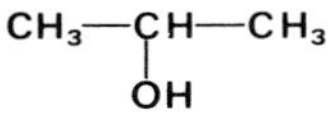 CH_3—CH—CH_3 with OH

    b) 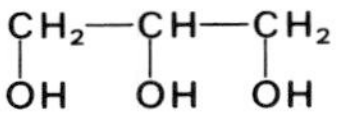 CH_2—CH—CH_2 with OH OH OH

    c) CH_2=CH—CH_2—OH

d)

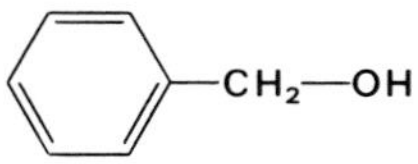

e)

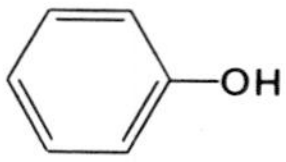

f) 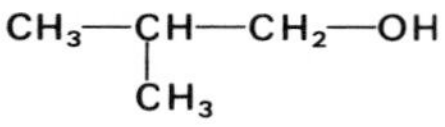 CH_3—CH—CH_2—OH with CH_3

g)

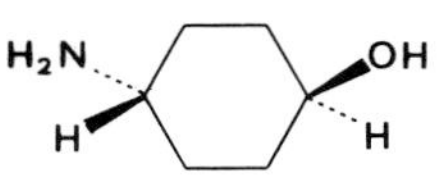

h) 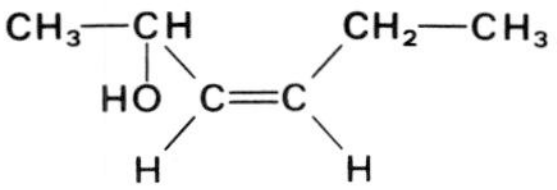 CH_3—CH ... CH_2—CH_3 (HO—C=C—H H)

i) 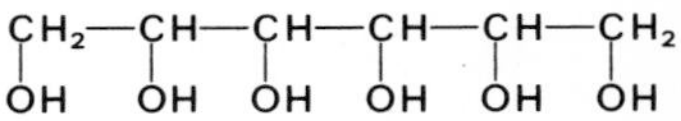 CH_2—CH—CH—CH—CH—CH_2 with OH OH OH OH OH OH

j)

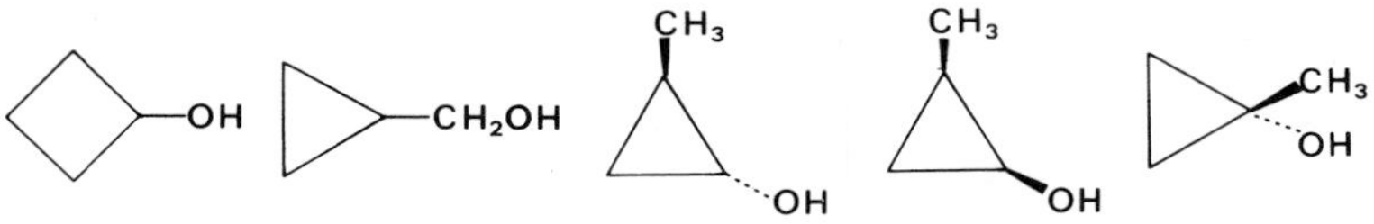

k)

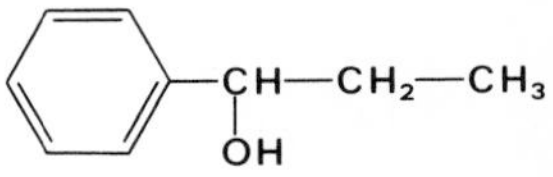

5.3

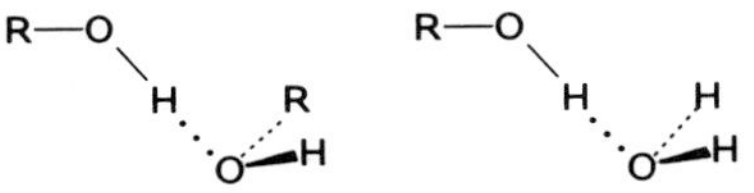

    The bonding hydrogen lies on the line connecting the two oxygens.

5.4 a) $(CH_3)_2CHCH_2$—$OH + H_2SO_4 \longrightarrow (CH_3)_2CHCH_2$—$\overset{+}{O}H_2 + HSO_4^-$

    b) $2\ (CH_3)_3C$—$OH + 2\ K^0 \longrightarrow 2\ (CH_3)_3C$—$O^-K^+ + H_2$

    c)

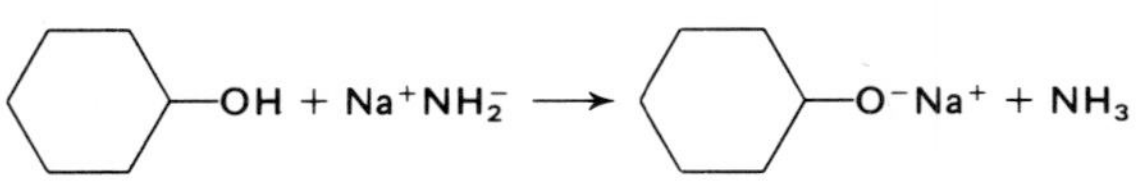

5.5 a) $CH_3CH_2CH_2\underset{\underset{\displaystyle OH}{|}}{\overset{\overset{\displaystyle CH_3}{|}}{C}}CH_2CH_3 + HCl \longrightarrow CH_3CH_2CH_2\underset{\underset{\displaystyle Cl}{|}}{\overset{\overset{\displaystyle CH_3}{|}}{C}}CH_2CH_3 + H_2O$

b)

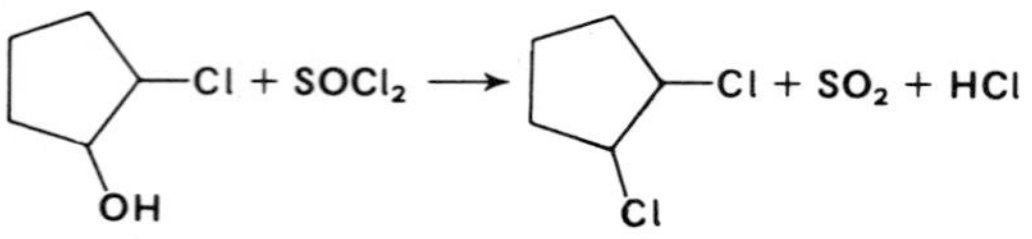

$+ SOCl_2 \longrightarrow$ —Cl $+ SO_2 + HCl$

c) $CH_3\underset{\underset{\displaystyle OH}{|}}{C}HCH_3 + HBr \longrightarrow CH_3\underset{\underset{\displaystyle Br}{|}}{C}HCH_3 + H_2O$

d) $3\ CH_3(CH_2)_4CH_2{-}OH + PBr_3 \longrightarrow 3\ CH_3(CH_2)_4CH_2{-}Br + P(OH)_3$

e)

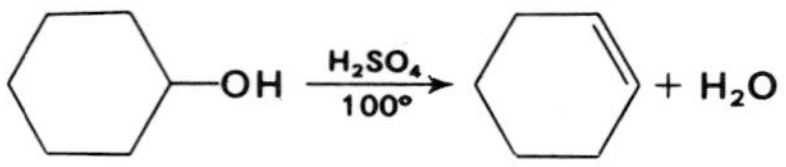

—OH $\xrightarrow[100°]{H_2SO_4}$ $+ H_2O$

5.6 a) $CH_3CH_2\overset{\overset{\displaystyle CH_3}{|}}{C}H - \overset{\overset{\displaystyle OH}{|}}{C}HCH_2CH_3$

b) $CH_3\overset{\overset{\displaystyle C_6H_5}{|}}{C}HCH_2CH_2{-}OH$

c)

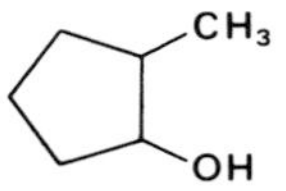

5.7 a) A: $CH_3CH_2CH_2CH_2CH_2{-}OH$, B: $CH_3CH_2CH_2CH_2CH_2{-}Br$

b) C: $CH_3CH_2CH_2\underset{\underset{\displaystyle OH}{|}}{C}HCH_3$, D: $CH_3CH_2CH_2\underset{\underset{\displaystyle Br}{|}}{C}HCH_3$

c) E: $CH_3CH_2CH_2CH_2CH_2{-}OH$, F: $CH_3CH_2CH_2CH_2CH_2{-}Cl$

5.8 a) *m*-ethylphenol   **b)** 3-methoxy-5-methylphenol   **c)** *p*-aminophenol

d) 2,4,6-trimethylphenol   **e)** 1,2-dihydroxybenzene (catechol)   **f)** 2-chloro-1,4-dihydroxybenzene

5.9 a) $C_6H_5{-}O^- + H_2CO_3 \longrightarrow C_6H_5{-}OH + HCO_3^-$ **b)** $CH_3CO_2H + HCO_3^- \longrightarrow CH_3CO_2^- + H_2CO_3$

5.10

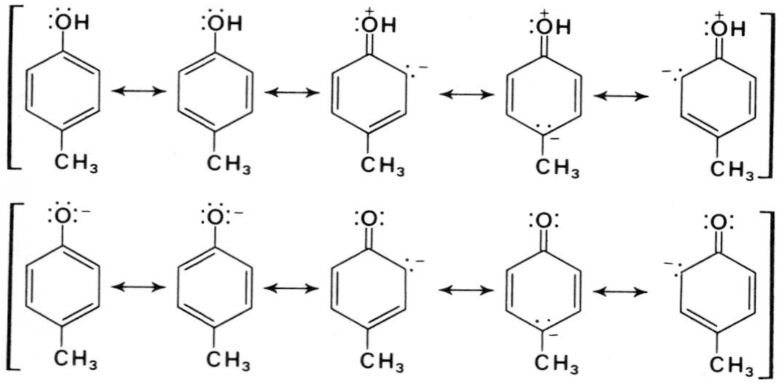

5.11 a) $CH_3(CH_2)_3CH{=}CH_2$; A: $CH_3(CH_2)_3\underset{\underset{\displaystyle O}{\diagdown\diagup}}{CH{-}CH_2}$, B: $CH_3(CH_2)_3\underset{\underset{\displaystyle OH}{|}}{C}H{-}CH_2{-}OH$

b)

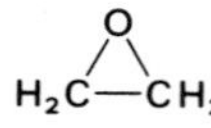

$H_2C\underset{\underset{\displaystyle O}{\diagup\diagdown}}{-}CH_2$; C: $HO{-}CH_2CH_2{-}OCH_3$

c) $CH_3CH{=}CHCH_3$; D: $CH_3\underset{\underset{\displaystyle O}{\diagdown\diagup}}{CH{-}CH}CH_3$, E: $CH_3\underset{\underset{\displaystyle OH}{|}}{C}H{-}\underset{\underset{\displaystyle SH}{|}}{C}HCH_3$

d)

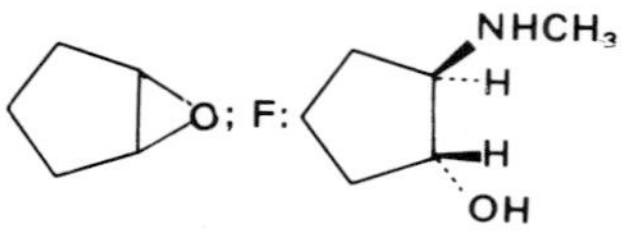

5.12 a) Rings are saturated except as follows: cholesterol B, one C=C; progesterone, testosterone, cortico-sterone, and aldosterone A, one C=C; estradiol and estrone A, aromatic.

b) cholesterol, progesterone, testosterone—4 substituents in same positions; estradiol and estrone—3; corticosterone and aldosterone—5

c) Carbon #3 in each compound carries the oxygen function: OH in cholesterol, estradiol, and estrone, =O in four other compounds.

d) carbons #13 and #17

e) hydroxyl to carbonyl, double bond from ring B to A, hydrocarbon chain on C-17 shortened to acetyl group

f) for testosterone—acetyl group at C-17 changed to OH; for corticosterone OH put on at C-11

g) hydrogen removed from unsaturated cyclic ketone to give phenol, C-10 methyl lost

5.13 a)

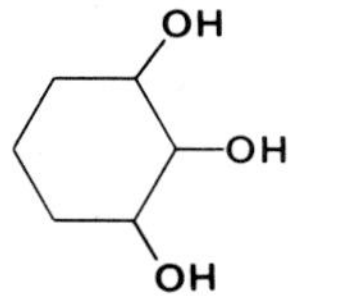

b) $CH_3CH_2CH=CHCH_2-OH$

c) $(CH_3)_3C-O^-K^+$

d)

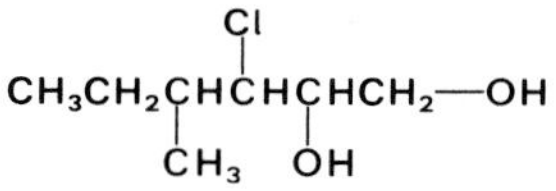

$CH_3CH_2CHCHCHCH_2-OH$ with Cl, CH₃, OH substituents

e)

f)

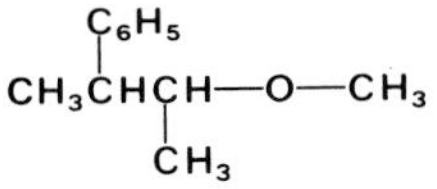

$CH_3CHCH-O-CH_3$ with C_6H_5 and CH_3

g)

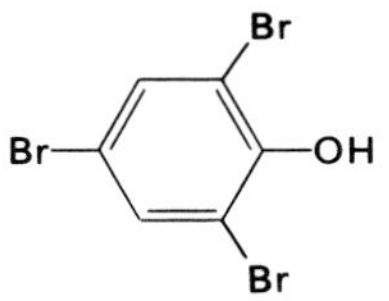

h)

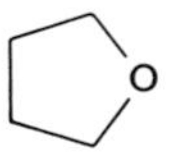

i)

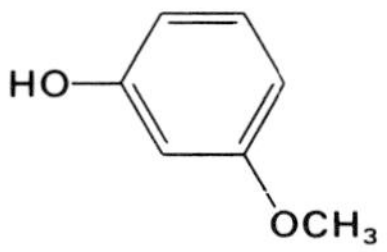

j)

k)

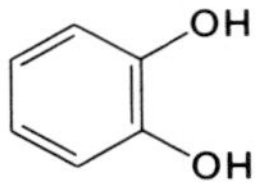

l)

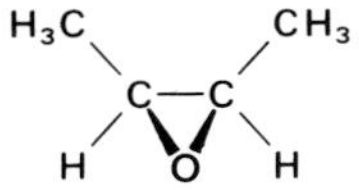

5.14 a)

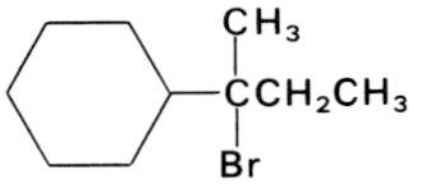

b) 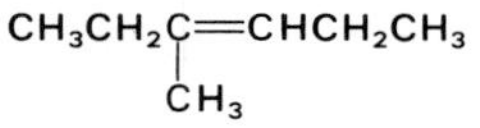 $CH_3CH_2C=CHCH_2CH_3$ and $CH_3CH=CCH_2CH_2CH_3$ (with CH_3)

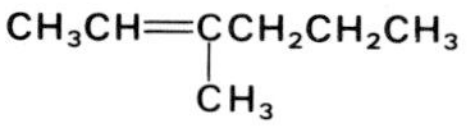

c) $(CH_3CH_2CH_2CH_2CH_2CH_2)_3B$

d) 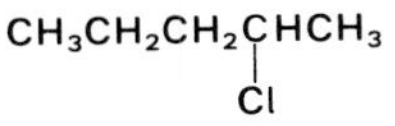 $CH_3CH_2CH_2CHCH_3$ with Cl

e) 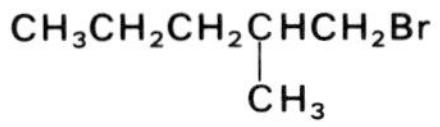 $CH_3CH_2CH_2CHCH_2Br$ with CH_3

f) 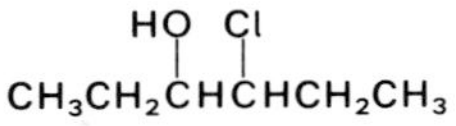 $CH_3CH_2CHCHCH_2CH_3$ with HO, Cl

g)

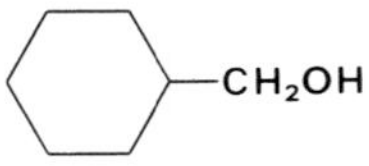

h) $CH_3CH_2CH_2CH_2-OH$

5.15 a)

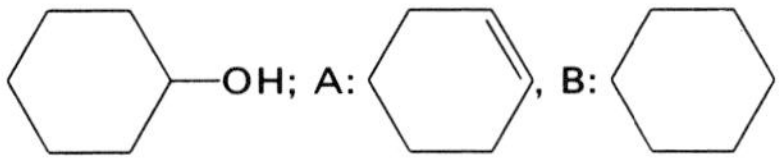

b) 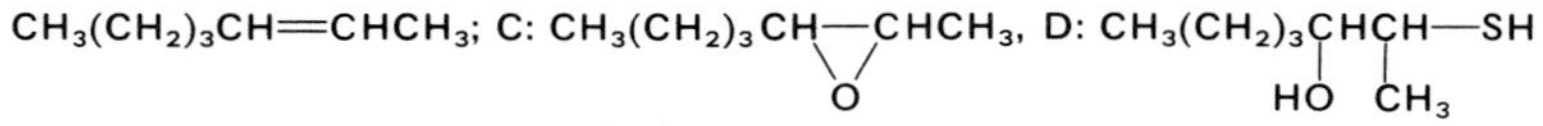 $CH_3(CH_2)_3CH=CHCH_3$; C: $CH_3(CH_2)_3CH-CHCH_3$ (with O); D: $CH_3(CH_2)_3CHCH-SH$ (with HO, CH_3)

c) $CH_3CH_2CHCH_3$ (with OH); E: $CH_3CH=CHCH_3$, F: $CH_3CH-CHCH_3$ (with OH, OH)

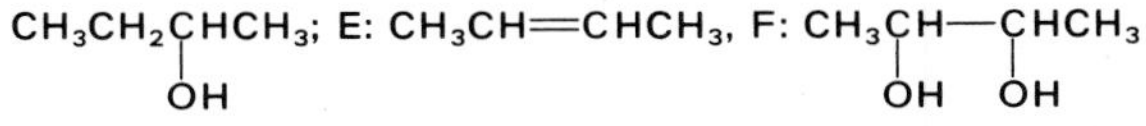

d) $CH_3(CH_2)_8CH=O$; G: $CH_3(CH_2)_8CH_2-OH$; H: $CH_3(CH_2)_8CH_2-Cl$

e) $CH_3CH_2CH_2C(CH_3){=}CH_2$; I: $(CH_3CH_2CH_2CH(CH_3)CH_2)_3B$; J: $CH_3CH_2CH_2CH(CH_3)CH_2{-}OH$

f) $CH_3C(C_6H_5)(OH)CH_3$; K: $CH_3C(C_6H_5){=}CH_2$, L: $CH_3CH(C_6H_5)CH_2{-}OH$

5.16

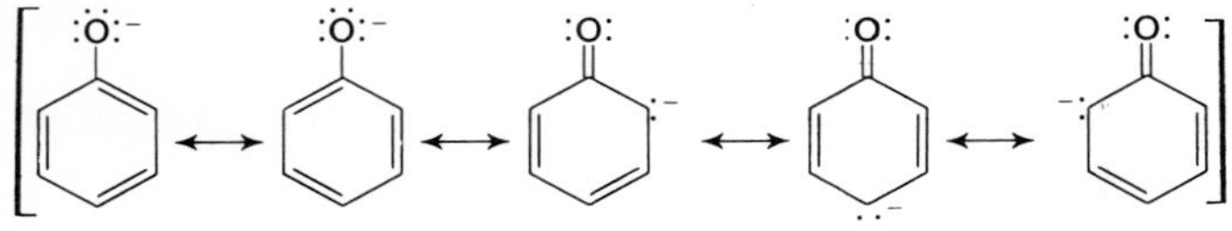

5.17 In **a** and **b** only. In **d**, Cl is not electrophilic enough, and in **e** and **f** the OH groups are too far apart.

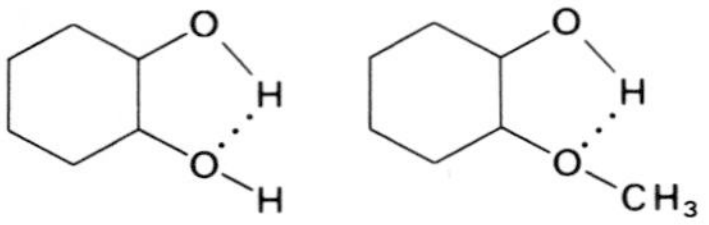

5.18 a) b) c)

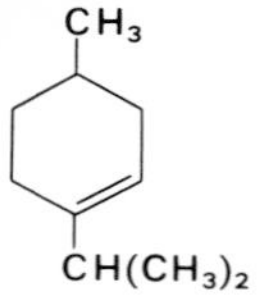

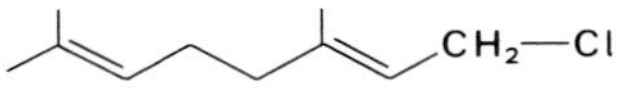

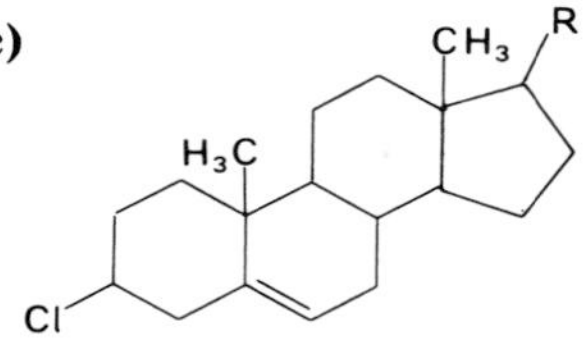

5.19 a) $CH_3CH_2C(CH_3){=}CH_2$, B_2H_6, H_2O_2, NaOH b) $CH_3CH_2C(CH_3){=}CH_2$ or $CH_3CH{=}C(CH_3)CH_3$, H_2O, H_2SO_4

c) , H_2O, H_2SO_4

5.20 a) $CH_3CH(OH)CH(CH_3)CH_2CH_2CH_3$, H_2SO_4 b) $HO{-}CH_2CH_2CH_2CH_3$, H_2SO_4

Chapter 6

6.1 a) $CF_3{-}CHBrCl$ b)   c)   d) $CH_3CHBrCH_2CBr_2CH_3$ e) CHI_3 f) $CH_2{=}CHCH_2{-}$

6.2 **a** and **b** secondary; **c** primary; **d** tertiary

6.3 a) $CH_3{-}I + CH_3CH(CH_3){-}O^-Na^+ \longrightarrow CH_3CH(CH_3){-}O{-}CH_3 + Na^+I^-$

b) $-O^-Na^+ + CH_3CH_2CH_2CH(Cl)CH_3 \longrightarrow C_6H_5{-}O{-}CH(CH_3)CH_2CH_2CH_3 + Na^+Cl^-$

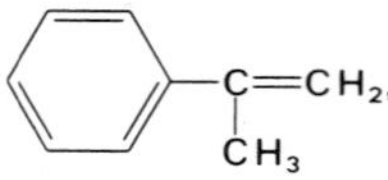

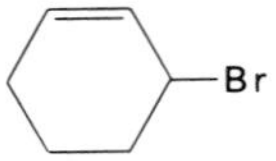

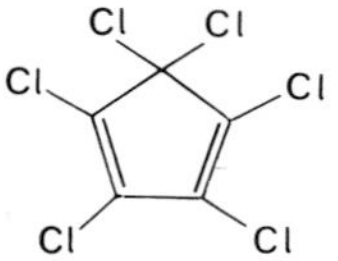

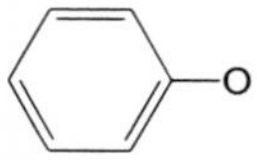

c) $CH_3CH_2CH_2{-}Br + Na^+CN^- \longrightarrow CH_3CH_2CH_2{-}CN + Na^+Br^-$

6.4

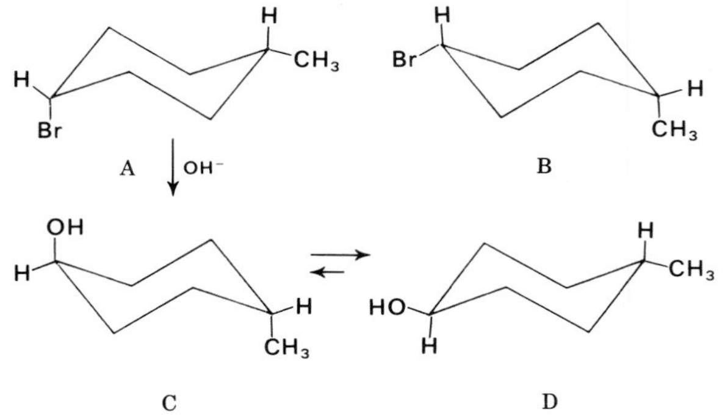

A ↓ OH⁻

B

Back-side attack on A from the top is easier than on B from inside ring. The *trans*-alcohol is formed in conformation C but converts to more stable conformation D.

C   D

6.5 CH_3—I > $CH_3CH_2CH_2CH_2$—Br > $CH_3\underset{CH_3}{CH}CH_2$—Br > $CH_3\underset{Br}{C}HCH_2CH_3$ > $(CH_3)_3C$—Br

6.6 Carbonium ion intermediate allows both axial and equatorial positions to be occupied by water or ethanol, thus product is a mixture of isomers.

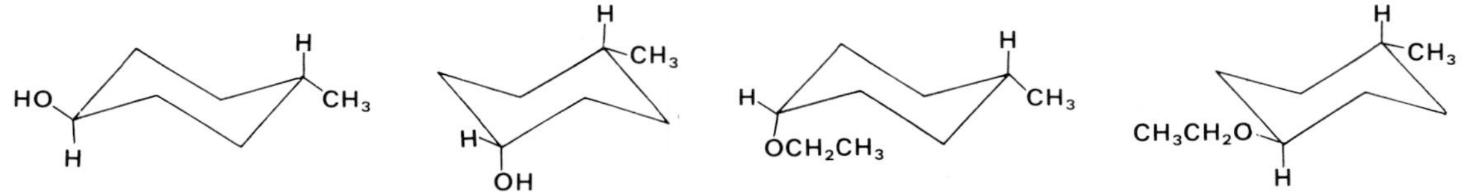

6.7 CH_2=CH—CH_2—Br > $(CH_3)_3C$—Br > $CH_3\underset{Br}{C}HCH_3$ > $CH_3CH_2CH_2$—Br > CH_3—Br

6.8 a) CH_3CH=$CHCH_3$ **b)** $CH_3\underset{}{\overset{CH_3}{C}}$=$CHCH_3$ **c)** $CH_3\overset{H_3C}{\underset{}{C}}$=$\overset{C_6H_5}{\underset{}{C}}CH_2CH_3$

6.9 a) A: H_3C—⬡—SO_2Cl, B: ⬡—CH_2OH, C: H_3C—⬡—$\overset{O}{\underset{O}{S}}$—O—$CH_2$—⬡

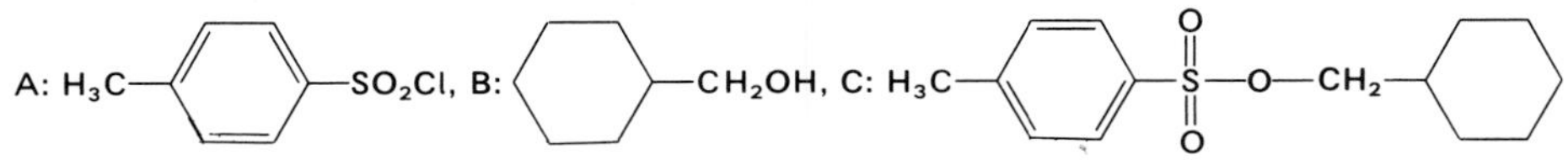

b) D: H_2C=⬡ + HO—$C(CH_3)_3$ + H_3C—⬡—$SO_3^-K^+$

c) E: CH_3OCH_2—⬡ + H_3C—⬡—SO_3H

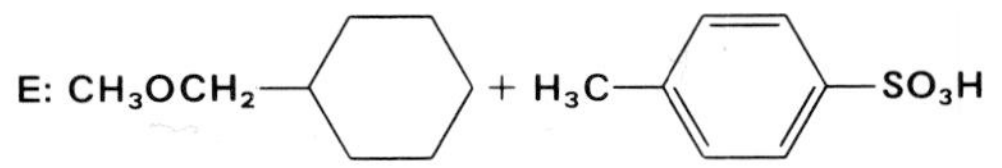

6.10 a) $CH_3\underset{}{\overset{CH_3}{C}}HCH_2$—O—$\overset{O}{\underset{O^-}{P}}$—OH   **b)** $CH_3\overset{CH_3}{\underset{}{C}}H\overset{CH_3}{\underset{}{C}}H$—O—$\overset{O}{\underset{O^-}{P}}$—O—$\overset{O}{\underset{O^-}{P}}$—OH   **c)** ⬡—$CH_2$—$OPO_3PO_3PO_3H^{3-}$

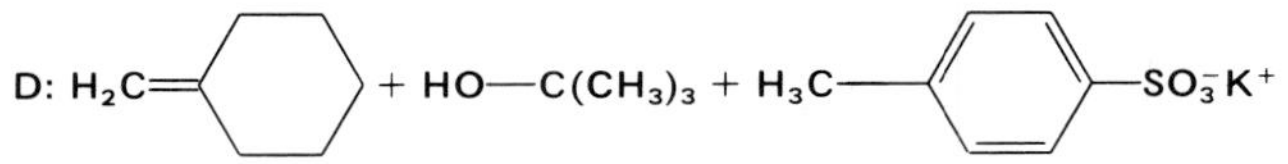

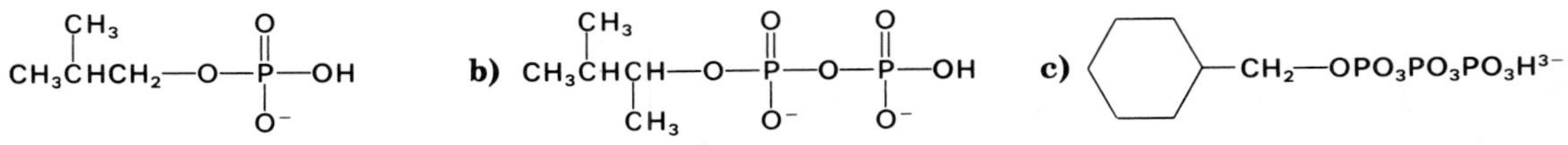

d) CH_3CH_2—O—$\overset{O}{\underset{O^-}{P}}$—O—$CH_2CH_3$

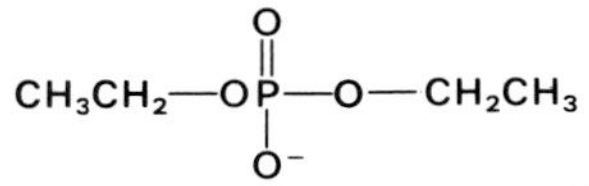

6.11

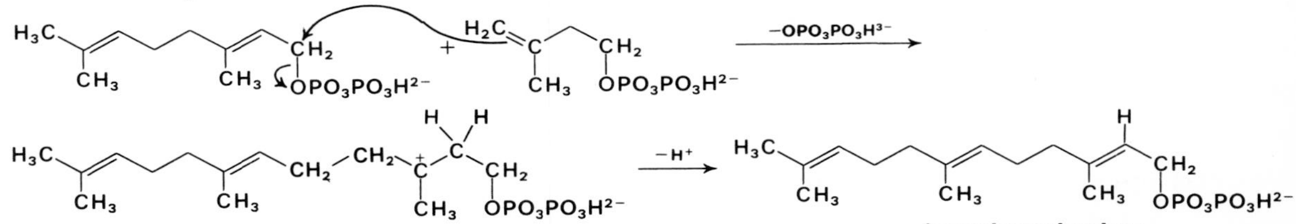

farnesyl pyrophosphate

6.12 a) $CH_3CH_2CH(CH_3)-SH + CH_3-O-SO_2-\langle\text{ring}\rangle-CH_3 \longrightarrow CH_3CH_2CH(CH_3)-S-CH_3 + HO_3S-\langle\text{ring}\rangle-CH_3$

b) $CH_3CH_2-S-\langle\text{cyclopentyl}\rangle + CH_3CH_2-I \longrightarrow (CH_3CH_2)_2\overset{+}{S}-\langle\text{cyclopentyl}\rangle \; I^-$

6.13 $^-O_2C\overset{\overset{+NH_3}{|}}{C}HCH_2CH_2-S-CH_3 + Ad-CH_2-OPO_3PO_3PO_3H^{3-} + H_2O \longrightarrow$

$HOPO_3^{2-} + HP_2O_7^{3-} + {}^-O_2C\overset{\overset{+NH_3}{|}}{C}HCH_2CH_2-\overset{\overset{CH_2-Ad}{|}}{\underset{+}{S}}-CH_3$

$^-O_2C\overset{\overset{+NH_3}{|}}{C}HCH_2CH_2-\overset{\overset{CH_2-Ad}{|}}{\underset{+}{S}}-CH_3 + H_2N-CH_2CH_2-OH \longrightarrow CH_3-\overset{+}{N}H_2-CH_2CH_2-OH + {}^-O_2C\overset{\overset{+NH_3}{|}}{C}HCH_2CH_2-\overset{\overset{CH_2-Ad}{|}}{S}$

6.14 a) $CH_3CH_2\overset{\overset{}{\underset{C_6H_5}{|}}}{C}HCH_2CH_2-Br + Na^+CN^- \longrightarrow CH_3CH_2\overset{\overset{}{\underset{C_6H_5}{|}}}{C}HCH_2CH_2-CN + Na^+Br^-$

b) $CH_3CH_2-O^-Na^+ + Br-CH_2CH=CH_2 \longrightarrow CH_3CH_2-O-CH_2CH=CH_2 + Na^+Br^-$

c) $CH_3CH_2\overset{\overset{}{\underset{Br}{|}}}{C}HCH_3 + Na^+OH^- \xrightarrow{\Delta} CH_3CH=CHCH_3 + Na^+Br^-$

d) $\langle\text{ring}\rangle-O^-Na^+ + (CH_3)_2CHCH_2Cl \longrightarrow \langle\text{ring}\rangle-O-CH_2CH(CH_3)_2 + Na^+Cl^-$

e) $\overset{\overset{CH_3CHOH}{|}}{\langle\text{ring}\rangle} + \overset{\overset{SO_2Cl}{|}}{\langle\text{ring}\rangle-CH_3} + \langle\text{pyridine}\rangle \longrightarrow \overset{\overset{C_6H_5\overset{\overset{CH_3}{|}}{C}H-O-SO_2}{|}}{\langle\text{ring}\rangle-CH_3} + \langle\text{pyridinium}\rangle N^+-Cl^- H$

f) $\langle\text{cyclopentyl}\rangle-O-SO_2-\langle\text{ring}\rangle-CH_3 + CH_3OH \longrightarrow \langle\text{cyclopentyl}\rangle-OCH_3 + HO_3S-\langle\text{ring}\rangle-CH_3$

g) $CH_3CH_2\overset{\overset{}{\underset{CH_3}{|}}}{C}HBr + Na^+HS^- \longrightarrow CH_3CH_2\overset{\overset{}{\underset{CH_3}{|}}}{C}HSH + Na^+Br^-$

6.15 a) $CH_3CH_2CH_2OH$; A: $CH_3CH_2CH_2Cl$, B: $CH_3CH_2CH_2OCH(CH_3)_2$

b) $CH_3CH_2\overset{\overset{}{\underset{CH_3}{|}}}{C}HCH_2CH_2OH$; C: $CH_3CH_2\overset{\overset{}{\underset{CH_3}{|}}}{C}HCH_2CH_2Cl$, D: $CH_3CH_2\overset{\overset{}{\underset{CH_3}{|}}}{C}HCH_2CH_2CN$

c) $CH_3CH_2CH_2CH=CH_2$; E: $CH_3CH_2CH_2CH_2CH_2-OH$, F: $CH_3CH_2CH_2CH_2CH_2-Br$

d) $(CH_3)_3COH$; G: $(CH_3)_3CO^-K^+$, H: $(CH_3)_3COCH_2-\langle\text{ring}\rangle$

e) $(CH_3)_2CHCl$; I: $(CH_3)_2CHSH$, J: $(CH_3)_2CHSCH_3$

f) $CH_3CHCH_2CH_2OH$; K: $CH_3CHCH_2CH_2OSO_2$—⬡—CH_3, L: $CH_3CHCH=CH_2 + K^+ {}^-SO_3$—⬡—$CH_3$
$\quad\;\; |$ $\qquad\qquad\;\; |$ $\qquad\qquad\qquad\qquad\qquad\qquad\qquad\qquad |$
$\quad\;\; CH_3$ $\qquad\qquad CH_3$ $\qquad\qquad\qquad\qquad\qquad\qquad\qquad\qquad CH_3$

g) ⬠—OH; M: ⬠—O⁻ Na⁺, N: ⬠—OH + $(CH_3)_2C=CH_2$

6.16 1) dehydration ⬡—CH_2CH_2OH $\xrightarrow[180°]{H_2SO_4}$ ⬡—$CH=CH_2 + H_2O$

2) dehydrohalogenation ⬡—CH_2CH_2OH $\xrightarrow[-H_2O]{+HBr}$ ⬡—CH_2CH_2Br $\xrightarrow[-Br^-]{NaOH\;\Delta}$

$\qquad\qquad\qquad\qquad\qquad\qquad\qquad\qquad\qquad\qquad\qquad\qquad\qquad\qquad$ ⬡—$CH=CH_2 + H_2O$

3) via sulfonate ester ⬡—CH_2CH_2OH $\xrightarrow[-HCl]{\text{tosyl chloride}}$ $CH_2CH_2OSO_2$ $\xrightarrow{NaOH}$ $CH=CH_2 + Na\overset{+}{O}\overset{-}{S}O_2$

$\qquad\qquad\qquad\qquad\qquad\qquad\qquad\qquad\qquad\qquad\qquad\quad$ ⬡ $\qquad$ ⬡ $\qquad\qquad\qquad$ ⬡ $\qquad$ ⬡
$\qquad\qquad\qquad\qquad\qquad\qquad\qquad\qquad\qquad\qquad\qquad\qquad\qquad\;\; CH_3 \qquad\qquad\qquad\qquad\qquad CH_3$

6.17 $\left[\text{⬡}—\overset{+}{C}H_2 \longleftrightarrow \text{⬡}—\overset{+}{C}H_2 \longleftrightarrow \text{⬡}=CH_2 \longleftrightarrow + \text{⬡}=CH_2 \longleftrightarrow \text{⬡}=CH_2\right]$

6.18 $\overset{+NH_3}{\underset{|}{}}$ $\overset{CH_2Ad}{\underset{|}{}}$ $\overset{+NH_3}{\underset{|}{}}$ $\overset{CH_2Ad}{\underset{|}{}}$

$^-O_2C\overset{|}{C}HCH_2CH_2$—S—$CH_3$ + $\overset{|}{O}PO_3PO_3PO_3H^{4-}$ ⟶ $^-O_2C\overset{|}{C}HCH_2CH_2$—$\overset{+}{S}$—$CH_3$ + HPO_4^{2-} + $HP_2O_7^{3-}$

$\overset{+NH_3}{\underset{|}{}}$ $\overset{CH_2Ad}{\underset{|}{}}$ $\qquad\qquad\qquad CONH_2 \qquad\qquad\qquad CONH_2$

$^-O_2C\overset{|}{C}HCH_2CH_2$—$\overset{|}{S}$—$CH_3$ + N⬡ $\qquad$ ⟶ $\quad CH_3$—$\overset{+}{N}$⬡ $\qquad + \; ^-O_2C\overset{|}{C}HCH_2CH_2$—S $\overset{+NH_3}{\underset{}{}}$ $\overset{AdCH_2}{\underset{}{}}$

—CH_2Ad = adenosine

6.19 CH_3OH $\xrightarrow{PBr_3}$ CH_3Br; ⬡—OH $\xrightarrow{Na°}$ ⬡—O⁻ Na⁺ $\xrightarrow{CH_3Br}$ ⬡—OCH_3

6.20 A good nucleophile which is also a strong base can attack either the hydrogen or the carbon. Base removes the hydrogen without electrons. The electrons push the halide out.

$\qquad\qquad$ + HB + X⁻ bimolecular (E_2)- $\qquad\qquad\qquad\qquad\qquad\qquad\qquad\qquad\qquad\qquad$ + HB unimolecular (E_1)-
$\qquad\qquad\qquad\qquad\qquad\qquad\quad$ one step $\qquad\qquad\qquad\qquad\qquad\qquad\qquad\qquad\qquad\qquad\qquad\qquad\qquad$ two step
primary halide with hydroxide ion heated $\qquad\qquad\qquad\qquad$ tertiary halide in aqueous sodium hydroxide

Ionization occurs first. The base removes the hydrogen from the carbonium ion, leaving a pair of electrons to form the double bond. The first step (ionization) is slow. Removal of the proton is fast.

6.21 $[CH_2=CH—\overset{+}{C}H_2 \longleftrightarrow \overset{+}{C}H_2—CH=CH_2]$

6.22 a) $(CH_3)_3\overset{+}{C}$ (tertiary) $> (CH_3)_2\overset{+}{C}H$ (secondary) $> \overset{+}{C}H_3$ (methyl)

b) $[(CH_3)_2C{=}CH\overset{+}{C}H_2 \longleftrightarrow (CH_3)_2\overset{+}{C}CH{=}CH_2] > [CH_3CH{=}CH\overset{+}{C}H_2 \longleftrightarrow CH_3\overset{+}{C}HCH{=}CH_2]$

$> [CH_2{=}CH\overset{+}{C}H_2 \longleftrightarrow \overset{+}{C}H_2CH{=}CH_2]$

c) $[CH_3CH{=}CH\overset{+}{C}H_2 \longleftrightarrow CH_3\overset{+}{C}HCH{=}CH_2]$ (allylic) $> CH_3\overset{+}{C}HCH_2CH_3$ (secondary) $> CH_2{=}CHCH_2\overset{+}{C}H_2$ (primary)

6.23 The primary chloride and protonated primary alcohol form primary carbonium ions with great reluctance, if at all; therefore the two-step substitution going through carbonium ions is extremely slow. For the one-step substitution mechanism, the nucleophile approaches most easily from the side of the carbon opposite the leaving group. If this approach is blocked by one large carbon group (*tert*-butyl) then direct substitution becomes very difficult and slow. The inertness of this structure earns it a special name—a *neopentyl* structure.

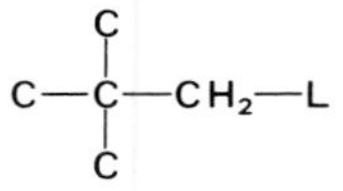

6.24 The stability of the allyl cation due to the resonance contributors allows the protonated primary alcohol to ionize readily under conditions for carbonium ion formation. Allyl chloride is a primary chloride and therefore reacts rapidly with strong nucleophiles.

Chapter 7

7.1 chiral—**a, b, e, f, g, k, m, o, p**

7.2 A B C D E H I K M O T U V W X Y

7.3 a)

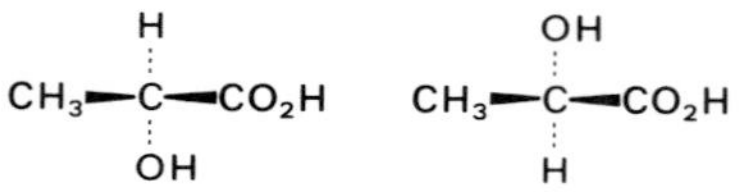

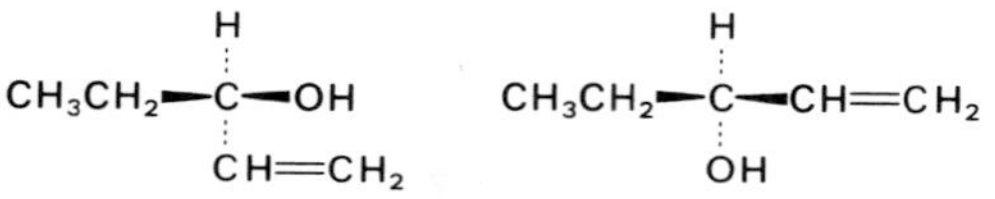

b)

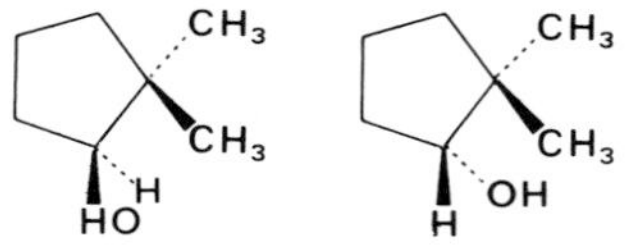

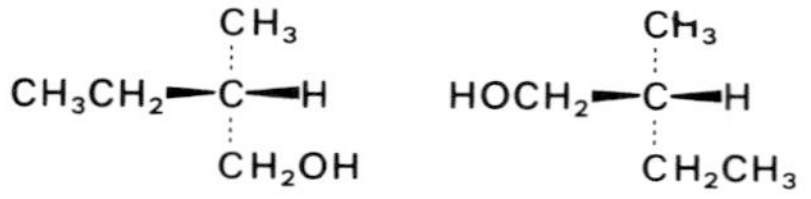

c)

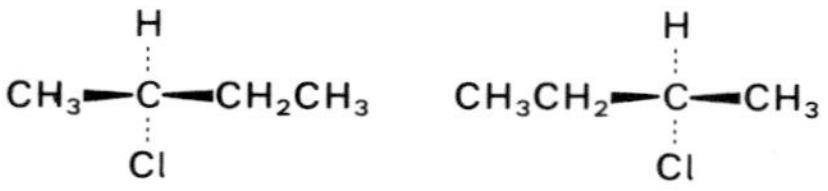

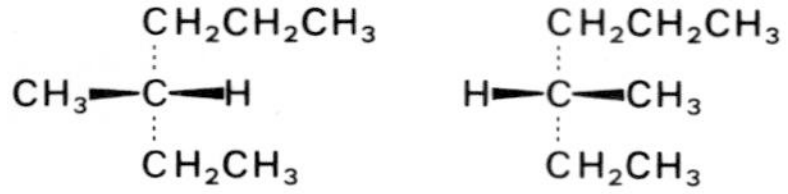

7.4 a, c, e, f:

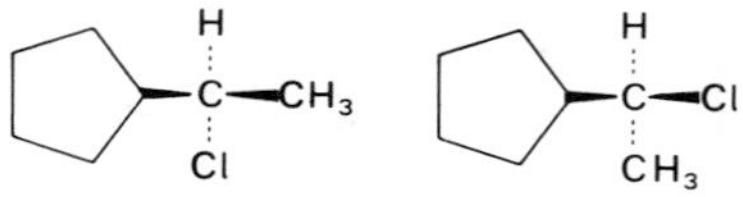

7.5 a) *R*

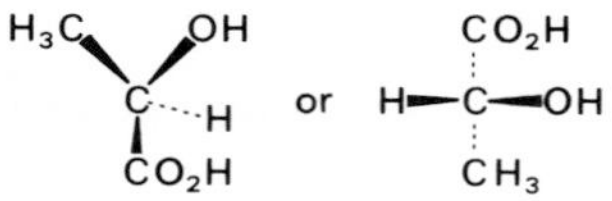

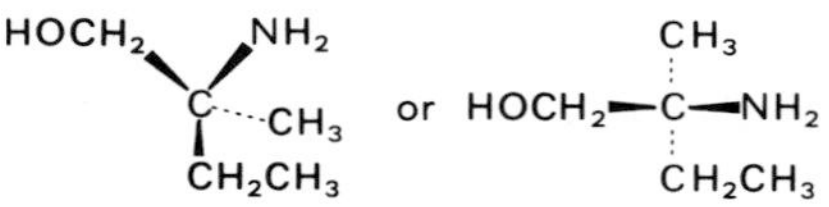

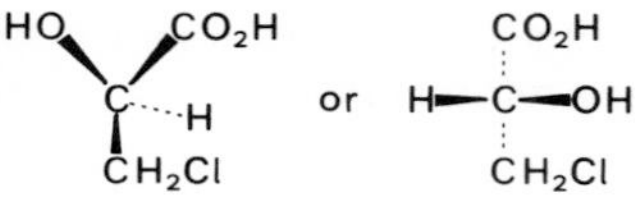

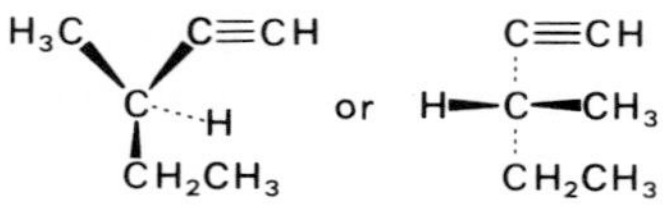

7.6 Rotation is due only to R compound in excess over racemic R, S.
$-120°/-157° = 76.4\%$ R in excess; 23.6% racemic, of which half is R or 11.8%;
total $R = 76.4 + 11.8 = 88.2\%$ of sample.

7.7 a) B: 3-(R)-phenyl-2-(S)-butanol; C: 3-(S)-phenyl-2-(S)-butanol; D: 3-(R)-phenyl-2-(R)-butanol

b) Notation of one enantiomer is exactly the opposite for each asymmetric carbon of the other enantiomer

c) Diastereoisomers must have at least one notation for a given asymmetric carbon identical and at least one notation opposite.

7.8 2^4 stereoisomers $= 16$, with eight racemic forms or pairs of enantiomers

7.9 a) two asymmetric centers and four stereoisomers

b) three centers and eight stereoisomers

c) two centers and four stereoisomers

7.10 $(\pm)$-amine $+ (-)$-tartaric acid $\rightarrow (+)(-)$-salt crystallized and filtered from solution; evaporation of solvent; $(-)(-)$-salt crystallized and filtered from solution; $(+)(-)$-salt $+$ NaOH$_{aq} \rightarrow (+)$-amine $+ (-)$-tartrate; $(+)$-amine extracted by organic solvent; $(-)(-)$-salt $+$ NaOH$_{aq} \rightarrow (-)$-amine $+ (-)$-tartrate; $(-)$-amine extracted by organic solvent

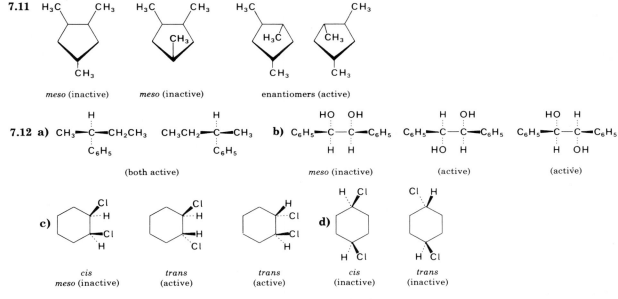

7.13 a) R **b)** S **c)** R left, S right **d)** R left, R right **e)** S left, S right

7.17 With HCl the reaction goes through a carbonium ion (S_N1) which allows Cl^- to enter from either side—racemic product. With tosyl chloride the tosyl ester is formed without breaking C—O bond, thus maintains the original configuration. Attack of Cl^- then is from back side (S_N2), product of inverted configuration. With $SOCl_2$ and pyridine the ester intermediate undergoes S_N2 substitution, giving product with inverted configuration.

7.18 a)

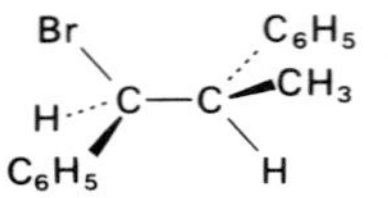

b)

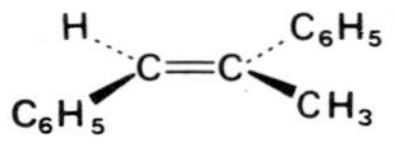

c) 1-(S)-2-(R)-enantiomer gives E-alkene (phenyls *trans*).

7.19 a) 2   **b)** 4   **c)** 0   **d)** 0   **e)** 8   **f)** 0   **g)** 2   **h)** 10   **i)** 8   **j)** 3

h)

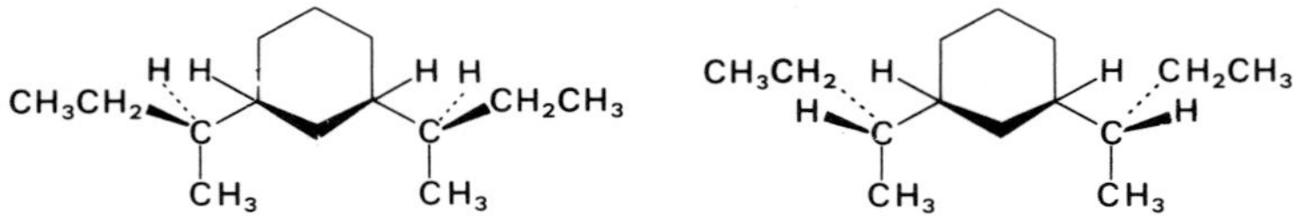

j)

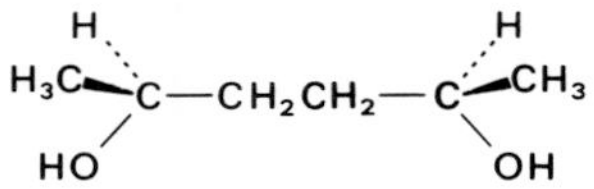

7.20 a) (S)

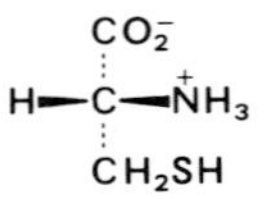

b) (R)

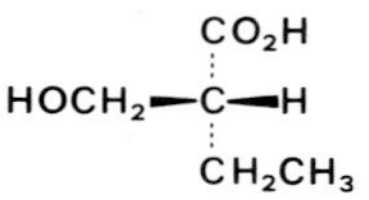

c) (RR)

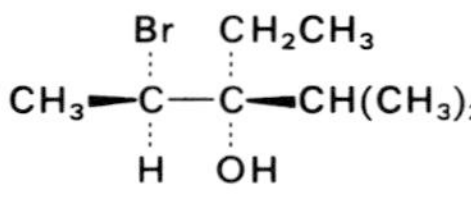

7.21 A:

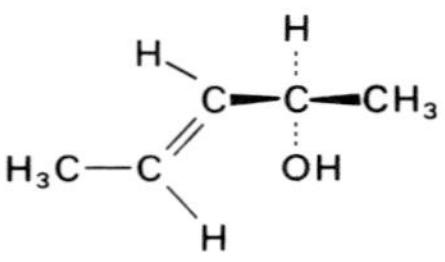

B:

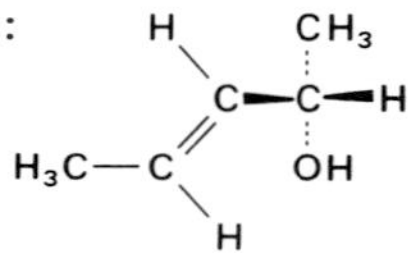

C:

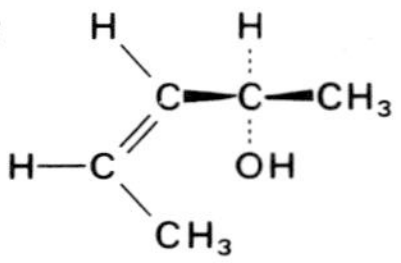

D:

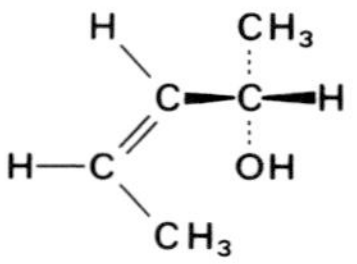

enantiomers: A and B, C and D; diastereoisomers: A and C or D;
B and C or D; *cis-trans* isomers: A and C, D and B

7.22

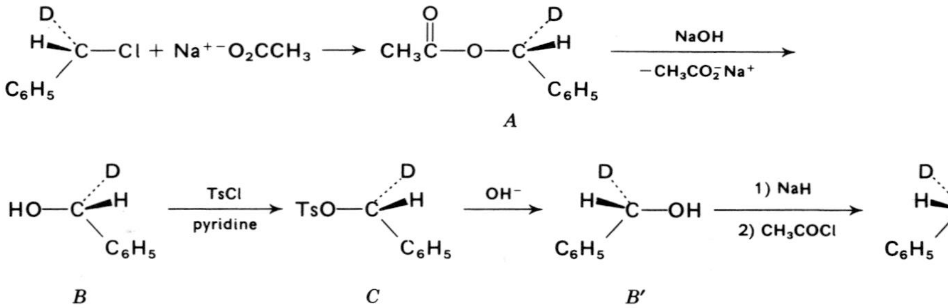

7.23 a) identical   **b)** enantiomers   **c)** identical (*meso*)
   d) diastereoisomers   **e)** identical (*meso*)   **f)** enantiomers

7.24

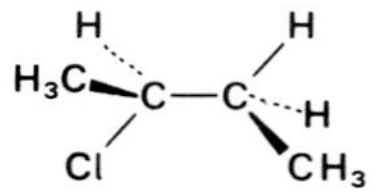

less stable
gives *cis*-alkene

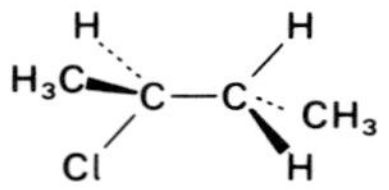

more stable
gives *trans*-alkene

Conformation on right has two methyl groups in anti or staggered positions. Less stable conformation has methyl gauche to **both** Cl and methyl.

7.25

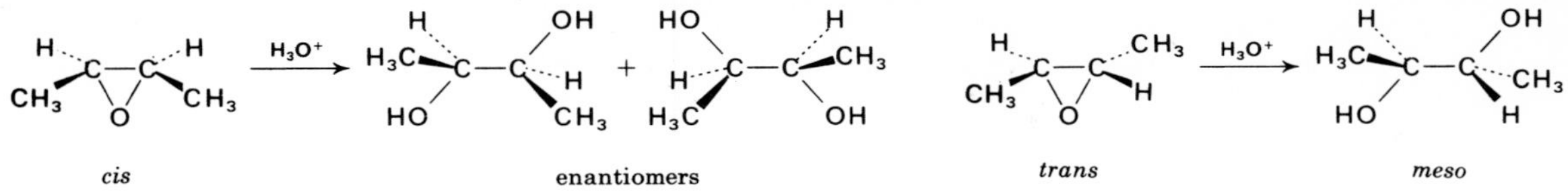

cis   enantiomers   *trans*   *meso*

In each case water attacks the protonated oxirane from the back side to invert the configuration of one carbon.

7.26

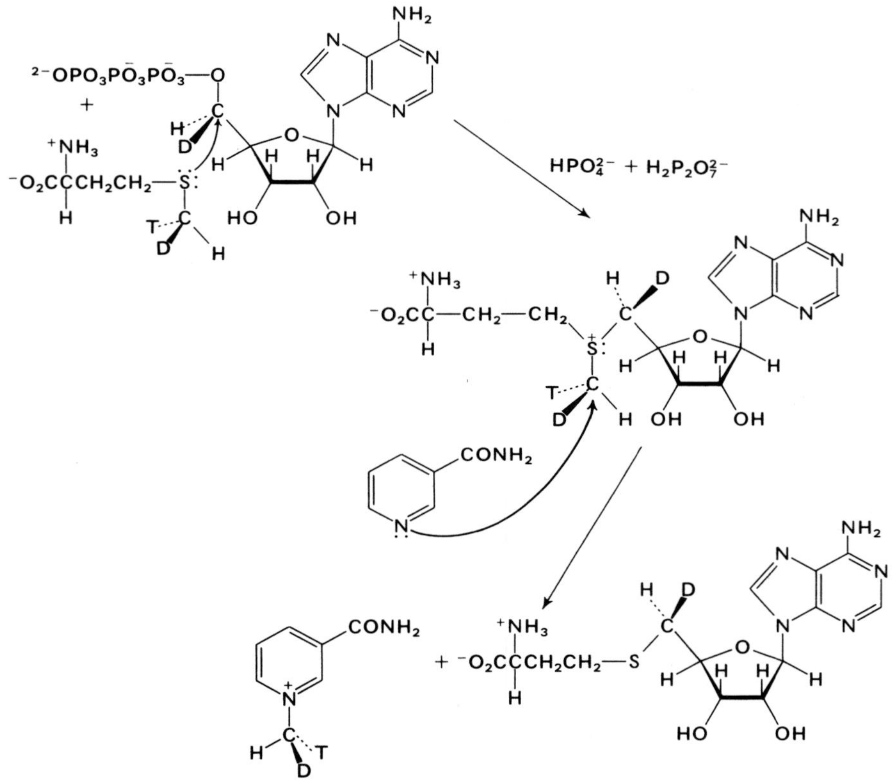

7.27 The nucleophilic substitution of Br^- on $(-)$-*sec*-butyl bromide is S_N2 and gives the inverted product $(+)$-*sec*-butyl bromide. When half of the original $(-)$-isomer has reacted, the compound exists as the racemic form $(\pm)$. Since the $(+)$ reacts equally well with Br^- as the $(-)$-isomer, the racemic form persists.

7.28 The nucleophilic substitution of Br^- on the optically active bromobutane is an S_N2 reaction. With the reaction of each molecule of ordinary optically active bromobutane with Br^{-*}, a molecule of radioactive bromobutane is produced of inverted configuration. As seen in Problem 7.27, the bromobutane is racemic when only half of the ordinary bromine has been replaced by Br^*.

Chapter 8

8.1 aspirin:

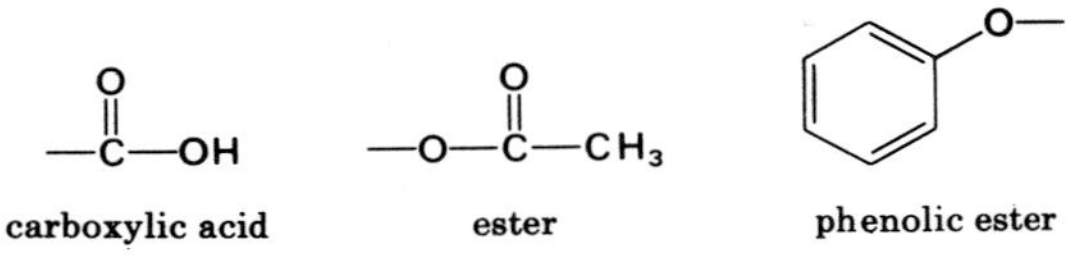

carboxylic acid   ester   phenolic ester

penicillin:

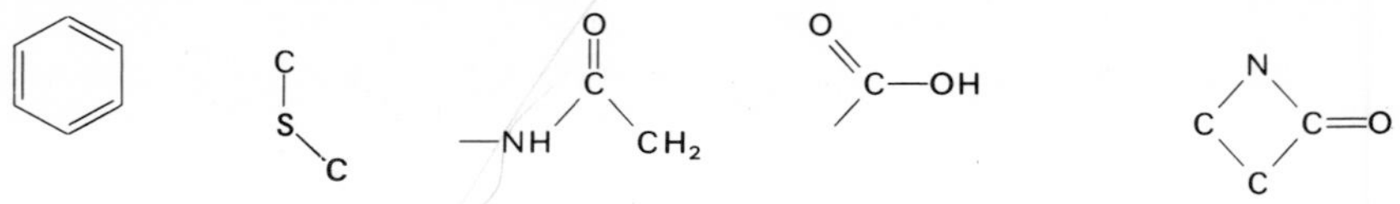

benzene ring   sulfide   amide   carboxylic acid   amide in 4-membered ring

vitamin C:

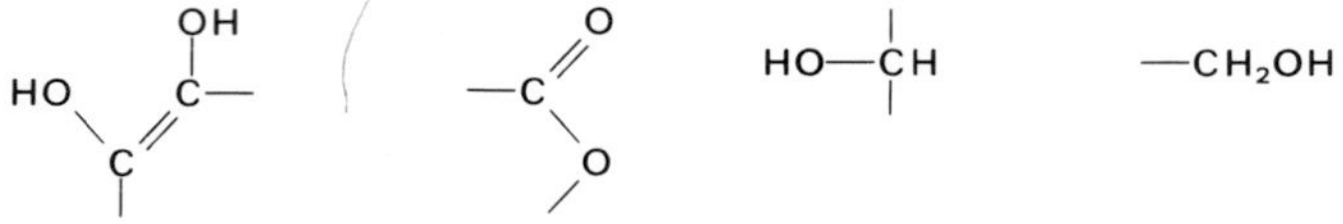

unsaturated alcohols (enols)   ester   secondary alcohol   primary alcohol

8.2 a) 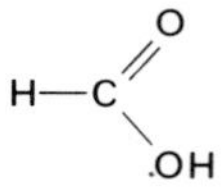 $H-C$ with $=O$ and $-OH$ **b)** Cl_2CH-C with $=O$ and $-OH$ **c)** CH_3-CH_2-CH-C with $=O$ and $-OH$, and CH_3CH_2 substituent

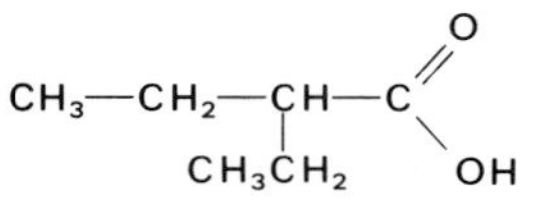

d) 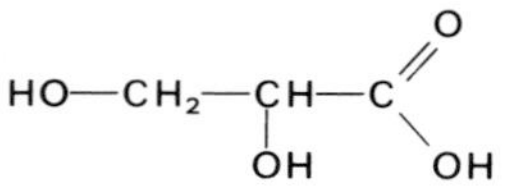 $HO-CH_2-CH-C$ with $=O$ and $-OH$, and OH substituent   **e)** 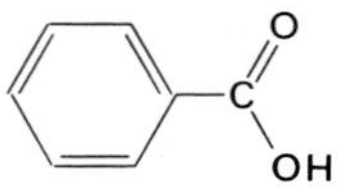 benzene ring$-C$ with $=O$ and $-OH$

8.3 $CH_3-CH_2-CH_2-CH-CH_2-CH_3$ with CO_2H substituent   2-ethylpentanoic acid—a simple example

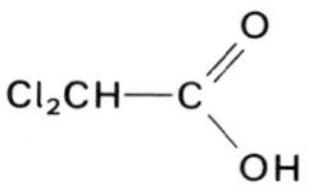

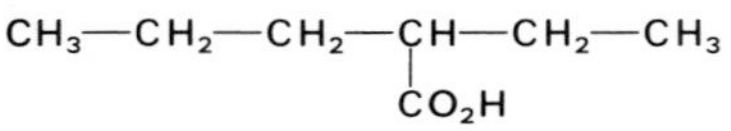

8.4 a) 2-methylpropanoic acid   **b)** 2-ethyl-3-hydroxybutanoic acid
c) 10,12-dihydroxyoctadecanoic acid   **d)** *cis*-3-phenylpropenoic acid
e) *trans*-2-pentenoic acid

8.5 a) 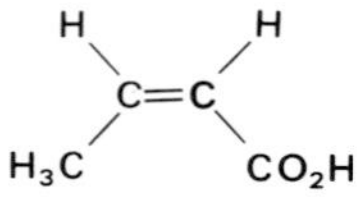 H_3C, H on $C=C$ with H, CO_2H **b)** $CH_3CH_2CH_2CH_2CHC-OH$ with $=O$ and CH_3CHCH_3 substituent   **c)** benzene ring$-CH_2CHC-OH$ with $=O$ and Br substituent

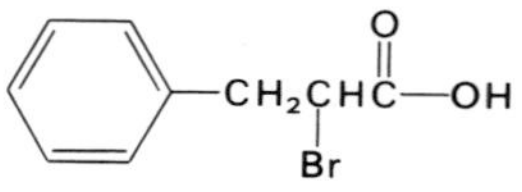

d) 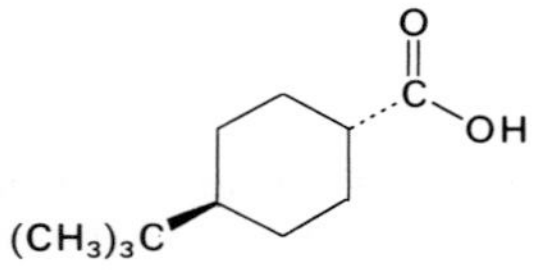 (cyclohexane with $(CH_3)_3C$ and $C(=O)OH$)   **e)** 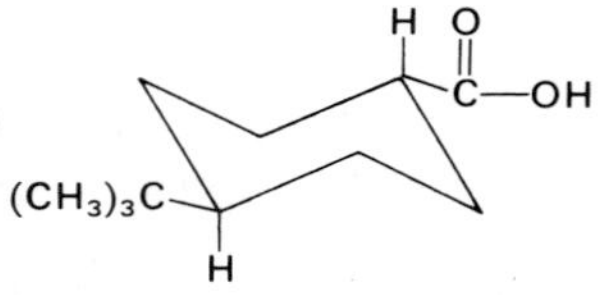 (cyclohexane with $(CH_3)_3C$, H, and $C-OH$ with $=O$)

8.6 $O=C$ (R)$-N(H)\cdots O=C$ (R)$-N(H)\cdots O=C$ (R)$-N(H)$

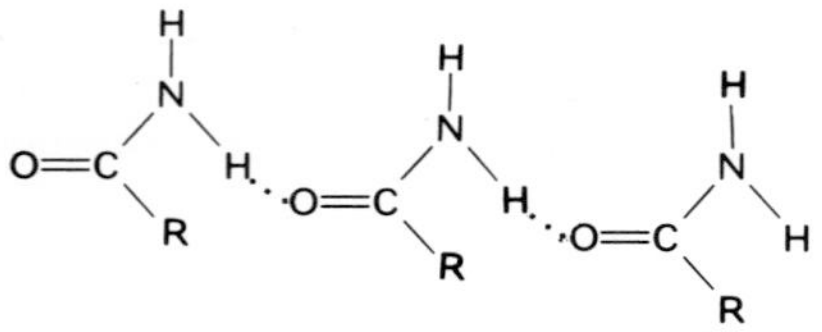

8.7 a) 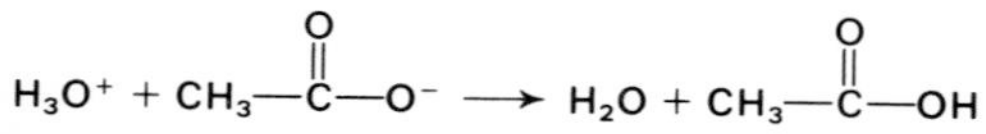 $H_3O^+ + CH_3-C(=O)-O^- \longrightarrow H_2O + CH_3-C(=O)-OH$ **b)** $H-C(=O)-OH + {}^-OCH_3 \longrightarrow H-C(=O)-O^- + HO-CH_3$

c) $CH_3CH_2CH_2C(=O)-OH + CH_3CH_2-NH_2 \longrightarrow CH_3CH_2CH_2C(=O)-O^- + CH_3CH_2\overset{+}{N}H_3$

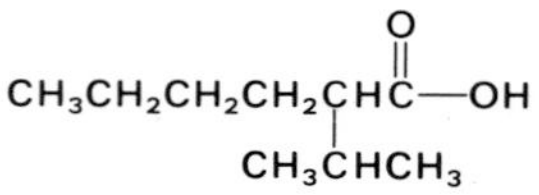

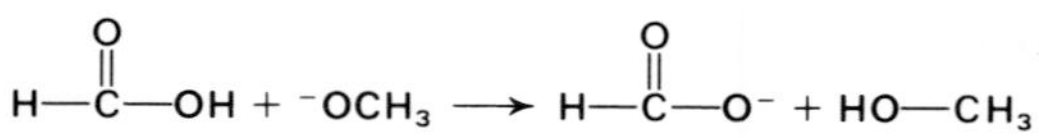

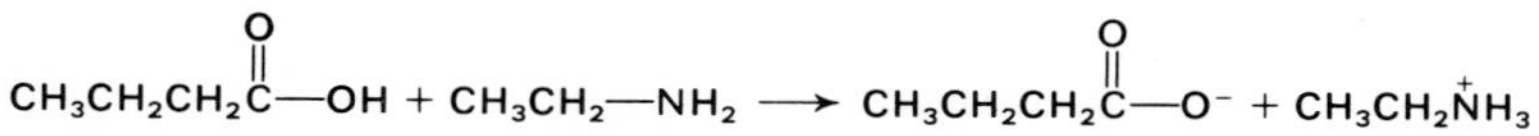

8.8 $OH^- > NH_3 > CH_3CH_2CO_2^- > CH_3CO_2^- > HCO_2^- > Cl_3CCO_2^- > F_3CCO_2^- > Br^-$

8.9 a) $CH_3CHFCO_2H > CH_3CHOHCO_2H > CH_3CHBrCO_2H > BrCH_2CH_2CO_2H$

b) $H_3\overset{+}{N}CH_2CO_2H > HOCH_2CO_2H > HSCH_2CO_2H$

8.10 a) $3\ C_6H_5-\overset{\overset{O}{\|}}{C}OH + PCl_3 \longrightarrow 3\ C_6H_5-\overset{\overset{O}{\|}}{C}-Cl + P(OH)_3$

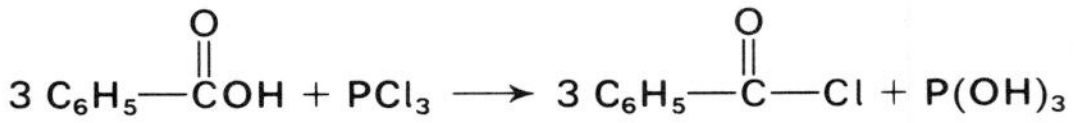

b) $CH_3CH_2CH_2\underset{\underset{OH}{|}}{C}H\overset{\overset{O}{\|}}{C}-OH + 2\ SOCl_2 \longrightarrow CH_3CH_2CH_2\underset{\underset{Cl}{|}}{C}H\overset{\overset{O}{\|}}{C}-Cl + 2\ SO_2 + 2\ HCl$

8.11 a)

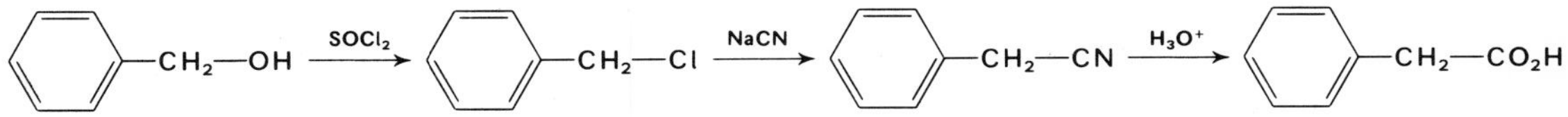

b) $(CH_3)_3C-Cl \xrightarrow[\text{dry ether}]{Mg} (CH_3)_3C-MgCl \xrightarrow{CO_2} (CH_3)_3C-\overset{\overset{O}{\|}}{C}-OMgCl \xrightarrow{H_3O^+} (CH_3)_3C-\overset{\overset{O}{\|}}{C}-OH$

c) $CH_3(CH_2)_3CH_2-OH \xrightarrow{Na_2Cr_2O_7,\ H_2SO_4} CH_3(CH_2)_3\overset{\overset{O}{\|}}{C}-OH$

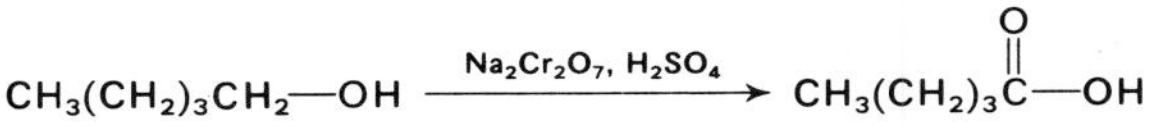

8.12 a) $CH_3\underset{\underset{CH_3}{|}}{C}HCH_2CH_2\overset{\overset{O}{\|}}{C}-Cl + HO-CH_2CH_3$ 4-methylpentanoyl chloride and ethanol

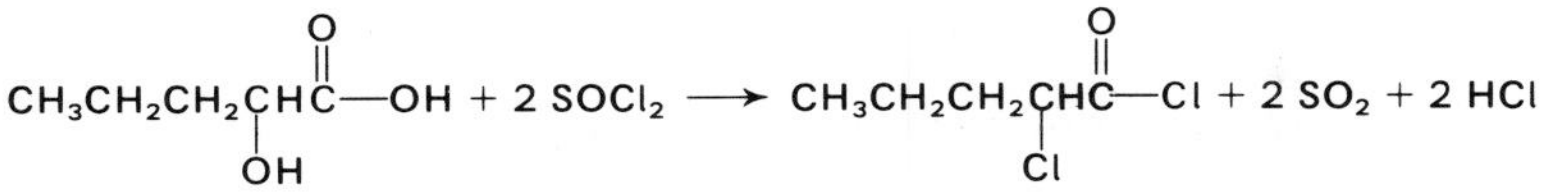

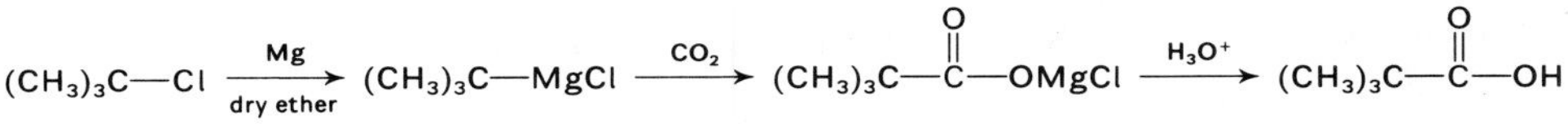

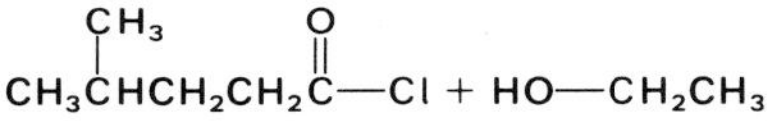

b)

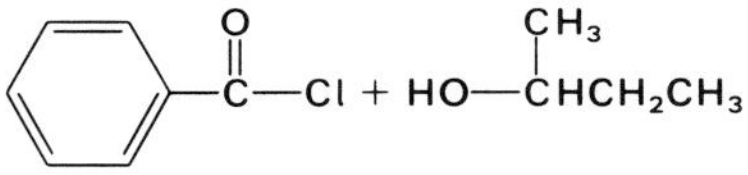

benzoyl chloride and *sec*-butyl alcohol

c) $CH_3\overset{\overset{O}{\|}}{C}-Cl + HO-$ acetyl chloride and cyclohexanol

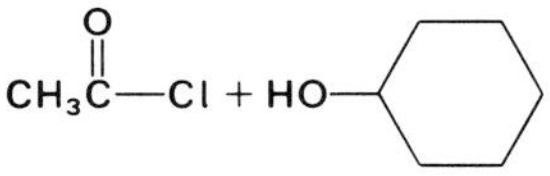

8.13 a) $CH_3(CH_2)_4\overset{\overset{O}{\|}}{C}-Cl + HO-$$\longrightarrow CH_3(CH_2)_4\overset{\overset{O}{\|}}{C}-O-$$+ HCl$

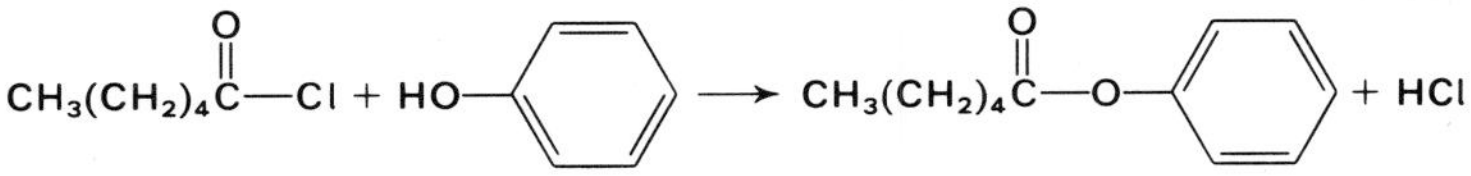

b) $CH_3(CH_2)_4\overset{\overset{O}{\|}}{C}-Cl + HO-C_6H_5 \longrightarrow CH_3(CH_2)_4-\underset{\underset{Cl}{|}}{\overset{\overset{-O}{|}}{C}}-\overset{\overset{H}{|}}{\underset{+}{O}}-C_6H_5$

$\longrightarrow CH_3(CH_2)_4\overset{\overset{O}{\|}}{C}-\underset{+}{O}-C_6H_5 + Cl^- \longrightarrow CH_3(CH_2)_4\overset{\overset{O}{\|}}{C}-O-C_6H_5 + HCl$

8.14 a) $CH_3CH_2CH_2CH_2\overset{\overset{O}{\|}}{C}-N(CH_2CH_3)_2 + (CH_3CH_2)_2\overset{+}{N}H_2$ **b)** $CH_3\underset{\underset{CH_3}{|}}{C}HCH_2\underset{\underset{CH_3}{|}}{C}HCH_2\overset{\overset{O}{\|}}{C}-NH-$$+ C_6H_5\overset{+}{N}H_3$

c) $\overset{\overset{O}{\|}}{C}-NH\underset{\underset{CH_3}{|}}{C}HCH_3 + CH_3\underset{\underset{CH_3}{|}}{C}H\overset{+}{N}H_2$

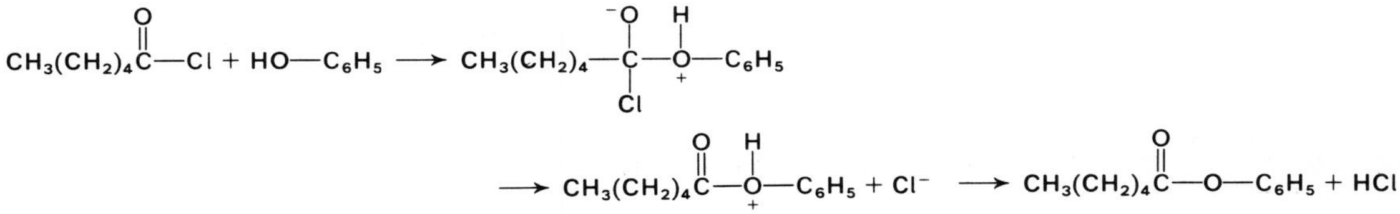

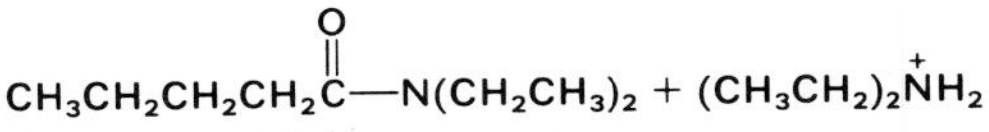

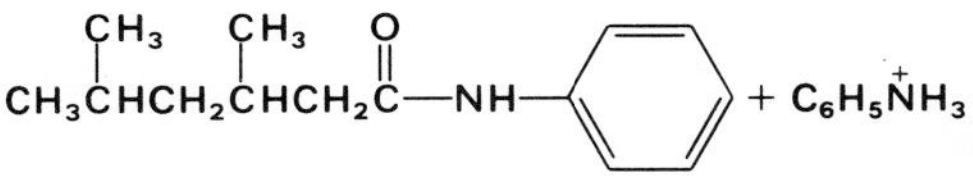

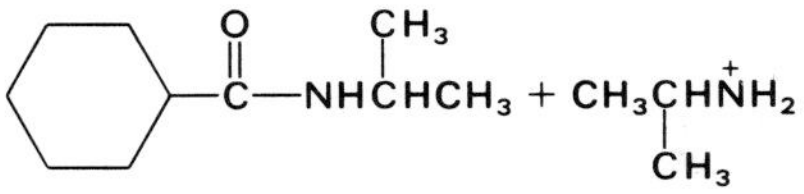

8.15 a)  $Cl-CH_2CH_2CH_2\overset{O}{\overset{\|}{C}}-Cl + HS-CH_2CH_2CH_2CH_3 \longrightarrow Cl-CH_2CH_2CH_2\overset{O}{\overset{\|}{C}}-S-CH_2CH_2CH_2CH_3 + HCl$

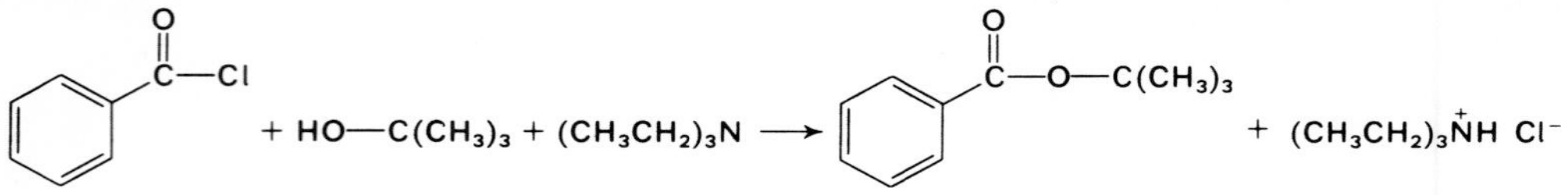

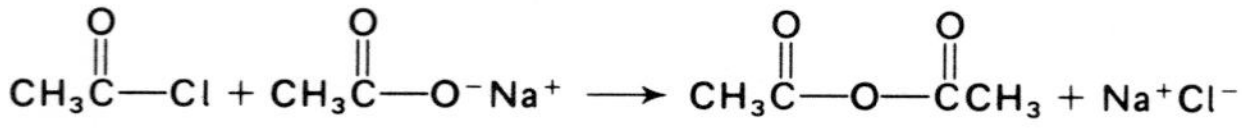

b) $+ HO-C(CH_3)_3 + (CH_3CH_2)_3N \longrightarrow$ $+ (CH_3CH_2)_3\overset{+}{N}H\ Cl^-$

c) $CH_3\overset{O}{\overset{\|}{C}}-Cl + CH_3\overset{O}{\overset{\|}{C}}-O^-Na^+ \longrightarrow CH_3\overset{O}{\overset{\|}{C}}-O-\overset{O}{\overset{\|}{C}}CH_3 + Na^+Cl^-$

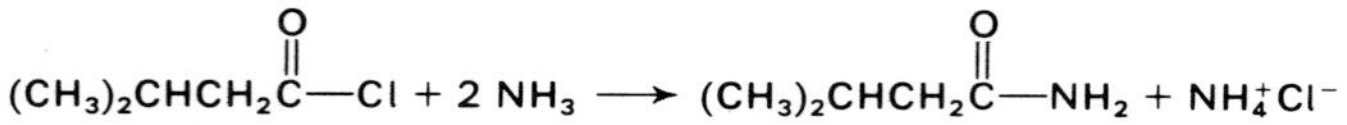

8.16 $CH_3\overset{O}{\overset{\|}{C}}-Cl + {}^-O-\overset{O}{\overset{\|}{C}}CH_3 \longrightarrow CH_3\overset{O^-}{\underset{Cl}{\overset{|}{C}}}-O-\overset{O}{\overset{\|}{C}}CH_3;\quad CH_3\overset{O^-}{\underset{Cl}{\overset{|}{C}}}-O-\overset{O}{\overset{\|}{C}}CH_3 \longrightarrow CH_3\overset{O}{\overset{\|}{C}}-O-\overset{O}{\overset{\|}{C}}CH_3 + Cl^-$

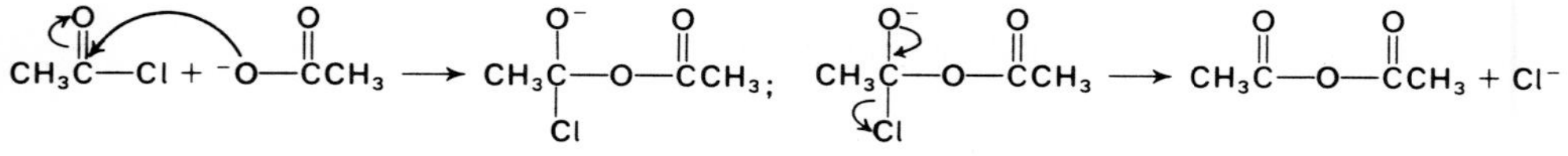

tetrahedral adduct intermediate

8.17 a) 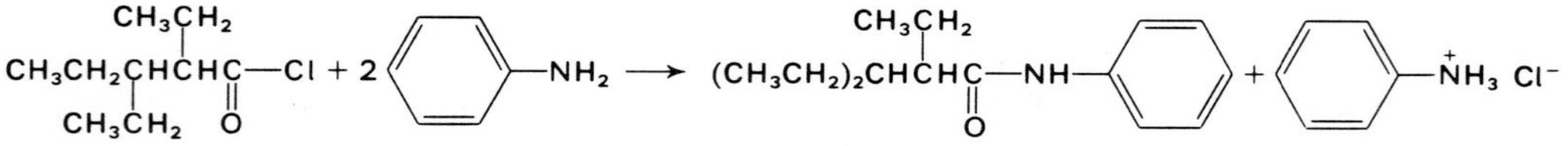 $CH_3CH_2\underset{CH_3CH_2}{\overset{CH_3CH_2}{\overset{|}{C}H}}\overset{O}{\underset{\|}{C}}-Cl + 2$ $-NH_2 \longrightarrow (CH_3CH_2)_2\overset{CH_3CH_2}{\overset{|}{C}H}\overset{O}{\underset{\|}{C}}-NH-$ $+$ $-\overset{+}{N}H_3\ Cl^-$

b) $(CH_3)_2CHCH_2\overset{O}{\overset{\|}{C}}-Cl + 2\ NH_3 \longrightarrow (CH_3)_2CHCH_2\overset{O}{\overset{\|}{C}}-NH_2 + NH_4^+Cl^-$

c) $CH_3CH_2\overset{O}{\overset{\|}{C}}-Cl + CH_3NHCH_2CH_2CH_2CH_2CH_3 \longrightarrow CH_3CH_2\overset{O}{\overset{\|}{C}}-\underset{CH_3}{\overset{|}{N}}-(CH_2)_4CH_3 + CH_3\overset{+}{N}H_2(CH_2)_4CH_3\ Cl^-$

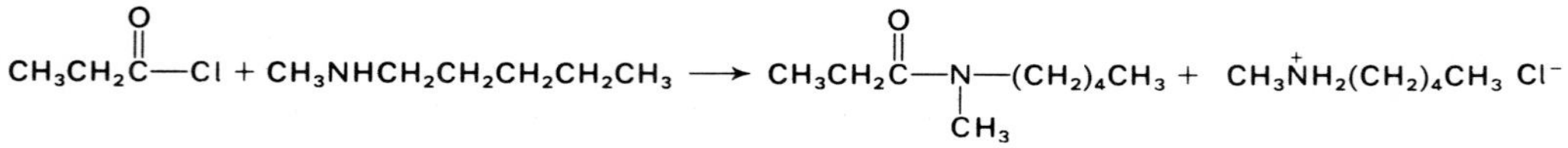

8.18

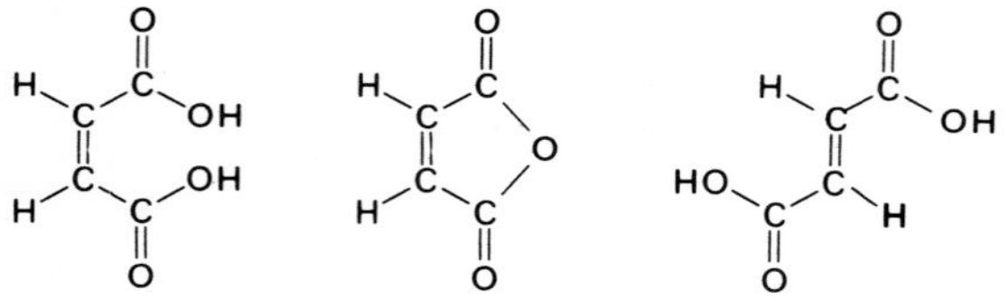

maleic acid   maleic anhydride   fumaric acid

Maleic acid has both carboxyls on the same side of the double bond, close enough to react with each other easily. Fumaric acid has carboxyls far apart on opposite sides of the double bond.

8.19 a) 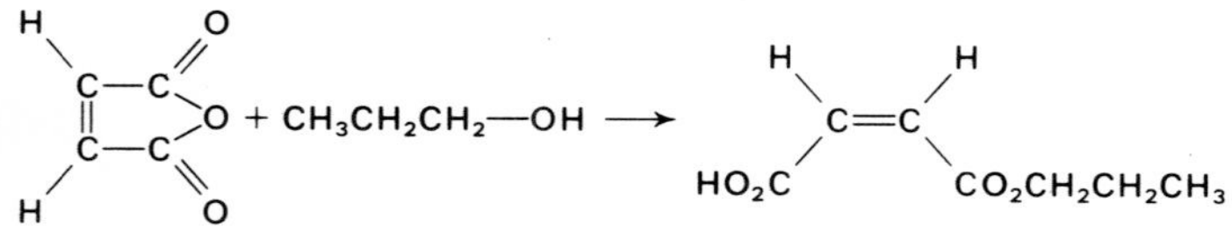 $+ CH_3CH_2CH_2-OH \longrightarrow$

b) 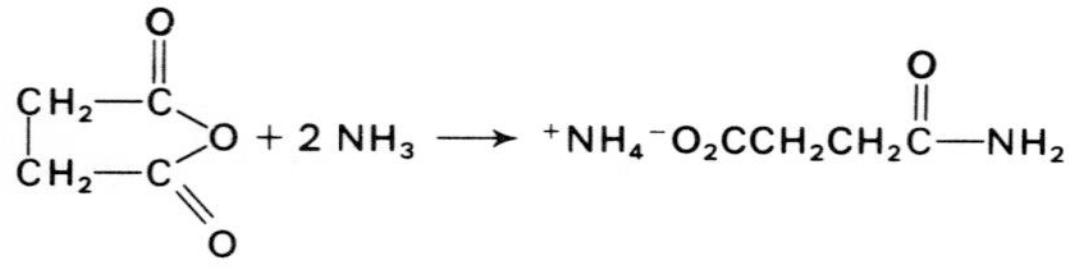 $+ 2\ NH_3 \longrightarrow {}^+NH_4{}^-O_2CCH_2CH_2\overset{O}{\overset{\|}{C}}-NH_2$

8.20 a) 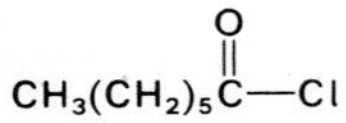 $CH_3(CH_2)_5\overset{O}{\overset{\|}{C}}-Cl$

b)

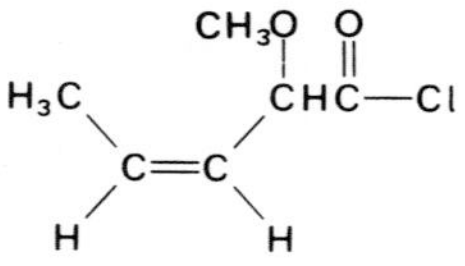

c) 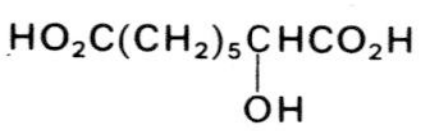 $HO_2C(CH_2)_5\underset{OH}{\overset{|}{C}}HCO_2H$

d) $HO_2CCH_2CO_2H$

e)

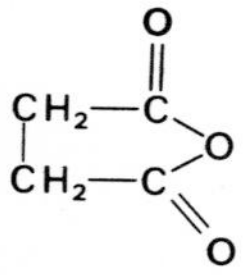

f)

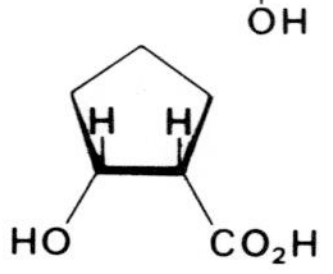

g) $CH_3(CH_2)_{16}CO_2H$ **h)** $CH_3C{\equiv}CCO_2H$ **i)** $HO_2C(CH_2)_3CO_2H$ **j)**

8.21 a) benzoyl chloride

b) *trans,trans*-4,7-decadienoic acid

c) 2-pyridinecarboxylic acid

d) fumaric acid (*trans*-butenedioic acid)

e) 2-ethylhexanedioic acid (α-ethyladipic acid)

f) glutaric anhydride

8.22 a)

b) $CH_3CH_2\underset{\underset{\textstyle CH_3}{|}}{CH}\overset{\overset{\textstyle O}{\|}}{C}OH + SOCl_2 \longrightarrow CH_3CH_2\underset{\underset{\textstyle CH_3}{|}}{CH}\overset{\overset{\textstyle O}{\|}}{C}{-}Cl + SO_2 + HCl$

c) $3\ CH_3CH_2\underset{\underset{\textstyle O}{\|}}{C}{-}OH + PCl_3 \longrightarrow 3\ CH_3CH_2\underset{\underset{\textstyle O}{\|}}{C}{-}Cl + P(OH)_3$

d)

e) $C_6H_5\underset{\underset{\textstyle O}{\|}}{C}{-}Cl + C_6H_5CO_2^-\,Na^+ \longrightarrow C_6H_5\underset{\underset{\textstyle O}{\|}}{C}{-}O{-}\underset{\underset{\textstyle O}{\|}}{C}C_6H_5 + Na^+Cl^-$

f) $CH_2{=}CH\underset{\underset{\textstyle O}{\|}}{C}{-}Cl + 2\ C_6H_5{-}NH_2 \longrightarrow CH_2{=}CH\underset{\underset{\textstyle O}{\|}}{C}{-}NHC_6H_5 + C_6H_5NH_3^+\ Cl^-$

g) $CH_3COCl + H_2O \longrightarrow CH_3CO_2H + HCl$

h) $CH_3\underset{\underset{\textstyle Br}{|}}{CH}{-}\underset{\underset{\textstyle O}{\|}}{C}{-}Cl +$ ⟨benzene⟩$-OH \longrightarrow CH_3\underset{\underset{\textstyle Br}{|}}{CH}{-}\underset{\underset{\textstyle O}{\|}}{C}{-}O{-}$⟨phenyl⟩$ + HCl$

8.23 a) $CH_3\overset{\overset{\textstyle O}{\|}}{C}{-}OH \xrightarrow{SOCl_2} \underset{A}{CH_3\overset{\overset{\textstyle O}{\|}}{C}{-}Cl} \xrightarrow{(CH_3)_2CH{-}NH_2} \underset{B}{CH_3\overset{\overset{\textstyle O}{\|}}{C}{-}NH{-}CH(CH_3)_2}$

b) ⟨cyclohexyl⟩$\overset{\overset{\textstyle O}{\|}}{C}{-}OH \xrightarrow{PCl_3} \underset{C}{⟨cyclohexyl⟩\overset{\overset{\textstyle O}{\|}}{C}{-}Cl} \xrightarrow{CH_3CH_2\overset{\overset{\textstyle CH_3}{|}}{CH}{-}OH} \underset{D}{⟨cyclohexyl⟩\overset{\overset{\textstyle O}{\|}}{C}{-}O{-}\overset{\overset{\textstyle CH_3}{|}}{CH}CH_2CH_3}$

c) $CH_3CH_2CH_2\underset{\underset{\textstyle CH_3CH_2}{|}}{CH}\overset{\overset{\textstyle O}{\|}}{C}{-}OH \xrightarrow{SOCl_2} \underset{E}{CH_3CH_2CH_2\underset{\underset{\textstyle CH_3CH_2}{|}}{CH}\overset{\overset{\textstyle O}{\|}}{C}{-}Cl} \xrightarrow{CH_3{-}SH} \underset{F}{CH_3CH_2CH_2\underset{\underset{\textstyle CH_3CH_2}{|}}{CH}\overset{\overset{\textstyle O}{\|}}{C}{-}S{-}CH_3}$

d)

$\xrightarrow{H_3O^+}$ (G) $\xrightarrow{140°}$ $(CH_3)_2CHCO_2H + CO_2$ (H)

e)

$$\underset{\text{CH}_2-\text{C}}{\overset{\text{CH}_2-\text{C}}{\big\langle}} \overset{\text{O}}{\underset{\text{O}}{\|}} \text{O} \xrightarrow{\text{CH}_3\text{OH}} \text{HO}_2\text{CCH}_2\text{CH}_2\overset{\text{O}}{\overset{\|}{\text{C}}}\text{O}-\text{CH}_3 \xrightarrow{\text{SOCl}_2} \text{ClC}\overset{\text{O}}{\overset{\|}{\text{C}}}\text{CH}_2\text{CH}_2\overset{\text{O}}{\overset{\|}{\text{C}}}\text{O}-\text{CH}_3$$

I   J

$$\xrightarrow{\text{C}_6\text{H}_5-\text{OH}} \text{C}_6\text{H}_5-\text{O}-\overset{\text{O}}{\overset{\|}{\text{C}}}\text{CH}_2\text{CH}_2\overset{\text{O}}{\overset{\|}{\text{C}}}-\text{OCH}_3 + \text{HCl}$$

K

8.24

+ H$_2$O

polymer

polymer

8.25 a) $\text{ClCH}_2\overset{\text{O}}{\overset{\|}{\text{C}}}-\text{OH} \xrightarrow{\text{SOCl}_2} \text{ClCH}_2\overset{\text{O}}{\overset{\|}{\text{C}}}-\text{Cl} \xrightarrow{\text{CH}_3\text{OH}} \text{ClCH}_2\overset{\text{O}}{\overset{\|}{\text{C}}}-\text{OCH}_3$

b) $\text{CH}_3(\text{CH}_2)_4\text{CH}_2-\text{OH} \xrightarrow{\text{CH}_3\text{COCl}} \text{CH}_3(\text{CH}_2)_5-\text{O}-\overset{\text{O}}{\overset{\|}{\text{C}}}\text{CH}_3$

c) $\text{CH}_2=\text{CHCO}_2\text{H} \xrightarrow{\text{PCl}_3} \text{CH}_2=\text{CH}\overset{\text{O}}{\overset{\|}{\text{C}}}-\text{Cl} \xrightarrow{\text{CH}_3\text{CH}_2-\text{SH}} \text{CH}_2=\text{CH}\overset{\text{O}}{\overset{\|}{\text{C}}}-\text{S}-\text{CH}_2\text{CH}_3$

d) $\text{C}_6\text{H}_5\text{CH}_2\text{CO}_2\text{H} \xrightarrow{\text{SOCl}_2} \text{C}_6\text{H}_5\text{CH}_2\overset{\text{O}}{\overset{\|}{\text{C}}}-\text{Cl} \xrightarrow{\text{NH}_3} \text{C}_6\text{H}_5\text{CH}_2\overset{\text{O}}{\overset{\|}{\text{C}}}-\text{NH}_2$

8.26 Both have four substituents on the ring and two double bonds in the side chains. E_2 has four asymmetric carbons and $F_{2\alpha}$ has five.

8.27 a) $\text{CH}_3(\text{CH}_2)_4\text{CH}_2-\text{OH} \xrightarrow{\text{SOCl}_2} \text{CH}_3(\text{CH}_2)_4\text{CH}_2-\text{Cl} \xrightarrow{\text{NaCN}} \text{CH}_3(\text{CH}_2)_4\text{CH}_2-\text{CN} \xrightarrow{\text{H}_3\text{O}^+} \text{CH}_3(\text{CH}_2)_4\text{CH}_2\text{CO}_2\text{H}$

b) $\text{CH}_3(\text{CH}_2)_4\text{CH}_2-\text{OH} \xrightarrow{\text{H}_2\text{SO}_4, \text{ heat}} \text{CH}_3(\text{CH}_2)_3\text{CH}=\text{CH}_2 \xrightarrow{\text{HCl}} \underset{\overset{|}{\text{Cl}}}{\text{CH}_3(\text{CH}_2)_3\text{CHCH}_3}$

$$\xrightarrow[\text{dry ether}]{\text{Mg}} \underset{\overset{|}{\text{CH}_3}}{\text{CH}_3(\text{CH}_2)_3\text{CH}-\text{MgCl}} \xrightarrow[\text{2) H}_3\text{O}^+]{\text{1) CO}_2} \underset{\overset{|}{\text{CH}_3}}{\text{CH}_3(\text{CH}_2)_3\text{CHCO}_2\text{H}}$$

Chapter 9

9.1 a) *sec*-butyl 2-chloropropanoate   **b)** isopropyl formate   **c)** isobutyl hexanoate

9.2 a) $\underset{\overset{|}{\text{CH}_3}}{\text{CH}_3\text{CHCH}_2}-\text{O}-\underset{\overset{\|}{\text{O}}}{\text{C}}-\text{C}_6\text{H}_5$ **b)** $\underset{\overset{|}{\text{CH}_3}}{\text{CH}_3(\text{CH}_2)_5\text{CH}}-\text{O}-\underset{\overset{\|}{\text{O}}}{\text{C}}\text{CF}_3$

c) $CH_3CH_2CH_2CHC$ (with $\overset{O}{\overset{\|}{C}}$, OH below) $-O-$ (phenyl ring)

d) $CH_3CH_2-O-\overset{O}{\underset{\|}{C}}-\underset{C_6H_5}{CHCH_2CH_3}$

9.3 The formation of the tetrahedral intermediate is very much more strongly hindered in the case of the substituted ester by the bulkiness of the *tert*-butyl group than is formation of the intermediate from the unsubstituted ester.

$CH_3\overset{H_3C}{\underset{H_3C}{\overset{|}{\underset{|}{C}}}}-\overset{O^-}{\underset{OH}{\overset{|}{\underset{|}{C}}}}-OCH_2CH_3$

9.4 a) $(CH_3CH_2)_2CHCH_2\overset{}{\underset{O}{\overset{\|}{C}}}-O-C(CH_3)_3 + (CH_3CH_2)_2NH \longrightarrow (CH_3CH_2)_2CHCH_2\overset{}{\underset{O}{\overset{\|}{C}}}-N(CH_2CH_3)_2 + (CH_3)_3C-OH$

b) $CH_3(CH_2)_8CH_2-O-\overset{O}{\overset{\|}{C}}CH_3 + CH_3-O^-Na^+ \longrightarrow CH_3(CH_2)_8CH_2-O^-Na^+ + CH_3-O-\overset{O}{\overset{\|}{C}}CH_3$

c) $CH_3(CH_2)_{16}\overset{O}{\overset{\|}{C}}-OCH_3 + Na^+OH^- \xrightarrow{\;C_2H_5OH\;} CH_3(CH_2)_{16}CO_2^-Na^+ + CH_3OH$

9.5 $CH_3(CH_2)_4\overset{O}{\overset{\|}{C}}-O-CH(CH_3)_2 + CH_3CH_2CH_2-NH_2 \longrightarrow CH_3(CH_2)_4\overset{O^-}{\overset{|}{\underset{+NH_2-CH_2CH_2CH_3}{C}}}-O-CH(CH_3)_2 \longrightarrow$

$CH_3(CH_2)_4\overset{O^-}{\underset{NH-CH_2CH_2CH_3}{\overset{|}{C}}}-\overset{H}{\overset{+}{O}}-CH(CH_3)_2 \longrightarrow CH_3(CH_2)_4\overset{O}{\overset{\|}{C}}-NHCH_2CH_2CH_3 + (CH_3)_2CH-OH$

9.6 a) $CH_3\overset{CH_3}{\overset{|}{C}}HCH_2\overset{CH_3}{\overset{|}{C}}HCH_2CH_2-OH + CH_3CH_2-OH$ **b)** $CH_3CH_2OH + CH_3(CH_2)_4-OH$

9.7 $CH_3(CH_2)_{14}\overset{O}{\overset{\|}{C}}-S-CoA + H_3C$ (steroid structure) $\longrightarrow HS-CoA + H_3C$ (steroid structure) $R = -\overset{CH_3}{\overset{|}{C}}H(CH_2)_3\overset{CH_3}{\overset{|}{C}}HCH_3$

9.8 Acid catalysis of transesterification should proceed easily in the same manner as hydrolysis of an ester. A thiol is a weaker base than water or alcohol and is itself less affected by the presence of acid than are either of those compounds. On the other hand, ammonia is a sufficiently good base that most of the catalyzing acid would react with ammonia instead of the carboxylic ester. If enough acid catalyst is used it will convert all the ammonia to ammonium ion, which is no longer a nucleophile.

9.9 $CH_3CH_2\overset{}{\underset{OH}{\overset{|}{C}}}(CH_3)_2 + H_3O^+ \xrightarrow{\;-H_2O\;} CH_3CH_2\overset{}{\underset{+OH_2}{\overset{|}{C}}}(CH_3)_2 \longrightarrow CH_3CH_2\overset{+}{C}(CH_3)_2 + H_2O$

$CH_3CH_2\overset{+}{C}(CH_3)_2 + H_2O \longrightarrow CH_3CH=C(CH_3)_2 + H_3O^+$

The catalyst H_3O^+ is regenerated to react with another alcohol molecule.

9.10 *R*-enantiomer.

9.11 11 asymmetric carbons; ring A to B junction *cis*, B to C *trans*, C to D *trans*.

9.12 Monoacylglycerols would be good for forming micelles since two polar and hydrogen-bonding OH groups are free. Diacylglycerols would be considerably less effective, with only one OH group, and triacylglycerols, with no hydrogen-bonding groups, would not form micelles.

9.13 a) AdCH$_2$—O—PO$_2$—O—$\overset{\text{O}}{\overset{\|}{\text{C}}}CH_3$

b) AdCH$_2$—$\overset{+}{\underset{\text{CH}_3}{\text{S}}}$—CH$_2CH_2$$\underset{^+\text{NH}_3}{\text{CHCO}_2^-}$

c) CH$_3$$\overset{\text{O}}{\underset{\|}{\text{C}}}$—S—CH$_2CH_3$

d) CH$_3$CO$_2^-$ + HO—$\overset{\text{O}^-}{\underset{\|}{\underset{\text{O}}{\text{P}}}}$—NHCH$_3$

e) CH$_3$$\overset{\text{O}}{\underset{\|}{\text{C}}}$—S(CoA) + AdCH$_2$—OPO$_3H^-$

9.14 a) H$\overset{\text{O}}{\overset{\|}{\text{C}}}$—O(CH$_2$)$_3CH_3$

b) CH$_3$CH$_2$CH$_2$$\underset{\text{CH}_3}{\text{CHCH}_2}$$\overset{\text{O}}{\underset{\|}{\text{C}}}$—OCH$_2CH_3$

c) ClCH$_2$$\overset{\text{O}}{\underset{\|}{\text{C}}}$—OCH$_2$CH(CH$_3$)$_2$

d) CH$_3$$\underset{\text{HO}}{\text{CH}}$$\overset{\text{O}}{\underset{\|}{\text{C}}}$—OCH(CH$_3$)$_2$

e) CH$_2$($\overset{\text{O}}{\underset{\|}{\text{C}}}$—OCH$_2CH_3$)$_2$

f) Na$^+$ $^-$O$_2$C$\underset{\text{HO}}{\text{CH}}$—$\underset{\text{OH}}{\text{CHCO}_2^-}$ Na$^+$

g) CH$_3$$\overset{\text{O}}{\underset{\|}{\text{C}}}$—S(CoA)

h) CH$_3$CH$_2$CH$_2$$\overset{\text{O}}{\underset{\|}{\text{C}}}$—S—C(CH$_3$)$_3$

i) CH$_3$CH$_2$CH$_2$CO$_2$—CH$_2$
CH$_3$CH$_2$CH$_2$CO$_2$—CH
CH$_3$CH$_2$CH$_2$CO$_2$—CH$_2$

j) CH$_3$(CH$_2$)$_{14}$$\overset{\text{O}}{\underset{\|}{\text{C}}}$—OCH$_2CH_3$

k) CH$_3$(CH$_2$)$_4$ —C=C— CH$_2$ —C=C— (CH$_2$)$_7$$\overset{\text{O}}{\overset{\|}{\text{C}}}OCH_3$ (with H, H H, H substituents)

l) (CH$_3$)$_3$C$\overset{\text{O}}{\underset{\|}{\text{C}}}$—S$\underset{}{\overset{\text{CH}_3}{\text{CH}}}CH_2CH_3$

m)

(CH$_3$)$_2$CH—[benzene ring]—$\overset{}{\underset{\text{O}}{\text{C}}}$—OCH$_2CH_3$

n) (CH$_3$)$_2$CHCH$_2$$\overset{}{\underset{\text{O}}{\text{C}}}$—O—C$_6H_5$

9.15 a) isopropyl propanethioate

b) *sec*-butyl formate

c) isobutyl 3-methylbutanoate

d) 4-chlorophenyl acetate

e) *tert*-butyl thiobenzoate

f) diethyl succinate

g) *sec*-butyl methyl nonanedioate

9.16 a) (CH$_3$)$_2$CH—O—$\overset{\text{O}}{\underset{\|}{\text{C}}}CH_3$ + Na$^+$OH$^-$ ⟶ (CH$_3$)$_2$CH—OH + CH$_3$$\overset{\text{O}}{\underset{\|}{\text{C}}}O^-Na^+$

b) CH$_3$$\underset{\text{CH}_3}{\text{CHCH}_2}$S—$\overset{\text{O}}{\underset{\|}{\text{C}}}$[benzene ring] + 2 K$^+$ + 2 OH$^-$ ⟶ CH$_3$$\underset{\text{CH}_3}{\text{CHCH}_2}S^-K^+$ + H$_2$O + [benzene ring]$\overset{\text{O}}{\overset{\|}{\text{C}}}$—O$^-K^+$

c) (CH$_3$)$_3$C—O—$\overset{}{\underset{\text{O}}{\text{C}}}$[benzene ring] + CH$_3O^-Na^+$ ⇌ (CH$_3$)$_3$C—O$^-$Na$^+$ + [benzene ring]$\overset{\text{O}}{\overset{\|}{\text{C}}}$—O—CH$_3$

d) CH$_3$CH$_2$$\underset{\text{CH}_3}{\text{CH}}$—O—$\overset{\text{O}}{\underset{\|}{\text{C}}}$—$\underset{\text{H}_3\text{C}}{\overset{\text{CH}_3}{\text{CH}}}\underset{\text{CH}_2\text{CH}_3}{\text{C}}CH_2CH_2CH_3$ + NH$_3$ ⟶ CH$_3$CH$_2$$\underset{\text{CH}_3}{\text{CH}}$—OH + CH$_3CH_2CH_2$$\underset{\text{CH}_3\text{CH}_2}{\overset{\text{H}_3\text{C}}{\text{C}}}$—$\overset{\text{CH}_3}{\text{CH}}$—$\overset{}{\underset{\text{O}}{\text{C}}}$—NH$_2$

e) CH$_2$—O—$\overset{\text{O}}{\underset{\|}{\text{C}}}$(CH$_2$)$_2CH_3$
CH—O—$\overset{\text{O}}{\underset{\|}{\text{C}}}$(CH$_2$)$_2CH_3$ + 3 K$^+$ + 3 OH$^-$ ⟶
CH$_2$—O—$\overset{\text{O}}{\underset{\|}{\text{C}}}$(CH$_2$)$_2CH_3$

CH$_2$OH
CHOH + 3 CH$_3$(CH$_2$)$_2$$\overset{\text{O}}{\underset{\|}{\text{C}}}O^-K^+$
CH$_2$OH

f)

$$\text{CH}_2\text{—O—}\overset{\displaystyle O}{\overset{\|}{\text{C}}}(\text{CH}_2)_{16}\text{CH}_3$$

$$\text{CH—O—}\overset{\displaystyle O}{\overset{\|}{\text{C}}}(\text{CH}_2)_{16}\text{CH}_3 + 2\ \text{Na}^+\text{OH}^- \longrightarrow$$

$$\text{CH}_2\text{—O—}\overset{\displaystyle O}{\underset{\underset{\text{O}^-}{|}}{\overset{\|}{\text{P}}}}\text{—OCH}_2\text{CH}_2\text{—}\overset{+}{\text{N}}(\text{CH}_3)_3$$

produces:

$$\begin{array}{l}\text{CH}_2\text{OH}\\ \text{CHOH} \qquad O\\ \text{CH}_2\text{—O—}\overset{\|}{\underset{\underset{O^-}{|}}{P}}\text{—OCH}_2\,\text{CH}_2\overset{+}{\text{N}}(\text{CH}_3)_3 \ + \ 2\ \text{CH}_3(\text{CH}_2)_{16}\overset{O}{\overset{\|}{\text{C}}}\text{—O}^-\,\text{Na}^+\end{array}$$

g) $\text{CH}_3\text{CH}_2\text{CH}_2\overset{O}{\overset{\|}{\text{C}}}\text{—S—CH}_2\text{CH}_3 + \text{CH}_3\text{CH}_2\text{CH}_2\text{—NH}_2 \longrightarrow \text{CH}_3\text{CH}_2\text{CH}_2\overset{O}{\overset{\|}{\text{C}}}\text{—NH—CH}_2\text{CH}_2\text{CH}_3 + \text{CH}_3\text{CH}_2\text{—SH}$

h) $\text{CH}_3\overset{}{\underset{\underset{\text{OH}}{|}}{\text{CH}}}\text{CH}_2\text{CH}_2\text{CO}_2\text{H} \xrightarrow{\text{H}^+} \text{CH}_3\text{—}$ (cyclic lactone) $=\text{O}$

9.17 a) A: SOCl_2 or PCl_3; B: $\text{C}_6\text{H}_5\text{—OH}$ **b)** C: PCl_3; D: $\text{CH}_3\text{CH}_2\text{—SH}$; E: HBr

c) F: CH_3OH, H_2SO_4; G: H_2O, H_2SO_4; H: CH_3COCl

9.18 The reactivity of a leaving group correlates with the acidity (pK_a) of H-L; the stronger H-L is as an acid, the better leaving group L$^-$ is.

9.19 a) The cyclic compound has the hydrocarbon portion of the molecule folded back all on one side and the electronegative oxygens both on one side of the molecule. The long compound has the methyl group on the opposite side of the oxygens from the ethyl group, and center of electronegativity is near the center of the molecule.

b) In the open-chain compound, the dipoles partially cancel; in the cycle they add to provide a higher dipole moment. The compound with the higher dipole moment exerts a greater intermolecular attraction and thus has a higher boiling point.

c) Unsaturated fatty acids and derivatives have lower melting points than do saturated fatty acids and derivatives.

d) Stearic acid, a saturated acid, easily forms an extended molecule. Linoleic acid, with two *cis* double bonds, has two predominant bends in the chain.

9.20 All of the compounds have a nonpolar long chain and a strongly polar or ionic end. They all serve as emulsifiers, being partly soluble in oil and partly soluble in water.

9.21 a) The carbonyl carbon in trifluoroacetate is much more electron-deficient and therefore much more reactive toward the nucleophilic hydroxide ion than in acetate ester. The three fluorines are strong electron-withdrawing groups, increasing the positive charge on the trifluoroacetate carbon over that of the unsubstituted ester.

b) The methyl groups in the 2 and 6 positions sterically hinder the approach of the hydroxide ion to the carbonyl carbon and the formation of the tetrahedral intermediate. The new C—O bond has little chance of forming and the hydrolysis of the ester by base is very slow.

9.22 Treatment of the ester with $^{18}\text{OH}^-$ and $\text{H}_2{}^{18}\text{O}$ would effect an exchange of unlabeled carbonyl oxygen to labeled carbonyl oxygen:

$$\text{C}_6\text{H}_5\overset{O}{\overset{\|}{\text{C}}}\text{—OCH}_2\text{CH}_3 + {}^{18}\text{OH}^- \xrightarrow[\text{H}_2{}^{18}\text{O}]{} \text{C}_6\text{H}_5\underset{\underset{{}^{18}\text{OH}}{|}}{\overset{\overset{\text{O}^-}{|}}{\text{C}}}\text{—OCH}_2\text{CH}_3 \longrightarrow$$

$$\text{C}_6\text{H}_5\underset{\underset{{}^{18}\text{O}^-}{|}}{\overset{\overset{\text{OH}}{|}}{\text{C}}}\text{—OCH}_2\text{CH}_3 \longrightarrow \text{C}_6\text{H}_5\overset{{}^{18}\text{O}}{\overset{\|}{\text{C}}}\text{—O—CH}_2\text{CH}_3 + \text{OH}^-$$

The $^{18}\text{OH}^-$ is made from sodium and $\text{H}_2{}^{18}\text{O}$:

$$\text{H}_2{}^{18}\text{O} + 2\ \text{Na}^0 \longrightarrow \text{H}_2 + 2\ \text{Na}^+ + 2\ {}^{18}\text{OH}^-$$

9.23 $CH_3CH_2CH_2C\overset{\text{O}}{\underset{\|}{C}}-O^- + AdCH_2OPO_2OPO_2OPO_3H^{-3} \longrightarrow CH_3CH_2CH_2C-\bar{O}PO_2OCH_2Ad + HP_2O_7^{-3}$

$CH_3CH_2CH_2\overset{\text{O}}{\underset{\|}{C}}-OP\bar{O}_2OCH_2Ad + CoA-SH \longrightarrow CH_3CH_2CH_2\overset{\text{O}}{\underset{\|}{C}}-S(CoA) + AdCH_2OPO_3H^-$

$CH_3CH_2CH_2\overset{\text{O}}{\underset{\|}{C}}-S(CoA) + CH_3CH_2-OH \longrightarrow CH_3CH_2CH_2\overset{\text{O}}{\underset{\|}{C}}-OCH_2CH_3 + CoA-SH$

9.24 $\overset{CH_3}{\underset{|}{R}CHCH_2CH_2\overset{O}{\underset{\|}{C}}}-S(CoA)$ formed as in Problem 9.23 above; R = tetracyclic nucleus of cholic acid, Sect. 9.14.

$\overset{CH_3}{\underset{|}{R}CHCH_2CH_2\overset{O}{\underset{\|}{C}}}-S(CoA) + H_2NCH_2CO_2^- \longrightarrow \overset{CH_3}{\underset{|}{R}CHCH_2CH_2\overset{O}{\underset{\|}{C}}}-NHCH_2CO_2^- + CoA-SH$

Chapter 10

10.1 a) One six-membered ring is aromatic (three double bonds); one five-membered ring has two double bonds

b) The two rings have two adjacent carbon atoms in common—called *fused rings*

c) one primary amine (on the aromatic ring); one secondary amine (in the five-membered ring); three unsaturated tertiary amines

10.2 seven rings—four six-membered, two five-membered, and one seven-membered

10.3 a) diisopropylamine   **b)** 2-aminooctane (2-octanamine)   **c)** 1-methylpiperidine

    d) *trans*-2-aminocyclohexanol   **e)** pyridine   **f)** benzylamine

10.4 a) $(CH_3CH_2CH_2CH_2)_3N$ **b)** 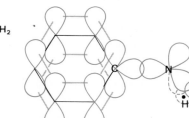   **c)** $(CH_3)_2\overset{+}{N}(CH_2CH_3)_2$

10.5 aniline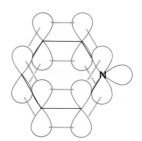

N has sp^3 hybrid orbitals. Orbital of unshared pair of electrons can overlap p orbitals of ring.

pyridine

N has sp^2 orbitals and p orbital. Electrons are in sp^2 orbital and do not overlap p orbitals of ring.

10.6 a) $NH_2^- > CH_3{-}O^- > CH_3CH_2CH_2{-}NH_2 > CH_3CO_2^-$

b) [structure: piperidine N—H] $>$ NH_3 $>$ [pyridine N] $>$ [aniline NH$_2$]   (according to information from Fig. 10.2)

10.7 a) $CH_3\overset{\underset{|}{CH_3}}{C}HCH_2{-}NH{-}\overset{\underset{\|}{O}}{C}CH(CH_3)_2$

b) $(CH_3)_2CHCH_2{-}\overset{+}{N}H_2{-}CH_2CH_3$ Br^-

c) [phenyl]$-NH{-}\overset{\overset{S}{\|}}{C}{-}NH{-}CH_2CH(CH_3)_2$

10.8 a) A: $(CH_3)_2CH\overset{\underset{\|}{O}}{C}{-}Cl$;   B: $(CH_3)_2CH\overset{\underset{\|}{O}}{C}{-}NHCH_3$;   C: $(CH_3)_2CHCH_2{-}NH{-}CH_3$

b) D: [cyclohexyl]$-CH_2{-}Cl$;   E: [cyclohexyl]$-CH_2{-}CN$;   F: [cyclohexyl]$-CH_2CH_2{-}NH_2$

10.9

imide / urea / imide — phenobarbital

imide / urea — caffeine

guanidine — guanine

10.10 a) butanamide

b) N-phenylacetamide

c) N-isopropyl-2-methylpropanamide

d) 3-pyridinecarboxamide (nicotinamide)

e) Z-3-ethyl-3-pentenamide

10.11 a) $H\overset{\underset{\|}{O}}{C}{-}NH_2$

b) [phenyl]$\overset{\overset{O}{\|}}{C}{-}N(CH_3)_2$

c) $CH_3\overset{\underset{|}{Cl}}{C}H\overset{\overset{O}{\|}}{C}{-}NHCH_3$

d) [phenyl]$-CH_2\overset{\underset{\|}{O}}{C}{-}NH_2$

e) $\underset{H}{\overset{H_3C}{>}}C{=}C\underset{\underset{\|}{\overset{O}{C}}NH_2}{\overset{H}{<}}$

10.12 a) $CH_3\overset{\overset{+NH_3}{|}}{C}HCO_2^-$

b) $\overset{\overset{+NH_3}{|}}{C}H_2CO_2H$

c) $CH_3{-}S{-}CH_2CH_2\overset{\overset{NH_2}{|}}{C}HCO_2^-$

d) $HO{-}CH_2\overset{\overset{+NH_3}{|}}{C}HCO_2^-$

e) $HS{-}CH_2\overset{\overset{+NH_3}{|}}{C}HCO_2^-$

f) [imidazole]$-CH_2\overset{\overset{+NH_3}{|}}{C}HCO_2^-$

10.13 a) $CH_3{-}S{-}CH_2CH_2\overset{\overset{+NH_3}{|}}{C}HCO_2H$, $HOCH_2\overset{\overset{+NH_3}{|}}{C}HCO_2H$, and $(CH_3)_2CH\overset{\overset{+NH_3}{|}}{C}HCH_2{-}OH$ designates the CO_2H terminal end unit, now changed to CH_2OH

b) $CH_3{-}S{-}CH_2CH_2CH$ [ring structure with HN, C=S, N, phenyl, C=O] and $H_3\overset{+}{N}\overset{\overset{}{\underset{|}{HOCH_2}}}{C}HCO{-}NH\overset{\overset{}{\underset{|}{CH(CH_3)_2}}}{C}HCO_2H$ designates NH_2 terminal end unit

10.14 Ser-Ala-Lys-Ser-His-Phe-Ala-Arg-Ala-Gly-Trp-Gly (full explanation of solution given in Study Guide).

10.15 $C_6H_5CH_2\!-\!O\!-\!\underset{O}{\overset{}{C}}\!-\!Cl + NH_2\underset{CH_3}{\overset{}{C}}HCO_2^- \longrightarrow C_6H_5CH_2\!-\!O\!-\!\underset{O}{\overset{}{C}}\!-\!NH\underset{CH_3}{\overset{}{C}}HCO_2H + Cl^-$

$C_6H_5CH_2\!-\!O\!-\!\underset{O}{\overset{}{C}}\!-\!Cl + NH_2CH_2CO_2^- \longrightarrow C_6H_5CH_2\!-\!O\!-\!\underset{O}{\overset{}{C}}\!-\!NHCH_2CO_2H + Cl^-$

$C_6H_5CH_2\!-\!O\!-\!\underset{O}{\overset{}{C}}\!-\!NH\underset{CH_3}{\overset{}{C}}HCO_2H + NH_2\underset{CH_2CH_2SCH_3}{\overset{}{C}}HCO_2CH_3 \xrightarrow{\text{DCC}} C_6H_5CH_2O\underset{O}{\overset{}{C}}\!-\!NH\underset{H_3C}{\overset{}{C}}H\underset{O}{\overset{}{C}}\!-\!NH\underset{CH_2CH_2SCH_3}{\overset{}{C}}HCO_2CH_3 \xrightarrow{\text{HBr}}$

$NH_2\underset{H_3C}{\overset{}{C}}H\underset{O}{\overset{}{C}}\!-\!NH\underset{CH_2CH_2SCH_3}{\overset{}{C}}HCO_2CH_3 \xrightarrow[\text{DCC}]{C_6H_5CH_2OCONHCH_2CO_2H.} C_6H_5CH_2O\underset{O}{\overset{}{C}}\!-\!NHCH_2\underset{O}{\overset{}{C}}\!-\!NH\underset{H_3C}{\overset{}{C}}H\underset{O}{\overset{}{C}}\!-\!NH\underset{CH_2CH_2SCH_3.}{\overset{}{C}}HCO_2CH_3 \xrightarrow{\text{HBr}}$

$NH_2CH_2\underset{O}{\overset{}{C}}\!-\!NH\underset{H_3C}{\overset{}{C}}H\underset{O}{\overset{}{C}}\!-\!NH\underset{CH_2CH_2SCH_3}{\overset{}{C}}HCO_2CH_3 \xrightarrow{\text{H}_2\text{SO}_4,\ \text{H}_2\text{O}} {}^+NH_3CH_2\underset{O}{\overset{}{C}}\!-\!NH\underset{H_3C}{\overset{}{C}}H\underset{O}{\overset{}{C}}\!-\!NH\underset{CH_2CH_2SCH_3}{\overset{}{C}}HCO_2H$

10.16 Amino acid R groups: glycine H, alanine CH_3, lysine $NH_2(CH_2)_4\!-$, glutamic acid $HO_2CCH_2CH_2\!-$. The α-helix conformation depends upon the ability of the R groups to lie close together in the spiral. Neutral R groups offer no problem and polyalanine easily forms a helix. R groups having the same charge, however, repel each other and do not allow the helix to form. Thus at pH 7, the NH_2 groups of lysine exist as NH_3^+ and repel each other, and no helix forms. At pH 12 the end NH_2 groups exist as uncharged NH_2 groups and offer no resistance to helix formation. Similarly at pH 7 the end CO_2H groups of glutamic acid exist as Co_2^- groups and repel each other, while at pH 2 they are uncharged CO_2H groups, offering no resistance to helix formation. The H of glycine is very small and offers no steric hindrance for any conformation. When sterically possible with small R groups, the stretched conformation of the pleated sheet appears to be preferred.

10.17 a) $H_3\overset{+}{N}\!-\!\underset{CH_3}{\overset{\overset{\displaystyle CO_2^-}{|}}{\underset{|}{C}}}\!-\!H$ **b)** $H_3\overset{+}{N}CH_2CO_2^-$ **c)** $H_3\overset{+}{N}\!-\!\underset{CH_2OH}{\overset{\overset{\displaystyle CO_2^-}{|}}{\underset{|}{C}}}\!-\!H$ **d)** $H_3\overset{+}{N}\!-\!\underset{CH_2}{\overset{\overset{\displaystyle CO_2^-}{|}}{\underset{|}{C}}}\!-\!H$ (imidazole ring, NH) **e)** $H_3\overset{+}{N}\!-\!\underset{CH_2SH}{\overset{\overset{\displaystyle CO_2^-}{|}}{\underset{|}{C}}}\!-\!H$

10.18 a) S-alanine **c)** S-serine **d)** S-histidine **e)** R-cysteine

10.19 a) $(CH_3CH_2)_3\overset{+}{N}H\ Cl^-$ **b)** $CH_3\!-\!NH\!-\!$(phenyl) **c)** $CH_3CH_2\underset{C_6H_5}{\overset{}{C}}H\underset{O}{\overset{\overset{\displaystyle O}{\|}}{C}}\!-\!N(CH_2CH_3)_2$

d) (pyridinium) $\overset{+}{N}\!-\!CH_3\ Br^-$ **e)** $H_3\overset{+}{N}\underset{CH_3}{\overset{}{C}}H\underset{O}{\overset{\overset{\displaystyle O}{\|}}{C}}\!-\!NH\underset{HOCH_2}{\overset{}{C}}HCO_2^-$ **f)** $HO_2CCH_2CH_2\underset{{}^+NH_3}{\overset{}{C}}HCO_2^-$

10.20 a) urea **b)** guanidine **c)** succinimide **d)** nicotinamide **e)** adenine

10.21 a) $-\underset{O}{\overset{\overset{\displaystyle O}{\|}}{C}}\!-\!(CH_2)_4\!-\!\underset{O}{\overset{\overset{\displaystyle O}{\|}}{C}}\!-\!NH\!-\!(CH_2)_6\!-\!NH\!- \xrightarrow{\text{H}^+,\ \text{H}_2\text{O}} {}^+NH_3(CH_2)_6NH_3^+ + HO\!-\!\underset{O}{\overset{\overset{\displaystyle O}{\|}}{C}}(CH_2)_4\underset{O}{\overset{\overset{\displaystyle O}{\|}}{C}}\!-\!OH$

1,6-hexanediammonium ion   hexanedioic acid

$\downarrow \text{OH}^-$

b) $^-O\!-\!\underset{O}{\overset{\overset{\displaystyle O}{\|}}{C}}(CH_2)_4\underset{O}{\overset{\overset{\displaystyle O}{\|}}{C}}\!-\!O^- + NH_2(CH_2)_6NH_2$

hexanedioate ion   1,6-diaminohexane

10.22 The amine nitrogen is more basic than the amide nitrogen.

10.23 a) $CH_3CH_2CH_2CO_2^- + NH_3$

b) $CH_3CO_2H + C_6H_5\!-\!CH_2\overset{+}{N}H_3$

c) $CH_3(CH_2)_4\underset{\underset{CH_3}{|}}{C}HNH^- \ Li^+ + C_6H_6$

d) $H_3\overset{+}{N}CH_2CO_2H$

e) $CH_3\underset{\underset{NH_2}{|}}{C}HCO_2^-$

f) $CH_3(CH_2)_3CH_2\!-\!NH_2$

g) $CH_3CH_2CH_2CH_2\!-\!NH\!-\!\overset{\overset{S}{\|}}{C}\!-\!NH^-C_6H_5$

h) $C_6H_5CH_2NH\!-\!\overset{\overset{O}{\|}}{C}CH_2CH_3 + C_6H_{11}NH\!-\!\overset{\overset{O}{\|}}{C}\!-\!NHC_6H_{11}$

i) [pyridinium ring] $\overset{+}{N}CH_2CH_3 \ Br^-$

j) $CH_3(CH_2)_4NHCH_2CH_3$

10.24 a) A: $SOCl_2$ or PCl_3; B: NaCN; C: H_2 + Pt, or $LiAlH_4$
b) D: $SOCl_2$ or PCl_3; E: CH_3NH_2; F: $LiAlH_4$ followed by H_2O

10.25 a) Heat furnishes the energy to break the hydrogen bonds in the α-helix structure of the protein. With the hydrogen bonds broken, it is possible to pull the protein chain to its greatest length just as a coiled spring can be stretched by pulling.
b) The formation of hydrogen bonds gives off energy. The protein chain reverts to its helical structure because this is the conformation of lowest energy.

10.26 The isomers are diastereoisomers.

[Structure: L-valinyl-L-serine, (S)-valinyl-(S)-serine]
$H_3\overset{+}{N}\!-\!\underset{\underset{H}{|}}{\overset{\overset{(CH_3)_2CH}{|}}{C}}\!-\!\overset{\overset{O}{\|}}{C}\!-\!NH\!-\!\underset{\underset{H}{|}}{\overset{\overset{CH_2OH}{|}}{C}}\!-\!CO_2^-$

[Structure: L-valinyl-D-serine, (S)-valinyl-(R)-serine]
$H_3\overset{+}{N}\!-\!\underset{\underset{H}{|}}{\overset{\overset{(CH_3)_2CH}{|}}{C}}\!-\!\overset{\overset{O}{\|}}{C}\!-\!NH\!-\!\underset{\underset{CH_2OH}{|}}{\overset{\overset{H}{|}}{C}}\!-\!CO_2^-$

L-valinyl-L-serine   L-valinyl-D-serine
(S)-valinyl-(S)-serine   (S)-valinyl-(R)-serine

10.27 The strong inductive effect of the positively charged $-NH_3^+$ group of protonated glycine withdraws electrons from the O—H group, increasing its tendency to ionize. The conjugate base formed is the zwitterion, which is neutral, charges cancelling, and is more stable than the anion of a carboxylic acid, such as acetic acid. The carboxyl group CO_2H of protonated glycine or of acetic acid is a stronger acid than is the ammonium ion of the zwitterion of glycine or of protonated glycine.

10.28 $C_6H_5\!-\!\underset{\underset{OH}{|}}{C}H\!-\!\underset{\underset{\underset{\underset{C_6H_5 \quad O}{\diagdown\!\diagup}}{C}}{NH}}{C}H\!-\!CH_3 \underset{OH^-}{\overset{H^+}{\rightleftarrows}} C_6H_5\!-\!\underset{\underset{\underset{\underset{C_6H_5 \quad O}{\diagdown\!\diagup}}{C}}{O}}{C}H\!-\!\underset{\underset{^+NH_3}{|}}{C}H\!-\!CH_3$

The amide is more stable than the ester under neutral or basic conditions and an ester is easily converted to an amide by treatment with an amine. Under acidic conditions, however, the amine exists as an ammonium ion which is not a nucleophile. Just as amides are hydrolyzed under acidic conditions by water, so can they be converted in a nonreversible reaction to an ester and ammonium ion by treatment with an alcohol. The proximity of the alcohol in the molecule makes this reaction easier than it would be if the two functional groups were in separate molecules.

10.29

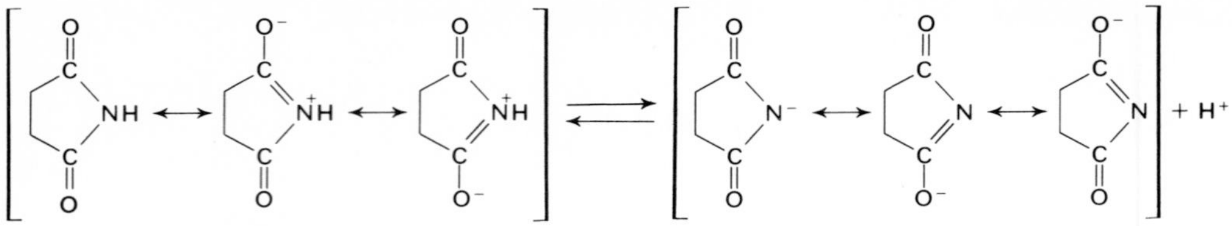

succinimide

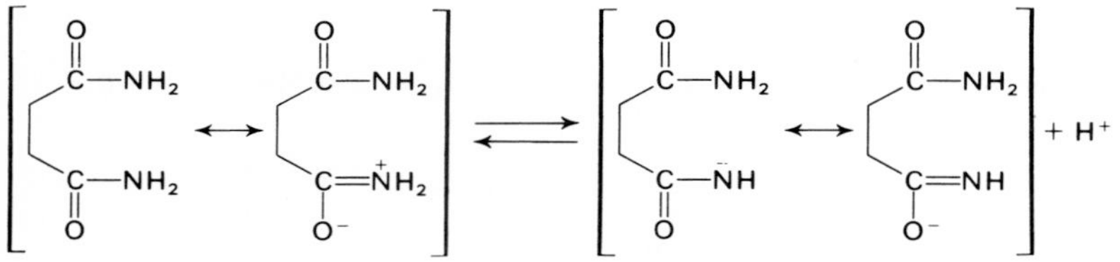

succinamide

The two resonance structures of the imide group of succinimide which have a positive charge on the N make the H on nitrogen more acidic. The three resonance structures of the imide anion with the negative charge distributed over two carbonyl oxygens and one nitrogen stabilize this anion much more than the anion of the diamide is stabilized, by distribution of the negative charge onto one oxygen and one nitrogen.

10.30

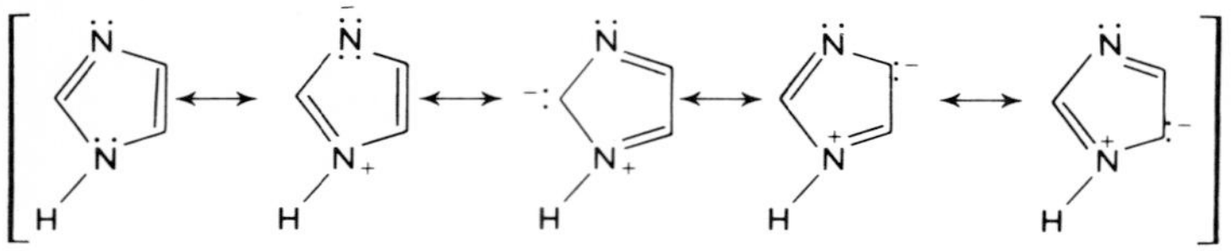

2 imidazole

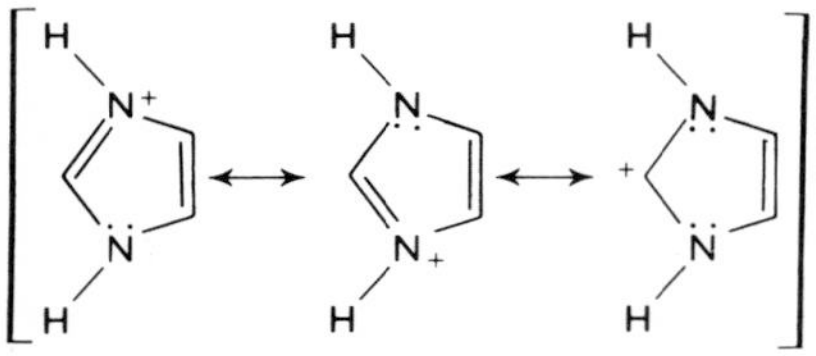

3 conjugate acid of imidazole

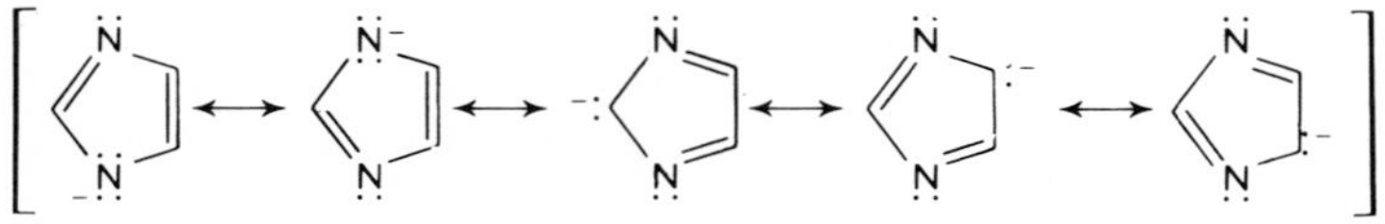

4 conjugate base of imidazole

Chapter 11

11.1 a) camphor—bicyclic ketone; two five-membered rings; three methyl substituents

   b) vanillin—a benzene ring with three substituents: an aldehyde, a phenolic OH, and a methyl ether (a methoxyl) group

c) pyridoxal—a pyridine ring with four substituents: an aldehyde, a phenolic OH, a hydroxymethyl, and a methyl group

d) glyceraldehyde—a straight-chain aldehyde, a secondary and a primary hydroxyl group

e) glucose—a straight-chain aldehyde, four secondary hydroxyls, and one primary hydroxyl group

11.2 a) 2,2-dimethylpropanal **b)** 2-isopropylpentanal

c) 4-bromo-1-phenyl-2-pentanone **d)** 4-bromo-2-*sec*-butylbenzaldehyde

e) dichloroacetaldehyde **f)** cyclobutanone

11.3 a) 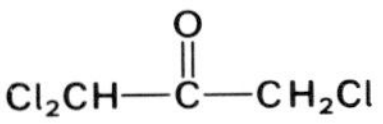 **b)** 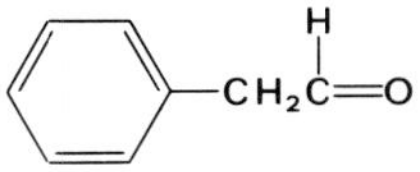 **c)** 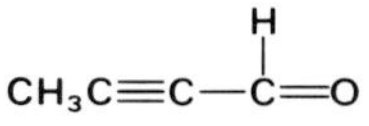

d) 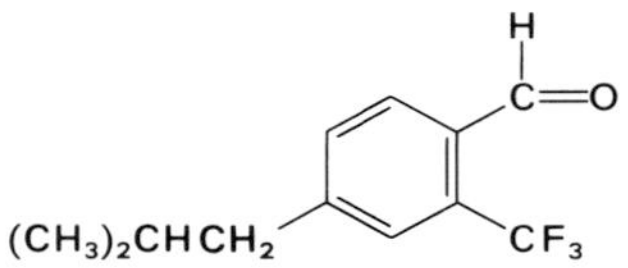 **e)** 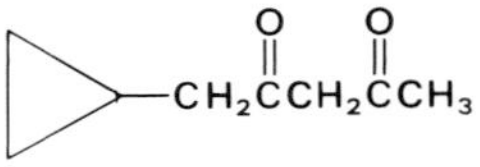 **f)** 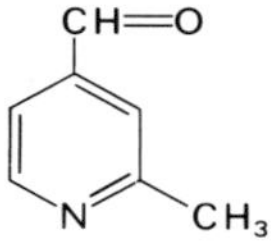

11.4 a) 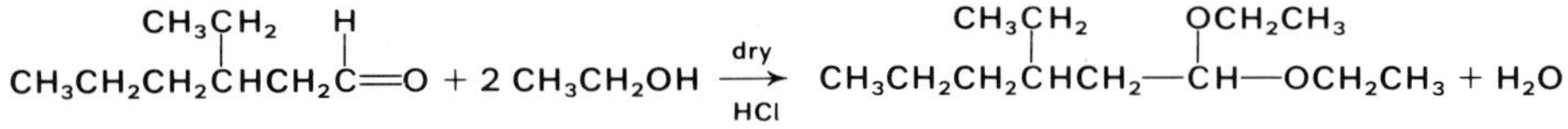

b) $CH_3CH_2CH_2CH_2CH(OCH_3)_2 + H_2O \xrightarrow{HCl} CH_3CH_2CH_2CH_2CH{=}O + 2\,CH_3OH$

11.5 a) 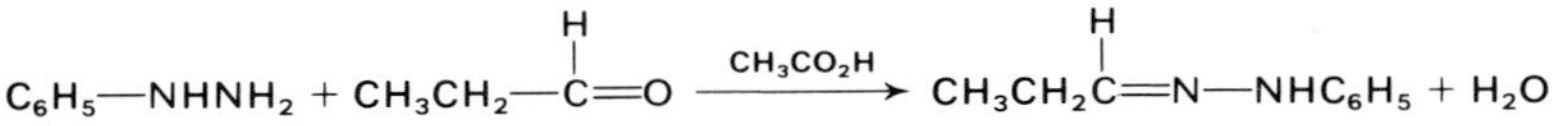

b) 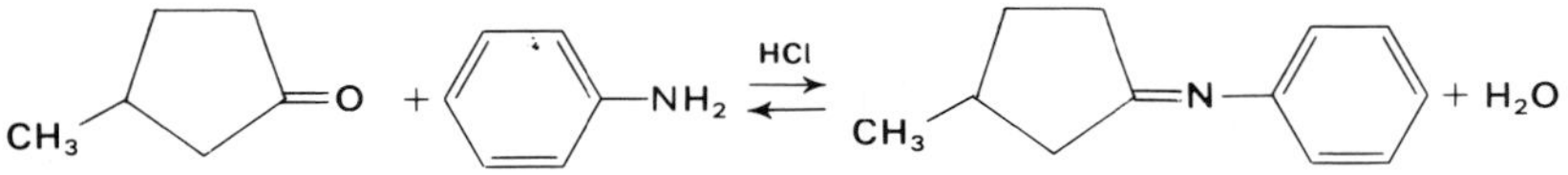

c) 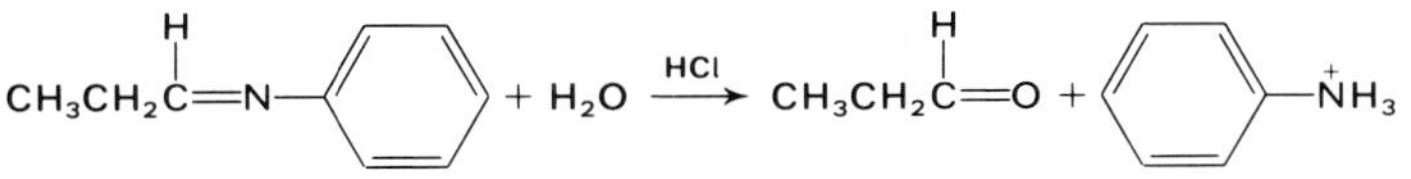

d) 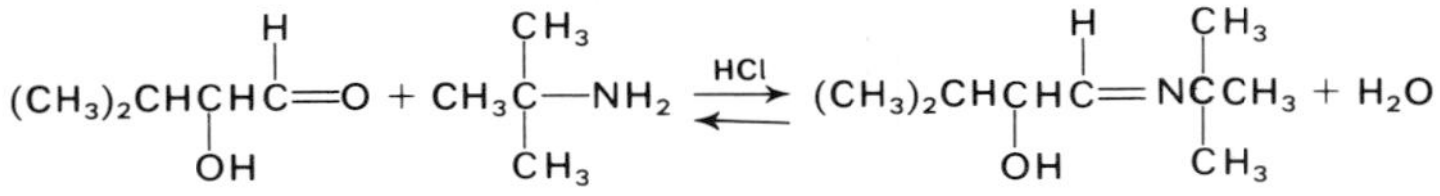

e) 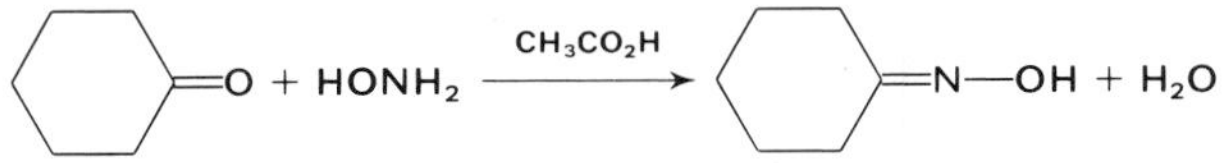

11.6 a) There would be no reaction, since HCN is a very weak acid and does not ionize enough to give enough CN^- nucleophile for reaction.

b) No addition product is formed. Addition of CN^- to give a tetrahedral intermediate anion is followed by loss of CN^- to give starting material, since there is no source of protons to form a stable, neutral product.

11.7 a) $CH_3CH_2CH_2CH{=}O$ **b)** 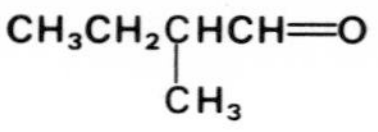 $CH_3CH_2\underset{\underset{\displaystyle CH_3}{|}}{C}HCH{=}O$ **c)** $CH_3\underset{\underset{\displaystyle CH_3}{|}}{C}HCH_2CH{=}O$

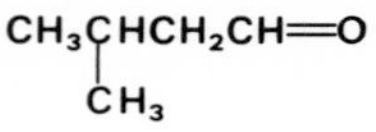

d) 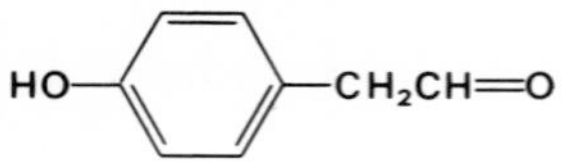 HO—⟨benzene ring⟩—$CH_2CH{=}O$ **e)** $C_6H_5CH_2CH{=}O$ **f)** $^-O_2CCH_2CH_2CH{=}O$

11.8 a) 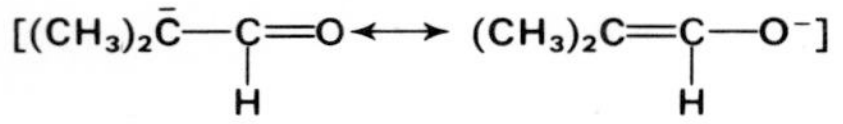 $[(CH_3)_2\bar{C}{-}\underset{\underset{\displaystyle H}{|}}{C}{=}O \longleftrightarrow (CH_3)_2C{=}\underset{\underset{\displaystyle H}{|}}{C}{-}O^-]$

b) 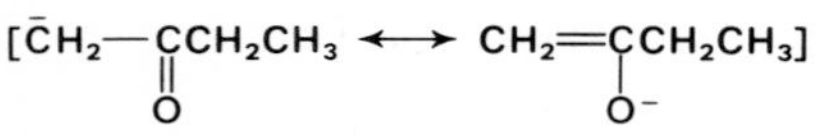 $[\bar{C}H_2{-}\underset{\underset{\displaystyle O}{||}}{C}CH_2CH_3 \longleftrightarrow CH_2{=}\underset{\underset{\displaystyle O^-}{|}}{C}CH_2CH_3]$ more stable

$[CH_3\underset{\underset{\displaystyle O}{||}}{C}\,\bar{C}HCH_3 \longleftrightarrow CH_3\underset{\underset{\displaystyle O^-}{|}}{C}{=}CHCH_3]$

c) $[\bar{C}H_2{-}CH{=}CH{-}\underset{\underset{\displaystyle H}{|}}{C}{=}O \longleftrightarrow CH_2{=}CH{-}\bar{C}H{-}\underset{\underset{\displaystyle H}{|}}{C}{=}O \longleftrightarrow CH_2{=}CH{-}CH{=}\underset{\underset{\displaystyle H}{|}}{C}{-}O^-]$

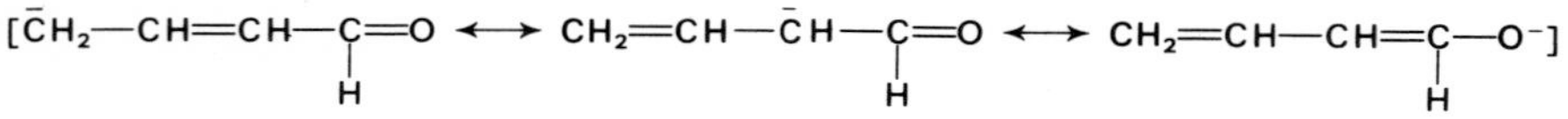

d)

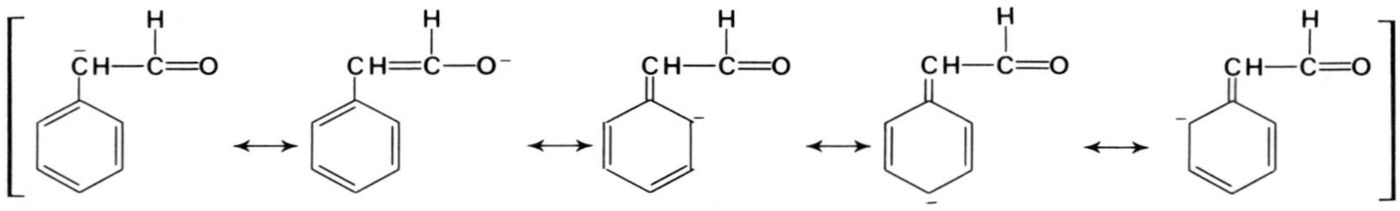

11.9 a) 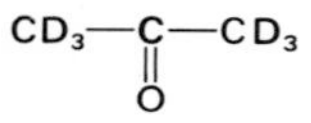 $CD_3{-}\underset{\underset{\displaystyle O}{||}}{C}{-}CD_3$ **b)** 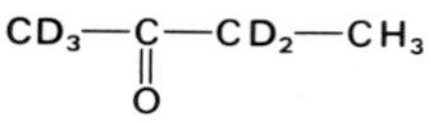 $CD_3{-}\underset{\underset{\displaystyle O}{||}}{C}{-}CD_2{-}CH_3$ **c)**

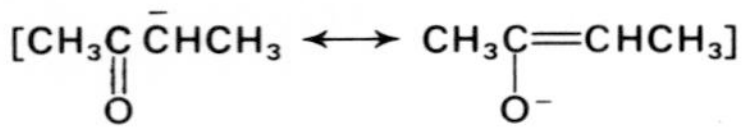

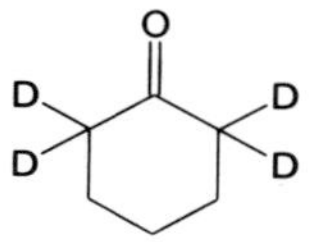

d) 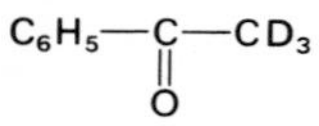 $C_6H_5{-}\underset{\underset{\displaystyle O}{||}}{C}{-}CD_3$ **e)** 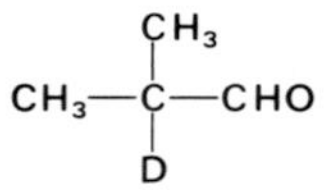 $CH_3{-}\underset{\overset{\displaystyle CH_3}{|}}{\underset{\underset{\displaystyle D}{|}}{C}}{-}CHO$ **f)**

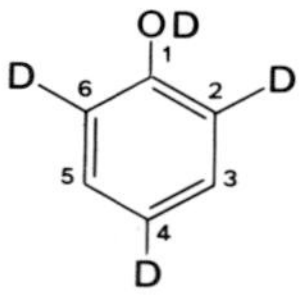

11.10 a) 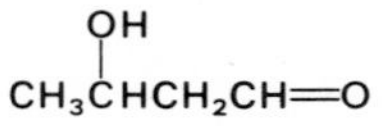 $CH_3\overset{\displaystyle OH}{\overset{|}{C}}HCH_2CH{=}O$ and

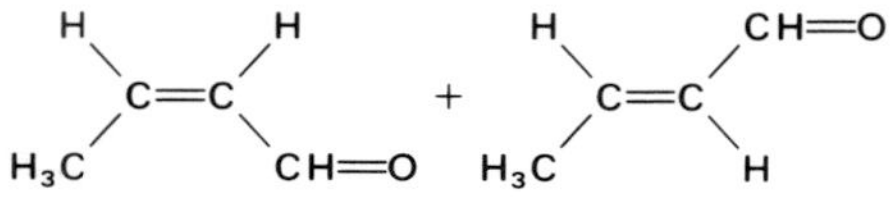

b) 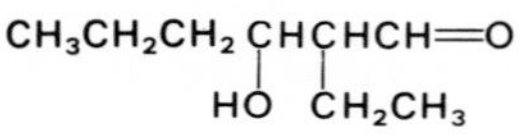 $CH_3CH_2CH_2\underset{\underset{\displaystyle HO}{|}}{C}H\underset{\underset{\displaystyle CH_2CH_3}{|}}{C}HCH{=}O$ and

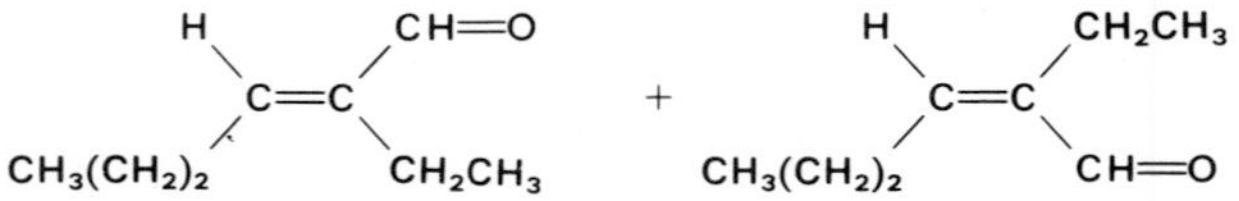

11.11 a)

$$H, CH{=}O \quad C{=}C \quad H_3C, H \qquad \text{and} \qquad H_3C, CH{=}O \quad C{=}C \quad H, H$$

b)

$$H, CH{=}O \quad C{=}C \quad C_6H_5CH_2, C_6H_5 \qquad \text{and} \qquad C_6H_5CH_2, CH{=}O \quad C{=}C \quad H, C_6H_5$$

c)

$$H, CH{=}O \quad C{=}C \quad C_6H_5, H \quad , \quad C_6H_5, CH{=}O \quad C{=}C \quad H, H \quad , \quad \text{and the two structures listed under } \mathbf{a}$$

d)

$$H, CHO \quad C{=}C \quad H_3C, CH_3 \quad , \quad H_3C, CHO \quad C{=}C \quad H, CH_3 \quad , \quad H, CHO \quad C{=}C \quad CH_3CH_2, H \quad ,$$

$$CH_3CH_2, CHO \quad C{=}C \quad H, H \quad , \quad H, CHO \quad C{=}C \quad CH_3CH_2, CH_3 \quad , \quad CH_3CH_2, CHO \quad C{=}C \quad H, CH_3 \quad ,$$

and the two structures listed under **a**

e) $H_2C{=}CH{-}CHO$ and the two structures listed under **a**

11.12 a) H_2O catalyzed by H_2SO_4 and $HgSO_4$; $CH_3CH_2\overset{\overset{O}{\|}}{C}CH_3$

b) (1) $H_2O + H_2SO_4$, (2) $Na_2Cr_2O_7 + H_2SO_4$; [cyclohexanone structure]${=}O$

c) CrO_3-pyridine; $CH_3CH_2CH{=}O$

11.13 $CH_3(CH_2)_6\overset{\overset{OH}{|}}{C}HCH_2\overset{\overset{O}{\|}}{C}{-}S(CoA) +$ [pyridinium ring with CONH$_2$, N$^+$, R]

$\longrightarrow CH_3(CH_2)_6\overset{\overset{O}{\|}}{C}CH_2\overset{\overset{O}{\|}}{C}{-}S(CoA) +$ [dihydropyridine ring with CONH$_2$, N, R] $+ H^+$

oxidation of hydroxyester to ketoester

11.14 a) $C_6H_5\overset{\overset{CH_3}{|}}{C}H{-}N{=}\overset{\overset{CH_3}{|}}{C}CH_2CH_3 + {}^-OC(CH_3)_3 \longrightarrow$ $\left[C_6H_5\overset{\overset{CH_3}{|}}{C}{-}N{=}\overset{\overset{CH_3}{|}}{C}CH_2CH_3 \longleftrightarrow C_6H_5\overset{\overset{CH_3}{|}}{C}{=}N{-}\overset{\overset{CH_3}{|}}{C}CH_2CH_3 \right] + HOC(CH_3)_3$

$\longrightarrow C_6H_5\overset{\overset{CH_3}{|}}{C}{=}N{-}\overset{\overset{CH_3}{|}}{C}HCH_2CH_3 + {}^-OC(CH_3)_3$

b) $C_6H_5\overset{\overset{CH_3}{|}}{C}H{-}NH_2 + O{=}\overset{\overset{CH_3}{|}}{C}CH_2CH_3; \quad C_6H_5\overset{\overset{CH_3}{|}}{C}{=}O + NH_2{-}\overset{\overset{CH_3}{|}}{C}HCH_2CH_3$

11.15 a) acetophenone (methyl phenyl ketone) **b)** 1,3-cyclooctanedione

c) 2,4-dihydroxy-6-octenal **d)** 3-pentanone dimethyl acetal or 3,3-dimethoxypentane

e) trichloroacetaldehyde

f) 2-hydroxy-3-methylbutanonitrile

g) pyridoxal

h) methyl vinyl ketone or 3-buten-2-one

11.16 a) $Cl_3CC\overset{H}{\underset{OH}{|}}-OH$ **b)** (phenyl)$-\overset{N-OH}{C}CH_2CH_2CH_3$ **c)** $CH_3CH_2\overset{}{\underset{CH_3}{C}}HCH=N-$(phenyl)   **d)** $\begin{array}{c} H_3C \\ H_3C \end{array} C \begin{array}{c} O-CH_2 \\ O-CH_2 \end{array} CH_2$

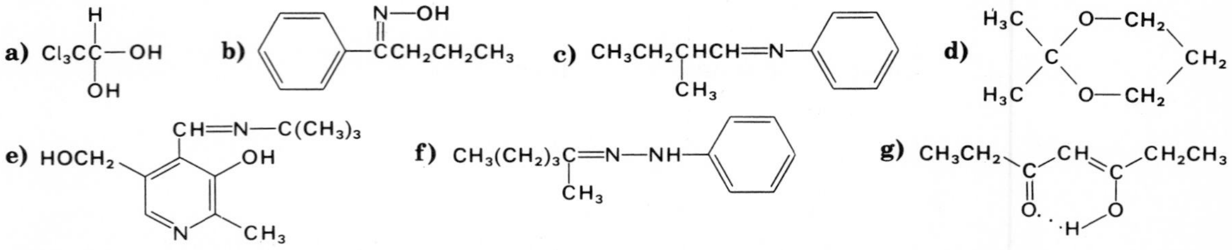

e) $HOCH_2$— pyridine ring with $CH=N-C(CH_3)_3$, OH, N, CH_3

f) $CH_3(CH_2)_3\overset{}{\underset{CH_3}{C}}=N-NH-$(phenyl)

g) $CH_3CH_2-\overset{CH}{C}\cdots\overset{CH_2CH_3}{C}$ with $O\cdots H\cdots O$

11.17 a) $CH_3CH=CHCH_2CH_2-\overset{}{\underset{OCH_3}{C}}-OCH_3$ **b)** $CH_2=O + HOCH_2CH_2OH$ **c)** C_6H_5CH (oxazolidine ring with O, N, H)   **d)** CH_3CH (thiazolidine ring with S, N, H)

e) (phenyl)$-CH_2\overset{}{\underset{NH_2}{C}}HCH\equiv N$ **f)** (phenyl)$-\overset{}{\underset{OH}{C}}HCH\equiv N$ **g)** cyclopentane with H and OH **h)** $CH_3(CH_2)_3\overset{}{\underset{OH}{C}}H-\overset{}{\underset{CH=O}{C}}HCH_2CH_2CH_3$

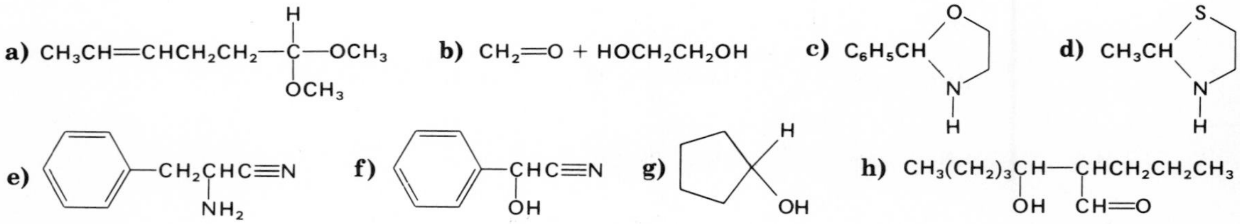

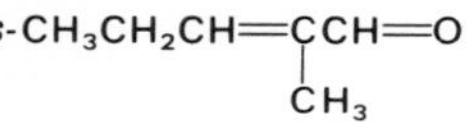

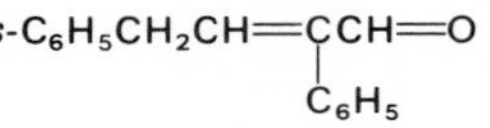

11.18 a and **d** are racemic products:

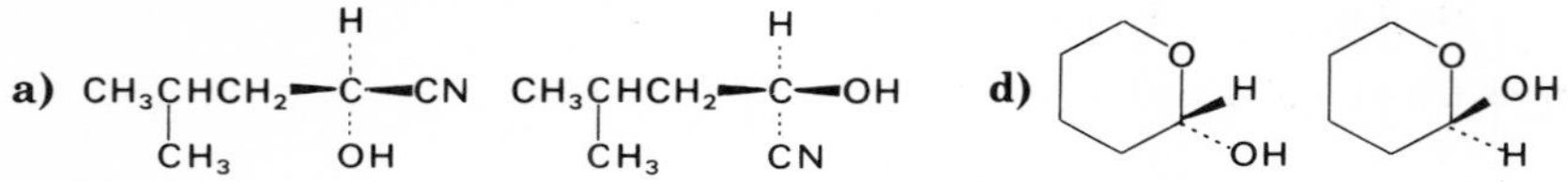

a) $CH_3CHCH_2-\overset{H}{\underset{OH}{C}}-CN$ $CH_3CHCH_2-\overset{H}{\underset{CN}{C}}-OH$ **d)** tetrahydropyran with H, OH; tetrahydropyran with OH, H

11.19 a) *cis* and *trans*-$CH_3CH_2CH=\overset{}{\underset{CH_3}{C}}CH=O$ **b)** *cis* and *trans*-$C_6H_5CH_2CH=\overset{}{\underset{C_6H_5}{C}}CH=O$

11.20 $\left[-\overset{\beta}{C}=\overset{\alpha}{C}-\overset{}{\underset{O}{C}}- \longleftrightarrow -C=C-\overset{+}{\underset{O^-}{C}}- \longleftrightarrow -\overset{+}{C}-C=\overset{}{\underset{O^-}{C}}- \right]$

    A   B   C

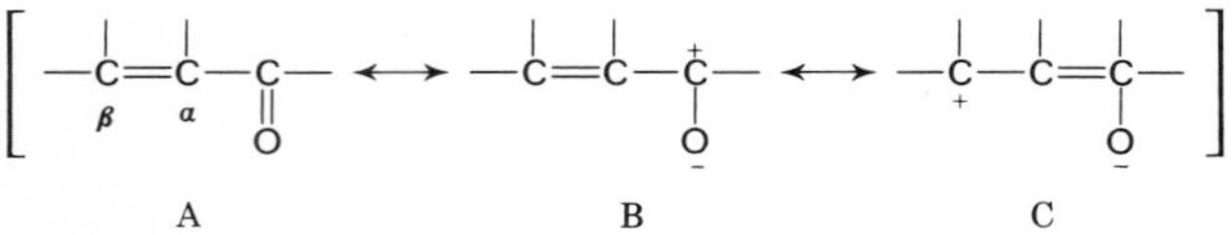

Resonance structure B has a positive charge on the carbonyl carbon and C has a positive charge on the β-carbon Nucleophiles attack both positive carbons.

11.21 a) $C_6H_5-CH=O + NH_3 + HCN \xrightarrow{NaOH} C_6H_5-\overset{}{\underset{NH_2}{C}}H-C\equiv N \xrightarrow{H_3O^+Cl^-}$

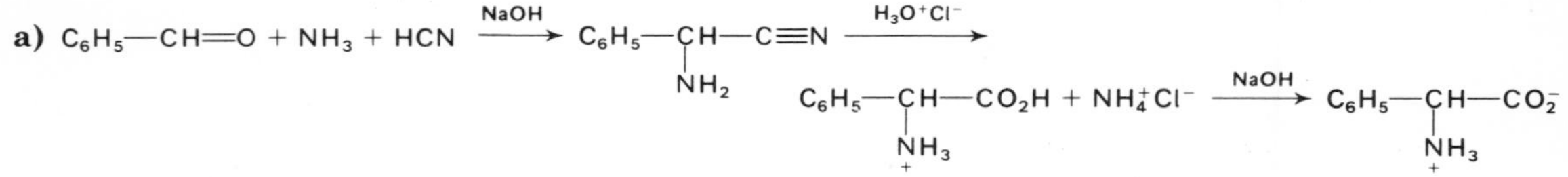

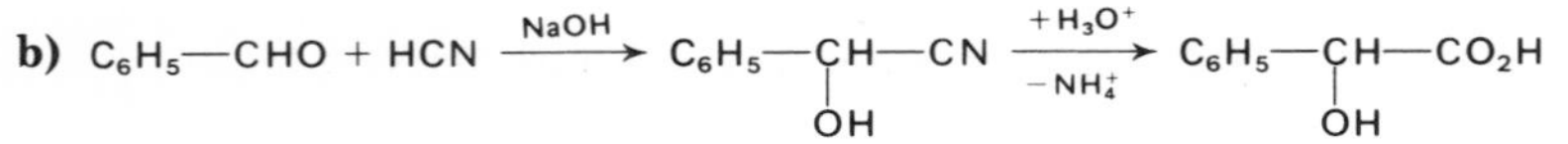

$C_6H_5-\overset{}{\underset{\overset{+}{N}H_3}{C}}H-CO_2H + NH_4^+Cl^- \xrightarrow{NaOH} C_6H_5-\overset{}{\underset{\overset{+}{N}H_3}{C}}H-CO_2^-$

b) $C_6H_5-CHO + HCN \xrightarrow{NaOH} C_6H_5-\overset{}{\underset{OH}{C}}H-CN \xrightarrow[-NH_4^+]{+H_3O^+} C_6H_5-\overset{}{\underset{OH}{C}}H-CO_2H$

11.22 a) $C_6H_5-\overset{}{\underset{O}{C}}\cdots\overset{CH}{\cdots}\overset{}{\underset{O}{C}}-C_6H_5$ with $\cdots H\cdots$ **b)** $H_3C-\overset{}{\underset{O}{C}}\cdots\overset{CH}{\cdots}\overset{}{\underset{O}{C}}-OCH_3$ with $\cdots H\cdots$ **c)** $C_6H_5-\overset{}{\underset{OH}{C}}=\overset{CH}{\cdots}\cdots C\equiv N:$

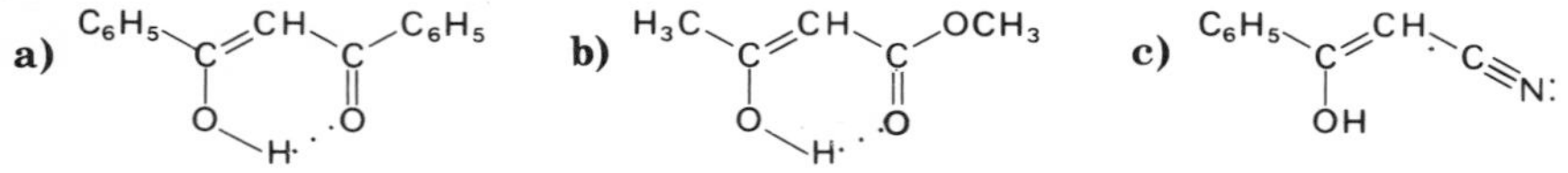

No internal hydrogen bond in **c** is possible because electron pair on N is too far away.

d)

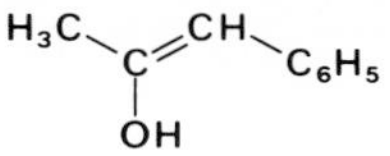

e)

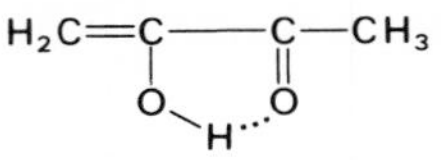

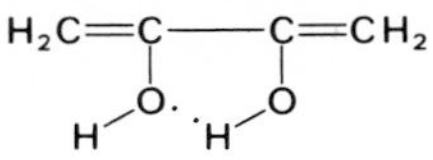

f) 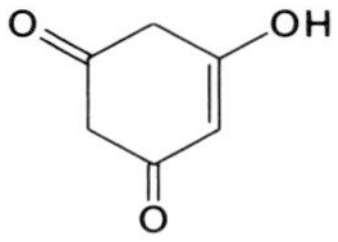 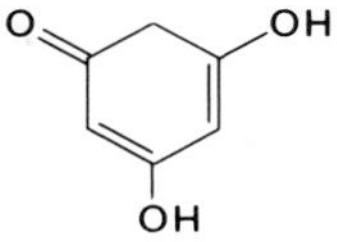 and

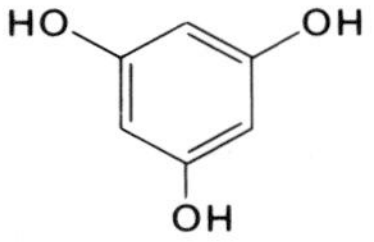

11.23 a)

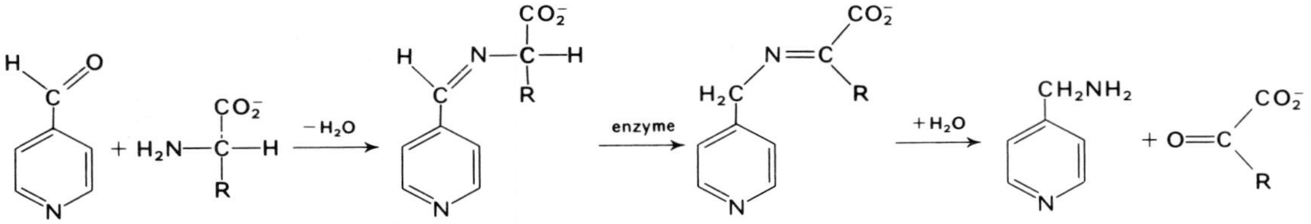

pyridoxal   glutamate ion $R = CH_2CH_2CO_2^-$ pyridoxamine   α-keto-
(substituents   glutarate ion
on ring of
pyridoxal
omitted)

b)

pyridoxamine   pyruvate   pyridoxal   alanine ion
(substituents   ion
omitted)

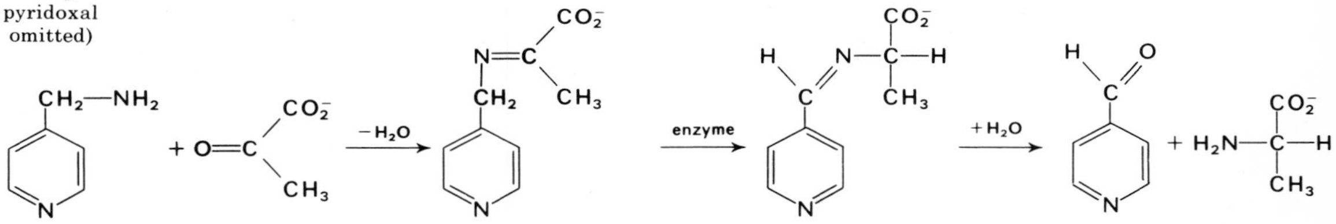

11.24 a) $C_{21}H_{39}O_{11}N_7$

b) ring 1: two guanidine substituents, three hydroxyls, one acetal
ring 2: a cyclic acetal, an acetal, an aldehyde
ring 3: a cyclic acetal, two hydroxyls, hydroxymethyl, methylamino

c) Every ring carbon is asymmetric: ring 1 has six, ring 2 has four and ring 3 has five.

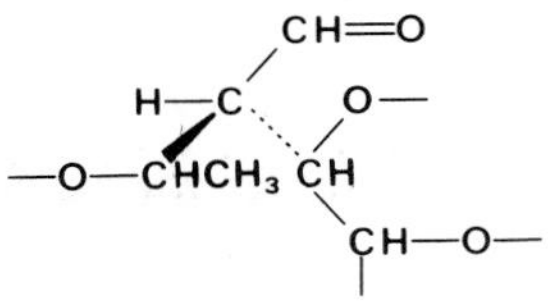

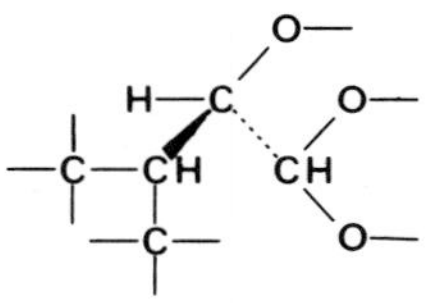

d)

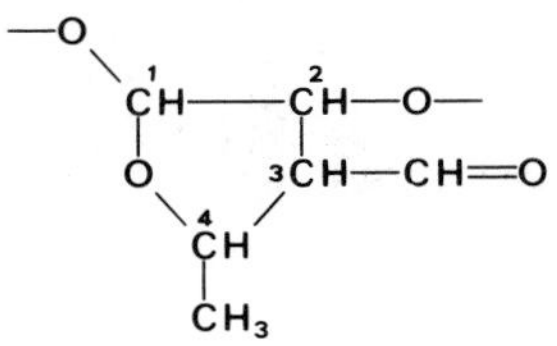

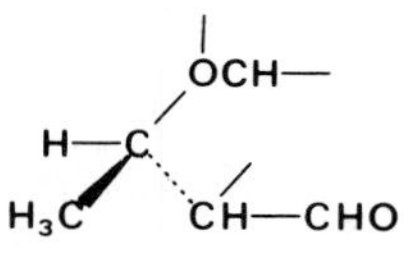

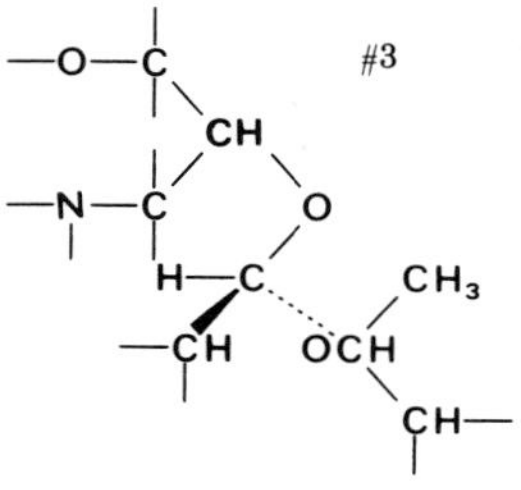

e) The two C=NH groups within the two guanidine groups on ring 1, and the amine group on ring 3.

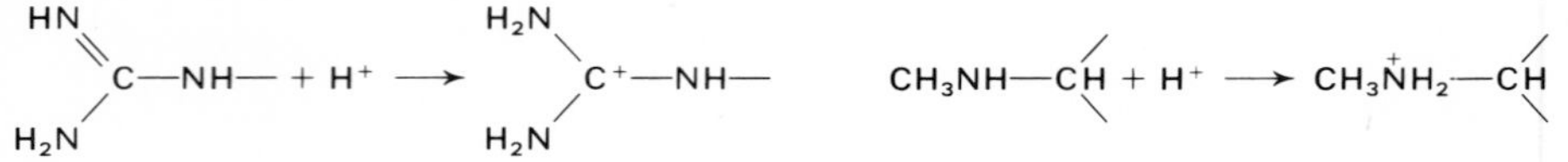

f) The reaction is hydrolysis of the two acetyl groups which connect rings 1 and 2 and rings 2 and 3.

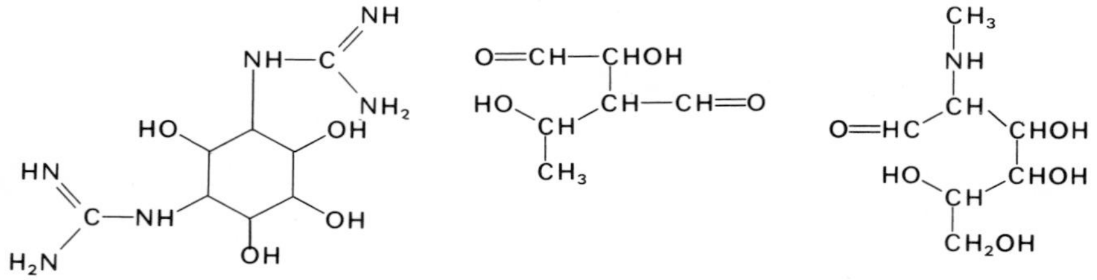

g) Racemization by enolization only would occur at the carbon adjacent to the free aldehyde group in ring 2.

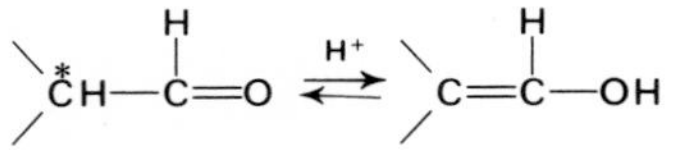

Hydrolysis of acetals would be followed by enolization of the two new free aldehyde groups and racemization of α-carbons.

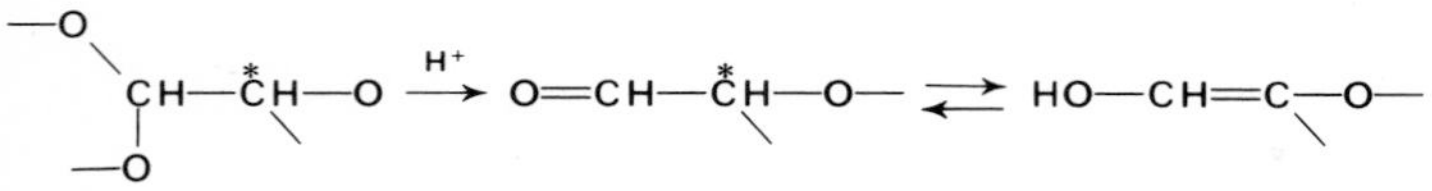

ring 2

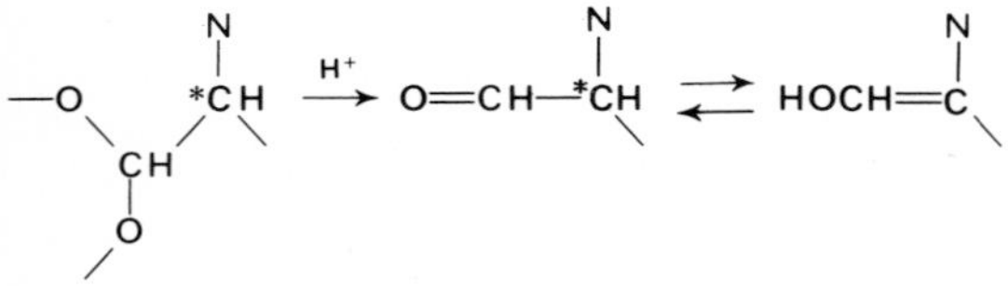

ring 3

Chapter 12

12.1 a) 2 **b)** 2 **c)** 1 **d)** 1

12.2 six inner and twelve outer hydrogens, three *cis* and six *trans* C=C

12.3

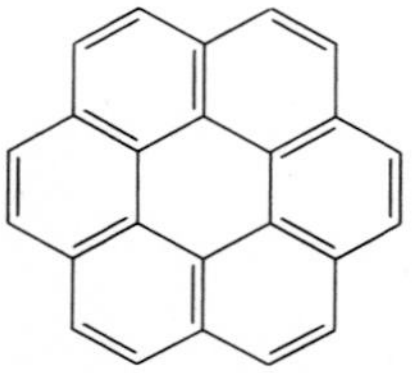

12.4 a)

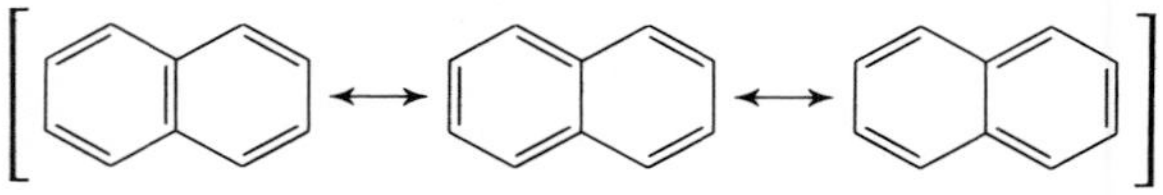

naphthalene

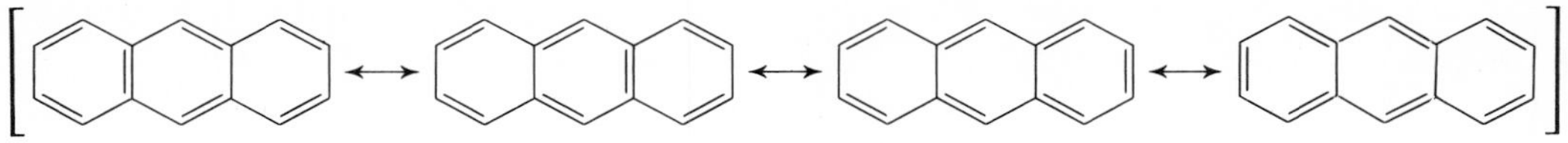

anthracene

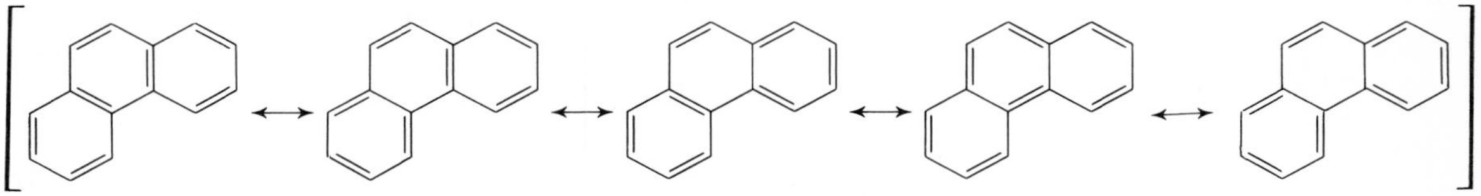

phenanthrene

b)

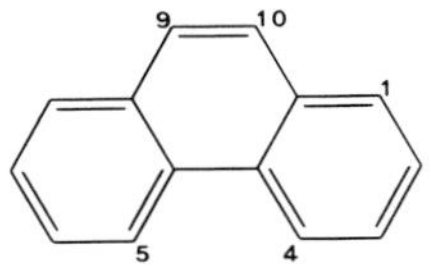

In four of the five resonance structures for phenanthrene, shown in the answer to Problem 12.4(a) above, the bond between C–9 and C–10 is double, while the bonds on each side are single. The C_9–C_{10} double bond is more susceptible to addition than others in the molecule because it is more nearly a true double bond, an alkene, and not fully aromatic.

12.5 a)

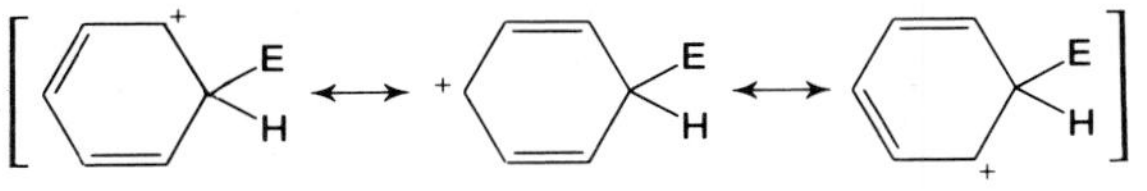

b)

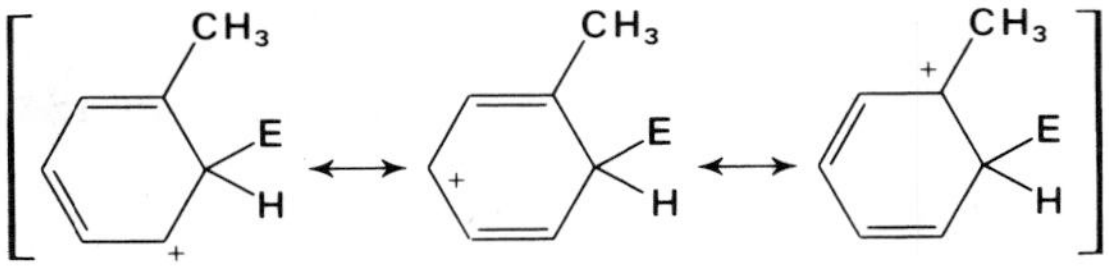

c)

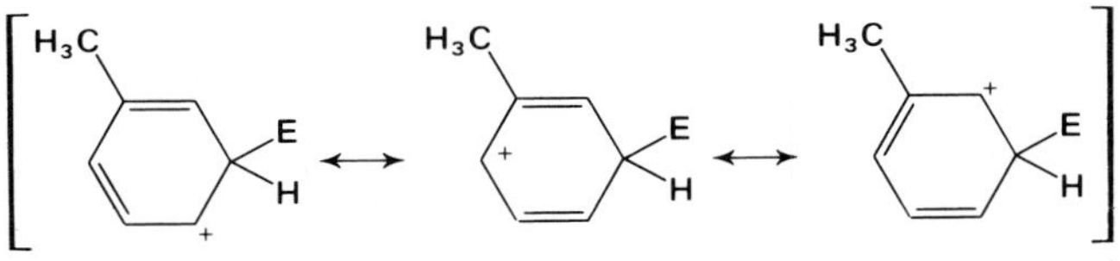

d)

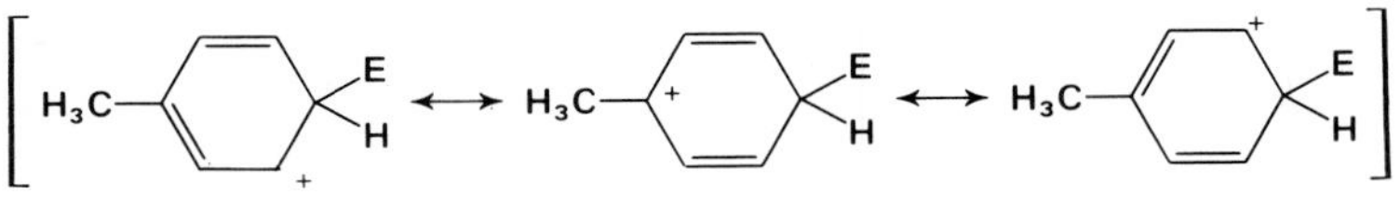

12.6 a)

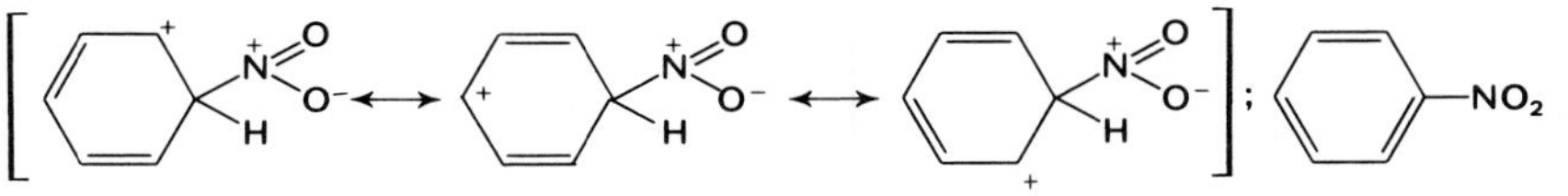

; NO_2 **nitrobenzene**

b) 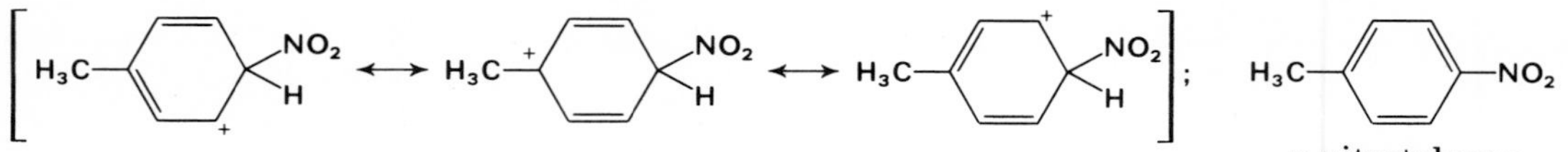 [resonance structures] ; p-nitrotoluene

H_3C—⬡—NO_2 p-nitrotoluene

c) 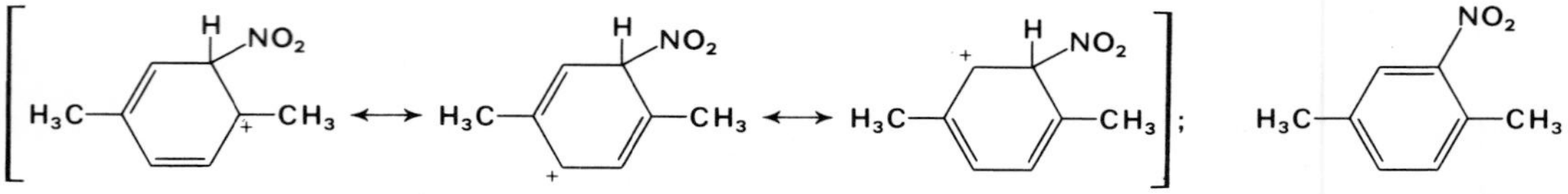 [resonance structures] ; 1,4-dimethyl-2-nitrobenzene

H_3C—⬡(NO_2)—CH_3 1,4-dimethyl-2-nitrobenzene

12.7 a) 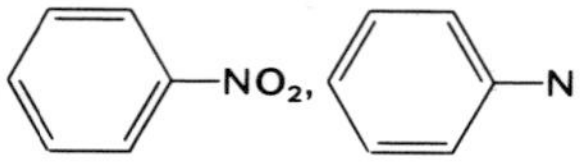 ⬡—NO_2, ⬡—NH_2, aniline   **b)**  H_3C—⬡—NO_2, H_3C—⬡—NH_2, p-methylaniline

c) 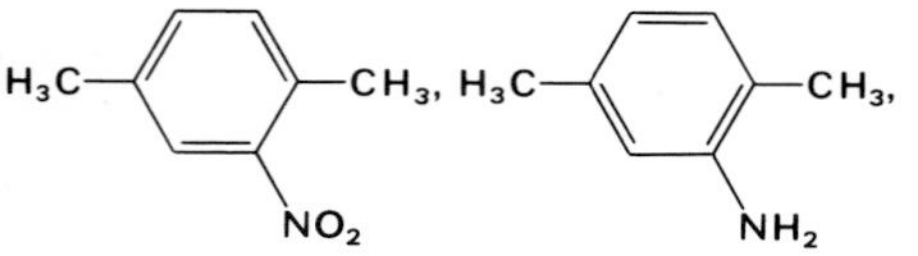 H_3C—⬡(NO_2)—CH_3, H_3C—⬡(NH_2)—CH_3, 2,5-dimethylaniline

12.8 a) 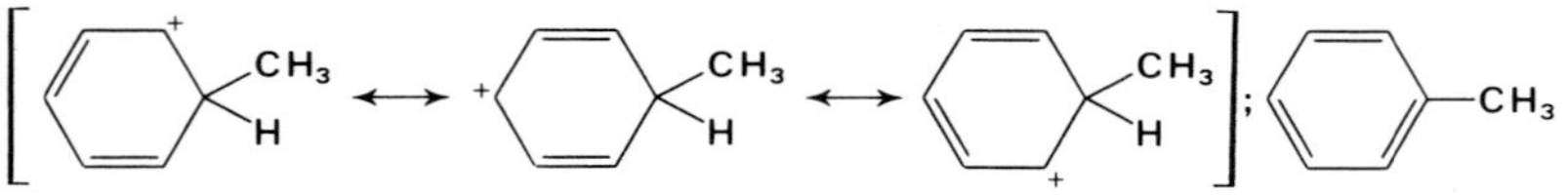 [resonance structures] ; ⬡—CH_3 toluene

b) 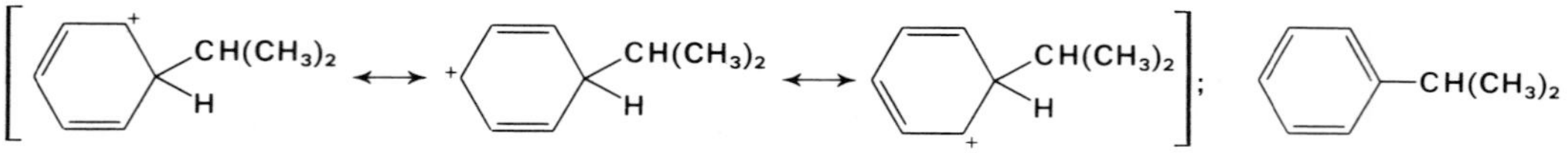 [resonance structures] ; ⬡—$CH(CH_3)_2$ isopropylbenzene

c) 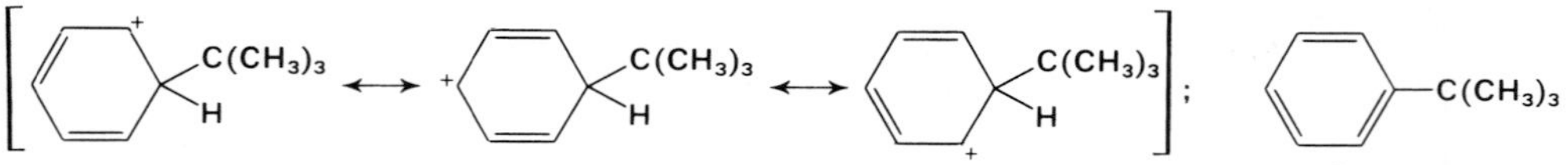 [resonance structures] ; ⬡—$C(CH_3)_3$ tert-butylbenzene

12.9 a) 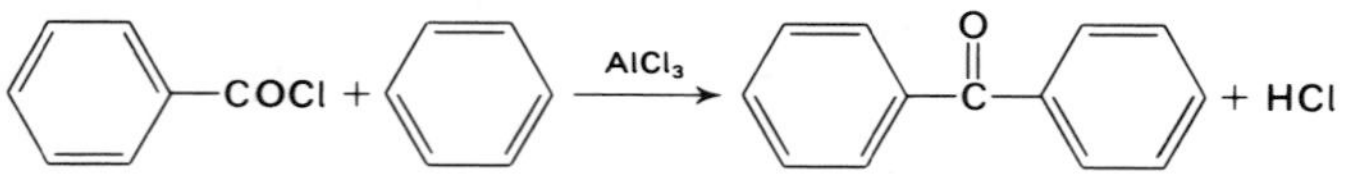 ⬡—COCl + ⬡ $\xrightarrow{AlCl_3}$ ⬡—CO—⬡ + HCl

benzoyl chloride   diphenyl ketone

b) 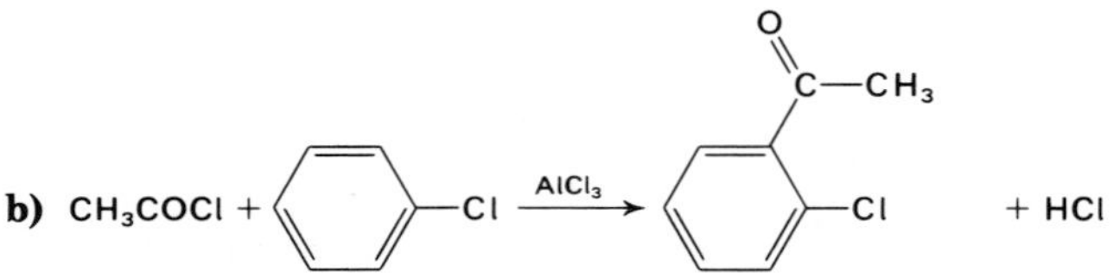 CH_3COCl + ⬡—Cl $\xrightarrow{AlCl_3}$ ⬡(CO—CH_3)—Cl + HCl

acetyl chloride   o-chloroacetophenone

c)

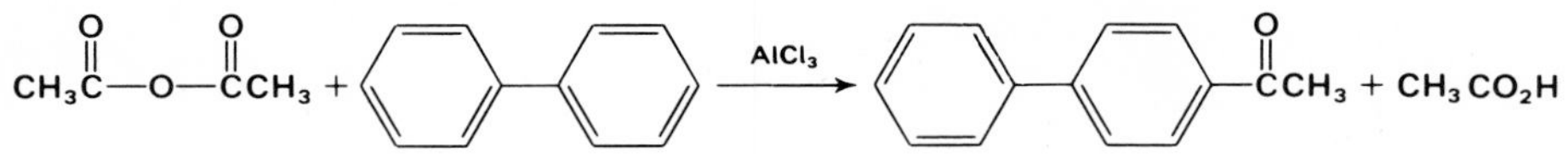

acetic anhydride   4-acetylbiphenyl

12.10 a)

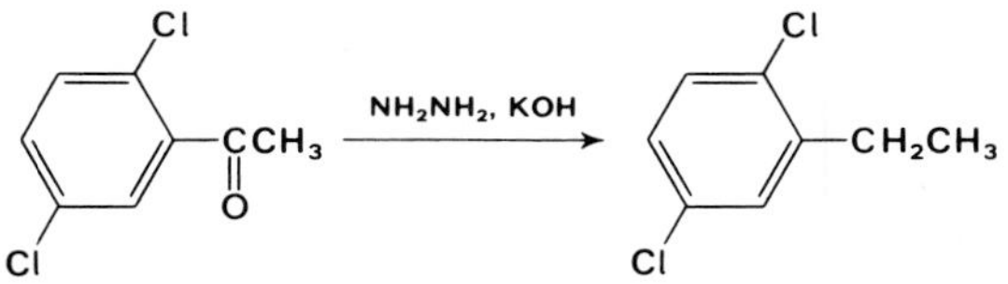

1,4-dichloro-2-ethylbenzene

b)

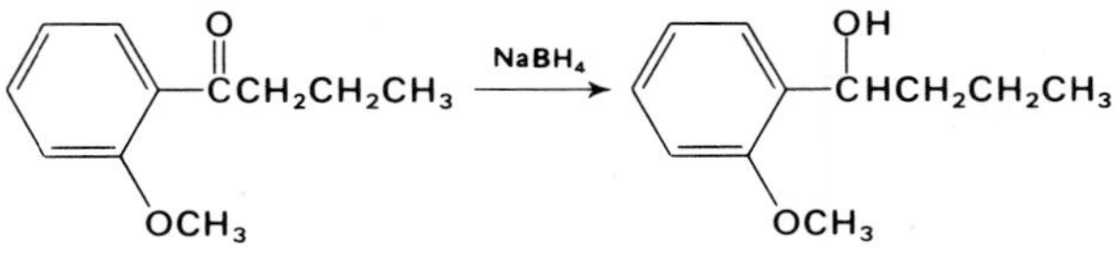

1-(o-methoxyphenyl)-1-butanol

12.11

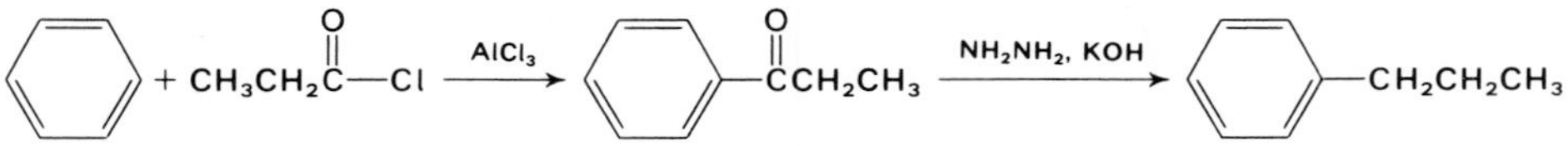

12.12 a)

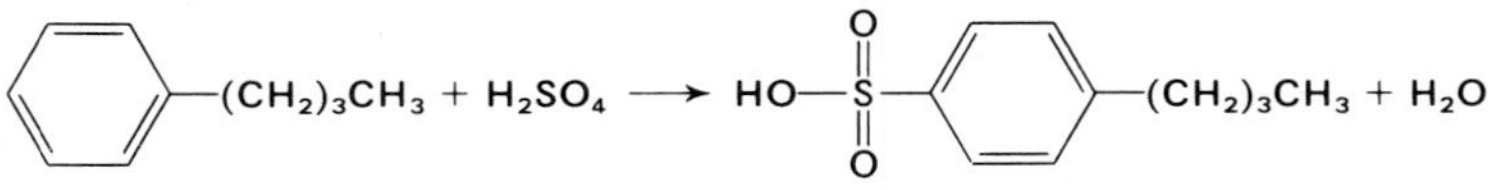

p-butylbenzenesulfonic acid

b)

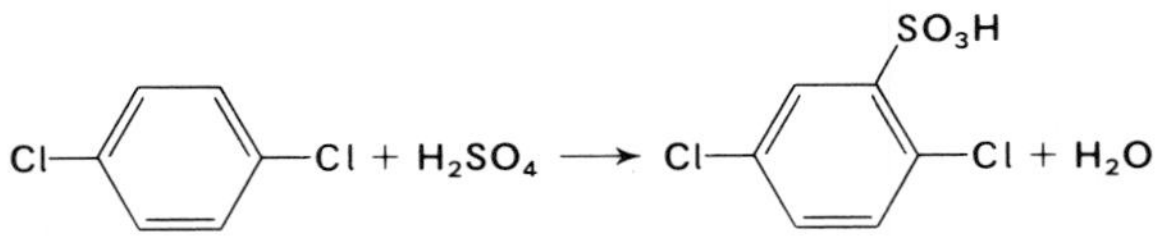

2,5-dichlorobenzenesulfonic acid

c)

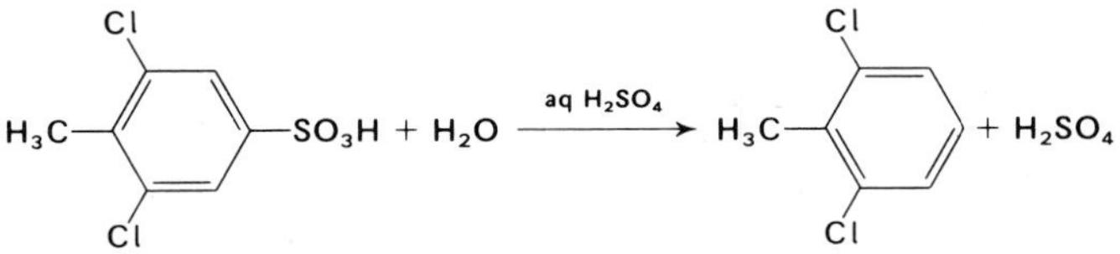

12.13

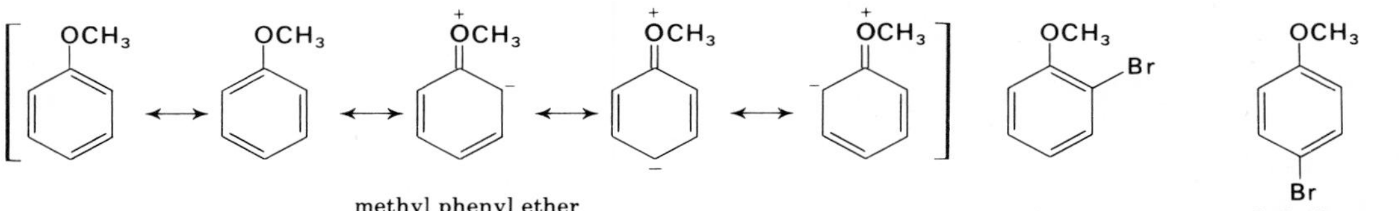

methyl phenyl ether   methyl o-bromo-   methyl p-bromo-
                                                 phenyl ether   phenyl ether

12.14 a)

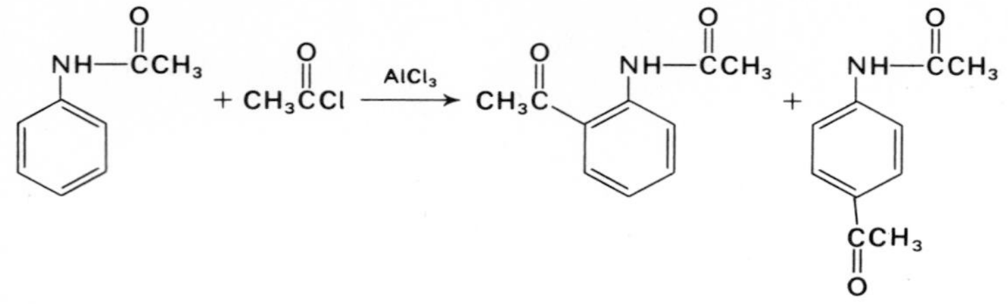

N-phenylacetamide   o-acetyl-   p-acetyl-
(acetanilide)   acetanilide   acetanilide

b)

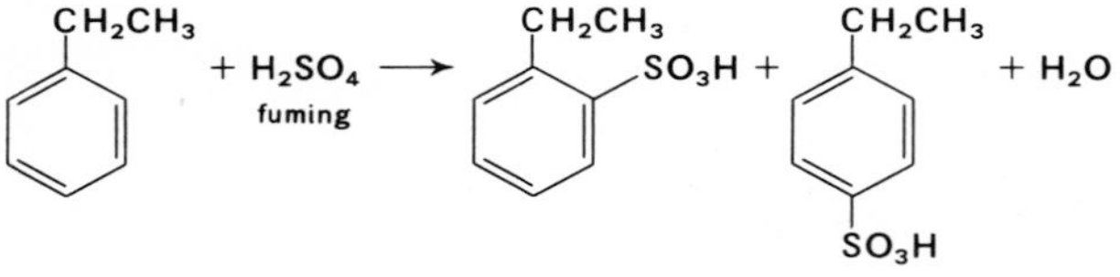

o-ethylbenzene-   p-ethylbenzene-
sulfonic acid   sulfonic acid

c)

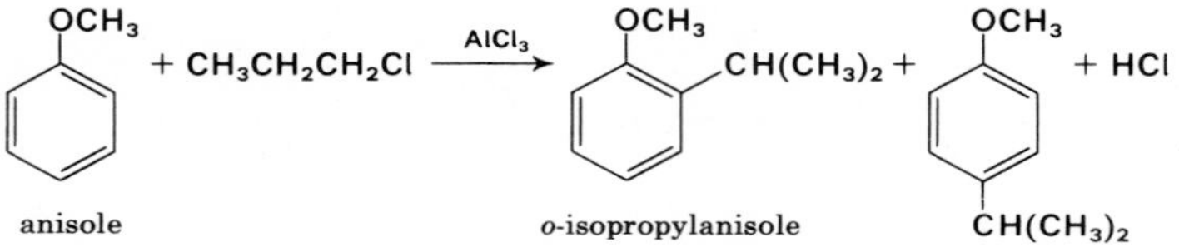

anisole   o-isopropylanisole

p-isopropylanisole

d)

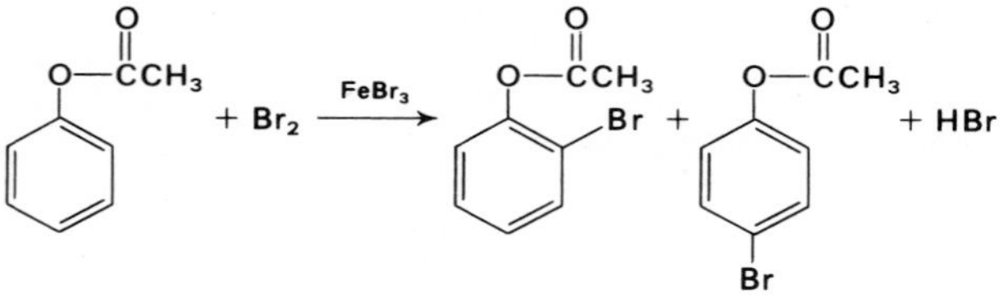

o-bromophenyl   p-bromophenyl
acetate   acetate

e)

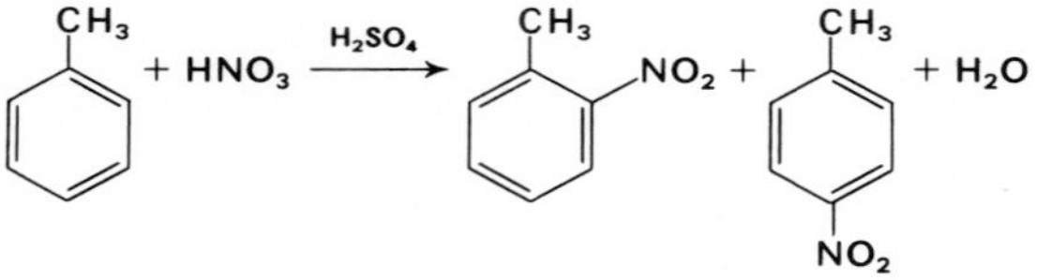

o-nitrotoluene   p-nitrotoluene

f)

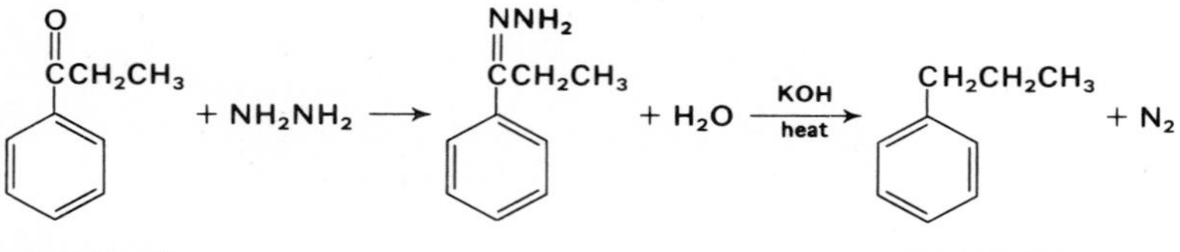

ethyl phenyl   propylbenzene
ketone

g)

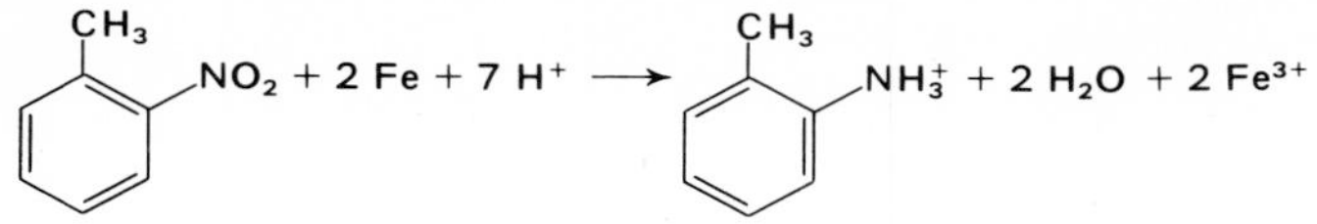

o-nitrotoluene (CH₃, NO₂) $+ 2\ Fe + 7\ H^+ \longrightarrow$ o-methylanilinium ion (CH₃, NH₃⁺) $+ 2\ H_2O + 2\ Fe^{3+}$

o-nitrotoluene   o-methylanilinium ion

12.15 a) 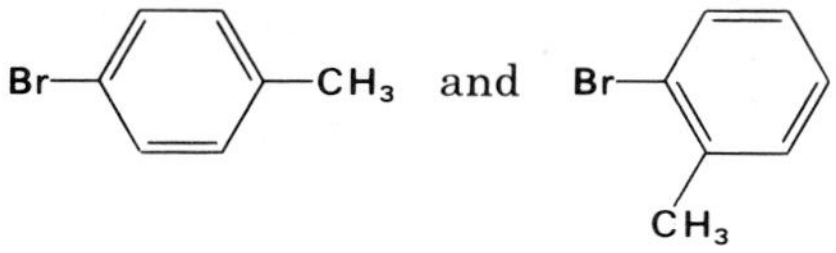 Br—⟨⟩—CH₃ and Br—⟨⟩(CH₃)

b) 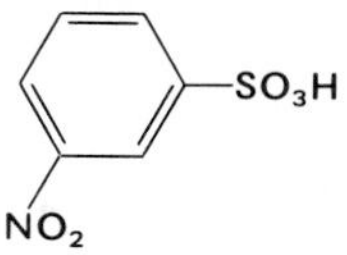 ⟨⟩—SO₃H, NO₂

c) 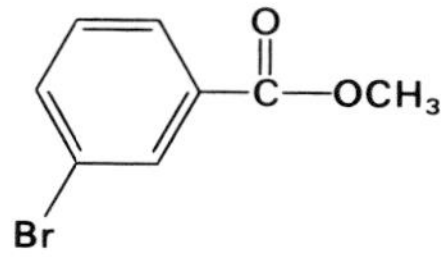 ⟨⟩—C(=O)—OCH₃, Br

d) 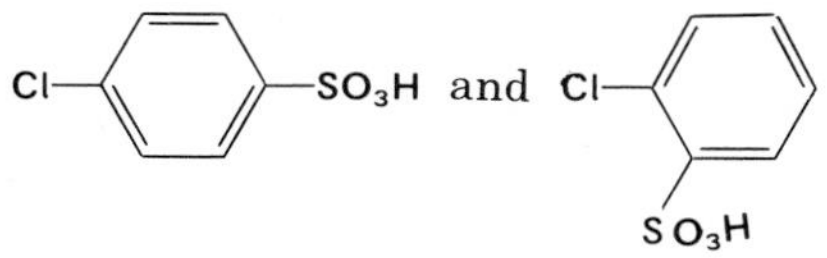 Cl—⟨⟩—SO₃H and Cl—⟨⟩—SO₃H

e) 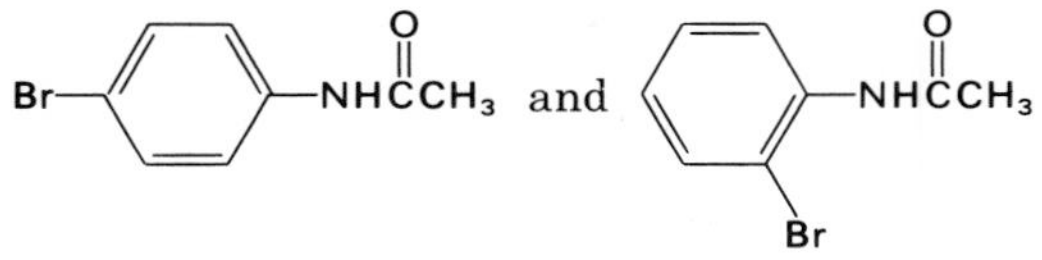 Br—⟨⟩—NHC(=O)CH₃ and ⟨⟩—NHC(=O)CH₃, Br

12.16 a) 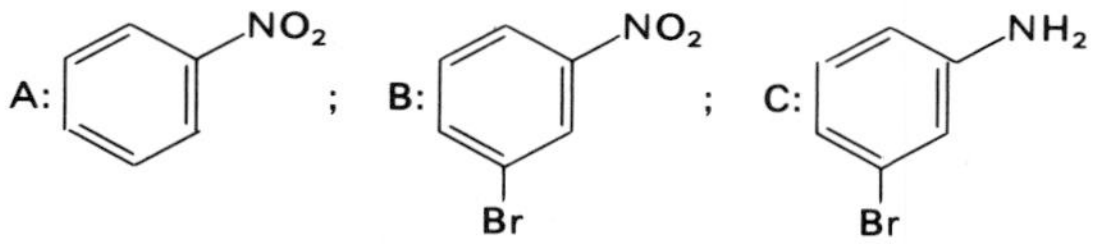 A: ⟨⟩—NO₂ ; B: ⟨⟩—NO₂, Br ; C: ⟨⟩—NH₂, Br

b) 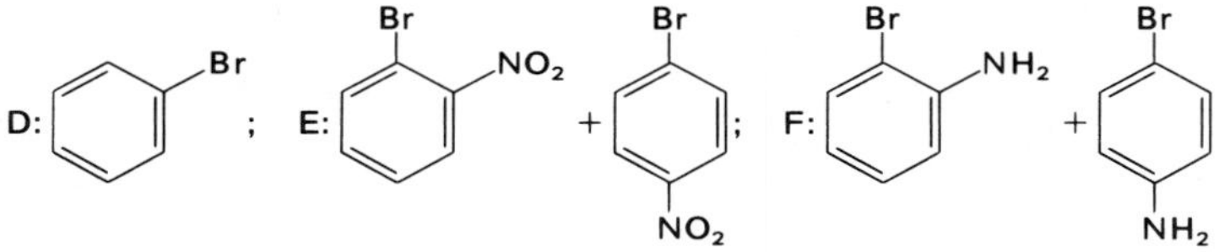 D: ⟨⟩—Br ; E: ⟨⟩(Br)—NO₂ + Br—⟨⟩—NO₂ ; F: ⟨⟩(Br)—NH₂ + Br—⟨⟩—NH₂

c) G: ⟨⟩—NH₂ ; H: ⟨⟩—NH—C(=O)CH₃ ; I: ⟨⟩(Br)—NHC(=O)CH₃ + ⟨⟩—NH—C(=O)CH₃(Br) ; J: ⟨⟩(Br)—NH₂ + ⟨⟩—NH₂(Br)

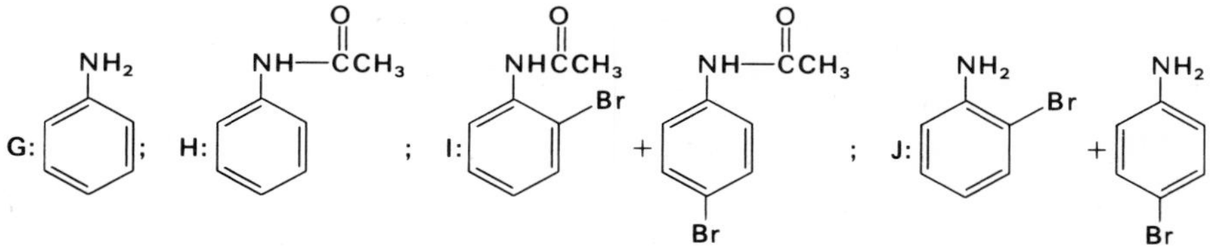

12.17 a) 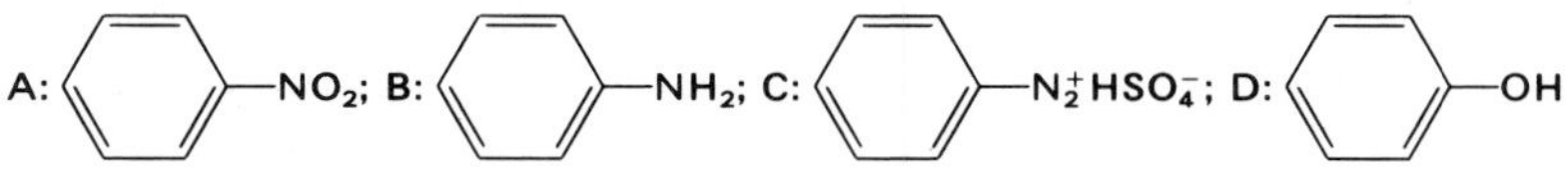 A: ⟨⟩—NO₂; B: ⟨⟩—NH₂; C: ⟨⟩—N₂⁺HSO₄⁻; D: ⟨⟩—OH

b) E: ⟨⟩—NO₂(Br); F: ⟨⟩—NH₂(Br); G: ⟨⟩—N₂⁺Br⁻(Br); H: ⟨⟩—Br(Br)

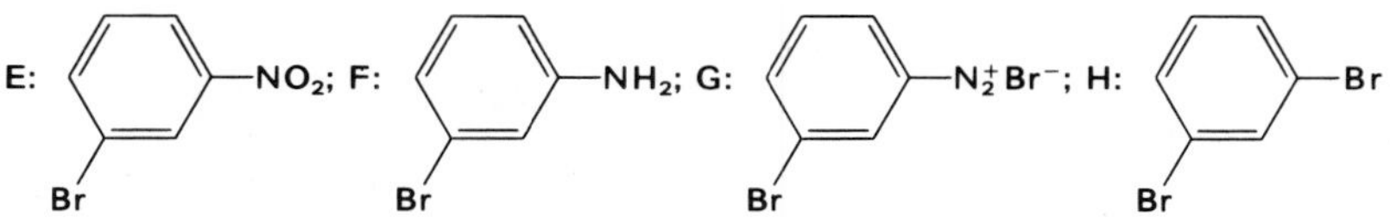

c) 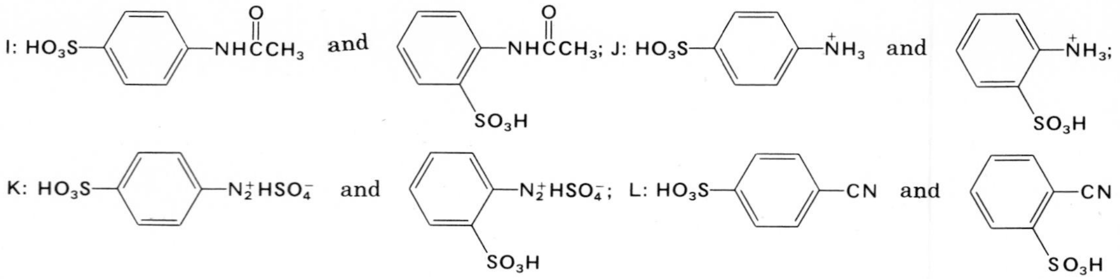 I: HO_3S—⬡—$NHCCH_3$ (with C=O) and ⬡(with SO_3H ortho)—$NHCCH_3$ (C=O); J: HO_3S—⬡—$\overset{+}{N}H_3$ and ⬡(with SO_3H ortho)—$\overset{+}{N}H_3$;

K: HO_3S—⬡—$N_2^+HSO_4^-$ and ⬡(with SO_3H ortho)—$N_2^+HSO_4^-$; L: HO_3S—⬡—CN and ⬡(with SO_3H ortho)—CN

12.18 one biphenyl and two naphthalene ring systems; amino, sodium sulfonate, and azo ($—N=N—$) groups

12.19 ⬡—$\overset{..}{N}=\overset{..}{N}$—⬡—$\overset{..}{N}(CH_3)_2$ ⟷ ⬡—N=N—⬡—$\overset{+}{N}(CH_3)_2$ ⟷

⬡—N—N—⬡—$\overset{+}{N}(CH_3)_2$ ⟷ ⬡—$\overset{..}{N}$—N—⬡—$\overset{+}{N}(CH_3)_2$ etc.

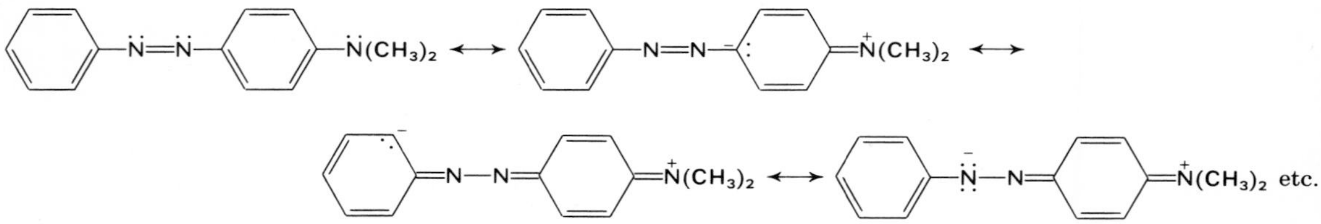

12.20 a) 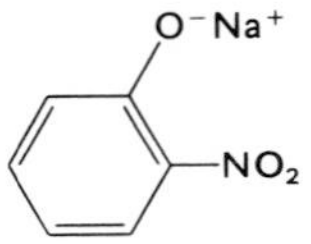 2-nitrophenol sodium salt (O^-Na^+ and NO_2 ortho)   **b)** 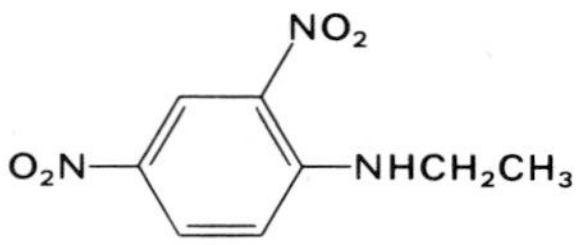 O_2N—⬡(NO_2, $NHCH_2CH_3$)   **c)** 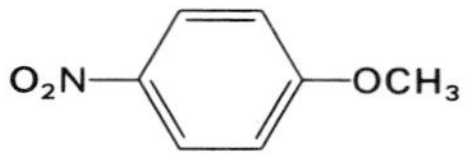 O_2N—⬡—OCH_3

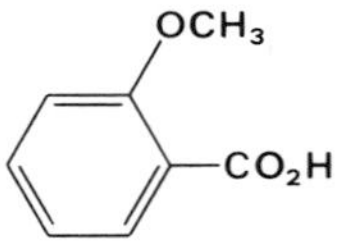

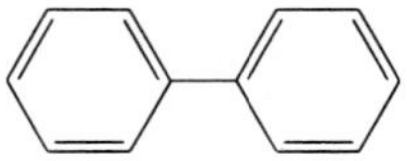

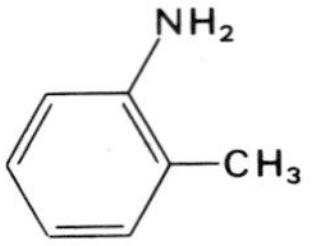

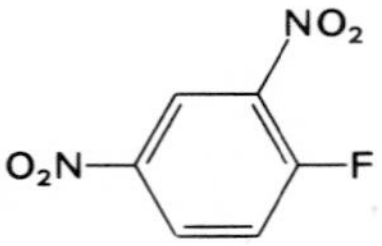

12.21 a) 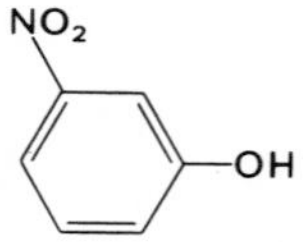 NO_2—⬡—OH (meta)   **b)** 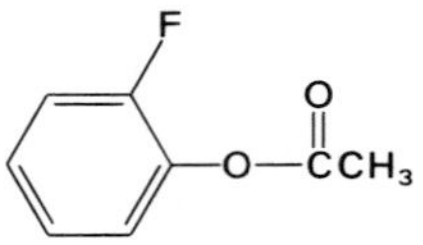 ⬡(F, O—$\overset{O}{C}CH_3$)   **c)** ⬡(OCH_3, CO_2H)

d) $H_3\overset{+}{N}$—⬡—SO_3^- **e)** ⬡—⬡ (biphenyl)   **f)** ⬡(NH_2, CH_3)

g) O_2N—⬡(NO_2, F)   **h)** Cl—⬡—Cl   **i)** 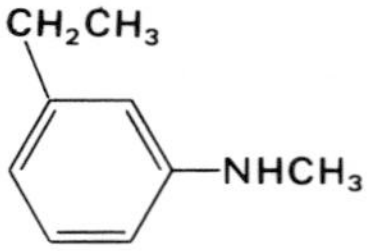 ⬡(CH_2CH_3, $NHCH_3$)

12.22 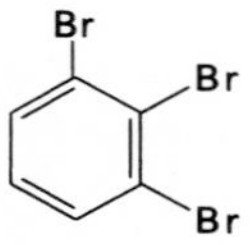 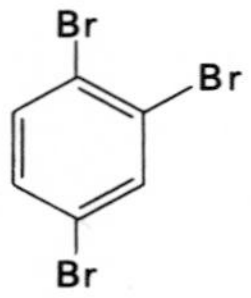 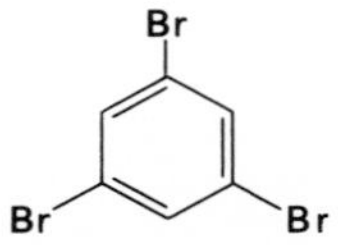 (a) tribromobenzene, (b) tribromobenzene, (c) 1,3,5-tribromobenzene

There are three tribromobenzene isomers. From *ortho*-dibromobenzene (a) and (b) are possible. From *meta* all three. From *para* only (b) is possible.

(a)   (b)   (c)

12.23

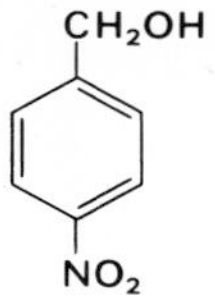

Structure I: para-CH₂OH with NO₂
Structure A: para-CH₂OH with NH₂
Structure B: para-CH₂OH with N₂⁺
Structure C: para-CH₂OH with CN
Structure D: para-CH₂OH with CO₂H

Labels: I 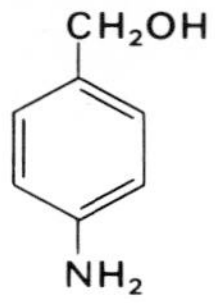   A 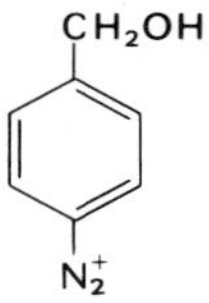   B 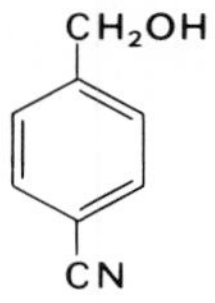   C 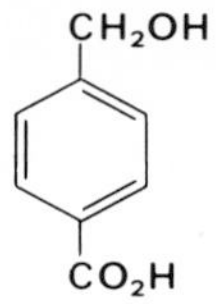   D

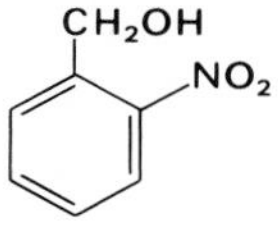

Structure II: ortho-CH₂OH with NO₂
Structure F: ortho-CH₂OH with NH₂
Structure G: ortho-CH₂OH with N₂⁺
Structure H: ortho-CH₂OH with CN
Structure J: lactone ring

Labels: II 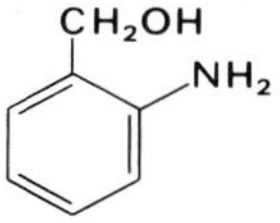   F 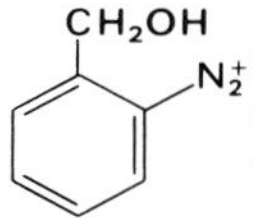   G 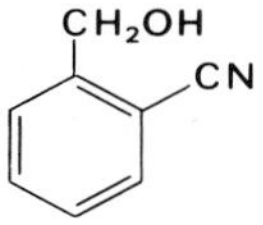   H 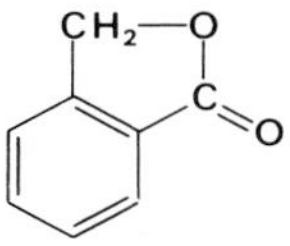   J

12.24 a)

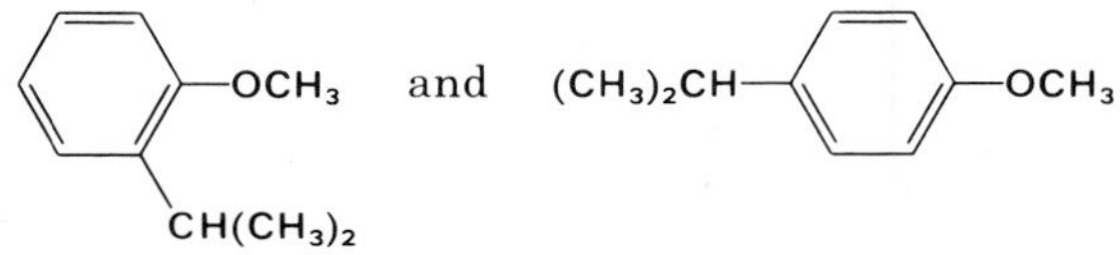

$-OCH_3$ with $CH(CH_3)_2$ and $(CH_3)_2CH-$ benzene $-OCH_3$

b)

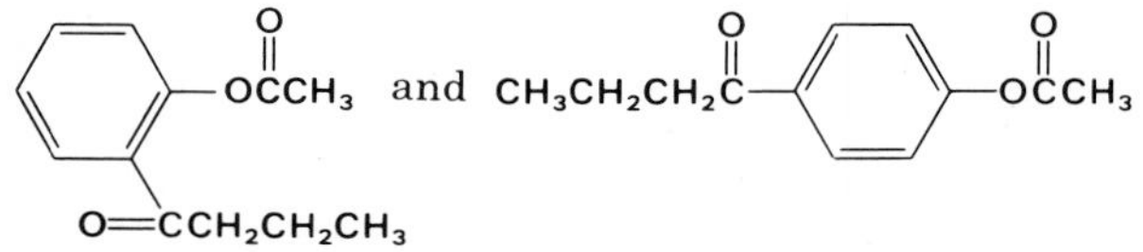

$-O\overset{O}{\overset{\|}{C}}CH_3$ with $O=\overset{}{C}CH_2CH_2CH_3$ and $CH_3CH_2CH_2\overset{O}{\overset{\|}{C}}-$ benzene $-O\overset{O}{\overset{\|}{C}}CH_3$

c) benzene $-\overset{O}{\overset{\|}{C}}-OH$ with NO_2

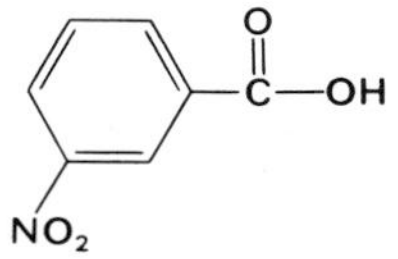

d) benzene $-\overset{O}{\overset{\|}{C}}CH_3$ with Br

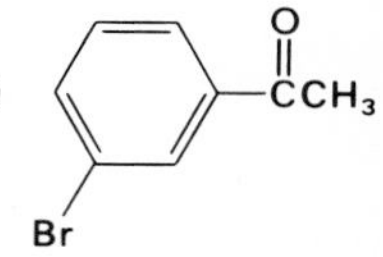

e) HO_3S- benzene $-NH\overset{O}{\overset{\|}{C}}CH_3$ and   benzene $-NH\overset{O}{\overset{\|}{C}}CH_3$ with SO_3H

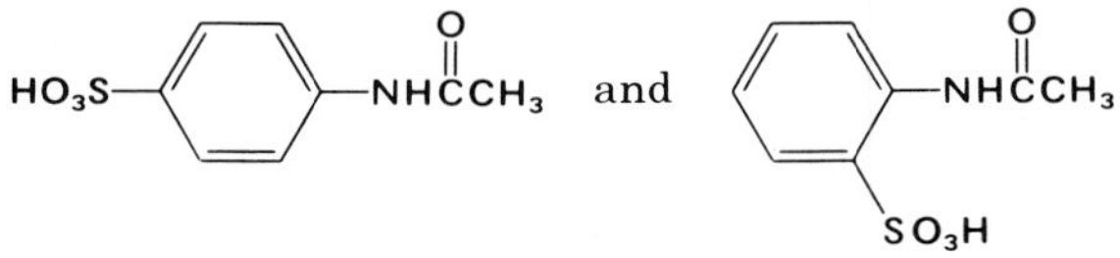

f) benzene $-NHCH_2CH_2CH_3$ with NO_2

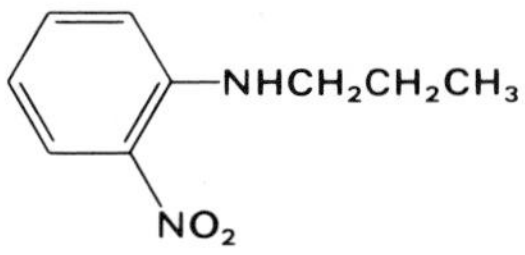

12.25 a) A: benzene $-\overset{O}{\overset{\|}{C}}CH_2CH_2CO_2H$;   B: benzene $-CH_2CH_2CH_2CO_2H$;   C: benzene $-CH_2CH_2CH_2\overset{O}{\overset{\|}{C}}Cl$;

D: ; E:

b) F: benzene $-NO_2$ with Br;   G: benzene $-NH_2$ with Br;   H: benzene $-N_2^+$ with Br;   I: benzene $-OH$ with Br

J: benzene $-OCH_3$ with Br;   K: benzene $-OCH_3$ with $MgBr$;   L: benzene $-OCH_3$ with CO_2H

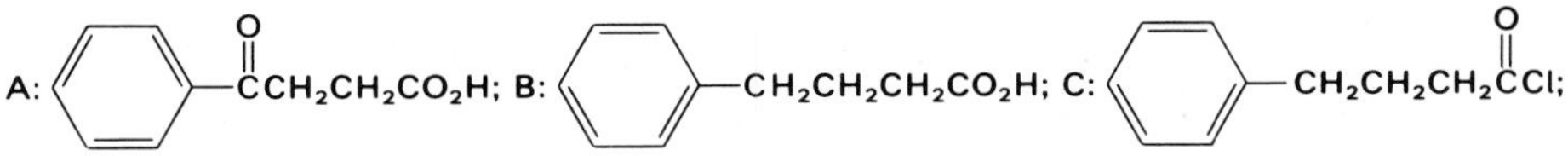

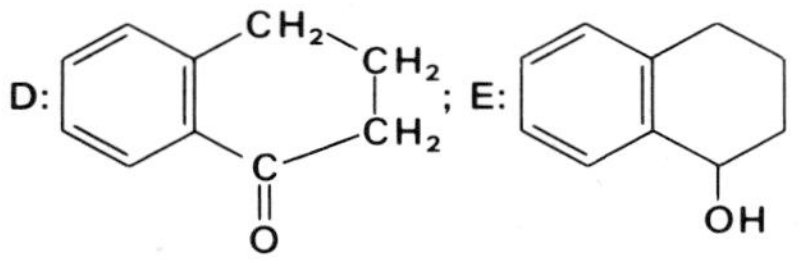

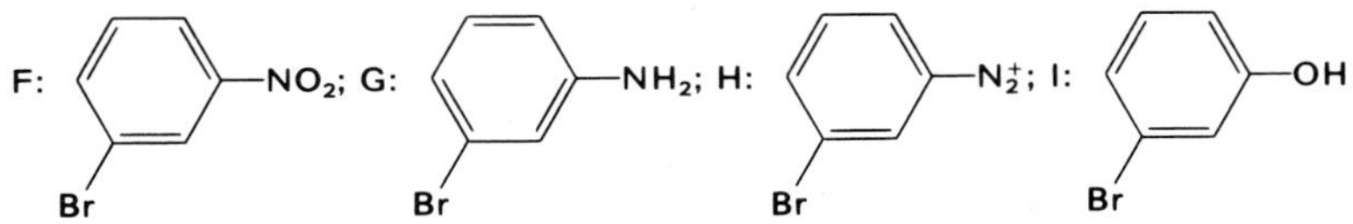

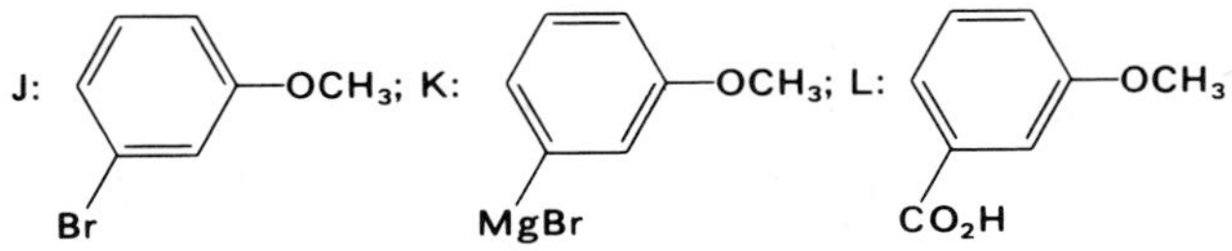

12.26 There are seven amide groups on side chains to the porphyrin ring structure and four nitrogens in the ring bonding to the metal; nine asymmetric carbons of which six are adjacent carbons.

12.27 a) 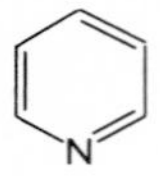 **b)** 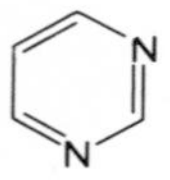 **c)** 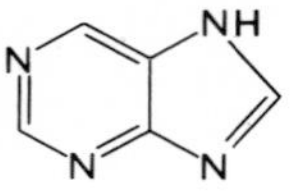 **d)** 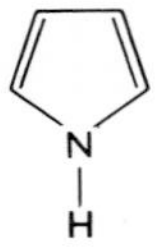 **e)**

12.28 a) nicotinamide and NAD$^+$ **b)** thymine, uracil, cytosine **c)** adenine, guanine
 d) chlorophyll, hemin **e)** histidine, vitamin B$_{12}$

12.29 a)

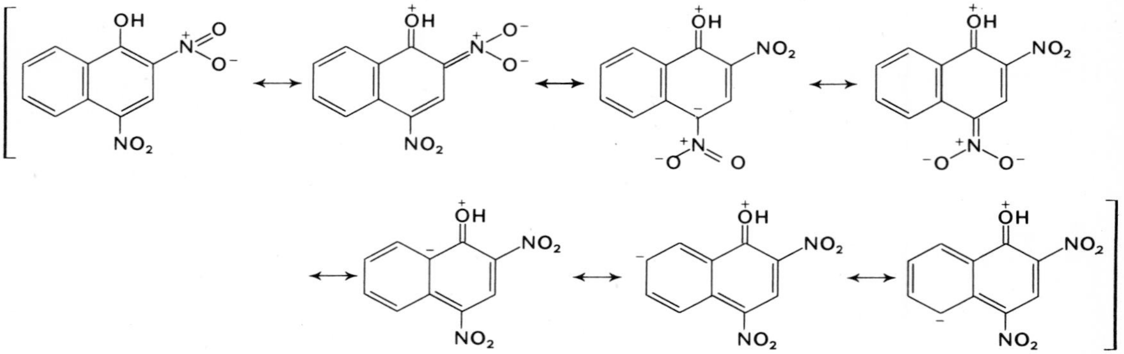

b)

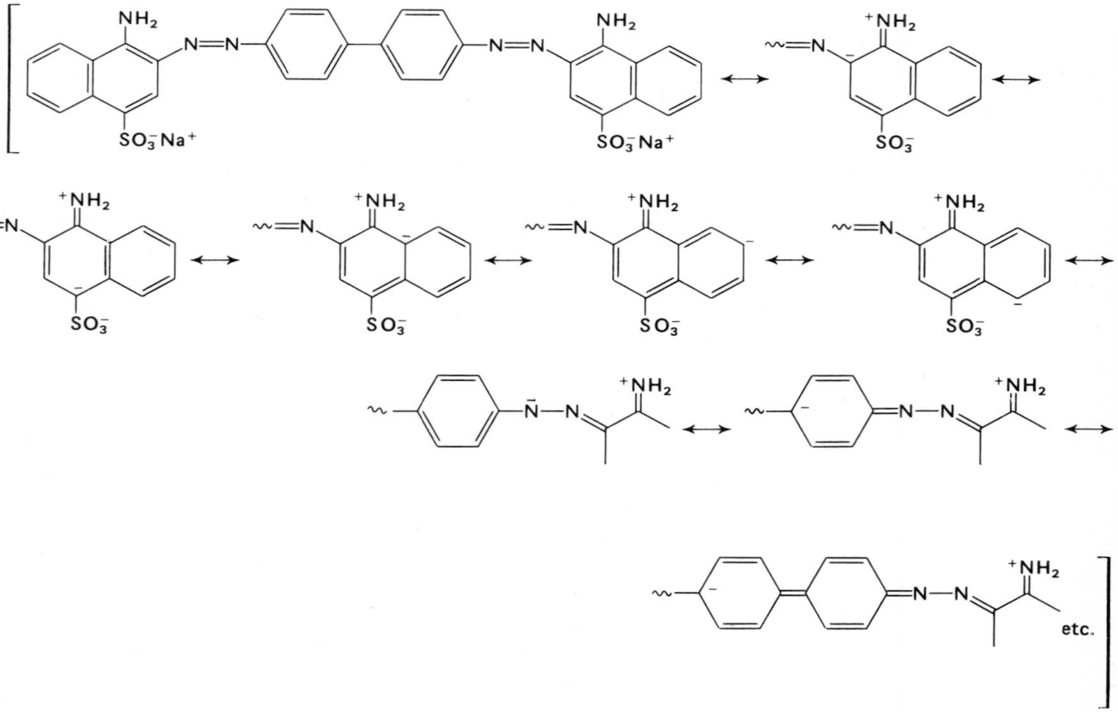

12.30 a) $(CH_3)_2C{=}CH_2 + H_2SO_4 \longrightarrow (CH_3)_2\overset{+}{C}CH_3 \ HSO_4^- \xrightarrow{\ C_6H_6\ }$

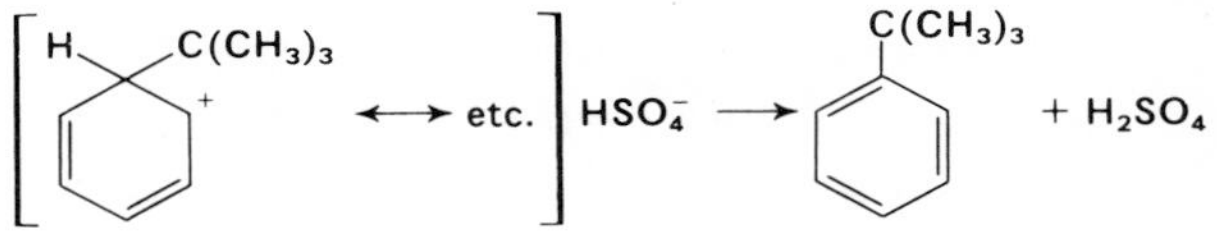

The addition reaction to form the carbonium ion follows Markownikoff's rule.

b)

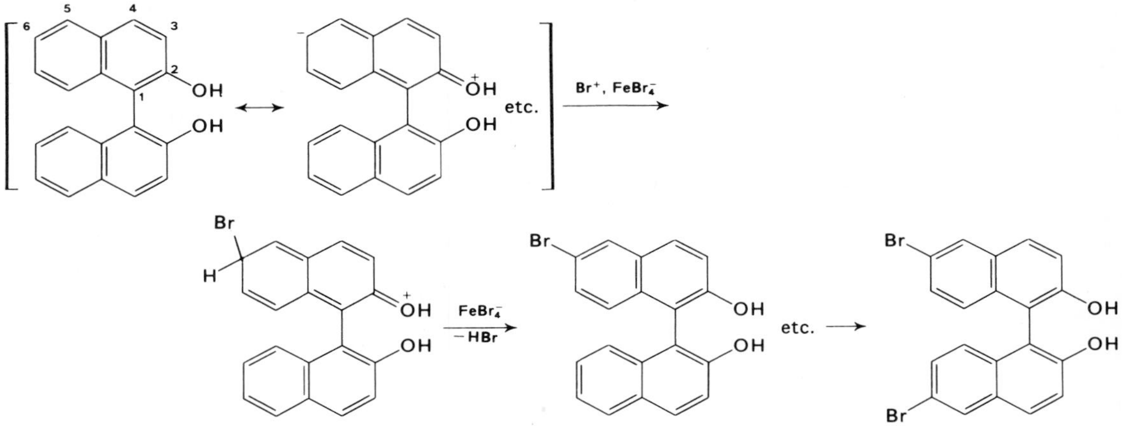

The 6,6'-positions are most activated, followed by 3,3'. These positions distribute negative charge by delocalization. The 8-position is hindered and difficult for a group to enter.

c) Attack at the two *ortho* positions is sterically blocked. Reaction intermediates for the *meta* attack provide resonance structures which allow the two positive charges to be farthest apart. Attack at the *para* position gives reaction intermediate with charges close together. *Meta* substitution predominates but reaction is slow; $-NH_3^+$ is a *meta*-director with deactivation.

d) In dilute acid there is enough free aniline to undergo the nitration reaction. The NH_2 group is *ortho-para*-directing with activation. With concentrated acid, most of the aniline is in the anilinium form, which is deactivating and *meta*-directing, as seen in part (c) above.

e) Methyl groups activate benzene rings toward further electrophilic substitution, and the toluene initially produced goes to xylenes. Acetyl group deactivates the ring toward further electrophilic attack.

f) Alkyl groups on benzene are usually *ortho-para*-directing. However the tert-butyl group is so large that attack on the *ortho* position is sterically hindered; *para*-product is major product.

Chapter 13

13.1 $HOCH_2\overset{*}{C}HOH\overset{*}{C}HOH\overset{*}{C}HOH\overset{*}{C}HOHCH{=}O$; $n = 4$; $2^n = 2^4 = 16$ stereoisomers

13.2 a)

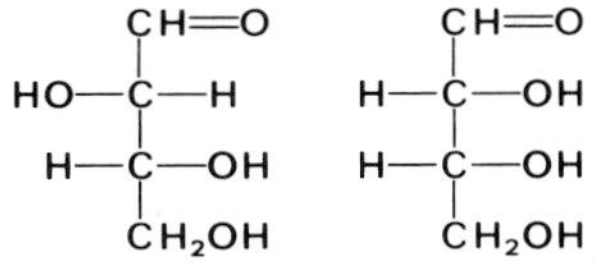

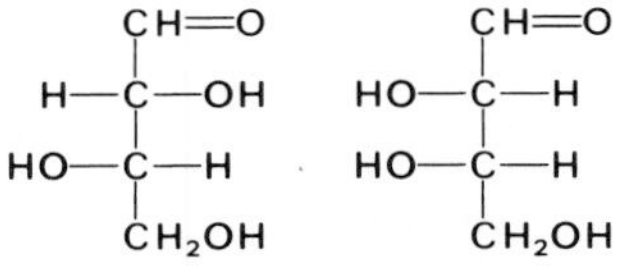

13.3 A and D are L-isomers; B and C are D-isomers; A and C are enantiomers. (Enantiomers of B and D not shown.)

13.4

CH=O   1) HCN   CO₂H   CO₂H   Na(Hg)   CH=O   CH=O
HO—C—H ——→   HO—C—H   +   H—C—OH   ——→   HO—C—H   +   H—C—OH
CH₂OH   2) H₃O⁺ HO—C—H   HO—C—H   HO—C—H   HO—C—H
                 CH₂OH   CH₂OH   CH₂OH   CH₂OH

L-glyceraldehyde   diastereoisomeric acids   diastereoisomeric L-aldotetroses

HNO₃ ↓   ↓ HNO₃

CO₂H   CO₂H
HO—C—H   H—C—OH
HO—C—H   HO—C—H
CO₂H   CO₂H

meso-tartaric acid   L-(+)-tartaric acid

13.5 Use D-erythrose to start the syntheses of both D-ribose and D-glucose; D-erythrose has both hydroxyls on the right side in the Fischer projection. Both D-ribose and D-glucose have the *lower* two hydroxyls on the right also.

13.6 a)

CH=O   1) HCN   CH=O   CH=O
—OH   2) H₃O⁺   —OH   HO—
—OH   3) Na(Hg)   —OH   —OH
CH₂OH   —OH   —OH
                   CH₂OH   CH₂OH

D-erythrose   D-ribose   D-arabinose

Oxidation of D-ribose by nitric acid gives a *meso*-dicarboxylic acid; D-arabinose gives an optically active one. (Simplified Fischer formulas show side that the hydroxyl is on for asymmetric carbons.)

b)

CH=O   1) HCN   CH=O   CH=O
HO—   2) H₃O⁺   —OH   HO—
—OH   3) Na(Hg) HO—   HO—
—OH   —OH   —OH
CH₂OH   —OH   —OH
                  CH₂OH   CH₂OH

D-arabinose   D-glucose   D-mannose

Oxidation of glucose by nitric acid gives an optically active dicarboxylic acid; mannose gives a *meso*-dicarboxylic acid.

13.7 $2^3 = 8$ aldopentoses

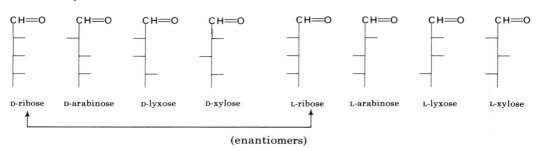

CH=O   CH=O   CH=O   CH=O   CH=O   CH=O   CH=O   CH=O

D-ribose D-arabinose D-lyxose D-xylose   L-ribose L-arabinose L-lyxose L-xylose

(enantiomers)

13.8 α-anomer ($[\alpha]_D = 112°$) present as 35%; β-anomer ($[\alpha]_D = 19°$) as 65%; in equilibrium of $[\alpha]_D = 52°$

13.9

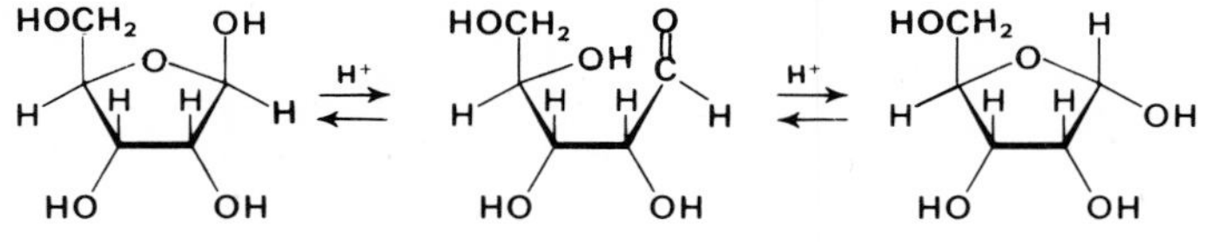

β   ⇌ (H⁺) ⇌   ⇌ (H⁺) ⇌   α

13.10

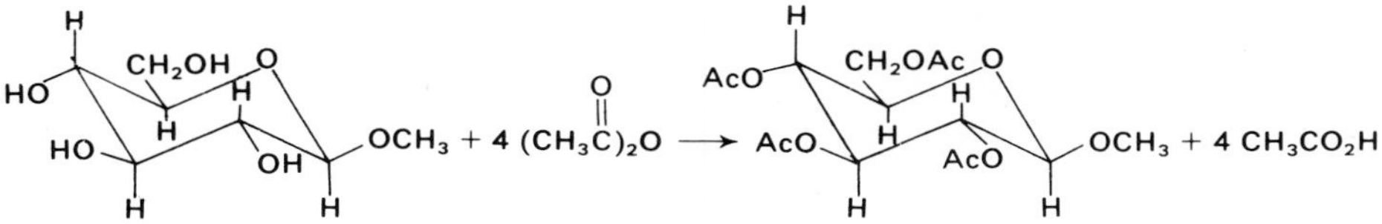

$$\text{HO} \cdots \text{CH}_2\text{OH} \cdots \text{OCH}_3 + 4\,(CH_3\overset{O}{C})_2O \longrightarrow \text{AcO} \cdots \text{CH}_2\text{OAc} \cdots \text{OCH}_3 + 4\,CH_3CO_2H$$

13.11

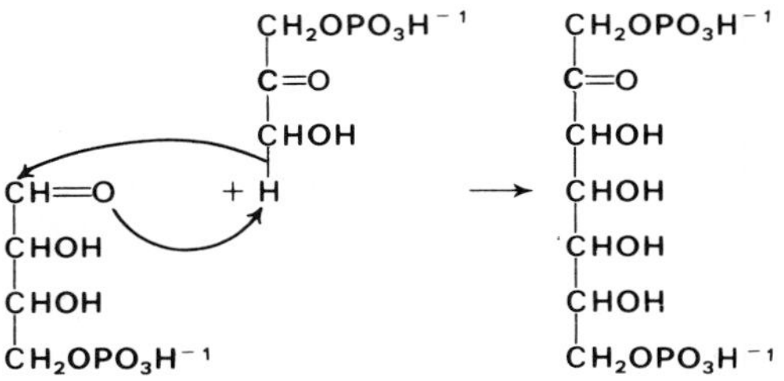

$$\begin{array}{l} CH_2OPO_3H^{-1} \\ C{=}O \\ CHOH \end{array} \quad + \quad \begin{array}{l} CH{=}O \\ CHOH \\ CHOH \\ CH_2OPO_3H^{-1} \end{array} \quad +\,H \longrightarrow \quad \begin{array}{l} CH_2OPO_3H^{-1} \\ C{=}O \\ CHOH \\ CHOH \\ CHOH \\ CHOH \\ CH_2OPO_3H^{-1} \end{array}$$

13.12

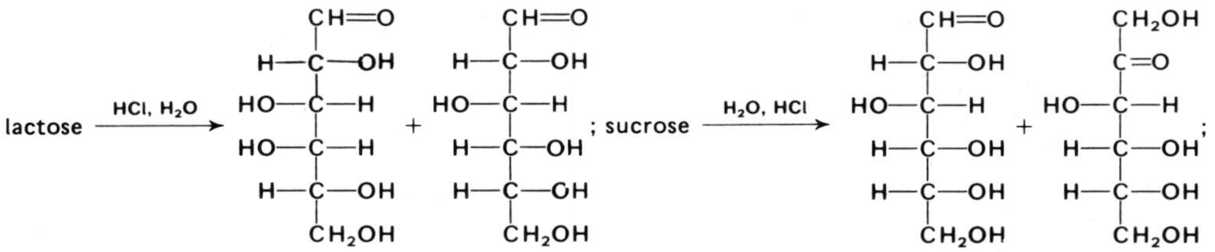

lactose $\xrightarrow{\text{HCl, H}_2\text{O}}$

CH=O
H—C—OH
HO—C—H
HO—C—H
H—C—OH
CH₂OH
D-galactose

+

CH=O
H—C—OH
HO—C—H
H—C—OH
H—C—OH
CH₂OH
D-glucose

; sucrose $\xrightarrow{\text{H}_2\text{O, HCl}}$

CH=O
H—C—OH
HO—C—H
H—C—OH
H—C—OH
CH₂OH
D-glucose

+

CH₂OH
C=O
HO—C—H
H—C—OH
H—C—OH
CH₂OH
D-fructose

;

cellobiose $\xrightarrow{\text{H}_2\text{O, HCl}}$ 2 **D-glucose**

13.13

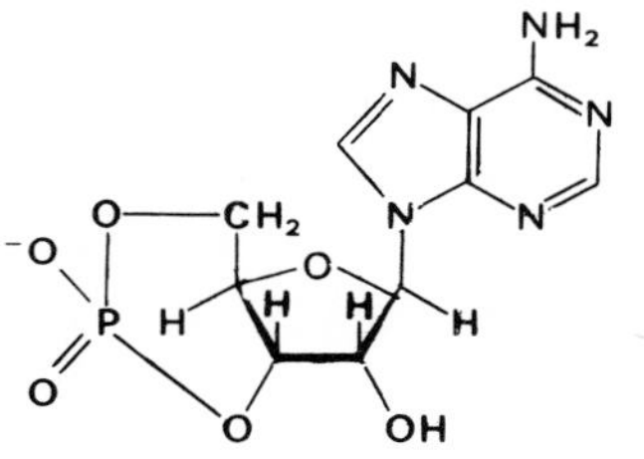

cyclic AMP

13.14

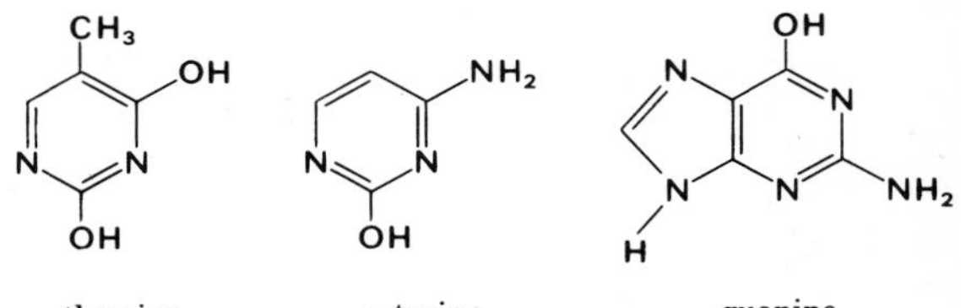

tautomeric forms of: **thymine**   **cytosine**   **guanine**

13.15

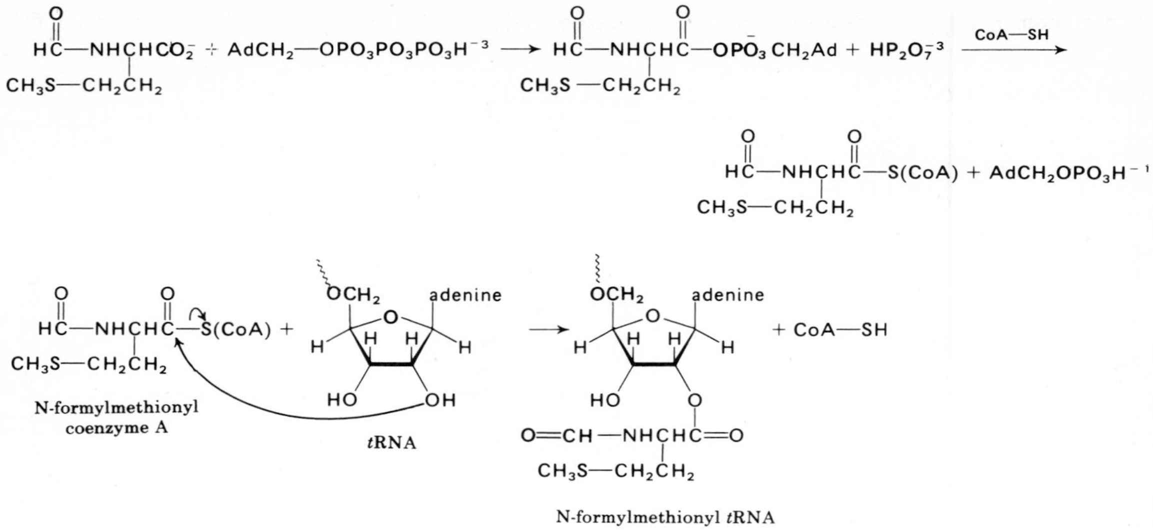

$$HC{-}NHCHCO_2^- + AdCH_2{-}OPO_3PO_3PO_3H^{-3} \longrightarrow HC{-}NHCHC{-}OPO_3^-CH_2Ad + HP_2O_7^{-3} \xrightarrow{\text{CoA}-\text{SH}}$$

$$HC{-}NHCHC{-}S(CoA) + AdCH_2OPO_3H^{-1}$$

N-formylmethionyl coenzyme A + *t*RNA ⟶ N-formylmethionyl *t*RNA + CoA—SH

13.16

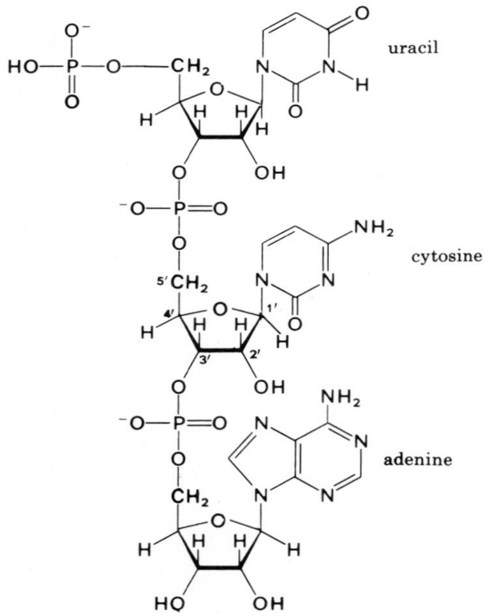

uracil

cytosine

adenine

13.17

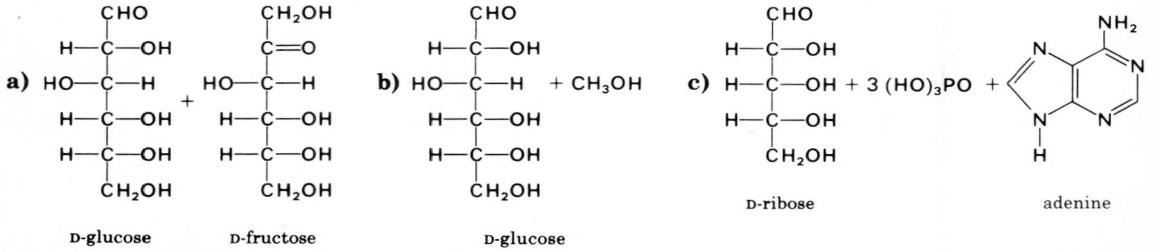

a) D-glucose + D-fructose

b) D-glucose + CH_3OH

c) D-ribose + 3 $(HO)_3PO$ + adenine

13.18 a)

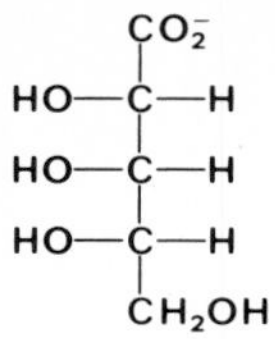

$$
\begin{array}{c}
CO_2^- \\
HO-C-H \\
HO-C-H \\
HO-C-H \\
CH_2OH
\end{array}
$$

b)

$$
\begin{array}{c}
CH_2OH \\
H-C-OH \\
HO-C-H \\
H-C-OH \\
H-C-OH \\
CH_2OH
\end{array}
+
\begin{array}{c}
CH_2OH \\
HO-C-H \\
HO-C-H \\
H-C-OH \\
H-C-OH \\
CH_2OH
\end{array}
$$

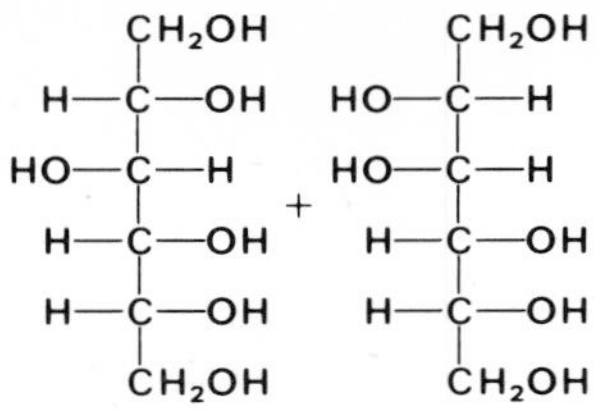

c)

$$
\begin{array}{c}
CO_2H \\
H-C-OH \\
H-C-OH \\
H-C-OH \\
CO_2H
\end{array}
$$

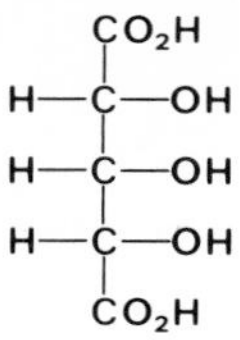

13.19 a)

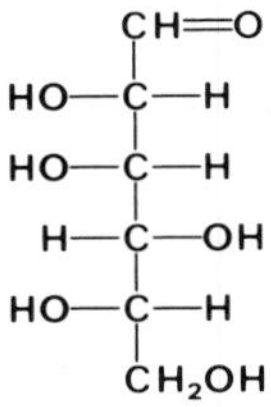

$$
\begin{array}{c}
CH=O \\
HO-C-H \\
HO-C-H \\
H-C-OH \\
HO-C-H \\
CH_2OH
\end{array}
$$

an L-aldohexose

b)

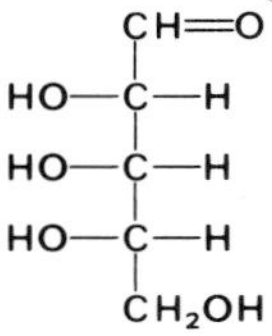

$$
\begin{array}{c}
CH=O \\
HO-C-H \\
HO-C-H \\
HO-C-H \\
CH_2OH
\end{array}
$$

L-ribose

13.20 a)

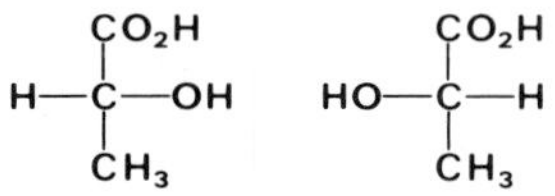

$$
\begin{array}{c}
CO_2H \\
H-C-OH \\
CH_3
\end{array}
\qquad
\begin{array}{c}
CO_2H \\
HO-C-H \\
CH_3
\end{array}
$$

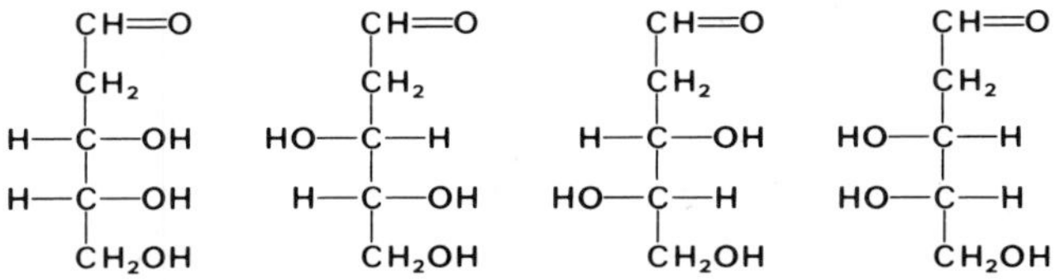

b)

$$
\begin{array}{c}
CH=O \\
CH_2 \\
H-C-OH \\
H-C-OH \\
CH_2OH
\end{array}
\quad
\begin{array}{c}
CH=O \\
CH_2 \\
HO-C-H \\
H-C-OH \\
CH_2OH
\end{array}
\quad
\begin{array}{c}
CH=O \\
CH_2 \\
H-C-OH \\
HO-C-H \\
CH_2OH
\end{array}
\quad
\begin{array}{c}
CH=O \\
CH_2 \\
HO-C-H \\
HO-C-H \\
CH_2OH
\end{array}
$$

c)

$$
\begin{array}{c}
CHO \\
H-C-OH \\
CH_2OCH_3
\end{array}
$$

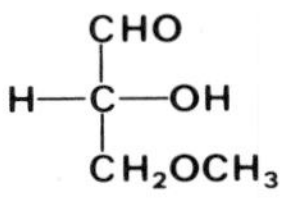

d)

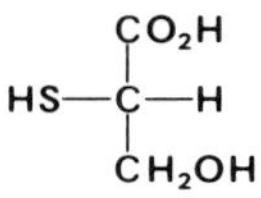

$$
\begin{array}{c}
CO_2H \\
HS-C-H \\
CH_2OH
\end{array}
$$

e)

$$
\begin{array}{c}
CH_2OH \\
| \\
| \\
CH_2OH
\end{array}
\quad
\begin{array}{c}
CH_2OH \\
| \\
| \\
CH_2OH
\end{array}
\quad
\begin{array}{c}
CH_2OH \\
| \\
| \\
CH_2OH
\end{array}
\quad
\begin{array}{c}
CH_2OH \\
| \\
| \\
CH_2OH
\end{array}
$$

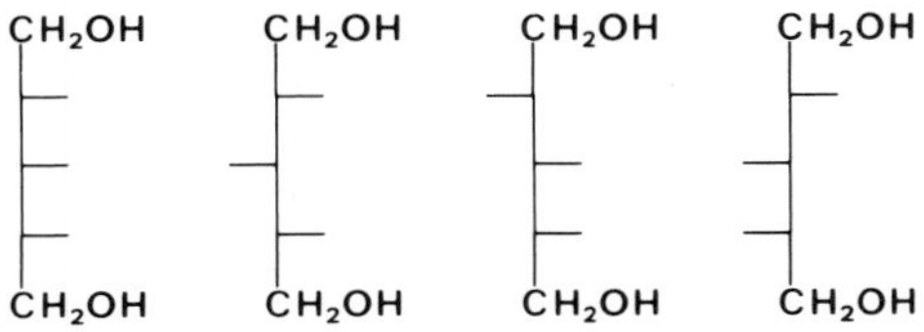

meso *meso* enantiomers

f)

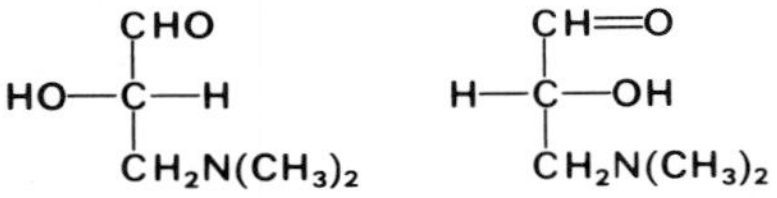

$$
\begin{array}{c}
CHO \\
HO-C-H \\
CH_2N(CH_3)_2
\end{array}
\qquad
\begin{array}{c}
CH=O \\
H-C-OH \\
CH_2N(CH_3)_2
\end{array}
$$

13.21 a)

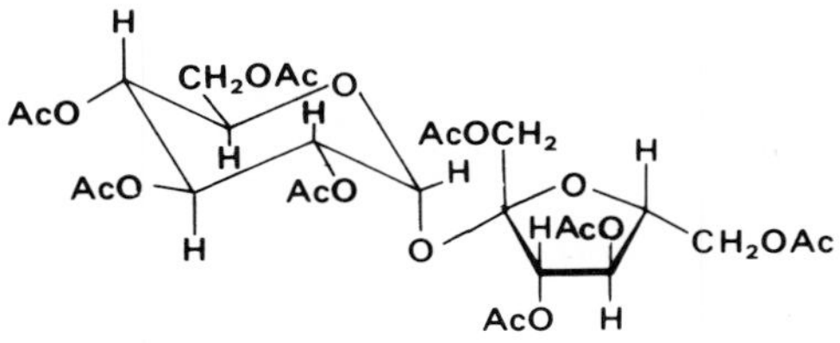

b)

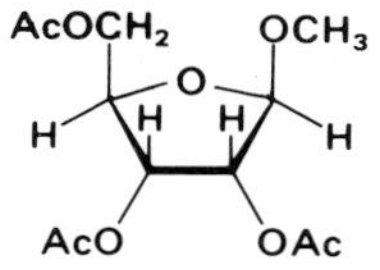

c)

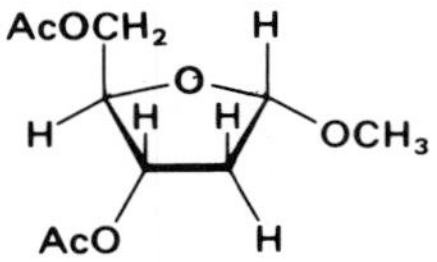

13.22 A → T → G; G → T → A; A → G → T; T → G → A; T → A → G; G → A → T

13.23

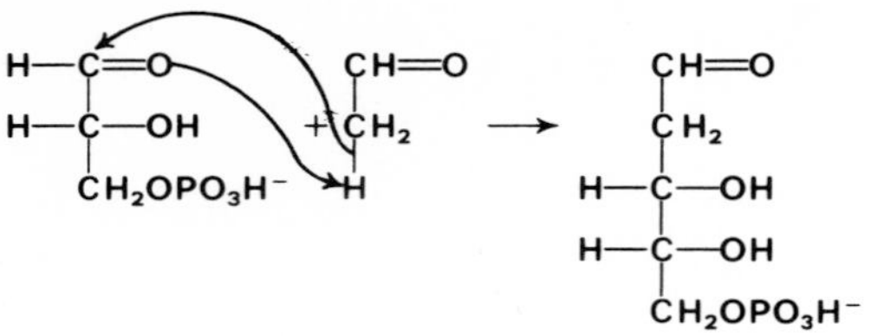

13.24

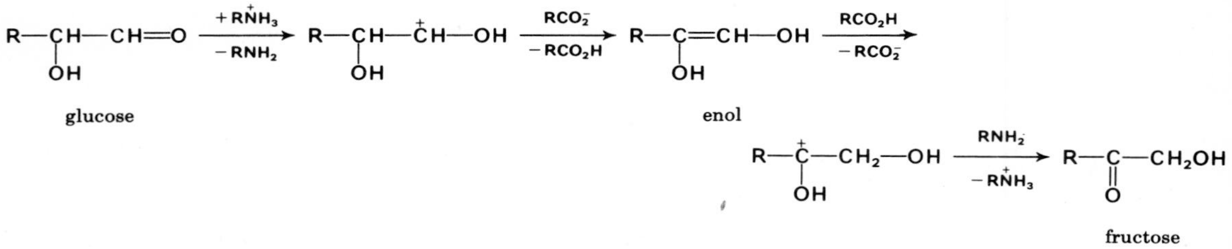

$$R-\overset{\underset{\displaystyle OH}{|}}{CH}-CH=O \xrightarrow[-RNH_2]{+R\overset{+}{N}H_3} R-\overset{\underset{\displaystyle OH}{|}}{CH}-\overset{+}{CH}-OH \xrightarrow[-RCO_2H]{RCO_2^-} R-\overset{\underset{\displaystyle OH}{|}}{C}=CH-OH \xrightarrow[-RCO_2^-]{RCO_2H}$$

glucose   enol

$$R-\overset{\underset{\displaystyle OH}{|}}{\overset{+}{C}}-CH_2-OH \xrightarrow[-R\overset{+}{N}H_3]{RNH_2} R-\overset{\underset{\displaystyle O}{\|}}{C}-CH_2OH$$

fructose

13.25 reduce $HC=O$ to CH_2OH; oxidize CH_2OH to CO_2H; dehydrogenation to dienol

Chapter 14

14.1 a)

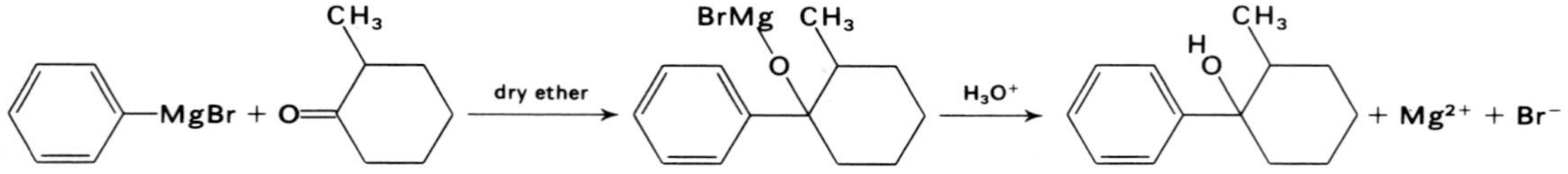

b)

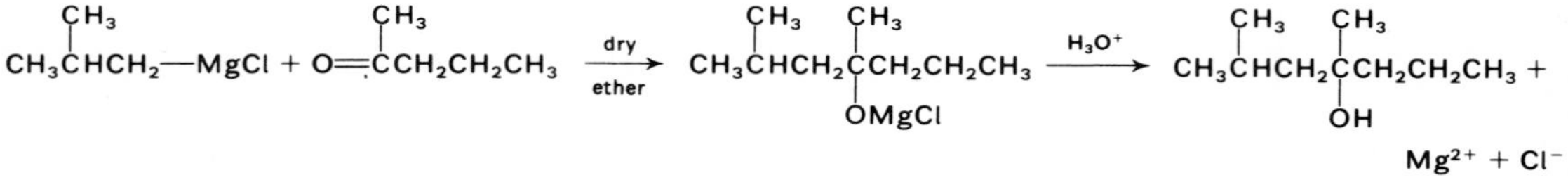

c)

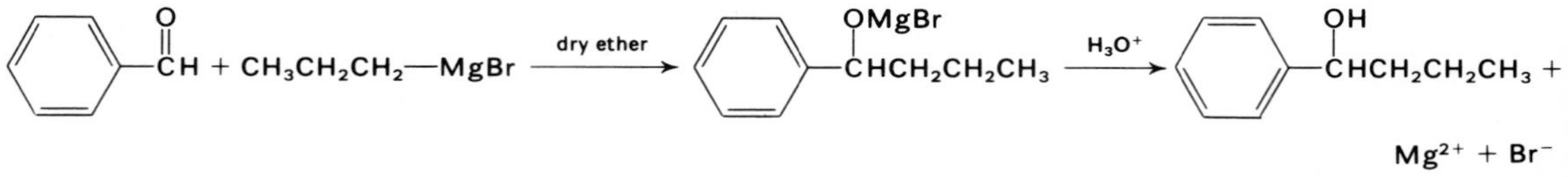

d)

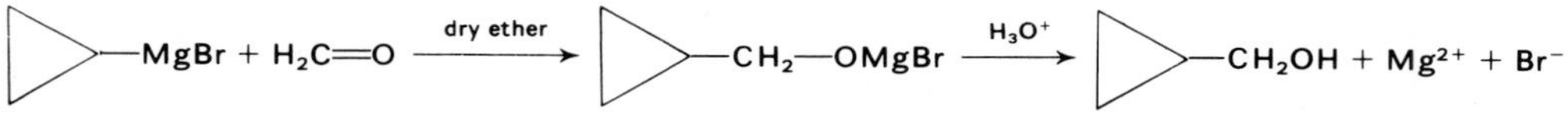

14.2 a)

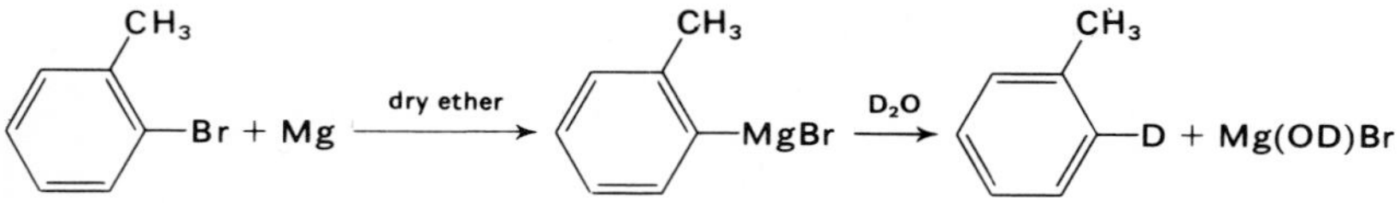

b)

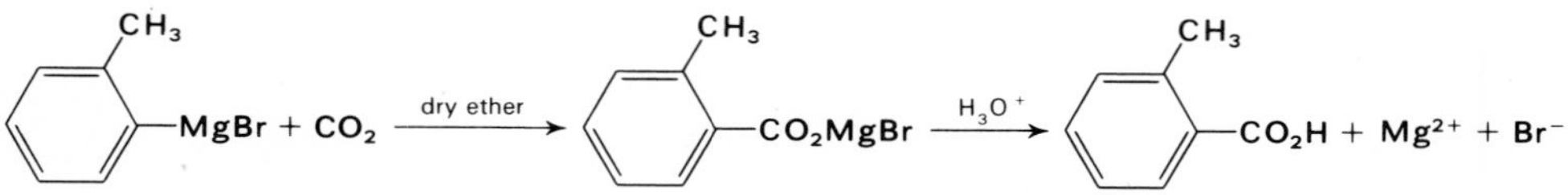

14.3 a) $CH_3-\overset{\displaystyle OH}{\underset{\displaystyle CH_2CH(CH_3)_2}{\underset{|}{\overset{|}{C}}}}-CH_2CH(CH_3)_2 + CH_3CH_2OH$ **b)** ![phenyl]$-\overset{\displaystyle OH}{\underset{\displaystyle (CH_2)_3CH_3}{\underset{|}{\overset{|}{C}}}}-(CH_2)_3CH_3 + CH_3OH$

14.4 $CH_3\overset{\displaystyle CH_3}{\underset{|}{C}}=CHCH_2CH_2\overset{\displaystyle CH_3}{\underset{|}{C}}=CH-\overset{\displaystyle OH}{\underset{|}{C}H}-CH_2\overset{\displaystyle O}{\overset{\|}{C}}CH_3$ ketone carbanion adding to aldehyde C=O

14.5 The ethoxide ion $CH_3CH_2-O^-$ is a stronger base, by a factor of 100, than is methoxide ion CH_3O^-. The greater strength of the base causes the acid–base equilibrium which forms the first carbanion to be shifted further in favor of forming the ester carbanion.

14.6 The 1,3-diketone is a much stronger acid ($pK_a = 11$) than ethanol ($pK_a = 18$). Thus the proton transfer between diketone and ethoxide ion goes nearly to completion of formation of the carbanion, which is stabilized by charge delocalization to two oxygens.

14.7 a) $CH_3CH_2CH_2\overset{\displaystyle O}{\overset{\|}{C}}\overset{\displaystyle O}{\underset{\displaystyle CH_3CH_2}{\underset{|}{C}}H}\overset{\|}{C}OC_2H_5$ **b)** ![cyclopentanone ring]$-\overset{\displaystyle O}{\overset{\|}{C}}-OC_2H_5$

14.8 A: $CH_3(CH_2)_2CH=CH-\overset{\displaystyle O}{\overset{\|}{C}}-S(CoA)$;   B: $CH_3(CH_2)_2\overset{\displaystyle OH}{\underset{|}{C}}HCH_2\overset{\displaystyle O}{\overset{\|}{C}}-S(CoA)$;   C: $CH_3(CH_2)_2\overset{\displaystyle O}{\overset{\|}{C}}CH_2\overset{\displaystyle O}{\overset{\|}{C}}-S(CoA)$;

D: $CH_3(CH_2)_2\overset{\displaystyle O}{\overset{\|}{C}}-S(CoA)$;   E: $CH_3\overset{\displaystyle O}{\overset{\|}{C}}-S(CoA)$;   F: $CH_3CH=CH\overset{\displaystyle O}{\overset{\|}{C}}-S(CoA)$;   G: $CH_3\overset{\displaystyle OH}{\underset{|}{C}}HCH_2\overset{\displaystyle O}{\overset{\|}{C}}-S(CoA)$;

H: $CH_3\overset{\displaystyle O}{\overset{\|}{C}}CH_2\overset{\displaystyle O}{\overset{\|}{C}}-S(CoA)$

14.9 A: $CH_3\overset{\displaystyle O}{\overset{\|}{C}}\overset{\displaystyle O}{\overset{\|}{C}}HC-OC_2H_5$;   B: $CH_3\overset{\displaystyle O}{\overset{\|}{C}}\overset{\displaystyle O}{\underset{\displaystyle CH_2CH_2CH_3}{\underset{|}{C}}H}\overset{\|}{C}-OC_2H_5$;   C: $CH_3\overset{\displaystyle O}{\overset{\|}{C}}\underset{\displaystyle CH_2CH_2CH_3}{\underset{|}{C}}HCO_2^-$; D: $CH_3\overset{\displaystyle O}{\overset{\|}{C}}CH_2CH_2CH_2CH_3 + CO_2$

14.10 $R-CH_2\overset{\displaystyle CO_2CH_3}{\underset{\displaystyle CH_3}{\underset{|}{\overset{|}{C}}}}-CH_2\overset{\displaystyle \cdot}{\underset{\displaystyle C_6H_5}{\underset{|}{C}H}} + CH_2=\overset{\displaystyle CO_2CH_3}{\underset{\displaystyle CH_3}{\underset{|}{\overset{|}{C}}}} \longrightarrow R-CH_2\overset{\displaystyle CO_2CH_3}{\underset{\displaystyle CH_3}{\underset{|}{\overset{|}{C}}}}-CH_2\overset{\displaystyle }{\underset{\displaystyle C_6H_5}{\underset{|}{C}H}}-CH_2\overset{\displaystyle CO_2CH_3}{\underset{\displaystyle CH_3}{\underset{|}{\overset{|}{C}}}}\cdot \xrightarrow{C_6H_5CH=CH_2}$

$R-CH_2\overset{\displaystyle CO_2CH_3}{\underset{\displaystyle CH_3}{\underset{|}{\overset{|}{C}}}}-CH_2\overset{\displaystyle }{\underset{\displaystyle C_6H_5}{\underset{|}{C}H}}-CH_2\overset{\displaystyle CO_2CH_3}{\underset{\displaystyle CH_3}{\underset{|}{\overset{|}{C}}}}-CH_2\overset{\displaystyle }{\underset{\displaystyle C_6H_5}{\underset{|}{C}H}}\cdot$

14.11 a) $[CH_3CH_2\overset{\displaystyle H}{\underset{|}{\overset{|}{C}}}H-\overset{\displaystyle H}{\underset{|}{C}}=O \longleftrightarrow CH_3CH_2CH=\overset{\displaystyle H}{\underset{|}{C}}-O^-]$ **b)** $[\bar{C}H_2-\overset{\displaystyle O}{\overset{\|}{C}}-OCH_3 \longleftrightarrow CH_2=\overset{\displaystyle O^-}{\underset{|}{C}}-OCH_3]$

c) $[\bar{C}H_2-\overset{\overset{\displaystyle O}{\|}}{C}-NHCH_3 \longleftrightarrow CH_2=\overset{\overset{\displaystyle O^-}{|}}{C}-NHCH_3]$ 　　　**d)** $[CH_3\bar{C}H-C\equiv N \longleftrightarrow CH_3CH=C=N^-]$

e) $[CH_3O-\overset{\overset{\displaystyle O}{\|}}{C}-\bar{C}H-\overset{\overset{\displaystyle O}{\|}}{C}-OCH_3 \longleftrightarrow CH_3O-\overset{\overset{\displaystyle O^-}{|}}{C}=CH-\overset{\overset{\displaystyle O}{\|}}{C}-OCH_3 \longleftrightarrow CH_3O-\overset{\overset{\displaystyle O}{\|}}{C}-CH=\overset{\overset{\displaystyle O^-}{|}}{C}-OCH_3]$

f) $[CH_3\overset{\overset{\displaystyle O}{\|}}{C}-\bar{C}H_2 \longleftrightarrow CH_3\overset{\overset{\displaystyle O^-}{|}}{C}=CH_2]$

g) $[CH_3CH_2\overset{\overset{\displaystyle O}{\|}}{C}-\bar{C}H-\overset{\overset{\displaystyle O}{\|}}{C}-OCH_3 \longleftrightarrow CH_3CH_2\overset{\overset{\displaystyle O^-}{|}}{C}=CH-\overset{\overset{\displaystyle O}{\|}}{C}-OCH_3 \longleftrightarrow CH_3CH_2\overset{\overset{\displaystyle O}{\|}}{C}-CH=\overset{\overset{\displaystyle O^-}{|}}{C}-OCH_3]$

h) $[\bar{C}H_2\overset{\overset{\displaystyle O}{\|}}{C}-\overset{\overset{\displaystyle O}{\|}}{C}-OCH_3 \longleftrightarrow CH_2=\overset{\overset{\displaystyle ^-O}{|}}{C}-\overset{\overset{\displaystyle O}{\|}}{C}-OCH_3]$

i) $[CH_3\overset{\overset{\displaystyle O}{\|}}{C}-\bar{C}H-\overset{\overset{\displaystyle O}{\|}}{C}CH_3 \longleftrightarrow CH_3\overset{\overset{\displaystyle O^-}{|}}{C}=CH-\overset{\overset{\displaystyle O}{\|}}{C}CH_3 \longleftrightarrow CH_3\overset{\overset{\displaystyle O}{\|}}{C}-CH=\overset{\overset{\displaystyle O^-}{|}}{C}-CH_3]$

14.12 a) $CH_3\overset{\overset{\displaystyle CH_3}{|}}{C}H-MgCl + CH_3(CH_2)_3\overset{\overset{\displaystyle O}{\|}}{C}H \xrightarrow{\text{dry ether}} CH_3\overset{\overset{\displaystyle CH_3}{|}}{C}H-\overset{\overset{\displaystyle OMgCl}{|}}{C}H(CH_2)_3CH_3 \xrightarrow{H_3O^+}$

$CH_3\overset{\overset{\displaystyle CH_3}{|}}{C}H-\overset{\overset{\displaystyle OH}{|}}{C}H(CH_2)_3CH_3 + Mg^{++} + Cl^-$

b) $2\ CH_3CH_2CH_2\overset{\overset{\displaystyle O}{\|}}{C}-OC_2H_5 + C_2H_5O^-Na^+ \longrightarrow CH_3CH_2CH_2\overset{\overset{\displaystyle O}{\|}}{C}-\overset{\overset{\displaystyle \ }{\underset{\underset{\displaystyle CH_3CH_2}{|}}{C}}}{}-\overset{\overset{\displaystyle O}{\|}}{C}-OC_2H_5Na^+ + 2\ C_2H_5OH$

$\xrightarrow{H_3O^+} CH_3CH_2CH_2\overset{\overset{\displaystyle O}{\|}}{C}\overset{\overset{\displaystyle \ }{\underset{\underset{\displaystyle CH_3CH_2}{|}}{C}H}}{}\overset{\overset{\displaystyle O}{\|}}{C}-OC_2H_5 + Na^+Cl^-$

c) $CH_3\overset{\overset{\displaystyle CH_3}{|}}{C}HCH_2\overset{\overset{\displaystyle O}{\|}}{C}H \xrightarrow[\text{heated}]{NaOH} CH_3\overset{\overset{\displaystyle CH_3}{|}}{C}HCH_2CH=\overset{\overset{\displaystyle CH_3}{|}}{\underset{\underset{\displaystyle (CH_3)_2CH}{|}}{C}}-CH=O + H_2O$

d) $2\ CH_3MgI + CH_3(CH_2)_3\overset{\overset{\displaystyle O}{\|}}{\underset{\underset{\displaystyle CH_3CH_2CH_2}{|}}{C}H C}-OC_2H_5 \xrightarrow{\text{dry ether}} CH_3(CH_2)_3\overset{\overset{\displaystyle CH_3}{|}}{\underset{\underset{\displaystyle CH_3CH_2CH_2}{|}}{C}H-C}-OMgI + C_2H_5OMgI \xrightarrow{H_3O^+}$

$CH_3(CH_2)_3\overset{\overset{\displaystyle CH_3}{|}}{\underset{\underset{\displaystyle CH_3CH_2CH_2}{|}}{C}H-C}-OH + C_2H_5OH + 2\ Mg^{++} + 2\ I^-$

e) $CH_2(CO_2C_2H_5)_2 + Na^+{}^-OC_2H_5 \longrightarrow Na^+{}^-CH(CO_2C_2H_5)_2 + C_2H_5OH \xrightarrow{C_6H_5CH_2Cl} C_6H_5CH_2CH(CO_2C_2H_5)_2 + Na^+Cl^-$

14.13 a) $CH_2=CHCH_2CO_2H$ 　　　**b)** $CH_3(CH_2)_3\overset{\overset{\displaystyle OH}{|}}{C}HC_6H_5$ 　　　**c)** $CH_3\overset{\overset{\displaystyle HO}{|}}{C}H\text{—}\langle\text{cyclopentane ring}\rangle$ with CH_3

d) CH_3CH_2—$\overset{\displaystyle OH}{\underset{\displaystyle CH_2CH_3}{C}}$—$CH_2CH_2$—$\overset{\displaystyle OH}{\underset{\displaystyle CH_2CH_3}{C}}$—$CH_2CH_3 + 2\ CH_3OH$

e) $CH_3(CH_2)_3\overset{\displaystyle OH}{\underset{\displaystyle CH_2CH_2CH_2}{CH}}CHCH=O$

f) $C_6H_5CH_2\overset{\displaystyle O}{\overset{\|}{C}}\overset{\displaystyle \underset{\displaystyle C_6H_5}{|}}{CH}\overset{\displaystyle O}{\overset{\|}{C}}$—$OC_2H_5$

g) $CH_3CH_2\overset{\displaystyle CH_3}{\underset{\displaystyle |}{C}}(CO_2C_2H_5)_2$

h) $H_2C\overset{\displaystyle H_2C-CH_2}{\underset{\displaystyle H_2C-\underset{\displaystyle CO_2C_2H_5}{CH}}{\diagup\diagdown}}C=O\quad +\ C_2H_5OH$

i) ⬡—$CH_2CH_2CH_2CH_2CO_2H + CO_2 + 2\ C_2H_5OH$

14.14 a) $CH_3CH_2CH_2\overset{\displaystyle O}{\overset{\|}{C}}$—$OR + CH_3MgI$; or $CH_3CH_2CH_2\overset{\displaystyle O}{\overset{\|}{C}}CH_3 + CH_3MgI$; or $CH_3\overset{\displaystyle O}{\overset{\|}{C}}CH_3 + CH_3CH_2CH_2MgCl$

b) $CH_3\overset{\displaystyle \underset{\displaystyle CH_3}{|}}{CH}MgCl + H_2C=O$;

c) $CH_3CH_2CH_2\overset{\displaystyle O}{\overset{\|}{C}}CH_3 + CH_3CH_2MgCl$; or $CH_3CH_2CH_2MgCl + CH_3CH_2\overset{\displaystyle O}{\overset{\|}{C}}CH_3$

or $CH_3CH_2CH_2\overset{\displaystyle O}{\overset{\|}{C}}CH_2CH_3 + CH_3MgI$

d) ⬡—$MgBr + CO_2$

e) $CH_3CH_2CH=O + CH_3CH_2CH_2MgCl$; or $CH_3CH_2CH_2CH=O + CH_3CH_2MgCl$

14.15 a) A: $CH_3CH_2\overset{\displaystyle OH}{\underset{\displaystyle CH_3}{CH}}CHCH=O$; B: $CH_3CH_2CH=\overset{\displaystyle \underset{\displaystyle CH_3}{|}}{C}CH=O$; C: $CH_3CH_2CH_2\overset{\displaystyle \underset{\displaystyle CH_3}{|}}{CH}CH_2OH$; D: $CH_3CH_2CH_2\overset{\displaystyle \underset{\displaystyle CH_3}{|}}{CH}CO_2H$

b) E: $CH_3CH_2CH_2CH_2OH$; F: $CH_3CH_2CH=CH_2$;

G: $CH_3CH_2\overset{\displaystyle \underset{\displaystyle Br}{|}}{CH}CH_3$; H: $CH_3CH_2\overset{\displaystyle \underset{\displaystyle CH_3}{|}}{CH}$—$MgBr$; I: $CH_3CH_2CH_2\overset{\displaystyle \underset{\displaystyle OH}{|}}{CH}$—$\overset{\displaystyle \underset{\displaystyle CH_3}{|}}{CH}CH_2CH_3$;

J: $CH_3CH_2CH_2\overset{\displaystyle \underset{\displaystyle CH_3}{|}}{\overset{\displaystyle Cl}{CH}}CHCH_2CH_3$; K: $CH_3CH_2CH_2\overset{\displaystyle \underset{\displaystyle CH_3}{|}}{\overset{\displaystyle MgCl}{CH}}CHCH_2CH_3$; L: $CH_3CH_2CH_2\overset{\displaystyle \underset{\displaystyle CH_3CH_2CHCH_3}{|}}{CH}CO_2H$

c) M: $CH_3CH_2CH_2OH$; N: $CH_3CH_2CH_2Cl$; O: $CH_3CH_2CH_2MgCl$; P: $(CH_3CH_2CH_2)_2\overset{\displaystyle \underset{\displaystyle OH}{|}}{C}CH_3$

d) Q: $C_6H_5CH_2\overset{\displaystyle O}{\overset{\|}{C}}OC_2H_5$; R: $C_6H_5CH_2\overset{\displaystyle O}{\overset{\|}{C}}\overset{\displaystyle \underset{\displaystyle C_6H_5}{|}}{CH}\overset{\displaystyle O}{\overset{\|}{C}}OC_2H_5$; S: $C_6H_5CH_2\overset{\displaystyle OH}{\underset{\displaystyle C_6H_5}{CH}}\overset{\displaystyle \underset{\displaystyle}{|}}{CH}\overset{\displaystyle O}{\overset{\|}{C}}OC_2H_5$

14.16 a)

$CH_2OPO_3H^-$
$|$
$C=O$
$|$
H—C—OH
$|$
H—C—OH
$|$
H—C—OH
$|$
$CH_2OPO_3H^-$

Remember there are only two possible configurations for #3 carbon.

b) The stereospecificity is determined by the enzyme which catalyzes the reaction. The aldol reaction is base-catalyzed, with the enzyme removing the deuterium. The enzyme must also hold the glyceraldehyde in position on the same side of the dihydroxyacetone as the deuterium. After the deuterium is removed, the carbanion can attach itself to the carbonyl carbon immediately. Thus the stereochemical courses for both the carbanion and the carbonyl are fixed.

c) The (ii) enantiomer could not replace the dihydroxyacetone with the same enzyme because the CH_3 group occupies the position of the hydrogen which should be removed. If there is no hydrogen in that position no reaction can occur. However, in (i), the hydrogen occupies the correct position and this enantiomer could substitute for dihydroxyacetone provided there is space in the enzyme pocket for the CH_3 group.

14.17

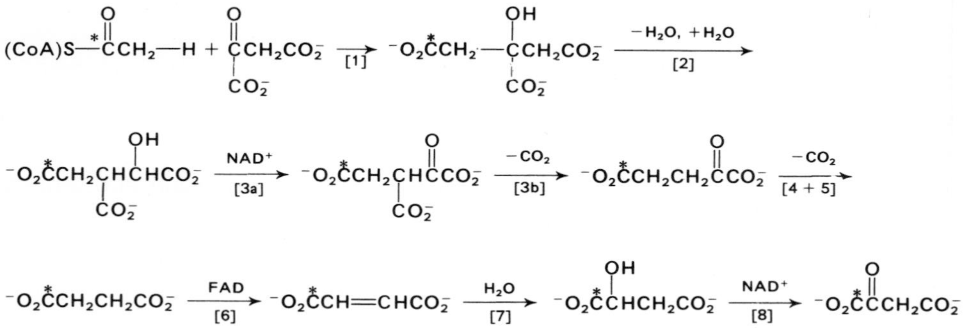

$C* = {}^{14}C$

According to this scheme, the two molecules of CO_2 lost in the cycle are the two original CO_2^- groups of the oxaloacetate ion. Thus the carboxylate carbon of the incoming acetyl ester is not lost until the second cycle; the methyl group of the incoming acetyl group becomes a carboxyl carbon in a subsequent cycle.

Chapter 15

15.1 A: $CH_3\overset{O}{\overset{\|}{C}}CH_3$ with λ_{max} at 280 nm due to $C{=}O$ B: $CH_2{=}CH{-}CH_2OH$, $C{=}C$ absorption below 210 nm

15.2 C:  conjugated diene indicated by λ_{max} at 256 nm;

D:  unconjugated diene absorption below 210 nm

15.3 a) $CH_3{-}O{-}CH_3$ **b)** $CH_3{-}CCl_2{-}CH_3$ **c)** $CH_3\overset{}{\underset{\overset{\|}{O}}{C}}CH_3$ **15.4 a)** 3 **b)** 3 **c)** 2

15.5 ↑↑↑ ↑↑↓ ↑↓↓ ↓↓↓   1:3:3:1 relative areas
   ↑↓↑ ↓↑↓
   ↓↑↑ ↓↓↑

15.6 a) CH₃—O—CH₂—CH₃ *a*—singlet (3H); *b*—quartet (2H); *c*—triplet (3H)
  a   b   c

b) 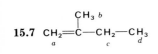 *a*—singlet (1H); *b*—septet (1H); *c*—doublet (6H)

c) Cl₂CH—CH—CHCl₂ *a*—doublet (2H); *b*—triplet (1H)
       a   b   a
       (with Cl on central carbon)

15.7 CH₂=C—CH₂—CH₃
         a | c   d
            CH₃ *b*

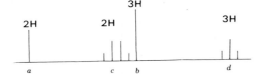

15.8 a) CH₃CH₂CH₂CH=CH₂ or CH₃CH₂CH=CHCH₃
  b) CH₃CH₂CH₂C≡CH

15.9 (CH₃)₂CH—O—CH₃ (OH band not present in IR, therefore no R—OH.)

15.10 The IR spectra would show strong C=O and OH absorption bands for both compounds. The NMR spectra would show —CH=O signal about δ 9 and —CO₂H signal about δ 12. Acetic acid would have two singlets while hydroxyacetaldehyde would have three signals—singlet, doublet, and triplet.

15.11 alcohol; saturated (absence of =CH and C=C bands), therefore a cyclic alcohol such as cyclohexanol

15.12 eliminated: CH₃CH=O (absence of strong C=O band); CH₂=CH—OH (absence of strong OH band); probable structure with one NMR signal and no strong high frequency IR bands other than CH: H₂C—CH₂ with O bridging

15.13 A:B = 3:2

15.14 (CH₃)₂CH—C(=O)—CH(CH₃)₂ two NMR signals: septet (2H) and doublet (12H). Treatment with NaOD in D₂O causes exchange of α-protons by deuterons giving (CH₃)₂CDCOCD(CH₃)₂, which would have one singlet. No exchange of CH₃ protons occurs.

15.15 CH₃—CH₂—C(=O)—O—CH₃ methyl propanoate (Note relative positions of CH₃C(=O)— and O—CH₃ in NMR spectrum of methyl acetate in Fig. 15.4.)

15.16
12 outer H at δ 8.9, similar to benzene;
6 H's inside ring at δ −1.9, more shielded than TMS

15.17 HCl salt indicates an amino group, *p*-aminobenzoate C₇H₆O₂N; therefore remaining portion has C₆H₁₄N.

IR—1700 cm⁻¹—aromatic ester Ar—C(=O)—OR

NMR—triplet and quarter at δ 1.0 and 2.6 suggest two identical ethyl groups, such as —N(CH$_2$CH$_3$)$_2$; δ 4.2 singlet due to NH$_2$; δ 6.5–8.0 multiplet aromatic H's; δ 4.35 triplet and δ 2.75 triplet must be O—CH$_2$—CH$_2$—N

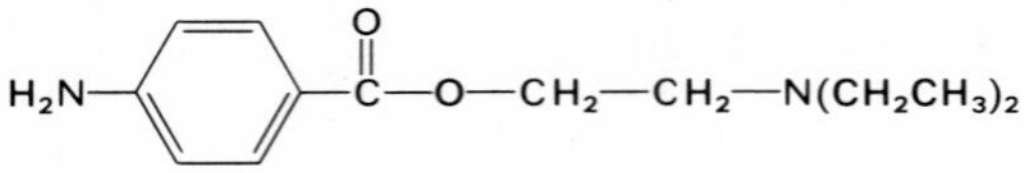

H$_2$N—⟨aromatic ring⟩—C(=O)—O—CH$_2$—CH$_2$—N(CH$_2$CH$_3$)$_2$

novocaine

Index